Active and Non-Linear Electronics

Thomas F. Schubert, Jr.
University of San Diego

Ernest M. Kim
University of San Diego

 John Wiley & Sons, Inc.
New York • Chichester • Brisbane • Toronto • Singapore

ACQUISITION EDITOR Steven Elliot
MARKETING MANAGER Debra Riegert
SENIOR PRODUCTION EDITOR Cathy Ronda
DESIGN SUPERVISOR Ann Marie Renzi
MANUFACTURING MANAGER Mark Cirillo
ILLUSTRATION Rosa Bryant
ILLUSTRATORS Wellington Studios
TEXT DESIGN Hudson River Studio
COVER DESIGN Kenny Beck
COVER ART Roy Wieman

This book was set in 10 x 12 Palatino by University Graphics and
printed and bound by Donnelley/Willard. The cover was printed by Lehigh Press.

Recognizing the importance of preserving what has been written, it is a
policy of John Wiley & Sons, Inc. to have books of enduring value published
in the United States printed on acid-free paper, and we exert our best
efforts to that end.

The paper on this book was manufactured by a mill whose forest management programs include
sustained yield harvesting of its timberlands. Sustained yields harvesting principles ensure that
the number of trees cut each year does not exceed the amount of new growth.

PSpice® is a registered trademark of MicroSim Corporation in the United States and other countries
throughout the world. All other MicroSim product names, including MicroSim™, are trademarks of
MicroSim Corporation. Representations of PSpice® and other MicroSim products are used by
permission of and are proprietary to MicroSim Corporation. Copyright 1994 MicroSim Corporation.

Library of Congress Cataloging in Publication Data:
Schubert, Thomas (Thomas F.)
    Active and non-linear electronics / Thomas Schubert, Jr., Ernest
Kim
        p. cm.
    Includes bibliographical references.
    ISBN 0-471-57942-4 (cloth : alk. paper)
    1. Electronic circuit design. 2. Electric circuits, Nonlinear—
Design. 3. Electric networks, Active—Design. I. Kim, Ernest M.
II. Title.
TK7867.S285   1996
621.3815—dc20   95-40850
                                                           CIP

Printed in the United States of America

10 9 8 7 6 5 4 3 2 1

PREFACE

Our purpose in writing this book is to provide a gradual and logical presentation of active and non-linear electronics. The focus is on the principles necessary to understand, analyze, and design electronic circuitry using currently available technologies. To accomplish that goal a graded approach is utilized. Discussions typically progress from the simple to the complex: this progression helps build a solid foundation of knowledge before advanced concepts and designs are presented. New material is introduced only when needed. For example, transistors are initially modeled in their simplest forms: complexity is added to the models when further development of analysis and design is demanded. We believe that this presentation results in the solid understanding of fundamental principles and prepares the reader for future professional activities and advanced study.

Electronic circuit analysis and circuit design are dynamic processes that require carefully nurtured development. Within this pedagogical framework, the primary heuristics in this book are as follows:

- To provide a strong foundation in the basics of electronics so that the student can understand current and future technologies.
- To reflect current industrial and professional practices
- To develop the student's ability to design

INTRODUCTION

This text has been designed primarily for use as an upper division course in electronics for electrical engineering students. Typically such a course spans a full academic year consisting of two semesters or three quarters. As such, the main body of material can easily be covered: additional material in the fourth section of the text provides a good introduction to advanced topics. An accompanying laboratory course is highly recommended. Secondary applications of this text include use in a one-semester electronics course for engineers or as a reference for practicing engineers.

Prerequisite course material in mathematics, physics, and electrical circuits is used extensively in this text. Students should have a working knowledge of algebra, electrical circuit analysis techniques, elementary calculus, and computer usage. Although prior knowledge of a circuit simulator (i.e., SPICE) and a computerized math tool (i.e., MathCAD) can be helpful, each, with the use of appropriate references, can be easily assimilated by the student through the many examples in the text. Similarly, a prior course in solid-state device physics is suggested, but it is not necessary for obtaining a working knowledge of the principles of electronics developed here.

Whenever possible, real or realistic device parameters are used in examples and exercises so that laboratory test or computer simulation can verify the analysis. Device parameters and macromodels for use with popular versions of SPICE are provided. Standard component values (e.g., capacitors and resistors) are used throughout the book.

ORGANIZATION

This book is divided into four main sections. The first section, *Quasistatic, Large-Signal Analysis* comprises four chapters describing the basic operation of each of the four fundamental building blocks of modern electronics: operational amplifiers, semiconductor diodes, bipolar junction transistors, and field effect transistors. Attention is focused on the student's obtaining a clear understanding of each of the devices when they are operated in equilibrium. Ideas fundamental to the study of electronic circuits are also developed in this section at a basic level to lessen the possibility of misunderstandings at a higher level. The difference between linear and non-linear operation is explored through the use of a variety of circuit examples, including amplifiers constructed by using operational amplifiers and elementary digital logic gates constructed with various transistors types.

The second section, *Linear Amplifiers—Small-Signal Analysis,* consists of four chapters that describe the fundamentals of amplifier action. Beginning with a review of two-port analysis, Chapter 5 introduces the modeling of the response of transistors to AC signals. Basic one-transistor amplifiers are extensively discussed. Chapter 6 expands the discussion to multiple transistor amplifiers. Chapter 7 concludes the discussion of simple amplifiers with an examination of power amplifiers. The discussion defines the limits of small-signal analysis and explores the realm where these simplifying assumptions are no longer valid. Chapter 8 concludes the section with the first of two chapters on the significant topic of feedback.

The third section, *Frequency Dependence,* consists of five chapters and is an in-depth look at the variation of circuit performance as a function of signal frequency. Chapter 9 introduces the frequency domain through a review of ideal filters and Bode diagrams. A discussion of the design of active filters using operational amplifiers emphasizes the significance of capacitors and resistors in frequency response. Chapter 10 explores the frequency response of transistor amplifiers. It is at this point that complex models of transistors are introduced to the discussion so that proper high-frequency modeling can be accomplished. Chapter 11 is the second chapter on feedback: here the effects of feedback on frequency response are explored in depth. Stabilization against amplifier oscillation is examined thoroughly. Chapter 12 expands on linear feedback theory to present linear oscillators. Chapter 13 delves into the non-linear domain to explore non-linear oscillators and waveshaping.

The fourth section, *Application-Specific Electronics,* is generally beyond the range of course for which this text is primarily designed. We believe, however, that no electronics text can be complete without these chapters. Chapter 14 explores the design of circuitry to provide safe, stable power to the circuits covered in the first three sections. Chapter 15 examines the circuitry necessary for signal communication. Chapter 16 expands on the elementary digital logic gates developed in Chapters 3 and 4, describing the non-linear world of digital electronics. Operational limitations such as the speed of operation of digital gates is explored extensively.

A typical first semester course in electronic circuits could consist of sequential coverage of the material in Chapters 1 to 6. For a single semester course targeting computer engineering students, a more detailed introduction to digital electronics may be desirable. For these students, the sequence of chapter coverage could be Chapters 1 to 4, 16, and 5. A second semester course in electronic circuits begins with either Chapter 6 or 7 and continues through Chapters 11 or 12. Chapter 12 can be replaced with any of the chapters in Section 4 at the instructor's discretion. From our experience teaching from the manuscript, significant time should be allocated for Chapters 10 and 11. Active filter design, Chapter 9, is often limited to Butterworth filters, with selected examples of Chebyshev filters and elliptical filters as variations.

FEATURES

Worked-Out Examples

Worked-out example problems provide valuable guidance to the reader. Each chapter contains several worked-out examples that are carefully chosen to illustrate the fundamental concepts in the chapter. Initial examples typically focus on analysis techniques whereas later examples tend to explore design possibilities.

Summary Design Examples

Each chapter is concluded with a summary design example that presents a realistic design problem related to the concepts discussed in the chapter. The intent of the summary design examples is to aid and amplify students' understanding of the circuit principles that are described in the chapter and to provide a glimpse into the design process.

Chapter Summaries

As an aid to students, each chapter ends with a set of concluding remarks summarizing the material presented in the chapter and putting it in perspective. Whenever possible, tables summarizing results have been provided as a quick reference aid.

Homework Problems

Homework problems are provided at the end of each chapter. Each chapter within the primary focus of the text has a minimum of 40 problems: there are more than 600 homework problems in total. These are divided into three basic categories:

- analysis,
- focused design, and
- design problems.

Analysis problems focus on determination of specific properties of a given electronic situation. Focused design problems provide an introduction into the realm of design by limiting the range of possible solutions while still keeping the basic properties of engineering design. Design problems open the range of possibilities while exploring more typical constraints. As an aid, the focused design and design problems are identified with the icons shown below:

Focused design   Design

Focused and open-ended design problems are used to introduce the concept of the complete design cycle—a process that spans the analysis of the options and design of the selected approach to verification by using SPICE simulations and, if necessary, redesign.

Computer Usage

Examples, summary design examples, and homework problems depend extensively on computer simulation as either an aid to finding a solution or as a verification of the solution obtained. Two basic simulation tools have been used throughout the text:

- a circuit simulator, and
- a computerized math tool.

While PSpice has been used as the primary circuit simulator, an effort has been made to have solutions that are transportable to other forms of SPICE. MathCAD has been used as the primary computerized math tool. Use of each of these computer aids is explored through the examples presented within the chapters. As an aid to the reader, a [icon] is included in the text wherever SPICE is used to illustrate an example.

Color Usage

Color has been used throughout the book for emphasis. In example problems, color highlights the desired results. In the problems, color is used to identify individual problems that require SPICE simulation. In circuit diagrams, color typically differentiates between the given circuit topology and the defined quantities that are necessary for analysis and design. Graphical representations are typically the result of computer simulation: the careful choice of colors used in this book is designed to accentuate the significant information.

DEVELOPMENT PROCESS

In Class Usage

Throughout the development of this book, we used drafts of the manuscript as the primary text in the electronic circuit and design sequence at the University of San Diego. Careful scrutiny by undergraduate students provided us with invaluable feedback concerning clarity and presentation style as well as accuracy. The following students provided invaluable help in the development of this work: Christine Bridewell, Scott Denton, Maureen Feiner, Jorge Geremia, Barbara Hammack, Douglas Harder, Kiyoshi Kanzawa, Brian Kuehnert, Brandon M. Knaggs, Daniel Leuthner, Thomas Mack, Michael Mahan, Eric Malek, Michael Malone, Mauricio Martinez, Keith Resch, Romeo H. Rodriguez, John Simbulan, Mary Joy Sotic, Dorothy Sze, Langford Wasada, and Kevin Wensley. We thank them all for their help and patience.

Reviews

Professors from a broad range of universities throughout the United States carefully reviewed each draft of this manuscript. Two initial reviews were limited to the formative sections of the text: the two final reviews spanned the entire text. All suggestions by the reviewers were carefully considered and incorporated whenever possible.

ACKNOWLEDGMENTS

In the development of any textbook, it seems that an infinite number of people provide an incalculable amount of guidance and help. While our thanks goes out to all those who helped us, we can only mention a few of our many benefactors here. Special thanks go to Lynn Cox, who sparked our interest in writing an electronics text and facilitated our initial negotiations, Robert L. Mertz for his help concerning feedback stability and, of course, to the editorial staff at John Wiley & Sons, most notably Steven Elliot, Deborah Riegert, Cathy Ronda, Ann Renzi, and Rosa Bryant.

We thank the following reviewers for their suggestions and encouragement throughout the development of this text:

Kay D. Baker	Utah State University
Martin Peckerar	University of Maryland
Jerome K. Butler	Southern Methodist University
Hollace L. Cox	University of Louisville
Milton E. Hamilton	California State Polytechnic University—Pomona
J. Pineda de Gyrez	Texas A&M University
Bryen E. Lorenz	Widener University
Bahram Nabet	Drexel University
Robert D. Strattan	University of Tulsa
Shirshak K. Dhali	Southern Illinois University at Carbondale
Jack R. Smith	University of Florida
Charles G. Nelson	California State University—Sacramento
Jiann-Shiun Yuan	University of Central Florida
Robert K. Feeney	Georgia Institute of Technology
William L. Kuriger	University of Oklahoma
Seth Wolpert	University of Maine
H. J. Allison	Oklahoma State University
Peter Aronhime	University of Louisville

While all reviewer suggestions were carefully considered, the final judgments concerning organization, emphasis, and pedagogy were made by the authors. We will be pleased to receive comments from readers and will attempt to acknowledge all such communications.

Thomas F. Schubert, Jr., Ph.D., P.E.
Ernest M. Kim, Ph.D., P.E.

Contents

Section 2 Linear Amplifiers: Small-Signal Analysis 247

Active and Non-Linear Electronics

3,356,858
LOW STAND-BY POWER COMPLEMENTARY
FIELD EFFECT CIRCUITRY
Frank M. Wanlass, Mountain View, Calif., assignor to
Fairchild Camera and Instrument Corporation, Syosset,
N.Y., a corporation of Delaware
Filed June 18, 1963, Ser. No. 288,786
3 Claims. (307–88.5)

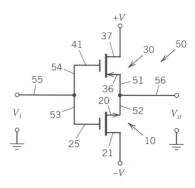

2. A circuit comprising a first insulated gate field-effect transistor having a substrate of one conductivity type, an insulating layer adjacent said substrate, a metal layer upon said insulating layer opposite said substrate, and a channel region adjacent said insulating layer having a conductivity type opposite said substrate upon application of a first predetermined bias voltage to said metal layer, thereby forming a PN junction between said substrate and said channel region, source and drain electrodes on opposite ends of said channel region, a second insulated gate field-effect transistor having a substrate of said opposite conductivity type, an insulating layer adjacent said substrate, a metal layer upon said insulating layer opposite said substrate, and a channel region adjacent said insulating layer having said one conductivity type upon application of a second predetermined bias voltage to said metal layer of polarity opposite to said first predetermined bias voltage, source and drain electrodes on opposite ends of said channel region, one of said source and drain electrodes of said first insulated gate field-effect transistor being coupled to one of said source and drain electrodes of said second insulated field-effect transistor, a means for applying a fixed bias voltage across the other of said source and drain electrodes of said first and second insulated field-effect transistors, and a means for applying a voltage signal of a single predetermined polarity to both of said metal layers of said first and second insulated gate field-effect transistors adapted to render said first insulated gate field-effect transistor conductive between its source and drain electrodes while rendering said second insulated gate field-effect transistor non-conductive between its source and drain electrodes, whereby said second non-conductive insulated gate field-effect transistor is employed as the active load for said first conductive insulated gate field-effect transistor.

Quasistatic, Large-Signal Analysis

It is expected that the reader of this text is familiar with the common passive elements of linear circuit analysis (resistors, inductors, capacitors, and transformers) as well as the idealized linear active elements (independent and dependent voltage and current sources). As is implied in the title, *Active and Non-Linear Electronics*, the field of electronics makes great use of active elements that do not fall into either of the above categories. These active elements may behave in either a linear or a non-linear fashion depending on their circuit application.

The study of electronic circuit behavior traditionally begins with three active semiconductor electronic elements:

- Semiconductor Diode
- Bipolar Junction Transistor (BJT)
- Field-Effect Transistor (FET)

To this trio of fundamental devices has been added an additional electronic circuit building block, the operational amplifier (OpAmp). While the OpAmp is composed of tens of transistors (usually either BJTs or FETs, but sometimes both) and often a few diodes, its easily understood terminal properties, high use in industry, and commercial availability make it a good companion for study with the other devices.

Quasistatic analysis explores the potentially non-linear action of each of these four elements (or any other similar element) in a variety of applications. The fundamental assumption in this exploration is that voltage and current transitions take place slowly and that the circuit is always in equilibrium: hence the term *quasistatic*.

The authors have chosen to begin the study of electronics with a chapter on the OpAmp for several reasons, among which are the following:

- In most simple applications, the OpAmp behaves in a near-ideal fashion.
- Typical analysis of OpAmp circuitry provides a good review of basic circuit analysis techniques.
- Discussion of the OpAmp provides a good framework for understanding of electronic circuitry.

While many readers will find much in this chapter on OpAmps a review, the chapter presents several concepts fundamental to the study of electronic circuitry. Most significant among these concepts are the following:

- Undistorted amplification
- Gain
- Device modeling
- Conditions under which device models, particularly linear models, fail

Of particular importance is the concept that a device with extremely complex interior working mechanisms can be modeled simply by its terminal characteristics.

The remaining three chapters in this section present the semiconductor diode, the BJT, and the FET. Each chapter follows the same basic framework and has the same goals:

- To present each device through real experimental data and through theoretical functional relationships
- To use the above presented relationships to observe the action of the device in relatively simple circuits
- To devise a progression of realistic piecewise-linear models for the devices. The theoretical basis for each model is presented and the appropriate use of these models is explored. Only when a model fails to properly predict device behavior will new, more complex models be introduced. This simple-to-complex route provides for progressively more detailed analysis using the newly introduced models.
- To use realistic applications to demonstrate the usefulness of the device models
- To provide a solid foundation for the linear and non-linear modeling and applications found in later sections

Upon completing Section I, the reader will have a good foundation in the operation of these four basic active, non-linear devices. The fundamental regions of operation for each device will have been explored: both linear and non-linear device models will be available for further investigations.

CHAPTER 1

OPERATIONAL AMPLIFIERS AND APPLICATIONS

The operational amplifier (commonly referred to as the OpAmp) is one of the primary active devices used to design low- and intermediate-frequency analog electronic circuitry: Its importance is surpassed only by the transistor. OpAmps have gained wide acceptance as electronic building blocks that are useful, predictable, and economical. Understanding OpAmp operation is fundamental to the study of electronics.

The name *operational amplifier* is derived from the ease with which this fundamental building block can be configured, with the addition of minimal external circuitry, to perform a wide variety of linear and non-linear circuit functions. Originally implemented with vacuum tubes and now as small, transistorized integrated circuits, OpAmps can be found in applications such as signal processors (filters, limiters, synthesizers, etc.), communication circuits (oscillators, modulators, demodulators, phase-locked loops, etc.), analog/digital converters (both A/D and D/A), and circuitry performing a variety of mathematical operations (multipliers, dividers, adders, etc.).

The study of OpAmps as circuit building blocks is an excellent starting point in the study of electronics. The art of electronics circuit and system design and analysis is founded on circuit realizations created by interfacing building block elements that have specific terminal characteristics. OpAmps, with near-ideal behavior and electrically good interconnection properties, are relatively simple to describe as circuit building blocks.

Circuit building blocks, such as the OpAmp, are primarily described by their terminal characteristics. Often this level of modeling complexity is sufficient and appropriately uncomplicated for electronic circuit design and analysis. However, it is often necessary to increase the complexity of the model to simplify the analysis and design procedures. These models are constructed from basic circuit elements so that they match the terminal characteristics of the device. Resistors, capacitors, and voltage and current sources are the most common elements used to create such a model: An OpAmp can be described at a basic level with two resistors and a voltage-controlled voltage source.

OpAmp circuit analysis also offers a good review of fundamental circuit analysis techniques. From this solid foundation, the building block concept is explored and expanded throughout this text. With the building block concept, all active devices are treated as functional blocks with specified input and output characteristics derived from the device terminal behavior. Circuit design is the process of interconnecting active building blocks with passive components to produce a wide variety of desired electronic functions.

1.1 BASIC AMPLIFIER CHARACTERISTICS

One of the fundamental characteristics of an amplifier is its gain.[1] Gain is defined as the factor that relates the output to the input signal intensities. Referring to Figure 1.1-1, a time-dependent input signal $x(t)$ is introduced to the "black box" that represents an amplifier and another time-dependent signal $y(t)$ appears at the output.

In actuality, $x(t)$ can represent either a time-dependent or a time-independent signal. The output of a good amplifier, $y(t)$, is of the same functional form as the input with two significant differences: The magnitude of the output is scaled by a constant factor A and the output is delayed by a time t_d. This input–output relationship can be expressed as

$$y(t) = Ax(t - t_d) + \alpha, \tag{1.1-1}$$

where

A = amplifier gain
α = output direct current (DC) offset
t_d = time delay between input and output signals

The signal is "amplified" by a factor of A. Amplification is a ratio of output signal level to the input signal level. The output signal is amplified when $|A| > 1$. For $|A| < 1$, the output signal is said to be attenuated. If A is a negative value, the amplifier is said to invert the input. Should $x(t)$ be sinusoidal, inversion of a signal is equivalent to a phase shift of 180°: Negative A implies the output signal is ±180° out of phase with the input signal. For time-varying signals, it may be convenient to find the amplification (ratio) by comparing either the root-mean-squared (RMS) values or the peak values of the input and output signals. Good measurement technique dictates that amplification is found by measuring the input and output RMS values since peak values may, in many instances, be ambiguous and difficult to quantify.[2] Unfortunately, in many practical instances, RMS or power meters are not available dictating the measurement of peak amplitudes. The delay time is an important quantity that is often overlooked in electronic circuit analysis and design.[3] The signal encounters delay between the input and output of an amplifier simply because it must propagate through a number of the internal components of the amplifying block.

In Figure 1.1-1, $x(t)$ and $y(t)$ are time-dependent signals. Depending on the amplifier, $x(t)$ and $y(t)$ can be either current or voltage signals. Every amplifier draws power from a power supply, typically in the form of current from a DC voltage source. As will be shown in later sections of this text, the maximum possible output signal level is determined largely by the power supply voltage and current limitations. For instance, assume that the amplifier in Figure 1.1-1 is powered by a DC voltage source with output equal to V_{cc}. If the output signal $y(t)$ is a voltage signal,

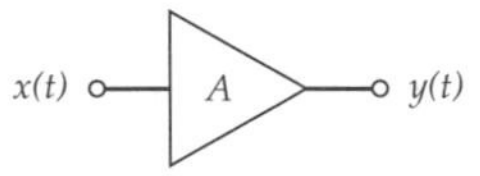

$x(t)$ ○—▷ A ▷—○ $y(t)$

Figure 1.1-1
Black box representation of amplifier with input $x(t)$ and output $y(t)$.

---

[1] Other amplifier specifications of interest include input and output impedances, power consumption, frequency response, noise factor, mean time to failure (MTTF), and operational temperature range. An understanding of the basis for these specifications and their impact on design will be developed in the chapters that follow. The discussion in this chapter will, for the most part, be restricted to gain and time-domain effects.

[2] Peak values are also strongly affected by the presence of noise.

[3] The implication of delay time is addressed in the transistor amplifier time-domain analysis portion of Section III of this text.

the maximum output voltage attainable under ideal conditions for the gain block is V_{cc}.[4] The phenomenon of limiting output voltage levels to lie within the limits set by the power supplies is called *saturation*. Should the power supply be unable to provide sufficient current to the gain block, the output will also be limited, although in a manner that is not as simple as in saturation.

In order to discuss terminal characteristics of commercially available OpAmps, a specific amplifier must be selected. The μA741 (or LM741, MC1741) is a good choice since it is the most commonly used and studied OpAmp available. The prefixes μA, OP, LM, and MC designate the manufacturer of the integrated circuit (IC): μA represents Fairchild Semiconductor, OP is used by Linear Technologies, LM by National Semiconductor, and MC by Motorola Semiconductors. The specification sheets for three OpAmps can be found in the Appendices. In many instances, one or two letters follow the numerical designation of the IC. These letters indicate the package type or size and package material. For example, a μA741CP is a 741 IC manufactured by National Semiconductor that is in a commercial-grade plastic standard eight-lead dual-in-line package (DIP) or MINI DIP. Other manufacturers, such as Texas Instruments, manufacture the μA741CP using the Fairchild part designation.

Other common OpAmps include the OP-27, LF411, and LM324. The OP-27 and LF411 OpAmps have specifications that are similar to the μA741 and come in selected packages. The OP-27 and LF411 OpAmps are, like the μA741, dual-power rail amplifiers; that is, the amplifier usually operates with both positive and negative power supply voltages. The LM324, on the other hand, is a single supply amplifier; it requires a positive voltage and a common reference (ground).

Figure 1.1-2 shows a top view of a μA741CP package with the terminal designations. The terminals of interest are the following:

- Inverting input (pin 2)
- Noninverting input (pin 3)
- Output (pin 6)
- Positive power supply (designated $+V_{cc}$, pin 7)
- Negative power supply (designated $-V_{cc}$, pin 4)

The offset null pins (1 and 5) are used to compensate for minor fabrication imperfections as well as degradation due to aging. Although commonly left disconnected by the circuit designer, these pins are sometimes utilized in applications that require the amplification of very small level signals. The μA741 OpAmp is a compensated amplifier. The performance implications of compensated and uncompensated amplifiers are related to frequency response and stability: It will be discussed in detail in Section III.

A conventional simplified OpAmp schematic representation is shown in Figure 1.1-3. This representation shows two input terminals, designated $(-)$ and $(+)$, cor-

Figure 1.1-2
Top view of μA741CP package with pin numbers.

---

[4] Note that uppercase letters are used for DC signals and lowercase letters for time-varying signals. Lowercase letters with lowercase subscripts are used for alternating current (AC) signals. Lowercase letters with uppercase subscripts are used for AC signals with DC components.

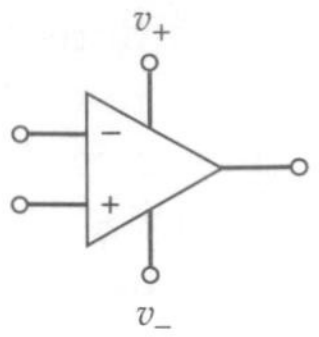

Figure 1.1-3
OpAmp schematic representation.

responding to the inverting and noninverting inputs, respectively, the output terminal, and the positive and negative power supply terminals labeled V^+ and V^-, respectively. Not shown are the offset null pins. Unless used, these pins are usually not included in schematic representations.

Notice that the schematic symbol of the OpAmp does not have a ground pin. In many ways, the lack of a ground pin on the OpAmp is the key to its operation. Ideally, only the differential voltage between the two input pins affect the output voltage of an OpAmp. A ground reference is provided external to the chip package.

1.2 MODELING THE OPAMP

Terminal voltages and currents are used to characterize OpAmp behavior. In order to unify all discussions of OpAmp circuitry, it is necessary to define appropriate descriptive conventions. All voltages are measured relative to a common reference node (or ground) that is external to the chip, as is shown in Figure 1.2-1. The voltage between the inverting pin and ground is denoted as v_1: the voltage between the noninverting pin and ground is v_2. The output voltage referenced to ground is denoted as v_o. Power is typically applied to an OpAmp in the form of two equal-magnitude supplies, denoted V_{CC} and $-V_{CC}$, which are connected to the V^+ and V^- terminals of the OpAmp, respectively.

The reference current directions are shown in Figure 1.2-1. The direction of current flow is always into the nodes of the OpAmp. The current into the inverting input terminal is i_1; current into the noninverting input terminal is i_2; current into the output terminal is i_o; and the currents into the positive and negative power supply terminals are $I_c{}^+$ and $I_c{}^-$, respectively.

The voltage and current constraints inherent to the input and output terminals of an OpAmp must be understood prior to connecting external circuit elements. The OpAmp is considered as a building block element with specific rules of operation. A short discussion of these rules of operation follows.

The terminal voltages are constrained by the following relationships[5]:

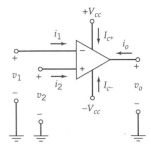

Figure 1.2-1
Terminal voltages and currents.

---

[5] For introductory purposes, the time delay factor t_d has been assumed to be zero. Delay time considerations will be discussed at length in Section III of this text.

$$v_o = A(v_2 - v_1) \tag{1.2-1}$$

and

$$-V_{CC} \leq v_o \leq V_{CC}. \tag{1.2-2}$$

The first of the two voltage constraints states that the output voltage is proportional to the difference between the noninverting and inverting terminal inputs, v_2 and v_1, respectively. The proportionality constant A is the called the *open-loop gain*, whose significance will be detailed later. The second voltage constraint states that the output voltage is limited to the power supply rails.[6] That is, v_o must lie between $\pm V_{CC}$. If the output reaches the limiting values, the amplifier is said to be saturated. In reality, the amplifier saturates at voltages slightly shy of $\pm V_{CC}$ due to device characteristics within the OpAmp. So long as $|v_o| < V_{CC}$, the amplifier is operating in the linear region. Between the limiting values lies the "linear region" where the output voltage is related to the input voltage by the proportionality constant A. Figure 1.2-2a shows an idealized voltage transfer characteristic[7] of an OpAmp. A more realistic voltage transfer characteristic is shown in Figure 1.2-2b, where the amplifier response gradually tapers toward saturation at higher input voltages due to the characteristics of the circuit design internal to the chip.

From the data sheet for the μA741C in the Appendix, the typical open-loop gain (designated as large-signal voltage gain) A is 200,000. It is reasonable to assume that all OpAmps have very large voltage gain and that a first-order approximation of the voltage gain is

$$A \approx \infty. \tag{1.2-3}$$

In the 741 OpAmp, the absolute maximum supply voltages (V_{CC} and $-V_{CC}$) are ± 18 V. Therefore, the output cannot exceed ± 18 V. Knowing the maximum value of the output voltage v_o and the typical A, the maximum difference between v_2 and v_1 is found to be

$$(v_2 - v_1)_{max} = \frac{18 \text{ V}}{200,000} = 0.09 \text{ mV}.$$

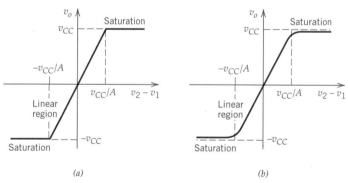

Figure 1.2-2

Transfer characteristics of (a) ideal OpAmp and (b) more realistic "soft-limiting" OpAmp transfer characteristic.

---

[6] The power supply voltages are commonly called *rails*: They limit the output voltages of a functional electronic block.

[7] A *transfer characteristic* is a graphical representation of the output as a function of the input. In this instance, the voltage input–output relationship is shown. A *transfer function* is usually a mathematical description of the output as a function of the input.

In most OpAmp applications this voltage level can be considered negligible. Therefore, in the linear region of operation, the input voltages are assumed to be equal:

$$v_1 \approx v_2. \tag{1.2-4a}$$

The input terminals of an OpAmp exhibit high input resistance: The μA741 typically has an input resistance of 2 MΩ. The input resistance of an ideal OpAmp is approximated as infinite.

$$R_i \approx \infty. \tag{1.2-5}$$

Within the linear region of operation the maximum current flowing between the two input terminals is given by

$$i_{input(max)} = \frac{(v_2 - v_1)_{max}}{2 \text{ M}\Omega} = \frac{0.09 \text{ mV}}{2 \text{ M}\Omega} = 45 \text{ pA}.$$

The input currents i_1 and i_2 are extremely small and are considered to be approximately zero:

$$i_1 = i_2 \approx 0. \tag{1.2-4b}$$

The two relationships given in Equations 1.2-4a and 1.2-4b together form what is referred to as a "virtual short" between the inverting and noninverting input terminals of the OpAmp. A virtual short implies that two terminals act in a *voltage sense* as if they were *shorted*, but *no current flows between the terminals*.

Kirchhoff's current law (KCL) can be used to sum the currents of an OpAmp. Since the input currents are very small, the resulting relationship is

$$i_o = -(I_{C^+} + I_{C^-}). \tag{1.2-6}$$

Equation 1.2-6 indicates that although the input currents are negligible, the output current is substantial. That is, $i_o \neq 0$.

For completeness, the typical output resistance of a 741 OpAmp is 75 Ω. All OpAmps have low output resistance: Ideal OpAmps are considered to have zero-value output resistance:

$$R_o \approx 0. \tag{1.2-7}$$

Equations 1.2-1 to 1.2-7 define the *ideal OpAmp model*. These defining properties are summarized in Table 1.2-1.

The SPICE macromodel[8] for the μA741C OpAmp yields a value of A of approximately 195,000. The input and output impedances are complex and vary with input signal frequency. The real part of the input impedance dominates with a value of approximately 996 kΩ. The output impedance is essentially 50 Ω at frequencies above 100 Hz. The component parameters specified in the SPICE macromodel of the

---

[8] A SPICE macromodel is a complex subcircuit model of a device intended to correctly model all terminal performance characteristics of a device. The SPICE macromodel for the μA741C OpAmp was provided by MicroSim Corp. and can be found in the Appendices.

TABLE 1.2-1 OpAmp Characteristic Property Value

Property	Typical OpAmp Value	Ideal OpAmp Value
Gain, A	>200,000	∞
Input resistance, R_i	>2 MΩ	∞
Output resistance, R_0	<75 Ω	0
Input voltage difference, $v_2 - v_1$	<0.1 mV	0 (virtual short)
Input current, i_1 or i_2	<50 pA	0 (virtual short)
Output voltage limits	$|v_o| < V_{CC}$	$|v_o| \leq V_{CC}$

μA741C OpAmp and the typical specifications found in the data sheet both lie within the acceptable range of parameter values found in manufactured components. Therefore, using either specification will yield acceptable results when designing a circuit using the μA741C.

The voltage and current constraints given are the parameters used to describe an *ideal* OpAmp model. By attaching external components to an OpAmp, a functional circuit can be designed. The significance of this exercise is to demonstrate the power of modeling active elements (in this case an OpAmp) in terms of its current and voltage characteristics at the input and output ports. Although an understanding of the internal operation of the active device is desirable, circuits can be designed using the device's input and output port parameters. The terminal characteristics of the active device in conjunction with Kirchhoff's current and voltage laws are used to analyze the circuit.

A simple application of an OpAmp is the unity-gain buffer, which is primarily used to isolate electronic signals. A unity-gain buffer or voltage follower is shown in Figure 1.2-3. The circuit can be analyzed using the voltage and current constraints given by Equations 1.2-4a and 1.2-4b, respectively,

$$v_1 \approx v_2$$

and

$$i_1 = i_2 \approx 0.$$

The input signal voltage v_i is equal to v_2 since negligible current flows into the positive input terminal of the OpAmp ($i_2 = 0$): There is no voltage drop across R_s. Owing to the virtual short between the inverting and noninverting terminals, the noninverting terminal voltage is also at the same voltage level, v_i, that is,

$$v_i = v_2 = v_1.$$

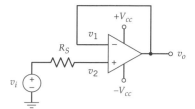

Figure 1.2-3
Unity-gain buffer.

Since the inverting terminal is directly connected to the output of the OpAmp,

$$v_o = v_1 = v_i,$$

or

$$\frac{v_o}{v_i} = 1. \tag{1.2-8}$$

The voltage follower is called a unity-gain *buffer* since it is an ideal impedance transformer. The input impedance of the voltage follower is very high and its output impedance is, for all practical purposes, zero. Verification of these impedance characteristics provides a useful exercise in the study of OpAmp properties.

The input and output resistances of the voltage follower can be found by using the simplified equivalent circuit for the OpAmp. The simplified equivalent circuit, shown in Figure 1.2-4, differs from the ideal OpAmp model in that the equivalent circuit is a lumped-parameter (resistors and sources) model of the OpAmp in the linear region of operation. It is a functional equivalent of the OpAmp that is not the actual circuitry in the OpAmp chip but behaves functionally as an OpAmp to external circuitry (the principle is much the same as Thévenin or Norton equivalent circuits). The simplified OpAmp equivalent model shown here assumes frequency-independent behavior; that is, the response of the amplifier is independent of signal frequency.[9]

The current and voltage constraints in Equations 1.2-4a and 1.2-4b assumed in the ideal OpAmp model must be discarded when using the simplified equivalent model. Equation 1.2-1 is no longer valid due to the nonzero output resistance R_o.

The voltage follower circuit is analyzed using the simplified equivalent circuit shown in Figure 1.2-5. In the model, $R_i = 2$ MΩ, $R_o = 75$ Ω, and $A = 200{,}000$. Let $R_s = 1$ kΩ.

The current i is found by Kirchhoff's voltage law (KVL),

$$i = \frac{v_i - A(v_2 - v_1)}{R_i + R_o + R_s}, \tag{1.2-9}$$

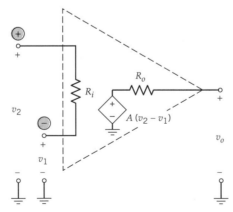

Figure 1.2-4
Simplified equivalent circuit model of OpAmp.

---

[9] In reality this frequency independence is not true. The frequency-dependent nature of OpAmps will be discussed in Chapter 9.

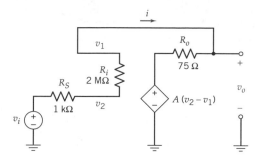

Figure 1.2-5
Voltage follower analysis with
simplified equivalent OpAmp model.

but i is also given by

$$i = \frac{v_2 - v_1}{R_i}.$$ (1.2-10)

Since the output terminal of a voltage follower is connected to the negative input terminal of the OpAmp,

$$v_o = v_1.$$ (1.2-11)

Equating Equations 1.2-9 and 1.2-10, substituting v_i for v_2, and solving for v_o/v_i yields the expression for voltage gain for the voltage follower:

$$\frac{v_o}{v_i} = 1 - \frac{R_i}{R_s + R_o + R_i(1 + A)}.$$ (1.2-12)

It can easily be seen that large values of the OpAmp gain A will lead to a voltage gain closely approximating unity. The parameter values for this example, $R_i = 2\,\text{M}\Omega$, $R_o = 75\,\Omega$, $R_s = 1\,\text{k}\Omega$, and $A = 200{,}000$, result in

$$\frac{v_o}{v_i} = 1 - 2.6875 \times 10^{-9} \approx 1.$$ (1.2-13)

The input and output resistances of the voltage follower can be calculated using the simplified equivalent model. Thévenin equivalent resistances are found using the standard "test source" method. To calculate the input resistance, a test source v_t is used to excite the circuit shown in Figure 1.2-6. Note that v_t is applied by adding R_s

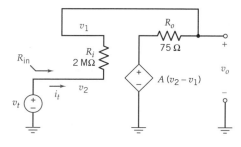

Figure 1.2-6
Using test source to find Thévenin
equivalent input resistance R_in.

to the Thévenin resistance found at v_2. The Thévenin equivalent input resistance R_{in} is the ratio of the test voltage over the current delivered by the test source,

$$R_{in} = \frac{v_t}{i_t}. \tag{1.2-14}$$

Using KVL, the current delivered by the test voltage, v_t, is

$$i_t = \frac{v_t - A(v_2 - v_1)}{R_i + R_o}. \tag{1.2-15}$$

But

$$v_2 = v_t, \tag{1.2-16}$$

and knowing the voltage drop across R_i,

$$v_1 = v_t - i_t R_i. \tag{1.2-17}$$

Substituting Equations 1.2-16 and 1.2-17 into 1.2-15 yields

$$i_t = \frac{v_t - A i_t R_i}{R_i + R_o}. \tag{1.2-18}$$

The Thévenin input resistance R_{in} is found by rearranging Equation 1.2-18,

$$R_{in} = \frac{v_t}{i_t} = R_i(1 + A) + R_o. \tag{1.2-19}$$

For the given typical parameter values (R_i = 2 MΩ, A = 200,000, and R_o = 75 Ω), the input resistance can be calculated to be the very large value R_{in} = 400 × 10^9 Ω. It is reasonable to assume that the input resistance of a unity-gain buffer, R_{in}, is, for all practical purposes, infinite.

EXAMPLE 1.2-1 Determine the output resistance of an OpAmp voltage follower.

SOLUTION

To find the output resistance R_{out}, a test voltage source is connected to the output of the voltage follower to find the Thévenin equivalent resistance at the output. Note also that all independent sources must be zeroed. That is, all independent voltage sources are short circuited and all independent currents are open circuited. The circuit used to find R_{out} is shown in Figure 1.2-7.[10]

To find the Thévenin equivalent output resistance, a test voltage source v_t is connected at the output. The circuit draws i_t source current. The input at v_2 has been short circuited to ground to set independent sources to zero.

---

[10]In this example the resistance of the source connected to the input was considered to be zero; that is, R_s = 0. When the resistance of the source is not zero, it appears in series with the input resistance of the OpAmp, R_i. The effect of considering nonzero source resistance in the results of this example is insignificant.

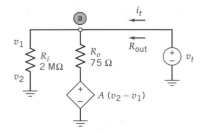

Figure 1.2-7
Simplified equivalent circuit for finding R_{out} of OpAmp voltage follower.

Using the node voltage method of analysis by summing all currents flowing into node a yields

$$i_t + \frac{v_2 - v_1}{R_i} + \frac{A(v_2 - v_1) - v_t}{R_o} = 0. \tag{1.2-20}$$

But

$$v_2 = 0 \quad \text{and} \quad v_1 = v_t. \tag{1.2-21}$$

Then Equation 1.2-20 is simplified to

$$i_t = \frac{R_o + (A + 1)R_i}{R_o R_i} v_t, \tag{1.2-22}$$

and R_{out} is given as

$$R_{out} = \frac{v_t}{i_t} = \frac{R_o R_i}{R_o + (A + 1)R_i}. \tag{1.2-23}$$

A typical value for the output resistance of the unity-gain buffer, R_{out}, can be calculated using the given typical OpAmp values of $R_i = 2 \text{ M}\Omega$, $R_o = 75 \text{ }\Omega$, and $A = 200{,}000$. Here, R_{out} is found to be 375 $\mu\Omega$, which in most circuit applications can be considered essentially zero.

A SPICE simulation to determine the output resistance of a unity-gain buffer using the circuit of Figure 1.2-7 is shown in Example 1.2-2. Solution 1 of Example 1.2-2 uses the **.TF** command in conjunction with a source voltage to yield the gain and input and output resistances. Although the **.TF** command is adequate for this particular example, a more general approach is to use a test source at the output, as shown in Solution 2. This is particularly important when the input and output resistances are complex and are dependent on frequency. For instance, the input and output impedances of real OpAmps are complex and frequency dependent.

EXAMPLE 1.2-2 Find the output resistance for a unity-gain buffer using SPICE and the simplified OpAmp equivalent circuit.

SOLUTION 1

Following is the SPICE solution using the **.TF** command with a source voltage:

```
**** 07/15/92 11:08:35 ******** PSpice 5.0A (Sep 1991) ******** ID# 67642 ****
*Gain, Rin, and Rout using .TF -- 15 Jul 92

****      CIRCUIT DESCRIPTION

*******************************************************************************
```

```
.OPTION NOPAGE
V1    1    0    1V
Ri    1    2    2MEG
Ro    2    3    75
E1    3    0    1    2    200K
*E is the designation of a voltage controlled voltage source.
.TF     V(3)    V1
*Finding the transfer fuction V(3)/V1.
.END

    ****     SMALL SIGNAL BIAS SOLUTION      TEMPERATURE =   27.000 DEG C

  NODE    VOLTAGE      NODE    VOLTAGE     NODE    VOLTAGE     NODE    VOLTAGE
(    1)    1.0000  (    2)    1.0000  (    3)    1.0000

     VOLTAGE SOURCE CURRENTS
     NAME            CURRENT
     V1            -2.500E-12
     TOTAL POWER DISSIPATION  2.50E-12   WATTS

   ****     SMALL-SIGNAL CHARACTERISTICS
     V(3)/V1 =   1.000E+00
     INPUT RESISTANCE AT V1 = 4.000E+11
     OUTPUT RESISTANCE AT V(3) =  0.000E+00
         JOB CONCLUDED
         TOTAL JOB TIME          1.16
```

Using the **.TF** command yields $R_{in} = 400 \times 10^9 \ \Omega$ and $R_{out} = 0 \ \Omega$. The input resistance found here exactly matches the calculations using Equation 1.2-19. The output resistance is functionally the same value as found in Example 1.2-1.

SOLUTION 2

Following is the SPICE solution using a test source at the output terminal of the equivalent model:

```
*Gain, Rin, and Rout using a test source - 15 Jul 92.
.OPTION NOPAGE
Ri    1    0    2MEG
Ro    1    2    75
E1    2    0    0    1    200K
Vt    1    0    AC    1V
.AC    DEC    10    10    10K
.PROBE
.END
```

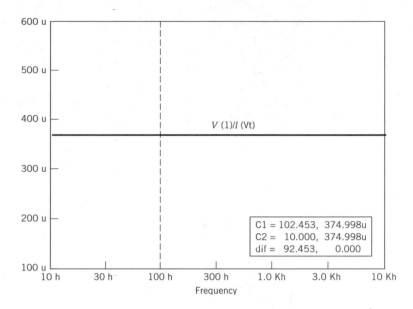

Using a test source at the output of the equivalent model yields $R_{out} = 375\ \mu\Omega$, which exactly matches the hand-calculated results of Example 1.2-1 and is, for all practical purposes, zero.

1.3 BASIC APPLICATIONS OF THE OPAMP

Although the OpAmp can be used in infinite circuit configurations, several configurations have become basic electronic building blocks. These commonly found OpAmp circuit configurations are the inverting amplifier, summing amplifier, noninverting amplifier, difference amplifier, integrator, and differentiator. All five configurations can be analyzed using the voltage and current constraints and the ideal model of the OpAmp discussed in Sections 1.1 and 1.2.

1.3.1 Inverting Amplifier

The inverting amplifier configuration shown in Figure 1.3-1 amplifies and inverts the input signal in the linear region of operation. The circuit consists of a resistor R_s in series with the voltage source v_i connected to the inverting input of the OpAmp. The noninverting input of the OpAmp is short circuited to ground (common). A resistor R_f is connected to the output and provides a negative-feedback path to the

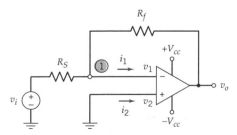

Figure 1.3-1
Inverting amplifier configuration.

inverting input terminal.[11] Because the output resistance of the OpAmp is nearly zero, the output voltage v_o will not depend on the current that might be supplied to a load resistor connected between the output and ground.

For most OpAmps, it is appropriate to assume that their characteristics are approximated closely by the ideal OpAmp model of Section 1.2. Therefore, analysis of the inverting amplifier can proceed using the voltage and current constraints of Equations 1.2-4a and 1.2-4b,

$$v_1 = v_2$$

and

$$i_1 = i_2 \approx 0.$$

Since v_2 is connected to the common, or ground, terminal,

$$v_2 = 0.$$

Node 1 is said to be a virtual ground due to the virtual short circuit between the inverting and noninverting terminals (which is grounded) as defined by the voltage constraint

$$v_1 = v_2 = 0. \tag{1.3-1}$$

The node voltage method of analysis is applied at node 1,

$$0 = \frac{v_i - v_1}{R_s} + \frac{v_o - v_1}{R_f} + i_1. \tag{1.3-2}$$

By applying Equation 1.3-1, obtained from the virtual short circuit, and the constraint on the current i_2 as defined in Equation 1.2-4b, Equation 1.3-2 is simplified to

$$0 = \frac{v_1}{R_s} + \frac{v_o}{R_f}. \tag{1.3-3}$$

Solving for the voltage gain v_o/v_i yields

$$\frac{v_o}{v_i} = -\frac{R_f}{R_s}. \tag{1.3-4}$$

Notice that the voltage gain is dependent only on the ratio of the resistors external to the OpAmp, R_f and R_s. The amplifier increases the amplitude of the input signal by this ratio. The negative sign in the voltage gain indicates an inversion in the signal.

The output voltage is also constrained by the supply voltages V_{CC} and $-V_{CC}$,

$$|v_o| < V_{CC}.$$

Using Equation 1.3-4, the maximum resistor ratio R_f/R_s for a given input voltage v_i is given by

---

[11]Further detailed discussion of feedback theory and the implication of negative feedback is found in Chapter 8. Analysis of the OpAmp circuits in this chapter will rely on standard circuit analysis techniques.

$$\frac{R_f}{R_S} < \left| \frac{V_{CC}}{v_i} \right|. \tag{1.3-5}$$

The input resistance of the inverting amplifier can be readily determined by applying the voltage constraint $v_1 = v_2 = 0$. Therefore, the resistance that the signal source v_i encounters is simply R_s due to the virtual short to ground at the inverting terminal. Then R_s must be large for a high input resistance. If R_s is large, R_f must be very large to achieve large gain, R_f/R_s. In some instances, R_f may be prohibitively high.[12] Therefore in most applications, the input resistance of the inverting amplifier is low to moderate.

EXAMPLE 1.3-1 For the circuit shown in Figure 1.3-2, find the gain and i_o. If $v_i = 2 \sin \omega_o t$ volts, what is the output? What input voltage amplitude will cause the amplifier to saturate?

SOLUTION

The output voltage v_o is independent of the load resistor R_L because of the low output resistance of the OpAmp. Therefore, the gain of the amplifier is given by

$$\frac{v_o}{v_i} = -\frac{R_f}{R_S}$$

$$= -\frac{47 \text{ k}\Omega}{10 \text{ k}\Omega} = -4.7.$$

Using KCL at node a,

$$0 - i_o + i_f + i_l.$$

The currents i_f and i_l are given by

$$i_l = -\frac{v_o}{22,000} = -\frac{4.7 \times 2 \sin \omega_o t}{22,000}$$

and

$$i_f = -\frac{v_o}{47,000} = -\frac{4.7 \times 2 \sin \omega_o t}{47,000}.$$

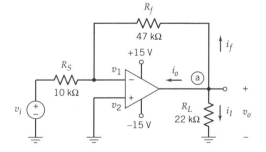

Figure 1.3-2
Inverting amplifier with load resistor R_L.

---

[12]The nonideal characteristics of OpAmps place constraints on externally connected elements. A discussion of these constraints is found in Section 1.5.

Solving for i_o yields

$$i_o = \left(\frac{1}{47,000} + \frac{1}{22,000}\right) \times 4 \times 4.7 \sin \omega_o t$$

$$i_o = 1.255 \sin \omega_o t \quad \text{mA}.$$

For an input voltage signal $v_i = 2 \sin \omega t$ volts,

$$v_o = \frac{v_o}{v_i} (2 \sin \omega t)$$

$$= -4.7(2 \sin \omega t)$$

$$= -9.4 \sin \omega t \quad \text{V}.$$

For operation in the linear region of the amplifier, the input amplitude must not exceed

$$|v_i| < \frac{+V_{CC}}{|R_f/R_s|} < \frac{15}{4.7} = 3.19 \text{ V}.$$

Input signal amplitudes greater than or equal to 3.19 will cause the amplifier to saturate.

1.3.2 Summing Amplifier

The output voltage of a summing amplifier is an inverted, amplified sum of the input voltages. A summing amplifier can theoretically have a large number of input voltages. Figure 1.3-3 shows a summing amplifier with three inputs, v_{i1}, v_{i2}, and v_{i3}.

Using the node voltage method of analysis by summing the current entering node 1,

$$0 = \frac{v_{i1} - v_1}{R_1} + \frac{v_{i2} - v_1}{R_2} + \frac{v_{i3} - v_1}{R_3} + \frac{v_o - v_1}{R_f}. \tag{1.3-6}$$

The virtual short between input terminals of the OpAmp leads to

$$v_1 = v_2 = 0.$$

Therefore, Equation 1.3-6 simplifies to

$$0 = \frac{v_{i1}}{R_1} + \frac{v_{i2}}{R_2} + \frac{v_{i3}}{R_3} + \frac{v_o}{R_f}. \tag{1.3-7}$$

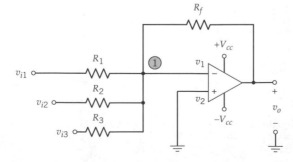

Figure 1.3-3
Summing amplifier with three input signals.

Solving for the output voltage v_o yields

$$v_o = -\left(\frac{R_f}{R_1} v_{i1} + \frac{R_f}{R_2} v_{i2} + \frac{R_f}{R_3} v_{i3}\right). \tag{1.3-8}$$

The output voltage is an inverted sum of scaled input voltages.

A particularly useful case occurs when $R_1 = R_2 = R_3 = R_s$. In this case, Equation 1.3-8 is simplified to

$$v_o = -\frac{R_f}{R_s}(v_{i1} + v_{i2} + v_{i3}). \tag{1.3-9}$$

The number of input signal voltages may be increased to meet the requirements of the application. For n input signals,

$$v_o = -\frac{R_f}{R_s}\sum_{j=i}^{n}(v_i)_j. \tag{1.3-10}$$

EXAMPLE 1.3-2 Two voltage signals

$$v_{i1} = 2\cos(\omega_o t + 25°) \text{ V}$$

and

$$v_{i2} = 1.5\cos(\omega_o t - 35°) \text{ V}$$

are added by the summing amplifier in Figure 1.3-4.

Find the output voltage v_o.

SOLUTION

Since $R_1 = R_2$, the expression for the output voltage is

$$v_o = -\frac{R_f}{R_1}(v_{i1} + v_{i2}).$$

The input voltages in this example are sinusoids of the same frequency. Therefore, the two input voltages can be combined using phasor representation,

$$\boldsymbol{V_i} = \boldsymbol{V_{i1}} + \boldsymbol{V_{i2}}.$$

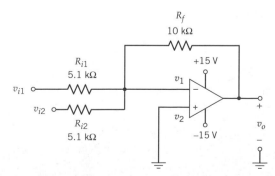

Figure 1.3-4
Summing amplifier with two input voltages.

The sum of the voltages is

$$v_i = v_{i1} + v_{i2}$$
$$= 2\cos(\omega_o t + 25°) + 1.5\cos(\omega_o t - 35°).$$

The sum of the voltages in phasor representation is given as

$$\mathbf{V}_i = \mathbf{V}_{i1} + \mathbf{V}_{i2}$$
$$= 2\angle 25° + 1.5\angle{-35°}$$
$$= (1.81 + j0.845) + (1.23 - j0.860)$$
$$= 3.04 - j0.015$$
$$= 3.04\angle{-0.285°}\ \text{V}.$$

The output voltage in phasor notation is

$$\mathbf{V}_o = -\frac{R_f}{R_1}\mathbf{V}_i$$
$$= -\frac{10\ \text{k}\Omega}{5.1\ \text{k}\Omega}(3.04\angle{-0.285°})$$
$$= -5.96\angle{-0.285°}\ \text{V}.$$

In time-domain notation, the output voltage is

$$v_o = -5.96\cos(\omega_o t - 0.285°)\ \text{V}.$$

Note that the resulting output voltage requires the use of the phase of the two input signals. The output voltage in this case is the amplifier gain, $-R_f/R_1 = -1.96$, multiplied by phasor sum of the two input voltages. This example demonstrates that proper attention to the phase and frequency of the input signals is required when designing and analyzing circuits.

1.3.3 Noninverting Amplifier

A noninverting amplifier is shown in Figure 1.3-5, where the source is represented by v_s and a series resistance R_s.

The analysis of the noninverting amplifier in Figure 1.3-5 assumes an ideal

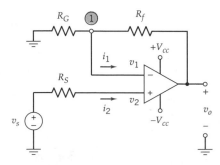

Figure 1.3-5
Noninverting amplifier configuration.

OpAmp operating within its linear region. The voltage and current constraints at the input to the OpAmp yield the voltage at node 1,

$$v_1 = v_2 = v_s,$$

since $i_1 = i_2 = 0$. Using the node voltage method of analysis, the sum of the currents flowing into node 1 is

$$0 = \frac{0 - v_1}{R_G} + \frac{v_o - v_1}{R_f}. \tag{1.3-11}$$

Solving for the output voltage v_o using the voltage constraints $v_1 = v_s$ yields

$$v_o = v_i\left(1 + \frac{R_f}{R_G}\right). \tag{1.3-12}$$

The gain of the noninverting amplifier is given by

$$\frac{v_o}{v_i} = 1 + \frac{R_f}{R_G}. \tag{1.3-13}$$

Unlike the inverting amplifier, the noninverting amplifier gain is positive. Therefore, the output and input signals are ideally in phase. The amplifier will operate in its linear region when

$$1 + \frac{R_f}{R_G} < \left|\frac{V_{CC}}{v_S}\right|. \tag{1.3-14}$$

Note that, like the inverting amplifier, the gain is a function of the external resistors R_f and R_G.

1.3.4 Difference Amplifier

The output voltage signal of a difference amplifier is proportional to the difference of the two input voltage signals. A schematic of a difference amplifier is shown in Figure 1.3-6.

By assuming an ideal OpAmp operating in the linear region, the current constraints can be used to yield the voltages at nodes 1 and 2 as a simple voltage division at the noninverting input:

$$v_1 = v_2 = v_b\left(\frac{R_D}{R_C + R_D}\right). \tag{1.3-15}$$

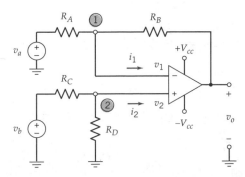

Figure 1.3-6
Difference amplifier with input voltages v_a and v_b.

The node voltage method of analysis is used to determine the output voltage v_o with respect to the input voltages v_a and v_b,

$$0 = \frac{v_a - v_1}{R_A} + \frac{v_o - v_1}{R_B}.$$ (1.3-16)

Solving for the output voltage v_o yields

$$v_o = \frac{R_B}{R_A}(v_1 - v_a) + v_1.$$ (1.3-17)

Substituting Equation 1.3-15 into 1.3-17 provides the output voltage as a function of the input voltages,

$$v_o = \frac{R_D}{R_C + R_D}\left(\frac{R_B}{R_A} + 1\right)v_b - \frac{R_B}{R_A}v_a.$$ (1.3-18)

The expression for the output voltage in Equation 1.3-18 can be simplified for the particular case where the resistor ratios are given by

$$\frac{R_A}{R_B} = \frac{R_C}{R_D}.$$ (1.3-19)

By applying the ratio of Equation 1.3-19, the output voltage in Equation 1.3-18 is reduced to a scaled difference of the input voltages,

$$v_o = \frac{R_B}{R_A}(v_b - v_a).$$ (1.3-20)

The difference amplifier is commonly used in circuits that require comparison of two signals to control a third (or output) signal. For instance, v_a could be a voltage reading representing temperature from a thermistor (a resistor that changes values with temperature) circuit and v_b a reference voltage representing a temperature setting. The output of the difference amplifier would then be the deviation of the measured temperature from the reference temperature setting.

EXAMPLE 1.3-3 The difference amplifier in Figure 1.3-7 has an input voltage $v_a = 3$ V. What values of v_b will result in operation in the linear region?

SOLUTION

The limits on the output voltage are determined by te power supply rail voltages. In this example, the supply voltages are +15 and −15 V. Therefore, the output voltage must be

$$-15 \text{ V} < v_o < +15 \text{ V}.$$

Since $R_A/R_B = R_C/R_D$, the input voltage v_b from Equation 1.3-20 can be calculated,

$$v_b = \frac{R_A}{R_B}v_o + v_a.$$

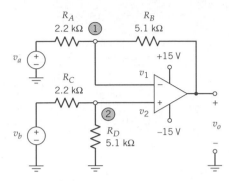

Figure 1.3-7
Difference amplifier of Example 1.3-3.

Substituting $R_A = 2.2$ kΩ and $R_B = 5.1$ kΩ into the equation for v_b yields, for the upper and lower limits of the output voltage,

$$v_o = +15 \text{ V:} \qquad v_b = 9.47 \text{ V}$$

and

$$v_o = -15 \text{ V:} \qquad v_b = -3.47 \text{ V}.$$

Then the input voltage range for v_b to ensure linear operation of the amplifier is

$$-3.47 \text{ V} < v_b < 9.47 \text{ V}.$$

The SPICE circuit file and simulation results are given below. The simplified model of the OpAmp is used. The voltage v_b is swept from -15 to 15 V. The output voltage is at node 4. Note that the voltage at node 4, $V(4)$, extends well above 15 V and below -15 V. The excursion occurs because the simplified OpAmp model assumes operation in the linear region. Therefore, care must be taken when using this model to take into account the limits on the output voltage. As shown in the PROBE output, the range of values for v_b is from -3.47 to 9.47 V, as indicated by the cursors C1 and C2 for output voltages of -15 and 15 V, respectively.

```
.OPTION NOPAGE
*
*OpAmp model
*
Ri    2    5    2MEG
Ro    3    4    75
E1    3    0    5    2    200K
*
*External resistors
*
RA    1    2    2.2K
RB    2    4    5.1K
RC    6    5    2.2K
RD    5    0    5.1K
*
*Input voltage sources
*
Va    1    0    3V
Vb    6    0    -15
*
```

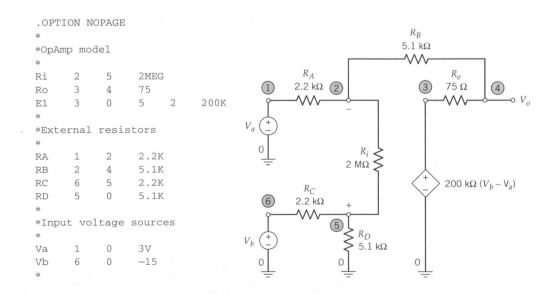

```
*Sweep the voltage input vb from −15 V to 15 V in increments of 1 V.
*
.DC     Vb     −15      15     1
.PROBE
.END
```

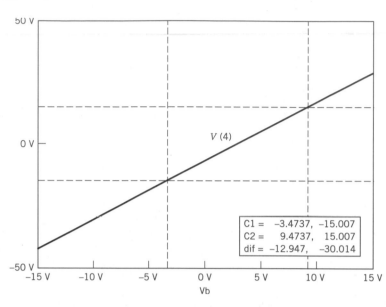

1.3.5 Integrator

The integrator is commonly used in signal generation or processing applications. The name of the circuit is accurately descriptive; the integrator performs an integration operation on the input signal. An integrator is shown in Figure 1.3-8. The circuit is similar to the inverting amplifier with the feedback resistor R_f replaced by a capacitor C.

With an ideal OpAmp operating in its linear region, the node voltage method of analysis can be applied at node 1 using the OpAmp voltage and current constraints,

$$0 = \frac{v_i - v_1}{R} + C \frac{d}{dt}(v_o - v_1). \tag{1.3-21}$$

But $v_1 = 0$ due to the virtual ground, so

$$0 = \frac{v_i}{R} + C \frac{dv_o}{dt}. \tag{1.3-22}$$

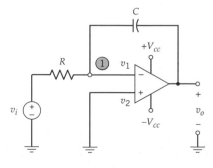

Figure 1.3-8
Integrator circuit.

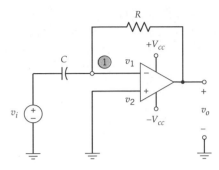

Figure 1.3-9
Differentiator circuit.

Solving for v_o yields

$$v_o = -\frac{1}{RC} \int v_i \, dt.$$

(1.3-23)

Equation 1.3-23 shows that the output voltage of an integrator circuit is a product of the reciprocal of the RC time constant and the integral of the inverted input signal.

1.3.6 Differentiator

If the capacitor and resistor positions in the integrator schematic are switched, the circuit performs a differentiation operation on the input signal. The resulting circuit is shown in Figure 1.3-9.

The analysis of the differentiator circuit is similar to the integrator. Apply the node voltage method of analysis at node 1 assuming ideal OpAmp characteristics to yield

$$0 = C \frac{d(v_i - v_1)}{dt} + \frac{v_o - v_1}{R}.$$

(1.3-24)

Using the voltage constraint $v_1 = v_2 = 0$, Equation 1.3-24 simplifies to

$$0 = C \frac{dv_i}{dt} + \frac{v_o}{R}.$$

(1.3-25)

The output voltage v_o is therefore given by

$$v_o = -RC \frac{dv_i}{d_t}.$$

(1.3-26)

Equation 1.3-26 shows that the output voltage of a differentiator circuit is a product of the RC time constant and the derivative of the inverted input signal.

1.4 DIFFERENTIAL AMPLIFIERS

A differential amplifier is any two-input amplifier that has an output proportional to the difference of the inputs. The defining equation for a differential amplifier is then

$$y_o = A(x_{i1} - x_{i2}),$$

(1.4-1)

where the output y_o and the inputs $\{x_i\}$ could be either voltages or currents. Previous discussions in this chapter have explored two differential amplifiers: the difference amplifier (shown in Figure 1.4-1) and the basic OpAmp itself. Each of these two examples has an output voltage that is proportional to the difference of two input voltages. In the case of the difference amplifier, the output expression was derived to be

$$v_o = \frac{R_B}{R_A}(v_{i2} - v_{i1}) \tag{1.4-2}$$

if the resistor values were chosen that

$$\frac{R_A}{R_B} = \frac{R_C}{R_D}. \tag{1.4-3}$$

Ideally this amplifier (or any differential amplifier) is sensitive only to the difference in the two input signals and is completely insensitive to any common component of the two signals. That is, if the difference in inputs remains constant, the output should not vary if the average value of the two inputs changes. Unfortunately, a differential amplifier rarely meets this goal, and the output has a slight dependence on the average of the input signals. The output for this type of imperfect differential amplifier is given by

$$v_o = A_{DM}(v_{i2} - v_{i1}) + A_{CM}[\tfrac{1}{2}(v_{i2} + v_{i1})], \tag{1.4-4}$$

where A_{DM} = amplification of input signal difference $v_2 - v_1$

A_{CM} = amplification of input signal average $\tfrac{1}{2}(v_2 + v_1)$

The *quality* of a differential amplifier is displayed in its ability to amplify the differential signal while suppressing the common signal. A measure of this quality is the ratio of the differential gain to the amplification of the average (or common) part of the input signals. The measure of quality is named the *common-mode rejection ratio* (CMRR) and is usually expressed in decibels (dB). The defining equation for CMRR is

$$\text{CMRR} = 20 \log_{10} \left| \frac{A_{DM}}{A_{CM}} \right|. \tag{1.4-5}$$

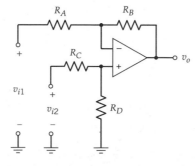

Figure 1.4-1

Difference amplifier. Shows OpAmps without power $\{\pm V_{CC}\}$ connections explicitly indicated. Omission is intentional and common practice: Power connections are to be assumed whenever not explicitly indicated. OpAmps require external power connections in order to operate correctly.

Unfortunately, usual analysis procedures do not produce an expression in the form of Equation 1.4-4: The differential-mode gain A_{DM} and the common-mode gain A_{CM} are not the usual results of analysis. A more typical result of analysis procedures is

$$v_o = A_1 v_{i1} + A_2 v_{i2}. \tag{1.4-6}$$

A conversion between the two output expressions can be obtained by realizing that

$$v_{i1} = \tfrac{1}{2}(v_{i2} + v_{i1}) - \tfrac{1}{2}(v_{i2} - v_{i1}) \tag{1.4-7a}$$

and

$$v_{i2} = \tfrac{1}{2}(v_{i2} + v_{i1}) + \tfrac{1}{2}(v_{i2} - v_{i1}). \tag{1.4-7b}$$

If Equations 1.4-7a and 1.4-7b are combined with Equation 1.4-6, the result is

$$v_o = \tfrac{1}{2}(A_2 - A_1)(v_{i2} - v_{i1}) + (A_1 + A_2)[\tfrac{1}{2}(v_{i2} + v_{i1})]. \tag{1.4-8}$$

The conclusions easily drawn from Equations 1.4-8 and 1.4-4 are

$$A_{DM} = \tfrac{1}{2}(A_2 - A_1) \quad \text{and} \quad A_{CM} = (A_1 + A_2). \tag{1.4-9}$$

A good differential amplifier has $A_1 \approx -A_2$: The differential-mode gain will be large and the common-mode gain small. The CMRR will be large for a good differential amplifier: As an example, the μA741 OpAmp has a typical CMRR of 90 dB with a guaranteed minimum CMRR of 70 dB.[13]

EXAMPLE 1.4-1 The difference amplifier of Figure 1.4-1 is constructed with an ideal OpAmp and 1% tolerance resistors of nominal values 2.2 kΩ and 5.1 kΩ. The resistors were measured and found to have the following resistance values:

$$R_A = 2.195 \text{ k}\Omega, \qquad R_C = 2.215 \text{ k}\Omega,$$
$$R_B = 5.145 \text{ k}\Omega, \qquad R_D = 5.085 \text{ k}\Omega.$$

Determine the gain of the differential amplifier and its CMRR.

SOLUTION

The design gain of this amplifier is

$$A = \frac{R_B}{R_A} = \frac{5.1 \text{ k}\Omega}{2.2 \text{ k}\Omega} = 2.318.$$

However, this value is based on the assumption of equal resistor ratios:

$$\frac{R_A}{R_B} = \frac{R_C}{R_D}.$$

---

[13]CMRR is dependent on external circuitry as well as on the fundamental properties of an OpAmp. The condition under which the μA741 measurements is made is that the output resistance of the source (and any series resistance between the source and the OpAmp) must be less than 10 kΩ.

The quality of this difference amplifier depends strongly on whether the resistor ratio described in Equation 1.4-3 is *exactly* valid. In this case, $0.42663 \neq 0.43559$. Typical resistor variation leads to the conclusion that Equation 1.4-3 is slightly in error and the more exact expression for the output voltage as a function of the inputs is more correct:

$$v_o = \frac{R_D}{R_C + R_D}\left(\frac{R_B}{R_A} + 1\right)v_{i2} - \frac{R_B}{R_A}\,v_{i1}.$$

This input–output transfer function is of the general form of Equation 1.4-6:

$$v_o = A_1 v_{i1} + A_2 v_{i2},$$

where

$$A_2 = \frac{R_D}{R_C + R_D}\left(\frac{R_B}{R_A} + 1\right) \quad \text{and} \quad A_1 = -\frac{R_B}{R_A}.$$

The common-mode and differential-mode gains can now be evaluated using the correct gain expression and Equations 1.4-9:

$$A_2 = \frac{5.085}{2.215 + 5.085}\left(\frac{5.145}{2.195} + 1\right) = 2.329$$

$$\text{and} \quad A_1 = -\frac{5.145}{2.195} = -2.344.$$

Therefore

$$A_{\text{DM}} = 2.337 \quad \text{and} \quad A_{\text{CM}} = -0.01464.$$

The CMRR is then obtained using Equation 1.4-5:

$$\text{CMRR} = 20\log_{10}\left|\frac{A_{\text{DM}}}{A_{\text{CM}}}\right| = 20\log|-159.6| = 44.06 \text{ dB}.$$

The ratio of differential-mode gain to common-mode gain is about 160: This is only a moderately good differential amplifier. A good circuit designer would notice that the resistors in this example were paired in the worst possible manner. If the physical resistors used for R_A and R_C were exchanged, the resistor ratios in each gain path would be more nearly exact: $0.4305 \approx 0.4317$. Continuing with the gain calculations for this new configuration leads to

$$A_2 = \frac{5.085}{2.195 + 5.085}\left(\frac{5.145}{2.215} + 1\right) = 2.321$$

$$\text{and} \quad A_1 = \frac{-5.145}{2.215} = -2.323$$

with

$$A_{\text{DM}} = 2.322 \quad \text{and} \quad A_{\text{CM}} = -0.00186$$

which results in

$$\text{CMRR} = 20 \log_{10} \left| \frac{A_{\text{DM}}}{A_{\text{CM}}} \right| = 20 \log |-1248.0| = 61.92 \text{ dB}.$$

Rearranging the resistors has brought an improvement in the CMRR of almost 18 dB: The ratio of gains has been improved by a factor of about 7.8. Obviously care must be taken in the choice and placement of element values to provide the optimum amplifier.

Differential amplifiers are not restricted to circuits with single OpAmps. It is possible to construct a differential amplifier without any OpAmps, while many differential amplifiers have three or more OpAmps as essential elements. Figure 1.4-2 shows the basic schematic representation of an instrumentation amplifier using two OpAmps. Instrumentation amplifiers are high-performance voltage amplifiers that are primarily used for the initial amplification of signals from a variety of types of transducers. They are available packaged as a single item in a DIP package or can be realized with discrete components. Packaged instrumentation amplifiers usually have greater control on the factors contributing to the CMRR and are often the advantageous choice for the circuit designer.

The basic topology of this particular instrumentation amplifier is that of an inverting amplifier connected in series with a summing amplifier. The extra resistors at the positive terminals of each OpAmp, R_2 and R_2', serve no obvious function if the OpAmps are considered to be ideal: Their function is to reduce the effects of input parameter variations that are nonideal properties of OpAmps.[14] The inversion of v_{i1} prior to summation allows for the output of the amplifier to be a multiple of the difference of the two inputs. This particular circuit topology also allows, with appropriate external resistor choices, for large (on the order of 100 V) input voltages. Analysis of this amplifier begins, as usual, with the assumption that each OpAmp is near ideal: The inputs are virtually shorted, the gain is infinite, the input resistance is infinite, and the output resistance is zero. Since no current flows into the inputs of either OpAmp, there is no voltage drop across the resistors, R_2 and R_2'. The OpAmp positive input terminals are therefore at ground potential: Each OpAmp circuit acts in the same manner as if its positive terminal were directly connected to ground. The zero output resistance of the first OpAmp implies that the remainder of the circuit does not affect its output, and the voltage at node a is obtained from the gain equation for an inverting amplifier:

$$v_a = -\frac{R_3}{R_1} v_{i1}. \tag{1.4-10}$$

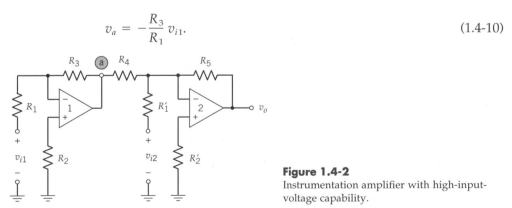

Figure 1.4-2
Instrumentation amplifier with high-input-voltage capability.

---

[14]The effects of input parameter variations on OpAmp performance is discussed in Section 1.5.

The zero output resistance of the first OpAmp circuit also implies that it acts as a perfect voltage source input to the summing amplifier of the second OpAmp. The output of the summing amplifier is then given by

$$v_o = -\frac{R_5}{R_4} v_a - \frac{R_5}{R_1'} v_{i2} \tag{1.4-11}$$

or, with the results of Equation 1.4-10, by

$$v_o = \frac{R_5 R_3}{R_4 R_1} v_{i1} - \frac{R_5}{R_1'} v_{i2}. \tag{1.4-12}$$

If proper resistor value choices are made, this instrumentation amplifier becomes a true differential amplifier. The necessary restriction on the resistor values to create a differential amplifier is

$$\frac{R_1 R_4}{R_3} = R_1'. \tag{1.4-13}$$

Additional decisions are made for good design.[15] It is often important to load the input sources equally. The input resistance at each input to the instrumentation amplifier is therefore set to the same value: $R_1' = R_1$. The restriction of Equation 1.4-13 then requires that $R_4 = R_3$. With these restrictions, the final idealized output relationship for this instrumentation amplifier is

$$v_o = \frac{R_5}{R_1} (v_{i1} - v_{i2}), \tag{1.4-14}$$

which is of the general form for a differential amplifier.

EXAMPLE 1.4-2 Determine the common-mode and differential-mode gains and the CMMR for the instrumentation amplifier of Figure 1.4-2 with the following resistor values:

$$R_1 = 50.15 \text{ k}\Omega, \qquad R_1' = 49.80 \text{ k}\Omega, \qquad R_5 = 49.85 \text{ k}\Omega$$

$$R_2 = 8.215 \text{ k}\Omega, \qquad R_2' = 8.250 \text{ k}\Omega,$$

$$R_3 = 10.05 \text{ k}\Omega, \qquad R_4 = 9.965 \text{ k}\Omega.$$

SOLUTION

In order to determine CMRR, it is necessary to use the results of Equation 1.4-12 to determine the gains. The input–output relationship is given by

$$v_o = \frac{(49.85)(10.05)}{(9.965)(50.15)} v_{i1} - \frac{49.85}{49.80} v_{i2} = 1.0025 v_{i1} - 1.0010 v_{i2}.$$

---

[15]Additional design guidelines are discussed in Section 1.5.

The differential-mode gain and the common-mode gain are calculated using Equation 1.4-9 and are found to be

$$A_{DM} = 1.00175 \quad \text{and} \quad A_{CM} = 0.001493.$$

CMRR is calculated using Equation 1.4-5 and is given by

$$\text{CMRR} = 20 \log_{10} \left| \frac{A_{DM}}{A_{CM}} \right| = 20 \log|671.07| = 56.54 \text{ dB}.$$

1.5 NONIDEAL CHARACTERISTICS OF OPAMPS

In this section the most significant limitations of the nonideal OpAmp are discussed. A fundamental understanding of these nonideal properties allows the electronics designer to choose circuit topologies and parameter values so that the performance of real, practical circuitry closely approximates the ideal case. The concept of an *ideal* OpAmp has allowed the use of simplified circuit analysis techniques to determine the performance of OpAmp circuits and concentration on the design philosophy behind the various OpAmp circuit topologies. The ideal OpAmp was defined with the following properties:

- Infinite voltage gain
- Infinite input resistance
- Zero output resistance
- Output independent of power source characteristics
- Properties independent of input frequency

A number of nonideal characteristics have been considered briefly in prior sections of this chapter:

- Output saturation
- Finite input resistance
- Finite voltage gain
- Nonzero output resistance

These characteristics will be further discussed along with the following additional nonideal characteristics:

- Input parameter variations
- Output parameter limitations
- Supply and package related parameters

In addition, the performance of an OpAmp is dependent on the frequency of the input signals. In many low-frequency applications this frequency dependence is not significant: OpAmps are commonly used in the audio frequency range and beyond without significant distortion. A discussion of frequency-dependent behavior and its close relative, slew rate, is beyond the scope of this section; a discussion of the frequency dependence of OpAmp circuit performance can be found in Section 9.8 of this text.

1.5.1 Finite Gain, Finite Input Resistance, and Nonzero Output Resistance

The properties of finite gain, finite input resistance, and nonzero output resistance were observed in previous discussions concerning using an OpAmp to create a unity-gain buffer in Section 1.2. The effect of these nonideal properties was approached through a simple equivalent model of the OpAmp, as shown in Figure 1.2-4. That approach will be continued here. While the effects due to these properties may vary slightly with OpAmp application, a demonstration of their typical effects on circuit performance will be illustrated using the basic inverting amplifier configuration.

An inverting amplifier can be constructed using an OpAmp as shown in Figure 1.5-1. While the power supply, $\pm V_{CC}$, is not shown in this figure, its presence is assumed. Analysis of this circuit begins with replacing the OpAmp with its simple equivalent circuit, as shown in Figure 1.5-2. Here the OpAmp model is enclosed within the dashed box with an input voltage $v_+ = v_2 - v_1$.

Equations expressing Kirchhoff's Current Law at the input and output terminals of the OpAmp are the first step to obtaining an expression for the voltage gain:

$$\frac{v_i - (-v_+)}{R_S} + \frac{v_+}{R_i} + \frac{v_o - (-v_+)}{R_f} = 0 \tag{1.5-1}$$

and

$$\frac{Av_+ - v_o}{R_o} - \frac{v_o}{R_l} + \frac{(-v_+) - v_o}{R_f} = 0. \tag{1.5-2}$$

Solving for v_+ in Equation 1.5-2 yields

$$v_+ = v_o\left(\frac{G_o + G_l + G_f}{AG_o - G_f}\right), \tag{1.5-3}$$

where the subscript quantities $\{G_x\}$ are conductances corresponding to the resistances with the same subscript, $\{R_x\}$. For example,

$$G_f = \frac{1}{R_f} \quad \text{and} \quad G_o = \frac{1}{R_o}.$$

Equations 1.5-3 and 1.5-1 can now be combined to obtain an expression for the voltage gain:

$$\frac{v_o}{v_i} = -\frac{G_S}{G_f + \dfrac{(G_o + G_l + G_f)(G_S + G_i + G_f)}{(AG_o - G_f)}}. \tag{1.5-4}$$

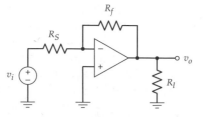

Figure 1.5-1
Simple inverting amplifier.

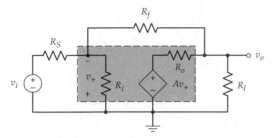

Figure 1.5-2
Inverting amplifier equivalent circuit.

The expression for voltage gain using the ideal OpAmp model in Section 1.3 was

$$\frac{v_o}{v_i} = -\frac{R_f}{R_S} = -\frac{G_S}{G_f}. \tag{1.5-5}$$

Consideration of the nonideal characteristics of an OpAmp has added complexity to the gain function and increases the magnitude of the denominator of the expression: The overall gain is reduced. Good circuit design practices imply that near-ideal performance is the desired goal. With *appropriate* external element choices, the gain function can approach the ideal. A first obvious design choice to decrease the size of the additional term in the denominator of Equation 1.5-4 is to make the parallel combination of R_l and R_f large with respect to the output impedance of the OpAmp, R_o. This choice of resistors is equivalent to making $G_o \gg (G_l + G_f)$, which allows for the simplification of the gain expression to

$$\frac{v_o}{v_i} \approx -\frac{G_S}{G_f + (1/A)(G_S + G_i + G_f)} \tag{1.5-6}$$

$$= -\frac{R_f \parallel \{A \cdot (R_S \parallel R_i \parallel R_f)\}}{R_S}.$$

If A is large, the parallel combination of R_f and $A \cdot (R_S \parallel R_i \parallel R_f)$ will be very close to R_f in value and the gain of the circuit will be near that of the idealized case.

EXAMPLE 1.5-1 Assume the circuit parameters for an inverting amplifier are

$$R_f = 47 \text{ k}\Omega, \qquad R_S = 10 \text{ k}\Omega, \qquad R_l = 22 \text{ k}\Omega,$$

and those for a nonideal OpAmp are

$$R_i = 2 \text{ M}\Omega, \qquad R_o = 75 \ \Omega, \qquad A = 200,000.$$

Determine the voltage gain of the amplifier and compare to the ideal gain.

SOLUTION
The admittances are first calculated:

$$G_f = 21.28 \ \mu S, \qquad G_i = 0.500 \ \mu S, \qquad G_S = 100.0 \ \mu S,$$
$$G_o = 13.33 \text{ mS}, \qquad G_l = 45.46 \ \mu S.$$

Equation 1.5-4 becomes

$$\frac{v_o}{v_i} = \frac{100 \ \mu S}{21.28 \ \mu S + \dfrac{(13.33 \ mS + 45.46 \ \mu S + 21.28 \ \mu S)(100 \ \mu S + 0.5 \ \mu S + 21.28 \ \mu S)}{(200{,}000 \cdot 13.33 \ mS - 21.28 \ \mu S)}}$$

$$= -\frac{100 \ \mu S}{21.28 \ \mu S + 611.96 \ pS} = -4.699865.$$

This result corresponds to a -0.0029% change in value from the ideal case (-4.7). Obviously, proper choices lead to near-ideal performance. Equation 1.5-6 could have also been used with the given set of circuit parameter values (the resistor values fit the necessary restriction): It yields similar results (-4.699865).

The input resistance of an inverting amplifier using a nonideal OpAmp can be obtained using many of the results from the gain calculations previously derived. In order to simplify the process, R_S is removed from the circuit, as shown in Figure 1.5-3, and the Thévenin input resistance of the remaining circuit is calculated: The input resistance of the total amplifier will be

$$R_{in} = R_S + R_{th} \tag{1.5-7}$$

Calculation of the Thévenin input resistance begins with determination of the two currents i_i and i_f:

$$i_i = \frac{-v_+}{R_i} \tag{1.5-8}$$

and

$$i_f = \frac{-v_+ - v_o}{R_f}. \tag{1.5-9}$$

Equation 1.5-3 can be combined with Equation 1.5-9 to eliminate v_o:

$$i_f = -v_+\left[G_f\left(1 + \frac{AG_o - G_f}{G_o + G_l + G_f} \right) \right], \tag{1.5-10}$$

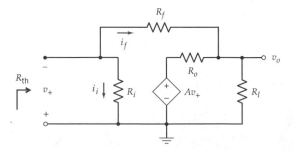

Figure 1.5-3
Inverting amplifier equivalent circuit with R_S removed.

which leads to the Thévenin resistance

$$R_{\text{th}} = \frac{-v_+}{i_i + i_f} \tag{1.5-11}$$

$$= \frac{1}{G_i + G_f\{1 + [(AG_o - G_f)/(G_o + G_l + G_f)]\}}$$

and the total input resistance

$$R_{\text{in}} = R_S + R_{\text{th}} \tag{1.5-12}$$

$$= R_S + \frac{1}{G_i + G_f\{1 + [(AG_o - G_f)/(G_o + G_l + G_f)]\}}$$

The input resistance has been increased by the quantity R_{th}. In order to make the nonideal performance mirror the ideal approximations, *appropriate* external element choices can be made. Once again, if the resistors are chosen so that $R_l \parallel R_f \gg R_o$, the expression for the input resistance reduces to

$$R_{\text{in}} \approx R_S + \frac{1}{G_i + G_f(1 + A)} = R_S + \left(R_i \parallel \frac{R_f}{1 + A}\right). \tag{1.5-13}$$

If Equation 1.5-13 is to be a reasonable approximation to the idealized expressions $R_{\text{in}} = R_S$, it is necessary that R_f be limited in magnitude. If R_{in} is not to vary by more than a few ohms from the ideal value, a reasonable choice for the maximum value of R_f is

$$R_f < 1 + A \quad \Omega. \tag{1.5-14}$$

EXAMPLE 1.5-2 Given the circuit of Example 1.5-1, determine the input resistance of the inverting amplifier and compare to the ideal case.

SOLUTION

Equation 1.5-12 yields

$$R_{\text{in}} = 10 \text{ k}\Omega + 0.2362 \text{ }\Omega$$

The appropriate choices for external resistor values have been made for Equation 1.5-13 to be valid. It yields a value for the input resistance of

$$R_{\text{in}} = 10 \text{ k}\Omega + 0.2350 \text{ }\Omega$$

The results are deviations of less than 0.0024% from the ideal value of 10 kΩ.

Calculation of the output resistance of an inverting amplifier using a nonideal OpAmp is accomplished using Thévenin techniques. The input source is set to zero, the load is removed (output resistance calculations rarely include the load), and the

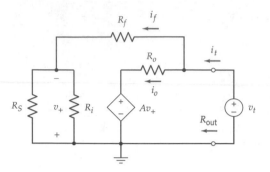

Figure 1.5-4
Inverting amplifier adjustments for R_{out} calculations.

output is driven by a voltage source as shown in Figure 1.5-4. The ratio of the driving voltage v_t to the driving current i_t gives the output resistance R_{out}.

The driving current is the sum of the currents through R_f and R_o:

$$i_t = i_o + i_f \tag{1.5-15}$$

The current through R_f is given by

$$i_f = \frac{v_t}{R_f + R'}, \tag{1.5-16}$$

where

$$R' = R_S \parallel R_i.$$

The current through R_o is given by

$$i_o = \frac{v_t - Av_+}{R_o}, \tag{1.5-17}$$

where v_+ can be obtained through a simple voltage division:

$$v_+ = -\frac{R'}{R' + R_f} v_t. \tag{1.5-18}$$

Combining Equations 1.5-17 and 1.5-18 with 1.5-16 and 1.5-14 gives the total driven current,

$$i_t = \frac{R_o + R_f + (1 + A)R'}{R_o(R_f + R')} v_t \tag{1.5-19}$$

and finally the output resistance,

$$R_{out} = \frac{v_t}{i_t} = \frac{R_o(R_f + R')}{R_o + R_f + (1 + A)R'} \tag{1.5-20}$$

$$= \frac{R_o(R_f + R_S \parallel R_i)}{R_o + R_f + (1 + A)(R_S \parallel R_i)}.$$

Reasonable assumptions (such as have been previously described) on the choice of external resistor values lead to an approximation of the output resistance expression:

$$R_{\text{out}} \approx \frac{R_o(R_f + R_S)}{(1 + A)(R_S)}. \tag{1.5-21}$$

EXAMPLE 1.5-3 Given the circuit of Example 1.5-1, determine the output resistance of the inverting amplifier and compare to the ideal case.

SOLUTION

Equation 1.5-20 yields

$$R_{\text{out}} = 0.00215 \ \Omega.$$

The external resistor simplification parameters leading to the use of Equation 1.5-21 have been met; therefore,

$$R_{\text{out}} \approx 0.00214 \ \Omega.$$

The idealized value of $R_{\text{out}} = 0$ seems justified for practical circuitry given appropriate choices of external components.

It has been shown that finite input impedance, finite gain, and nonzero output resistance do have an effect on the performance of an inverting amplifier. Still, it is possible to have near-ideal performance if appropriate choices on the external circuitry are imposed. A reasonable set of restrictions on the external components has been shown to be as follows:

- $R_l \parallel R_f \gg R_o$
- $R_f < (1 + A)\Omega$

If resistances connected to the inputs and output of an OpAmp circuit obey these general restrictions, the OpAmp circuit performance will be near ideal. Any nonideal variations can be detected with a computer simulation; good circuit design practice always includes simulation.

1.5.2 Input Parameter Variations

When an ideal OpAmp has zero output, one expects that the input will have the following properties:

- The voltage difference at the input will be zero.
- Each input current will be zero.

Real OpAmps have input voltage differences and currents that vary from the ideal. These differences are described by the quantities *input offset voltage, input bias current,* and *input offset current.*

Input Offset Voltage

The input offset voltage V_{OS} is defined as the difference in voltage between the OpAmp input terminals when the output voltage is zero.[16] This voltage difference is due to slightly different properties of the input circuitry at each of the input terminals. For typical OpAmps the offset voltage is a few millivolts or less and can often be nulled with an external three-terminal potentiometer connected between the offset null terminals of the OpAmp with the middle terminal of the potentiometer connected either to ground or one of the supply voltage terminals (the connections vary with OpAmp type and manufacturer). If the offset voltage is not nulled, it appears as an additional input voltage in series with the true inputs to the OpAmp (see Figure 1.5-5). The input offset voltage is also a function of temperature: The nulling circuitry may need to be adjusted as OpAmp temperature varies. The offset voltage and its variation with temperature place a lower limit on the magnitude of DC voltages that can act as inputs to an OpAmp without erroneous circuit operation.

Input Bias and Offset Current

The input circuitry of an OpAmp also draws a small amount of current: This nonzero input current is in variation to the ideal OpAmp assumptions. Input bias current is defined as the average of the two input currents when the output of the OpAmp is zero volts:

$$I_{\text{bias}} = \tfrac{1}{2}(I_{\text{in1}} + I_{\text{in2}}). \tag{1.5-22}$$

The magnitudes of the input bias current of an OpAmp lies in the range of a few picoamperes to tens of nanoamperes depending on the type of input circuitry. BJT input stages tend to have larger bias currents while FET input stages have smaller bias currents. In many applications the effect of a balanced bias current (both input currents equal) can be eliminated through the use of external circuit elements. Slightly different properties of the input circuitry at each of the input terminals, particularly due to random manufacturing variations, creates a more serious problem. Input offset current is defined as the difference between the input currents:

$$I_{OS} = I_{\text{in1}} - I_{\text{in2}}. \tag{1.5-23}$$

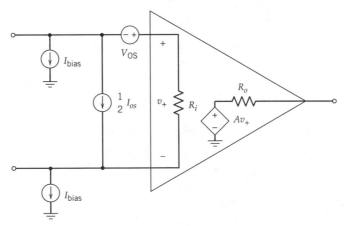

Figure 1.5-5
Equivalent circuit for OpAmp including input offset voltage and current, input bias current, finite input and nonzero output resistance, and finite voltage gain.

---

[16]An alternate (but entirely equivalent) definition is: V_{OS} is the differential input voltage that must be applied to drive the output voltage to zero.

Typical variation of input circuitry bias current is approximately 5% of the mean value. This means that the mismatch in the two currents is random (among OpAmps) and cannot be compensated by a fixed external resistor.

An equivalent circuit diagram of an OpAmp including input offset voltage and current, input bias current, finite input and nonzero output resistance, and finite voltage gain is shown in Figure 1.5-5. The effects of finite input and nonzero output resistance and finite voltage gain have been discussed at length: The effects of input bias current, input offset current, and input offset voltage can be analyzed in a similar fashion. It should be noted that manufacturer specifications on these parameters denote the maximum *magnitude* of the parameter—the parameter can be either positive or negative for a particular OpAmp.

1.5.3 Output Parameter Limitations

The maximum output voltage swing was described in Section 1.2 as slightly shy of $\pm V_{CC}$ due to device characteristics within the OpAmp. Manufacturers guarantee the minimum value of this parameter using a graph of output voltage swing as a function of supply voltage for a specified load resistance (often 2 or 10 kΩ). Variation in the load resistance will also alter the maximum output voltage swing. Manufacturers often provide a graph of this variation as well.

The maximum output current is often specified through a graph. For protection purposes, many OpAmps have a current-limiting circuit in the output stage that will limit the maximum output current to a specified value. A μA741 OpAmp (which has a current-limiting output stage) can source, or sink, approximately 25 mA.

1.5.4 Package and Supply Related Parameters

There are several other limitations on the operation of OpAmps that relate to the manufacturer's package and the power supply to which the OpAmp is connected. The limitations can be found in OpAmp data sheets. The primary limitations are as follows:

- Power dissipation
- Operating temperature range
- Supply voltage range
- Supply current
- Power supply rejection ratio

All OpAmps have a limitation on the maximum amount of power that can be dissipated safely on a continuous basis. Power dissipation is package dependent with ceramic packages having the highest rating. Metal and plastic packages have lower ratings with plastic the lowest. Typical values are in the 100–500 mW range.

OpAmps are guaranteed to operate within specifications provided the temperature of the package is within the operating temperature range specification. Commercial-grade devices have a temperature range of 0 to +70°C, the range for industrial-grade devices is −25° to +85°C, and military-grade devices operate from −55° to +125°C.

The supply voltages $\pm V_{CC}$ have *maximum* and *minimum* values for proper operation of the OpAmp. Typical maxima are in the range of $\pm$18 to $\pm$22 V, but specialized units may operate at much higher levels. Minima are typically about $\pm$5 V but may range as low as $\pm$2 V. As has been mentioned before, the output voltage swing must lie within the rails set by the supply voltage. Specialized OpAmps can operate with a one-sided supply: Typically the negative power supply terminal is grounded and the other power supply terminal is connected to $+V_{CC}$.

The supply current is defined as the current that an OpAmp draws from the power supply when the OpAmp output is zero. This is a particularly important parameter in battery-operated applications.

Variations in the supply voltage $\pm V_{CC}$ can feed through to the output, typically through offset voltage variation. The ratio of the change in offset voltage to the change in power supply voltage is defined as the power supply rejection ratio (PSRR):

$$\text{PSRR} = \frac{\Delta V_{OS}}{\Delta V_{CC}}.$$

PSRR can be expressed in microvolts per volt (μV/V) or in decibels, where

$$\text{PSRR}|_{dB} = 20 \log \frac{\Delta V_{OS}}{\Delta V_{CC}}.$$

If OpAmps are used with a high-performance voltage regulator, the error due to PSRR can essentially be eliminated in OpAmp applications.

While the list of nonideal OpAmp properties may seem large, each property contributes but a small error that, with careful choices of circuit topology and circuit element value, can be nearly eliminated. There is insufficient space in a text of this nature to investigate all effects in all possible circuits. While the demonstrations have been kept to a minimum, it is hoped that the reader has developed a "feel" for the most important effects and a sense of how to compensate for them. The good circuit designer should keep all of these second-order effects in mind and act appropriately.

1.6 CONCLUDING REMARKS

The OpAmp has been described in this chapter as a highly useful device with near-ideal terminal characteristics, summarized in Table 1.6-1. Many common OpAmp applications can be described using only these characteristics and simple circuit analysis techniques.

In well-designed OpAmp applications, the properties of the application depend most strongly on the circuit elements external to the OpAmp rather than on the

TABLE 1.6-1 IDEALIZED OPAMP CHARACTERISTICS

Property	Ideal OpAmp Value
Gain, A	∞
Input resistance, R_i	∞
Output resistance, R_O	0
Input voltage difference, $v_2 - v_1$	0 (virtual short)
Input current, i_1 or i_2	0 (virtual short)
Output voltage limits	$|v_o| \le V_{CC}$

OpAmp itself. In order to preserve this primary dependence on the external circuit elements, certain design restrictions have been presented. In general, these restrictions relate to the resistance values connected to the terminals of the OpAmp:

- Resistors connected to the output should be large with respect to the output impedance.
- Resistors connected to the input should be less than $1 + A$ ohms.

Additional design restrictions concerning frequency response will be discussed in Chapter 9.

While a variety of linear applications have been examined in this chapter, the possibilities for circuitry using OpAmps extend far beyond what has been shown here. Additional OpAmp linear applications and many nonlinear applications will be examined in later chapters. Later chapters will also investigate components used in the internal design of several OpAmp types and will shed light on nonideal characteristics and the limitations these characteristics impose on OpAmp usage.

SUMMARY DESIGN EXAMPLE

In order to investigate the low-frequency volt–ampere (V–I) relationship of a two-terminal electronic device, it is often desirable to display the V–I relationship on the screen of an oscilloscope. A typical experimental circuit diagram for such a display is shown below: It consists of the series connection of a low-frequency function generator, a resistor, and the device under test (DUT).

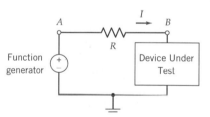

The voltage across the DUT is given by V_B and will serve as one of the inputs to the oscilloscope. The other input to the oscilloscope is the loop current. The most economical method for measuring this current is given by (current probes are quite expensive)

$$I = \frac{V_{AB}}{R}.$$

The location of the ground node in this circuit presents a problem. Safety regulations require that one terminal of the output of most function generators be at ground potential; similarly, one terminal of the input to most oscilloscopes is at ground potential. These ground connections do not pose a problem in measuring the voltage V_B, but measuring the voltage V_{AB} is difficult. The differential input mode to most oscilloscopes can solve this difficulty in measurement, but this mode cannot usually be invoked simultaneously with the required x–y display mode.

The obvious solution to the measurement problem is an external differential amplifier with inputs at nodes A and B and an output to one of the oscilloscope channels. Design such a differential amplifier.

Solution:

A list of specifications is necessary for good design. The connection of the differential amplifier across the resistor R must not significantly disturb the measurements. It should have very high input resistance: $R_{in} > 1$ MΩ matches the input resistance of most oscilloscopes. Similarly, the output of the differential amplifier should have low output resistance so that an accurate measurement can be made; $R_{out} < 100$ Ω is adequate. The amplifier differential gain should be either unity or 10 so that a mix of oscilloscope probes can be utilized. The CMRR should be high.

If a low-frequency function generator is used, OpAmps can be used for the realization of the differential amplifier. The differential amplifier of Figure 1.4-1 can easily be designed to meet all the specifications except input resistance. If the resistors are chosen to be sufficiently large to meet input resistance requirements, the ideal OpAmp approximations will fail. Therefore it is necessary connect unity-gain buffers in series with each input. The following circuit topology is therefore chosen:

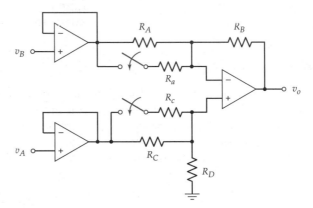

The input and output resistance of this circuit automatically meets specifications using all common, commercial OpAmps. The requirement for two distinct values of the differential gain is accomplished with a double-pole, single-throw switch. When both indicated switches (each is a pole of the actual switch) are open, the differential gain is unity; when both are closed, the differential gain is 10. After these topological design decisions, all that remains in the design is the choice of resistor values.

For unity magnitude gain in the v_B path, $R_A = R_B$ (Equation 1.3-18). For a gain magnitude of 10 in that path,

$$R_A \parallel R_a = 0.1R_B = 0.1R_A \Rightarrow R_A = 9R_a.$$

Assuming the above results, unity magnitude gain in the v_A path implies $R_C = R_D$ (Equation 1.3-19). Similarly, a gain magnitude of 10 in that path implies

$$R_C \parallel R_c = 0.1R_D = 0.1R_C \Rightarrow R_C = 9R_c.$$

While many choices will fulfill these ratios with adequate accuracy, one reasonable choice is

$$R_A = R_B = R_C = R_D = 90.9 \text{ k}\Omega,$$

$$R_a = R_c = 10.1 \text{ k}\Omega.$$

Both these resistor values are available as 0.5% resistors. Resistors with such small tolerances will ensure a high CMRR.

REFERENCES

Ghausi, M. S., *Electronic Devices and Circuits: Discrete and Integrated,* Holt, Rinehart and Winston, New York, 1985.

Gray, P. R., and Meyer, R. G., *Analysis and Design of Analog Integrated Circuits,* 2nd. Ed., John Wiley & Sons, Inc., New York, 1984.

Millman, J., *Microelectronics, Digital and Analog Circuits and Systems,* McGraw-Hill Book Company, New York, 1979.

Nilsson, J. W., *Electric Circuits,* 3rd. Ed., Addison-Wesley Publishing Co., Reading, 1989.

Soclof, S., *Analog Integrated Circuits,* Prentice-Hall, Inc., Englewood Cliffs, 1985.

Wojslaw, C. F., and Moustakas, E. A., *Operational Amplifiers,* John Wiley & Sons, Inc., New York, 1986.

PROBLEMS

1-1. A sinusoidal input is applied to the input of a linear amplifier. The input and the output voltage signals are displayed on the screen of an oscilloscope as in the figure. The oscilloscope vertical scale is set a 2 V/div. and the horizontal scale is set at 1 ms/div. It is known that the gain of the amplifier is greater than unity. Find the following:

a. Frequency of signals

b. Gain of amplifier

c. Delay time

d. Phase shift

e. Mathematical expression for input and output voltages

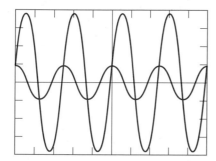

1-2. A sinusoidal input is applied to the input of a linear amplifier. The input and the output voltage signals are displayed on the screen of an oscilloscope in its *x–y* display mode as shown. The input signal is displayed on the horizontal axis and the output on the vertical scale. Both input amplifiers for the oscilloscope are adjusted so that the scales are 1 V/div. It is known that the output signal lags the input signal by a phase angle between zero and 90°. Determine:

a. Gain of amplifier

b. Phase shift

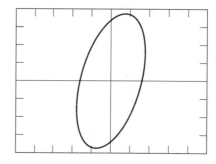

1-3. In order to measure the output resistance of an amplifier, an engineer connects a variable resistor to the amplifier output. The engineer then carefully measures the peak-to-peak (p–p) output voltage for several settings of the variable resistor. The following experimental data result:

MEASUREMENT NUMBER	RESISTOR VALUE, Ω	OUTPUT VOLTAGE (P–P), V
1	20	1.68
2	50	3.20
3	100	4.57
4	200	5.82
5	500	6.96

Determine the output resistance of the amplifier.

1-4. In order to measure the input resistance of a linear amplifier, a voltage source is connected in series with a 1-kΩ resistor and the input terminal of the amplifier. The voltage at each end of the resistor is displayed on an oscilloscope as shown, with the vertical scales set at 1 V/div. The signals have zero-voltage offset. What is the input resistance of the amplifier?

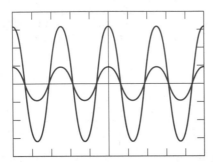

1-5. In order to measure the input resistance and voltage gain of a linear amplifier, a voltage source is connected in series with a true RMS ammeter and the input terminal of the amplifier. The input and output signals are displayed on an oscilloscope. Peak-to-peak readings of the input and output voltages are found to be 1.8 and 6.7 V_{p-p}, respectively. Each has zero-voltage offset. The ammeter reads 1 mA RMS. Determine the input resistance and voltage gain of the amplifier.

1-6. The design specifications for a simple inverting amplifier require an input resistance of 10 kΩ and a voltage gain $A = -6.2$:

a. Prepare a design using an ideal OpAmp.

b. Assume a real OpAmp with the following properties is used:

- $R_i = 2$ MΩ
- $A = 200,000$
- $R_o = 75$ Ω

What error in the gain and input resistance will result due to the nonideal properties of the OpAmp?

1-7. The design specifications for a simple inverting amplifier require an input resistance of 10 kΩ and a voltage gain $A = -6.2$:

a. Prepare a design using an ideal OpAmp.

b. Determine the maximum error in the gain and input resistance that will result due to resistor variation if:

- 5% resistors are used
- 1% resistors are used

1-8. An ideal OpAmp can be modeled using the **SPICE** statement

```
Eopamp   101  0  103  102  10MEG
```

where the nodes are related to the OpAmp shown. Explain why this one statement can be used as a very simple model of an OpAmp. Draw appropriate circuit diagrams.

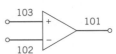

1-9. Design a noninverting amplifier with a voltage gain of approximately 12 with an input resistance of 12 kΩ. Use **SPICE** to confirm the result using a simple model of the OpAmp.

1-10. For the circuit shown determine the voltage gain v_o/v_i and the input resistance R_{in}.

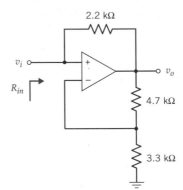

1-11. Determine the current through the load resistor R_L as a function of the input voltage v_i for the given circuit.

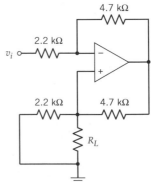

 1-12. Design an operational amplifier circuit to meet the following specifications:

a. $v_o = 3.0v_2 - 5.0v_1$

b. $R_{i1} = 15$ kΩ (input resistance seen by source v_1)

c. $R_{i2} = 25$ kΩ (input resistance seen by source v_2)

1-13. In an attempt to create a noninverting, summing amplifier, the circuit topology shown is chosen. Complete the design so that the output voltage is given by

$v_o = 5v_{i1} + 3v_{i2}$.

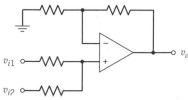

1-14. Determine the expression of the output voltage for the circuit shown for

$v_{i1} = \cos(\omega_o t)$

and

$v_{i2} = 1.5 \cos(\omega_o t + 30°)$.

Use **SPICE** to confirm the result.

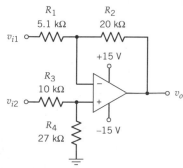

1-15. For the circuit shown, determine the range of values of a and b for the output to remain in the linear region of operation when

$v_{i1} = a \cos(\omega_o t)$

and

$v_{i2} = b \cos(\omega_o t + 45°)$

for $a = 0.3b$. Use **SPICE** to confirm the result.

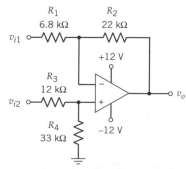

1-16. In applications where the output of an amplifier is no longer ground referenced (as in a

D'Arsonval meter), an amplifier like that shown may be used. Determine the current gain $A_i = i_i/i_o$. Assume ideal OpAmp characteristics. Use **SPICE** to confirm the result.

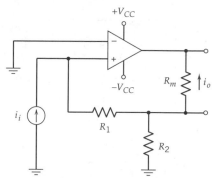

1-17. Determine the voltage gain and output resistance of the amplifier shown. Use a simplified equivalent model of an OpAmp for the analysis. Use **SPICE** to confirm the result.

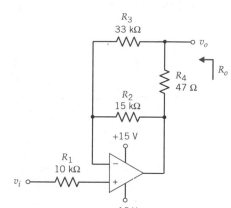

1-18. A voltage v_i to current i_o converter circuit is shown. Complete the design of the circuit by determining $R_4 - R_1$ so that

$i_o = v_i$ mA.

Let $2R_1 = 2R_2 = R_3 + R_4 = 1020$ Ω.

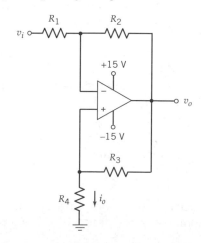

1-19. For the amplifier shown, confirm that v_o is half of v_i. Find the output resistance R_o. Use **SPICE** to confirm the result.

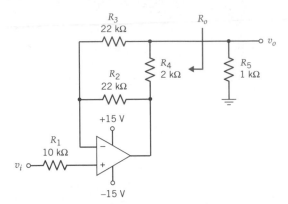

1-20. Design an OpAmp circuit to implement the equation

$v_o = v_1 - 5.1v_2.$

1-21. A difference amplifier with a gain of 10 is to be designed.

a. Prepare a design that will not have a saturated output with input voltages of -0.5 and $+0.2$ V.

b. Determine the error in gain for a real OpAmp with the following specifications:

$R_i = 2 \text{ M}\Omega, \quad A = 200{,}000, \quad R_o = 75 \ \Omega.$

1-22. A differential amplifier has a differential gain of 92 dB and a CMRR of 80 dB. Find the magnitude of the differential-mode output $v_{o(DM)}$ and the common-mode output $v_{c(CM)}$ if:

a. $v_1 = 1.6 \ \mu\text{V}$ and $v_2 = 2 \ \mu\text{V}$

b. $v_1 = -1.6 \ \mu\text{V}$ and $v_2 = 2 \ \mu\text{V}$

1-23. a. Design an OpAmp differential amplifier with a gain of 67 and a minimum input resistance of 22 kΩ for each input.

b. For an OpAmp with CMRR = 67 with a maximum common-mode signal of 0.08 V, find the differential input signal for which the differential-mode output is greater than 90 times the common-mode output.

1-24. A differential amplifier is constructed with an ideal OpAmp and resistors of nominal values 1.0 and 4.7 kΩ (i.e., $R_A \approx R_C \approx 1.0$ kΩ and $R_B \approx R_D \approx 4.7$ kΩ).

a. What is the worst-case common-mode gain using resistors with 5% tolerance?

b. What is the CMRR for that case?

c. Repeat parts a and b for 0.5% tolerance resistors.

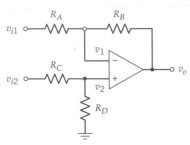

1-25. A differential amplifier shown has a differential-mode gain $A_{DM} = 5000$ and a CMRR of 56 dB. Let $v_o = v_{o(CM)} + v_{o(DM)} = 1.2$ V. Construct a graph of v_{i2} vs. v_{i1} showing the locus of all possible inputs that provide this output. Compare significant graph points to resistor ratios. Assume that $v_{i2} \geq v_{i1}$ and the outputs add and maintain $|v_{i2}| \leq 5$ V.

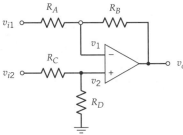

1-26. For the differential amplifier shown, the common-mode and differential-mode inputs are $v_{CM} = 150$ mV and $v_{DM} = 25$ mV, respectively. Let $A_{DM} = 10{,}000$ and $A_{CM} = 3$. Assume that $v_{o(DM)}$ and $v_{o(CM)}$ add. Find:

a. CMRR in dB

b. v_1

c. v_2

d. $v_{o(CM)}$

e. $v_{o(DM)}$

f. v_o

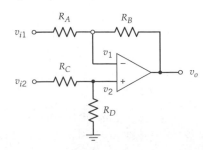

1-27. Determine the output voltage v_o as a function of the input voltage v_i for the given circuit. (*Hint:* This is an integrator of some sort.)

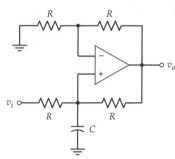

1-28. Assuming that the OpAmp for the circuit shown is ideal and the capacitor is uncharged, find:

a. $v_o(t)$ for $t > 0$

b. Time for output voltage $v_o(t)$ to reach 3V

Use **SPICE** to confirm the results.

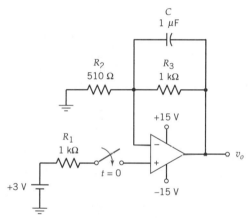

1-29. For the circuit shown, graph the output signal for the square-wave input signal as shown. Show amplitude and time scales. Simulate the circuit using **SPICE**.

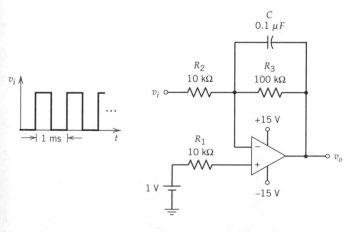

1-30. Given the attached circuit constructed with ideal OpAmps:

a. Determine the output voltage as a function of the input voltage in the given circuit.

b. At what input voltages will the output saturate?

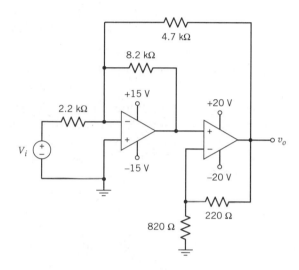

1-31. Determine the output voltage v_o as a function of the two input voltages v_{i1} and v_{i2} for the given circuit.

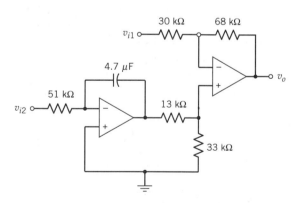

1-32. Design an inverting amplifier with voltage gain ≈ -5.1 and an input resistance ≈ 10 kΩ using a μA741 OpAmp. The μA741 has the following typical performance parameters:

Input resistance	2 MΩ
Voltage gain	200 kV/V
Output resistance	75Ω

Determine the variation from the ideal design goals due to the nonideal properties of the μA741 OpAmp.

1-33. Design an inverting amplifier with voltage gain ≈ -5.6 and an input resistance ≈ 10 kΩ using an OP27 OpAmp. The OP27 has the following typical performance parameters:

Input resistance	2 GΩ
Voltage gain	1.5 MV/V
Output resistance	70 Ω

Determine the variation from the ideal design goals due to the nonideal properties of the OP27 OpAmp.

1-34. Find an expression for the input resistance of the noninverting amplifier shown in Figure 1.3-5 if a nonideal OpAmp is used. What conditions must be met for the input resistance to be approximated by the ideal?

1-35. Find an expression for the output resistance of the noninverting amplifier shown in Figure 1.3-5 if a nonideal OpAmp is used. What conditions must be met for the output resistance to be approximated by the ideal?

1-36. In high-gain applications, it is useful to replace the unity-gain buffers of the circuit described in the Summary Design Example with an alternate input buffer. In the circuit shown, the new input buffer increases the differential-mode gain without altering the common-mode gain; an improvement in the CMRR of the total circuit results. Determine expressions for the common-mode and differential-mode gain of the new input buffer stage. *Note:* In applications where there are two outputs, the gain quantities are defined as follows:

$$A_{DM} = \frac{\{v_a - v_b\}}{\{v_A - v_B\}}, \qquad A_{CM} = \frac{\frac{1}{2}\{v_a + v_b\}}{\frac{1}{2}\{v_A + v_B\}}.$$

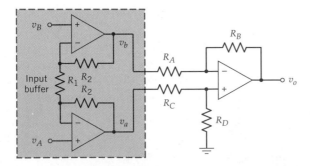

1-37. For the circuit shown in the preceding problem, the CMRR of the input buffer is 20 dB while the CMRR of the differential amplifier is 54 dB. What is the CMRR of the total amplifier (consisting of the buffer in series with the differential amplifier). Present theoretical validation of the results.

1-38. The circuit shown is a variable-gain difference amplifier. Determine an expression for the gain as a function of the fixed resistors and the variable resistor R_V.

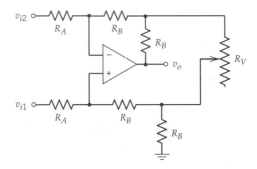

1-39. The circuit shown is a variable-gain difference amplifier. Determine an expression for the gain as a function of the fixed resistors and the variable resistor R_V.

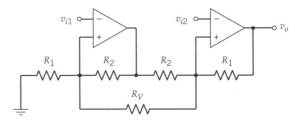

1-40. For the amplifier shown, determine the values of the load resistor R_L that will lead to gain that deviates from the ideal value by -0.01%. Assume the OpAmp has the following properties:

$$A_v = 500{,}000, \qquad R_i = 1 \text{ M}\Omega, \qquad R_o = 75 \ \Omega$$

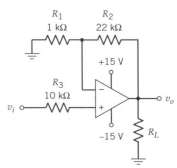

CHAPTER 2

DIODE CHARACTERISTICS AND CIRCUITS

Simple electronic circuit elements can be divided into two fundamental groups by their terminal characteristics:

- Linear devices: devices that can be described by linear algebraic equations or linear differential equations
- Non-linear devices: those devices that are described by non-linear equations

Resistors, capacitors, and inductors are examples of passive circuit elements that are basically linear.[1] Operational amplifiers, when functioning within certain operational constraints (as described in Chapter 1), are linear, active devices.

The diode is the most basic of the non-linear electronic circuit elements. It is a simple two-terminal device whose name is derived from the vacuum tube technology device with similar characteristics: a tube with two electrodes (*di* = two; *ode* = path), the anode and the cathode. Vacuum tube devices have largely been superseded in electronic applications by semiconductor junction diodes. This chapter will restrict its discussion to semiconductor diodes, diode characteristics, and simple electronic diode applications.

2.1 BASIC FUNCTIONAL REQUIREMENTS OF IDEAL DIODE

There are many applications in electronic circuitry for a one-way device: that is, a device that provides zero resistance to current flowing in one direction but infinite resistance to current flowing in the opposite direction. Protection against misapplied currents or voltages, converting alternating current (AC) into direct current (DC), demodulating Amplitude-Modulated (AM) radio signals, and limiting voltages to specified maximum or minima are but a few of the many possible applications. While a device with such ideal characteristics may be impossible to manufacture, its study is still instructive and, in many applications, ideal devices closely approximate real devices and provide insight into real-device circuit operation.

---

[1] All linear electronic devices can become nonlinear if input currents or voltages are allowed to become too large. Thermal effect, dielectric breakdown, magnetic saturation, and other physical phenomena can cause non-linearities in device transfer characteristics. Still, devices that are categorized as linear have a region of operation, usually specified by the manufacturer, in which the transfer characteristics are extremely linear.

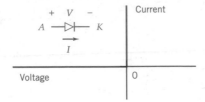

Figure 2.1-1
Volt–ampere transfer relationship for ideal diode.

An ideal diode is a true one-way electronic device. Its volt-ampere (V–I) transfer relationship is shown in Figure 2.1-1.[2] The two terminals of a such a diode retain the names first used in the vacuum tube diode:

A—the anode (Greek, *ana* = up + *hodos* = way

K—the cathode (Greek, *kata* − down + *hodos* = way)

Analytically the transfer relationship can be described as

$$I = 0 \quad \text{for } V < 0$$
$$V = 0 \quad \text{for } I \geq 0$$

(2.1-1)

It is important to notice that the definition of the sign convention of Figure 2.1-1 is extremely important. For many devices that are linear (e.g., a resistor), reversing the polarity of both the voltage and current yields the same V–I relationship (Ohm's law) as long as the passive sign convention[3] is followed. Reversing the polarity of the voltage and current (still keeping the passive sign convention) can yield, in general, a drastically different V–I relationship for a non-linear device. Thus the sign convention takes special significance.

The functional relationships of Equation 2.1-1 are a piecewise linearization of the V–I transfer relationship for an ideal diode and lead to two piecewise linear models that are often used to replace a diode for analysis purposes:

$$A \text{——————} K \qquad I \geq 0$$
$$A \cdot \qquad\qquad \cdot K \qquad V < 0$$

The two linear models are as follows:

- A *short circuit* when the applied *current* is *positive*
- An *open circuit* when the applied *voltage* is *negative*

While at first it is not always obvious which model will accurately predict the state of a diode in an electronic circuit, analysis using one model will produce results

---

[2] Ideal diodes will symbolically be shown using the symbol in Figure 2.1-1: the triangle portion of the symbol will be empty. Real diodes will be shown with a triangle that is filled, as seen in Figure 2.2-1.

[3] The passive sign convention allows consistent equations to be written to characterize electronic devices. In two-terminal devices, it simply states that, when describing the device, positive reference current enters the positive-voltage reference node and exits the negative-voltage node.

consistent with model assumption: The other model will produce a result that contradicts the assumptions upon which that model is based.

EXAMPLE 2.1-1 For the simple ideal diode circuit shown, determine the current in the diode if:

(a) $V_S = 1$ V
(b) $V_S = -1$ V

SOLUTION

(a) $V_S = 1$. Choosing the short-circuit model to replace the diode, the current I is found to be

$$I = \tfrac{1}{100} = 10 \text{ mA}.$$

If the open-circuit model is used to replace the diode, the current is found to be

$$I = \frac{1}{100 + \infty} = 0 \text{ mA}.$$

In the first case (the short circuit) the diode current (here, the same as I) is within the restrictions of the model assumptions ($I \geq 0$) and there is no contradiction to that model's assumptions. In the second case the diode voltage violates the second model assumptions ($V = 1$ violates the model assumption $V < 0$). Thus the diode appears to act as a short circuit and the true value of the current given by

$$I = 10 \text{ mA}.$$

(b) If the diode is replaced by its short-circuit model, the current is calculated to be

$$I = -\tfrac{1}{100} = -10 \text{ mA}.$$

This result violates the defining constraint for the model ($I > 0$). Therefore the open-circuit model must apply:

$$I = \frac{-1}{100 + \infty} = 0 \text{ mA}.$$

Here the voltage across the diode is -1 V, which fulfills the defining assumption for the model. Consequently the diode current is zero valued.

While the studying the action of an ideal diode often provides useful insight into the operation of many electronic circuits, real diodes have a more complex V–I relationship. The fundamental operation of a real semiconductor diode in its conducting and nonconducting regions is discussed in Section 2.2. When large reverse voltages are applied to a real diode (in what should be the far extremes of nonconducting region), the diode will enter a region of reverse conduction (the Zener region) due to one or more of several mechanisms. This sometimes useful, sometimes destructive region of reverse conduction is discussed in Section 2.7.

2.2 SEMICONDUCTOR DIODE VOLT–AMPERE RELATIONSHIP

Semiconductor diodes are formed with the creation of a *p–n* junction. This junction is a transition region between a semiconductor region that has been injected (doped) with acceptor atoms (a *p* region) and one that has been injected with donor atoms (an *n* region).[4] The *p* region becomes the anode and the *n* region becomes the cathode of a semiconductor diode. Semiconductor diodes are real diodes and have *V–I* relationships that are in many ways similar to the *V–I* relationship for an ideal diode. There are, however, distinct differences:

- In the nonconducting region (when the *p–n* junction is reverse biased) the diode current is not exactly zero: The diode exhibits a small reverse leakage current.
- The diode requires a small positive voltage to be applied before it enters the conducting region (when the *p–n* junction is forward biased). When in the conducting region the diode has a nonzero dynamic resistance.
- For large input voltages and/or currents the diode enters breakdown regions. In the forward direction, power dissipation restrictions lead to thermal destruction of the diode. In the reverse direction, the diode will enter first a Zener region of conduction and then thermal destruction.

The similarities between the semiconductor diode and the ideal diode allow the semiconductor diode to be used for the applications mentioned in Section 2.1. The differences mean circuit designers and engineers must be careful to avoid an oversimplification of the analysis of diode circuitry. Other differences allow for a few applications not possible with an ideal diode.

In the region near the origin of the *V–I* relationships for a semiconductor diode, the *V–I* curve can be described analytically by two equivalent expressions:

$$I = I_S\left[\exp\left(\frac{qV}{\eta \, kT}\right) - 1\right] = I_S\left[\exp\left(\frac{V}{\eta \, V_t}\right) - 1\right] \tag{2.2-1a}$$

or

$$V = \eta V_t \ln\left(\frac{I}{I_S} + 1\right). \tag{2.2-1b}$$

The physical constants fundamental to the diode *V–I* relationship are given by

$$q = \text{electronic charge } (160 \times 10^{-21} \text{ C})$$

$$k = \text{Boltzmann's constant } (13.8 \times 10^{-24} \text{ J/K})$$

---

[4] Discussions of the atomic semiconductor physics that lead to a *p–n* junction forming a diode are not within the scope of this electronics text. The authors suggest several texts in semiconductor physics and electronic engineering materials at the end of this chapter for those readers interested in these aspects of physical electronics.

$$V_t = \text{voltage equivalent temperature of diode}$$

$$= kT/q \approx T/11{,}600 \approx 26 \text{ mV at room temperature } (\approx 300 \text{ K})$$

It is difficult to describe the non-linear behavior of this mathematical expression for the diode V–I relationship throughout the entire range of possible values. However, discussion of the behavior in its two extremes is useful. In the *strongly reverse-biased region*, that is, when

$$V \ll -V_t$$

the exponential term of Equation 2.2-1 is much smaller than unity and the diode current is very nearly constant, with the value

$$I \approx -I_S.$$

The nonzero value of the current implies that a reverse leakage current of value I_S is present for the diode when it is in its nonconducting region. This leakage is very small, typically in the range of a few hundredths of a nanoampere to several nano-amperes.

When the diode is in its *strongly forward biased region* ($V \gg V_t$), the current experiences an exponential growth and the diode appears to have near-zero *dynamic resistance*. Dynamic resistance is defined as the *incremental* change in voltage with respect to an *incremental* change in current. For the diode the dynamic resistance is given by

$$r_d = \frac{\partial V}{\partial I} = \frac{\eta V_t}{I_S + I} = \frac{\eta V_t}{I_S} \exp\left(-\frac{V}{\eta V_t}\right) \tag{2.2-2}$$

In the strongly forward biased and strongly reversed biased regions the dynamic resistances are then, respectively

$$r_d \approx 0 \quad \text{when } V \gg V_t \text{ or equivalently } I \gg I_S$$

and

$$r_d \approx \infty \quad \text{when } V \ll -V_t \text{ or equivalently } I \approx -I_S.$$

These dynamic resistance values are, asymptotically, the values for forward and reverse resistance of an ideal diode. Thus the behavior of a real diode is similar to that of an ideal diode.

A plot of the V–I relationship for a typical diode with $I_S = 1$ nA and $\eta = 1, 2$ at a temperature of 300 K appears in Figure 2.2-1. Notice that the basic shape of this relationship is similar to that of the ideal diode with the following exceptions:

- In the reverse-biased region, the current is not exactly zero; it is instead a small leakage value I_S.

- The forward-biased region exhibits real, nonzero resistance. The vertical portion of the curve is displaced to the right. There appears to be a voltage at which the diode begins to conduct. This voltage is often called a *threshold*

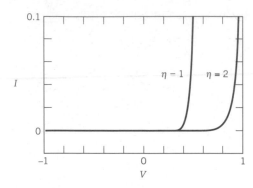

Figure 2.2-1
Typical diode *V–I* relationships.

voltage, which for silicon diodes lies in the range of 0.6–0.9 V. Threshold voltage will be more thoroughly discussed in Section 2.5.

The quantity η is an empirical scaling constant that for typical devices lies in the range

$$1 \le \eta \le 2.$$

This scaling constant, η, is dependent on the semiconductor material, the doping levels of the p and n regions, and the physical geometry of the diode. Typical germanium diodes have a scaling constant near unity while silicon diodes have $\eta \approx 2$.

It is important to note the temperature dependence of Equation 2.2-1. While the temperature dependence of V_t is evident, the temperature dependence of I_S is not explicit. It can be shown through basic principles of semiconductor physics that I_S is strongly temperature dependent. In silicon, I_S approximately doubles for every 6 K increase in temperature in the temperature range near 300 K (room temperature). That is,

$$I_S(T_2) = I_S(T_1) \times 2^{(T_2 - T_1)/6}. \tag{2.2-3}$$

Other semiconductor materials exhibit similar variation of I_S with temperature. A graphical demonstration of the change in the diode *V–I* characteristic with temperature is given in Figure 2.2-2.

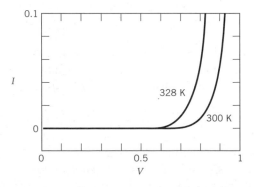

Figure 2.2-2
Diode *V–I* Characteristics at two temperatures.

EXAMPLE 2.1-1 The saturation current of a silicon diode is 2.0 nA and the empirical scaling constant $\eta = 2$. Calculate the following:

(a) At room temperature (300 K) the diode current for the following voltages: $-5, -1, 0.5,$ and 0.9 V

(b) The above quantities if the temperature is raised to 55°C

SOLUTION

(a) From Equation 2.1-1, at room temperature,

$$I = I_S\left[\exp\left(\frac{V}{\eta\,V_t}\right) - 1\right] = 2 \times 10^{-9}\left[\exp\left(\frac{V}{2\times 0.026}\right) - 1\right]$$

$$= 2 \times 10^{-9}\left[\exp\left(\frac{V}{0.052}\right) - 1\right].$$

Substituting the values of V yields

$$
\begin{array}{lll}
V = -5 & \qquad & I = 2.00 \text{ nA}\\
= -1 & & = 2.00 \text{ nA}\\
= 0.5 & & = 29.98 \text{ } \mu A\\
= 0.9 \text{ V,} & & = 65.72 \text{ mA.}
\end{array}
$$

(b) If the temperature is raised to 55°C, the equivalent temperature is 328 K. This is a change in temperature of 28 K. Thus

$$V_t = \frac{328}{11,600} = 28.28 \text{ mV}$$

and

$$I_S(328 \text{ K}) = 2^{(28/6)}\,I_S(300 \text{ K}) = 50.8 \text{ nA.}$$

Then

$$I = 50.8 \times 10^{-9}\left[\exp\left(\frac{V}{2\times 0.02828}\right) - 1\right]$$

$$= 50.8 \times 10^{-9}\left[\exp\left(\frac{V}{0.05656}\right) - 1\right].$$

Substituting the values of V yields

$$
\begin{array}{lll}
V = -5 & \qquad & I = 50.80 \text{ nA}\\
= -1 & & = 50.80 \text{ nA}\\
= 0.5 & & = 351.2 \text{ } \mu A\\
= 0.9 \text{ V,} & & = 414.4 \text{ mA.}
\end{array}
$$

Significant differences occur. Notice also that the warm diode with 0.9 V across it is now dissipating 0.38 W while the cool diode dissipates only 0.059 W in the same circumstances. Heating leads to increased power dissipation, which in turn leads to heating: This cyclic process can lead to thermal runaway and eventual destruction of the diode.

2.3 DIODE AS CIRCUIT ELEMENT

Due to the non-linear nature of the V–I relationship for a semiconductor diode, analysis techniques for circuits containing diodes are complex. This section deals with exact analytical solutions and lays the foundation for the graphical techniques of Section 2.4. Simplified piecewise linear modeling techniques that allow for the use of linear analysis techniques are discussed in Section 2.5.

One of the most simple circuits involving a diode is shown in Figure 2.3-1. This circuit consists of an independent voltage source, a resistor, and a diode in series. Simple Thévenin extensions of this circuit show that the discussion presented here can be extended to any linear circuit connected in series with a single diode.

Kirchhoff's voltage Law taken around the closed loop yields

$$V - IR - V_d = 0, \tag{2.3-1}$$

where

$$V_d = \eta V_t \ln\left(\frac{I}{I_S} + 1\right), \tag{2.3-2}$$

which leads to the non-linear equation

$$V - IR - \eta V_t \ln\left(\frac{I}{I_S} + 1\right) = 0 \tag{2.3-3}$$

that must be solved for I. Equation 2.3-2 is the diode V–I relationship expressed as the voltage across the diode as a function of the diode current.

2.3.1 Numerical Solutions

A simple closed-form solution for Equation 2.3-3 does not exist. The best technique for solution is usually a structured numerical search. Structured searches can be easily performed using mathematical software packages such as MathCAD, TKSolver, or similar programs. Another common technique uses programmable calculators with built-in numerical equation solvers (root finders).

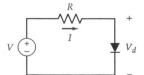

Figure 2.3-1
Simple diode circuit.

EXAMPLE 2.3-1 For the circuit of Figure 2.3-1, assume the values

$$V = 5 \text{ V}, \qquad R = 1 \text{ k}\Omega$$

and diode parameters

$$I_S = 2 \text{ nA}, \qquad \eta = 2.$$

Find, at room temperature, the diode current, the voltage across the diode, and the power dissipated by the diode.

SOLUTION

MathCAD can determine the solution through the use of a "solve block" as shown:

```
T. F. Schubert & E. M. Kim
Circuit Parameters
V: = 5        Is: = 2 · 10⁻⁹  Vt: = 0.026
R: = 1000 η: = 2
Guess Values         Vd: = 1   I: = 0.001
Solve Block
Given
```

$$V_d = \eta \cdot V_t \cdot \ln\left(\frac{I}{I_S} + 1\right)$$

$$V - I \cdot R - V_d = 0$$

$$a: = \text{Find}(I, V_d)$$

$$a = \begin{pmatrix} 4.242 \cdot 10^{-3} \\ 0.758 \end{pmatrix}$$

Similar techniques applied to a hand-held calculator yield

$$I = 4.242 \text{ mA}, \qquad V_d = 0.758 \text{ V}.$$

2.3.2 Simulation Solutions

In addition to mathematical equation solving computer programs, there exists a variety of electronic circuit simulators that perform similar solutions more efficiently. Most common among these simulators is SPICE[5] and its many derivatives. These simulation programs easily solve the type of problem described in Example 2.3-1 and with little effort can provide solutions to more complex problems. A simple extension to Example 2.3-1 with a time-varying voltage source is described below.

EXAMPLE 2.3-2 Assume the voltage source given in Figure 2.3-1 is a time-dependent voltage source:

$$V = 2.0 + 4 \sin[2\pi(80)t].$$

Determine the diode current as a function of time.

---

[5] SPICE, System Program with Integrated Circuit Emphasis, was developed at the University of California at Berkeley. The authors use one of its derivatives PSpice, developed by MicroSim Corporation. PSpice is available in a low-cost version from MicroSim to run on both IBM-PC compatible and Macintosh type computers as well as in a professional version, which will run on a variety of large and small computers.

SOLUTION (USING PSPICE)

First an input data file describing the circuit must be created. A typical data file describing the circuit of interest follows:

```
**** 02/23/92 11:42:22 ******** Evaluation PSpice (September 1991) ***********
Typical Input Information - T.F. Schubert & E.M. Kim - date
 ****      CIRCUIT DESCRIPTION
****************************************************************************

*the first line is a title - it is good practice to include authors and date
R1 1 2 1k
V1 1 0 SIN(2 4 80 0 0 0)
D1 2 0 DIODE1
.MODEL DIODE1 D (IS=2E-9 N=2)
.TRAN (0.00025 0.0125)
.PRINT TRAN V(1) I(V1)
.PROBE
.OPTIONS NOPAGE
.END

 ****      Diode MODEL PARAMETERS

            DIODE1
        IS   2.000000E-09
         N   2


 ****      INITIAL TRANSIENT SOLUTION       TEMPERATURE =  27.000 DEG C
 NODE    VOLTAGE       NODE    VOLTAGE      NODE    VOLTAGE    NODE    VOLTAGE
 (    1)    2.0000  (    2)      .6927

     VOLTAGE SOURCE CURRENTS
     NAME         CURRENT
     V1          -1.307E-03
     TOTAL POWER DISSIPATION  2.61E-03   WATTS


 ****      TRANSIENT ANALYSIS            TEMPERATURE = 27.000 DEG C
   TIME         V(1)          I(V1)
   0.000E+00    2.000E+00    -1.307E-03
   2.500E-04    2.501E+00    -1.792E-03
   5.000E-04    2.993E+00    -2.272E-03
   7.500E-04    3.470E+00    -2.739E-03
   1.000E-03    3.924E+00    -3.185E-03
   1.250E-03    4.348E+00    -3.602E-03
   1.500E-03    4.734E+00    -3.984E-03
   1.750E-03    5.077E+00    -4.323E-03
   2.000E-03    5.372E+00    -4.614E-03
   2.250E-03    5.614E+00    -4.853E-03
   2.500E-03    5.798E+00    -5.036E-03
   2.750E-03    5.923E+00    -5.159E-03
   3.000E-03    5.986E+00    -5.221E-03
   3.250E-03    5.986E+00    -5.221E-03
   3.500E-03    5.923E+00    -5.159E-03
   3.750E-03    5.798E+00    -5.036E-03
   4.000E-03    5.614E+00    -4.853E-03
   4.250E-03    5.372E+00    -4.614E-03
   4.500E-03    5.077E+00    -4.323E-03
   4.750E-03    4.734E+00    -3.983E-03
```

```
****      TRANSIENT ANALYSIS          TEMPERATURE = 27.000 DEG C
  TIME         V(1)           I(V1)
  5.000E-03    4.347E+00    -3.602E-03
  5.250E-03    3.924E+00    -3.185E-03
  5.500E-03    3.470E+00    -2.739E-03
  5.750E-03    2.993E+00    -2.272E-03
  6.000E-03    2.500E+00    -1.792E-03
  6.250E-03    2.000E+00    -1.308E-03
  6.500E-03    1.499E+00    -8.316E-04
  6.750E-03    1.007E+00    -3.832E-04
  7.000E-03    5.297E-01    -7.581E-05
  7.250E-03    7.595E-02    -1.126E-06
  7.500E-03   -3.475E-01     1.784E-09
  7.750E-03   -7.339E-01     2.001E-09
  8.000E-03   -1.077E+00     2.001E-09
  8.250E-03   -1.372E+00     2.001E-09
  8.500E-03   -1.614E+00     2.002E-09
  8.750E-03   -1.798E+00     2.002E-09
  9.000E-03   -1.923E+00     2.002E-09
  9.250E-03   -1.986E+00     2.002E-09
  9.500E-03   -1.986E+00     2.002E-09
  9.750E-03   -1.923E+00     2.002E-09
  1.000E-02   -1.798E+00     2.002E-09
  1.025E-02   -1.614E+00     2.002E-09
  1.050E-02   -1.372E+00     2.001E-09
  1.075E-02   -1.077E+00     2.001E-09
  1.100E-02   -7.337E-01     2.001E-09
  1.125E-02   -3.473M-01     1.985E-09
  1.150E-02    7.615E-02    -6.345E-08
  1.175E-02    5.300E-01    -7.182E-05
  1.200E-02    1.007E+00    -3.874E-04
  1.225E-02    1.500E+00    -8.323E-04
  1.250E-02    2.000E+00    -1.307E-03
```

With the above input file, output takes the following form:

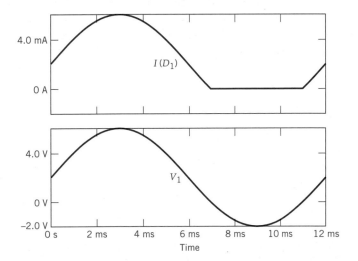

2.4 LOAD LINES

While numerical techniques are often quite useful to solve electronic circuit problems, they require the use of calculators or computers and in many circumstances provide no insight into the operation of the circuit. It is also necessary to know the parameters of the non-linear circuit elements with a reasonable degree of accuracy. Diode $V–I$ curves are often obtained through experimental procedures and only then are the parameters derived from the experimental data. A more direct approach to working with non-linear elements that has traditionally been taken is the use of graphical techniques. The process of constructing solutions through these graphical techniques often provides useful insight into the operation of the circuit.

2.4.1 Graphical Solutions to Static Circuits

Graphical solutions to the simple diode circuit of Figure 2.3-1 involves the use of the graphical representation of the diode $V–I$ relationship. The graph of the $V–I$ relationship can be obtained from Equation 2.3-2 or determined experimentally. The other elements in the circuit are then combined to create another relationship between the diode current and voltage. In the simple case of a DC Thévenin equivalent source driving a single diode, this additional relationship can be easily obtained by rewriting Equation 2.3-1 to become

$$V_d = V - IR. \tag{2.4-1}$$

Equation 2.4-1 is called a load line and gives its name to this type of graphical analysis: *load line analysis*. Notice that a plot of the load line crosses the horizontal (V_d) axis at the value of the Thévenin voltage source and the vertical (I) axis at V_d/R. The slope of the load line is the negative of the inverse of the Thévenin resistance ($-1/R$).

The two relationships involving the diode current and voltage (the diode $V–I$ relationship and the load line) can now be plotted on the same set of axes. The intersection of the two curves yields the value of the diode current and voltage for the particular values of the Thévenin source. For circuits involving static (DC) sources the intersection set of values is called the *quiescent point* or *Q point* of the circuit. Figure 2.4-1 demonstrates the load line technique applied to the circuit of

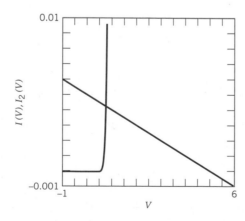

Figure 2.4-1
Load line analysis applied to Example 2.3-1

Figure 2.3-1 using the diode V–I relationship using diode and circuit parameters of Example 2.3-1. The Q point for this example appears to be

$$V_d \approx 0.76 \text{ V}, \qquad I \approx 0.0042 \text{ A},$$

which compares nicely with the analytic solution previously obtained.

If the Thévenin source voltage V is changed, a new load line must be plotted. This new load line will have the same slope $(-1/R)$ and intersects the V_d axis at the new value of the Thévenin voltage. The diode current and voltage, a new Q point, can be obtained by finding the intersection of this new load line and the diode V–I characteristic.

2.4.2 Graphical Solutions to Circuits with Time-Varying Sources

The load line technique outlined above can be expanded to provide solutions to circuits with time-varying sources. Here the Thévenin equivalent voltage source will be time varying. Graphical load line analysis is performed as above for several instants of time $\{t_i\}$. At each instant of time the Thévenin voltage $\{v_i\}$ is known, a load line can be plotted, and values for the diode voltage and/or current can be obtained. These intersection values $\{Q_i\}$ obtained from the load line analysis are then plotted against the time variables $\{t_i\}$ to obtain the time-varying output of the circuit.

EXAMPLE 2.4-1 Load lines apply to time-varying sources. For the circuit shown, assume values

$$V = 2 + 4 \sin[2\pi(80t)], \qquad R = 1 \text{ k}\Omega$$

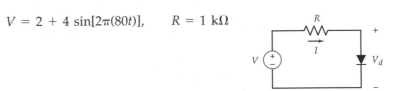

and diode parameters

$$I_S = 2 \text{ nA}, \qquad \eta = 2.$$

At room temperature, find the diode current as a function of time.

SOLUTION

The diode current is plotted as a function of the diode voltage. On this plot the time-dependent source voltage is also plotted as a function of time with the voltage axis corresponding to the diode voltage axis and the time axis parallel to the diode current axis. The time-varying voltage is sampled and a load line is created for each sample value. The Q point for each of these lines is determined and a plot of current as a function of time is created. The graphical representation of these steps is shown in Figure 2.4-2.

The same principles can be applied to other non-linear devices and circuits.

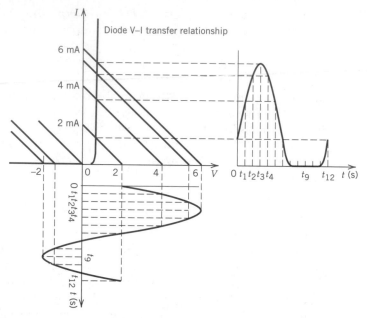

Figure 2.4-2
Load line analysis applied to time-varying sources.

2.5 SIMPLIFIED PIECEWISE LINEAR MODELS OF DIODE

The previous sections of this chapter have treated a diode as either an extremely simple device (the ideal diode) or as a complex non-linear represented by non-linear equations or curves. While each of these treatments has its place in the analysis of electronic circuits, it is often useful to find a technique in the middle ground between these two extremes. One such technique involves the regional linearization of circuit element *V–I* characteristics.

When this technique is applied to a diode, there are two basic regions:

- The region when the diode is basically conducting (forward bias)
- The region when the diode is basically nonconducting (reverse bias)

In Example 2.1-1 the ground work for the two-region linearization was laid with the boundary between the two regions being at the origin of the ideal diode *V–I* transfer relationship. With real diodes the transition between the two regions lies at a positive diode voltage, hereafter called the *threshold voltage* V_γ.

2.5.1 Forward-Bias Modeling

In the region where the diode is conducting, a good linear model of the diode is a straight line tangent to the diode *V–I* relationship at a *Q* point. The slope of this line is the dynamic resistance of the diode at the *Q* point, which was derived in Section 2.2:

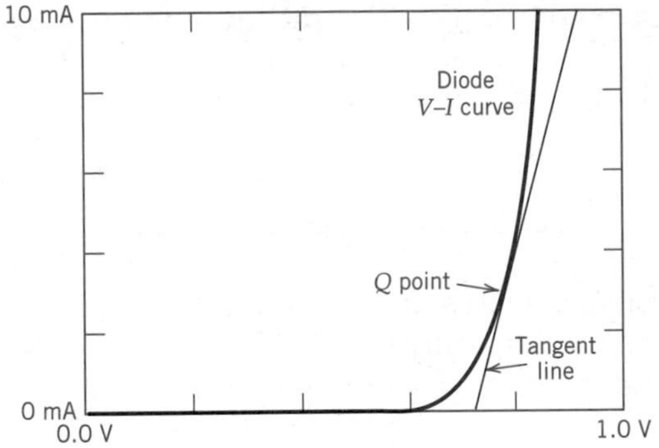

Figure 2.5-1
Modeling diode with tangent line at Q point

$$r_d = \frac{\partial V}{\partial I} = \frac{\eta V_t}{I_S + I} = \frac{\eta V_t}{I_S} \exp\left(-\frac{V}{\eta V_t}\right). \qquad (2.5\text{-}1)$$

The value of the threshold voltage can be derived by finding the intersection of this tangent line with the diode voltage axis:

$$V_\gamma = V - I r_d = V - \eta V_t\left[1 - \exp\left(-\frac{V}{\eta V_t}\right)\right] \qquad (2.5\text{-}2)$$

Figure 2.5-1 demonstrates this principle, showing a line tangent to a diode curve at the Q point of

$$I = 3 \text{ mA}, \qquad V = 0.776 \text{ V}.$$

EXAMPLE 2.5-1 Determine a linear forward-bias model for a diode with the parameters

$$I_S = 1 \text{ nA}, \qquad \eta = 2$$

near the region where the diode current is 3 mA.

SOLUTION
The diode voltage at this Q point is given by

$$V_d = \eta V_t \ln\left(\frac{I}{I_S} + 1\right),$$

$$V = 2(0.026)\ln\left(\frac{3 \text{ mA}}{1 \text{ nA}} + 1\right) = 0.776 \text{ V}.$$

The diode dynamic resistance can then be calculated as

$$r_d = \frac{\eta V_t}{I + I_S}$$

$$= \frac{2(0.026)}{1 \text{ nA} + 3 \text{ mA}} = 17.33 \text{ } \Omega.$$

And finally the threshold voltage is calculated as

$$V_\gamma = V - I r_d$$
$$= 0.776 - (3 \text{ mA})(17.33 \text{ } \Omega) = 0.724 \text{ V}.$$

The linear forward-bias model of the diode therefore has a V–I relationship

$$V = V_\gamma + Ir_d = 0.724 + 17.33\ I$$

as shown in Figure 2.5-1.

The linear model of a forward-biased diode can simply be modeled with linear circuit elements as a voltage source (typically shown as a battery) in series with a resistor. The voltage source takes the value V_γ and the resistor becomes r_d. The approximate model is shown in Figure 2.5-2. Care should be taken as to the polarity of the voltage source: Forward-biased diodes experience a voltage drop, as is shown in the figure.

While this technique gives an accurate approximation of the diode V–I characteristic about a Q point, it is not always clear what Q point should be chosen. In practice, one usually chooses a Q point as follows:

- By using an ideal diode model to get an approximate Q point for static cases
- By choosing a Q point that approximately bisects the expected range of diode currents within the application of interest

If ideal diode modeling is excessively difficult or if the range of diode currents is not easily determined, then less accurate approximations must be made. Figure 2.5-3 is a plot of the threshold voltage as a function of diode quiescent current for a typical silicon diode with reverse saturation current of 1 nA. The threshold voltage increases sharply for small diode quiescent currents and then becomes relatively constant at a value between 0.7 and 0.8 V: approximate values should lie in that range for this silicon diode. The dynamic resistance of this diode as a function of threshold voltage is given in Figure 2.5-4. This resistance, while not constant, has value of only a few ohms (here seen to be $27 > r_d > 4$). A reasonable guess at the dynamic resistance, without prior knowledge of the diode state might be $r_d \approx 15\ \Omega$ for this diode (obviously, diodes with different defining parameters will have other dynamic resistance values). Approximate models create an error in the calculation of solution but allow for the use of simple linear algebraic solution techniques.

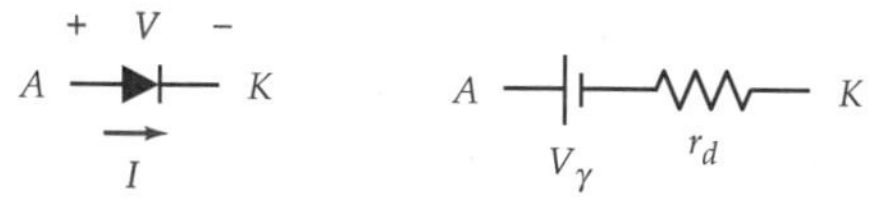

Figure 2.5-2
Linear modeling of forward-biased diode.

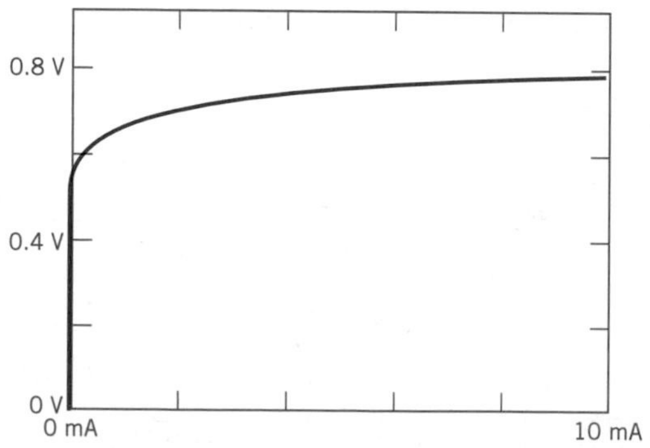

Figure 2.5-3 Threshold voltage V_γ as function of diode current.

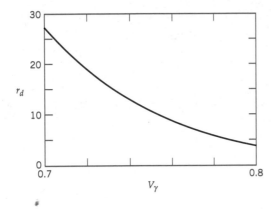

Figure 2.5-4
Diode resistance r_d as function of threshold voltage.

EXAMPLE 2.5-2 Assume the diode of Example 2.5-1 is connected in series with a 4-V source and a resistance of 820 Ω so that the diode is forward biased.

(a) Calculate the diode current and voltage with an approximate diode model.
(b) Calculate the diode current and voltage using the model derived in Example 2.5-1

SOLUTION

(a) Using an ideal model for the diode, one would expect the diode current to be somewhat less than 5 mA. Since curves for the diode threshold voltage as a function of diode current exist, choose $V_\gamma = 0.74$ V. Again, since these curves exist, Figure 2.5-4 implies a diode resistance $r_d = 13\ \Omega$ for that threshold voltage.

The diode V–I relationship can now be approximted by

$$V = V_\gamma + r_d I$$
$$= 0.74 + 13I.$$

The load line derived from the other circuit elements is given by

$$V = V_S - RI$$
$$= 4 - 820I.$$

Simple linear algebraic techniques applied to two equations with two unknowns lead to a solution of

$$V = 0.791\ \text{V}, \quad I = 3.91\ \text{mA}.$$

If the diode curves for threshold voltage and resistance did not exist, other approximate values could be chosen. For example, $V_\gamma = 0.7$ and $r_d = 15\ \Omega$ lead to

$$V = 0.759\ \text{V}, \quad I = 3.95\ \text{mA}.$$

Figure 2.5-5
Simplified forward-bias diode model.

(b) With the model derived in Example 2.5-1, the diode V–I relationship is given by

$$V = 0.724 + 17.33I.$$

The load line is the same as in part (a). Similar linear algebraic techniques give the solution

$$V = 0.792 \text{ V}, \qquad I = 3.91 \text{ mA}.$$

Notice that all the approximate models give solutions that are within $\approx 1\%$ in the diode current and $\approx 5\%$ in the diode voltage. Numerical solution of this problem as outlined in Example 2.3-1 (using the theoretical non-linear diode V–I relationship) give the exact solution to be

$$V = 0.789 \text{ V}, \qquad I = 3.92 \text{ mA}.$$

The models have all given results within $\approx 0.8\%$ in current and $\approx 4\%$ in voltage of the exact theoretical values.

When accuracy of the order seen in Example 2.5-2 is not necessary, a diode can be represented as simply a voltage drop of approximately V_γ. This model shown in Figure 2.5-5 assumes that the small series resistance is negligible with respect to the circuit Thévenin resistance as seen by the diode and forms an intermediate linear model between the ideal diode model and the two-element linear model. Use of this model leads to inaccuracies much larger than seen in previous linear models. It does, however, simplify circuit analysis greatly.

EXAMPLE 2.5-3 Assume the diode of Example 2.5-1 is connected in series with a 4-V source and a resistance of 820 Ω so that the diode is forward biased.

Calculate the diode current and voltage using a simplified forward-bias diode model.

SOLUTION

The diode voltage for this simple model is just a voltage source V_γ. For this simple model choose the approximate value

$$V_\gamma = 0.7.$$

Kirchhoff's voltage law applied to the loop yields

$$V_\gamma = V_S - RI.$$

Thus

$$0.7 = 4 - 820I$$

and

$$I = 4.02 \text{ mA}.$$

This approximate solution has an error of approximately 2.7%. The diode voltage (which was just guessed at) is in error by approximately 11.3%.

2.5.2 Reverse-Bias Modeling

In the region where the diode is basically not conducting, there are several possible linear models from which to choose. Each of these is based on the principle that the diode has a small leakage current that is fairly constant in the reverse-bias region; that is, when the diode voltage is between a few negative multiples of V_t and the Zener breakdown voltage,[6] the diode current is constant at $-I_S$. The two common models are a current source of value $-I_S$ or a large resistor. These models are shown in Figure 2.5-6.

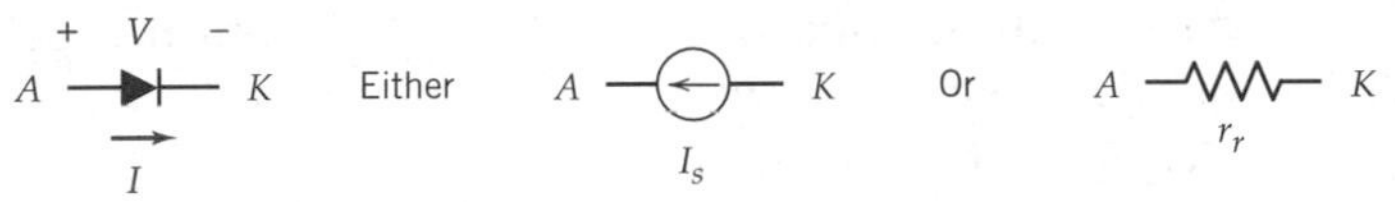

Figure 2.5-6
Linear reverse-bias diode models.

The value of the reverse resistance r_r for the second model can be approximated by one of two techniques: (1) using Equation 2.5-1 to determine the dynamic resistance about some Q point or (2) assuming that the diode achieves its true reverse saturation current at the Zener breakdown voltage. While method 1 allows for an exact dynamic resistance a some point, it is often difficult to choose the proper Q point for a particular application. Method 2 is easier to calculate but is less accurate at any point and underestimates the dynamic resistance for large reverse voltages.

EXAMPLE 2.5-4 Assume the diode of Example 2.5-1 (with Zener breakdown occurring at a voltage of -25 V) is connected in series with a 4V source and a resistance of 820 Ω so that the diode is *reverse biased*.
Calculate the diode current and voltage using:

(a) Current source model for a reverse-biased diode

(b) Resistor model for a reverse-biased diode

SOLUTION

(a) If the diode is replaced by a 1 nA current source, then all circuit elements carry 1 nA of current and the diode current is -1 nA. The voltage across the resistor is given by

$$V_r = (1 \text{ nA})(820 \text{ } \Omega) = 820 \text{ nV}.$$

---

[6] When a large negative voltage is applied to a semiconductor diode (i.e., a voltage that exceeds some reverse threshold voltage called the Zener voltage), the diode enters a region of reverse conduction. The Zener conduction region of semiconductor diodes is discussed thoroughly in Section 2.7.

Kirchhoff's voltage law applied to the loop gives the resulting diode voltage

$$V_d = V_r - 4 = -3.99999918 \approx -4.00 \text{ V}.$$

(b) The reverse resistance can be approximated as

$$r_r \approx \frac{|-25 \text{ V}|}{1 \text{ nA}} = 25 \text{ G}\Omega.$$

The diode voltage and current are given as

$$V_d = \frac{25 \text{ G}\Omega}{25 \text{ G}\Omega + 820 \text{ }\Omega} (-4 \text{ V}) = -3.999999869 \approx -4 \text{ V},$$

$$I_d = \frac{-4 \text{ V}}{25 \text{ G}\Omega + 820 \text{ }\Omega} = 160 \text{ pA}.$$

Comments: Each solution yields diode voltage solutions that are extremely close—essentially all the source voltage appears across the reverse biased diode. The current values vary by more than a factor of 6, but it must be remembered that these are extremely small values that are difficult to verify experimentally. Qualitatively, the two solutions are the same.

Numerical solutions of the type outlined in Example 2.3-1 suffer from lack of precision capability and, due to the almost zero slope of the *V–I* relationship, often cannot converge to a solution. In this particular problem MathCAD is unable to effectively search for solutions if the source voltage is more negative than about −1 V.

2.6 DIODE APPLICATIONS

Typical applications of diodes are considered in this section. The diode circuits studied are as follows:

- Limiter or clipping circuit
- Full- and half-wave rectifiers
- Peak detector
- Clamping or DC restoring circuit
- Voltage multiplier
- Diode logic gates
- Superdiode

All of the circuits in this section perform some form of wave-shaping operation on the input signal to yield a desired output. The clipping circuit "truncates" the input to some desired value beyond which the signal is not to exceed. Full- and half-wave rectifiers pass only the signals of the desired polarity (positive or negative amplitude) and are commonly used in DC power supply designs. The peak detector follows only the maximum amplitudes of an incoming signal and is commonly used in AM radio receivers in communications applications. Clamping circuits perform a level shifting operation on the input waveform and are used to measure the duty cycle of a pulse waveform. Clamping circuits are commonly used to detect information carried on pulse-width-modulated signals (i.e., the information of interest is represented by increasing or decreasing the pulse width of a pulse waveform) by retrieving the DC component of the modulated signal. Voltage multipliers perform an integer multiplication on the input signal to yield a higher output voltage. Diode logic gates are simple circuits for performing Boolean operations. The superdiode is a combination of an OpAmp and diode that eliminates the undesirable diode threshold voltage and dynamic resistance characteristics. Diodes in a circuit can, in most instances, be replaced by superdiodes to design precision circuits.

2.6.1 Limiter or Clipping Circuit

Diodes are often used in wave-shaping applications. In particular, when used with a DC voltage in series with the diode, the output signal can be limited to the reference voltage level of the DC voltage source. Examples of clipping circuits are shown in Figure 2.6-1.

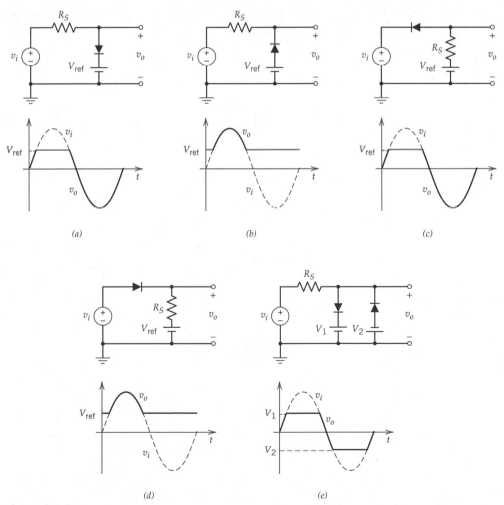

Figure 2.6-1
Diode clipping circuits.

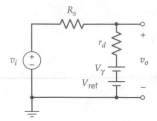

Figure 2.6-2
Simplified equivalent circuit of clipping circuit of Figure 2.6-1a for $v_i > V_\gamma + V_{ref}$.

The simplified forward-bias diode model of Figure 2.5-5 can be used to analyze clipping circuits.

The circuit of Figure 2.6-1a will be used as an example of this analysis. When the input voltage $v_i \leq V_\gamma + V_{ref}$, the diode is reverse biased (or off).[7] Therefore, the diode can be thought of as an open circuit. The output voltage in this case follows the input voltage,

$$v_o = v_i.$$

When the voltage $v_i > V_\gamma + V_{ref}$, the diode is forward biased (or on). Using the piecewise linear model of the forward-biased diode, a simplified equivalent circuit of the clipping circuit of Figure 2.6-1a is developed in Figure 2.6-2.

The output voltage v_o of the clipping circuit when the diode is forward biased is found by analyzing the circuit in Figure 2.6-2 using superposition and voltage division,

$$v_o = \frac{r_d}{R_s + r_d} v_i + \frac{R_s}{R_s + r_d} (V_\gamma + V_{ref}). \qquad (2.6\text{-}1)$$

If $r_d \ll R_s$, then the output voltage is held at a constant value:

$$v_o = V_\gamma + V_{ref}. \qquad (2.6\text{-}2)$$

The input–output voltage relationships for the five diode clipper circuits are given in Table 2.6-1.

EXAMPLE 2.6-1 For the clipping circuit shown, find the output waveform v_o for the input voltage.

$$v_i = 5 \sin \omega_o t.$$

---

[7] The diode resistance r_r under reverse-bias conditions is assumed to be much larger than the series resistance R_S in the derivation ($r_r \gg R_S$).

TABLE 2.6-1 INPUT–OUTPUT VOLTAGE RELATIONSHIPS FOR DIODE CLIPPINGS CIRCUITS

Clipping Circuits of Figure 2.6-1	Output Voltage v_o		Simplified Output Voltage v_o	
(a)	$v_o = \dfrac{r_d}{R_s + r_d} v_i + \dfrac{R_s}{R_s + r_d}(V_\gamma + V_{ref})$	$v_i > V_\gamma + V_{ref}$	$v_o \approx V_\gamma + V_{ref}$	$v_i > V_\gamma + V_{ref}$
	$v_o = v_i$	$v_i \leq V_\gamma + V_{ref}$	$v_o = v_i$	$v_i \leq V_\gamma + V_{ref}$
(b)	$v_o = \dfrac{r_d}{R_s + r_d} v_i + \dfrac{R_s}{R_s + r_d}(V_{ref} - V_\gamma)$	$v_i < V_{ref} - V_\gamma$	$v_o \approx V_{ref} - V_\gamma$	$v_i < V_{ref} - V_\gamma$
	$v_o = v_i$	$v_i \geq V_{ref} - V_\gamma$	$v_o = v_i$	$v_i \geq V_{ref} - V_\gamma$
(c)	$v_o = V_{ref}$	$v_i + V_\gamma > V_{ref}$	$v_o = V_{ref}$	$v_i + V_\gamma > V_{ref}$
	$v_o = \dfrac{r_d}{R_s + r_d} V_{ref} + \dfrac{R_s}{R_s + r_d}(v_i + V_\gamma)$	$v_i + V_\gamma \leq V_{ref}$	$v_o \approx v_i + V_\gamma$	$v_i + V_\gamma \leq V_{ref}$
(d)	$v_o = V_{ref}$	$v_i - V_\gamma < V_{ref}$	$v_o = V_{ref}$	$v_i - V_\gamma < V_{ref}$
	$v_o = \dfrac{r_d}{R_s + r_d} V_{ref} + \dfrac{R_s}{R_s + r_d}(v_i - V_\gamma)$	$v_i - V_\gamma \geq V_{ref}$	$v_o \approx v_i - V_\gamma$	$v_i - V_\gamma \geq V_{ref}$
(e)	$v_o = \dfrac{r_d}{R_s + r_d} v_i + \dfrac{R_s}{R_s + r_d}(V_1 + V_\gamma)$	$v_i > V_\gamma + V_1$	$v_o \approx V_\gamma + V_1$	$v_i > V_\gamma + V_1$
	$v_o = \dfrac{r_d}{R_s + r_d} v_i - \dfrac{R_s}{R_s + r_d}(V_2 + V_\gamma)$	$v_i < -V_2 - V_\gamma$	$v_o \approx -V_2 - V_\gamma$	$v_i > -V_2 + V_\gamma$
	$v_o = v_i$ $-V_\gamma - V_2 \leq v_i \leq V_\gamma + V_1$		$v_o = v_i$ $V_\gamma - V_2 \leq v_i \leq V_\gamma + V_1$	

The diode has the following characteristics:

$$r_d = 15 \ \Omega, \qquad V_\gamma = 0.7 \ \text{V}, \qquad r_r \to \infty.$$

SOLUTION 1:

Since $R_s \gg r_d$, the simplified voltage result from Table 2.6-1 can be constructed. That is,

$$v_o = V_\gamma + V_{\text{ref}}, \qquad v_i > V_\gamma + V_{\text{ref}},$$

$$v_o = v_i, \qquad v_i \leq V_\gamma + V_{\text{ref}}.$$

So, when $v_i > 2.7$ V, $v_o = 2.7$ V. When $v_i < 2.7$ V, $v_o = 5 \sin \omega_o t$.

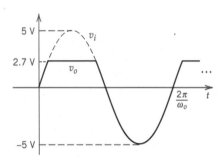

Figure 2.6-3
Output waveform for Example 2.6-1.

SOLUTION 2:

Transfer Function Analysis

A solution can be constructed from a transfer function analysis of the circuit as shown below. A transfer function defines the input–output relationship of a circuit. In this case, the transfer function is described in Table 2.6-1.

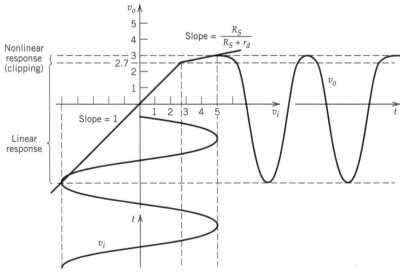

Figure 2.6-4
Transfer function solution for Example 2.6-1.

SOLUTION 3:

SPICE Solution

In order to use the diode parameters provided, an ideal diode in series with a diode dynamic resistance r_d and diode voltage V_d is constructed. In SPICE, an ideal diode is modeled with $n = 0.01$ with a very small reverse saturation current $I_s = 100$ pA. So, the model statement used for an ideal diode is

```
.model ideal_diode D(Is = 100pA n = 0.01)
```

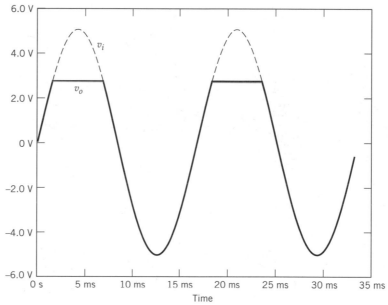

Figure 2.6-5
SPICE solution for Example 2.6-1

2.6.2 Half-Wave Rectifiers

One of the most common diode applications is the conversion of power from AC to DC for use as power supplies. Today's power supplies involve sophisticated design principles that will be detailed in Chapter 14. However, the basic principles of converting from AC to DC power can be explored in this section.

Figure 2.6-6 shows a half-wave rectifier circuit. The circuit is so named because it only allows current from the positive half cycle of the input to flow through the load resistor R. Figure 2.6-6 is identical to the clipping circuit of Figure 2.6-1d with zero reference voltage.

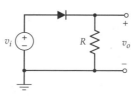
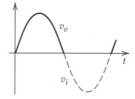

Figure 2.6-6
Half-wave rectifier with output voltage waveform.

If v_i is a sinusoidal voltage with peak voltage V_m and radian frequency ω,

$$v_i(t) = V_m \sin \omega t,$$

the average voltage V_{dc} across the load R is given by

$$V_{dc} = \frac{1}{T} \int_0^T v_o(t) \, dt, \tag{2.6-3}$$

where T is the period of the sinusoid.

Since the diode is OFF in the interval $\frac{1}{2}T \le t \le T$, output voltage is[8]

$$v_o(t) = \begin{cases} V_m \sin \omega t - V_\gamma, & 0 \le t < \frac{1}{2}T, \\ 0, & \frac{1}{2}T \le t \le T. \end{cases} \tag{2.6-3a}$$

Substituting Equation 2.6-3a into 2.6-3 to solve for V_{dc} yields

$$\begin{aligned}
V_{dc} &= \frac{1}{T} \int_0^{T/2} (V_m \sin \omega t - V_\gamma) \, dt \\
&= \frac{V_m}{\omega T} (-\cos \omega t) \Big|_0^{T/2} - \frac{V_\gamma}{T}(t) \Big|_0^{T/2} \\
&= \frac{-V_m}{\omega T} \left(\cos \frac{\omega T}{2} - 1 \right) - \frac{1}{2} V_\gamma.
\end{aligned} \tag{2.6-4}$$

But $\omega = 2\pi/T$, so Equation 2.6-4 simplifies to

$$V_{dc} = \frac{V_m}{\pi} - \frac{V_\gamma}{2}. \tag{2.6-5}$$

Recall that the "effective," or RMS, voltage quantifes the amount of energy delivered to a resistor in T seconds. The use of RMS comes from the desire to compare the ability of a sinusoid to deliver energy to a resistor with the ability of a DC source. The RMS value of any period waveform $v_o(t)$ is defined as

$$V_{rms} = \left[\frac{1}{T} \int_0^T v_o^2(t) \, dt \right]^{1/2}. \tag{2.6-6}$$

The output RMS voltage for the half-wave rectifier with an output waveform defined by Equation 2.6-3a is

$$\begin{aligned}
V_{rms} &= \left[\frac{1}{T} \int_0^{T/2} (V_m \sin \omega t - V_\gamma)^2 \, dt \right]^{1/2} \\
&= \left[\frac{1}{T} \int_0^{T/2} (V_m^2 \sin^2 \omega t - 2V_\gamma V_m \sin \omega t + V_\gamma^2) \, dt \right]^{1/2} \\
&= \left(\frac{V_m^2}{4} - \frac{2V_\gamma V_m}{\pi} + \frac{V_\gamma^2}{2} \right)^{1/2}.
\end{aligned} \tag{2.6-7}$$

---

[8] In reality, the interval over which the diode is on is slightly less than $\frac{1}{2}T \le t \le T$ due to the diode threshold voltage V_γ. The analysis assumes that $V_m \gg V_\gamma$. If V_m is very small, the analysis becomes more complex.

If $V_\gamma \ll V_m$, then Equation 2.6-7 reduces to

$$
\begin{aligned}
V_{\text{ms}} &= \left[\frac{1}{T} \int_0^{T/2} (V_m \sin \omega t)^2 \, dt \right]^{1/2} \\
&= \frac{1}{\sqrt{2}} \frac{V_m}{\sqrt{2}} = \frac{V_m}{2}.
\end{aligned}
\tag{2.6-8}
$$

The efficiency of rectification is defined as

$$
\eta = \frac{P_{\text{dc}}}{P_{\text{ac}}},
\tag{2.6-9}
$$

where P_{ac} and P_{dc} are AC and DC powers, respectively. For the half-wave rectifier in Figure 2.6-6 the efficiency is

$$
\eta = \frac{P_{\text{dc}}}{P_{\text{ac}}} = \frac{(V_m/\pi)^2/R}{(V_m/2)^2/R} = \frac{4}{\pi^2} \Rightarrow 40.6\%.
\tag{2.6-10}
$$

The result of Equation 2.6-10 is for an ideal half-wave rectifier and represents the maximum efficiency attainable. In real systems, the efficiency will be lower due to power losses in the resistor and diode.

In order to produce a DC voltage from a half-wave rectifier, a large capacitor is placed in parallel to the load resistor. The capacitor must be large enough so that the RC time constant of the capacitor and load resistor is large compared to the period of the output waveform. This has the effect of "smoothing" the output waveform. Clearly, an efficient filter is required to eliminate any ripple in the output waveform.

In many rectifier applications, it is desirable to transformer couple the input voltage source to the rectifier circuit. This method is commonly used in the design of power supplies where there is a requirement to "step down" the AC input voltage to a lower DC voltage. For example, a 15V peak half-wave rectified voltage can be derived from a 120 V AC (household power is defined in RMS volts) source through the use of a transformer. Transformers also provide isolation of the circuit from the household power line, providing protection from the possibility of shock from those lines.

The turns ratio N_p/N_s (primary winding over the secondary windings) determines the stepdown ratio. For example, if the voltage input at the primary is 120 V AC (RMS), which is approximately 170 V peak, a transformer turns ratio of 11:1 (actually 11.3:1) is required to yield a 15V peak half-wave-rectified signal. If the coefficient of coupling is nearly 1.0, implying no loss occurs in the transformer, the inductance ratio of the primary coil, L_p, and the secondary coil, L_s, is given by

$$
\frac{L_p}{L_s} = \left(\frac{N_p}{N_s}\right)^2.
\tag{2.6-11}
$$

EXAMPLE 2.6-2 For the circuit in Figure 2.6-7, determine the inductance of the secondary coil for a transformer turns ration of 11:1. What is the peak output voltage? Assume that the coefficient of coupling is $k = 1$ and that the diode threshold voltage V_γ of the 1N4148 is 0.76 V.

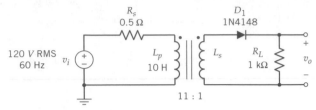

Figure 2.6-7
Transformer-coupled input to half-wave rectifier.

SOLUTION

Apply Equation 2.6-11 to find the secondary inductance:

$$L_s = \frac{L_p N_s^2}{N_p^2} = \frac{(10)(1^2)}{11^2} = 82.6 \text{ mH.}$$

The peak output voltage is

$$v_o = \frac{(v_{I \text{ peak}})}{N_p/N_s} - V_d = \frac{120\sqrt{2}}{11} - 0.76 = 14.64 \text{ V peak.}$$

In order to simulate the circuit using SPICE, several items must be added to the circuit. The SPICE circuit is shown in Figure 2.6-8.

Observe that in the original circuit (Figure 2.6-7), the output did not reference to a common (ground) point. A large isolation resistor at the secondary facilitates the reference to ground at the secondary. The model statement for the 1N4148 is

```
.model D1N4148 D(Is=0.1pA Rs=16 CJO=2p Tt=12n Bv=100 Ibv=0.1p)
```

Many of the SPICE model parameters are useful for frequency analysis that is used in the latter chapters. However, two parameters are of interest at this time:

Is = reverse saturation current

Bv = reverse breakdown voltage

The reverse breakdown voltage of the diode must not be exceeded. If the breakdown voltage is exceeded, the diode may suffer catastrophic failure.

The SPICE circuit model and the output voltage waveform are shown in Figure 2.6-9.

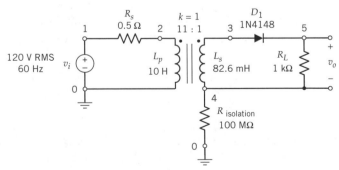

Figure 2.6-8
SPICE circuit topology.

```
Input transformer coupled half-wave rectifier
vi       1        0        SIN(0 169V 60Hz)
Rs       1        2        0.5
*transformer
Lp       2        0        10
Ls       3        4        82.6m
*the coupling coefficient of the transformer:
Kxfrmr   Lp       Ls       0.99
Riso     4        0        100M
D1       3        5        D1N4148
RL       5        4        1K
.model D1N4148 D(IS=0.1pA Rs=16 CJO=2p Tt=12n Bv=100 Ibv=0.1p)
.TRAN 1us 50ms 0 0.05m
.PROBE
.END
```

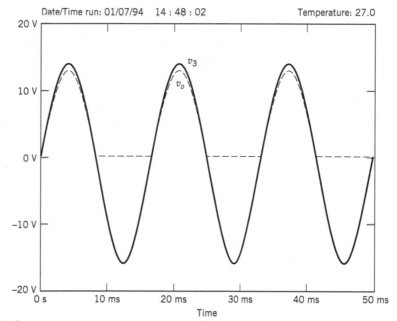

Figure 2.6-9
SPICE solution for Figure 2.6-6.

If the transformer is assumed to have a coupling coefficient of nearly 1.0, then it may be replaced by one of the equivalent circuits of the ideal transformer shown in Figure 2.6-10.

2.6.3 Full-Wave Rectifiers

Removing the ripple from the output of a half-wave rectifier may require a very large capacitance. In many instances, the capacitor required to reduce the ripple on the half-wave rectified output voltage to the desired design specification may be prohibitively large.

A full-wave rectifier circuit can be used as a more efficient way to reduce ripple on the output voltage. A center-tapped input transformer-coupled full-wave rectifier is shown in Figure 2.6-11. Each half of the transformer with the associated diode acts

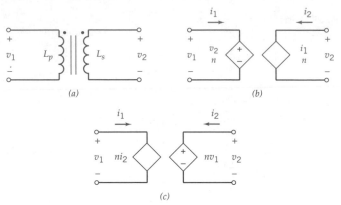

Figure 2.6-10
(*a*) ideal transformer; (*b*, *c*) equivalent circuits.

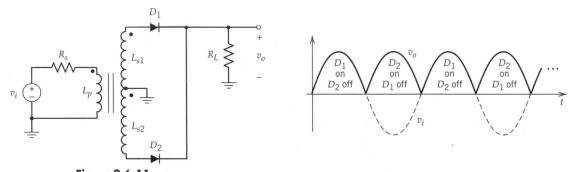

Figure 2.6-11
Full-wave rectifier with center-tapped transformer.

as a half-wave rectifier. The diode D_1 conducts when the input $v_i > V_\gamma$ and D_2 conducts when the input $v_i < V_\gamma$. Note that the secondary winding is capable of providing twice the voltage drop across the load resistor. Additionally, the input to the diodes and the output share a common ground between the load resistor and the center tap.

An isolation transformer is not required to design a full-wave rectifier. If ground isolation is not required, only a center-tapped well-coupled coil is required, as shown in Figure 2.6-12.

An alternate configuration for a full-wave rectifier exists with an addition of two diodes. In the alternate configuration, called a *bridge rectifier*, shown in Figure 2.6-13, the source and load do not share an essential common terminal. Additionally, the

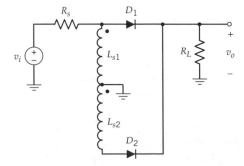

Figure 2.6-12
Full-wave rectifier without isolation transformer.

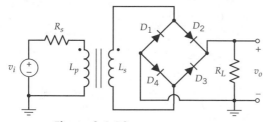 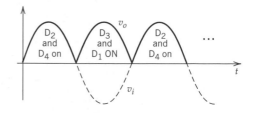

Figure 2.6-13
Bridge rectifier with input transformer.

secondary transformer does not require a center tap and provides a voltage only slightly greater than half that of the secondary in Figure 2.6-11.

In the bridge rectifier circuit, diodes D_2 and D_4 are ON for the positive half cycle of the voltage across the secondary of the transformer. Diodes D_3 and D_1 are off in the positive half cycle since their anode voltages are less than the cathode voltages. This is due to the voltage drop across the ON diodes and the load resistor. In the negative half cycle of the voltage across the secondary of the transformer, diodes D_3 and D_1 are ON, with D_2 and D_4 OFF. In both half cycles, the current through the load resistor is in the same direction. Therefore, in each half cycle, the output voltage appears in the same polarity.

From Equation 2.6-3, the output DC voltage of a full-wave rectifier circuit is twice that of the half-wave rectifier since its period is half that of the half-wave rectifier circuit,

$$[V_{dc}]_{\text{full wave}} = \frac{1}{T} \int_0^T v_o(t) \, dt$$

$$= 2[V_{dc}]_{\text{half wave}} \qquad (2.6\text{-}12)$$

$$= \frac{2V_m}{\pi}.$$

Similarly, the RMS output voltage of a full-wave rectifier is found by applying Equation 2.6-6,

$$V_{\text{ms}} = \left[\frac{1}{T} \int_0^T v_o^2(t) \, dt \right]^{1/2}$$

$$= \frac{V_m}{\sqrt{2}}. \qquad (2.6\text{-}13)$$

The maximum possible efficiency of the full-wave rectifier is significantly greater than that of the half-wave rectifier since power from both positive and negative cycles are available to produce a DC voltage,

$$\eta_{\text{full wave}} = \frac{P_{dc}}{P_{ac}} = \frac{(2V_m/\pi)^2/R_L}{(V_m/\sqrt{2})^2/R_L}$$

$$= \frac{8}{\pi^2} = 2\eta_{\text{half wave}} \Rightarrow 81.2\%. \qquad (2.6\text{-}14)$$

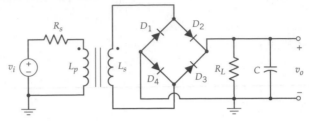

Figure 2.6-14
Filtered full-wave rectifier circuit.

In order to produce a DC source from the output of a full-wave rectifier, a capacitor is placed in parallel to the load resistor, as shown in Figure 2.6-14. The RC time constant must be long with respect to $\frac{1}{2}T$ to smooth out the output waveform.

Let t_1 and t_2 be the time between two adjacent peaks of the filtered rectified voltage as shown in Figure 2.6-15. Then the output voltage between t_1 and t_2 is

$$v_o = V_m e^{(t-t_1)/R_L C}, \qquad t_1 \leq t \leq t_2. \tag{2.6-15}$$

The peak-to-peak ripple is defined as

$$\begin{aligned} v_r &= v_o(t_1) - v_o(t_2) \\ &= V_m(1 - e^{(t_2-t_1)/R_L C}). \end{aligned} \tag{2.6-16}$$

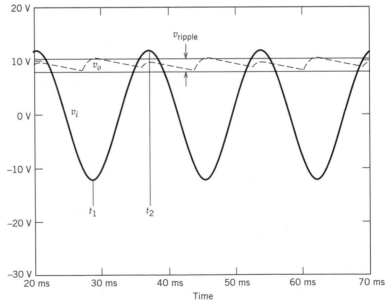

Figure 2.6-15
Full-wave rectified voltage with ripple.

If $R_L C \gg (t_2 - t_1)$, then the exponential approximation can be used,

$$e^{-x} \approx 1 - x \quad \text{for } |x| \ll 1. \tag{2.6-17}$$

Applying Equation 2.6-17 to 2.6-16, a good approximation for the peak-to-peak ripple voltage can be derived,

$$v_r(\text{peak-to-peak}) \approx V_m \frac{t_2 - t_1}{R_L C}. \tag{2.6-18}$$

Since $t_2 - t_1 \approx \frac{1}{2}T = 1/2f_0$, where f_0 is the frequency of the input signal and T is the period of that signal, the peak-to-peak ripple voltage for a full-wave rectifier circuit is

$$v_r(\text{peak-to-peak})_{\text{full wave}} \approx \frac{V_m}{2f_0 R_L C}. \tag{2.6-19}$$

The DC component of the output signal is

$$
\begin{aligned}
V_{O,\text{dc}} = V_{\text{dc}} &= V_m - \tfrac{1}{2}v_r \\
&= V_m \left(1 - \frac{1}{4f_0 R_L C} \right).
\end{aligned} \tag{2.6-20}
$$

For a half-wave rectifier, $t_2 - t_1 \approx T = 1/f_0$, since only half the cycle of the input signal is passed. Therefore, the peak-to-peak ripple voltage of a half-wave rectifier is

$$v_r(\text{peak-to-peak})_{\text{half wave}} \approx \frac{V_m}{f_0 R_L C}. \tag{2.6-21}$$

In both the half- and full-wave rectifiers, the DC voltage is less than the peak rectified voltage.

EXAMPLE 2.6-3 Consider the full-wave rectifier circuit of Figure 2.6-14 with $C = 47$ µF and transformer winding ration of 14:1. If the input voltage is 120 V AC (RMS) at 60 Hz, what is the load resistor value for a peak-to-peak ripple less than 0.5 V? What is the output DC voltage?

SOLUTION

Since the transformer turns ratio is 14:1, the voltage across the secondary is

$$V_m = \frac{120\sqrt{2}}{14} = 12.1 \text{ V}.$$

From Equation 2.6-19 for peak-to-peak ripple,

$$R \leq \frac{V_m}{2f_0 C v_r} = \frac{12.1}{2(60)(47 \times 10^{-6})(0.5)}.$$

$$\leq 4.29 \text{ k}\Omega.$$

The DC voltage at the output is found by using Equation 2.6-20,

$$V_{dc} \leq V_m - \tfrac{1}{2} v_r$$

$$\leq 12.1 - \tfrac{1}{2}(0.5)$$

$$\leq 11.9 \text{ V.}$$

2.6.4 Peak Dectector

One of the first applications of the diode was a "detector" in radio receivers that retrieved information from AM radio signals. The AM signal consists of a radio-frequency "carrier" wave that is at a high frequency and varies in amplitude at an audible frequency. The detector circuit, shown in Figure 2.6-16, is similar to a half-wave rectifier. The RC time constant is approximately the same as the period of the carrier so that the output voltage can follow the variation in amplitude of the input.

2.6.5 Clamping or DC Restoring Circuits

Diode circuits can be designed to clamp a voltage so that the output voltage is shifted to never exceed (or fall below) a desired voltage. A clamping circuit is shown in Figure 2.6-17. In the steady state, the input waveform is shifted by an amount that makes the peak voltage equal to the value of V_{ref}. The waveform is shifted and "clamped" to V_{ref}. The clamping circuit allows shifting of the waveform without a

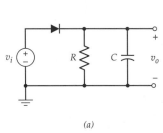

(a)

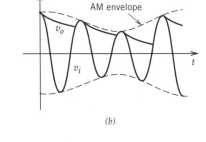

(b)

Figure 2.6-16
Peak detector and associated waveforms.

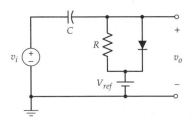

Figure 2.6-17
Clamping circuit.

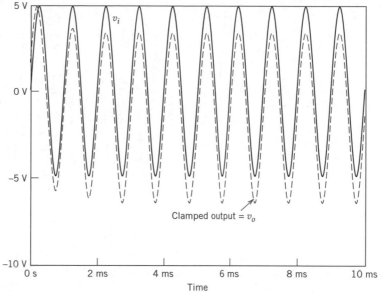

Figure 2.6-18
Input and ouput waveforms for clamping circuit.

priori knowledge of the input wave shape. In Figure 2.6-17, the capacitor charges to a value equal to the difference between the peak input voltage and the reference voltage of the clamping circuit, V_{ref}. The capacitor then acts like a series battery whose value is the voltage across the capacitor, shifting the waveform to the value shown in Figure 2.6-18.

The clamping circuit configuration in Figure 2.6-17 clamps the *maximum* to the reference voltage. If the diode is reversed, the circuit will clamp the minimum voltage of the signal to the reference voltage.

2.6.6 Voltage Multiplier

Diode circuits may be used as voltage doublers as shown in Figure 2.6-19. The circuit is a clamper found by C_1 and D_1 and a peak rectifier formed by D_1 and C_2. With a peak input signal V_m, the clamping section yields the waveforms shown in Figure 2.6-20. The positive voltage is clamped to 0 V. Across diode D_1, the negative peak reaches $-V_m$ due to the charge stored in capacitor C_1. The voltage stored in C_1 is $V_{C1} = V_m$ corresponding to the maximum negative input voltage. Therefore, the voltage across diode D_1 is

$$v_{D1} = V_m\sin \omega t - V_{C1}. \qquad (2.6\text{-}22)$$

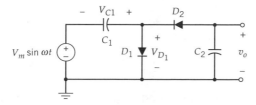

Figure 2.6-19
Diode voltage doubler circuit.

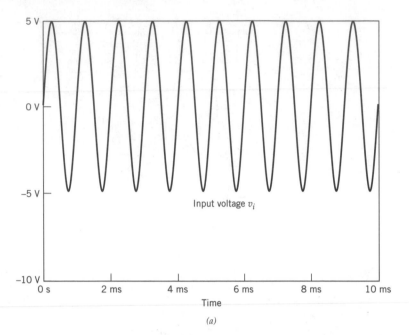

(a)

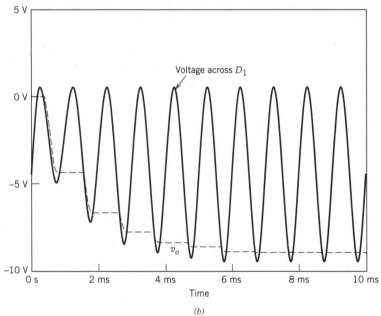

(b)

Figure 2.6-20
Output signal from voltage doubler circuit.

The section consisting of D_2 and C_2 is a peak rectifier. Therefore, the output voltage v_o across C_2 is, after some time, the DC voltage shown in Figure 2.6-19b,

$$v_o = -(V_m + V_{C1}). \tag{2.6-23}$$

By adding more capacitor and diode sections, higher multiples of the input voltage are achievable.

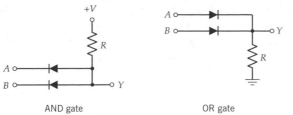

Figure 2.6-21
Diode logic AND and OR gates.

AND gate   OR gate

2.6.7 Diode Logic Gates

Diodes together with resistors can be used to perform logic functions. Figure 2.6-21 shows diode AND and OR gates.

In the AND gate, when either input is connected to ground, the diode in series with that input is forward biased. The output is then equal to one forward-biased diode voltage drop above ground, which is interpreted as logic 0. When both inputs are connected to $+V$, both diodes are zero biased, yielding an output voltage of $+V$, which is interpreted as logic 1. The Boolean notation for the circuit is

$$Y = A \cdot B.$$

In the OR gate, if one or both of the diodes are connected to $+V$, the diodes will conduct, clamping the output voltage to a value equal to $+V - V_\gamma$, or logic 1. Therefore, the Boolean notation for the circuit is

$$Y = A + B.$$

2.6.8 Superdiode

Figure 2.6-22 shows a precision half-wave rectifier using a superdiode. The super diode consists of an OpAmp and a diode. The operation of the circuit is as follows: For positive v_i, the output of the OpAmp will go positive, causing the diode to conduct. This in turn closes the negative-feedback path, creating an OpAmp voltage followers. Therefore,

$$v_o = v_i, \qquad v_i \geq 0. \tag{2.6-24}$$

The slope of the voltage follower transfer function is unity.

For $v_i < 0$, the output voltage of the OpAmp follows the input and goes negative. The diode will not conduct since it is reverse biased. Therefore, no current flows through the resistor R, and

$$v_o = 0, \qquad v_i < 0. \tag{2.6-25}$$

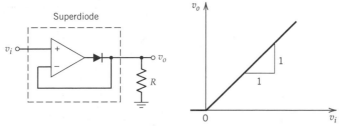

Figure 2.6-22
Precision rectifier using superdiode.

The advantage of the superdiode is the very small turn-on voltage exhibited and ideal transfer function for positive v_i.

Precision circuits (clamper, peak detector, etc.) can be designed using the super-diode in place of regular diodes to eliminate dynamic resistance and diode threshold voltage effects. The superdiode is commonly used in small-signal applications and is not used in power circuits.

2.7 ZENER DIODES AND APPLICATIONS

Diodes that are designed with adequate power dissipation capabilities to operate in breakdown are called Zener diodes and are commonly used as voltage reference or constant-voltage devices.

The two mechanisms responsible for the breakdown characteristics of a diode are avalanche breakdown and Zener breakdown. Avalanche breakdown occurs at high voltages (≥ 10 V) where the charge carriers acquire enough energy to create secondary hole–electron pairs that act as secondary carriers. This chain reaction causes avalanche breakdown of the diode junction and a rapid increase in current at the breakdown voltage. Zener breakdown occurs in the heavily doped p and n regions on both sides of the diode junction and occurs when the externally applied potential is large enough to create a large electric field across the junction to force bound electrons from the p-type material to *tunnel* across to the n-type region. A sudden increase in current is observed when sufficient external potential is applied to produce the required ionization energy for tunneling.

Regardless of the mechanism for breakdown, the breakdown diodes are usually called Zener diodes. The symbol and characteristic curve of a low-voltage (referring to the breakdown voltage) Zener diode are shown in Figure 2.7-1. The forward-bias characteristic is similar to conventional p–n junction diodes. The reverse-bias region

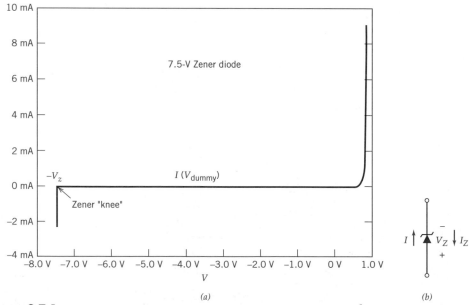

Figure 2.7-1

(*a*) Characteristic curve of Zener diode; (*b*) Zener diode symbol.

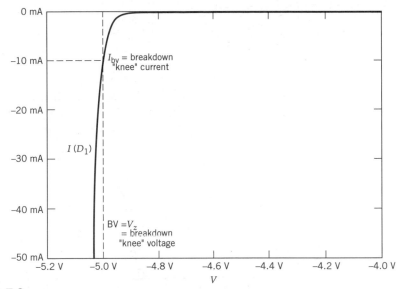

Figure 2.7-2
Reverse breakdown knee voltage B_v and corresponding current I_{bv}.

depicts the breakdown occurring at V_Z that is nearly independent of diode current. A wide range of Zener diodes are commercially available over a wide range of break-down voltages and power ratings to 100 W.

Changes in temperature generally cause a shift in the breakdown voltage. The temperature coefficient is approximately +2 mV/°C for Zener breakdown. For avalanche breakdown, the temperature coefficient is negative.

The simplified SPICE model of a Zener diode is identical to that of the conventional diode with the addition of the reverse breakdown "knee" voltage B_v and the corresponding reverse breakdown "knee" current I_{bv}. The relationship between B_v and I_{bv} is shown in the reverse-bias portion of the Zener diode characteristic curve of the Zener diode in Figure 2.7-2. To obtain a steeper reverse breakdown characteristic, a higher breakdown current I_{bv} may, in general, be used without incurring significant errors. The Zener diode model statement for Figure 2.7-2 is

```
.model zener_diode D(Is=10pA N=2 Bv=5 Ibv=10m)
```

Both B_v and I_{bv} are positive quantities. If I_{bv} is large, the reverse breakdown curve is steeper.

The dynamic resistance of the Zener diode in the reverse breakdown region,r_Z, is the slope of the diode curve at the operating reverse-bias current. Since the reverse current increases rapidly with small changes in the diode voltage drop, r_Z is small (typically 1–15 Ω). The Zener diode piecewise linear model and its simplified version are shown in Figure 2.7-3.

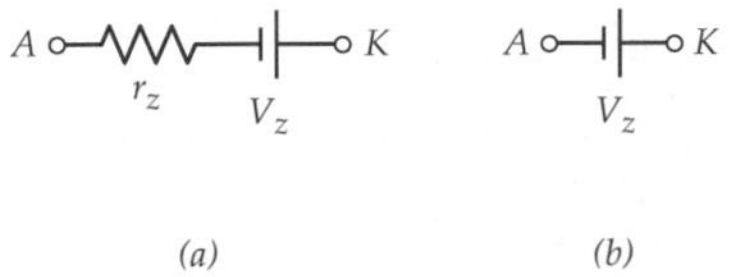

$$A \circ\!\!-\!\!\wedge\!\!\wedge\!\!\wedge\!\!-\!\!\dashv\!\!\vdash\!\!-\!\!\circ K \qquad A \circ\!\!-\!\!\dashv\!\!\vdash\!\!-\!\!\circ K$$
$$r_z \qquad V_z \qquad\qquad\quad V_z$$

Figure 2.7-3
(a) Zener diode piecewise linear model; (b) simplified Zener diode model.

(a) $\qquad\qquad\qquad$ (b)

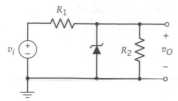

Figure 2.7-4
Zener diode voltage reference circuit.

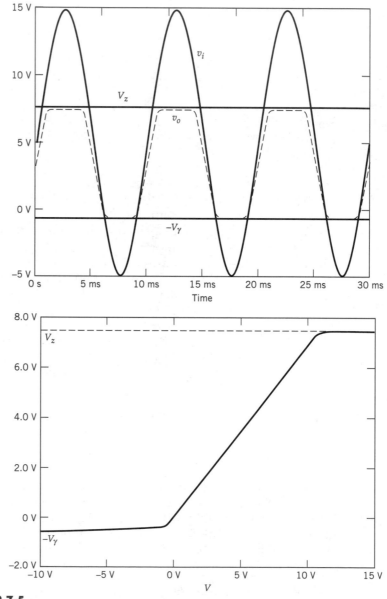

Figure 2.7-5
(a) input and output waveforms; (b) transfer characteristic of Zener voltage reference shown in Figure 2.7-4.

A typical application of Zener diodes is in the design of the voltage reference circuit shown in Figure 2.7-4.

To simplify the analysis, first consider the circuit without the Zener diode. The output voltage is simply that of a voltage divider,

$$v_{O, \text{ no diode}} = \frac{R_2}{R_1 + R_2} v_I \tag{2.7-1}$$

where the input voltage v_I and output voltage v_O have both AC and DC components. Replace the Zener diode in the circuit. If the breakdown voltage V_Z is greater than $v_{O, \text{ no diode}}$, then the Zener diode is operating in the reverse-bias region between the Zener knee and 0 V or in the forward-bias region. Therefore, the Zener diode operates as a conventional diode,[9] and

$$v_O = \begin{cases} -V_\gamma, & v_I \le \dfrac{R_1 + R_2}{R_2}(-V_\gamma) \\[2mm] v_{O, \text{ no diode}}, & v_I > \dfrac{R_1 + R_2}{R_2}(-V_\gamma) \end{cases} \tag{2.7-2}$$

If the breakdown voltage V_Z is less than $v_{O, \text{ no diode}}$, then the Zener diode operates in the breakdown region beyond V_Z and forces

$$v_O = V_Z \quad \text{when } v_I \ge \frac{R_1 + R_2}{R_2}(V_Z). \tag{2.7-3}$$

Figure 2.7-5 shows the input, output, and transfer characteristics of a Zener voltage reference circuit.

The Zener voltage reference circuit clamps the output voltage to the Zener breakdown voltage. The current through the Zener diode forces the voltage drop across resistor R_1 to $v_O - v_I$ since V_Z is invariant over a wide range of currents I_Z and the dynamic resistance in the breakdown region is negligible compared to R_2. Here, R_1 must be chosen to limit I_Z to safe operating values as specified by the diode manufacturer.

EXAMPLE 2.7-1 The Zener diode in the circuit shown has a working range of current for proper regulation,

$$5 \text{ mA} \le I \le 50 \text{ mA},$$

and a Zener voltage

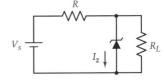

$$V_z = 50 \text{ V}.$$

(a) If the input voltage V_s varies from 150 to 250 V and $R_L = 2.2 \text{ k}\Omega$, determine the range of values for the resistor R to maintain regulation.

(b) If R is chosen as the midpoint of the range determined in part (a), how much variation in the load resistance R_L is now possible without losing regulation?

---

[9] A piecewise simplified linear model of a diode is used in this example.

SOLUTION

(a) The current through the load is given by

$$I_L = \frac{50 \text{ V}}{2.2 \text{ k}\Omega} = 22.74 \text{ mA}.$$

The current through the resistor R must lie in the range of the load current plus the diode current:

$$5 \text{ mA} + 22.74 \text{ mA} \leq I \leq 50 \text{ mA} + 22.74 \text{ mA},$$

$$27.74 \text{ mA} \leq I \leq 72.74 \text{ mA},$$

but this current is also dependent on the source voltage, the Zener voltage, and the resistance R:

$$I = \frac{V_s - V_z}{R} = \frac{V_z - 50}{R}$$

or

$$R = \frac{V_s - 50}{I}.$$

Since regulation must occur for both extremes of the source voltage, R must be the intersection of the limits determined by the above equation:

$$\frac{V_{s,\text{ min}} - 50}{I_{\text{max}}} \leq R \leq \frac{V_{s,\text{ min}} - 50}{I_{\text{min}}},$$

$$\frac{V_{s,\text{ max}} - 50}{I_{\text{max}}} \leq R \leq \frac{V_{s,\text{ max}} - 50}{I_{\text{min}}},$$

or, after applying the intersection of the ranges,

$$\frac{V_{s,\text{ max}} - 50}{I_{\text{max}}} \leq R \leq \frac{V_{s,\text{ min}} - 50}{I_{\text{min}}},$$

$$\frac{250 - 50}{72.74} \leq R \leq \frac{150 - 50}{27.74} \Rightarrow 2.75 \text{ k}\Omega \leq R \leq 3.60 \text{ k}\Omega.$$

(b) The midpoint of the above range is $R = 3.175 \text{ k}\Omega$. The resistor current is given by

$$I = \frac{V_s - 50}{R} \Rightarrow 31.5 \text{ mA} \leq I \leq 63 \text{ mA}.$$

Since regulation must hold for all values of R, the load current must lie in the intersection of the possible ranges of the resistor current minus the diode current:

$$3.15 - 50 \text{ mA} \leq I_L \leq 31.5 - 5 \text{ mA}, \qquad 63 - 50 \text{ mA} \leq I_L \leq 63 - 5 \text{ mA}.$$

Thus,

$$13 \text{ mA} \leq I_L \leq 26.5 \text{ mA}.$$

Since

$$R_L = \frac{V_Z}{I_L},$$

the range of R_L is found to be

$$3.85 \text{ k}\Omega \geq R_L \geq 1.93 \text{ k}\Omega.$$

Another typical application of Zener diodes occurs in AC–DC conversion. Recall that the output of a filtered rectifier has residual voltage ripple. By adding a Zener diode voltage reference circuit at the output of a rectifier and filter circuit, as shown in Figure 2.7-6, the residual voltage ripple at the output can be eliminated. The resistors R_1 and R_L must be carefully selected to yield the desired voltage output. If R_1 is chosen to be much less than R_L, and if the voltage across the capacitor is greater than V_Z, then the output voltage will be clamped at V_Z.

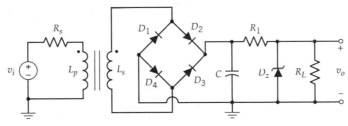

Figure 2.7-6
Full-wave rectifier circuit with output voltage reference.

 EXAMPLE 2.7-2 Consider the full-wave rectifier circuit with a Zener clamp shown in Figure 2.7-6. If the input voltage is 120 V AC (RMS) at 60 Hz, transformer winding ratio is 14:1, $C = 47$ μF, $R_L = 3.8$ kΩ, and $R_1 = 220$ Ω, show that the output is clamped to 5 V if a Zener diode with $V_Z = 5$ V is used.

SOLUTION

This is the same problem as Example 2.6-3. The resistance for the circuit without R_1 and the Zener diode was found to be less than 4.29 kΩ. The DC output voltage was determined to be less than or equal to 11.9 V. Since $V_Z < 11.9$ V, the resulting output voltage is clamped to 5 V. To confirm the result, a SPICE simulation of the circuit is performed. The SPICE output file and the input and outputs at various points on the circuit follow.

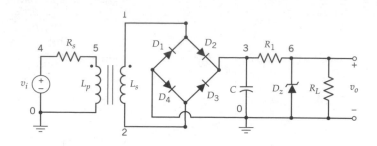

```
*******************************PSpice*********************************
Filtered full-wave rectifier with Zener regulator
****   CIRCUIT DESCRIPTION
***********************************************************************
V1    4    0    SIN(0 169V 60Hz)
Rs    4    5    0.5
*transformer
LP    5    0    10
LS    1    2    51m
Kxfrmr    Lp    Ls    0.999
*main circuit
D1    0    1    new_diode
RL    6    0    3.8k
R1    3    6    220
DZ    0    6    zener_diode
C     3    0    47u
D2    1    3    new_diode
D3    2    3    new_diode
D4    0    2    new_diode
.model new_diode D(Is=10pA n=2)
.model zener_diode D(Is=10pA n=2 Bv=5 Ibv=20m)
.TRAN    lus  70ms 0   100us
.PROBE
.END
*******************************PSpice*********************************
 ****   Diode MODEL PARAMETERS
***********************************************************************
        new_diode    zener_diode
     IS  10.000000E-12 10.000000E-12
      N  2            2
     BV               5
     IBV              .02
*******************************PSPICE*********************************
**** INITIAL TRANSIENT SOLUTION TEMPERATURE = 27.000 DEG C
***********************************************************************
NODE VOLTAGE NODE VOLTAGE NODE VOLTAGE NODE VOLTAGE
(  1) 1.546E-18     ( 2) 1.546-18     ( 3)-527.6E-21    ( 4) 0.0000
(  5)   0.0000      ( 6)-517.4E-21


VOLTAGE SOURCE CURRENTS
NAME    CURRENT
VI    0.000E+00


TOTAL POWDER DISSIPATION 0.00E+00 WATTS
JOB CONCLUDED


TOTAL JOB TIME 27.63
```

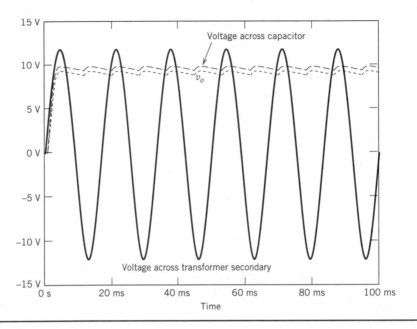

A design trade-off exists between using a large-capacitor filter to remove the output voltage ripple or a smaller capacitor and a Zener diode clamp. Large capacitors require large "real estate," or space, on a printed circuit board and height above the circuit board. However, capacitor filters only dissipate energy when in the charging or discharging cycles.

Although the Zener diode is small in size and requires substantially less circuit volume than large capacitors, the diode must be able to dissipate from zero to the maximum current delivered to the load. If the Zener diode is to carry large currents over much of the operating cycle, the power dissipation is high and a large capacitor filter may be preferred. The Zener diode may be preferred if the power dissipation in the diode can be limited.

2.8 OTHER COMMON DIODES AND APPLICATIONS

In the previous sections of this chapter, the conventional p–n junction diode and the Zener diode and their applications were introduced. Although the conventional p–n junction diode and the Zener diode are the most common diode types used in electronic design, other types of diodes are designed into certain electronic applications. Some different types of diodes include the tunnel diode, backward diode, Schottky barrier diode, Varactor diode, p–i–n diode, IMPATT diode, TRAPATT diode, BARRITT diode, solar cell, photodiode, light-emitting diode, and semiconductor laser diode.

Unfortunately, it is far beyond the scope of this book to discuss all of the different types listed above. However, four common types of diodes are presented in this section for discussion:

- Tunnel diode
- Schottky-barrier diode
- Photodiode
- Light-emitting diode

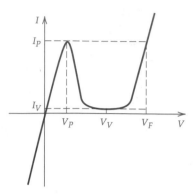

Figure 2.8-1
Voltage–current characteristic of tunnel diode.

Figure 2.8-2
Tunnel diode symbol.

2.8.1 Tunnel Diode

The tunnel diode (also called the Esaki diode, after the L. Esaki, who announced the new diode in 1958) V–I characteristic is shown in Figure 2.8-1. The figure shows that the tunnel diode is an excellent conductor in the reverse direction. Figure 2.8-2 is the circuit symbol for the tunnel diode.

For small forward voltages (on the order of 50 mV in germanium), the resistance is on the order of 5 Ω. At the peak current I_P corresponding to the voltage V_P, the slope of the characteristic curve is zero. As the voltage increases beyond V_P, the current also decreases. The tunnel diode characteristic curve in this region exhibits a *negative dynamic resistance* between the peak current I_P and the minimum or valley current I_V. At the valley voltage V_V corresponding to the valley current I_V, the slope of the characteristic curve is again zero. Beyond V_V, the curve remains positive. At the peak forward voltage V_F, the current again reaches the value of I_P.

Since it is difficult to manufacture silicon tunnel diodes with a high ratio of peak-to-valley current I_P/I_V, most commercially available tunnel diodes are made from germanium (Ge) or gallium arsenide (GaAs). Table 2.8-1 summarizes some of the static characteristics of these devices.

The operating characteristics of the tunnel diode are highly dependent on the load line of the circuit in which the diode is operating. Some load lines may intersect the tunnel diode characteristic curve in three places: the regions between O and V_P, between V_P and V_V, and beyond V_V. This multivalued feature makes the tunnel diode useful in high-speed pulse circuit design. High-frequency (microwave) oscillators are often designed so that the tunnel diode is based in its negative dynamic resistance region.

TABLE 2.8-1 TYPICAL TUNNEL DIODE PARAMETERS

Parameter	Ge	GaAs	Si
I_P/I_V	8	15	3.5
V_P, V	0.055	0.15	0.065
V_V, V	0.35	0.5	0.42
V_F, V	0.5	1.1	0.7

Figure 2.8-3
Schottky diode symbol.

2.8.2 Schottky-Barrier Diode

The Schottky-barrier diode (or simply Schottky diode) is a metal–semiconductor diode. The circuit symbol of the Schottky diode is shown in Figure 2.8-3. Metal–semiconductor diodes are formed by bonding a metal (usually aluminum or platinum) to n- or p-type silicon. Metal–semiconductor diode voltage–current characteristics are very similar to conventional p–n junction diodes and can be described by the diode equation with the exception that the threshold voltage V_γ is in the range from 0.3 to 0.6 V. The physical mechanisms of operation of the conventional p–n junction diode and the metal–semiconductor diode are not the same.

The primary difference between metal–semiconductor and p–n junction diodes is in the charge storage mechanism. In the Schottky diode, the current through the diode is the result of the drift of majority carriers. This is in contrast to the current in the p–n junction diode, which is the result of a diffusion of minority carriers. There is no minority carrier storage in the Schottky diode, allowing the switching time from forward to reverse bias to be very short compared to the p–n junction diode.

Therefore, Schottky diodes are often used in integrated circuits for high-speed switching applications. The Schottky diode is easy to fabricate on integrated circuits because of its construction. The low-noise characteristics of the Schottky diode is ideal for the detection of low-level signals like those encountered in radio-frequency electronics and radar detection applications.

2.8.3 Photodiode

The photodiode converts optical energy to electric current. The circuit symbol of the photodiode is shown in Figure 2.8-4.

In order to make this energy conversion, the photodiode is reverse biased. Intensifying the light on the photodiode induces hole–electron pairs, which increases the magnitude of the diode reverse saturation current. The useful output of the photodiode is *photocurrent*, which for all practical purposes is proportional to the light intensity (in watts) on the device. The proportionality constant is called the responsivity R, which is usually given in amperes per watt and is dependent on the wavelength of the light. Figure 2.8-5 shows a photodiode characteristic curve.

If the intensity of the light on the photodiode is constant, the photodiode can be modeled as a constant current source so long as the voltage does not exceed the avalanche voltage. Naturally, the photocurrent will vary with varying input light intensity. Since the photocurrent can be very small, an electronic amplifier is used in many applications to both boost the signal level and to convert from a current to a voltage output. For example, in optical fiber communication receivers, the average intensity of a time-varying infrared light on the photodiode can be significantly less than 1 μW. Taking a typical photodiode responsivity for fiber-optic application of 0.7 A/W, 1 μW of light will produce 0.7 μA of average current. This low-level signal must be amplified by electronic amplifiers for processing by other electronic circuits to retrieve the transmitted information.

Figure 2.8-4
Photodiode symbol.

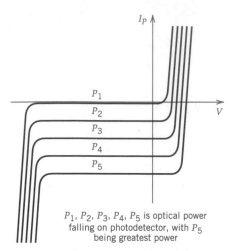

P_1, P_2, P_3, P_4, P_5 is optical power falling on photodetector, with P_5 being greatest power

Figure 2.8-5

Characteristic curve of photodiode.

2.8.4 Light-Emitting Diode

The light-emitting diode (LED) converts electric energy to optical energy (light). The LEDs are used for displays and are used as the light source for low-cost fiber-optic communication transmitters. By appropriate doping, the emission wavelength of the LED can be varied from the near infrared (<2 μm) to the visible (400–780 nm). The symbol for the LED, shown in Figure 2.8-6, is similar to that of the photodiode, except that the direction of the arrows represents light being emitted.

When the LED is conducting, its diode voltage drop is about 1.7 V. The intensity of the light emitted from the LED is proportional to the current through the diode and is characterized by the so-called light intensity–current (L–I) curve shown in Figure 2.8-7a. The LED also has a current–voltage relationship depicted in Figure 2.8-7b.

When using the LED in a circuit, a series current-limiting resistor is used to prevent destruction should large currents flow through the LED. The magnitude of the current-limiting resistor is calculated by limiting the current though the LED to a desired level I_{OP}, less than the maximum operating current with a diode threshold voltage of V_γ. For example, in Figure 2.8-8, if the diode threshold voltage is 1.7 V

Figure 2.8-6

Light-emitting diode symbol.

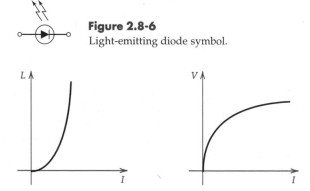

(a)   (b)

Figure 2.8-7

(a) The L–I characteristic curve of LED;
(b) LED I–V characteristic curve.

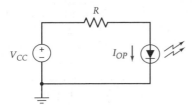

Figure 2.8-8
Simple LED driving circuit.

and $I_{OP} = 10$ mA provides a satisfactory optical output, the current-limiting resistor is

$$R = \frac{V_{CC} - V_\gamma}{I_{OP}}$$

$$= \frac{5 - 1.7}{10 \times 10^{-3}} = 330 \ \Omega.$$

2.9 CONCLUDING REMARKS

The semiconductor diode has been described in this chapter as the most basic non-linear electronic device. It is a two-terminal device that provides small resistance when currents flow through the diode from the anode to the cathode and extremely large resistance to currents in the reverse direction. Large reverse voltages force the diode into breakdown and the dynamic resistance again becomes small. Diode applications utilize the characteristic properties of the diode in one or more of these three regions of operation:

- Forward-bias region ($V \geq V_\gamma$)
- Reverse-bias region ($V_\gamma > V \geq V_Z$)
- Zener region ($V < V_Z$)

While nearly exact, all-region, analytic expressions for the diode V–I relationship were presented, it was shown that piecewise linear approximate expressions for the

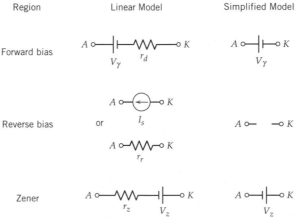

Figure 2.9-1
Regional linear models of diode.

diode *V–I* relationship can, in many cases, provide an adequate representation of the performance of the diode. Other applications allow for even further simplification. These linear models and their simplified versions are summarized in Figure 2.9-1.

The diode was shown to be useful in a number of applications determined by the surrounding circuitry. The Zener diode, tunnel diode, Schottky diode, photodiode, and LED were introduced as diodes with properties that are necessary for specialized electronic applications. In later chapter, additional linear and nonlinear diode applications will be examined.

SUMMARY DESIGN EXAMPLE

Many electronic applications have a need for backup electrical power when there is a primary power failure. This backup may come in the form of very large value capacitors (1 F or greater), battery systems, or generators powered by an internal combustion engine. One such electronic application operates with the following power requirements:

| Allowable input voltage | $5\text{ V} < V_{CC} < 9\text{ V}$ |
| Current draw | $100\text{ mA} < I < 500\text{ mA}$ |

The primary power source V_{CC} provides a nominal voltage of 8 ± 1 V. Design a backup power system that will provide adequate auxiliary power when the primary power source fails (i.e., $V_{CC} = 0$). The maximum duration of primary power failure is 2 h.

Solution

The total power requirement for this system lies in the range

$$0.5\text{ W} < P_T < 4.5\text{ W}.$$

Capacitive backup is a poor choice for this system: Even if the largest available capacitors are used, it will only last a few seconds. Conversely, motorized generators are inappropriately large for this system: a 0.04-hp motor would be more than adequate to power a 5 W generator. Of the three given choices, the best for this application is a rechargeable battery backup.

The design goals for the backup system include the following

- The battery backup powers the system only during primary power failure.
- The battery backup is recharged at a controlled rate during normal operation.
- The battery backup is protected against excessive recharging.

All design goals can be met with a network composed of diodes and a resistor. The proposed design is as shown. During normal operation the system state is:

- D_1, on; D_2, off; D_z, either off or in the Zener region. Resistor R controls the rate of changing of the battery and D_z prevents overcharging the battery.

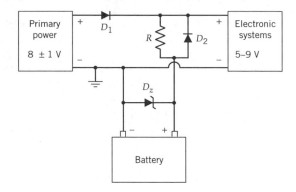

During backup operation the system state is:

- D_1, off; D_2, on; D_z, off. Here, D_1 blocks current discharge to the primary power source and D_2 provides a low-impedance path to the load.

The properties of the system components must be specified:

1. The battery voltage must lie within acceptable ranges to both provide auxiliary power and receive current for recharge. Auxiliary power voltage must be greater than the minimum voltage required by the load *plus V_γ* and smaller than the minimum voltage provided by the primary power source *minus V_γ*:

    $$5 + V_\gamma < \text{battery voltage} < 7 - V_\gamma.$$

    A 6 V battery is a good choice for this system. The battery must also have a capacity so that backup will be provided during the entire time of primary power failure. Battery capacity is measured in the product of current and time (i.e., ampere-hours). This system requires a battery with a capacity greater than 1 A-hr:

    $$\text{Battery capacity} > (\text{maximum current})(\text{time})$$
    $$= (500 \text{ mA})(2 \text{ h}) = 1 \text{ A-hr.}$$

2. Diodes D_1 and D_2 must have sufficiently large current ratings. In each case the diode must be capable of carrying a minimum of 500 mA continuously.

3. Battery recharging current must be limited. Every battery type has a recommended rate of charge in order to maintain proper operation for a long life. A typical value for a small battery such as the one specified here is approximately 20 mA. The resistor R can then be determined from the nominal values for the primary power and battery voltages:

    $$R = \frac{8 - 6 \text{ V}}{20 \text{ mA}} = 100 \text{ }\Omega.$$

    The power rating is given by the maximum voltage squared divided by the resistance:

$$P_R > \frac{(9-6)^2}{100} = 0.03 \text{ W} \Rightarrow \text{a } \frac{1}{4} \text{ W resistor will suffice.}$$

4. The Zener diode provides protection against overcharging the battery. As batteries are overcharged, the battery voltage increases. Depending on the type of battery chosen, a maximum voltage will determine the Zener voltage. Maximum recharging current determines the capacity of the Zener diode. A typical overvoltage for small batteries is approximately 0.2 V, which leads to

$$V_Z = 6.2 \text{ V.}$$

The maximum recharging current is 30 mA. This corresponds to a maximum power dissipation in the Zener diode of 0.19 W. A $\frac{1}{4}$ W diode will suffice.

Summary component list

- One 6 V, rechargeable battery with at least 1 A-hr capacity
- Two power diodes with at least a 500-mA rating
- One $\frac{1}{4}$ W, 100 Ω resistor
- One 6.2 V, $\frac{1}{4}$ W Zener diode

REFERENCES

Ghausi, M. S., *Electronic Devices and Circuits: Discrete and Integrated*, Holt, Rinehart and Winston, New York, 1985.

Millman, J., *Microelectronics,* Digital and Analog Circuits and Systems, McGraw-Hill Book Company, New York, 1979.

Colclaser, R. A., Neaman, D. A., and Hawkins, C. F., *Electronic Circuit Analysis: Basic Principles*, John Wiley & Sons, New York, 1984.

Mitchell, F. H and Mitchell, F. H., *Introduction to Electronics*, Second Edition, Prentice-Hall, Englewood Cliffs, NJ, 1992.

Savant, C. J., Roden, M. S., and Carpenter, G. L., *Electronic Design: Circuit and Systems*, Benjamin/Cummings Publishing Company, Redwood City, 1991.

Sedra A. S. and Smith, K. C., *Microelectronic Circuits, Second Edition*, Holt, Rinehart and Winston, New York, 1987.

Tuinenga, P. W., *SPICE: A Guide to Circuit Simulation & Analysis Using PSpice, Second Edition*, Prentice-Hall, Englewood Cliffs, NJ, 1992.

Thorpe, T. W., *Computerized Circuit Analysis with SPICE*, John Wiley & Sons, New York, 1992.

Colclaser, R. A. and Diehl-Nagle, S., *Materials and Devices for Electrical Engineers and Physicists*, McGraw-Hill, New York, 1985.

Streetman, B. G., *Solid State Electronic Devices, Second Edition*, Prentice-Hall, Englewood Cliffs, 1990.

Navon, D. H., *Semiconductor Micro-devices and Materials*, Holt, Rinehart & Winston, New York, 1986.

Grebene, A. B., *Bipolar and MOS Analog Integrated Circuit Design,* John Wiley & Sons, New York, 1984.

Mayer, W. and Lau, S. S., *Electronic Materials Science: For Integrated Circuits in Si and GaAs*, Macmillan, New York, 1990.

■ PROBLEMS

2-1. A silicon diode has a reverse saturation current of 0.1 nA and an empirical scaling constant $\eta = 2$. Assume operation at room temperature.

a. At what diode voltage will the reverse current attain 99% of the saturation value?

b. At what diode voltage will the forward current attain the same magnitude?

c. Calculate the forward currents at diode voltages of 0.5–0.8 V in 0.1 V increments.

2-2. A silicon diode ($\eta = 2$) at room temperature conducts 1 mA when 0.6 V is applied across its terminals.

a. Determine the diode reverse saturation current.

b. What will the diode current be if 0.7 V is applied across it?

2-3. Experimental data at 25°C indicates that the forward biased current I_D flowing through a diode is 2.5 μA with a voltage drop across the diode V_D of 0.53 V and $I_D = 1.5$ mA at $V_D = 0.65$ V. Determine:

1. η

2. I_S

3. I_D at $V_D = 0.60$ V

2-4. For the diode in the above problem, what is the diode voltage drop V_D at:

a. $I_D = 1.0$ mA at 50°C

b. $I_D = 1.0$ mA at 0°C

2-5. A silicon diode has a reverse saturation current of 1 nA and an empirical scaling constant $\eta = 2$. Assume operation at room temperature.

a. A positive voltage of 0.6 V is applied across the diode. Determine the diode current.

b. What voltage must be applied across the diode to increase the diode current by a factor of 10?

c. What voltage must be applied across the diode to increase the diode current by a factor of 100?

2-6. A diode is operating in a circuit in series with a constant current source of value I. What change in the voltage across the diode will occur if an identical diode is placed in parallel with the first? Assume $I \gg I_S$. What if two identical diodes are placed in parallel with the first?

2-7. A silicon diode has a reverse saturation current of 1 nA and an empirical scaling constant $\eta = 1.95$. Determine the percentage change in diode current for a change of temperature from 27 to 43°C for diode voltages of:

a. -1 V

b. 0.5 V

c. 0.8 V

2-8. A diode at room temperature has 0.4 V across its terminals when the current through it is 5 mA and 0.42 V when the current is twice as large. What values of the reverse saturation current and empirical scaling constant allow this diode to be modeled by the diode equation?

2-9. At room temperature, a diode with $\eta = 2$ has 2.5 mA flowing through it with 0.6 V across its terminals.

a. Find V_d when $I_D = 10$ mA.

b. Determine the reverse saturation current.

c. The diode is connected in series with a 3 V DC source and a resistance of 200 Ω. Find I_D if the diode is operating in the forward-bias region.

2-10. Let the reverse saturation current of a diode equal 15 nA and $\eta V_t = 25.6$ mV. If $I_D = 5$ mA, find:

a. V_d

b. V_d/I_D

c. r_d

If I_D varies over the range 4.8 mA $\leq I_D \leq$ 5.2 mA, determine the range of:

d. V_d

e. r_d

2-11. When 20 A current is initially applied to a silicon diode of a particular characteristic, the voltage across the diode is $V_D = 0.69$ V at room temperature. With so much current flowing through the diode, the power dissipation raises the operating temperature of the device. The increased temperature eventually causes the diode voltage to stabilize at $V_D = 0.58$V.

a. What is the temperature rise in the diode?

b. How much power is dissipated in the diode at the two operating conditions mentioned?

c. What is the temperature rise per watt of power dissipation?

2-12. A silicon diode with parameters

$$I_S = 5 \text{ nA}, \qquad \eta = 2$$

is placed in the given circuit.

a. Determine the diode current and voltage.

b. How much power is dissipated in each circuit element?

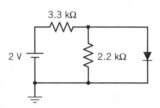

2-13. Determine the current I in the given circuit if the diode is described by

$$I_S = 3 \text{ nA}, \qquad \eta = 1.9.$$

Hint: Find the Thévenin equivalent of the total circuit connected to the diode terminals.

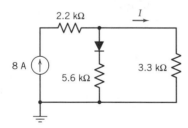

2-14. Find the values of I and V for the circuits shown. Assume that the diodes are ideal.

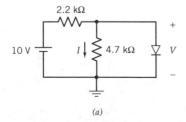

(a)

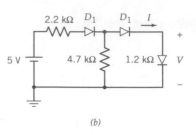

(b)

2-15. Find the values of all currents I and voltage V for the circuits shown. Assume that the diodes are ideal.

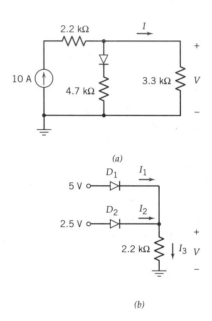

(a)

(b)

2-16. The reverse saturation current for the silicon diode operating room temperature in the circuit shown is $I_S = 15$ pA.

a. Sketch v_o as a function of time in milliseconds for v_i.

b. Repeat part a if $v_i(t)$ is a 2-V peak-to-peak signal with the same period as the square wave in part a.

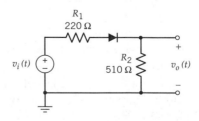

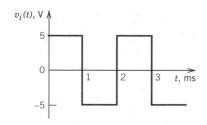

2-17. The circuit shown includes a diode with the specifications

$$\eta = 2 \quad \text{and} \quad I_S = 15 \text{ pA}$$

operating at room temperature. Let

$$v_i(t) = 0.1 \sin \omega_0 t \quad \text{V}.$$

    a. Determine the quiescent $[v_i(t) = 0]$ current in the diode.

    b. Find the dynamic resistance r_d of the diode.

    c. Find the output voltage v_o.

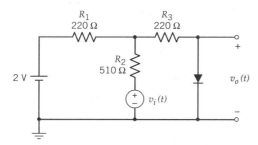

2-18. A silicon diode has a reverse saturation current of 8 nA and an empirical scaling constant $\eta = 2$. Find the diode current in the given circuit as a function of time:

    1. Using load line techniques

    b. Using SPICE

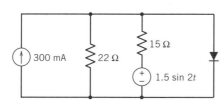

2-19. The four-diode circuits given use identical diodes described by $\eta = 2$ and $I_S = 1.5$ pA.

    a. For the circuit shown in part (*a*) of the figure, determine the value of the current source to obtain an output voltage $V_O = 2.6$ V.

    b. Find the change in output voltage for the circuit in part (*b*). How much current is drawn away by the load?

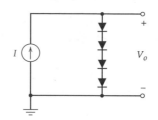

(*a*)

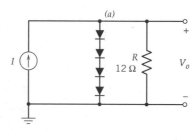

(*b*)

2-20. A silicon diode with parameters
$$I_S = 5 \text{ nA}, \qquad \eta = 2$$
is carrying a forward current of 1 mA. Find:

    a. The diode forward dynamic resistance

    b. The diode threshold voltage

    c. A linear model of the diode at that operating point

2-21. For the logarithmic amplifier shown (let $v_D \gg \eta V_t$ and assume ideal OpAmps):

    a. Find the expression for the output voltage in terms of the input voltage.

    b. Find the expression for the output voltage in terms of the input voltage for the antilogarithmic amplifier.

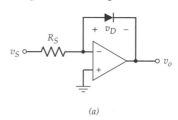

(*a*)

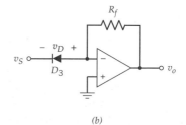

(*b*)

2-22. The diodes in the circuit are modeled by a simple model:

$$V_\gamma = 0.7, \qquad r_d = 0, \qquad r_r = \infty.$$

Sketch the transfer characteristic for $-25 \leq v_i \leq 25$ V. In each region indicate which diode(s) are on. Also indicate all slopes and voltage levels on the sketch.

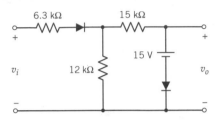

2-23. For the given circuits, plot the output voltage against the input voltage. Assume silicon diodes.

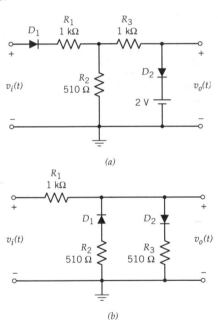

(a)

(b)

2-24. Design a half-wave unregulated power supply to provide an output DC voltage of 10 V with a peak-to-peak ripple voltage of 0.1 V. Assume a 120-V, 60-Hz supply. Use transformer coupling. Provide both analytical design and SPICE outputs. Piecewise linear models of diodes may be used for your analytical design.

2-25. Design an unregulated half-wave rectifier power supply with transformer input coupling that has an input of 120 V_{RMS} at 60 Hz and requires a maximum DC output voltage of

17 V and a minimum of 12 V. The power supply will provide power to an electronic circuit that requires a constant current of 1 A. Determine the circuit configuration, transformer winding ratio, and capacitor size. Assume no losses by the transformer and a diode $V_\gamma = 0.7$ V. Use SPICE to confirm the operation of the circuit.

2-26. Design a full-wave bridge rectifier to provide an output DC voltage of 10 V with a peak-to-peak ripple voltage of 0.1 V. Assume a 120 V, 60 Hz supply. Use transformer coupling. Provide only the analytical solution. Piecewise linear models of diodes may be used for your design.

2-27. Given the accompanying diode circuit. Assume diodes with the following properties:

$$V_\gamma = 0.7 \qquad r_d = 0, \Omega \qquad r_r = \infty \ \Omega.$$

a. What range of values for V_{CC} will produce *both* the following design goals?

- If $V_1 = 25$ V and $V_2 = 0$ V, then $V_o = 3$ V.
- If $V_1 = V_2 = 25$ V, then $V_o \geq 10$ V.

b. Choose V_{CC} to be the midpoint of the range calculated in part a. Calculate the diode currents for $V_1 = V_2 = 25$ V.

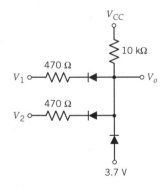

2-28. The diodes in the circuit are modeled by a simple model:

$$V_\gamma = 0.7 \text{ V} \qquad r_d = 0 \ \Omega \qquad r_r = \infty \ \Omega.$$

Sketch the transfer characteristic for $0 \leq v_i \leq 30$ V. In each region indicate which diode(s) are on. Also indicate all slopes and voltage levels on the sketch.

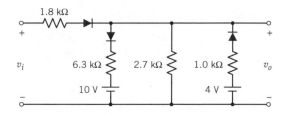

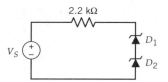

	I_S	η	V_Z
D_1	2 μA	1.8	30 V
D_2	3 μA	2.0	25 V

2-29. Sketch the transfer characteristic with $-40 \text{ V} \le V_i \le 40 \text{ V}$ for the circuit shown. On the sketch indicate all significant voltage levels and slopes. In each region indicate the state of each diode. The diodes have the following approximate properties:

$$r_d \approx 0 \ \Omega \qquad V_\gamma \approx 0.7 \text{ V} \qquad r_r \approx \infty \ \Omega.$$

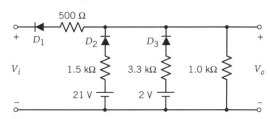

2-30. Design a circuit that clamps a signal to 5 V and clips it below -5 V. Plot the output when the input signal is $v_i(t) = 10 \cos[2\pi(1000)t]$ volts. Piecewise linear models of diodes may be used for your analytical design.

2-31. Analyze the voltage tripler–quadrupler shown. Assume $v_i(t)$ is sinusoidal.

a. Calculate the maximum voltage across each capacitor.

b. Calculate the peak inverse voltage of each diode.

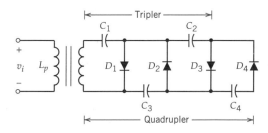

2-32. Given the circuit and diode data:

a. Calculate the diode currents and voltages if $V_S = 5$ V.

b. Calculate the diode currents and voltages if $V_S = 50$ V.

2-33. A simple Zener diode voltage regulator is under design. The design constraints are:

Supply voltage	$V_S = 150$ V
Zener voltage	$V_Z = 50$ V
Zener current range	5 mA $\le I_Z \le 50$ mA

a. Determine the value of the regulator resistor R so that voltage regulation is maintained for a load current in the range 1 mA $\le I_L \le I_{L,\,max}$. What is the maximum load current $I_{L,\,max}$?

b. What power rating is necessary for the resistor R?

c. If the variable load of part a is replaced by a 2 kΩ resistor and the regulator is constructed using the resistor R calculated in part a, over what range of supply voltages will regulation be maintained?

2-34. A load draws between 20 and 40 mA at a voltage of 20 V DC. The available DC power varies between 100 and 140 V DC.

a. Design a voltage regulator using the following components:

- Resistors: any size, any number

- Zener diode: regulates for a Zener current of 5–50 mA

b. The value engineering department has found an "equivalent" Zener diode that can be purchased for 70% of the cost of your original diode. The characteristics for this diode are:

- Regulates for a Zener current of 3–42 mA

What effect on your design will the substitution have? Is a redesign necessary?

2-35. A load draw exists between 100 and 400 mA at 5 V. It is receiving power from an unregulated power supply that can vary between 7.5 and 10 V. Design a Zener diode voltage regulator using a 5 V Zener diode that regulates for diode currents between 50 and 1.1 A. Show the circuit diagram for the completed design. Be sure to specify the power rating of any necessary resistors.

2-36. Design a wall socket power converter that plugs into a standard electrical wall socket and provides an unregulated DC voltage of 6 V to a portable compact disk player. For maximum conversion efficiency, a transformer-coupled full-wave bridge rectifier is required. The output ripple of the converter is specified to be less than 10% and must be able to provide 0.25 A of constant current. Assume no losses by the transformer and diode $V_\gamma = 0.7$ V.

    a. Determine the circuit configuration, transformer winding ratio, and capacitor size. Use SPICE to confirm the operation of the circuit.

    b. A 6 V Zener diode is used to clamp the output of the wall-mounted converter. Adjust the circuit parameters to ensure stable 6 V output from the converter. Use SPICE to confirm the operation of the circuit.

2-37. Each diode in the given circuit is described by a linearized volt–ampere characteristic with dynamic foward resistance r_d and threshold voltage V_γ. The values for these parameters are listed in the table. The voltage source has value

$V = 200$ V.

    a. Find the diode currents if $R = 20$ kΩ.

    b. Find the diode currents if $R = 4$ kΩ.

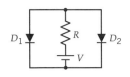

Diode	D_1	D_2
r_d	20	10
V_γ	0.2	0.6
r_r	∞	∞

2-38. For the circuit shown:

    a. Find the equation for the DC load line and plot on the diode characteristic curve provided.

    b. Find the AC load line equation and plot on the diode characteristic curve.

    c. Find v_o for $\omega_o = 2\,\pi(1000)$ rad/s.

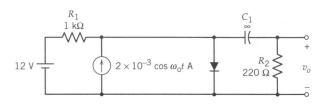

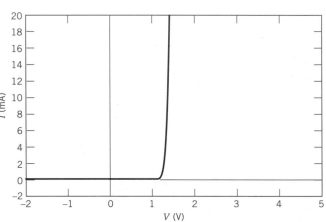

2-39. For the circuit shown:

    a. Find the equation for the DC load line and plot on the diode characteristic curve of the preceding problem.

    b. Find the AC load line equation and plot on the diode characteristic curve.

    c. Find v_o for $\omega_o = 2\pi(1000)$ rad/s.

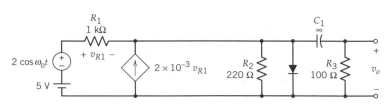

2-40. For the circuit shown:

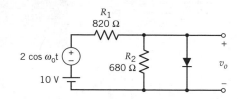

a. Find the equation for the DC load line and plot on the diode characteristic curve provided.

b. Find v_o for $\omega_o = 2\pi(1000)$ rad/s.

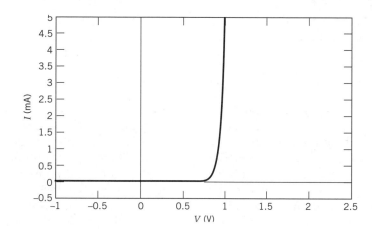

CHAPTER 3

BIPOLAR JUNCTION TRANSISTOR CHARACTERISTICS

The Bipolar Junction Transistor (BJT) is perhaps the most basic of three-terminal semiconductor devices. It can be found, for example, as a vital component in digital and analog integrated circuits, audio and other frequency range amplifiers, radio electronics, and electronic control devices with a wide range of applications. The BJT is an active device that is highly non-linear and, along with applications in non-linear circuitry, plays an important part in many linear electronic applications. The apparent contradiction of a non-linear device being useful in linear applications is placated by a region of BJT operation that is nearly linear. Non-linear BJT operation typically involves a transition between BJT operating regions.

Bipolar junction transistors are constructed with two *p–n* junctions sharing a common region, identified as the base region. This common region, lying between two regions of the complementary doping, causes the two diode-like *p–n* junctions to become coupled.[1] The base region may be doped as either a *p* region or an *n* region: The two types of BJTs formed are identified as *npn* or *pnp*, respectively.

Before proceeding with technical descriptions of the operation of a BJT, it is necessary to define appropriate descriptive conventions. The two circuit symbols for BJTs are shown in Figure 3-1. The three terminals of a BJT are uniquely defined by the circuit symbols and are identified as follows:

- *C*—the collector
- *B*—the base
- *E*—the emitter

The ordering of the characterizing letters *npn* or *pnp* indicates the doping type of the collector, base, and emitter regions, respectively. The two types of BJTs have unique symbolic representation characterized by the direction of the arrow on the emitter terminal, which indicates the direction current would flow if the base–emitter junction were forward biased. The current entering each terminal is identified with the subscript of the terminal: The positive direction for all currents is *into* the device (i.e., I_B is the current entering the base of the transistor, as is drawn in Figure 3-1). The BJT terminal voltage differences will be identified with standard double-sub-

---

[1] Extensive discussions of the semiconductor physics that lead to coupled *p–n* junctions forming a BJT are not within the scope of this electronics text. The authors suggest several texts in semiconductor physics and electronic engineering materials at the end of this chapter for those readers interested in these aspects of physical electronics.

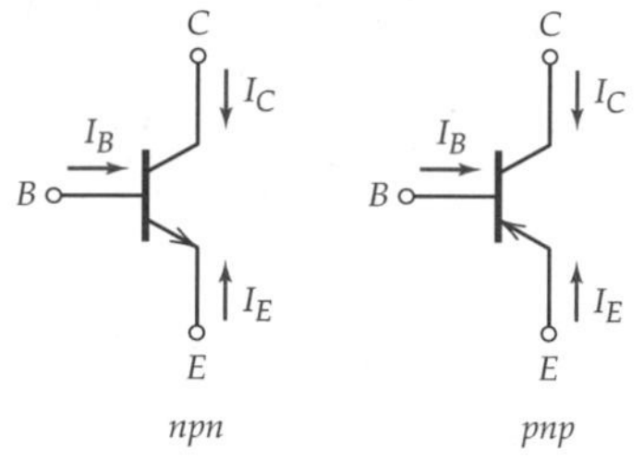

Figure 3-1
Bipolar junction transistor circuit symbols.

script notation: voltage at first subscript with the second subscript as reference. For example, V_{BE} is the voltage at the base terminal with the emitter terminal taken as reference ($V_{BE} = V_B - V_E$).

As is true of all chapters in Section I, the focus of this chapter is on quasistatic (low-frequency), large-signal analysis. Section II of this text will focus on small-signal linear applications (e.g., amplifiers, etc.) of BJTs. Section III will explore the higher frequency ranges.

This chapter begins with a discussion of the principles of BJT operation: The non-linear characteristics of BJT operation are explored through the non-linear Ebers–Moll equations. Quantitative results are obtained through graphic techniques and analytic characterization using SPICE and MathCAD. In order to simplify the analysis of BJT operation, four simple, linear models for the BJT, one for each of its four regions of operation, are derived. Digital logic gates provide a good example of circuitry effectively using BJT regional transitions: The operational characteristics of three BJT logic gate families are explored.

In order to use the BJT as a linear device, operation must be restricted to single-region operation. Amplifiers, the most common BJT linear devices, operate with BJTs biased into a linear region using a variety of circuit topologies. These bias circuits are explored with close interest on two significant design criteria: the establishment of a fixed quiescent operating point and the stability of that operating point to variation in the BJT characteristic parameters.

The Summary Design Example explores a non-linear use of a BJT as a controlled switch in a Zener diode voltage regulator circuit. Such usage can greatly increase circuit efficiency while reducing component cost.

3.1 BJT *V–I* RELATIONSHIPS

Much like the semiconductor diode, the BJT can be described empirically by a set of experimentally derived curves or theoretically by a set of equations. Because there are three terminals and the action of the two *p–n* junctions are coupled, a single *V–I* relationship (as was possible for the diode) is not applicable to the BJT: A *set* of curves or equations is necessary.

A set of empirical curves for a typical *npn* BJT is shown in Figures 3.1-1(*a–d*). While only the first (*a* and *b*) or last (*c* and *d*) pair of curves are necessary for complete description of the BJT, both pairs are shown for completeness. These characteristic curves are grouped into two categories:

Common-base characteristics

Input characteristics: In common-base configurations, the emitter terminal is the input and the base–emitter junction is the primary control. Figure 3.1-1*a* is a

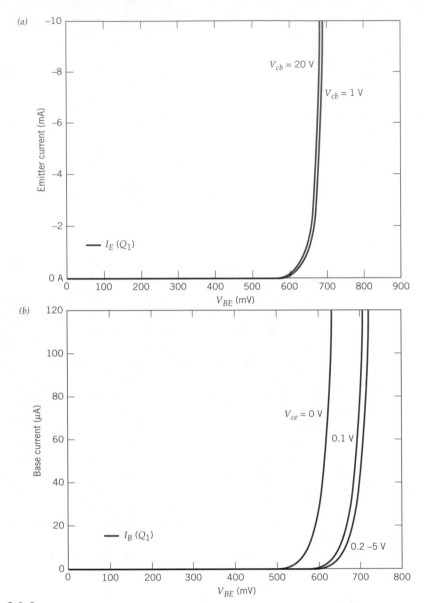

Figure 3.1-1
2N2222A input and output characteristics.

plot of the input (emitter) current as a function of the control (base–emitter) voltage with the output (base–collector) voltage as a parameter.

Output characteristics: The output for this configuration is at the collector terminal. Figure 3.1-1*b* shows the output (collector) current as a function of the output (collector–base) voltage with the input (emitter) current as a parameter.

Common-emitter characteristics

Input characteristics: In common-emitter configurations, the base terminal is the input and the base–emitter junction is the primary control. Figure 3.1-1*c* is a plot of the input (base) current as a function of the control (base–emitter) voltage with the output (collector–emitter) voltage as a parameter.

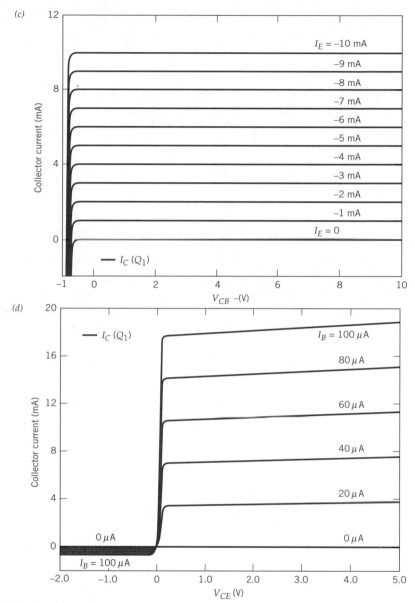

Figure 3.1-1 (continued)

Output characteristics: The output for this configuration is also at the collector terminal. Figure 3.1-1*d* shows the output (collector) current as a function of the output (collector-emitter) voltage with the input (base) current as a parameter.

One of the most accurate and simple theoretical characterizations of the BJT are the Ebers–Moll equations, initially derived by J. J. Ebers and J. L. Moll.[2] These equations relate the collector and emitter currents to the base–collector junction and base–

[2] J. J. Ebers and J. L. Moll, "Large Signal Behavior of Junction Transistors," *Proceedings of the IRE*, Vol. 42, No. 12, 1954, pp. 1761–1772.

emitter junction voltages. The third BJT current, base current, may then be calculated using KCL applied to the BJT as a whole:

$$I_B = -(I_C + I_E). \tag{3.1-1}$$

The collector–emitter voltage may be derived from the base–collector and base–emitter voltages by applying KVL around the BJT terminals:

$$V_{CE} = V_{BE} - V_{BC}. \tag{3.1-2}$$

The Ebers–Moll equations for an *npn* BJT[3] are given by

$$I_E = -I_{ES}(e^{V_{BE}/\eta V_t} - 1) + \alpha_R I_{CS}(e^{V_{BC}/\eta V_t} - 1) \tag{3.1-3a}$$

$$I_C = -I_{CS}(e^{V_{BC}/\eta V_t} - 1) + \alpha_F I_{ES}(e^{V_{BE}/\eta V_t} - 1) \tag{3.1-3b}$$

where V_t is the voltage equivalent temperature defined in Chapter 2, η is an empirical scaling constant that depends on geometry, material, and doping levels,[4] and I_{ES} and I_{CS} are the emitter and collector *p–n* junction saturation currents for a specified temperature. As with diode leakage currents, I_{ES} and I_{CS} have temperature variation similar to *p–n* junction temperature variations (described in Section 2.1): The saturation currents roughly double with every 6 K increase in temperature.

The quantities α_F and α_R have particular significance. In order to understand this significance, it is best to perform two mental experiments upon the equations. As a first test, set $V_{BE} \gg \eta V_t$ and $V_{BC} \ll -\eta V_t$, that is, strongly forward bias the base–emitter junction while the base–collector junction is strongly reverse biased. The terms with I_{CS} in Equations 3.1-3 become negligible and I_E and I_C are related by α_F:

$$I_C \approx -\alpha_F I_E. \tag{3.1-4}$$

Here, α_F is therefore identified as the DC collector–emitter current gain.

Figures 3.1-1*b,d* graphically demonstrate this relationship for two different circuit connections of a BJT. It can be seen from the curves that the quantity α_F is very close to, but slightly smaller than, unity. Equation 3.1-1 can be substituted into Equation 3.1-4 to get the ratio of the collector current to the base current:

$$I_C = \frac{\alpha_F}{1 - \alpha_F} I_B = \beta_F I_B \quad \text{where} \quad \beta_F = \frac{\alpha_F}{1 - \alpha_F}. \tag{3.1-5}$$

Figure 3.1-1*d* graphically demonstrates this relationship. It can be seen that these bias conditions lead to a region where an approximate *linear relationship* exists among the

---

[3] The *pnp* BJTs are constructed with the *p–n* junctions in the opposite direction from *npn* BJTs. Therefore the polarities of the junction voltages and currents must be reversed. The Ebers–Moll equation for *pnp* BJTs are

$$I_E = I_{ES}(e^{V_{EB}/\eta V_t} - 1) - \alpha_R I_{CS}(e^{(V_{CB}/\eta V_t} - 1),$$

$$I_C = I_{CS}(e^{V_{CB}/\eta V_t} - 1) - \alpha_F I_{ES}(e^{V_{EB}/\eta V_t} - 1).$$

[4] The Ebers–Moll equations are written here with a single value of η. In fact, it is possible for each of the *p–n* junctions to have an individual value of this scaling constant. In practice, the values are nearly identical—hence the use of a single value here. SPICE and most other circuit emulators allow for the possibility of individual junction values of η and require the user to input both values should a change from the default value of unity be desired.

BJT currents. This linear region is identified as the *forward-active* region and is defined by forward-biased base–emitter and reverse-biased base–collector junctions.

If the opposite biasing scheme is used, set $V_{BC} \gg \eta V_t$ and $V_{BE} \ll -\eta V_t$, that is, strongly forward bias the base–collector junction while the base–emitter junction is strongly reverse biased. The terms with I_{ES} in Equations 3.1-3 become negligible and I_E and I_C are now related by α_R:

$$I_E \approx -\alpha_R I_C, \tag{3.1-6}$$

where α_R is identified as the DC collector–emitter reverse current gain.

A relationship similar to Equation 3.1-5 can be determined relating the emitter and base currents:

$$I_E = \frac{\alpha_R}{1 - \alpha_R} I_B = \beta_R I_B \quad \text{where} \quad \beta_R = \frac{\alpha_R}{1 - \alpha_R}. \tag{3.1-7}$$

Figure 3.1-1*d* also demonstrates this relationship ($V_{CE} < 0$). Here the bias conditions again lead to an approximate linear relationship among the BJT currents. This linear region is identified as the *inverse-active* region and is defined by forward-biased base–collector and reverse-biased base–emitter junctions. Notice that β_R is typically much smaller than β_F. While it is *possible* for a BJT to have the forward and reverse values of β to be nearly identical, BJTs are typically designed for optimal performance in the forward-active region: This design process leads to the significantly larger value for β_F than for β_R. Manufacturing conditions lead to a relationship between the forward and reverse current gain expressions:

$$\alpha_R I_{CS} = \alpha_F I_{ES} = I_S, \tag{3.1-8}$$

where I_S, the *transistor saturation current*, is a constant for any particular BJT.

It should be noted that Equations 3.1-4 and 3.1-5 and 3.1-5 3.1-7 apply to *different biasing conditions* and cannot be valid simultaneously: Each set of bias conditions has its own application.

There are two other regions of particular interest in the operation of a BJT. The first of these occurs when both base–collector and base–emitter junctions are reverse biased. Under these bias conditions the emitter and collector currents become (simplifying Equation 3.1-3)

$$I_E = \approx I_{ES} - \alpha_R I_{CS} \tag{3.1-9}$$

and

$$I_C \approx I_{CS} - \alpha_F I_{ES}. \tag{3.1-10}$$

The currents become basically junction leakage currents. This region of BJT operation corresponds to two *p–n* junctions that are reverse biased or turned off. The region is called the *cutoff* region.

The last region of interest occurs when both *p–n* junctions are forward biased. This region is highly non-linear and is shown in Figures 3.1-1*b,d* by the convergence of curves near the origin of the horizontal axis. In this region the relationship between the BJT currents is not clear; however, the terminal base–emitter and base–collector

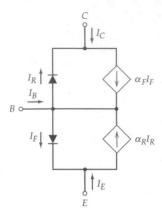

Figure 3.1-2
Ebers–Moll Model of *npn* BJT.

voltages correspond to voltage across forward-biased *p–n* junctions and therefore lie in the range of approximately 0.6–0.9 V. This region is called the *saturation* region.

The Ebers–Moll equations can be easily converted into a circuit model consisting of a pair of diodes and a pair of dependent current-controlled current sources. This highly useful model is shown for an *npn* BJT[5] in Figure 3.1-2.

Notice that KCL applied to the emitter and collector nodes produces the following equations:

$$I_E = -I_F + \alpha_R I_R, \tag{3.1-11}$$

$$I_C = -I_R + \alpha_F I_F, \tag{3.1-12}$$

where the currents in the diodes are given by

$$I_F = I_{ES}(e^{V_{BE}/\eta V_t} - 1), \tag{3.1-13}$$

$$I_R = I_{CS}(e^{V_{BC}/\eta V_t} - 1). \tag{3.1-14}$$

Substitution of these diode equations (as derived in Chapter 2) into Kirchhoff's equations at the nodes produces the usual form of the Ebers–Moll equations, as seen in Equation 3.1-3 (exercise left to the reader).

The Ebers–Moll model and the related Ebers–Moll equations accurately predict the behavior of a BJT throughout its regions of operation at low frequencies. The model does not include power dissipation restrictions or the possibility of the reverse breakdown of either junction due to excessive reverse voltages being applied. Addition of capacitors across the junctions as described in Chapter 10 expands the model to include high-frequency effects.

3.2 THE BJT AS A CIRCUIT ELEMENT

The operation of a BJT in a simple circuit can be derived in much the same manner as the operation of a diode in a simple circuit. There are several choices; among the most obvious are the following:

---

[5] The Ebers–Moll model for a *pnp* BJT takes exactly the same form as that of an *npn* BJT except the two diodes are *reversed in direction*. The dependent current sources and their controlling currents keep the *same polarity* even though forward biased and reverse biased now imply currents in the opposite direction.

- Use the Ebers–Moll set of equations and obtain a numerical solution.
- Use the empirical V–I curves and obtain a graphical solution.
- Use computer simulation to obtain a solution.

The choice of solution technique depends strongly on the complexity of the circuit in which the BJT is operating.

EXAMPLE 3.2-1 A BJT with the parameters

$$\alpha_F = 0.995, \qquad \alpha_R = 0.95, \qquad \eta = 1, \qquad I_S = 0.1 \text{ fA}$$

is connected as shown.

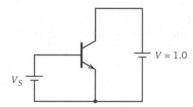

Determine the collector, base, and emitter currents for the following values of the source voltage V_S: 0.4, 0.6, 0.7, 0.8, 0.9, and 1.0 V.

SOLUTION

The simple connection of the voltage sources leads directly to the determination of the base–emitter and base–collector junction voltages. Thus, using the Ebers–Moll equations seems the most direct method of solution. The Ebers–Moll equations are

$$I_E = -I_{ES}(e^{V_{BE}/\eta V_t} - 1) + \alpha_R I_{CS}(e^{V_{BC}/\eta V_t} - 1),$$
$$I_C = -I_{CS}(e^{V_{BC}/\eta V_t} - 1) + \alpha_F I_{ES}(e^{V_{BE}/\eta V_t} - 1).$$

Here, the following substitutions due to the circuit connections can be made:

$$V_{BE} = V_S, \qquad V_{BC} = V_S - 1.0, \qquad I_{ES} = \frac{I_S}{\alpha_F} = 0.101 \text{ fA},$$

$$I_{CS} = \frac{I_S}{\alpha_R} = 0.105 \text{ fA}.$$

Here, I_B can be calculated by applying KCL to the BJT:

$$I_B = -(I_C + I_E).$$

Direct substitution of these values into the Ebers–Moll equations yields the following values:

V_S	I_C	I_E	I_B
0.4	480.2 pA	−482.7 pA	2.413 pA
0.6	1.052 μA	−1.058 μA	5.289 nA
0.7	49.27 μA	−49.51 μA	247.6 nA
0.8	2.306 mA	−2.318 mA	11.59 μA
0.9	108.0 mA	−108.5 mA	542.5 μA
1.0	5.054 A	−5.079 A	25.40 mA

Notice that when V_S is small (i.e., $V_S \approx 0.4$ V), the collector current is very small. When V_S increases by 50% to 0.6 V, the current jumps by a factor of more than 2000: The base–emitter junction of the BJT has become forward biased while the base–collector junction has remained reverse-biased, and the BJT has transitioned from the cutoff region to the forward-active region. For 0.6 V $\leq V_S \leq 1.2$ V, the BJT remains in the forward-active region and the base and collector currents are related by

$$I_C = \frac{\alpha_F}{1 - \alpha_F} I_B = \frac{0.995}{1 - 0.995} I_B = 199 I_B.$$

Notice that in the forward-action region, small changes in the base current produce significantly larger changes in the collector (and consequently the emitter) current.

EXAMPLE 3.2-2 A BJT with the parameters

$$\alpha_F = 0.995, \qquad \alpha_R = 0.95, \qquad \eta = 1, \qquad I_S = 0.1 \text{ fA}$$

is connected as shown, with the following resistor values:

$$R_b = 3.3 \text{ k}\Omega, \qquad R_c = 220 \ \Omega.$$

Determine the collector, base, and emitter currents for the following values of the source voltage V_S: 0.4, 0.6, 0.7, 0.8, 0.9, 1.0, and 1.2 V.

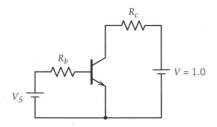

SOLUTION 1 (EBERS–MOLL EQUATIONS)

Here the Ebers–Moll equations are not sufficient to determine the currents. Two additional equations, dependent on the circuit topology and parameters, are needed.

Loop equations around base–emitter and base–collector loops yield

$$V_{BE} = V_S - I_B R_b,$$

$$V_{BC} = V_S - I_B R_b - (1.0 - I_C R_C).$$

Combining these two equations with Ebers–Moll equations and searching for solutions is, by hand, quite complex. Realistically, a computer search for solutions is the only practical method of solution. A Computer search solution using MathCAD for the above circuit can be performed as follows:

```
Solution of Simple npn BJT with Two Resistors—T.F. Schubert & E.M. Kim
Defining BJT and Circuit Parameters
```

$$I_s := 0.1 \cdot 10^{-15} \quad \alpha_F := 0.995 \quad \alpha_R := 0.95 \quad V_t := 0.026 \quad R_b := 3300$$

$$I_{es} := \frac{I_s}{\alpha_F} \quad I_{cs} := \frac{I_s}{\alpha_R} \quad \eta := 1 \quad V_q := 1.0 \quad R_c := 220$$

Guess values:

$$I_c := 0.002 \quad I_e := -0.0021 \quad V_{be} := 0.8 \quad V_{bc} := -0.2$$

Given (solve block):

$$V_{be} = V_S + (I_e + I_c) \cdot R_b \qquad V_{bc} = V_{be} - (1 - I_c \cdot R_c)$$

$$I_e = -I_{es} \cdot \left(e^{\frac{V_{be}}{\eta \cdot V_t}} - 1 \right) + \alpha_R \cdot I_{cs} \cdot \left(e^{\frac{V_{bc}}{\eta \cdot V_t}} - 1 \right)$$

$$I_c = -I_{cs} \cdot \left(e^{\frac{V_{bc}}{\eta \cdot V_t}} - 1 \right) + \alpha_F \cdot I_{es} \cdot \left(e^{\frac{V_{be}}{\eta \cdot V_t}} - 1 \right)$$

$$x := \text{Find} (I_c, I_e, V_{be}, V_{bc})$$

$$x = \begin{bmatrix} 4.29171 \cdot 10^{-3} \\ -4.34639 \cdot 10^{-3} \\ 0.81956 \\ 0.76374 \end{bmatrix} \qquad I_B := -(x_0 + x_1) \quad I_B = 5.468 \cdot 10^{-5}$$

Repeated use of this MathCAD program yield a set of results for the various input voltage values:

V_S	I_C	I_E	I_B
0.4	480.2 pA	−482.6 pA	2.413 pA
0.6	1.052 μA	−1.057 μA	5.285 nA
0.7	47.78 μA	−48.03 μA	240.1 nA
0.8	1.125 mA	−1.131 mA	5.654 μA
0.9	4.088 mA	−4.114 mA	25.62 μA
1.0	4.292 mA	−4.346 mA	54.68 μA
1.2	4.405 mA	−4.764 mA	113.6 μA

Notice that the addition of a resistor on the collector of the BJT creates a transition from the forward-active region to the saturation region. In the saturation region of a BJT the base and collector currents no longer have a constant linear relationship. For example, in this circuit the collector-to-base current relationships are as follows:

V_S	I_C/I_B	V_{CE}
0.4	199	1.000
0.6	199	1.000
0.7	199	0.989
0.8	199	0.7525
0.9	160	0.1005
1.0	78.5	0.0558
1.2	38.6	0.0349

The transition to the saturation region is signaled by a change in the ratio of the collector current to the base current. Saturation occurs when

$$\frac{I_C}{I_B} < \beta_F,$$

which appears to occur in this circuit application when V_S increases to a value larger than something slightly less than 0.9 V.

The last column in the table reports the voltage at the collector of the BJT. Notice that for small values of the input voltage ($V_S \le 0.6$), the collector voltage is essentially the collector supply voltage (1.0 V): for large values of the input voltage ($V_S > 0.9$), the collector voltage nears zero. This property of a near-constant output voltage value for a range of input values has special significance in digital applications circuitry. These applications are discussed at length in Section 3.5.

SOLUTION 2 (GRAPHICAL TECHNIQUES)

The empirical curves for the BJT coupled with load line techniques provide a more direct form of solution. The two supplemental equations derived in the first solution are actually the equations for load lines (one in the base–emitter loop and one in the collector–emitter loop):

$$V_{BE} = V_S - I_R R_b, \tag{3.2-1}$$

$$V_{BC} = V_S - I_B R_b - (1.0 - I_C R_C). \tag{3.2-2}$$

The V–I relationships for this BJT are shown in Figures 3.2-1a,b.[6]

---

[6] The curves for this fictitious BJT were generated using PSpice with the following input file (the BJT parameter assignments are explained in Solution 3 of this example):

```
Transistor Curve Traces for Example 3.2-2—T. F. Schubert & E. M. Kim
IB 0 1 DC 0
VCE 2 0 DC 1
Q1 2 1 0 NPNSPEC
.MODEL NPNSPEC NPN (IS=0.1E-15 BF=199 BR=19 NF=1 NR=1)
*For each curve choose one of the following ''.DC'' lines-comment the other
*The following line produces the output transistor curves
.DC DEC VCE 0.001 1 10 LIN IB 0 200U 5U
*The following line produces the input transistor curves
*.DC DEC IB 0.1U 400U 10 DEC VCE 0.01 1 2
.PROBE
.OPTIONS NOPAGE TNOM=28.6
.END
```

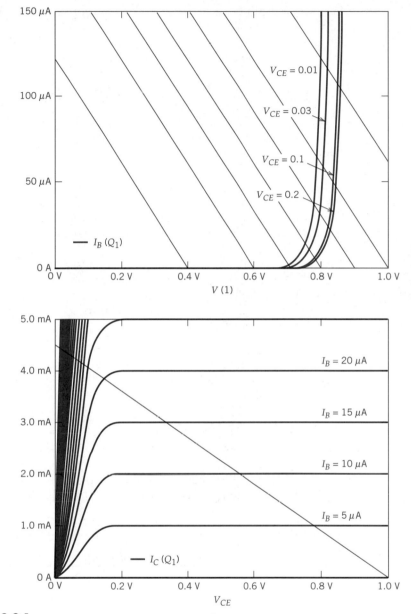

Figure 3-2.1

Transistor curve traces for Example 3.2-2.

Solution takes the form of plotting the load lines on the transistor curves: Equation 3.2-1 is plotted on the output curves (Figure 3.2-1b), and Equation 3.2-2 is plotted as a series of parallel lines on the input transistor curves (Figure 3.2-1a), one line for each of the specified values of V_S. Unfortunately, the load lines cross several of the transistor parameter curves. It is the task of the circuit analyst/designer to obtain a solution that is consistent with all constraints. For example, take the case where $V_S = 0.9$ V. Looking at the input curves, one concludes that the base current must approximately lie in the range

$$20 \ \mu A \leq I_B \leq 40 \ \mu A.$$

From the output curves, using those values of I_B, one then concludes that V_{CE} must lie in the range

$$0.08 \text{ V} \leq V_{CE} \leq 0.15 \text{ V}.$$

This set of restrictions reduces the range of allowable base currents to

$$23 \text{ }\mu\text{A} \leq I_B \leq 27 \text{ }\mu\text{A},$$

which in turn reduces the range of V_{CE} to

$$0.09 \text{ V} \leq V_{CE} \leq 0.11 \text{ V}.$$

Continuing along this path, one progresses to a final solution for $V_S = 0.9$. This solution takes the form of the following voltage and current values:

$$V_{CE} \approx 0.1 \text{ V},$$

$$I_B \approx 25 \text{ }\mu\text{A},$$

$$I_C = \frac{1.0 - V_{CE}}{220} \approx \frac{1.0 - 0.1}{220} = 4.1 \text{ mA}.$$

While reading values off graphs will introduce some margin of error, this result is very close to that of Solution 1: Other values of V_S will similarly produce results equivalent to those previously obtained.

SOLUTION 3 (SPICE SIMULATION)

This problem is particularly suited for DC analysis using SPICE. Of particular importance is the modeling of the BJT to suit the Ebers–Moll parameters given. SPICE uses a model for BJTs that in its most simple formulation becomes the Ebers–Moll model. The parameters necessary for input for this problem are as follows:

Parameter	SPICE Variable	Value
I_S	IS	0.1×10^{-15}
η	NF	1
	NR	1
β_F	BF	$\dfrac{\alpha_F}{1 - \alpha_F} = \dfrac{0.995}{1 - 0.995} = 199$
β_R	BR	$\dfrac{\alpha_R}{1 - \alpha_R} = \dfrac{0.95}{1 - 0.95} = 19$

When comparing simulation techniques, it is also necessary to make sure all other equation parameters are identical. The MathCAD simulation used the common approximation $V_t = 0.026$ V. This approximation is quite valid for room temperature ≈ 300 K; it corresponds to $\approx 28.6°C \approx 301.7$ K. The output file for the SPICE simulation is shown in Figure 3.2-2. Notice the results are within $\pm 0.11\%$ of those calculated using the Ebers–Moll equations solved directly with MathCAD.

```
*********************Evaluation PSpice****************************
T. F. Schubert & E. M. Kim

**** CIRCUIT DESCRIPTION
*********************************************************
VIB 11 0 DC 0
RB 11 1 3300
VCE 22 0 DC 1
RC 22 2 220
Q1 2 1 0 NPNSPEC
.MODEL NPNSPEC NPN (IS=0.1E-15 BF=199 BR=19 NF=1 NR=1)
.DC VIB LIST 0.4 0.6 0.7 0.8 0.9 1.0 1.2
.PRINT DC IC(Q1) IE(Q1) IB(Q1) V(1) V(1,2)
.OPTIONS NOPAGE TNOM=28.6
.END

**** BJT MODEL PARAMETERS

     NPNSPEC
     NPN
  IS 100.000000E-18
  BF 199
  NF 1
  BR 19
  NR 1

****  DC TRANSFER CURVES    TEMPERATURE = 28.600 DEG C
VIB         IC(Q1)      IE(Q1)       IB(Q1)      V(1)       V(1,2)
4.000E-01   4.807E-10   -4.831E-10   2.381E-12   4.000E-01  -6.000E-01
6.000E-01   1.050E-06   -1.056E-06   5.279E-09   6.000E-01  -3.998E-01
7.000E-01   4.769E-05   -4.793E-05   2.396E-07   6.992E-01  -2.903E-01
8.000E-01   1.124E-03   -1.129E-03   5.646E-06   7.814E-01   2.857E-02
9.000E-01   4.088E-03   -4.114E-03   2.560E-05   8.155E-01   7.149E-01
1.000E+00   4.292E-03   -4.346E-03   5.466E-05   8.196E-01   7.638E-01
1.200E+00   4.387E-03   -4.500E-03   1.136E-04   8.251E-01   7.902E-01

     JOB CONCLUDED

     TOTAL JOB TIME    2.09
```

Figure 3.2-2
SPICE simulation output.

3.3 REGIONS OF OPERATION IN BJTs

Bipolar junction transistor operation has been seen to fall into four basic regions of useful operation. The regions are described by the state of bias of the two p–n junctions within the transistor. The four possible combinations and the corresponding region names are shown in Figure 3.3-1. Briefly, the four regions of operation are as follows:

1. The cutoff region is defined by both base–emitter and base–collector junctions being reverse biased. Reverse biasing both junctions reduces all current in a BJT to small leakage values in the picoampere to nanoampere range: The BJT essentially looks like an open circuit. Applications for this region are primarily in the switching and digital logic areas.

2. The saturation region is defined by both junctions being forward biased. Here the possibility exists for large current flow between the collector and

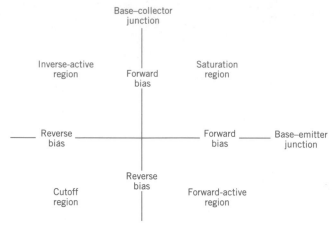

Figure 3.3-1
Four basic regions of BJT operation.

emitter terminals with minimal dynamic resistance ($V_{CE} \approx 0$). Applications again fall in the switching and digital logic areas.

3.   The forward-active region is defined as a forward-biased base–emitter junction and a reverse-biased base–collector junction. The BJTs operating in this region are characterized by a relatively constant collector current to base current ratio. The region is most commonly used for amplification with the parameters α_F and β_F describing the amplification.

4.   The inverse-active region is the direct opposite of the forward-active region: The base–collector junction is forward biased and the base–emitter junction is reverse biased. Here emitter current is a multiple of base current with α_R and β_R describing the amplification. While it is certainly possible to manufacture a BJT with amplification in the inverse-active region as large as that of the forward-active region, most BJTs are optimized for forward amplification, resulting in much smaller values for α and β in this region. The region is rarely used with the notable exception of the input stages of the transistor–transistor logic (TTL) family of digital logic gates.

In addition to these four regions, there is an additional region of severe consequences: the breakdown region. A BJT enters the breakdown region when one or both of the p–n junctions are sufficiently reverse biased so that a Zener-like breakdown occurs. Transistors are not manufactured to withstand extended use in the breakdown region and typically will exhibit catastrophic thermal runaway and destruction.

Manufacturers list maximum voltages that can be safely applied to the junctions and maximum power limitations in order to ensure that the devices are operated safely.

3.4 MODELING BJT IN ITS REGIONS OF OPERATION

The Ebers–Moll model for BJTs is a flexible, but rather complex, model that can be used in all four useful regions of operation. It can, however, be simplified within each region to provide a set of elementary BJT models, one for each region. The derivation of each of the individual models is described below and a summary table

Single-region model equivalent circuits (*npn* voltage polarities shown) (*pnp voltage* polarities are reversed)	Characteristics of models	
	npn	*pnp*
1. Cutoff region model 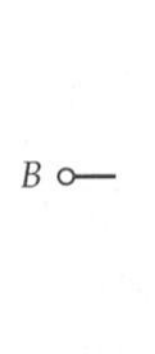	$I_C = I_B = I_E = 0$ $V_{BE} < V_{BE(on)}$ $V_{BC} < V_{BC(on)}$	$I_C = I_B = I_E = 0$ $V_{BE} > -V_{BE(on)}$ $V_{BC} > -V_{BC(on)}$
2. Saturation region model 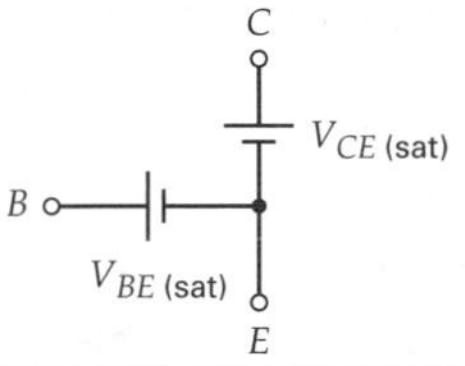	$I_B > 0$ $I_C > 0$ $I_C < \beta_F I_B$ $V_{BE} > V_{BE(on)}$	$I_B < 0$ $I_C < 0$ $|I_C| < |\beta_F I_B|$ $V_{BE} < -V_{BE(on)}$
3. Forward-active model 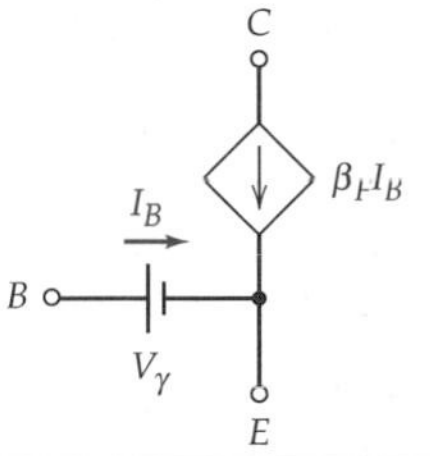	$I_C = \beta_F I_B$ $I_B > 0$ $V_{BE} > V_{BE(on)}$ $V_{CE} > V_{CE(sat)}$	$I_C = \beta_F I_B$ $I_B < 0$ $V_{BE} < -V_{BE(on)}$ $V_{CE} < -V_{CE(sat)}$
4. Inverse-active model 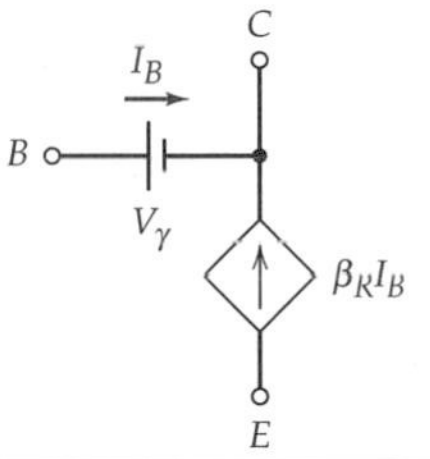	$I_E = \beta_R I_B$ $I_B > 0$ $V_{BC} > V_{BE(on)}$ $V_{EC} > V_{CE(sat)}$	$I_E = \beta_R I_B$ $I_B < 0$ $V_{BC} < -V_{BE(on)}$ $V_{EC} < -V_{CE(sat)}$

Figure 3.4-1
Bipolar junction transistors: linear models.

listing each model and conditions to test the validity of each model appears in Figure 3.4-1.[7]

1. The *cutoff region* is defined by both base–emitter and base–collector junctions being reverse biased. Reverse-biased *p–n* junctions provide extremely high impedance and low leakage currents; the low currents in the junctions reduce the dependent current sources of the Ebers–Moll model to near-zero value.

---

[7] In Figure 3.4-1 the voltage source polarities are correct for *npn* BJTs. The circuit diagrams for *pnp* BJTs are identical with the polarity of the voltage sources reversed. One common error in making this change involves the polarity of the dependent current sources: The polarity (or direction) of the dependent current sources *stays the same* for both types of BJT. The currents flow in opposite directions, but the *ratio*, as indicated by the direction of the dependent current sources, does not change polarity.

Therefore, a simple model of a BJT in cutoff is three terminals with open circuits between. The typical turn-on voltage (the voltage at which a junction becomes forward biased) for a silicon BJT p–n junction is

$$V_{BE(on)} = V_{BC(on)} = 0.6 \text{ V}$$

2. The *saturation region* is defined by both p–n junctions being forward biased. Forward-biased junctions can be modeled by a voltage source in series with a small resistance. The model most commonly used ignores this small resistance and models the BJT with two constant-voltage sources. Rather than show a voltage source connected from the base to each of the other terminals, it is most common to model the BJT with a voltage source from the base to the emitter and another from the emitter to the collector: This second voltage source will be a small-value source since it models the difference in forward-bias voltage for the two junctions. Typically one junction will tend to be more forward biased than the other (i.e., the base–emitter junction is often more forward biased than the base–collector junction) due to the different currents passing through each junction. For silicon transistors typical model values are

$$V_{BE(sat)} = 0.8 \text{ V}, \quad V_{CE(sat)} = 0.2\text{V}.$$

3. The *forward-active region* is defined as a forward-biased base–emitter junction and a reverse-biased base–collector junction. Here the forward-biased base–emitter junction is modeled by a voltage source and the reverse-biased junction by an open circuit. The current through the forward-biased base–emitter junction activates the dependent current source between the collector and base in the Ebers–Moll model while the other current source is inactive (its controlling current is near-zero value). Typical values for silicon BJTs are

$$V_{\gamma} = 0.7 \text{ V}.$$

4. The *inverse-active region* is the opposite of the forward-active region. The base–collector junction is modeled by a voltage source and a dependent current source is connected between the emitter and base.

When one complex model is separated into a group of simple models, several difficulties can occur. The most prevalent of these difficulties is the choice of the proper model for the circumstances in question. Whenever a circuit element is replaced by a model that is only correct for one region of operation of the circuit element, the assumptions upon which the model rests much be tested in order to verify the validity of the replacement. Experience leads to the proper choice of current model on the first guess; incorrect guesses, when tested for validity, give clues as to which is the correct model to choose next.

EXAMPLE 3.4-1 Assume as given the circuit shown with element values

$$V_{bb} = 2 \text{ V}, \quad V_{cc} = 10 \text{ V}, \quad R_b = 22 \text{ k}\Omega, \quad R_e = 100 \text{ }\Omega,$$
$$R_c = 2 \text{ k}\Omega$$

and a silicon BJT with

$$\alpha_F = 0.99.$$

Determine the region of operation for the BJT and the base and collector currents.

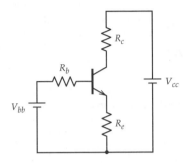

SOLUTION

First a region of operation must be chosen. It seems clear that the inverse-active region can clearly be eliminated: The base must be at higher voltage than the collector with positive base and negative collector currents. Also easily eliminated is the cutoff region: With no current flowing in the BJT, the base–emitter and base–collector junctions would clearly be forward biased (a violation of the assumptions for that region). The choice clearly lies between the forward-active region and the saturation region. The correct choice is not obvious.

As a first try, assume the BJT is operating in the saturation region.

ATTEMPT 1 (SATURATION)

Replace the BJT with its saturation region model (indicated in Figure 3.4-1) and then calculate the terminal currents for the BJT.

Around the left mesh, KVL gives

$$V_{bb} - R_b I_B - 0.8 - R_e(I_B + I_C) = 0.$$

Around the right mesh, the equation is

$$V_{cc} - R_c I_C - 0.2 - R_e(I_B + I_C) = 0.$$

Inserting the circuit values yields two equations in the two unknowns, I_B and I_C:

$$22{,}100 I_B + 100 I_C = 1.2, \qquad 100 I_B + 2100 I_C = 9.8.$$

The solutions are

$$I_B = 33.19 \ \mu\text{A} \quad \text{and} \quad I_C = 4.665 \ \text{mA}.$$

These solutions must now be checked to see if they are consistent with the saturation region model of the BJT. From Figure 3.4-1, the *npn* saturation region characteristics of interest are

$$I_B > 0, \qquad I_C > 0, \qquad I_C < \beta_F I_B.$$

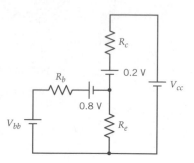

The base and collector currents are positive, but the ratio of these currents is

$$\frac{I_C}{I_B} = \frac{4.665 \text{ mA}}{33.19 \text{ μA}} = 140.6 > \beta_F = \frac{\alpha_F}{1 - \alpha_F} = \frac{0.99}{1 - 0.99} = 99.$$

Thus the basic assumptions of the saturation region model have been violated. The verification of assumptions has indicated that, in all likelihood, the collector current to base current ratio is not smaller than β_F—a sign that the BJT must be in the forward-active region.

ATTEMPT 2 (FORWARD ACTIVE)

Replace the BJT with its forward-active region model (indicated in Figure 3.4-1) and then calculate the terminal currents for the BJT.

Around the left mesh, KVL produces (notice that the current out of the emitter is the base current plus the collector current)

$$V_{bb} - I_B R_b - V_\gamma - (1 + \beta_F)I_B R_e = 0.$$

With the circuit element values and the value for V_γ (0.7) inserted, the only unknown is I_B. The solution of the equation is

$$I_B = 40.6 \text{ μA}.$$

Then I_C can be calculated as

$$I_C = \beta_F I_B = 4.02 \text{ mA}.$$

The appropriate region verification checks are

$$I_B > 0, \qquad V_{CE} > V_{CE(sat)} = 0.2.$$

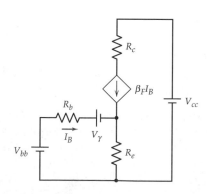

The base current has the correct sign: V_{CE} can be calculated by using KVL around the right mesh:

$$V_{cc} - I_C R_c - V_{CE} - (I_C + I_B)R_e = 0,$$

which yields

$$V_{CE} = 1.54 \text{ V}.$$

The forward-active region verifications have been shown to be valid; thus the currents calculated for the forward-active region are the correct values.

3.5 DIGITAL ELECTRONICS APPLICATIONS

Many digital electronics applications are based upon a BJT changing from one region of operation to another. The simple regional models of BJT operation are sufficient to understand the basic operation of several logic gate types.[8]

In digital systems there are only a few basic logic operations that must typically be performed. The most common of these operations are NOT, AND, and OR. A common property of these operations is that a variety of inputs produce an output in binary form; that is, an output that exists in one of two possible states. Typical electronic circuitry assigns to each logic value (i.e., 1 or 0, HIGH or LOW, ON or OFF) a specific range of voltage values. As an example, the transistor-transistor logic (TTL) circuit family, one of the basic logic circuit families that will be discussed in this section, assigns two logic voltage ranges:

Low-voltage range: $0 \text{ V} \leq V_L \leq 0.8 \text{ V}$

High-voltage range: $2.0 \text{ V} \leq V_H \leq 5.5 \text{ V}$

One of the major goals of an electronic circuit that will perform a logic operation is to provide a constant-value output that is invariant to this form of voltage variation on the input. The simple circuit of Example 3.2-2 showed this property of virtually invariant output voltage to a range of input voltage values. A circuit with similar properties is shown in Figure 3.5-1.

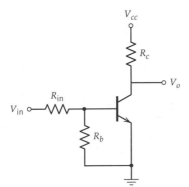

Figure 3.5-1
Simple logic inverter.

---

[8] Gate speed and some of the more advanced digital circuitry topics will be discussed in Chapter 16.

3.5.1 Logic Inverter Circuit

If one assumes that V_{cc} is a positive voltage, the operation of the circuit can be described as follows:

1. For small values of V_{in}, the BJT is in the cutoff region. The base and collector currents are near zero, and the output voltage V_o is basically the same value as V_{cc}.

2. As V_{in} increases, the base–emitter voltage on the BJT will increase until the BJT turns on and the BJT enters the forward-active region. This transition of regions will occur at an input voltage

$$V_{in} = \frac{R_b + R_{in}}{R_b} V_{BE(on)}. \tag{3.5-1}$$

As the input voltage continues to increase, the output voltage will decrease steadily until it approaches $V_{ce(sat)}$:

$$V_o = 5 - 50 I_B (2.2 \text{ k}\Omega),$$

where

$$I_B = \frac{V_{in} - V_{BE(on)}}{R_{in}} - \frac{V_{BE(on)}}{R_b}$$

3. When the output voltage reaches $V_{ce(sat)}$, the BJT will enter the saturation region, and further increases in the input voltage will result in negligible changes in the output voltage.

This simple circuit forms the basis of a NOT gate (also known as a logic inverter): Low-value input voltages become high output voltages and vice versa. The BJT switches between the cutoff and the saturation regions to form the two logic levels. The high-output region (BJT in cutoff) is described as point 1 above, and the low region (BJT in saturation) is described as point 3. A transition region (BJT in forward-active region) is described as point 2 and serves as a buffer to isolate the two regions.

EXAMPLE 3-5.1 The circuit of Figure 3-5.1 has the circuit element values

$$V_{cc} = 5 \text{ V}, \qquad R_{in} = 5.6 \text{ k}\Omega, \qquad R_b = 15 \text{ k}\Omega, \qquad R_c = 2.2 \text{ k}\Omega$$

and a silicon BJT with $\beta_F = 50$.

Determine the voltage transfer relationship for $0 \text{ V} \leq V_{in} \leq 5 \text{ V}$.

SOLUTION

The voltage transfer relationship is best shown with a plot of V_o as a function of V_{in} as the desired result.

It is best to begin at one end of the input range, for example, at $V_{in} = 0$. Clearly both BJT p–n junction are reverse biased and the transistor is in the cutoff region. With the transistor off, $V_o = 5 \text{ V}$. As the input voltage increases, the BJT will remain

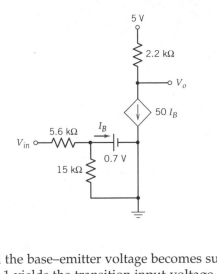

in the cutoff region until the base–emitter voltage becomes sufficiently large to turn the BJT on. Equation 3.5-1 yields the transition input voltage

$$V_{in(1)} = \frac{R_b + R_{in}}{R_b} V_{BE(on)} = \frac{15k\Omega + 5.6k\Omega}{15k\Omega} (0.6) = 0.824 \text{ V.}$$

Section 1 of the transfer relationship can be drawn (seen in Figure 3.5-2) with the above derived result.

When $V_{in} > V_{in(1)}$, the BJT is in the forward-active region. The circuit can then be redrawn using the forward-active model of the BJT.

The circuit-defining equations are

$$I_B = \frac{V_{in} - 0.7}{5.6k\Omega} - \frac{0.7}{15k\Omega}$$

$$= 178.6V_{in} - 171.7 \ \mu\text{A}$$

and

$$V_o = 5 - 2.2k\Omega(50I_B)$$

or

$$V_o = -19.65V_{in} + 23.88.$$

The input-output transfer relationship for this region of operation is of particular interest. Notice that the output voltage is a *multiple* of the input voltage (in this case, −19.65) with a DC offset (in this case, 23.88 V). Small variations of the input voltage create larger variations at the output! This important property, known as amplification, often associated with the forward-active region of a BJT, is discussed thoroughly in Chapter 5.

The BJT forward-active portion of the transfer relationship is valid until the BJT enters the saturation region, that is, when $V_o = V_{CE(sat)} = 0.2$ V. Solving for the input voltage that corresponds to the transition to the BJT saturation region yields

$$0.2 = -19.65V_{in(2)} + 23.88 \Rightarrow V_{in(2)} = 1.205 \text{ V.}$$

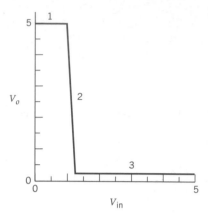

Figure 3.5-2
Voltage transfer relationship for Example 3.5-1 (obtained using three linearized models of BJT operation): (1) cutoff; (2) forward active; (3) saturation.

Section 2 of the transfer relationship can be plotted (shown in Figure 3.5-2) using the above relationships.

For $V_{in} > V_{in(2)} = 1.205$ V, the BJT is in saturation and $V_o \approx 0.2$ V. That result is shown in section 3 of the transfer relationship. It should be noted that the simple linearized models for BJTs have greatest error near the transition between regions. In this example, the errors are greatest near $V_{in(1)} \approx 0.83$ V. Here, because of the extreme steepness of the curve in region 2, region 1 ends at $V_{in(1)} = 0.824$ V and region 2 begins (the output voltage cannot exceed V_{cc} therefore when $V_o = 5$) at $V_{in} = 0.960$ V. Experience tells us to fill in the gap with a smooth curve.

Notice in the transfer relationship shown in Figure 3.5-2 that the two constant voltage levels are quite wide compared to the very rapid transition region. These characteristics are quite desirable in digital logic circuitry.

The logic inverter circuit serves a useful purpose in digital circuitry: It performs the NOT operation. In order to perform the more complex operations such as AND and OR it is necessary to increase the complexity of the circuitry. As a first example, the addition of diodes to the input of a BJT leads to the common Diode–Transistor Logic (DTL) family of gates. The most basic circuit of the DTL family is the NAND (NOT–AND) gate. It is formed by a connection of a diode logic AND and a transistor logic inverter. In Figure 3.5-3, this gate is shown with two input diodes, D_{1a} and D_{1b}. Additional inputs could be implemented with additional similarly connected diodes.

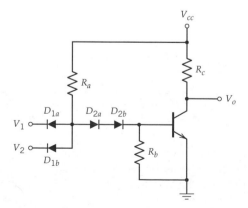

Figure 3.5-3
A DTL NAND Gate.

3.5.2 Diode–Transistor Logic Gate

The operation of this circuit can be described as follows[9]:

1. For small values of *either* input voltage (or *both* input voltages), the corresponding input diode D_1 will turn on. The voltage at the anode of the input diodes then becomes

    $$V_{\text{anode}} = V_{\text{in}} + V_\gamma.$$

    If V_{anode} is sufficiently small, that is, if

    $$V_{\text{anode}} < V_{BE(\text{on})} + 2V_\gamma,$$

    or equivalently,

    $$V_{\text{in}} < V_\gamma + V_{CE(\text{on})},$$

    the BJT will be in the cutoff region and the output voltage

    $$V_o = V_{cc}.$$

2. If the minimum value of *both* inputs is V_{in}, and as V_{in} increases beyond the constraints of the above region of operation, i.e., when

    $$V_{\text{anode}} > 2V_\gamma + V_{BE(\text{on})}$$

    or equivalently,

    $$V_{\text{in}} > V_\gamma + V_{BE(\text{on})},$$

    the BJT enters the forward-active region. The output voltage will steadily decrease until the BJT enters the saturation region. The forward-active region of the BJT will end when the input diodes *both* turn off, that is, when

    $$V_{in} > V_{BE(\text{sat})} + 2V_\gamma - V_\gamma = V_{BE(\text{sat})} + V_\gamma.$$

    As in the logic inverter circuit, this region is very narrow: The BJT is in the forward-active region for only a small range ($V_{BE(\text{sat})} - V_{BE(\text{on})} \approx 0.2$ V) of input voltage values.

3. Once the input diodes turn off, the input is essentially disconnected from the circuit. Further increases in the value of V_{in} produce no change in V_o, which remains at

    $$V_o \approx V_{CE(\text{sat})} = 0.2 \text{ V}.$$

---

[9] Once again V_{CC} is assumed to be a positive voltage. Often V_{CC} is chosen to be $\approx +5$ V, but there are several varieties of DTL circuitry that use other positive values.

To summarize the operation of this gate:

> One or more inputs low $\Rightarrow$ HIGH output $= V_{cc}$
>
> All inputs high $\Rightarrow$ LOW output $= V_{CE(sat)}$

This behavior forms a logic NAND gate.

It has been shown that input voltages are not necessarily a particular value: They can vary over a range of values without changing the output of the logic circuit:

> LOW input range: $0 \leq V_{iL} \leq V_{BE(on)} + V_{\gamma}$
>
> HIGH input range: $V_{BE(sat)} + V_{\gamma} \leq V_{iH} \leq V_{cc}$

The susceptibility of a logic gate to noise is dependent on these input level range and the nominal logic level that is expected. For this particular logic circuit the nominal logic levels are

$$\text{Nominal LOW} = V_L = V_{CE(sat)},$$
$$\text{Nominal HIGH} = V_H = V_{cc}.$$

The noise voltage at an input that will cause the circuit to function improperly is the *noise margin*. This noise margin typically is different for each logic level and is defined as the difference between the edge of the level range and the nominal level. The DTL NAND gate under consideration has noise margins of

$$\text{NM(LOW)} = V_{BE(on)} + V_{\gamma} - V_{CE(sat)},$$
$$\text{NM(HIGH)} = V_{BE(sat)} + V_{\gamma} - V_{cc}.$$

It should be noted that the magnitude of NM(LOW) is significantly smaller than the magnitude of NM(HIGH): The LOW input signal is much more susceptible to noise than the HIGH input signal. Moving the transition voltage nearer to the center of the range of input values can equalize the noise rejection properties between HIGH and LOW inputs. This equalization of noise margin magnitudes (rarely done in DTL gates) can be accomplished by inserting additional diodes in series with D_{2a} and D_{2b}, or, as in a closely related logic family, the High-Threshold Logic (HTL) family, the two diodes can be replaced by a reverse-biased Zener diode with an appropriate Zener voltage greater than $2V_{\gamma}$. The Zener diode allows an increase in the transition voltage values without an increase in component quantity. The HTL family is particularly useful in high-noise environments.

Another important factor when considering using logic gates is the quantity of gates of similar properties that can be connected in parallel to the output. The number of gates that can be driven by a single gate, without changing the value of the output voltages, is called the *fan-out* of the gate. The load that "slave" gates (the gates being driven) apply to the "master" gate (the gate driving the slaves) takes the form of a load current. When the input to a slave gate is HIGH, the input diodes to the slave are off: The slave gates draw no current and present no load to the master gate. When the input to a slave gate is LOW, current flows out of the slave gate into the output of the master gate: Sufficient current added into the collector of the master gate BJT will force the BJT out of saturation and therefore change the output voltage level.

EXAMPLE 3.5-2 For the circuit of Figure 3.5-3 assume the following circuit parameters:

$$V_{cc} = 5 \text{ V}, \qquad R_a = 3.9 \text{ k}\Omega, \qquad R_b = 5.6 \text{ k}\Omega, \qquad R_c = 2.2 \text{ k}\Omega.$$

Also assume silicon diodes and a silicon BJT with $\beta_F = 50$.
Determine the fan-out of the NAND gate.

SOLUTION

It has already been determined which quantities are significant in fan-out computations for this DTL logic gate (only low slave inputs present a load):

- Input current for a slave gate with low input
- The BJT base and collector currents for a master gate with low output

First calculate the input current for a slave gate with LOW input. The LOW output of the master gate will form the input voltage for the slave gate. Therefore,

$$V_{in} \approx V_{CE(sat)} = 0.2 \text{ V}.$$

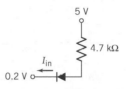

With this input voltage, the diodes D_2 are both off, and the input current to the slave gate can be obtained as

$$I_{in} = \frac{5 - 0.7 - 0.2}{4700} = 872 \text{ }\mu\text{A}.$$

The worst scenario of all the current exiting only one of the slave input diodes has been considered.

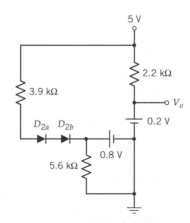

The BJT base and collector (no-load) currents for the master gate with LOW output can be determined as

$$I_B = \frac{5 - 0.8 - 2(0.7)}{3900} - \frac{0.8}{5600} = 575 \ \mu A,$$

$$I_{C(nl)} = \frac{5 - 0.2}{2200} = 2.182 \ mA.$$

For the master gate BJT to remain in saturation it is necessary that the total collector current be less than β_F times the base current. This total collector current is the sum of the no-load current and the input currents from N slave gates:

$$I_C = I_{C(nl)} + NI_{in} < \beta_F I_B.$$

or

$$2.182 \ mA + N(872 \ \mu A) < 50(575 \ \mu A).$$

Therefore

$$N < 30.5.$$

Only integer numbers of gates can be driven. Thus,

$$\text{Fan-out} = \mathbf{30} \text{ gates.}$$

3.5.3 Transistor–Transistor Logic Gate

The back-to-back arrangements in DTL logic gates of the input diodes and D_{2a} indicate that a possible replacement by an *npn* BJT might be possible. In order to accommodate multiple inputs, the transistor can be fabricated as a multiple-emitter structure as shown in Figure 3.5-4 (here shown as Q_1 with three emitters). This multiple-emitter transistor will not function exactly the same as the diodes in a DTL logic

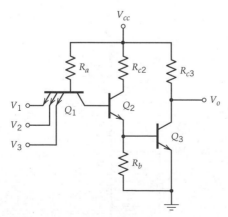

Figure 3.5-4
A TTL NAND Gate.

gate but will switch between the various modes of BJT operation as described in earlier sections of this chapter. When diode D_{2b} of the DTL gate is also replaced by a BJT (Q_2 in Figure 3.5-4), the resultant circuit is a full transistor implementation of a NAND gate. The family of logic gates to which this all-transistor NAND belongs is called *transistor–transistor logic*, or TTL.

The operation of this basic logic circuit can be described as follows:

1. For small values of *any one or more* input voltage, the input transistor Q_1 base–emitter junction will be forward biased. Since currents coming out of the base of Q_2 (this current is also the collector current of Q_1) are negligible, $I_{C1} < \beta_F I_{B1}$ and Q_1 is in saturation. The voltage at the base of Q_2 is given by

    $$V_{B2} = V_{\text{in}} + V_{CE(\text{sat})}.$$

    If this voltage is sufficiently small, that is, if

    $$V_{B2} < V_{BE(\text{on})2} + V_{\gamma3},$$

    or equivalently,

    $$V_{\text{in}} < V_{BE(\text{on})2} + V_{\gamma3} - V_{CE(\text{sat})1},$$

    the output transistor Q_3 will be in the cutoff region and the output voltage

    $$V_o = V_{cc}.$$

2. Let the minimum value of *both* inputs be V_{in}. As V_{in} increases beyond the constraints of the above region of operation, i.e., when

    $$V_{B2} > V_{BE(\text{on})2} + V_{\gamma3},$$

    or equivalently

    $$V_{\text{in}} > V_{BE(\text{on})2} + V_{\gamma3} - V_{CE(\text{sat})1},$$

    the transistor Q_1, with both its p–n junctions forward biased, begins to have current flowing out of its collector. This outward-flowing current brings Q_2 into the forward-active region, allowing current to flow through R_b. As V_{in} is increased farther, sufficient current flows through R_b to bring Q_3 into the active and finally the saturation region. Much of the action in this region is internal to the gate circuitry and not visible at the output. The output voltage transitions from HIGH to LOW over a very small range on input voltages typically of width 0.2 V or less.

3. Finally, when V_{in} increases sufficiently, that is, when

    $$V_{\text{in}} > V_{BE(\text{sat})3} + V_{BE(\text{sat})2} + V_{CE(\text{sat})1} = 2V_{BE(\text{sat})} + V_{CE(\text{sat})},$$

first Q_3 and then Q_2 have their base–emitter junctions sufficiently forward biased to enter the saturation region. As the bias on the base–emitter junction of Q_1 becomes less negative, Q_1 transitions through the saturation region (smaller inputs have the base–emitter junction more strongly forward biased; larger inputs force the base–collector junction to be more strongly forward biased) to the inverse-active region. It should be noted that Q_1 in the inverse-active region implies a current load on the input source. That load can certainly be present; however, Q_1 could also be in the saturation region (with the base–collector junction more strongly forward biased) to achieve the same output voltage. Here, Q_1 self-limits the amount of current that it draws to no more than is available from the source. The output voltage for HIGH inputs is

$$V_o \approx V_{CE(\text{sat})} = 0.2 \text{ V}.$$

To summarize the operation of this gate:

One or both inputs low $\Rightarrow$ HIGH output $= V_{cc}$ (Q_1, saturation; Q_2 and Q_3, cutoff)

Both inputs high $\Rightarrow$ LOW output $= V_{CE(\text{sat})}$ (Q_1, inverse active; Q_2 and Q_3, saturation)

This behavior forms a logic NAND gate.

EXAMPLE 3.5-3 Determine the fan-out of the TTL gate of Figure 3.5-4 with the following circuit parameters:

$$V_{cc} = 5 \text{ V}, \qquad R_a = 3.9 \text{ k}\Omega, \qquad R_b = 1.0 \text{ k}\Omega,$$
$$R_{c2} = 1.5 \text{k}\Omega, \qquad R_{c3} = 3.9 \text{ k}\Omega.$$

Assume we are using silicon BJTs with the properties

$$\beta_F = 50, \qquad \beta_R = 2.$$

SOLUTION

As in the DTL gate, it is necessary to find the following quantities[10]:

- Input current for a slave gate with low input
- Master gate output BJT base and collector currents when the master gate has a low output

---

[10]While a high input to a slave gate does draw current, it does not effect the proper operation of the slave gate. It does, however, draw down the output voltage of the master gate. In extreme cases the input transistor of the slave gate will enter the inverse-saturation (base–collector junction more strongly forward biased) region due to a limitation on the current available: The slave gate will self-limit the amount of current that it draws from the master gate.

First calculate the input current for a slave gate with low input. The low output of the master gate will form the input voltage for the slave gate. Therefore

$$V_{in} \approx V_{CE(sat)} = 0.2 \text{ V}.$$

Transistor Q_2 is off and Q_1 is in saturation; therefore,

$$I_{in} = \frac{5.0 - 0.2 - 0.8}{3900} = 1.026 \text{ mA}.$$

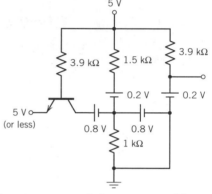

The worst scenario of all the current exiting only one of the emitters of the slave input transistor is considered.

The master gate output BJT collector and base currents can be determined with the following process:

$$I_{B1} = \frac{5 - (0.8 + 0.8 + 0.7)}{3900}$$

$$= 692 \text{ μA}.$$

If Q_1 is in the inverse-active region, its emitter current is given by

$$I_{E1} = \beta_R I_{B1}$$

$$= 2(692 \text{ μA}) = 1.384 \text{ mA}.$$

If the master gate is driven by other gates of the same type, it is unreasonable to assume that this large current is entering the emitter of Q_1 (it would draw the input voltage below zero). For fan-out calculations it is safer to assume the worst-case scenario where the input current is approximately zero. Under that scenario

$$I_{B2} \approx I_{B1} = 692 \text{ μA}.$$

The collector current of Q_2 is

$$I_{C2} = \frac{5 - 0.8 - 0.2}{1500} = 2.667 \text{ mA}$$

and the base current of Q_3 is therefore

$$I_{B3} = I_{B2} + I_{C2} - \frac{0.8}{1000} = 2.559 \text{ mA}.$$

The no-load collector current of Q_3 in the master gate is

$$I_{C3(nl)} = \frac{5 - 0.2}{3900} = 1.231 \text{ mA}.$$

The fan-out can now be calculated from

$$I_{C3} < \beta_F I_{B3}$$

or

$$I_{C3(nl)} + N(I_{in}) < \beta_F I_{B3}$$

or

$$1.231 \text{ mA} + N(1.026 \text{ mA}) < 50 \, (2.559 \text{ mA}) \Rightarrow N < 123.5.$$

The fan-out of this gate is 123 gates of similar construction.

3.5.4 Emitter-Coupled Logic Gate

Another common logic gate family is Emitter-Coupled Logic (ECL). A simple two-input ECL gate is shown in Figure 3.5-5.

The operation of this gate can be described as follows:

1. If V_1 and V_2 are both very negative (near -5.2 V), both Q_1 and Q_2 are in the cutoff region. Here, Q_3 is in the forward-active region, which allows the following calculations:

$$-I_{E3} = \frac{-1.15 - 0.7 - (-5.2)}{1200} = 2.792 \text{ mA},$$

$$I_{C3} = -\alpha_F I_{E3} = \frac{50}{1 + 50} \, (2.792 \text{ mA}) = 2.737 \text{ mA},$$

$$V_{C3} = 0 - I_{C3}(330) = -0.903 \text{ V},$$

$$V_o = V_{C3} - V_\gamma = -1.60 \text{ V}.$$

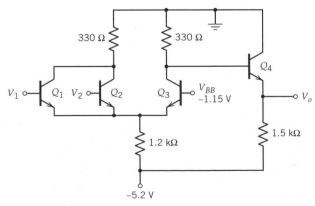

Figure 3.5-5
Simple ECL OR gate.

2. As either V_1 or V_2 increases to within $V_\gamma - V_{BE(on)} \approx 0.1$ V of V_{BB}, the corresponding input transistor will begin to turn on. Since the current in R_b must remain relatively constant, Q_3 will supply R_b an ever-decreasing portion of that current. Current I_{C3} will decrease and V_o will therefore increase. This process will continue until Q_3 enters the cutoff region. This linear region forms the basis for an amplifier type to be discussed in Chapter 6.

3. When either V_1 or V_2 becomes $V_\gamma - V_{BE(on)} \approx 0.1$ V greater than V_{BB}, Q_3 will enter the cutoff region. By considering Q_4 disconnected from Q_3, V_o can be calculated as follows:

$$330 I_B + 0.7 + 1500(\beta_F + 1)I_B = 5.2$$

$$I_B = 58.6 \ \mu A,$$

$$V_o = 0 - 330 I_B - V_\gamma = -0.72 \text{ V}.$$

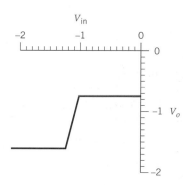

The gate operates as an OR gate with the following logic levels:

$$V_H = -0.72 \text{ V}, \ V_L = -1.60 \text{ V}.$$

The voltage transfer relationship for this gate is given in Figure 3.5-6.

Figure 3.5-6
Voltage transfer Relationship for ECL OR gate.

3.6 BIASING THE BJT

In the previous section, transistors were used in digital (non-linear) circuits. In the digital applications considered, the transistors operated in one of two states that resulted in an output of either a logic 1 or a logic 0: The transistors transitioned between the saturation and cutoff regions. In linear applications (e.g., in the design of linear amplifiers) the transistors are *biased* to operate in only the forward-active region of operation. The transistor is biased at a quiescent operating point, which is commonly called the Q point, based on the *DC conditions* of the transistor. The Q point is determined by the transistor characteristics and the applied external currents and voltages. It is commonly described by four DC quantities:

- Two transistor terminal voltages V_{BE} and V_{CE}
- Two transistor currents I_B and I_C

Once the Q point is established, a time-varying excursion of the input signal (e.g., a base current) will cause an output signal (collector voltage or current) of the same waveform. The amplifier design and analysis techniques in Section II of this book may be employed to determine the gain of the circuit.

If the output waveform is not a reproduction of the input signal (e.g., the waveform is clipped on one side), the Q point is unsatisfactory and must be relocated. The selection of the Q point in the forward-active region is also subject to the various transistor ratings that limit the range of useful operation. Manufacturers' specification sheets for transistors list the maximum collector dissipation (sometimes listed as maximum power dissipation) $P_{C,max}$, maximum collector current $I_{C,max}$, maximum collector–emitter voltage V_{CEO}, maximum emitter–base voltage V_{EBO}, and maximum collector–base voltage V_{CBO}.

A graphical representation of the operating limits of the transistor due to maximum power dissipation $P_{C,max}$, maximum collector current $I_{C,max}$, and maximum collector–emitter voltage V_{CEO} is shown in Figure 3.6-1. The safe operating region is not shaded and lies below the so-called maximum power dissipation hyperbola defined by $P_{C,max}$, where

$$P_{C,max} = I_{C,max}V_{CEO}. \tag{3.6-1}$$

In an increased temperature environment, the maximum power dissipation hyperbola encroaches into the safe operating region. Therefore, the transistor load line and the operating point should lie well within the unshaded safe operating region to avoid the possibility of transistor thermal failure.

To help in the design and analysis of transistor biasing circuits, a few key terminal current relationships are reiterated here. The direction of transistor terminal current flow is defined in Figure 3-1: All currents flow into the transistor. Three important transistor terminal current relationships from the previous sections of this chapter are

$$I_C = \beta_F I_B, \qquad I_C = -\alpha_F I_E, \qquad I_B = -(I_C + I_E).$$

From these equations, the emitter current I_E can be related to the collector current I_C and the base current I_B.

To find the relationship between the base and the emitter currents, KCL is applied to the BJT:

$$I_B = -(I_C + I_E).$$

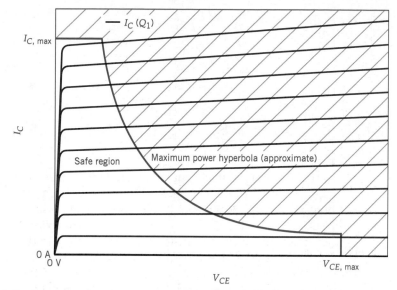

Figure 3.6-1
Safe operating region defined by maximum power dissipation hyperbola.

Rearranging the equation,

$$-I_E = I_B + I_C.$$

By substituting $I_C = \beta_F I_B$ for I_C in the equation above,

$$-I_E = I_B(\beta_F + 1), \tag{3.6-2}$$

or

$$I_B = \frac{-I_E}{\beta_F + 1}. \tag{3.6-3}$$

Using Equation 3.1-4,

$$I_C = -\alpha_F I_E.$$

and

$$\beta_F = \frac{\alpha_F}{1 - \alpha_F} \Rightarrow \alpha_F = \frac{\beta_F}{\beta_F + 1},$$

the expression for the collector current is

$$I_C = -\frac{\beta_F}{\beta_F + 1} I_E. \tag{3.6-4}$$

Figure 3.6-2
Early voltage of BJT.

In real transistors, the output characteristic curves in the forward-active region slope slightly upward, increasing I_C with V_{CE} for a constant I_B. The slope of the curve is determined by the BJT early voltage V_A. The *Early voltage* is that voltage that is the point of intersection of the $I_C = 0$ line and the extended line from the characteristic curves in the forward-active region, with values typically in the 75–100 V range. Figure 3.6-2 provides a pictorial definition of the BJT early voltage.

The **.model** statement in the SPICE model of the BJT can be altered to include an early voltage. For instance, the model statement for a BJT with $V_A = 75$ V,

```
.model NPXEX NPN(BF=200 VA=75)
```

is used to create the load line analysis plot in Figure 3.6-3.

Using resistive networks, several common BJT biasing methods that achieve the desired Q point will be discussed in this section. Current source biasing methods for achieving the desired Q point will be discussed in Chapter 6. For linear applications, the transistor Q point must be established in the forward-active region.

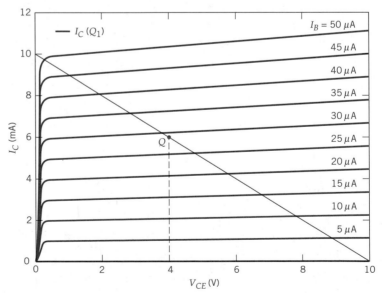

Figure 3.6-3
Load line analysis plot for BJT with early voltage of 75 V.

3.6.1 Fixed-Bias Circuit

One method of biasing a transistor to operate at a desired Q point is illustrated in Figure 3.6-4a, which shows the fixed-bias circuit (sometimes called the base-bias circuit). It is convenient to use the forward-active model of the npn BJT as shown in Figure 3.6-4b for analyzing bias circuits.

The collector–emitter voltage $V_{CE} = V_C - V_E$ is equal to the power supply voltage minus the voltage drop across the collector resistor R_C. That is,

$$V_{CE} = V_{CC} - I_C R_C, \tag{3.6-5}$$

where $\qquad V_{CE}$ = DC collector–emitter voltage

$\qquad\qquad V_{CC}$ = collector supply voltage

$\qquad\qquad I_C$ = DC collector current

$\qquad\qquad R_C$ = load resistance seen from collector

Kirchhoff's voltage law applied to the base–emitter loop yields the following expression for the base current:

$$I_B = \frac{V_{CC} - V_{BE}}{R_B}. \tag{3.6-6}$$

Substituting $I_C = \beta_F I_B$ and Equation 3.6-6 into Equation 3.6-5, the collector–emitter voltage becomes

$$V_{CE} = V_{CC} - \beta_F \frac{V_{CC} - V_{BE}}{R_B} R_C. \tag{3.6-7}$$

The Q point is defined by I_C and V_{CE} for a specified I_B.

The Q point may also be found through graphical methods. The base–emitter voltage is determined by performing a load line analysis on the common emitter input characteristic of the BJT. The slope of the input load line is $-1/R_B$ from Equation 3.6-6. The load line intersects the I_B axis at V_{CC}/R_B when $V_{BE} = 0$ and intersects

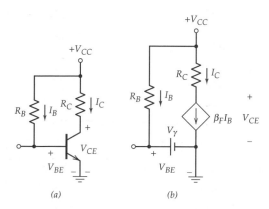

(a) (b)

Figure 3.6-4
(a) Fixed-bias circuit; (b) equivalent circuit using npn BJT model for forward-active region.

the V_{BE} axis when at V_{CC}. Note that relative to the range of possible V_{BE} values of the load line, the intersection of the load line and the characteristic curves for V_{CE} = 0 V and $V_{CE} = V_{CC}$ are approximately the same. Therefore, it is common practice to use the approximation $V_{BE} = V_{\gamma}$ (and therefore V_{BE} = 0.7 V for silicon BJTs). Unless a more accurate determination of V_{BE} is required, this approximate value will be used for the remainder of this book.

A load line representing R_C is superimposed on the transistor output characteristic, as shown in Figure 3.6-5b. As in the load line analysis performed for diode circuits in Chapter 2, the slope of the input load line is the negative of the inverse of the load resistance $(-1/R_C)$, as is evident in Equation 3.6-1. The intersection of this

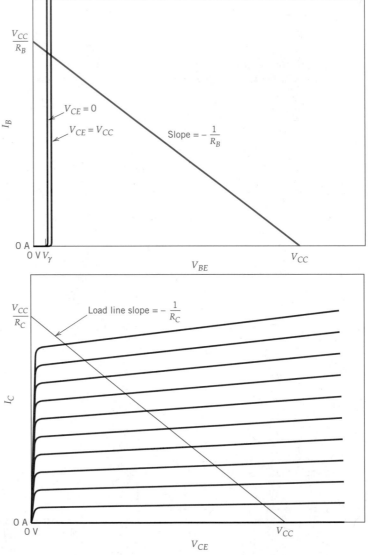

Figure 3.6-5

(a) Input load line analysis; (b) output load line analysis for fixed-bias transistor circuit.

load line with the common emitter transistor output characteristic curve for the desired I_B determines the Q point. This Q point is identified by the resultant transistor collector current and the collector–emitter voltage.

EXAMPLE 3.6-1 Complete the design of the fixed-bias transistor circuit shown by determining R_C and R_B for a Q point of $I_C = 6$ mA and $V_{CE} = 4$ V. The transistor forward current gain is $\beta_F = 200$ with a negligible β_R. Let $V_{BE} = 0.7$ V.

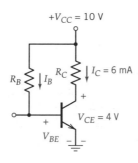

$+V_{CC} = 10$ V

R_B $\quad I_B$ $\quad R_C$ $\quad I_C = 6$ mA

$V_{CE} = 4$ V

V_{BE}

SOLUTION 1 (ANALYTICAL)

The load resistor R_C is calculated by applying KVL to the collector–emitter loop,

$$R_C = \frac{V_{CC} - V_{CE}}{I_C}$$

$$= \frac{10 - 4}{6 \times 10^{-3}} \Rightarrow R_C - 1 \text{ k}\Omega.$$

The base resistor R_B is calculated by applying KVL to the base–emitter loop,

$$R_B = \frac{V_{CC} - V_{BE}}{I_B} = \frac{V_{CC} - V_{BE}}{I_C/\beta_F}$$

$$= \frac{(10 - 0.7)/6 \times 10^{-3}}{200} \Rightarrow R_B = 310 \text{ k}\Omega.$$

SOLUTION 2 (GRAPHICAL)

In this example, the BJT was assumed to have a flat I_C-versus-V_{CE} curve for each I_B; that is, I_C was invariant with changes in V_{CE}. The variation in I_C with V_{CE} in the forward-active region is commonly described by a parameter called the Early voltage. No variation in I_C (a common first-order approximation) is modeled by infinite early voltage. The BJT output characteristic curve for this example can be generated in SPICE with a netlist similar to Example 3.3-1, Solution 2. The BJT is designated in the SPICE netlist as

```
Qname collector_node base_node emitter_node model_name
```

The *npn* transister **.model** statement for this example is as follows:

```
.model NPXEX NPN(BF=200)
```

The default values of the transport saturation current (IS), Early voltage (VA), reverse beta (BR), and forward current emission coefficient (NF) were used. The default values for these parameters are

```
IS = 1E-16
VA = ∞
BR=1
NF=1
```

First, the load line representing R_C must be found. This is accomplished by first establishing the desired Q point on the transistor output characteristics. The desired Q point is located knowing V_{CEQ}, I_{CQ}, and I_{BQ}, where the additional subscript Q denotes the voltages and currents at the Q point. From the graph, $I_{BQ} = 30$ μA. Therefore, the base resistance is

$$R_B = \frac{V_{CC} - V_{BE}}{I_B}$$

$$= \frac{10 - 0.7}{30 \times 10^{-6}} \Rightarrow R_B = 310 \text{ k}\Omega.$$

From the Q point, a straight line that intersects the V_{CE} axis at $V_{CE} = V_{CC}$ is drawn. The line is extended from the Q point to intersect the I_C axis to complete the load

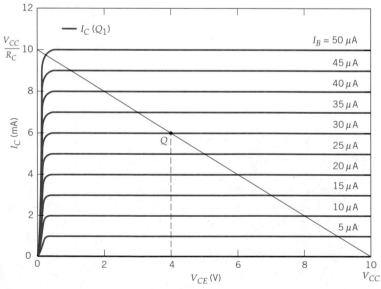

Figure 3.6-6
Load line analysis of Example 3.6-1.

line. These two points on the load line establish the extremes of the transistor operation. That is, when $V_{CE} = V_{CC}$, all of the voltage from the power supply is dropped across the transistor (collector and emitter). Therefore, there is no voltage drop across the load resistor R_C corresponding to zero current flowing through the resistor ($I_C = 0$). When the load line intersects the I_C axis, $V_{CE} = 0$. Therefore, the transistor (collector–emitter) acts as if it were a short circuit and the voltage drop across the load resistor equals the power supply voltage V_{CC}. From Equation 3.6-5, the current flowing through the resistor when $V_{CE} = 0$ is V_{CC}/R_C. The load line analysis of the fixed-bias circuit is shown in Figure 3.6-6. The slope of the line (actually, the negative inverse slope of the line) is the desired load resistor R_C, which in this case is 1 kΩ.

In the fixed-bias circuit, the base current is essentially fixed by the value of V_{CC} and R_B. The collector current is determined by I_B and β_F. Because β_F varies widely from one transistor to another and with temperature change, the collector current also varies widely with these changes. In Section 3.7, the fixed-bias circuit will be shown to be one of the worst ways to bias a transistor from the standpoint of stability of the Q point.

Several variations of the fixed-bias circuit are commonly used. These variations increase the stability of the Q point, making the circuit less susceptible to variations in performance due to changes in transistor parameters.

In the fixed-bias circuit with emitter feedback, shown in Figure 3.6-7, an emitter resistor is added to the basic fixed-bias circuit.

The collector current is found by applying KVL to the base–emitter loop,

$$V_{CC} = I_B R_B + V_{BE} + I_{EE} R_F,$$

(3.6-8)

where $I_{EE} = -I_E$, so $I_{EE} = [(\beta_F + 1)/\beta_F]I_C$. Substituting

$$I_{EE} = \frac{\beta_F + 1}{\beta_F} I_C \quad \text{and} \quad I_B = \frac{I_C}{\beta_F}$$

into Equation 3.6-8 yields

$$V_{CC} = V_{BE} + I_C\left(\frac{R_B}{\beta_F} + \frac{\beta_F + 1}{\beta_F} R_E\right).$$

(3.6-9)

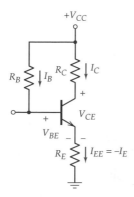

Figure 3.6-7
Fixed-bias circuit with emitter feedback.

Equation 3.6-9 yields the collector current with respect to the power supply voltage, the transistor β_B, and the external resistors,

$$I_C = \frac{V_{CC} - V_{BE}}{R_B/\beta_F + [(\beta_F + 1)/\beta_F]R_E}. \tag{3.6-10}$$

The relationship between V_{CE} and I_C is found by applying KVL to the collector–emitter loop,

$$\begin{aligned}
V_{CC} &= I_C R_C + V_{CE} + I_{EE} R_E \\
&= V_{CE} + I_C\left(R_C + \frac{\beta_F + 1}{\beta_F} R_E\right)
\end{aligned} \tag{3.6-11}$$

Solving Equation 3.6-11 to solve for V_{CE} yields

$$\begin{aligned}
V_{CE} &= V_{CC} - I_C R_C - I_{EE} R_E \\
&= V_{CC} - I_C\left(R_C + \frac{\beta_F + 1}{\beta_F} R_E\right)
\end{aligned} \tag{3.6-12}$$

The slope of the load line for the fixed-bias circuit with emitter resistor on the BJT output characteristic curve is

$$\text{Slope of load line} = -\left(R_C + \frac{\beta_F + 1}{\beta_F} R_E\right)^{-1}. \tag{3.6-13}$$

If $\beta_F \gg 1$, then the slope of the load line is simplified as

$$\text{Slope of load line} \approx -(R_C + R_E)^{-1}. \tag{3.6-14}$$

The load line described by Equation 3.6-12 is superimposed on the transistor output characteristics, as shown in Figure 3.6-8. As usual, the Q point is determined by the collector current and the collector–emitter voltage for a given base current of the transistor.

To draw the load line on the transistor output characteristics, the two extremes of the load line are determined. From Equation 3.6-12, the point at which the load line intersects the I_C axis occurs when $V_{CE} = 0$, so that

$$I_{C,V_{CE}=0} = \frac{V_{CC}}{R_C + [(\beta_F + 1)/\beta_F]R_E} \tag{3.6-15}$$

The other extreme of the load line occurs when the transistor collector–emitter voltage drops all of the voltage provided by the power supply. If all of the supply voltage is dropped across the transistor, the collector current must be zero ($I_C = 0$). Therefore the intersection of the load line with the V_{CE} axis must occur at V_{CC}.

The fixed-bias circuit with collector feedback is shown in Figure 3.6-9. The circuit is similar to the fixed-bias circuit with the exception that the base resistor is connected

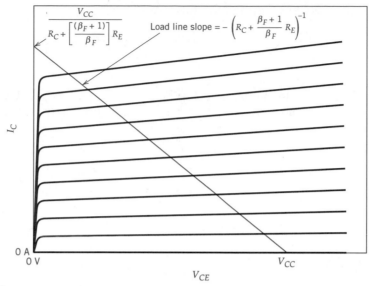

Figure 3.6-8
Load line analysis for fixed-bias emitter feedback transistor circuit.

directly to the collector of the BJT. The result is that the current provided by the power supply, I_{CC}, is not equal to the BJT collector current I_C. The current provided by the power supply is

$$I_{CC} = I_C + I_B. \tag{3.6-16}$$

The collector current is found by applying KVL to the base–emitter loop,

$$
\begin{aligned}
V_{CC} &= I_{CC}R_C + I_B R_B + V_{BE} \\
&= (I_C + I_B)R_C + I_B R_B + V_{BE} \\
&= I_C\left(\frac{\beta_F + 1}{\beta_F}\, R_C + \frac{R_B}{\beta_F}\right) + V_{BE}.
\end{aligned}
\tag{3.6-17}
$$

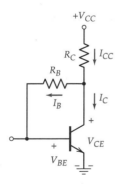

Figure 3.6-9
Fixed-bias circuit with collector feedback.

Solving for the collector current,

$$I_C = \frac{\beta_F}{\beta_F + 1} \frac{V_{CC} - V_{BE}}{R_C + R_B/(\beta_F + 1)}. \tag{3.6-18}$$

To find the relationship between V_{CE} and I_C, the KVL equation for the collector–emitter loop, which is identical to that of the fixed-bias circuit, is found,

$$
\begin{aligned}
V_{CC} &= I_{CC}R_C + V_{CE} \\
&= (I_C + I_B)R_C + V_{CE} \tag{3.6-19} \\
&= \frac{\beta_F + 1}{\beta_F} I_C R_C + V_{CE},
\end{aligned}
$$

or

$$V_{CE} = V_{CC} - I_C \frac{\beta_F + 1}{\beta_F} R_C. \tag{3.6-20}$$

From Equation 3.6-20, the slope of the load line for the fixed-bias circuit with collector feedback is

$$\text{Slope of load line} = -\frac{\beta_F}{(\beta_F + 1)R_C}. \tag{3.6-21}$$

The point at which the load line that intersects the I_C axis occurs when $V_{CE} = 0$, so from Equation 3.6-20,

$$I_{C,V_{CE}=0} = \frac{V_{CC}}{[(\beta_F + 1)/\beta_F]R_C} = \frac{V_{CC}}{R_C} \frac{\beta_F}{\beta_F + 1}. \tag{3.6-22}$$

The other extreme of the load line occurs when the transistor collector–emitter voltage drops all of the voltage provided by the power supply. If all of the supply voltage is dropped across the transistor, the collector current must be zero ($I_C = 0$). Therefore the intersection of the load line with the V_{CE} axis must occur at V_{CC}. The BJT output characteristic with the load line for the fixed-bias circuit with collector feedback is shown in Figure 3.6-10.

The last variation of base biasing presented is the fixed-bias circuit with collector and emitter feedback shown in Figure 3.6-11.

Applying KVL to the base–emitter loop,

$$
\begin{aligned}
V_{CC} &= I_{CC}R_C + I_B R_B + V_{BE} + I_{EE}R_E \\
&= (I_C + I_B)R_C + I_B R_B + V_{BE} + I_{EE}R_E \tag{3.6-23} \\
&= V_{BE} + I_C\left[\frac{\beta_F + 1}{\beta_F}(R_C + R_E) + \frac{R_B}{\beta_F}\right].
\end{aligned}
$$

The collector current is found by rearranging Equation 3.6-23,

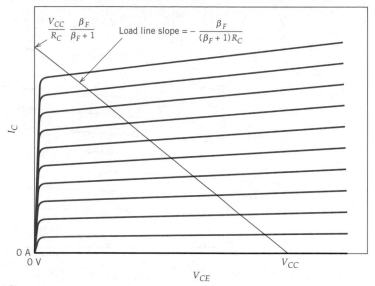

Figure 3.6-10
Load line analysis for fixed-bias collector feedback transistor circuit.

$$I_C = \frac{\beta_F}{\beta_F + 1} \frac{V_{CC} - V_{BE}}{R_C + R_E + R_B/(\beta_F + 1)}. \tag{3.6-24}$$

The relationship between V_{CE} and I_C is found by applying KVL to the collector–emitter loop,

$$
\begin{aligned}
V_{CC} &= I_{CC}R_C + V_{CE} + I_{EE}R_E \\
&= I_C \frac{\beta_F + 1}{\beta_F}(R_C + R_E) + V_{CE}.
\end{aligned}
\tag{3.6-25}
$$

By rearranging Equation 3.6-25, the expression for V_{CE} for the fixed-bias circuit with collector and emitter feedback is found,

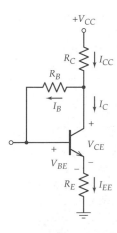

Figure 3.6-11
Fixed-bias circuit with collector and emitter feedback.

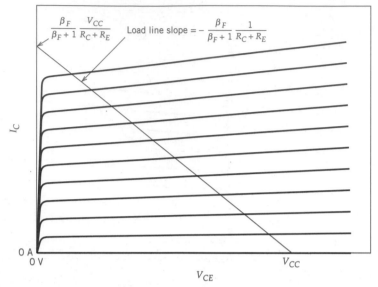

Figure 3.6-12
Load line analysis for fixed-bias collector and emitter feedback transistor circuit.

$$V_{CE} = V_{CC} - I_C \frac{\beta_F + 1}{\beta_F}(R_C + R_E). \tag{3.6-26}$$

The slope of the load line for the fixed-bias circuit with collector and emitter feedback is

$$\text{Slope of load line} = -\frac{\beta_F}{\beta_F + 1} \frac{1}{R_C + R_E}. \tag{3.6-27}$$

The point at which the load line that intersects the I_C axis occurs when $V_{CE} = 0$, so from Equation 3.6-26,

$$I_{C,V_{CE}=0} = \frac{V_{CC}}{[(\beta_F + 1)/\beta_F](R_C + R_E)} = \frac{\beta_F}{\beta_R + 1} \frac{V_{CC}}{R_C + R_E}. \tag{3.6-28}$$

The other extreme of the load line occurs when the transistor collector–emitter voltage drops all of the voltage provided by the power supply. If all of the supply voltage is dropped across the transistor, the collector current must be zero ($I_C = 0$). Therefore the intersection of the load line with the V_{CE} axis must occur at V_{CC}. The BJT output characteristic with the load line for the fixed-bias circuit with collector feedback is shown in Figure 3.6-12.

3.6.2 Emitter Bias Circuit (with Two Power Supplies)

The emitter bias circuit is often used when two power supplies (positive and negative) are available. In this configuration, shown in Figure 3.6-13 the collector current can easily be made to be essentially independent of β_F, making the circuit less sensitive to variations in β_F due to temperature or transistor replacement.

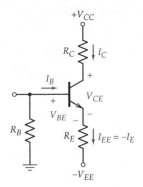

Figure 3.6-13
Emitter bias of transistor with two power supplies.

To find I_C, the KVL equation for the base–emitter loop is found,

$$V_{EE} = V_{BE} + I_B R_B + I_{EE} R_E$$
$$= V_{BE} + I_C \left(\frac{R_B}{\beta_F} + \frac{\beta_F + 1}{\beta_F} R_E \right).$$

(3.6-29)

Rearranging Equation 3.6-29 yields the expression for the collector current,

$$I_C = \frac{V_{EE} - V_{BE}}{R_B/\beta_F + [(\beta_F + 1)/\beta_F] R_E}$$
$$= \frac{\beta_F}{\beta_F + 1} \frac{V_{EE} - V_{BE}}{R_E + R_B/(\beta_F + 1)}.$$

(3.6-30)

The relationship between V_{CE} and I_C is found by the knowledge that

$$V_{CE} = V_C - V_E.$$

(3.6-31)

The collector voltage with respect to ground is

$$V_C = V_{CC} - I_C R_C.$$

(3.6-32)

The emitter voltage with respect to ground is found by applying KVL to the base-emitter loop,

$$V_E = - (V_{BE} + I_B R_B) = - \left(V_{BE} + I_C \frac{R_B}{\beta_F} \right).$$

(3.6-33)

By substituting Equations 3.6-32 and 3.6-33 into 3.6-31, the equation relating V_{CE} and I_C is

$$V_{CE} = (V_{CC} - I_C R_C) - \left[- \left(V_{BE} + I_C \frac{R_B}{\beta_F} \right) \right]$$
$$= V_{CC} + V_{BE} - I_C \left(R_C - \frac{R_B}{\beta_F} \right).$$

(3.6-34)

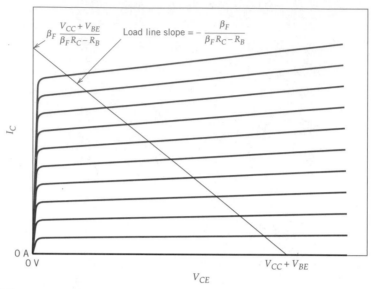

Figure 3.6-14
Load line analysis for two-supply emitter bias transistor circuit.

The slope of the load line for the fixed-bias circuit with collector and emitter feedback is

$$\text{Slope of load line} = -\frac{\beta_F}{\beta_F R_C - R_B}. \tag{3.6-35}$$

The point at which the load line that intersects the I_C axis occurs when $V_{CE} = 0$, so from Equation 3.6-34,

$$I_{C,V_{CE}=0} = \beta_F \frac{V_{CC} + V_{BE}}{\beta_F R_C - R_B}. \tag{3.6-36}$$

The other extreme of the load line occurs when the transistor is cut off ($I_C = 0$). By setting $I_C = 0$ in Equation 3.6-34, the intersection of the load line with the V_{CE} axis occurs at $V_{CC} + V_{BE}$. The BJT output characteristic with the load line for the fixed-bias circuit with collector feedback is shown in Figure 3.6-14.

EXAMPLE 3.6-2 Find the operating point (V_{CE}, I_B, and I_C) for the emitter-biased circuit with two power supplies shown. Let $V_{BE} = 0.84$V and $\beta_F = 200$.

SOLUTION 1 (ANALYTICAL)

From Equation 3.6-30 the collector current is

$$I_C = \frac{V_{EE} - V_{BE}}{\dfrac{R_B}{\beta_F} + [(\beta_F + 1)/\beta_F]R_E} = \frac{15 - 0.84}{\dfrac{10{,}000}{200} + (201/200)1000}$$

$$\Rightarrow I_C = 13.4 \text{ mA}.$$

The base current is then

$$I_B = \frac{I_C}{\beta_F} = \frac{13.4 \times 10^{-3}}{200}$$

$$\Rightarrow I_B = 67 \ \mu A.$$

From Equation 3.6-34 the collector–emitter current is

$$V_{CE} = V_{CC} + V_{BE} - I_C \left(R_C - \frac{R_B}{\beta_F} \right)$$

$$= 15 + 0.84 - (13.4 \times 10^{-3}) \left(1000 - \frac{10{,}000}{200} \right)$$

$$\Rightarrow V_{CE} = 3.1 \ V.$$

SOLUTION 2 (SIMULATION)

In the simulation, the BJT Early voltage is assumed to be 75V. The simulation uses the **.OP** command, which instructs the simulation to display the results of the operating point of the active devices (in this case the BJT) in the **.out** file. The output file and the node designations used for the netlist are shown below. The Q point values are highlighted in color.

```
*******************************PSpice*****************************************
Emitter Bias With Two Power Supplies
****   CIRCUIT DESCRIPTION
*****************************************************************************
.OPTION NOPAGE
VCC   1    0    15V
RC    1    2    1K
RE    3    5    1K
RB    4    0    10K
VEE   5    0    -15V
Q1    2    4    3    NPN3_6_2
.MODEL NPN3_6_2    NPN(BF=200 VA=75)
*Find the operating point of the transistor NPN3_6_2
.OP
.END
```

```
****   BJT MODEL PARAMETERS

      NPN3_6_2
       NPN
   IS 100.000000E-18
   BF 200
   NF 1
   VAF 75
   BR 1
   NR 1

****   SMALL SIGNAL BIAS SOLUTION   TEMPERATURE= 27.000 DEG C
  NODE VOLTAGE NODE VOLTAGE NODE VOLTAGE NODE VOLTAGE
  ( 1) 15.0000 ( 2) 1.5588 ( 3) −1.4935 ( 4) −.6528
  ( 5) −15.0000

  VOLTAGE SOURCE CURRENTS
  NAME     CURRENT

  VCC −1.344E-02
  VEE 1.351E-02
   TOTAL POWER DISSIPATION 4.04E-01 WATTS

****   OPERATING POINT INFORMATION   TEMPERATURE= 27.000 DEG C

**** BIPOLAR JUNCTION TRANSISTORS

  NAME        Q1
  MODEL       NPN3_6_2
  IB          6.53E-05
  IC          1.34E-02
  VBE         8.41E-01
  VBC        −2.21E+00
  VCE         3.05E+00
  BETADC      2.06E+02
  GM          5.20E-01
  RPI         3.96E+02
  RX          0.00E+00
  RO          5.74E+03
  CBE         0.00E+00
  CBC         0.00E+00
  CBX         0.00E+00
  CJS         0.00E+00
  BETAAC      2.06E+02
  FT          8.27E+18

     JOB CONCLUDED

     TOTAL JOB TIME    1.15
```

The simulation is in good agreement with the analytical solution.

3.6.3 Self-Bias Circuit (Emitter Bias with One Power Supply)

In many instances, two power supply voltages are not available to the designer to implement the emitter bias circuit in Section 3.6.2. In this case, a modified emitter bias configuration called the self-bias circuit shown in Figure 3.6-15 is used.

The analysis of the self-bias circuit of Figure 3.6-15 is facilitated by replacing the circuit to the left between the base and ground terminals with its Thévenin equivalent. The Thévenin equivalent of the circuit attached to the base of the transistor is shown in Figure 3.6-16. The self-bias circuit with the simplified base circuit using the Thévenin equivalent is shown in Figure 3.6-17.

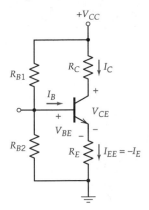

Figure 3.6-15
Self-bias circuit.

The analysis of the diagram of Figure 3.6-17 is similar to that of the fixed-bias circuit. Applying KVL on the base–emitter loop,

$$V_{BB} = I_B R_B + V_{BE} + I_{EE} R_E$$

$$= V_{BE} + I_C \left(\frac{R_B}{\beta_F} + \frac{\beta_F + 1}{\beta_F} R_E \right). \qquad (3.6\text{-}37)$$

Solving for the collector current,

$$I_C = \frac{V_{BB} - V_{BE}}{\dfrac{R_B}{\beta_F} + [(\beta_F + 1)/\beta_F]R_E}$$

$$= \frac{\beta_F}{\beta_F + 1} \frac{V_{BB} - V_{BE}}{R_E + R_B/(\beta_F + 1)} \qquad (3.6\text{-}38)$$

$$= \frac{\beta_F}{\beta_F + 1} \frac{V_{CC} R_{B2}/(R_{B1} + R_{B2}) - V_{BE}}{R_E + R_{B1} R_{B2}/[(R_{B1} + R_{B2})(\beta_F + 1)]}.$$

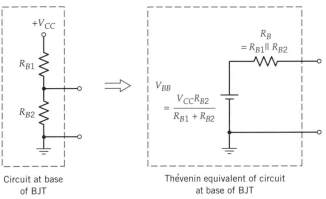

Circuit at base
of BJT

Thévenin equivalent of circuit
at base of BJT

Figure 3.6-16
Thévenin equivalent circuit at base of BJT.

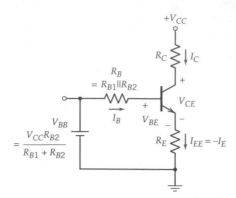

Figure 3.6-17

Simplification of self-bias circuit through use of Thévenin's theorem.

The equation relating V_{CE} to I_C (Equation 3.6-11, repeated as Equation 3.6-39 below) and the load line analysis are identical to the fixed-bias circuit with emitter feedback,

$$V_{CE} = V_{CC} - I_C R_C - I_{EE} R_E$$

$$= V_{CC} - I_C \left(R_C + \frac{\beta_F + 1}{\beta_F} R_E \right). \tag{3.6-39}$$

EXAMPLE 3.6-3 For the self-bias circuit shown, find V_{CE} and I_B for $I_C = 4$ mA. Complete the design by finding R_{B1}. Let $\beta = 200$ and $V_\gamma = 0.7$ V.

To find V_{CE}, apply Equation 3.6-39,

$$V_{CE} = V_{CC} - I_C R_C - I_{EE} R_E$$

$$= V_{CC} - I_C \left(R_C + \frac{\beta_F + 1}{\beta_F} R_E \right)$$

$$= 12 - (4 \times 10^{-3}) \left[1000 + \left(\frac{201}{200} \right) 510 \right]$$

$$\Rightarrow V_{CE} = 5.95 \text{ V}.$$

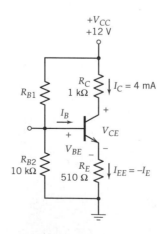

The base current is

$$I_B = \frac{I_C}{\beta_F} = \frac{4 \times 10^{-3}}{200}$$

$$\Rightarrow I_B = 20 \ \mu A.$$

To find R_{B1}, the Thévenin equivalent circuit (Figure 3.6-17) at the base of the BJT is used. Applying KVL to the base–emitter loop (Equation 3.6-37),

$$V_{BB} = I_B R_B + V_{BE} + I_{EE} R_E$$

$$= V_{BE} + I_C \left(\frac{R_B}{\beta_F} + \frac{\beta_F + 1}{\beta_F} R_E \right)$$

or

$$\frac{V_{CC} R_{B2}}{R_{B1} + R_{B2}} = V_{BE} + I_C \left(\frac{R_{B1} R_{B2}/(R_{B1} + R_{B2})}{\beta_F} + \frac{\beta_F + 1}{\beta_F} R_E \right).$$

Solving for R_{B1},

$$R_{B1} = \frac{V_{CC} R_{B2} - (V_{BE} + [(\beta_F + 1)/\beta_F] I_C R_E) R_{B2}}{\{V_{BE} + [(\beta_F + 1)/\beta_F] I_C R_E\} + I_B R_{B2}}$$

$$\Rightarrow R_{B1} = 31.4 \ k\Omega \approx 33 \ k\Omega \quad \text{(common value)}.$$

3.6.4 Biasing *pnp* Transistors

When *pnp* transistors are used, the polarity of all DC sources must be reversed. For operation in the forward-active region, the *pnp* transistor emitter voltage is greater than the collector voltage. The *pnp* characteristic curves are shown in Figure 3.6-18. Note that the polarity of the base and collector currents are negative in relation to the convention chosen in Figure 3.1. Because of the polarity reversal of the terminal voltages, it is customary to use positive values of V_{EC} and V_{EB} in the DC analysis and design of *pnp* transistor circuits.

Consider the *pnp* BJT fixed-bias circuit with collector feedback shown in Figure 3.6-19. The collector power supply voltage is a negative value, $-V_{CC}$.

The KVL expression for the base–emitter loop is

$$V_{CC} = V_{EB} - I_B R_B + I_{CC} R_C, \tag{3.6-40}$$

where $I_{CC} = -I_B - I_C$ and $I_C = \beta_F I_B$.

By substituting the current relationships above into Equation 3.6-40,

$$V_{CC} = V_{EB} - \frac{I_C}{\beta_F} R_B - I_C \left(1 + \frac{1}{\beta_F} \right) R_C. \tag{3.6-41}$$

By rearranging Equation 3.6-41, the expression for the collector current is found,

$$I_C = \frac{\beta_F}{\beta_F + 1} \frac{V_{EB} - V_{CC}}{R_C + R_B/(\beta_F + 1)}. \tag{3.6-42}$$

To find V_{EC}, KVL is used in the collector-emitter loop,

$$V_{EC} = V_{CC} - I_{CC} R_C. \tag{3.6-43}$$

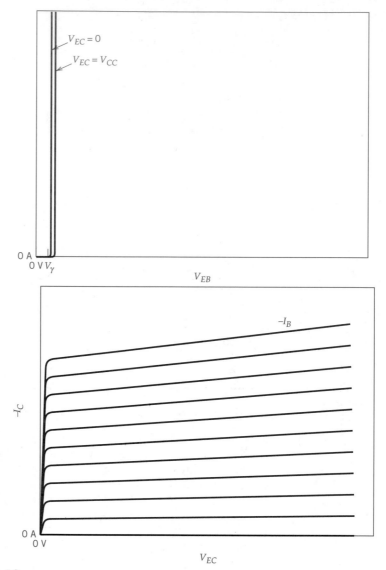

Figure 3.6-18

(a) Input and (b) output characteristic curves of *pnp* BJT.

Substituting $I_{CC} = -I_B - I_C$ and $I_C = \beta_F I_B$ into Equation 3.6-43 yields the expression for V_{EC},

$$V_{EC} = V_{CC} + I_C \frac{\beta_F + 1}{\beta_F} R_C. \tag{3.6-44}$$

The load line derived from Equation 3.6-44 is in the fourth quadrant of the BJT output characteristic curve. Since the characteristic curve is plotted with the first quadrant with the ordinate axis representing $-I_C$, the slope of the load line for the *pnp* BJT fixed-bias circuit with collector feedback is

$$\text{Slope of load line} = -\frac{\beta_F}{(\beta_F + 1)R_C}. \tag{3.6-45}$$

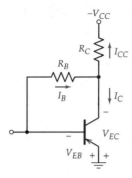

Figure 3.6-19
The *pnp* BJT fixed-bias circuit with collector feedback.

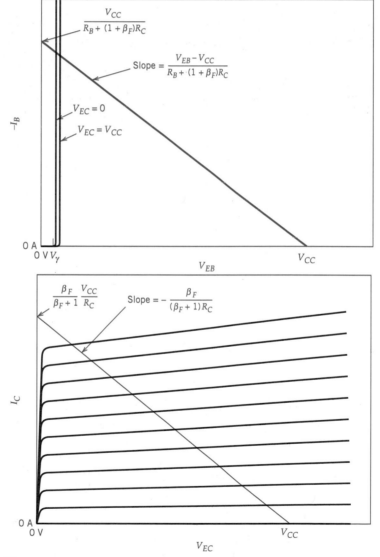

Figure 3.6-20
(*a*) Input and (*b*) output load line analysis for *pnp* BJT fixed-bias circuit with collector feedback.

The point at which the load line intersects the I_C axis occurs when $V_{EC} = 0$, so from Equation 3.6-44,

$$I_{C,V_{EC}=0} = -\frac{\beta_F}{\beta_F + 1}\frac{V_{CC}}{R_C}.$$

(3.6-46)

The other extreme of the load line occurs when the transistor is cut off ($I_C = 0$). By setting $I_C = 0$ in Equation 3.6-44, the intersection of the load line with the V_{EC} axis occurs at V_{CC}. Like the *npn* transistor, the emitter–base voltage is assumed to be approximately equal to V_γ as is evident in the input load line analysis shown in Figure 3.6-20. The BJT output characteristic with the load line for the fixed-bias circuit with collector feedback is shown in Figure 3.6-20*b*.

Similar analysis can be performed for the other biasing configurations discussed. The SPICE **.model** statement for a *pnp* transistor is

```
.model model_name PNP([parameter=value]. . .)
```

EXAMPLE 3.6-4 Complete the design of the *pnp* self-bias circuit shown. Let $V_{EB} = 0.7$ V, $\beta_F = 200$, and $V_{EC} = 7$ V. What is the collector current I_C?

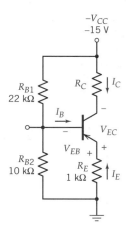

SOLUTION

The Thévenin equivalent voltage V_{BB} and the Thévenin equivalent resistance at the base of the circuit is

$$V_{BB} = \frac{-V_{CC}R_{B2}}{R_{B1} + R_{B2}} = -4.69 \text{ V}$$

and

$$R_B = \frac{R_{B1}R_{B2}}{R_{B1} + R_{B2}} = 6.875 \text{ k}\Omega.$$

The collector current is found by applying KVL on the base-emitter loop,

$$I_C = \frac{\beta_F}{\beta_F + 1} \frac{V_{EB} + V_{BB}}{R_E + R_B/(\beta_F + 1)}$$

$$= \frac{200}{210} \frac{0.7 - 4.69}{1000 + \dfrac{6875}{201}} \Rightarrow I_C = -3.84 \text{ mA}.$$

The collector resistance R_C is calculated by applying KVL on the collector-emitter loop,

$$V_{CC} = I_E R_E + V_{EC} - I_C R_C$$

$$= \frac{\beta_F + 1}{\beta_F} I_C R_E + V_{EC} - I_C R_C.$$

Rearranging the above equation yields a solution for R_C,

$$R_C = \frac{V_{EC} - V_{CC} - [(\beta_F + 1)/\beta_F] I_C R_E}{I_C}$$

$$= \frac{7 - 15 - 201/200(-3.84 \times 10^{-3})(1000)}{-3.84 \times 10^{-3}}$$

$$\Rightarrow R_C = 1.08 \text{ k}\Omega \approx 1 \text{ k}\Omega \quad \text{(common value)}.$$

3.7 BIAS STABILITY

The quiescent operating point (Q point) of a BJT in the forward-active region is dependent on the reverse saturation current, base-emitter voltage, and current gain of the transistor. This Q point of a BJT circuit can change due to variations in operating temperature or parameter variations that occur when interchanging individual transistors with slightly different characteristics in the biasing circuit. A stable Q point is desirable for the following reasons:

- That the transistor will operate over a specified range of DC voltages and currents is ensured.
- The desired amplifier gain, and input and output resistances, which are all dependent on the bias condition, are achieved.
- The maximum power hyperbola is not violated.

As an example of the parameter variation, an input transistor characteristic in Figure 3.7-1 shows a decrease in V_γ (for $V_{CE} = 0$) for a rise in operating temperature from 27 to 50°C. For this particular example, the change in V_γ is approximately -70 mV.

Figure 3.7-2 shows the variation in the *npn* BJT output characteristic curve with load line for a rise in operating temperature from 27 to 50°C. At elevated temperatures the change in the collector current, ΔI_C, increases for higher base currents. Therefore, for a constant I_B, the Q point of the BJT is shifted by some increment of

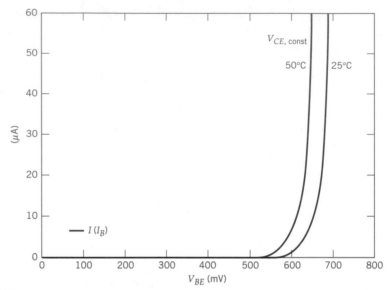

Figure 3.7-1
The *npn* BJT input characteristic curves for $V_{CE} = 0$ for operating temperatures of 27 and 50°C.

both I_C and V_{CE}. The change in the output characteristic is caused by increases in both the transistor β_F and the reverse saturation current.

In order to quantitatively determine the variation in quiescent conditions, it is necessary to examine the Ebers–Moll equations 3.1-3a and 3.1-3b in the forward-active region:

$$I_E = -I_{ES}(e^{V_{BE}/\eta V_t} - 1) + \alpha_R I_{CS}(e^{(V_{BE}-V_{CE})/\eta V_t} - 1),$$
$$I_C = -I_{CS}(e^{(V_{BE}-V_{CE})/\eta V_t} - 1) + \alpha_F I_{ES}(e^{V_{BE}/\eta V_t} - 1).$$

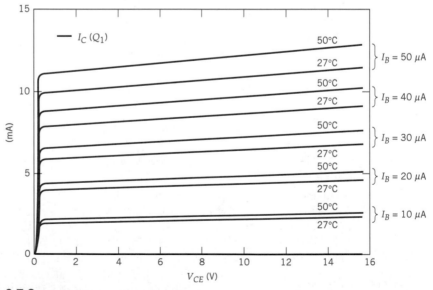

Figure 3.7-2
The *npn* BJT output characteristic curves with load line for operating temperatures of 27 and 50°C.

The collector current is found as a function of the emitter current by solving for $I_{ES}(e^{V_{BE}/\eta V_t} - 1)$ from Equation 3.1-3a,

$$I_{ES}(e^{V_{BE}/\eta V_t} - 1) = -I_E + \alpha_R I_{CS}(e^{(V_{BE}-V_{CE})/\eta V_t} - 1). \tag{3.7-1}$$

Substitution of Equation 3.7-2 into Equation 3.1-3b yields I_C as a function of I_E:

$$\begin{aligned} I_C &= -I_{CS}(e^{(V_{BE}-V_{CE})/\eta V_t} - 1) \\ &\quad + \alpha_F[-I_E + \alpha_R I_{CS}(e^{(V_{BE}-V_{CE})/\eta V_t} - 1)] \\ &= -\alpha_F I_E + (\alpha_F \alpha_R - 1)I_{CS}(e^{(V_{BE}-V_{CE})/\eta V_t} - 1). \end{aligned} \tag{3.7-2}$$

When the BJT is in the forward-active region, $V_{BE} - V_{CE} \ll -\eta V_t$. Then $I_{CS}(e^{(V_{BE}-V_{CE})/\eta V_t} - 1) \approx -I_{CS}$. Therefore, the collector current for an *npn* BJT in the forward-active region is

$$I_C = -\alpha_F I_E + (1 - \alpha_F \alpha_R)I_{CS}, \tag{3.73-a}$$

$$I_C = -\alpha_F I_E + I_{CO}, \tag{3.7-3b}$$

where the reverse saturation current I_{CO} is defined as

$$I_{CO} \approx (1 - \alpha_F \alpha_R)I_{CS}. \tag{3.7-4}$$

The collector current of Equation 3.7-3 can be written in terms of β_F as

$$I_C = \beta_F I_B + (\beta_F + 1)I_{CO}. \tag{3.7-5}$$

Equation 3.7-5 will be used to determine the dependence of the collector current of an *npn* BJT to changes in the reverse saturation current for different biasing arrangements.

A set of three stability factors is used to quantify the variation in the collector current with respect to the reverse saturation current, base–emitter voltage, and β_F. The stability factors are

$$\begin{aligned} S_I &= \frac{\partial I_C}{\partial I_{CO}} \approx \frac{\Delta I_C}{\Delta I_{CO}}, \\ S_V &= \frac{\partial I_C}{\partial V_{BE}} \approx \frac{\Delta I_C}{\Delta V_{BE}}, \\ S_\beta &= \frac{\partial I_C}{\partial \beta_F} \approx \frac{\Delta I_C}{\Delta \beta_F}. \end{aligned} \tag{3.7-6}$$

The stability of the bias configuration is quantified with respect to the collector current for the following reasons:

- The collector current is dependent on the base current and the collector–emitter voltage and determines the output signal of a BJT amplifier.
- Small variations in β_F, I_{CO}, and ΔV_{BE} can result in a large change in I_C.

The total incremental change in I_C for small changes in I_{CO}, V_{BE}, and β_F is

$$\Delta I_{CT} = S_I \Delta I_{CO} + S_V \Delta V_{BE} + S_\beta \Delta \beta_F. \tag{3.7-7}$$

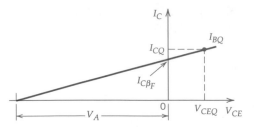

Figure 3.7-3
Actual Q point of BJT taking into account Early voltage V_A.

Equation 3.7-7 clearly shows that in order to keep the change in I_C small, the magnitude of the stability factors must also be kept small.

An accurate method of determining β_F at the operating point would be to include the effects of the Early voltage. Increase in accuracy is attained by using the slope of the output characteristics caused by the early voltage V_A shown in Figure 3.7-3.

The Q point given the base current I_{BQ} is I_{CQ}. The SPICE output file provides β_F for $V_{CE}= 0$ V. The collector current using I_{BQ} and β_F provided by SPICE yields the y-axis intercept, $I_{C\beta_F} = \beta_F I_{BQ}$. The collector-emitter voltage for a fixed-bias circuit is defined in Equation 3.6.1-1,

$$V_{CEQ} = V_{CC} - I_{CQ}R_C. \tag{3.7-8}$$

Using the expression for the equation of a line, the collector current can be written as

$$I_{CQ} = \frac{I_{C\beta_F}}{V_A} V_{CEQ} + I_{C\beta_F}. \tag{3.7-9}$$

Solving for I_{CQ},

$$I_{CQ} = \frac{(I_{C\beta_F}/V_A)V_{CC} + I_{C\beta_F}}{1 + (I_{C\beta_F}/V_A)R_C}. \tag{3.7-10}$$

In order to demonstrate the effect of circuit element values on quiescent point stability, the fixed-bias and self-bias circuits are examined in the remainder of this section.

3.7.1 Fixed-Bias Circuit Stability

For the fixed-bias circuit shown in Figure 3.6-4, the reverse saturation current stability factor S_I is found by applying Equation 3.7-5 to the expression for the base current in Equation 3.6-6:

$$I_B = \frac{V_{CC} - V_{BE}}{R_B}. \tag{3.7-11}$$

Substituting Equation 3.7-11 into Equation 3.7-5 yields

$$I_C = \beta_F \frac{V_{CC} - V_{BE}}{R_B} + (\beta_F + 1)I_{CO}. \tag{3.7-12}$$

The reverse saturation stability factor is

$$S_I = \frac{\partial I_C}{\partial I_{CO}}$$

$$= \frac{\partial}{\partial I_{CO}}\left[\beta_F \frac{V_{CC} - V_{BE}}{R_B} + (\beta_F + 1)I_{CO}\right] \tag{3.7-13}$$

$$= \beta_F + 1.$$

It is apparent from Equation 3.7-13 that S_I is very large. Therefore, a small change in I_{CO} leads to a large change in I_C.

The base–emitter voltage stability factor S_V is found by using Equation 3.7-11,

$$S_V = \frac{\partial I_C}{\partial V_{BE}}$$

$$= \frac{\partial}{\partial V_{BE}}\left(\beta_F \frac{V_{CC} - V_{BE}}{R_B}\right) \tag{3.7-14}$$

$$= -\frac{\beta_F}{R_B}.$$

The β_F stability factor S_β for the fixed bias circuit is

$$S_\beta = \frac{\partial I_C}{\partial \beta_F}$$

$$= \frac{\partial}{\partial \beta_F}\left(\beta_F \frac{V_{CC} - V_{BE}}{R_B} + (\beta_F + 1)I_{CO}\right) \tag{3.7-15}$$

$$= \frac{V_{CC} - V_{BE}}{R_B} + I_{CO} \approx \frac{I_C}{\beta_F}.$$

The change in collector current due to a change in β_F is found by using S_β,

$$\Delta I_C = S_\beta \, \Delta\beta_F$$

$$= \frac{I_{CQ1}}{\beta_{FQ1}} \Delta\beta_F, \tag{3.7-16a}$$

or

$$\frac{\Delta I_C}{I_{CQ1}} = \frac{\Delta\beta_F}{\beta_{FQ1}}, \tag{3.7-16b}$$

where I_{CQ1} and β_{FQ1} are the collector current and β_F at the known Q point.

Equation 3.7-16b states that there is a one-to-one correspondence between a percentage change in β_F to I_C. Therefore, the fixed-bias circuit is not a stable biasing arrangement. The desire is to reduce the percentage change in I_C for a given change in β_F.

EXAMPLE 3.7-1 For the fixed-bias transistor circuit shown, find the collector current at 50°C. Assume $V_{BE} \approx 0.7$ V.

A 2N2222 BJT is used with the following parameters:

BF $\equiv \beta_F = 255$

BR $\equiv \beta_R = 6$

VA $\equiv$ Early voltage $= 75$

IS $\equiv$ transport saturation current $= I_S = 14.34 \times 10^{-15}$

$I_{CO} \approx I_S [(1 - \alpha_R \alpha_R)/\alpha_R] = I_S[(\beta_R + 1)/\beta_R - \beta_F/(\beta_F + 1)]$

XTB $\equiv$ forward and reverse β temperature coefficient $= 1.5$, which is typically used for small-signal BJT transistors, where

$$\beta_F(T_2) = \beta_F(T_1)\left(\frac{T_2}{T_1}\right)^{XTB}$$

$$\beta_R(T_2) = \beta_R(T_1)\left(\frac{T_2}{T_1}\right)^{XTB}$$

The default value of XTI $= 3$ (IS temperature coefficient) is used.

```
*********************************PSpice*********************************
BJT Input Transistor Curve
**** CIRCUIT DESCRIPTION
**********************************************************************
VCE   1    0    12V
IB    0    2    50u
Q1    1    2    0   Q2N2222
.model Q2N2222 NPN(IS=14.34E-15 BF=255 BR=6 VA=75 XTB=1.5)
.DC   IB   0 60u .2u   TEMP 27 50 23
.PROBE
.END
*********************************PSpice*********************************
BJT Input Transistor Curve
**** BJT MODEL PARAMETERS
**********************************************************************
         Q2N2222
         NPN
      IS 14.340000E-15
      BF 255
      NF  1
      VAF  75
      BR   6
      NR   1
      XTB 1.5
      JOB CONCLUDED
      TOTAL JOB TIME    30.70
```

Figure 3.7-4
Netlist for determining input characteristics of 2N2222 BJT.

SOLUTION

First determine I_B:

$$I_B = \frac{V_{CC} - V_{BE}}{R_B} = \frac{12 - 0.7}{750 \times 10^3} = 15 \ \mu A.$$

The parameter changes with respect to temperature must be found. Using SPICE, the variation in I_{CO}, V_{BE}, and β_F are found. To determine the variation in V_{BE}, the input characteristic curves are generated using the SPICE netlist in Figure 3.7-4. The netlist shown is used to generate the BJT input characteristic curves at 27 and 50°C

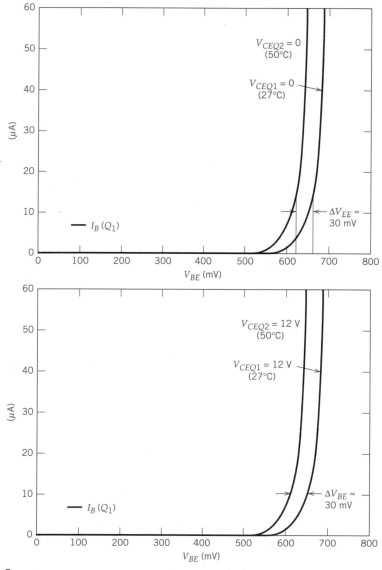

Figure 3.7-5

Input characteristic curves for 2N2222 BJT for (a) $V_{CE} = 0$ and (b) $V_{CE} = 12$ V, where $\Delta V_{BE} \approx -30$ mV.

```
*****************************PSpice********************************
BJT Output Transistor Curve (27 AND 50 C)
**** CIRCUIT DESCRIPTION
*******************************************************************

VCE   1    0     15V
IB    0    2     60 uA
Q1    1    2     0    Q2N2222
.model Q2N2222 NPN(IS=14.34E-15 BF=255 BR=6 VA=75 XTB=1.5)
.DC DEC VCE 0.001 15 10 IB 13.6u 15u 1.4u
.TEMP 27  50
.PROBE
.END

*****************************PSpice********************************
BJT Output Transistor Curve (27 AND 50 C)
**** BJT MODEL PARAMETERS
*******************************************************************
        Q2N2222
        NPN
    IS 14.340000E-15
    BF 255
    NF 1
    VAF  75
    BR   6
    NR   1
    XTB 1.5
*****************************PSpice********************************
BJT Output Transistor Curve (27 AND 50 C)
**** TEMPERATURE-ADJUSTED VALUES  TEMPERATURE= 50.000 DEG C
*******************************************************************
****BJT MODEL PARAMETERS
NAME      BF         ISE        VJE        CJE        RE         RB
          BR         ISC        VJC        CJC        RC         RBM
          IS         ISS        VJS        CJS
Q2N2222 2.849E+0 2  0.000E+00  7.095E-01  0.000E+00  0.000E+00  0.000E+00
        6.703E+00   0.000E+00  7.095E-01  0.000E+00  0.000E+00  0.000E+00
        3.796E-13   0.000E+00  7.095E-01  0.000E+00

    JOB CONCLUDED
    TOTAL JOB TIME   10.05
```

Figure 3.7-6
SPICE output file for 2N2222 BJT.

for $V_{CE} = 12$ V. By changing V_{CE} to 0 V, the other set of curves for $V_{CE} = 0$ V can be found. From the two graphs shown in Figures 3.7-5a,b, ΔV_{BE} is nearly identical for $V_{CE} = 0$ and $V_{CE} = 12$ V, where

$$\Delta V_{BE} \approx 30 \text{ mV}.$$

The output file for the SPICE output characteristics, shown in Figure 3.7-6, is used to determine the variation in I_{CO} and β_F.

Using load line analysis (Figure 3.7-7), the expected ΔI_C is slightly greater than 0.5 mA. From the SPICE output file in Figure 3.7-6, the variation in β_F is from 225 (27°C) to 285 (50°C) for $\Delta\beta_F = 30$. The variation in I_S is from 14.43×10^{-15} A (27°C) to 379.6×10^{-15} A (50°C). The variation in β_R is from 6 (27°C) to 6.7 (50°C) for $\Delta\beta_R = 0.7$. This corresponds to $\Delta I_{CO} = 55.54 \times 10^{-15}$ Δ.

From Equation 3.7-7 the total change in collector current is

$$\Delta I_{CT} = S_I \, \Delta I_{CO} + S_V \, \Delta V_{BE} + S_\beta \, \Delta \beta_F.$$

For a fixed-bias BJT circuit,

$$S_I = \beta_F + 1 = 255 + 1 = 256,$$

$$S_V = -\frac{\beta_F}{R_B} = -\frac{255}{750,000} = -0.36 \times 10^{-3} \text{ S},$$

$$S_\beta = \frac{V_{CC} - V_{BE}}{R_B} = \frac{12 - 0.7}{750,000} \approx 15 \ \mu\text{A}.$$

Therefore,

$$\Delta I_C = 256(55.54 \times 10^{-15}) + (-0.36 \times 10^{-3})(-0.030)$$
$$+ (15 \times 10^{-6})(30)$$
$$= 0.46 \text{ mA}.$$

The analytical result is in agreement with the load line result in Figure 3.7-7, where ΔI_C is slightly larger than 0.5 mA. The discrepancy is caused by the early voltage effect. Although β_F changes due to temperature, SPICE does not alter the early voltage (VA). Therefore, β_F increases as V_{CE} is increased for a given I_B.

For $I_B = 15 \ \mu\text{A}$ at 27°C and $I_{C\beta_F} = \beta_F I_{BQ} = 225(15 \times 10^{-6}) = 3.825$ mA, $I_{CQ} = 4.12$ mA. For the most accurate determination of I_{CQ}, SPICE should be used.

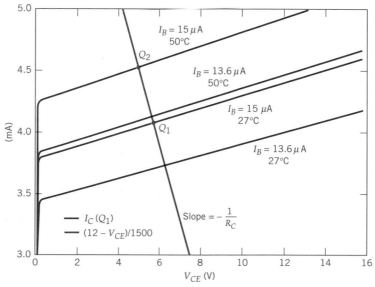

Figure 3.7-7
Output characteristic curves for 2N2222 BJT.

3.7.2 Self-Bias Circuit Stability

The bias stability factors for the self-bias circuit, shown in Figure 3.6-15, can be derived using the analysis for the fixed-bias circuit. Here, S_I for the self-bias circuit is found by applying Equation 3.7-5 to find the base and emitter currents:

$$I_B = \frac{I_C - (\beta_F + 1)I_{CO}}{\beta_F} \tag{3.7-17a}$$

and

$$I_{EE} = -I_E = \frac{\beta_F + 1}{\beta_F}(I_C - I_{CO}). \tag{3.7-17b}$$

The collector current is found by using the base–emitter loop equation of Equation 3.6-37,

$$V_{BB} = V_{BE} + I_B R_B + I_{EE} R_E. \tag{3.7-18}$$

Substituting Equations 3.7-17a and 3.7-17b into Equation 3.7-18 yields the expression for I_C,

$$I_C = \frac{\beta_F(V_{BB} - V_{BE}) + I_{CO}(\beta_F + 1)(R_B + R_E)}{R_B + (\beta_F + 1)R_E}. \tag{3.7-19}$$

Here, S_I is found by taking the derivative of Equation 3.7-19 with respect to I_{CO},

$$\begin{aligned}
S_I &= \frac{\partial I_C}{\partial I_{CO}} \\
&= \frac{\partial}{\partial I_{CO}}\left[\frac{\beta_F(V_{BB} - V_{BE}) + I_{CO}(\beta_F + 1)(R_B + R_E)}{R_B + (\beta_F + 1)R_E}\right] \\
&= \frac{(\beta_F + 1)(R_B + R_E)}{R_B + (\beta_F + 1)R_E}.
\end{aligned} \tag{3.7-20}$$

The term S_I can be reduced by choosing R_B as small as possible and R_E as large as possible. The range of values of R_B and R_E are limited by the required input resistance and the limitation of the power supply current output. That is, if a large R_E is chosen, the power supply must be able to provide a large collector current to the transistor to yield the required emitter voltage.

For the self-biased circuit S_V is also found by using the expression for the base–emitter loop of Equation 3.7-18. By rearranging Equation 3.7-18, the expression for I_C (for $I_{CO} \approx 0$) is found:

$$I_C = \frac{\beta_F(V_{BB} - V_{BE})}{R_B + (\beta_F + 1)R_E}. \tag{3.7-21}$$

From Equation 3.7-21 the base–emitter voltage stability factor is

$$
\begin{aligned}
S_V &= \frac{\partial I_C}{\partial V_{BE}} \\
&= \frac{\partial}{\partial V_{BE}} \frac{\beta_F(V_{BB} - V_{BE})}{R_B + (\beta_F + 1)R_E} \\
&= -\frac{\beta_F}{R_B + (\beta_F + 1)R_E}.
\end{aligned}
\tag{3.7-22}
$$

For maximum base–emitter stability, a large R_E is desirable.

The β_F stability factor is found by taking the derivative of Equation 3.7-21 with respect to β_F,

$$
\begin{aligned}
S_\beta &= \frac{\partial I_C}{\partial \beta_F} \\
&= \frac{\partial}{\partial \beta_F} \frac{\beta_F(V_{BB} - V_{BE})}{R_B + (\beta_F + 1)R_E} \\
&= (V_{BB} - V_{BE}) \frac{R_B + R_E}{[R_B + (\beta_F + 1)R_E]^2}.
\end{aligned}
\tag{3.7-23}
$$

Since S_β is a function of varying β_F, the expression does not indicate whether β_{FQ1} or β_{FQ2} should be used in Equation 3.7-23. This uncertainty is solved through the use of an alternate derivation of S_β. By taking finite differences rather than the derivative,

$$
S_\beta \approx \frac{I_{CQ2} - I_{CQ1}}{\beta_{FQ2} - \beta_{FQ1}} = \frac{\Delta I_C}{\Delta \beta_F}.
\tag{3.7-24}
$$

From the equation describing the collector current (Equation 3.6-38),

$$
\frac{I_{CQ2}}{I_{CQ1}} = \frac{\beta_{FQ2}}{\beta_{FQ1}} \frac{R_B + (\beta_{FQ1} + 1)R_E}{R_B + (\beta_{FQ2} + 1)R_E}.
\tag{3.7-25}
$$

Subtracting unity from both sides of Equation 3.7-25,

$$
\frac{I_{CQ2}}{I_{CQ1}} - 1 = \left(\frac{\beta_{FQ2}}{\beta_{FQ1}}\right) \frac{R_B + (\beta_{FQ1} + 1)R_E}{R_B + (\beta_{FQ2} + 1)R_E} - 1
\tag{3.7-26}
$$

yields an expression for ΔI_C as a function of $\Delta \beta_F$:

$$
\begin{aligned}
\Delta I_C = I_{CQ2} - I_{CQ1} &= I_{CQ1} \frac{\beta_{FQ2} - \beta_{FQ1}}{\beta_{FQ1}} \frac{R_B + R_E}{R_B + (\beta_{FQ2} + 1)R_E} \\
&= \frac{I_{CQ1}}{\beta_{FQ1}} \frac{R_B + R_E}{R_B + (\beta_{FQ2} + 1)R_E} \Delta \beta_F
\end{aligned}
\tag{3.7-27}
$$

TABLE 3.7-1 STABILITY FACTORS FOR BJT BIAS CIRCUIT CONFIGURATIONS.

Bias Configuration	S_I, I_C - Reverse Saturation Current Bias Stability Factor	S_V, I_C - Base-Emitter Voltage Bias Stability Factor	S_β, I_C - β Bias Stability Factor
Fixed-Bias	$\beta_F + 1$	$-\dfrac{\beta_F}{R_B}$	$\dfrac{V_{CC} - V_{BE}}{R_B}$
Fixed-Bias with Emitter Resistor	$\dfrac{(\beta_F + 1)(R_B + R_E)}{R_B + (\beta_F + 1)R_E}$	$\dfrac{-\beta_F}{R_B + (\beta_F + 1)R_E}$	$\dfrac{(V_{CC} - V_{BE})(R_B + R_E)}{[R_B + (\beta_{FQ1} + 1)R_E][R_B + (\beta_{FQ2} + 1)R_E]}$
Fixed-Bias with Collector Feedback	$\dfrac{(\beta_F + 1)(R_B + R_C)}{R_B + (\beta_F + 1)R_C}$	$\dfrac{-\beta_F}{R_B + (\beta_F + 1)R_C}$	$\dfrac{(V_{CC} - V_{BE})(R_B + R_C)}{[R_B + (\beta_{FQ1} + 1)R_C][R_B + (\beta_{FQ2} + 1)R_C]}$
Fixed-Bias with Collector and Emitter Feedback	$\dfrac{(\beta_F + 1)(R_B + R_C + R_E)}{R_B + (\beta_F + 1)(R_C + R_E)}$	$\dfrac{-\beta_F}{R_B + (\beta_F + 1)(R_C + R_E)}$	$\dfrac{(V_{CC} - V_{BE})(R_B + R_C + R_E)}{[R_B + (\beta_{FQ1} + 1)(R_C + R_E)][R_B + (\beta_{FQ2} + 1)(R_C + R_E)]}$
Emitter-Bias with Two Power Supplies	$\dfrac{(\beta_F + 1)(R_B + R_E)}{R_B + (\beta_F + 1)R_E}$	$\dfrac{-\beta_F}{R_B + (\beta_F + 1)R_E}$	$\dfrac{(V_{EE} - V_{BE})(R_B + R_E)}{[R_B + (\beta_{FQ1} + 1)R_E][R_B + (\beta_{FQ2} + 1)R_E]}$
Self-Bias	$\dfrac{(\beta_F + 1)(R_B + R_E)}{R_B + (\beta_F + 1)R_E}$	$\dfrac{-\beta_F}{R_B + (\beta_F + 1)R_E}$	$\dfrac{(V_{BB} - V_{BE})(R_B + R_E)}{[R_B + (\beta_{FQ1} + 1)R_E][R_B + (\beta_{FQ2} + 1)R_E]}$

Note: $\Delta I_{CT} = S_I \Delta I_{CO} + S_V \Delta V_{BE} + S_\beta \Delta \beta_F$

or

$$S_\beta = \frac{\Delta I_C}{\Delta \beta_F} = \frac{I_{CQ1}}{\beta_{FQ1}} \frac{R_B + R_E}{R_B + (\beta_{FQ2} + 1)R_E}$$

$$= \frac{(V_{BB} - V_{BE})R_B + R_E}{[R_B + (\beta_{FQ1} + 1)R_E][R_B + (\beta_{FQ2} + 1)R_E]}.$$

(3.7-28)

If a 1% change in I_C is desired for a 10% change in β_F, the ratio R_B/R_E can be determined for the self-bias circuit using Equation 3.7-27:

$$\frac{\Delta I_C}{I_{CQ1}} = \frac{\Delta \beta_F}{\beta_{FQ1}} \frac{R_B + R_E}{R_B + (\beta_{FQ2} + 1)R_E},$$

$$0.01 = 0.1 \frac{R_B + R_E}{R_B + (\beta_{FQ2} + 1)R_E}.$$

(3.7-29)

To achieve a bias stability of 1% change in I_C for a 10% percent change in β_F the ratio R_B/R_E is

$$\frac{R_B}{R_E} \leq \frac{0.1\beta_{F,\min} - 0.9}{0.9} = \frac{\beta_F}{9} - 1.$$

(3.7-30)

The ratio R_B/R_E in Equation 3.7-30 can be used as a rule of thumb when designing self-bias BJT circuits. It is evident from Equation 3.7-30 that increasing R_E results in increased stability. The value of R_E is determined by many factors, including the amplifier gain and maximum allowable swing, which will both be discussed in Chapter 5.

The bias stability factors for various *npn* BJT bias arrangements are shown in Table 3.7-1. In general, the addition of R_E to the bias network decreases all three stability factors with increases in the overall stability of the Q point of the BJT.

Since the β_F variation of a BJT is, in general, the dominant factor that determines Q-point stability, an approximate relationship for the change in collector current with respect to a change in β_F can be established. The change in collector current, ΔI_C, for a change in β_F for the different bias configurations is shown in Table 3.7-2.

EXAMPLE 3.7-2 Find the change in the collector current from 27 to 50°C for the self-bias BJT circuit of Example 3.6-3 shown. Assume a normal $V_{BE} = 0.7$ V. The circuit is designed for $I_B = 15$ μA. Compare the result to the fixed-bias circuit in Example 3.7-1.

A 2N2222 BJT is used with the following parameters (identical to Example 3.7-1):

BF ≡ β_F = 255

BR ≡ β_R = 6

VA ≡ Early voltage = 75 V

IS ≡ transport saturation current = I_S = 14.34E-15 A

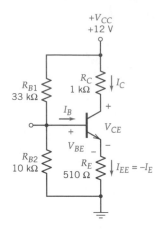

SOLUTION

The output file for the SPICE output characteristics, shown in Figure 3.7-6 in Example 3.7-1, is used to determine the variation in I_{CO} and β_F. The output characteristic of the BJT is shown with the circuit load line in Figure 3.7-8.

Unlike the fixed-bias circuit where the base current was invariant with parameter changes, in the self-bias circuit, the base current decreases with increasing β_F (the base current increases with decreasing β_F). In this example, at 27°C, the base current is

$$I_B = \frac{V_{BB} - V_{BE}}{R_B + (\beta_F + 1)R_C},$$

where

$$V_{BB} = \frac{V_{CC}R_{B2}}{R_{B1} + R_{B2}} = 2.79 \text{ V}, \qquad R_B = R_{B1} \parallel R_{B2} = 7.67 \text{ k}\Omega.$$

The base currents for 27°C($\beta_F = 255$) and 50°C ($\beta_F = 285$) are $I_{BQ1} = 15$ μA and $I_{BQ2} = 13.6$ μA, respectively.

Using the load line analysis shown in Figure 3.7-8, the expected ΔI_C is slightly greater than 0.056 mA. From the SPICE output file in Figure 3.7-6 in Example 3.7-1, the variation in β_F is from 255 (27°C) to 285 (50°C) for a $\Delta\beta_F = 30$. The variation in I_S is from 14.43×10^{-15} A (27°C) to 379.6×10^{-15} A (50°C). The variation in β_R is from 6 (27°C) to 6.7 (50°C) for $\Delta\beta_R = 0.7$. This corresponds to $\Delta I_{CO} = 55.54 \times 10^{-15}$ A.

From Equation 3.7-7, the total change in collector current is

$$\Delta I_{CT} = S_I \Delta I_{CO} + S_V \Delta V_{BE} + S_\beta \Delta\beta_F.$$

For a self-bias BJT circuit,

$$S_I = \frac{(\beta_F + 1)(R_B + R_E)}{R_B + (\beta_F + 1)R_E} = \frac{256(7670 + 510)}{7670 + 256(510)} = 15.2,$$

$$S_V = \frac{\beta_F}{R_B + (\beta_F + 1)R_E} = -\frac{255}{7670 + 256(510)} = -1.85 \text{ mS},$$

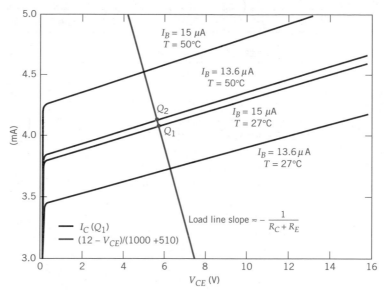

Figure 3.7-8
Output characteristic curves for 2N2222 BJT with load line.

$$S_\beta = \frac{(V_{BB} - V_{BE})(R_B + R_E)}{[R_B + (\beta_{FQ1} + 1)R_E][R_B + (\beta_{FQ2} + 1)R_E]}$$

$$= \frac{(2.79 - 0.7)(7670 + 510)}{[7670 + 256(510)][7670 + 286(510)]} = 0.81 \ \mu A.$$

Therefore,

$$\Delta I_C = 15.2(55.54 \times 10^{-15}) + (-1.85 \times 10^{-3})(-0.030)$$

$$+ (0.81 \times 10^{-6})(30)$$

$$= 79.8 \ \mu A.$$

The analytical result is in general agreement with the load line result in Figure 3.7-8, where ΔI_C is slightly larger than 0.056 mA.

Note that the collector current change due ΔI_{CO} is negligible and, in most cases, can be ignored.

The variation in collector current is 79.8 μA for the self-bias circuit, as opposed to 460 μA for the fixed-bias circuit. Therefore, the self-bias circuit has better bias stability than the fixed-bias arrangement.

 Some SPICE software packages allow for worst-case or Monte Carlo analysis of the circuit with variations in operating parameters. These packages are useful when analyzing bias stability when individual BJTs are replaced in a bias circuit without varying the operating temperature. For instance, in PSpice, the **.WCASE** command is used for worst-case analysis.

TABLE 3.7-2 CHANGE IN I_C FOR A CHANGE IN β

Bias Configuration	ΔI_C As a Function of $\Delta\beta_F$
Fixed-Bias	$\dfrac{I_{CQ1}\Delta\beta_F}{\beta_{FQ1}}$
Fixed-Bias with Emitter Resistor	$\dfrac{I_{CQ1}\Delta\beta_F}{\beta_{FQ1}} \dfrac{(R_B + R_E)}{[R_B + (\beta_{FQ2} + 1)R_E]}$
Fixed-Bias with Collector Feedback	$\dfrac{I_{CQ1}\Delta\beta_F}{\beta_{FQ1}} \dfrac{(R_B + R_C)}{[R_B + (\beta_{FQ2} + 1)R_C]}$
Fixed-Bias with Collector and Emitter Feedback	$\dfrac{I_{CQ1}\Delta\beta_F}{\beta_{FQ1}} \dfrac{(R_B + R_C + R_E)}{[R_B + (\beta_{FQ2} + 1)(R_C + R_E)]}$
Emitter-Bias with Two Power Supplies	$\dfrac{I_{CQ1}\Delta\beta_F}{\beta_{FQ1}} \dfrac{(R_B + R_E)}{[R_B + (\beta_{FQ2} + 1)R_E]}$
Self-Bias	$\dfrac{I_{CQ1}\Delta\beta_F}{\beta_{FQ1}} \dfrac{(R_B + R_E)}{[R_B + (\beta_{FQ2} + 1)R_E]}$

For instance, if β_F varies for an *npn* BJT from 200 to 220, the simple **.model** statement for PSpice is

```
.model device_name NPN(BF=200 DEV 20)
```

where DEV is the deviation from the nominal (BF=200).

To perform a worst-case analysis to determine the impact of the change in β_F on the collector current on the transistor Q_1, the netlist includes the following statement:

```
.WCASE DC IC(Q1) YMAX HI VARY DEV DEVICES Q
```

This statement performs a worst-case analysis to find the greatest difference from the nominal (YMAX), with BF = nominal BF + 20 (HI), varying the parameter denoted by DEV (VARY and DEV), on the transistor only (Q). For BF = nominal BF − 20, replace HI with LO.

A sample netlist in PSpice for finding the BJT output characteristic curve of an *npn* BJT is shown below:

```
Worst-Case Analysis with PSpice
VCE    1      0       15V
IB     0      2       60u
Q1     1      2       0      NPNBJT
```

```
.model  NPNBJT    NPN(BF=200 DEV 20 VA=75)
.DC DEC VCE 0.001 15 10 IB 0 50u 10u
.WCASE DC IC(Q1) YMAX HI VARY DEV DEVICES Q
.PROBE
.END
```

The resulting BJT output characteristic curve is shown in Figure 3.7-9.

3.8 CONCLUDING REMARKS

The BJT has been described in this chapter as a commonly used semiconductor device with four basic regions of operation: the saturation, foward-active, inverse-active, and cut-off regions. The BJT operation into these regions is controlled by the bias conditions on the transistor base–emitter and base–collector junctions. Typical applications lead to the use of the transistor base current and the collector–emitter voltage as more accessible quantities for region verification.

The Ebers–Moll transistor model was presented to quantify the current–voltage relationships of the BJT in all four regions of operation. As with the semiconductor diode, adequate representation of BJT performance can be obtained with piecewise linear approximations of the transistor characteristics. A set of simple linear models, one for each region of operation, was developed using transistor characteristics as expressed by the Ebers–Moll model and its corresponding set of equations.

The BJT logic gate applications have provided a good example of transistor circuitry using a variety of regions of operation. The two-state output necessary for binary gates is often achieved by the output BJT transitioning between the saturation or cutoff region. In addition, TTL gates provide an example of inverse-active region operation of the input BJT. Linear BJT applications utilize the forward-active region. Here, the operation of the transistor is nearly linear about a quiescent operating point (Q point) achieved with external bias circuitry. Several biasing circuits were devel-

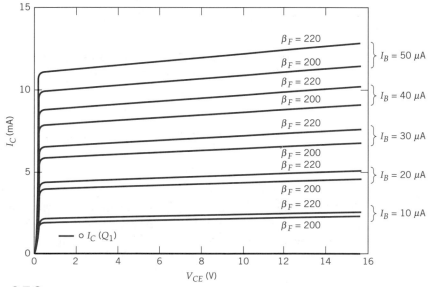

Figure 3.7-9
Output characteristic of *npn* BJT using netlist for varying BF.

oped, and the bias stability due to variations in transistor parameters was quantified. Based on the foundations developed in this chapter, additional linear and non-linear applications will be examined in later chapters.

SUMMARY DESIGN EXAMPLE

Bipolar junction transistors can often be used as a controlled current shunt to reduce power consumption in sensitive or expensive electronic components. A Zener diode voltage regulator is one device that can benefit from a shunting BJT. The basic topology of such a voltage regulator is shown in the accompanying figure.

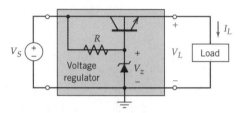

Without the use of a shunting BJT, the regulator resistor must be capable of carrying currents in excess of the load current. Similarly, the Zener diode also carries very large currents. The shunting BJT in this design topology reduces the current through both these devices, thereby reducing power consumption and component cost.

Design a Zener voltage regulator with a BJT shunt to meet the following design requirements:

Regulated load voltage	$V_L = 10V$
Load current	$0 \text{ A} \leq I_L \leq 5 \text{ A}$
Source voltage	$12 \text{ V} \leq V_S \leq 15 \text{ V}$

Determine the appropriate ratings for all components. Assume the minimum Zener current for proper regulation is 2 mA and a power BJT with typical forward current gain $\beta_F = 50$.

Solution

In order to maintain 10 V at the load, the diode Zener voltage must be

$$V_Z = V_L + V_\gamma = 10.7 \text{ V}.$$

The remainder of the design process for this circuit topology is similar to that of a simple Zener diode regulator except the current through the resistor R is the Zener current plus the base current of the BJT:

$$I_R = I_Z + \frac{I_I}{\beta_F + 1}.$$

The minimum design value for the resistor current I_R that will ensure regulation under all load and source variation is given by

$$I_{R(min)} = I_{Z(min)} + \frac{I_{L(max)}}{\beta_F + 1} = 2 \text{ mA} + \frac{5 \text{ A}}{51} = 100 \text{ mA}.$$

This minimum resistor current must occur with minimum source voltage V_S. The resulting maximum value for the resistor is therefore given by

$$R = \frac{12 \text{ V} - 10.7 \text{ V}}{100 \text{ mA}} = 13 \text{ } \Omega.$$

With this choice of resistor value, the maximum power dissipated in the resistor is given by

$$P_{R(max)} = \frac{(15 - 10.7)^2}{13} = 1.423 \text{ W}.$$

The maximum current through the Zener diode is given by the maximum resistor current less the minimum BJT base current:

$$I_{Z(max)} = \frac{15 - 10.7}{13} - \frac{I_{I(min)}}{\beta_F + 1} = 0.331 \text{ A}.$$

The maximum power dissipated by the Zener diode is 3.542 W. The BJT must be capable of an emitter current equal to the maximum load current (5 A) at a maximum V_{CE} (5 V). Thus the BJTs must be able to dissipate at least 25 W.

Total power consumption is given by

$$P_T = V_S \left(\frac{V_S - 10.7}{13} + \frac{\beta_F}{\beta_F + 1} I_L \right).$$

Thus, total power consumption ranges between 1.2 and 78.5 W depending on source and load conditions.

Summary of Required Components

Zener diode	$V_Z = 10.7$ V	$I_{Z(max)} > 0.331$ A	$P_{max} > 3.542$ W
Resistor	$R = 13$ Ω		$P_{max} > 1.432$ W
BJT	$\beta_F = 50$	$I_{c(max)} = 4.902$ A	$P_{max} > 25$ W

COMPARISON TO SIMPLE ZENER REGULATOR

A simple Zener diode regulator (without a shunting BJT) requires the following components:

Zener diode	$V_Z = 10$ V	$I_{Z(max)} > 12.6$ A	$P_{max} > 126$ W
Resistor	$R = 0.397$ Ω		$P_{max} > 63$ W

This simple design results in a power consumption of between 60.5 W ($V_S = 12$ V) and 190 W ($V_S = 15$ V) independent of the load current (maximum load power = 50 W). The difference in component specifications is striking: The BJT shunt provides for a more efficient and cost-effective solution to voltage regulation in this case.

REFERENCES

Antognetti, P. and Massobrio, G., *Semiconductor Device Modeling with SPICE,* Mc-Graw-Hill Book Company, New York, 1988.

Colclaser, R. A. and Diehl-Nagel, S., *Materials and Devices for Electrical Engineers and Physicists,* McGraw-Hill Book Company, New York, 1985.

Ghausi, M. S., *Electronic Devices and Circuits: Discrete and Integrated,* Holt, Rinehart and Winston, New York, 1985.

Gray, P. R., and Meyer, R. G., *Analysis and Design of Analog Integrated Circuits,* 3rd. Ed., John Wiley & Sons, Inc., New York, 1993

Malvino, A. P., *Transistor Circuit Approximations,* 2nd. Ed., McGraw-Hill Book Company, New York, 1973.

Millman, J., *Microelectronics, Digital and Analog Circuits and Systems,* McGraw-Hill Book Company, New York, 1979.

Millman, J. and Halkias, C. C., *Integrated Electronics: Analog and Digital Circuits and Systems,* McGraw-Hill Book Company, New York, 1972.

Sedra, A. S. and Smith, K. C., *Microelectronic Circuits,* 3rd. Ed., Holt, Rinehart, and Winston. Philadelphia, 1991.

Tuinenga, P., *SPICE: A Guide to Circuit Simulation and Analysis Using PSpice,* 2nd. Ed., Prentice Hall, Englewood Cliffs, NJ 1992.

PROBLEMS

3-1. A silicon *npn* BJT is described by the following parameters (remember $\eta = 1$ for silicon BJTs):

$$I_S = 1 \text{ fA}, \qquad \alpha_F = 0.922, \qquad \alpha_R = 0.94.$$

The BJT is operating at room temperature and its junctions are biased so that

$$V_{BC} = -1.2 \text{ V}, \qquad V_{BE} = 0.6 \text{ V}.$$

a. Determine the base, collector, and emitter currents.

b. The junction biasing is changed so that

$$V_{BC} = 0.6 \text{ V}, \qquad V_{BE} = -1.2 \text{V}.$$

Repeat part a.

3-2. A Silicon *pnp* BJT is described by the following parameters (remember $\eta = 1$ for silicon BJTs):

$$I_S = 1 \text{ fA}, \qquad \alpha_F = 0.995, \qquad \alpha_R = 0.91.$$

The BJT is operating at room temperature and its junctions are biased so that

$$V_{BC} = 0.4 \text{ V}, \qquad V_{BE} = 0.6 \text{ V}.$$

Determine the base, collector, and emitter currents.

3-3. A typical 2N4401 *npn* BJT is described by the following parameters:

$$I_S = 26 \text{ fA}, \qquad \alpha_F = 0.994, \qquad \alpha_R = 0.75.$$

a. Generate the output characteristic curve using PSpice. Increment the base current from 0 to 100 μA in 10-μA increments.

b. Generate the input characteristic curve using PSpice.

3-4. For the given circuit, determine the following transistor currents and voltages using load line analysis:

a. Collector current

b. Base current

c. Base–emitter voltage

d. Collector–emitter voltage

Assume the transistor is a 2N2222A *npn* BJT as is described in Figure 3.1-1.

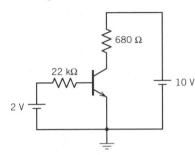

3-5. For the circuit shown, find the transistor currents and voltages using load line analysis:

a. Collector current

b. Base current

c. Base–emitter voltage

d. Collector–emitter voltage

Assume that the transistor is a 2N2222A *npn* BJT as shown in Figure 3.1-1.

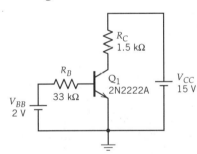

3-6. For the given circuit, determine the following transistor currents and voltages using load line analysis:

a. Collector current

b. Emitter current

c. Base–emitter voltage

d. Base–collector voltage

Assume the transistor is a 2N2222A *npn* BJT as is described in Figure 3.1-1.

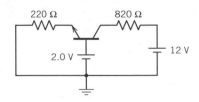

3-7. The transistor shown in the circuit has the characteristics given in Figure 3.1. Plot over the range $0 \text{ V} \le V_i \le 4 \text{ V}$, V_o versus V_i.

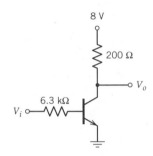

3-8. For the circuit shown, draw the transfer curve V_O vs. V_I for

$$-15 \text{ V} \le V_I \le 15 \text{ V} \quad \text{and} \quad \beta_F = 180.$$

Confirm the transfer curve using SPICE.

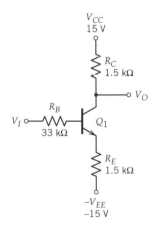

3-9. For the circuit shown:

a. Draw the transfer curve V_O vs. V_I for

$$-9 \text{ V} \le V_I \le 9 \text{ V}.$$

b. Determine the value(s) of V_I that will saturate the transistor.

c. Confirm the transfer curve using SPICE.

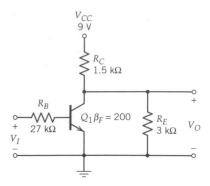

3-10. Calculate the collector and base currents in the silicon transistor shown. Assume β_F = 75. *Hint*: Make a Thévenin equivalent of the circuit connected to the base of the BJT.

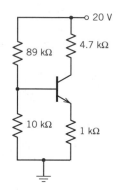

3-11. For the circuit shown, find the transistor currents and voltages using load line analysis:

   a. Collector current

   b. Base current

   c. Base–emitter voltage

   d. Collector–emitter voltage

Assume that the transistor is a 2N3906 *pnp* BJT. Use the *V–I* characteristics found in the Appendix.

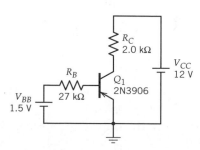

3-12. Determine the maximum value of the resistor R_b for which the transistor remains in saturation. Assume a silicon BJT with β_F = 150.

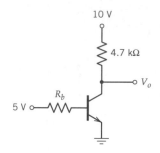

3-13. For the circuit shown, find the minimum V_{BB} for transistor saturation. Assume β_F = 210.

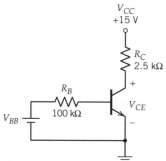

 3-14. For the circuit shown:

   a. Find V_{BB} for saturation. Use SPICE to confirm the result.

   b. Find V_{BB} and V_{CE} for operation in the forward-active region with $I_C = 0.5(I_C)_{\text{sat}}$. Use SPICE to confirm the result.

Assume the silicon BJT is described by β_F = 200 and has additional SPICE parameters IS = 6.7 fA and VA = 100 V.

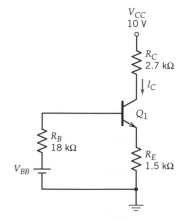

3-15. Complete the design of the circuit shown by finding R_B for $I_C = 11$ mA. Determine the following voltages and currents:

a. Base current

b. Base–emitter voltage

c. Collector–emitter voltage

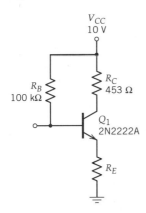

3-16. Complete the design of the fixed-bias BJT circuit shown, given $I_B = 60$ µA. Determine the resistance value for R_E. What is the transistor Q point? That is, find:

a. Collector–emitter voltage

b. Collector current

c. Base–emitter voltage

Assume that the transistor is a 2N2222A *npn* BJT as shown in Figure 3.1-1.

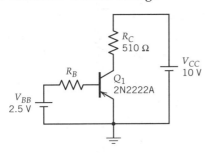

3-17. Complete the design of the self-biased BJT circuit shown given the following:

$$V_{BEQ} = -0.7 \text{ V}, \qquad \beta_F = 150, \qquad I_C = -2 \text{ mA}.$$

Determine the resistance value for R_{B2}. What is the transistor Q point? That is, find:

a. Base current

b. Collector–emitter voltage

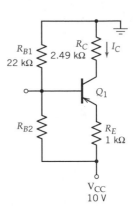

3-18. The simple logic inverter shown is constructed using a silicon BJT with $\beta_F = 75$.

a. What is the maximum input voltage v_i for which the output will be HIGH (≈ 5 V).

b. What is the minimum input voltage v_i for which the output will be LOW (< 0.2 V).

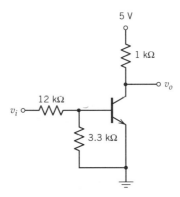

3-19. The DTL NAND Gate shown in Figure 3.5-3 is constructed using a silicon BJT described by $\beta_F = 75$ and the following circuit elements:

$$R_a = 3.6 \text{ k}\Omega, \qquad R_b = 6.2 \text{ k}\Omega, \qquad R_c = 1.8 \text{ k}\Omega.$$

Assume $V_{CC} = 5$ V.

Determine the fan-out of this gate.

3-20. Use SPICE to verify the operation of the TTL gate of Example 3.5-3.

3-21. Determine the fan-out of the simple ECL OR gate shown in Figure 3.5-5.

3-22. Assume as given the input circuit for

a TTL gate (with only one input) shown. The transistor parameters of interest are

$\beta_F = 200,$ $\beta_R = 6,$

$V_{BE(on)} = 0.6$ V, $V_{BE(sat)} = 0.8$ V, $V_{CE(sat)} = 0.2$ V.

Assume that Q_1 and Q_2 are identical.

a. Find V_{B1}, I_{B1}, I_{C1}, I_{E1}, V_{B2}, and V_O for input logic 1 (5 V) and logic 0 (0V).

b. Find the *fan-out* of the circuit.

c. Compare the results of parts a and b and comment on the potential design advantages of using the circuit analyzed in Example 3.5-3.

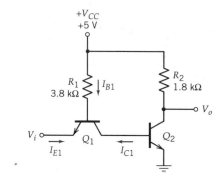

3-23. In an effort to reduce the power consumption for the TTL gate of Example 3.5-3, V_{CC} is reduced to 3.3 V.

a. Verify that the gate operates properly and calculate the noise margins and fan-out.

b. Compare average power consumption of the gate with $V_{CC} = 5$ V and $V_{CC} = 3.3$ V. *Hint*: Find the power supply currents for a 0 and a 1 output and then average them.

3-24. The circuit shown is a form of an HTL gate. Analytically determine the following gate properties:

a. Logic function performed

b. Logic levels

c. Fan-out

d. Noise margins

e. Average power consumption

Assume:

• Silicon diodes and BJTs
• $\beta_F = 100$
• $V_Z = 5.6$ V

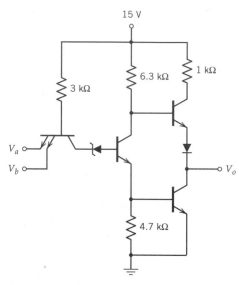

3-25. Complete the design of the self-biased BJT circuit shown given the input and output characteristic curves and

$I_C = 7$ mA, $V_{CE} = 3$ V.

a. What is I_B and V_{BE} at the Q point?

b. What is β_F?

c. Find the resistance values R_C and R_{B1}.

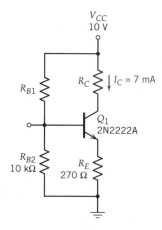

3-26. Complete the design of the BJT biasing circuit shown given the input and output characteristic curves and

$I_C = 3.5$ mA, $V_{CE} = 4$ V.

a. What is I_B and V_{BE} at the Q point?

b. What is β_F?

c. Find the resistance values R_C and R_B.

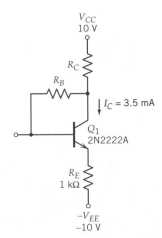

Assume that the transistor is a 2N3906 *pnp* BJT

a. What is I_B and V_{BE} at the Q point?

b. What is β_F?

c. Find the resistance values R_C and R_B.

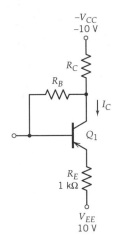

3-27. Complete the design of the circuit shown for $I_C = 5.3$ mA. Determine the resistance values for R_B. Find the transistor currents and voltages using load line analysis:

a. Collector–emitter voltage

b. Base current

c. Base–emitter voltage

Assume that the transistor is a 2N2222A *npn* BJT as shown in Figure 3.1-1.

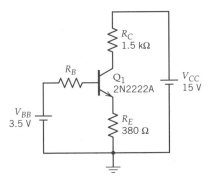

3-28. Complete the design of the *pnp* BJT biasing circuit shown to achieve the desired quiescent conditions

$$I_C = -4 \text{ mA}, \quad V_{CE} = -4 \text{ V}.$$

3-29. Complete the design of the circuit shown. Determine the resistance values for R_B. Find the transistor circuit and voltages using load line analysis if $V_{CE} = -5$ V:

a. Collector current

b. Base current

c. Base–emitter voltage

Assume that the transistor is a 2N3906 *pnp* BJT with V–I characteristics found in the Appendix.

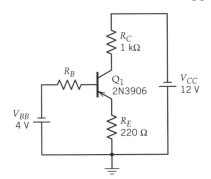

3-30. Complete the design of the circuit shown by determining the base bias voltage V_{BB} for $V_{CE} = 3$ V. Find all transistor terminal currents and voltages using load line analysis where applicable (the input and output characteristics are attached). Assume the BJT is a 2N2222A.

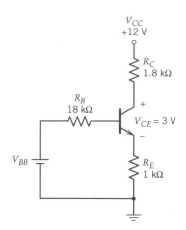

3-31. Design a fixed-bias circuit to achieve the following BJT Q point:

$$I_C = 4 \text{ mA}, \qquad V_{CE} = 8 \text{ V}.$$

Use a 20-V DC power supply and a silicon BJT with $\beta_F = 100$.

3-32. Design a self-bias circuit to achieve the following BJT Q point:

$$I_C = 4 \text{ mA}, \qquad V_{CE} = 8 \text{ V}.$$

Use a 20-V DC power supply and a silicon BJT with $\beta_F = 100$. Additional constraints on the bias resistors are

$$\frac{R_c}{R_e} = 5, \qquad R_{B1} \parallel R_{B2} = 15 \text{ k}\Omega.$$

3-33. For the self-bias circuit topology, determine the ratio R_B/R_E to achieve a bias stability that results in a 0.2% change in I_C for a 10% change in β_F.

3-34. Complete the design of the circuit shown so that a bias stability to achieve a 1% change in I_C for a 10% change in β_F operating over the temperature range of 15–50°C. The BJT SPICE parameters are BF = 100, IS = 1.1 fA, and VA = 120 V. Confirm the circuit stability using SPICE.

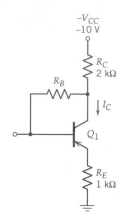

3-35. An *npn* BJT emitter bias configuration with two 15 V power supplies must be designed so that it achieves a bias stability of 1% change in I_C for the variation in β_F found in the transistor used to manufacture the circuit. The value of β_F due to component tolerances ranges from 170 to 230 at 25°C.

a. Design the circuit.

b. Simulate the circuit using SPICE to confirm that the design meets the stability requirement.

3-36. Design a self-bias circuit to accomplish the following design goal: as β_F of a silicon BJT varies between 80 and 200, the collector current lies within the range of 1.35–1.65 mA. Assume $V_{CC} = 15$ V, $R_C = 2.2$ kΩ, and no transistor variation in V_{BE} or I_{CO}.

3-37. A transistor with $\beta_F = 60$ and $V_{BE} = 0.8$ V is used in the self-bias circuit with $V_{CC} = 25$ V. The quiescent point is $I_C = 2.5$ mA and $V_{CE} = 15$ V. The transistor is replaced by another with $\beta_F = 200$ and $V_{BE} = 0.65$ V. It is desired that the effect of the change in β_F does not increase I_C by more than 0.1 mA and that the same should be true for the change in V_{BE}. Determine the resistor values to accomplish these design goals.

3-38. A silicion BJT with $\beta_F = 70$ and $V_{BE} = 0.75$ V produces a quiescent point of $I_C = 2$ mA and $V_{CE} = 10$ V when inserted into a self-bias circuit with a 20-V power supply. When the transistor is replaced by another silicon BJT with $\beta_F = 180$ and $V_{BE} = 0.67$ (no change in I_{CO}), the effect of each change increases I_C by 0.08 mA (final $I_C = 2.16$ mA). Determine the values of the four resistors in the self-bias configuration.

3-39. It is common to attempt to improve amplifier performance by connecting two transistors as shown. Assume identical silicon BJTs with $\beta_F = 100$. Determine resistor values to accomplish a quiescent condition or transistor Q_2 of

$$I_{C2} = 4 \text{ mA}, \qquad V_{CE2} = 8 \text{ V}.$$

with the design restrictions that

$$R_{CC} = 4R_{EE}, \qquad R_{B1}/R_{B2} = 22 \text{ k}\Omega.$$

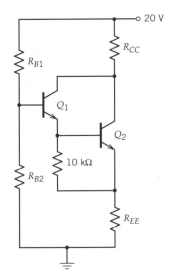

3-40. Design a Zener voltage regulator with BJT shunt to meet the following design requirements:

a. Regulated load voltage: $V_L = 5$V

b. Load current: $0 \text{ A} \leq I_L \leq 4 \text{ A}$

c. Source voltage: $12 \text{ V} \leq V_S \leq 16 \text{ V}$

Determine the appropriate ratings for all components. Assume the minimum Zener current for proper regulation is 1.5 mA and a power BJT with typical forward current gain, $\beta_F = 75$.

4

FIELD-EFFECT TRANSISTOR CHARACTERISTICS

In Chapter 3 BJTs were shown to be semiconductor devices that operate on carrier flow from the emitter to the base and then through to the collector. For example, *npn* BJTs are devices where the current flow from the collector to the emitter is regulated by the current injected into the base. Therefore, the BJT is a *current-controlled* three-terminal semiconductor device.

Field Effect Transistors (FETs) are semiconductor devices that employ a channel between the *drain* and the *source* to transport carriers. An adjacent controlling surface, called the *gate*, regulates the current flow through the drain–source channel. This channel is controlled by a voltage applied to the gate of the FET. Therefore, the FET can be described as a *voltage-controlled* three-terminal semiconductor device (see Figure 4-1). The physical properties of FETs make them suitable for amplification, switching, and other electronic applications.

The terminal characteristics of Junction Field Effect Transistors (JFETs) and Metal Oxide Semiconductor FETs (MOSFETs) are described in this chapter. Other types of FETs exist, but JFETs and MOSFET are the predominate FET types.[1] Metal–oxide–semiconductor FETs are used extensively in integrated circuits for digital applications: JFETs are most commonly found in analog applications. Terminal characterization of FETs is sufficient for electronic analysis and design: For discussions on the device physics of FETs, the reader is referred to the references.

Within each FET type, further categorization is based on two properties of the channel: channel doping and gate action on the channel. The FET channel may be fabricated from *n*- or *p-type* material. Therefore, the FETs are designated as *n*-channel JFETs, *p*-channel JFETs, *n*-channel MOSFETs (NMOSFETs) and *p*-channel MOSFETs (PMOSFETs). The equations governing the operation of the two channel types of FETs are identical with the exception that the current and voltages in the two types are of opposite polarities. That is, the *n*-channel positive current and voltages are replaced by negative current and voltages for *p*-channel devices. In addition to the two FET channel types, MOSFETs can either be depletion or enhancement mode MOSFETs. Junction JFETs are depletion mode devices. The terms *depletion* and *enhancement* refer to the action of the gate control voltage on the carriers in the channel and the channel itself.

Depletion mode devices can be thought of has having a normally open channel for charge carriers between the drain and the source. With the application of a po-

---

[1] Other FET types include the Metal Semiconductor FET (MESFET), Modulation-doped FET (MODFET), and Vertical MOSFET (VMOSFET). Analysis of JFETs and MOSFETs will allow for a general understanding of FET behavior that can be used with other FET devices.

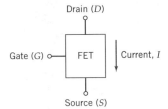

Figure 4-1

FET represented as three-terminal device.

tential of the proper polarity across the gate and source, the carriers in the channels are essentially "depleted," which "pinches off," or squeezes, the channel between the drain and source disallowing additional charge carriers (current) to flow through the channel. The point at which the current between the drain and the source is pinched off is regulated by decreasing gate voltage relative to the source for n-channel devices and increasing the voltage relative to the source for p-channel devices. For n-channel devices, decreasing the gate potential relative to the source squeezes the drain–source channel closed.

The drain–source channel in an enhancement mode device is described as normally closed. In the enhancement mode case, the application of a potential of the proper polarity between the gate and source terminals causes the drain–source channel to become enhanced, allowing additional carriers (current) to flow. For n-channel devices, increasing the gate potential relative to the source enhances the drain–source channel.

Whether n or p type, JFETs are depletion mode devices.[2] Depletion MOSFETs can operate either in the depletion mode or the enhancement mode depending on the gate potential relative to the source, allowing the drain–source channel to either open up to allow additional current flow (enhancement) or squeezing closed the channel and restricting further flow of current (depletion).

Like BJTs, FETs have different regions of operation. The regions are identified and characterized through the FET terminal characteristics. Simple circuits are developed that use the basic terminal characteristics of the various types of FETs in each region. Among the significant circuits analyzed and designed are a FET constant-current source, active resistive loads constructed with FETs, a complementary metal–oxide–semiconductor (CMOS) inverter, FET switches, and voltage variable resistors. SPICE modeling parameters for the various FETs and the effect of those parameters on circuit design and analysis are discussed.

4.1 JUNCTION FIELD-EFFECT TRANSISTORS

Junction field-effect transistors are either n-channel or p-channel devices. The terminal voltage and current relationships of n-channel JFETs are developed in this section. The p-channel JFET terminal voltage and current relationships are identical to that of the n-channel JFET with the exception that the polarities of the voltages and currents are reversed. A brief description of the p-channel JFET characteristics will be presented at the end of this section.

The circuit symbol for the n-channel JFET is shown in Figure 4.1-1a. The terminal

---

[2] Gallium arsenide (GaAs) enhancement mode JFETs exist. The GaAs enhancement JFETs operate in a similar manner to other enhancement mode devices. GaAs JFETs are currently not widely used. Unless specifically stated, JFETs are always thought of as being depletion mode devices in this text.

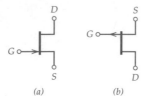

Figure 4.1-1
Circuit symbol for (*a*) *n*-channel JFET and (*b*) *p*-channel JFET.

(*a*)   (*b*)

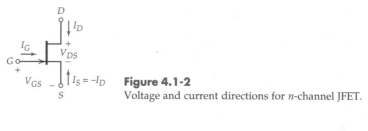

Figure 4.1-2
Voltage and current directions for *n*-channel JFET.

connection to the gate has an arrow whose direction indicates the channel type of device depicted. For *n*-channel JFETs the arrow points to the source on the gate terminal. For *p*-channel JFETs, the arrow points away from the source as shown in Figure 4.1-1*b*.

All FETs are three-terminal devices with the gate acting as the current-regulating terminal between the drain and the source. The voltage and current sign conventions for *n*-channel JFETs are shown in Figure 4.1-2.

When the gate–source junction is reverse biased in *n*-channel JFETs, the conductivity of the drain–source channel is reduced with decreasing gate to source voltage V_{GS}. The current through the drain–source channel is I_D. The source current is equal to $-I_D$,

$$I_S = -I_D. \tag{4.1-1}$$

4.1.1 The *n*-Channel JFET

Since JFETs normally operate with the gate junction reverse biased, the gate current is essentially zero,

$$I_G = 0. \tag{4.1-2}$$

The gate-to-source voltage that pinches off the drain–source channel is called the pinch-off voltage V_{PO},

$$V_{PO} = V_{GS}|_{I_D=0, V_{DS, \text{ small}}}. \tag{4.1-3}$$

For *n*-channel devices, V_{PO} is a negative voltage and is specific to the particular FET. If V_{GS} is positive (for *n*-channel JFETs), the gate junction is forward biased and the equations developed in this section do not apply.

The drain (common source) *n*-channel JFET characteristics are shown in Figure 4.1-3. The transfer characteristic is presented in Figure 4.1-4. The current and voltage reference directions were shown in Figure 4.1-2.

The different regions of operation of the *n*-channel JFET can be illustrated by selecting one curve from the output characteristics (e.g., the curve corresponding to $V_{GS} = 0$). For small values of V_{DS}, the channel allows current to readily flow with

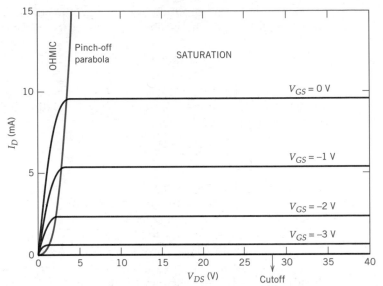

Figure 4.1-3
Output characteristics of n-channel JFET.

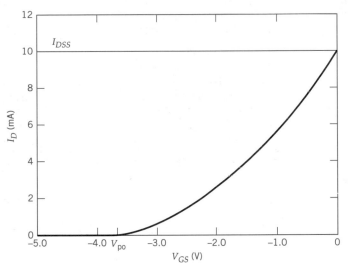

Figure 4.1-4
Transfer curve for n-channel JFET.

an initial resistance R_{DS}. The output curve in this region can be approximated by a straight line of slope R_{DS}^{-1}. This approximate linear relationship of I_D vs. V_{DS} leads to the descriptive name for this region of operation, the *ohmic region*. In the ohmic region, the JFET acts as a voltage variable resistor: The gate–source voltage V_{GS} controls the value of the equivalent resistance. As V_{GS} increases, the channel narrows, causing an increase in resistance. The result is a decrease in the slope of the characteristic curve.

With increasing V_{DS}, when V_{GS} is held constant, the channel resistance increases as evidenced by the leveling of I_D. At pinch-off the drain, current I_D remains almost constant due to low conductivity through the channel. The resistance of the device in this region is very high. The drain-to-source voltage at pinch-off is

$$V_{DS} \text{ (at pinch-off)} = V_{GS} - V_{PO}. \qquad (4.1\text{-}4)$$

The boundary between the *ohmic* and *saturation* region is the so-called pinch-off parabola (refer to Figure 4.1-3) defined by the relationship in Equation 4.1-4. The JFET is said to be in saturation when

$$V_{DS} > V_{GS} - V_{PO}. \qquad (4.1\text{-}5)$$

Saturation in FETs differs from saturation in BJTs. Field-effect transistors operating in the saturation region are analogous to BJTs in the forward-active region. As can be seen in the FET output characteristic of Figure 4.1-3, operation in saturation allows the drain current to be adjusted by varying a control voltage at the gate, in this case V_{GS}, independent of V_{DS}. Amplifiers designed using JFETs takes advantage of the small voltage variations in V_{GS} controlling the current flow through the device. JFETs can also be used as switches since large and abrupt changes in V_{GS} causes the current I_D to change from zero to a relatively large value.

The JFET does not require significant input gate current. Therefore, the input characteristic of the device provides limited and nearly useless information for circuit design. The transfer characteristic shown in Figure 4.1-4 is far more useful. Since the gate channel diode in JFETs must be reverse biased, only negative values of V_{GS} allow for operation in the ohmic and saturation regions.

From the transfer function, the current at $V_{GS} = 0$ is defined as I_{DSS}, where I_{DSS} is the drain current at $V_{GS} = 0$ at pinch-off ($V_{DS} = -V_{PO}$). Here, I_{DSS} is temperature dependent and decreases with increasing temperature.

The voltage and current relationships of n-channel JFETs can be described for the three most common regions of operation: ohmic, saturation, and cutoff. For n-channel JFETs operating in the ohmic or saturation regions, the following voltage and current conditions must hold:

$$I_D > 0, \qquad V_{PO} < V_{GS} \le 0, \qquad V_{DS} > 0.$$

The n-channel JFETs have the following characteristic parameters:

$$V_{PO} < 0, \qquad I_{DSS} > 0.$$

In the negative quadrant, where $I_D < 0$ and $V_{DS} < 0$, the drain–gate junction becomes forward biased causing the drain current to increase rapidly. The characteristics in this region are similar to diode characteristics (except that $I_D < 0$ and $V_{DS} < 0$) with the turn-on voltage determined by V_{GS}.

4.1.2 Ohmic Region

The ohmic region is that portion of the curve between $V_{DS} = 0$ and pinch-off on the output characteristic curve in Figure 4.1-3. The mathematical expression defining this region is

$$0 < V_{DS} \le V_{GS} - V_{PO}. \qquad (4.1\text{-}6)$$

The V–I relationship in this region is

$$I_D = I_{DSS} \left[2\left(\frac{V_{GS}}{V_{PO}} - 1\right) \frac{V_{DS}}{V_{PO}} - \left(\frac{V_{DS}}{V_{PO}}\right)^2 \right]. \tag{4.1-7}$$

For small values of V_{DS}, the drain current in Equation 4.1-7 is approximately,

$$I_D \approx 2I_{DSS}\left(\frac{V_{GS}}{V_{PO}} - 1\right) \frac{V_{DS}}{V_{PO}}. \tag{4.1-8}$$

If V_{GS} is held constant, Equation 4.1-8 is a linearly varying function of I_D and V_{DS}. Therefore the output resistance in this region is found by taking the derivative of Equation 4.1-8 with respect to V_{DS}:

$$R_{DS}^{-1} = \frac{\partial I_D}{\partial V_{DS}} = \frac{2I_{DSS}}{V_{PO}}\left(\frac{V_{GS}}{V_{PO}} - 1\right). \tag{4.1-9}$$

4.1.3 Saturation Region

The saturation region occupies the portion of the output characteristic curve of Figure 4.1-3 where $I_D > 0$ and to the right of the pinch-off parabola. That is,

$$V_{DS} \geq V_{GS} - V_{PO}. \tag{4.1-10}$$

The drain current is virtually independent of V_{DS} in this region:

$$I_D = I_{DSS}\left(1 - \frac{V_{GS}}{V_{PO}}\right)^2. \tag{4.1-11}$$

Equation 4.1-11 is called the transfer characteristic and is shown in Figure 4.1-4. The values of V_{PO} and I_{DSS} are specified by the manufacturers.

The expression for the pinch-off parabola can be derived from Equation 4.1-11 by substituting $V_{GS} = V_{DS} + V_{PO}$:

$$I_D = I_{DSS}\left(\frac{V_{DS}}{V_{PO}}\right)^2. \tag{4.1-12}$$

4.1.4 Cutoff Region

The JFET is said to be in the cutoff region when

$$V_{GS} < V_{PO}. \tag{4.1-13}$$

The drain current is zero when the JFET is cut off:

$$I_D = 0. \tag{4.1-14}$$

4.1.5 The p-Channel JFET

The voltage and current sign conventions for p-channel JFETs are illustrated in Figure 4.1-5. The arrow on the gate is leaving the device. Also note that the direction of the

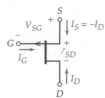

Figure 4.1-5
Voltage and current directions of p-channel JFETs.

drain current is opposite that for the n-channel JFET. In essence, all terminal voltages and currents have been reversed. The p-channel operating voltages and currents are given below:

$$V_{PO} > 0, \qquad V_{SD} > 0, \qquad -V_{PO} < V_{SG} \leq 0,$$
$$I_{DSS} < 0, \qquad I_D < 0.$$

The conditions that identifies the regions of operation are

$$0 < V_{SD} < V_{PO} + V_{SG} \quad \text{(ohmic region)},$$
$$V_{SD} \geq V_{PO} + V_{SG} \quad \text{(saturation region)}, \qquad (4.1\text{-}15)$$
$$V_{SD} = V_{PO} + V_{SG} \text{ (pinch-off parabola)}.$$

The output characteristic curve and the transfer characteristic for a p-channel JFET are shown in Figures 4.1-6 and 4.1-7.

The V–I relationship in the three regions can be found by using the mathematical expressions for n-channel JFETs in Equations 4.1-6–4.1-14. In the p-channel JFET in the ohmic and saturation regions, the gate-to-source and drain-to-source voltages are

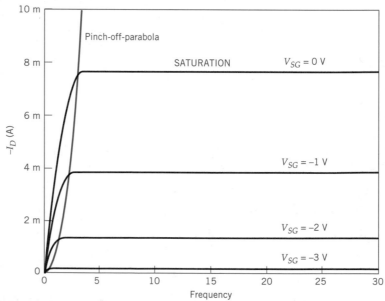

Figure 4.1-6
A p-channel JFET output characteristic curve.

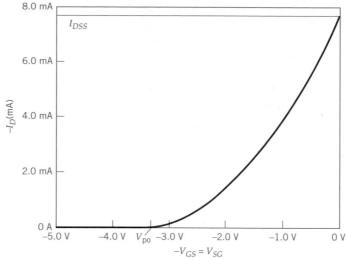

Figure 4.1-7
A p-channel JFET transfer curve.

of the opposite polarity to the n-channel JFET voltage. That is, in the p-channel JFET, the gate-to-source voltage is positive ($V_{GS} > 0$ or $V_{SG} < 0$), and the drain-to-source voltage is negative ($V_{DS} < 0$ or $V_{SD} > 0$).

In both p- and n-channel JFETs, the polarity of V_{PO} and I_D have opposite polarities.

EXAMPLE 4.1-1 An n-channel JFET has the following characteristics:

$$V_{PO} = -3.5 \text{ V}$$

and

$$I_{DSS} = 10 \text{ mA}.$$

Find the minimum drain-to-source voltage, V_{DS}, for the JFET to operate at in saturation for a gate-to-source voltage, $V_{GS} = -2$V. What is its resistance in the ohmic region?

SOLUTION
The condition for saturation is

$$V_{DS} \geq V_{GS} - V_{PO}.$$

Substituting yields

$$V_{DS} \geq -2 - (-3.5) = 1.5 \text{ V}.$$

In the ohmic region the resistance is

$$R_{DS}^{-1} = \frac{\partial I_D}{\partial V_{DS}} = \frac{2I_{DSS}}{V_{PO}}\left(\frac{V_{GS}}{V_{PO}} - 1\right).$$

Substituting $V_{GS} = -2$ V, $V_{PO} = -3.5$ V, and $I_{DSS} = 10$ mA yields an output resistance of

$$R_{DS} = 408 \ \Omega.$$

EXAMPLE 4.1-2 Given a p-channel JFET with the parameters

$$V_{PO} = 4 \text{ V}, \qquad I_{DSS} = -8 \text{ mA},$$

what is the drain current I_D for $V_{SG} = -3$ V and $V_{SD} = 2$ V?

SOLUTION
Determine operating region of the FET.

$$V_{PO} + V_{SG} = 4 - 3 = 1 \text{ V}.$$

But $V_{SD} = 2$ V so

$$V_{SD} \geq V_{PO} + V_{SG},$$

which indicates that the transistor is in *saturation*. Therefore, the drain current is

$$I_D = I_{DSS}\left(1 - \frac{V_{GS}}{V_{PO}}\right)^2.$$

Substituting and solving for I_D yields

$$I_D = -8 \times 10^{-3}\left(1 - \frac{-(-3)}{4}\right)^2 = -0.5 \text{ mA}.$$

4.2 METAL–OXIDE–SEMICONDUCTOR FIELD-EFFECT TRANSISTORS

Metal–oxide semiconductor field-effect transistors are widely used in integrated circuits. Because MOS devices can be fabricated in very small geometries and are relatively simple to manufacture, most very large scale integrated (VLSI) circuits are fabricated from MOS devices.

Metal–oxide–semiconductor FETs come in either of two types:

- Depletion type
- Enhancement type

Both types of MOSFETs are either n- or p-channel devices, commonly abbreviated NMOSFET and PMOSFET, respectively. The depletion MOSFET can operate in either the depletion or enhancement mode. The enhancement MOSFET operates in the enhancement mode only.

This section considers the depletion and enhancement-type MOSFETs separately.

4.2.1 Depletion-Type MOSFET

In this section, the terminal voltage and current relationships of depletion-type n-channel MOSFETs (NMOSFETs) will be developed. The depletion-type p-channel MOSFET (PMOSFET) terminal voltage and current relationships are identical to the NMOSFET with the exception that the polarities of the voltages and currents are reversed. The depletion MOSFET regions of operation are identical to those of the JFET. In fact, the $V–I$ relationships of the depletion MOSFET are identical in form to those of the JFET. Both types of transistors are depletion mode devices; that is, with the application of the appropriate polarity potential between the gate and the source (a negative potential for NMOSFETs), the carriers in the channel are depleted, which squeezes off the channel through which charge carriers flow. Therefore, the same types of transistor parameters are used to describe the $V–I$ characteristics of the depletion MOSFET as for the JFET. The operational characteristics of the depletion MOSFET does differ from the JFET by one important attribute, the depletion MOSFET can additionally operate in the enhancement mode by applying a positive gate–source potential, thus enhancing the channel through which the charge carriers flow.

The circuit symbol for the depletion NMOSFET is shown in Figure 4.2-1a. In addition to the three FET terminals of gate, drain, and source, the symbol depicts a terminal representing the substrate of the semiconductor (B for body) with an arrow pointing into the junction. It is common for the substrate to be electrically connected to the source: this connection does not affect the characteristics of the MOSFET. However, in integrated circuits using NMOSFETs, the substrate is commonly connected to the most negative supply voltage. With NMOSFETs connected in this fashion, it is guaranteed that the substrate is at signal ground. Unfortunately, there is some possibility that circuit performance may be compromised.

A simplified circuit symbol for the depletion-type NMOSFET is shown in Figure 4.2-2 with the standard voltage and current directions indicated. For NMOSFETs, the arrow points toward the source from the junction. The direction of the arrow corresponds to the direction of standard current flow toward the source terminal.

Figure 4.2-1

(a) Circuit symbol for depletion-type NMOSFET. (b) Depletion-type NMOSFET with the substrate connected to the source.

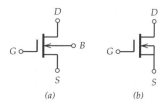

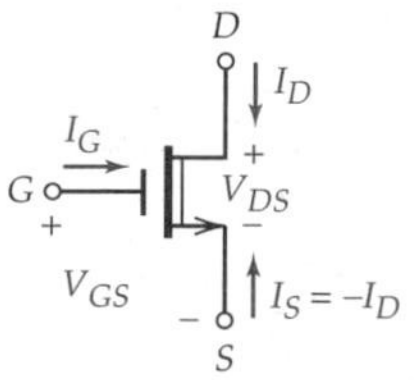

Figure 4.2-2

Simplified circuit symbol for depletion NMOSFET.

As in the n-channel JFET, the source current is equal to the negative of the drain current when the gate–source junction is reverse biased:

$$I_S = -I_D. \tag{4.2-1}$$

The depletion type MOSFET operates similarly to the JFET. Like the JFET, the MOSFET gate current is essentially zero:

$$I_G = 0. \tag{4.2-2}$$

As in the JFET, the pinch-off voltage is the gate-to-source voltage that depletes or squeezes off the drain–source channel. For NMOSFETs, V_{PO} is negative and is transistor specific.

Unlike the n-channel JFET, the depletion-type NMOSFET allows for enhancement mode operation with the application of a positive V_{GS}. Application of a positive gate-to-source potential "opens up" the drain–source channel by increasing the channel conductivity, allowing more current to flow. Enhancement mode operation of the depletion-type NMOSFET allows for drain currents in excess of I_{DSS}.

Typical depletion-type NMOSFET drain characteristics are shown in Figure 4.2-3. The transfer characteristic of the depletion NMOSFET is shown in Figure 4.2-4.

The depletion NMOSFET characteristic curves of Figure 4.2-3 and 4.2-4 are very similar to those of the n-channel JFET curves of Figures 4.1-3 and 4.1-4. The equations describing the curves for the depletion NMOSFET and the n-channel JFET are identical. As in the JFET, there are three common regions of operation: the ohmic, saturation, and cutoff regions.

For the depletion-type NMOSFET operating in the depletion mode ohmic or saturation regions, the following voltage and current conditions must hold:

$$I_D > 0, \qquad V_{PO} < V_{GS} \le 0, \qquad V_{DS} > 0.$$

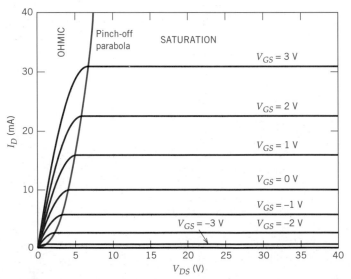

Figure 4.2-3
Depletion-type NMOSFET drain characteristics.

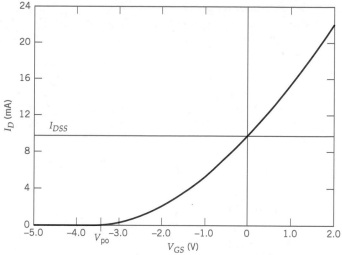

Figure 4.2-4
Depletion-type NMOSFET transfer characteristic.

For depletion-type NMOSFET operating in the enhancement mode, V_{GS} changes sign. The voltage and current conditions become

$$I_D > 0, \qquad V_{GS} > 0, \qquad V_{DS} > 0.$$

Depletion-type NMOSFETs have the following characteristic parameters:

$$V_{PO} < 0, \qquad I_{DSS} > 0.$$

As in the JFET operating in the negative quadrant, where $I_D < 0$ and $V_{DS} < 0$, the drain–gate junction becomes forward biased causing the drain current to increase rapidly. The characteristics in this region are similar to a diode (except that $I_D < 0$ and $V_{DS} < 0$) with the turn-on voltage determined by V_{GS}.

Ohmic Region
The ohmic region lies between $V_{DS} = 0$ and the pinch-off parabola in the drain characteristic curve. The mathematical expression defining this region is

$$0 < V_{DS} \le V_{GS} - V_{PO}. \tag{4.2-3}$$

The V–I relationship in this region is

$$I_D = I_{DSS}\left[2\left(\frac{V_{GS}}{V_{PO}} - 1\right)\frac{V_{DS}}{V_{PO}} - \left(\frac{V_{DS}}{V_{PO}}\right)^2\right]. \tag{4.2-4}$$

For small values of I_D, the drain current in Equation 4.2-4 is approximately

$$I_D \approx 2I_{DSS}\left(\frac{V_{GS}}{V_{PO}} - 1\right)\frac{V_{DS}}{V_{PO}}. \tag{4.2-5}$$

Saturation Region

The saturation region occupies the portion of the output characteristic curve of Figure 4.2-3 where $I_D > 0$ and to the right of the pinch-off parabola. That is,

$$V_{DS} \geq V_{GS} - V_{PO}. \tag{4.2-6}$$

The drain current is virtually independent of V_{DS} in this region and is described by the transfer characteristic (shown in Figure 4.2-4):

$$I_D = I_{DSS}\left(1 - \frac{V_{GS}}{V_{PO}}\right)^2. \tag{4.2-7}$$

The values of V_{PO} and I_{DSS} are specified by the manufacturers.

The expression for the pinch-off parabola can be derived from Equation 4.2-7 by substituting $V_{GS} = V_{DS} + V_{PO}$:

$$I_D = I_{DSS}\left(\frac{V_{DS}}{V_{PO}}\right)^2. \tag{4.2-8}$$

Cutoff Region

The depletion-type NMOSFET is said to be in the cutoff region when

$$V_{GS} < V_{PO}. \tag{4.2-9}$$

When the depletion-type NMOSFET is cut off,

$$I_D = 0. \tag{4.2-10}$$

EXAMPLE 4.2-1 For a depletion-type NMOSFET with $V_{PO} = -3$ V and $I_{DSS} = 8$ mA, what is the drain-to-source voltage required to saturate the transistor at $V_{GS} = +2$ V? What is the drain current?

SOLUTION

The pinch-off condition is

$$V_{DS} = V_{GS} - V_{PO}.$$

Substituting the values of the pinch-off voltage and the gate-source voltage,

$$V_{DS} = 2 - (-3) = 5 \text{ V}.$$

To find the drain current at pinch-off, the expression for the current in the saturation region is used:

$$I_D = I_{DSS}\left(1 - \frac{V_{GS}}{V_{PO}}\right)^2.$$

Substituting in the values for I_{DSS}, V_{GS}, and V_{PO} yields the drain current:

$$I_D = 8 \times 10^{-3} \left[1 - \frac{2}{(-3)} \right]^2 = 22.2 \text{ mA}.$$

Since $I_D > I_{DSS}$, the MOSFET is in the enhancement mode of operation.

Depletion-Type PMOSFET

The circuit symbol of the depletion-type PMOSFET is shown in Figure 4.2-5. All of the voltages and currents of the depletion-type PMOSFET are the opposite polarity to those of the NMOSFET. The pinch-off voltage of the depletion-type PMOSFET is greater than zero:

$$V_{PO} > 0. \tag{4.2-11}$$

The simplified circuit symbol of the depletion-type PMOSFET is shown in Figure 4.2-6. The current and voltage sign conventions are also shown.

The V–I relationship in the three regions of the depletion PMOSFET can be found by using the mathematical expressions for depletion NMOSFET in Equations 4.2-3– 4.2-10.

In both p- and n-channel depletion MOSFETs, the polarity of V_{PO} is always opposite from I_D.

For the depletion-type PMOSFET operating in the depletion mode-ohmic or saturation regions, the following voltage and current relationships must hold:

$$I_D < 0, \qquad -V_{PO} < V_{SG} \leq 0, \qquad V_{SD} > 0.$$

For the depletion-type PMOSFET operating in the enhancement mode, V_{SG} changes sign and the conditions become:

$$I_D < 0, \qquad V_{SG} > 0, \qquad V_{SD} > 0.$$

Note that

$$V_{SG} = -V_{GS}, \qquad V_{SD} = -V_{DS}.$$

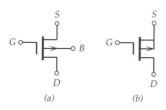

(a) (b)

Figure 4.2-5
(a) Circuit symbol of depletion-type PMOSFET. (b) Symbol for depletion-type PMOSFET with the substrate electrically connected to the source.

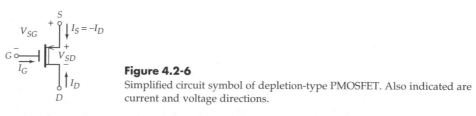

Figure 4.2-6
Simplified circuit symbol of depletion-type PMOSFET. Also indicated are current and voltage directions.

4.2.2 Enhancement-Type MOSFETs

The enhancement-type MOSFET is commonly used in integrated circuit design because of its ease of fabrication, small geometry, and low power dissipation. Without an applied voltage between the gate–source terminals, the drain–source channel is closed. With the application of a gate-to-source potential, the channel becomes enhanced to conduct carriers. The enhancement-type carrier conduction mechanism can be thought of as the antithesis of that of the depletion MOSFET.

An n-type channel layer (for the enhancement NMOSFET) is formed to conduct carriers from the drain to the source with the application of a positive V_{GS}. A p-type channel layer conducts carriers from the drain to the source with the application of a positive V_{SG} (negative V_{GS}) in enhancement PMOSFETs. The gate-to-source voltage that starts to form the drain–source channel is called the *threshold voltage* V_T. When V_{GS} is less than V_T (in NMOSFETs), I_D is zero since the drain–source channel does not exist. The value of V_T is dependent on the specific MOSFET device and commonly ranges in value from 1 to 5 V for enhancement NMOSFETs. Since $I_D = 0$ for $V_{GS} = 0$, the quantity I_{DSS} found in the depletion MOSFET and JFET is not pertinent to the enhancement-type MOSFET.

Unlike the depletion MOSFETs that operate in both enhancement and depletion modes, enhancement MOSFETs can only operate in the enhancement mode.

Enhancement-Type NMOSFET

The circuit symbol for the enhancement-type NMOSFET is shown in Figure 4.2-7. The symbol is similar to that of the depletion NMOSFET with the exception of the three short lines representing the junction area. As with the depletion-type NMOSFET, the symbol depicts a terminal representing the substrate with an arrow pointing into the junction.

A simplified circuit symbol for the enhancement-type NMOSFET is shown in Figure 4.2-8 with the applicable voltage and current sign conventions. The arrow on the circuit symbol points away from the junction to the source. The direction of the arrow corresponds to the direction of the current flow relative to the source terminal.

In the enhancement NMOSFET, the FET channel allows charge flow only when the gate–source potential, V_{GS}, is greater than some threshold voltage V_T. When $V_{GS} > V_T$,

$$I_S = -I_D, \qquad I_G = 0.$$

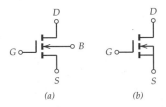

Figure 4.2-7
(*a*) Circuit symbol for enhancement-type NMOSFET. (*b*) Enhancement-type NMOSFET with the substrate connected to the source.

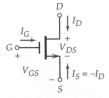

Figure 4.2-8
Simplified circuit symbol for enhancement NMOSFET.

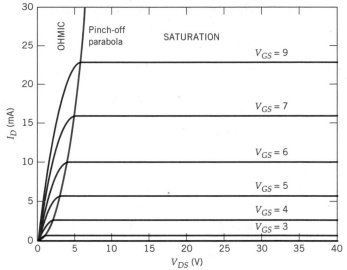

Figure 4.2-9
Enhancement-type NMOSFET drain characteristics.

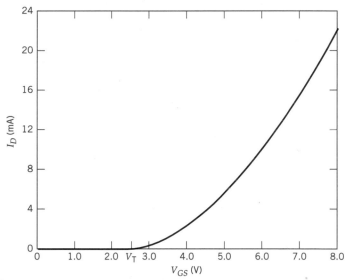

Figure 4.2-10
Enhancement-type NMOSFET transfer characteristic.

The enhancement NMOSFET characteristics are shown in Figure 4.2-9. The transfer characteristic of the enhancement NMOSFET is shown in Figure 4.2-10.

The threshold voltage V_T for the enhancement-type NMOSFET is a positive value. A positive V_{GS} greater than V_T allows current to flow through the FET by the formation of the drain–source channel. The NMOSFET is in the ohmic region when

$$0 < V_{DS} \le V_{GS} - V_T. \tag{4.2-12}$$

The pinch-off parabola, which delineated the boundary between the ohmic and saturation regions, is defined by

$$V_{DS} = V_{GS} - V_T. \tag{4.2-13}$$

The saturation region lies beyond the pinch-off parabola:

$$V_{DS} \geq V_{GS} - V_T. \tag{4.2-14}$$

For the enhancement-type NMOSFET operating in the ohmic or saturation regions,

$$I_D > 0, \qquad V_{GS} > V_T, \qquad V_{DS} > 0.$$

Enhancement-type NMOSFETs have a positive valued threshold voltage:

$$V_T > 0.$$

Ohmic Region

The ohmic region is that portion of the output characteristic curve between $V_{DS} = 0$ and the pinch-off parabola. The mathematical expression defining this region is

$$0 < V_{DS} \leq V_{GS} - V_T. \tag{4.2-15}$$

The V–I relationship in this region is

$$I_D = K[2(V_{GS} - V_T)V_{DS} - V_{DS}^2]. \tag{4.2-16}$$

The constant K is the transconductance factor in units of amperes/volt^2. This transconductance factor is determined by the geometry of the FET, gate capacitance per unit area, and the surface mobility of the electrons in the n channel.

For small values of V_{DS}, the drain current in Equation 4.2-16 is approximately

$$I_D = K[2(V_{GS} - V_T)V_{DS}]. \tag{4.2-17}$$

Equation 4.2-17 is a linearly varying function of I_D and V_{DS} for constant V_{GS}. The output resistance in this region is the derivative of Equation 4.2-17 with respect to V_{DS} for constant V_{GS}:

$$R_{DS}^{-1} = \frac{\partial I_D}{\partial V_{DS}} = 2\,K(V_{GS} - V_T). \tag{4.2-18}$$

By substituting Equation 4.2-13 into 4.2-16 the pinch-off parabolas is given by

$$I_D = KV_{DS}^2. \tag{4.2-19}$$

Saturation Region

The saturation region occupies the region of the output characteristic curve of Figure 4.2-9 where $I_D > 0$ and to the right of the pinch-off parabola. That is,

$$V_{DS} \geq V_{GS} - V_T. \tag{4.2-20}$$

The drain current is virtually constant with respect to V_{DS} for a given V_{GS} in this region due to high conductivity in the drain–source channel. The transfer characteristic is obtained by substituting $V_{GS} - V_T$ for V_{DS} in Equation 4.2-19:

$$I_D = K(V_{GS} - V_T)^2. \tag{4.2-21}$$

The transfer characteristic (Equation 4.2-20) is shown in Figure 4.2-10.

Cutoff Region

The cutoff region is defined as

$$V_{GS} < V_T. \tag{4.2-22}$$

In cutoff, the drain current is zero:

$$I_D = 0. \tag{4.2-23}$$

The FET is off in this region and does not conduct current. The region is used to implement the off state of a switch as described in Section 4.5.

The substrate potential affects the threshold current. In particular, for increasing negative substrate potential with respect to the source, the threshold voltage of the enhancement NMOSFET increases.

EXAMPLE 4.2-2 An enhancement-type NMOSFET with $V_T = 2$ V that conducts a current $I_D = 5$ mA for $V_{GS} = 4$ V and $V_{DS} = 5$ V. What is the value of I_D for $V_{GS} = 3$ V and $V_{DS} = 6$ V?

SOLUTION

First determine the region of operation at $V_{GS} = 4$ V and $V_{DS} = 5$ V.

$$V_{GS} - V_T = 4 \text{ V} - 2 \text{ V} = 2 \text{ V}.$$

But $V_{DS} = 5$ V, so

$$V_{DS} > V_{GS} - V_T,$$

implying that the FET is in the saturation region.

The unknown quantity is the transconductance factor K of Equation 4.2-21,

$$I_D = K(V_{GS} - V_T)^2.$$

Solving for K,

$$\begin{aligned}
K &= \frac{I_D}{(V_{GS} - V_T)^2} \\
&= \frac{5 \times 10^{-3}}{(4 - 2)^2} \\
&= 1.25 \times 10^{-3} \text{ A/V}^2.
\end{aligned}$$

With this information, the value of I_D for $V_{GS} = 3$ V and $V_{DS} = 6$ V can be determined. Again, the region of operation must be determined for the new operating parameters:

$$V_{GS} - V_T = 3 \text{ V} - 2 \text{ V} = 1 \text{ V}.$$

But $V_{DS} = 6$ V so,

$$V_{DS} > V_{GS} - V_T,$$

implying that the FET is again in the saturation region.

The current at I_D for $V_{GS} = 3$ V and $V_{DS} = 6$ V is

$$
\begin{aligned}
I_D &= K(V_{GS} - V_T)^2 \\
&= 1.25 \times 10^{-3} \, (3 - 2)^2, \\
&= 1.25 \text{ mA}.
\end{aligned}
$$

Enhancement Type PMOSFET

The circuit symbol for the enhancement-type PMOSFET is shown in Figure 4.2-11. The symbol is similar to that of the enhancement NMOSFET with the exception of the direction of the arrow on the body (substrate) terminal. As with the depletion-type PMOSFET, the symbol depicts a terminal representing the substrate with an arrow pointing into the junction.

A simplified circuit symbol for the enhancement-type PMOSFET is shown in Figure 4.2-12 with the voltage and current sign conventions. The arrow on the circuit symbol points toward the junction to the source. The direction of the arrow corresponds to the direction of the current flow relative to the source terminal.

All terminal voltages are of opposite polarity from that of the enhancement NMOSFET. The polarity of the threshold voltage is opposite that of the enhancement NMOSFET. The enhancement-type p-channel MOSFET voltages and currents have the following characteristics:

$$
\begin{aligned}
V_T < 0, \qquad K > 0, \qquad V_{SD} > 0, \\
V_{SG} > 0, \qquad I_D < 0.
\end{aligned}
$$

The conditions that identifies the regions of operation are

$$
\begin{aligned}
0 < V_{SD} < V_{SG} + V_T \quad &\text{(ohmic region)}, \\
V_{SD} \geq V_{SG} + V_T \quad &\text{(saturation region)}, \\
V_{SD} = V_{SG} + V_T \quad &\text{(pinch-off parabola)}.
\end{aligned} \tag{4.2-24}
$$

The enhancement PMOSFET characteristics are shown in Figure 4.2-13. The transfer characteristic of the enhancement PMOSFET is shown in Figure 4.2-14.

The threshold voltage V_T for the enhancement-type PMOSFET is a negative

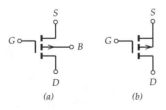

Figure 4.2-11
(*a*) Circuit symbol for enhancement-type PMOSFET. (*b*) Enhancement-type PMOSFET with the substrate connected to the source.

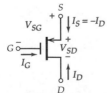

Figure 4.2-12
Simplified circuit symbol for enhancement PMOSFET.

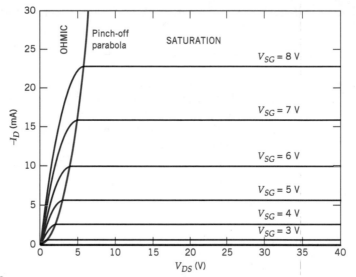

Figure 4.2-13
Enhancement-type PMOSFET drain characteristics.

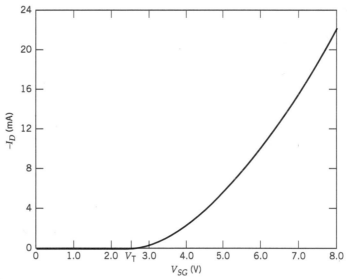

Figure 4.2-14
Enhancement-type PMOSFET transfer characteristic.

value. A positive V_{SG} greater than $|V_T|$ allows current to flow through the FET by the formation of the drain–source channel.

4.3 FET as Circuit Element

Using the terminal behavior of JFETs and MOSFETs, simple circuits using FETs can be developed. Field-effect transistors are used in circuits as constant-current devices, active loads, and voltage variable resistors, to name a few. Metal–oxide–semiconductor FETs are used extensively in digital circuits and have certain advantages over

BJT designs. The SPICE model for the JFET and MOSFET are also developed in this section.

4.3.1 FET SPICE Models

When trying to convert the FET parameters discussed in this chapter to that used in SPICE models, it rapidly becomes apparent that the correspondence between the two sets of parameters is not clear. The correspondence between the SPICE parameter names and the FET parameter names used in this book is accomplished with the equivalences presented in this section.

For modeling JFETs, the general form in the net list should be

```
Jname drain_node gate_node source_node model_name
```

The model forms are

```
.MODEL model_name NJF(parameter values)   for n-JFETs
.MODEL model_name PJF(parameter values)   for p-JFETs
```

The parameter values of interest are

VTO = threshold voltage = $\pm|V_{PO}|$

BETA = transconductance coefficient = $|I_{DSS}|/V_{PO}^2$

LAMBDA = $1/V_A$ = channel-length modulation

Here, V_A is similar to the "early" voltage of BJTs. The voltage V_A is that voltage (V_{DS} in the case n-JFETs) that is the point of intersection of the $I_D = 0$ line and the extended line from the characteristic curves in saturation (depicted in Figure 4.3-1).

For depletion mode JFETs, VTO < 1. For the extremely rare enhancement mode JFETs, VTO > 1. In this book, only depletion mode n- and p-channel JFETs are discussed.

An example of the .MODEL statement for an n-JFET is

```
.MODEL NJFET_TEST NJF(VTO=−4 BETA=625E-6 LAMBDA=2E-6)          .
```

The default values for the parameters are

$$\text{VTO} = -2 \text{ V} \qquad \text{BETA} = 1 \times 10^{-4} \text{ A/V}^2$$

$$\text{LAMBDA} = 0 \text{ V}^{-1}.$$

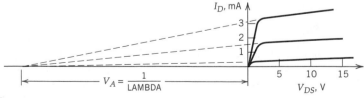

Figure 4.3-1
Channel-length modulation effect of FET.

For MOSFETs, the general form in the net list should be

```
Mname drain_node gate_node source_node substrate_node model_name
```

The substrate_node for *n*-channel devices is connected to the most negative voltage, whereas for *p*-channel devices the substrate_node is connected to the most positive voltage. The model forms are

```
.MODEL model_name NMOS(parameter values)  for n-MOSFETs
.MODEL model_name PMOS(parameter values)  for p-MOSFETs
```

The parameter values of interest are as follows:

VTO = zero-bias threshold voltage (V_{PO} for depletion MOSFETs, V_T for enhancement MOSFETS).

KP = transconductance coefficient ($|2I_{DSS}/V_{\text{PO}}^2|$ for depletion MOSFETs, $|2K|$ for enhancement MOSFETS).

LAMBDA ($= 1/V_A$) = channel-length modulation.

An example of the .MODEL statement for an depletion NMOSFET is

```
.MODEL MOS_TEST NMOS(VTO=-4 KP=1.25E-3 LAMBDA=2E-6)
```

The default values for the parameters are

$$\text{VTO} = 0 \text{ V} \qquad \text{LAMBDA} = 0 \text{ V}^{-1}.$$

4.3.2 FET as Voltage Variable Resistor

A voltage variable resistor (VVR) is a three-terminal device where the resistance between two of the terminals is controlled by a voltage on the third. In the ohmic region, FETs demonstrate a variation in resistance, R_{DS}, described by Equations 4.1-9 and 4.2-18:

$$R_{DS}^{-1} = \frac{\partial I_D}{\partial V_{DS}} = \frac{2I_{DSS}}{V_{\text{PO}}}\left(\frac{V_{GS}}{V_{\text{PO}}} - 1\right)$$

$$= \frac{\partial I_D}{\partial V_{DS}} = 2K(V_{GS} - V_T).$$

Written more conveniently as resistances, the drain–source resistance in the ohmic region of the *n*-JFET and depletion mode NMOSFET are

$$R_{DS} = \frac{V_{\text{PO}}^2}{2I_{DSS}(V_{GS} - V_{\text{PO}})}. \tag{4.3-1}$$

For the enhancement mode NMOSFET, the drain–source resistance is

$$R_{DS} = \frac{1}{2K(V_{GS} - V_T)}. \tag{4.3-2}$$

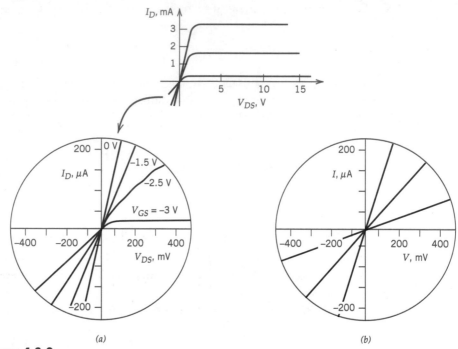

Figure 4.3-2

Comparison of (a) n-JFET and (b) resistor characteristics.

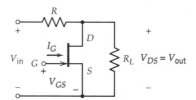

Figure 4.3-3

Simple voltage-controlled voltage divider.

The n-channel JFET is used as an example in this section. Operation of n-JFETs in the ohmic region implies that the drain–source voltage is held small. Figure 4.3-2 illustrates typical n-JFET and resistor characteristics. The slope of V_{DS}/I_D is a function of V_{GS}, and the drain–source resistance R_{DS} is controlled by V_{GS}.

Figure 4.3-3 shows a simple voltage-controlled voltage divider using an n-JFET VVR to provide a means of voltage controlling the divider ratio. As V_{GS} is changed from zero to V_{PO}, R_{DS} is changed.

The output voltage is

$$V_{\text{out}} = V_{\text{in}} \frac{R_L(1 + R_{DS}^{-1}R_L)^{-1}}{R + R_L(1 + R_{DS}^{-1}R_L)^{-1}}. \tag{4.3-3}$$

When V_{GS} approaches V_{PO}, R_{DS} approaches zero and the FET causes no signal attenuation. If $R_L \gg R_{DS}$, Equation 4.3-3 can be simplified to

$$V_{out} = V_{in} \frac{1}{1 + R/R_{DS}}.$$

(4.3-4)

Using Equation 4.1-9, the output voltage is expressed as a function of V_{GS}:

$$V_{out} = \frac{V_{in}}{1 + R/R_{DSO}^{-1}[(V_{PO} - V_{GS})/V_{PO}]},$$

(4.3-5)

where

$$R_{DSO} = \frac{\partial V_{GS}}{\partial I_D} \quad \text{at} \quad V_{GS}, V_{DS} = 0.$$

4.3.3 The n-JFET as Constant-Current Source

The ideal constant-current source would supply a given current to a load independent of the voltage across the load. In such a case, the output resistance of the source is infinite. The drain current of the JFET approaches saturation when operated with the gate–drain voltage greater than V_{PO}. Under this condition, the FET can be used as a constant-current source.

One way to force V_{GD} to be greater than V_{PO} is to tie the gate of the n-JFET directly to the source as shown in Figure 4.3-4.

In Figure 4.3-4, $V_{GS} = 0$, therefore in saturation, $V_{DS} = -V_{PO}$ and the drain current $I_D = I_{DSS}$. As long as the FET remains in saturation, it will provide a constant current flowing through the drain–source channel.

4.3.4 FET Inverter

The methods used to analyze FET circuits are similar to those used to solve BJT circuits in Chapter 3. Three choices are available to analyze FET circuits:

- Use the set of FET equations in Sections 4-1 and 4-2 along with additional circuit-dependent equations to obtain a numerical solution.
- Use empirical V–I curves and obtain a graphical solution.
- Use SPICE to obtain the solution.[3]

As in BJT circuit analysis, the choice of technique used to analyze the circuit is strongly dependent on the complexity of the FET circuit.

Figure 4.3-4
The n-JFET constant-current source.

---

[3] SPICE is the computer analysis tool of preference to the authors. Other computer simulation packages will produce similar results.

EXAMPLE 4.3-1 Consider an *n*-JFET inverter shown in Figure 4.3-5. The JFET characteristics are

$$I_{DSS} = 8 \text{ mA} \quad \text{and} \quad V_{PO} = -4 \text{ V}.$$

Find the drain current I_D and the drain–source voltage V_{DS}.

SOLUTION 1 (USING CHARACTERISTIC EQUATIONS FOR N-JFET)

In order to find V_{DS}, an additional equation dependent on circuit topology is required. The loop equation needed is

$$V_{DD} = I_D R_D + V_{DS}.$$

Assume that the JFET is operating in saturation. The drain current can then be calculated from Equation 4.1-11:

$$I_D = I_{DSS}\left(1 - \frac{V_{GS}}{V_{PO}}\right)^2$$

$$= 8 \times 10^{-3}\left(1 - \frac{-2}{-4}\right)^2$$

$$= 2 \text{ mA.}$$

The drain–source voltage can then be obtained:

$$V_{DS} = V_{DD} - I_D R_D$$

$$= 10 - (2 \times 10^{-3})(2200)$$

$$= 5.6 \text{ V.}$$

To confirm that the *n*-JFET is operating in saturation, the condition for saturation is checked:

$$V_{DS} \geq V_{GS} - V_{PO},$$

$$5.6 \geq -2 - (-4),$$

$$5.6 \geq 2.$$

The saturation region assumption is verified and the calculations are valid.

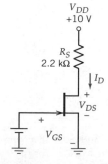

Figure 4.3-5
The *n*-JFET inverter in Example 4.3-1.

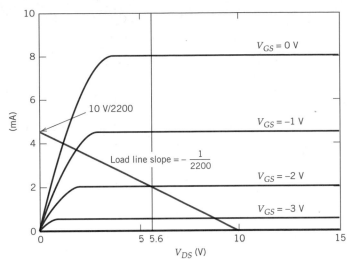

Figure 4.3-6
Load line solution for Example 4.3.4-1.

Solution 2 (Graphical Method)

The n-JFET output curves to perform a load line analysis of the circuit must first be plotted. In the load line method of analysis, the solution (operating point) of the n-JFET circuit is the intersection of the straight line representing the constant load (R_D) and the curve for the desired V_{GS}. The values of I_D and V_{DS} corresponding to the intersection of the load line and the V_{GS} curve is the solution. The graphical solution of this example is shown in Figure 4.3-6. The intersection of the load line and the $V_{GS} = -2$ V line occurs at $I_D = 2$ mA and $V_{DS} = 5.6$ V, confirming solution 1. Also note that the solution lies in the saturation region.

The graphical technique also provides solutions for this circuit with other values of the source connected to the gate of the n-JFET. The intersection of the load line and the plotted curves for these values of V_{GS} are

V_{GS} (V)	I_D (mA)	V_{DS} (V)
-4	0.0	10
-3	0.5	8.9
-2	2.0	5.6
-1	3.7	1.8
0	4.0	1.2

Notice that the intersections of the load line and the JFET curves for $V_{GS} = 0$ V and -1 V occur in the ohmic region.

If other gate–source voltage values are chosen, the analysis must be expanded in one of two ways:

- Additional JFET curves must be plotted.
- The solutions must be obtained by interpolation between the curves.

```
*****************************Evaluation
PSpice*******************************
SPICE analysis of n-JFET inverter
**** CIRCUIT DESCRIPTION
*****************************************************************************
VDD   1   0   10
VGS   2   0   -5
RD    1   3   2.2K
J1    3   2   0   NJFET1
.MODEL   NJFET1   NJF(VTO=-4 BETA=500E-6)
.DC LIN   VGS   -4.5   0   .5
.PRINT   DC   ID(J1)   V(3)
.END
****   Junction FET MODEL PARAMETERS
       NJFET1
       NJF
     VTO  -4
     BETA 500.000000E-06
****   DC TRANSFER CURVES        TEMPERATURE= 27.000 DEG C
VGS          ID(J1)       V(3)
-4.500E+00   1.451E-11   1.000E+01
-4.000E+00   1.401E-11   1.000E+01
-3.500E+00   1.250E-04   9.725E+00
-3.000E+00   5.000E-04   8.900E+00
-2.500E+00   1.125E-03   7.525E+00
-2.000E+00   2.000E-03   5.600E+00
-1.500E+00   3.125E-03   3.125E+00
-1.000E+00   3.741E-03   1.768E+00
-5.000E-01   3.911E-03   1.396E+00
 0.000E+00   4.011E-03   1.176E+00
       JOB CONCLUDED
       TOTAL JOB TIME     .77
```

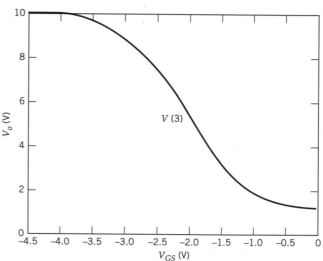

Figure 4.3-7
SPICE analysis of *n*-JFET inverting amplifier.

Figure 4.3-8
SPICE analysis: Output voltage vs. input voltage.

SOLUTION 3 (SPICE SOLUTION)

The SPICE solution is shown in Figure 4.3-7. The solution is in the form of the PSpice **.OUT** file. The calculated values are printed out as follows:

The SPICE results match those of the analytical and graphical approaches shown in Solutions 1 and 2. For the SPICE simulation, the output current and voltage have also been calculated for additional values of the voltage at the gate of the JFET. A plot of the resultant output voltage as a function of the input gate–source voltage is shown in Figure 4.3-8. Notice that for low values of input ($V_{GS} \approx -4$ V) the output is high ($V_{DS} \approx 10$ V); for high value of the input ($V_{GS} \approx 0$V) the output is low ($V_{DS} \approx 0$ V). The output forms a digital inverse of the input. In the region between these two extremes the output is roughly a negative multiple of the input: A circuit of this topology forms a small-signal inverting amplifier. Field-effect transistor amplifiers are explored in Chapter 5.

4.3.5 FET as Active Load

In some applications, an FET active load to replace the passive resistor used as a load in an inverting FET amplifier may be desirable. For instance, in integrated circuit fabrication, passive resistors consume large areas of the chip compared to transistors. Both enhancement and depletion NMOSFET active loads are commonly used. However, the load line characteristics are quite different between the two types of NMOSFETs.

An enhancement NMOSFET inverting amplifier with an enhancement NMOSFET load is shown in Figure 4.3-9. In this circuit, Q_1 acts as the load and Q_2 is the driver.

By studying the circuit, several interesting relationships can be found. The output voltage V_O is

$$V_O = V_{DS2}.$$

Also, since the gate of Q_1 is connected to its own drain,

$$V_{GS1} = V_{DS1}.$$

Because the Q_1 gate–source and drain–source voltages are equal, Q_1 is in saturation since the difference between the gate–source and threshold voltages will always be less than the drain–source voltage:

$$V_{DS} \geq V_{GS} - V_T.$$

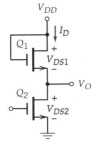

Figure 4.3-9
Enhancement NMOSFET inverting amplifier with enhancement NMOSFET active load.

The relationship between the two NMOSFETs is readily apparent using elementary circuit analysis. The sum of the voltages V_{DS1} and V_{DS2} must always equal the source voltage V_{DD}:

$$V_{DS1} + V_{DS2} = V_{DD}.$$

The load line created by Q_1 can be found by replacing V_{DS1} with $V_{DD} - V_{DS2}$:

$$
\begin{aligned}
I_{D1} &= K(V_{GS1} - V_T)^2 \\
&= K[(V_{DD} - V_{DS2}) - 2]^2.
\end{aligned}
\tag{4.3-6}
$$

The Q_1 load line superimposed on the Q_2 characteristic equation is shown in Figure 4.3-10. The curves were created using

$$K = \tfrac{1}{2}KP = 625 \times 10^{-6} \qquad V_T = 2 \text{ V}, \qquad V_{DD} = 10 \text{ V}.$$

The load line ends at

$$V_{DD} - V_{T1} = 8 \text{ V}, \qquad I_D = K(V_{DD} - V_T)^2 = 40 \text{ mA}.$$

If the load line is plotted using Equation 4.3-6, the load lines curves back up at $V_{DS} > V_{DD} - V_T$. In real circuits, the line ends at $V_{DS} = V_{DD} - V_T$.

Another common configuration is the enhancement NMOSFET driver with a depletion NMOSFET active load shown in Figure 4.3-11.

By studying the circuit in Figure 4.3-11, several circuit relationships are evident:

$$V_O = V_{DS2}, \qquad V_{GS1} = 0, \qquad V_{DD} = V_{DS1} + V_{DS2}.$$

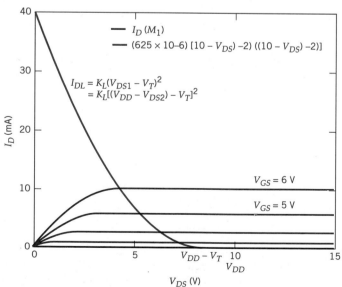

Figure 4.3-10

Enhancement NMOSFET inverting amplifier with enhancement NMOSFET load.

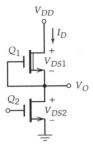

Figure 4.3-11
Enhancement NMOSFET inverting amplifier with depletion NMOSFET active load.

The depletion NMOSFET Q_1 is in saturation only when

$$V_{DS1} = V_{DD} - V_{DS2} \geq V_{GS1} - V_{PO} \geq -V_{PO}.$$

Since $V_{GS} = 0$, the current through Q_1 is I_{DSS} in the saturation region.

As V_{DS2} increases toward V_{DD}, V_{DS1} eventually becomes less than $-V_{PO}$, causing Q_1 to operate in the ohmic region. As Q_1 enters its ohmic region, the load line curves down toward V_{DD} following the characteristic equation of the ohmic region:

$$
\begin{aligned}
I_D &= I_{DSS}\left[2\left(\frac{V_{GS1}}{V_{PO}} - 1\right)\frac{V_{DS1}}{V_{PO}} - \left(\frac{V_{DS1}}{V_{PO}}\right)^2\right] \\
&= I_{DSS}\left[-2\frac{V_{DS1}}{V_{PO}} - \left(\frac{V_{DS1}}{V_{PO}}\right)^2\right] \\
&= I_{DSS}\left[-2\frac{V_{DD} - V_{DS2}}{V_{PO}} - \left(\frac{V_{DD} - V_{DS2}}{V_{PO}}\right)^2\right].
\end{aligned}
$$

(4.3-7)

The resulting graphical solution is shown in Figure 4.3-12.

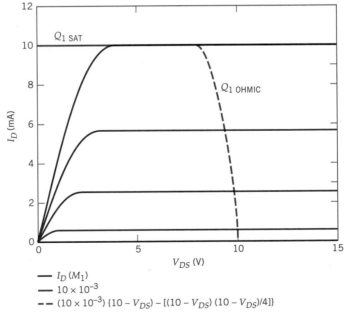

Figure 4.3-12
Enhancement NMOSFET inverting amplifier with depletion NMOSFET load.

4.3.6 CMOS Inverter

Complementary symmetry metal–oxide–semiconductor (CMOS) circuits are fabricated with both enhancement NMOSFETs and enhancement PMOSFETs on the same chip. The main advantage of CMOS technology is the ability to design circuits with essentially zero DC power dissipation. Power is dissipated only during switching transitions. The CMOS inverter is shown in Figure 4.3-13.

In Figure 4.3-13 the enhancement PMOSFET Q_1 is the load and the enhancement NMOSFET Q_2 is the driver.

Qualitatively, the operation of the CMOS inverter is simple. When $V_I = 0$, $V_{SG1} = V_{DD}$ and $V_{GS2} = 0$. When the gate–source voltage of the enhancement NMOSFET is zero, Q_2 is cut off. Therefore, V_O is V_{DD}.

When $V_I = V_{DD}$, $V_{GS2} = V_{DD}$ and $V_{SG1} = 0$. Since the source–gate voltage of the Q_1 enhancement PMOSFET is zero, Q_1 is cut off. Therefore, $V_O = 0$.

Note that at logic zero and logic one outputs ($V_O = 0$ and V_{DD}), there is no current flowing though the circuit. Therefore, the CMOS inverter only dissipates power when transitioning between the two logic levels.

A SPICE simulation clearly shows the transfer function of the CMOS inverter. The transfer function is shown in Figure 4.3-14.

Load line analysis of the CMOS circuit is similar to other active load analysis. The circuit yields the relationships

$$V_O = V_{DS2}, \qquad V_I = V_{GS1} = V_{GS2}.$$

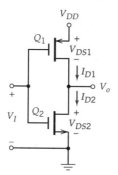

Figure 4.3-13
CMOS inverter.

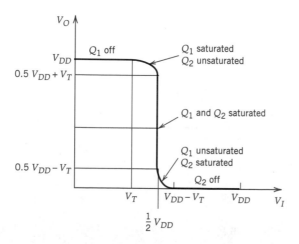

Figure 4.3-14
Transfer characteristics of CMOS inverter.

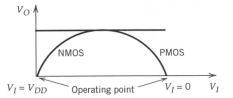

Figure 4.3-15
Load line analysis of CMOS inverter.

Using the condition for saturation,

$$V_{DS} \geq V_{GS} - V_T,$$

a graphical solution of the CMOS inverter can be found. The load lines are shown in Figure 4.3-15.

4.4 REGIONS OF OPERATION IN FETs

Field-effect transistor operation has been seen to fall into three regions of useful operation. The regions are described by the state of the drain–source channel controlled by the gate voltage. The three regions of interest are shown in Figure 4.4-1. Briefly, the three regions of operation are as follows:

1. The cutoff region is defined as that region where the gate voltage disallows charge flow in the drain–source channel. The FET essentially looks like an open circuit. Applications for this region are primarily in switching and digital logic circuits.

2. The ohmic region is defined by a gradual increase in charge flow in the drain–source channel, the rate of which is controlled by the gate voltage. Applications for this region include the use of FETs as voltage variable resistors.

3. The saturation region is defined by constant charge flow in the drain–source channel without regard to the drain–source voltage. The amount of constant-current flow is regulated by the gate voltage. The region is commonly used for amplification with the modulation of the gate voltage and for constant-current source applications.

In addition to these three regions, there is an additional region of severe consequences: the breakdown region. Field-effect transistors are not manufactured to with-

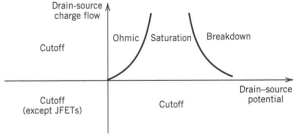

Figure 4.4-1
FET regions of operation.

stand extended use in the breakdown region and will exhibit catastrophic thermal runaway and destruction. To ensure proper operation of the devices, manufacturers' specifications on maximum voltages must be heeded. Due to the JFET fabrication process, the drain and source can, in most cases, be interchanged when the JFET is used as a circuit element without affecting the desired operating characteristics. This property (drain–source reciprocity) is not common to all types of FETs.

The characteristics of the FETs in the different regions are shown in Table 4.4-1.

4.5 FET AS ANALOG SWITCH

Both BJTs and FETs can be used as switches in a wide variety of analog electronic applications. Each type provides a great advantage over mechanical switches in both speed, reliability, and resistance to deterioration. Field-effect transistors are the more common choice due to the inherent symmetry of FETs and the undesirable offset voltage (V_{CE} at $I_C = 0$) that is present in BJTs. The offset voltage of BJTs, typically on the order of a few millivolts, can produce significant errors in the transmission of low-level analog signals. Another advantage of FET switches is the high gate input impedance and the consequent low load that the voltage control port presents to control electronics.

Among the many possible electronic applications of solid-state analog switches are the following:

- Sample-and-hold circuits
- Switchable gain amplifiers
- Switched-capacitor filters
- Digital-to-analog conversion
- Signal gating and squelch control
- Chopper stabilization of amplifiers

In addition, multiple switches connected to share a common output form a multiplexer: a common building block for analog and digital systems.

The basic function of a switch is to electrically isolate or connect two sections of a circuit. In order to accomplish that functionality, an ideal analog switch has the following design goals:

- In the ON state, it passes signals without attenuation or distortion.
- In the OFF state, signals are not passed.
- The transitions between the ON and OFF states are instantaneous.

In order to accomplish these design goals, real switches have the more realistic specifications:

- Very low ON resistance compared to other circuit resistances
- Very high OFF resistance compared to other circuit resistances
- Low leakage currents in the OFF state
- Low capacitive and/or inductive effects

TABLE 4.4-1 FET CHARACTERISTICS

FET Type	Ohmic Region	Saturation Region	Cutoff Region
n-JFET	$0 < V_{DS} \leq V_{GS} - V_{PO}$, $\quad I_D = I_{DSS}\left[2\left(\dfrac{V_{GS}}{V_{PO}} - 1\right)\dfrac{V_{DS}}{V_{PO}} - \left(\dfrac{V_{DS}}{V_{PO}}\right)^2\right]$	$V_{DS} \geq V_{GS} - V_{PO}$, $\quad I_D = I_{DSS}\left(1 - \dfrac{V_{GS}}{V_{PO}}\right)^2$	$V_{GS} < V_{PO}, I_D = 0$
p-JFET	$V_{SD} \leq V_{PO} + V_{SG}$, $\quad I_D = I_{DSS}\left[2\left(\dfrac{V_{GS}}{V_{PO}} - 1\right)\dfrac{V_{DS}}{V_{PO}} - \left(\dfrac{V_{DS}}{V_{PO}}\right)^2\right]$	$V_{SD} \geq V_{SG} + V_{PO}$, $\quad I_D = I_{DSS}\left(1 - \dfrac{V_{GS}}{V_{PO}}\right)^2$	$V_{SG} < -V_{PO}, I_D = 0$
Enhancement NMOSFET	$0 < V_{DS} \leq V_{GS} - V_T$, $\quad I_D = K[2(V_{GS} - V_T)V_{DS} - V_{DS}^2]$	$V_{DS} \geq V_{GS} - V_T$, $\quad I_D = K(V_{GS} - V_T)^2$	$V_{GS} < V_T, I_D = 0$
Depletion NMOSFET In depletion mode: $I_D > 0, V_{PO} < V_{GS} \leq 0, V_{DS} > 0$ In enhancement mode: $I_D > 0, V_{GS} > 0, V_{DS} > 0$	$0 < V_{DS} \leq V_{GS} - V_{PO}$, $\quad I_D = I_{DSS}\left[2\left(\dfrac{V_{GS}}{V_{PO}} - 1\right)\dfrac{V_{DS}}{V_{PO}} - \left(\dfrac{V_{DS}}{V_{PO}}\right)^2\right]$	$V_{DS} \geq V_{GS} - V_{PO}$, $\quad I_D = I_{DSS}\left(1 - \dfrac{V_{GS}}{V_{PO}}\right)^2$	$V_{GS} < V_{PO}, I_D = 0$
Enhancement PMOSFET	$V_{SD} \leq V_{SG} + V_T$, $\quad -I_D = K[2(V_{GS} - V_T)V_{DS} - V_{DS}^2]$	$V_{SD} \geq V_{SG} + V_T$, $\quad -I_D = K(V_{GS} - V_T)^2$	$V_{SG} < -V_T, I_D = 0$
Depletion PMOSFET In depletion mode: $I_D < 0, -V_{PO} < V_{SG} \leq 0, V_{SD} > 0$ In enhancement mode: $I_D < 0, V_{SG} > 0, V_{SD} > 0$	$0 < V_{SD} \leq V_{SG} + V_{PO}$, $\quad I_D = I_{DSS}\left[2\left(\dfrac{V_{GS}}{V_{PO}} - 1\right)\dfrac{V_{DS}}{V_{PO}}\left(\dfrac{V_{DS}}{V_{PO}}\right)^2\right]$	$V_{SD} \geq V_{SG} + V_{PO}$, $\quad I_D = I_{DSS}\left(1 - \dfrac{V_{GS}}{V_{PO}}\right)^2$	$V_{SG} < -V_{PO}, I_D = 0$

Field-effect transistors satisfy these specifications satisfactorily for many applications. Control signals applied to the gate–source port of an FET will vary the drain–source port between ON and OFF switch positions. Reasonably low ON resistance combined with extremely high OFF resistance make the FET ideal as a voltage controlled analog switch element. In Section 4.3.2 it was seen that the drain–source resistance, R_{DS}, for an FET is highly dependent on the gate–source voltage.[4] The drain–source resistance expression for depletion type FETs is

$$R_{DS} = \frac{V_{PO}^2}{2I_{DSS}(V_{GS} - V_{PO})}, \qquad V_{GS} \geq V_{PO}. \tag{4.5-1}$$

For enhancement-type FETs,

$$R_{DS} = \frac{1}{2K(V_{GS} - V_T)}, \qquad V_{GS} \geq V_T. \tag{4.5-2}$$

Variation in the control signal, V_{GS} can change this resistance from the range of a few ohms to many megohms. A simple application of an analog switch using a single FET is shown in Figure 4.5-1. Such simple analog switches typically use enhancement mode FETs, although it is possible to form a switch with depletion mode FETs.[5] In order to keep the switch in the OFF state for all values of the input voltage v_s, the control voltage v_c must be less than the minimum input signal value plus the threshold voltage of the FET:

$$v_{c(\text{off})} < v_{s(\text{min})} + V_T. \tag{4.5-3}$$

Similarly, to keep the switch in the ON state for all values of the input voltage, the control voltage must be greater than the maximum input signal value plus the threshold voltage of the FET[6]:

$$v_{c(\text{on})} > v_{s(\text{max})} + V_T. \tag{4.5-4}$$

Equations 4.5-3 and 4.5-4 provide expressions for the absolute limits of ON and OFF control voltages. In order to ensure good switch performance, it is necessary to provide a design margin beyond these absolute limits.

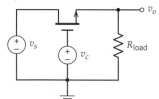

Figure 4.5-1
Simple FET analog switch application.

---

[4] For simplicity of presentation, only the n-channel FET equations are given. Shown are the equations for drain–source resistance for depletion mode and enhancement mode FETs, respectively.
[5] Enhancement mode FETs will be used in the discussions of this section.
[6] Care must be taken so that the FET gate breakdown voltage is not exceeded when choosing control voltages. Typical gate breakdown voltages are in excess of 25 V.

EXAMPLE 4.5-1 The FET analog switch of Figure 4.5-1 is used on the output of a OpAmp whose power supplies are set at ± 15 V DC. These voltages are chosen as the control voltages for the switch. If the threshold voltage for the FET is 2 V, what range of output voltages will be properly controlled by the switch?

SOLUTION

The minimum signal voltage is given by Equation 4.5-3:

$$v_{s(min)} > v_{c(off)} - V_T$$

or

$$v_{s(min)} > -15 \text{ V} - 2 \text{ V} = -17 \text{ V}.$$

But, for this case the signal voltage is limited by the output of the OpAmp to ≈ -15 V.
    The maximum signal voltage is given by Equation 4.5-4:

$$v_{s(max)} < v_{c(on)} - V_T$$

or

$$v_{s(max)} < 15 \text{ V} - 2V = +13 \text{ V}.$$

Thus the absolute maximum signal range is limited range to

$$-15 \text{ V} < v_s < 13 \text{ V}.$$

Good design practice would place tighter limits on the signal. These limits are functions of whatever additional design specifications may apply to the particular application of this switch.

One of the problems associated with the simple FET analog switch of Figure 4.5-1 is that the ON resistance of the switch is not constant as the input signal varies. Variation in ON resistance can be a serious limitation in some circuit applications as it can cause distortion of the analog signal. Variation in the input signal will cause the voltage at the source of the FET to vary. If the gate terminal of the FET is set at a specific control voltage level, the input signal variation therefore produces a variation in V_{GS} for the FET and consequently a variation in the drain–source resistance of the switch. Determination of the switch resistance as a function of input signal level requires the solution of two nonlinear equations. For this particular circuit the equations reduce to

$$R_{DS} = \frac{v_s}{i} - R_{load} \tag{4.5-5}$$

and

$$R_{DS} = \frac{1}{v_{c(on)} - iR_{load} - V_T}. \tag{4.5-6}$$

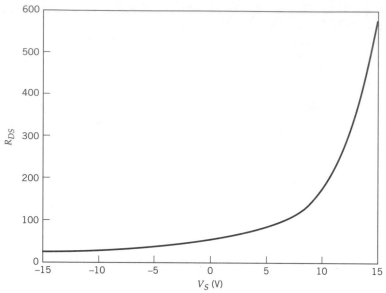

Figure 4.5-2
Switch resistance as a function of input signal.

Computer simulation or the application of load line techniques are perhaps the two best methods to calculate the variation of switch resistance with input. A typical graph[7] of the switch resistance as a function of input signal is given in Figure 4.5-2. Notice that while the switch resistance is fairly small throughout most of the input signal range (24 Ω < R_{DS} < 100 Ω as -15 V < v_s < 6.6 V), it increases dramatically as the level of the input signal approaches the positive control voltage. This drastic increase in switch resistance can greatly diminish the utility of such a simple switch. The OFF resistance of this switch is typically very high ($\geq$ 20 MΩ) and the OFF performance of such a switch is limited only by a very small leakage current in the FET.

A common approach to designing a better analog switch using FETs is shown in Figure 4.5-3. This switch is a parallel combination of a two switches: one con-

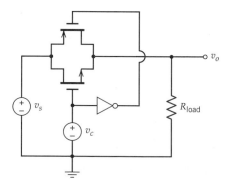

Figure 4.5-3
Parallel CMOS analog switch.

---

[7] The graph was obtained using PSpice using the circuit parameters of Example 4.5-1 with R_{load} = 1 kΩ and K = 1.5 mA/V².

structed with an *n*-channel FET, the other with a *p*-channel FET. The triangular symbol is an inverter: It reverses the control signal levels so that the *p*-channel FET will operate with the *n*-channel FET.

It has been shown that the *n*-channel FET in this configuration will have low ON resistance for inputs signals near the negative limits. The *p*-channel FET will have low ON resistance for inputs signals near the positive limits. Since the total ON resistance of this parallel switch is the parallel combination of the two individual switch resistances, the switch has nearly constant ON resistance throughout the range of possible input signals. Figure 4.5-4 demonstrates this near-constant resistance property.[8] In this example the switch has an ON resistance of approximately 22.2 Ω throughout most of the possible range of input voltage:

$$v_s \leq v_{c(on)} - V_T. \tag{4.5-7}$$

When the input voltage level exceeds the constraint of Equation 4.5-7, the input resistance drops slightly. In addition, the OFF resistance of this switch is extremely high: The OFF performance of this switch is limited by a very small leakage current through the FETs. The OFF resistance *suffers a significant degradation* when the limit of Equation 4.5-7 is exceeded: Input signals levels must be limited to these constraints.

The FET analog switches have specific limitations that are usually described in the specifications provided by the manufacturer. Among the most common of the limitations are:

- *Analog output leakage current*: the algebraic sum of currents from the power supplies, ground, input signal, and control signal through a OFF switch

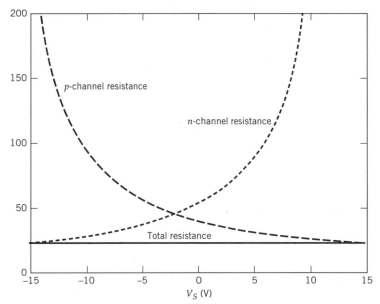

Figure 4.5-4
Complementary switch resistance as a function of input signal.

---

[8] Figure 4.5-4 was generated using PSpice and the same circuit and FET parameters as Figure 4.5-2.

- *Analog voltage range*: The range of analog voltage amplitudes with respect to ground over which the analog switch operates within the ON and OFF specifications
- *ON resistance and ON resistance variation*: the resistance of the switch over the analog voltage range
- *Output switching times*: the time it takes the switch to change states
- *Switch current*: the maximum amount of current that can be fed through the switch

In addition to the MOSFET switches described here, *n*-channel JFETs are used for analog switches. In order to maintain a depletion mode JFET switch in the ON state, a rather complicated electronic control circuit is necessary. This control circuitry is usually fabricated on the same semiconductor chip as the switch and consists of both bipolar and JFET devices. These switches have a very constant ON resistance over the entire analog signal range. The disadvantage of these switches is their relatively high cost.

4.6 BIASING THE FET

The selection of an appropriate quiescent operating point for a FET is determined by conditions similar to those for a BJT. Here, the quiescent conditions are the zero-input DC values of the FET drain current I_D and the terminal voltage differences V_{GS} and V_{DS}. In this section, several bias circuits for FETs are examined. While bias circuitry can be used to put the FET in any of its regions of operation, the focus of this section is the saturation region. The saturation region is, of course, significant as the region where FET amplification occurs. While the examples in this section use *n*-channel FETs, *p*-channel FETs are biased in a similar manner except for a change in polarity of the voltage supplies.

The emphasis in this section in on biasing using voltage supplies and resistors. It is also common to bias transistors using current sources. Biasing FETs for amplifier applications using current sources is discussed extensively in Chapter 6.

4.6.1 Source Self-Bias Circuit

The source self-bias circuit shown in Figure 4.6-1 is particularly useful in biasing JFETs and depletion mode FETs of other types. In this application, the external gate

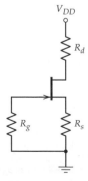

Figure 4.6-1
Source self-bias circuit.

resistor R_G serves to tie the voltage at the gate of the FET to ground (the gate current is essentially zero). This resistor is typically chosen to be a very large value (often on the order of a few megohms). For a specified drain current I_D, the gate–source voltage V_{GS} can be determined from the FET transfer characteristic in either analytic or graphical form. Then R_s is determined from V_{GS} and I_D:

$$R_s = -\frac{V_{GS}}{I_D} \tag{4.6-1}$$

EXAMPLE 4.6-1 An n-channel JFET with the following characteristics:

$$V_{PO} = -3.5 \text{ V} \quad \text{and} \quad I_{DSS} = 10 \text{ mA,}$$

placed in the source self-bias circuit of Figure 4.6-1. Quiescent conditions of

$$I_D = 5 \text{ mA} \quad \text{and} \quad V_{DS} = 5 \text{ V}$$

are desired. Determine the bias resistors necessary to establish the quiescent conditions if the power supply voltage is $V_{DD} = 15$ V.

SOLUTION

The quiescent conditions for the JFET distinctly imply that it is in saturation:

$$V_{DS} \geq V_{GS} - V_{PO}.$$

Therefore, Equation 4.1-11 is the determining factor in finding V_{GS}:

$$I_D = I_{DSS}\left(1 - \frac{V_{GS}}{V_{PO}}\right)^2.$$

Substitution of values yields

$$\frac{5}{10} = \left(1 - \frac{V_{GS}}{-3.5}\right)^2$$

or

$$V_{GS} = 3.5\left(\pm\frac{1}{\sqrt{2}} - 1\right) = -1.025 \text{ V} \quad \text{or} \quad -5.975 \text{ V}.$$

The value between 0 V and V_{PO} is the only valid solution to Equation 4.1-11: The other is a spurious result caused by taking the square root. Thus, $V_{GS} = -1.025$ V. The resistor R_s can now be determined:

$$R_s = -\frac{-1.025 \text{ V}}{5 \text{ mA}} = 205 \text{ }\Omega.$$

The resistor R_d is determined by writing a loop equation around the drain–source loop:

$$I_D R_s + V_{DS} + I_D R_d - V_{DD} = 0$$

or

$$R_d = \frac{V_{DD} - V_{DS}}{I_D} - R_s = \frac{15 - 5}{5 \text{ m}} - 205 = 1.795 \text{ k}\Omega.$$

The resistor R_g is typically chosen arbitrarily large:

$$R_g = 1 \text{ M}\Omega.$$

One drawback of the source self-bias circuit is that the quiescent conditions are sensitive to variation in the FET parameters V_{PO} and I_{DSS}. The restriction of Equation 4.6-1,

$$V_{GS} = -I_D R_s,$$

coupled with the FET characteristic equations potentially leads to wide variation in the quiescent condition. This variation is best described graphically as in Figure 4.6-2. In the circuit of Example 4.6-1, an FET with the following parameter variations is used:

$$8 \text{ mA} < I_{DSS} < 12 \text{ mA},$$
$$-4 \text{ V} < V_{PO} < -3 \text{ V}.$$

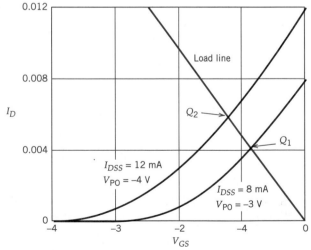

Figure 4.6-2
Change in quiescent point due to FET parameter variation.

If the design procedure results of Example 4.6-1 are used, the resultant ranges in quiescent conditions due to FET parameters variation can be read off Figure 4.6-2. They are

$$-1.20 \text{ V} < V_{GS} < -0.85 \text{ V},$$

$$4.13 \text{ mA} < I_D < 5.87 \text{ mA}.$$

The variation in FET parameters has caused a change in the two quiescent quantities of more than ±17% from the nominal quiescent conditions calculated in Example 4.6-1. In many cases it is necessary to control the quiescent point to a greater degree than is possible with the source self-bias circuit.

4.6.2 Fixed-Bias Circuit

In addition to sensitivity to FET parameter variation, the source self-bias circuit mandates overly restrictive constraints on the external bias resistors in many application. As will be seen in Chapter 5, the resistors R_d and R_s play important roles in determining amplifier gain and output impedance. It is a rare occurrence when quiescent conditions are more significant than these two amplifier performance factors. The restrictive nature of the source self-bias circuit is centered at the hold of the FET gate terminal at ground potential. Removing that restriction greatly improves the versatility of a bias scheme. The fixed-bias circuit of Figure 4.6-3 is a simple depletion mode FET bias circuit that allows the gate to be at voltages other than ground.

Since the gate current in the FET is essentially zero, the two gate resistors R_{g1} and R_{g2} set the voltage at the gate at any desired value between ground and V_{DD}:

$$V_G = \frac{R_{g2}}{R_{g1} + R_{g2}} V_{DD}. \tag{4.6-2}$$

Setting the FET gate voltage at values other than zero allows for a wider possible choice of source and drain resistances to accomplish specific quiescent conditions.

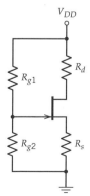

Figure 4.6-3
Fixed-bias circuit.

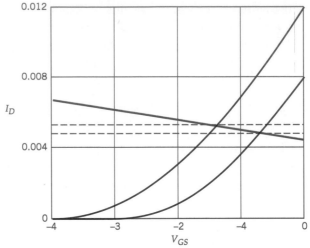

Figure 4.6-4
Change in quiescent point due to FET parameter variation: fixed-bias circuit.

Design choices for these resistors are based primarily on one or more of the following specific criteria:

- Amplification specifications
- Output resistance specifications
- Quiescent point stability

The first two of these design criteria will be discussed in Chapter 5: Quiescent point stability involves reducing the variation in drain current due to FET parameter variation. The relationship of drain current to the external resistors in a fixed-bias circuit is determined by the voltage across the drain resistor:

$$I_D = \frac{V_S}{R_s} = \frac{V_G - (V_G - V_S)}{R_s} = \frac{V_G - V_{GS}}{R_s}, \tag{4.6-3}$$

where V_G is given by Equation 4.6-2. The variation of I_D due to variation of the FET parameter, V_{GS}, is inversely proportional to the value of the source resistor R_S. While other design criteria may put upper limits on R_S, quiescent point stability indicates that a large value for R_s is desirable. The basic load line interpretation of this stability principle is described in Figure 4.6-4.

EXAMPLE 4.6-2   An n-channel JFET with the nominal characteristic parameters

$$V_{PO} = -3.5 \text{ V} \quad \text{and} \quad I_{DSS} = 10 \text{ mA}$$

is to be biased with quiescent conditions of

$$I_D = 5 \text{ mA} \quad \text{and} \quad V_{DS} = 5 \text{ V}.$$

The FET is subject to parameter variation:

$$-4 \text{ V} < V_{PO} < -3 \text{ V}, \qquad 8 \text{ mA} < I_{DSS} < 12 \text{ mA}.$$

Determine the bias resistors necessary to establish the quiescent conditions so that the drain current will not vary more than ±4% due to the FET parameter variation. The power supply is given to be $V_{DD} = 20$ V.

SOLUTION

The nominal quiescent conditions for the FET are identical to those of Example 4.6-1. Thus the value of V_{GS} is obtained in the same manner and is

$$V_{GS(\text{nominal})} = -1.025 \text{ V}.$$

The extreme values of V_{GS} can be obtained either graphically, as in Figure 4.6-4, or analytically using Equation 4.1-11:

$$I_D = I_{DSS}\left(1 - \frac{V_{GS}}{V_{PO}}\right)^2.$$

These extreme values are found to be

$$V_{GS(\text{min})} = -1.37 \text{ V} \qquad (I_{D(\text{max})} = 5.2 \text{ mA}),$$
$$V_{GS(\text{max})} = -0.676 \text{ V} \qquad (I_{D(\text{min})} = 4.8 \text{ mA}).$$

The minimum value of R_s that will satisfy the stability requirements can be calculated as the *maximum* value of

$$R_{s(\text{min})} = \max\left(\frac{\Delta V_{GS}}{\Delta I_D}\right).$$

For this particular case,

$$R_{s(\text{min})} = \max\left(\frac{+1.025 + 1.367}{0.0002} \text{ or } \frac{-1.025 + 0.676}{-0.0002}\right) = 1.75 \text{ k}\Omega.$$

If R_s is chosen to be this minimum value, the nominal voltage at the source and gate of the FET are

$$V_s = I_{D(\text{nominal})}R_s = 8.75 \text{ V}, \qquad V_G = V_S + V_{GS(\text{nominal})} = 7.725 \text{ V}.$$

The gate voltage can be obtained with a combination of a large variety of resistors. One pair that will satisfy the minimal design goals for this case is

$$R_{g1} = 1.62 \text{ M}\Omega, \qquad R_{g2} = 1.02 \text{ M}\Omega.$$

The remaining resistor, R_d, can be determined from the drain–source voltage requirement as

$$R_d = \frac{V_{DD} - V_{DS}}{I_D} - R_s = \frac{20 - 5}{5 \text{ m}} - 1750 = 1.25 \text{ k}\Omega.$$

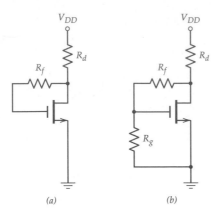

Figure 4.6-5
Additional bias circuits for enhancement mode FETs.

(a)   (b)

4.6.3 Biasing Enhancement Mode FETs

The fixed-bias circuit presented in the preceding section is also useful in biasing enhancement mode FETs. The only significant difference in the procedure necessary to determining resistance values to properly bias the FET is in the initial determination of the quiescent conditions. In enhancement FETs V_{GS} carries the opposite sign as for depletion FETs, and a different relationship between V_{GS} and I_D exists. The proper expression for the FET gate–source voltage in the saturation region is given by Equation 4.2-21:

$$I_D = K(V_{GS} - V_T)^2.$$

Bias stability can be achieved by following the procedure outlined in Example 4.6-2.

The source self-bias circuit of Figure 4.6-1 is not useful for enhancement FETs. However, several other possibilities for biasing this type of FET exist. In addition to the fixed-bias circuit of Figure 4.6-3, two other common bias circuits for enhancement mode FETs are given in Figure 4.6-5. Both circuits offer some bias stabilization, with the circuit of Figure 4.6-5b offering superior stability and flexibility in resistor value choice. The connection in these two circuits between the gate and drain through resistor R_f provides a signal path between input and output when the FET is used in amplifier applications. This path provides advantages and disadvantages that will be discussed in Chapter 8, Feedback Amplifier Principles.

4.7 CONCLUDING REMARKS

The FET has been described in this chapter as a highly useful device with three basic regions of operation: the saturation, ohmic, and cutoff regions. Entrance into each of these regions is controlled by two voltages: the gate–source voltage and the drain–source voltage.

The large-signal characterization of an FET is non-linear. While it can be modeled through local linearization, these linear models are highly dependent on the point of linearization. Therefore, unlike the BJT, the FET cannot easily be described with a unique, simple linear model in each of its regions of operation. The design engineer must rely on a set of non-linear analytic expressions for FET description.

Switching applications, including logic gate applications, are achieved with the

FET in transitioning between the cutoff and ohmic region. Linear applications take place in the saturation region. While a variety applications have been examined in this chapter, the possibilities for circuitry using FETs extend far beyond what has been shown here. Many additional FET linear applications and several non-linear applications will be examined in later chapters. Additional restrictions placed on FET circuit design by frequency response limitations will be discussed in Chapter 9.

SUMMARY DESIGN EXAMPLE

In order to provide bidirectional current to a DC motor, the motor is connected to an "H-driver" circuit. The basic topology of such a circuit contains four switches as shown.

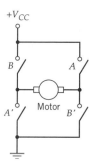

The H-driver switches operate in counteracting pairs: Only one pair is closed at a time. When the A–A' pair is closed, current flows through the motor from right to left (in the diagram): Closing the B–B' pair reverses the current flow.

It is necessary to design a circuit to control a small DC motor bidirectionally. Control signals will be provided by standard CMOS logic levels (high $\approx$ 5 V, low $\approx$ 0 V). The voltage and current ratings for the motor are as follows:

Voltage: 3–15 V

Maximum current: 300 mA

Design an H driver to provide proper power to the motor when control signals are applied.

Solution

The switches in an H driver can be either mechanical or solid state. Because of the small currents being switched and the complexity necessary for control of mechanical switches, solid-state switching is a good choice for this design. Field-effect transistor switches are the preferred choice in most solid-state switching applications.

The best FET H driver is essentially formed from two counteracting CMOS gates (with appropriate, high-current FETs) and a buffer FET. The basic topology is shown in the accompanying diagram.

Each complementary pair of FETs (Q_A–$Q_{B'}$ and Q_B–$Q_{A'}$) acts as a counteracting switch, and the interconnection ensures that Q_A, $Q_{A'}$ and Q_B, $Q_{B'}$ act simultaneously. The buffer FET, Q_C, provides 0 V and $+ V_{CC}$ to ensure accurate switching. The switching FETs must be enhancement mode FETs with a threshold voltage significantly less than the minimum applied voltage and with maximum current capability in excess of 300 mA. The buffer FET Q_C must also be a similar enhancement

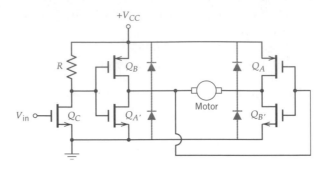

mode FET, but with smaller current capability. Resistance R should be small compared to the input resistance of $Q_{A'}$ and Q_B but large to avoid wasting power.

A summary of appropriate design choices is as follows:

Q_A, Q_B: p-channel, enhancement MOSFETs, $V_T \approx -1$ V, $I_{D(\text{max})} \approx -0.5$ A

$Q_{A'}$, $Q_{B'}$: n-channel, enhancement MOSFETS, $V_T \approx 1$ V, $I_{D(\text{max})} \approx 0.5$ A

Q_C: n-channel, enhancement signal MOSFET, $V_T \approx 1$ V

R: 100 kΩ

It is common when switching loads that have a significant inductive component (i.e., motors) to shunt the output of a switch with a reverse-biased diode (shaded in the diagram). This diode provides a current path during the switching transition. For such a small motor, these diodes are probably not necessary. Many switching FETs have diodes incorporated into their design.

REFERENCES

Analog IC Data Book, Precision Monolithics, Inc., Santa Clara, 1990.

Integrated Circuits Data Book, Siliconix Inc., Santa Clara, 1988

Baliga, B. J. and Chen, D. Y., editors, *Power Transistors: Device Design and Applications*, IEEE Press, New York, 1984.

Colclaser, R. A. and Diehl-Nagle, S., *Materials and Devices for Electrical Engineers and Physicists*, McGraw-Hill Book Company, New York, 1985.

Ghausi, M. S., *Electronic Devices and Circuits: Discrete and Integrated*, Holt, Rinehart and Winston, New York, 1985.

Gray, P. R., and Meyer, R. G., *Analysis and Design of Analog Integrated Circuits*, 3rd. Ed., John Wiley & Sons, Inc., New York, 1993.

Horowitz, P., and Hill, W., *The Art of Electronics*, 2nd. Ed. Cambridge University Press, Cambridge, 1992.

Millman, J., *Microelectronics, Digital and Analog Circuits and Systems*, McGraw-Hill Book Company, New York, 1979.

Sedra, A. S. and Smith, K. C., *Microelectronic Circuits*, 3rd. Ed., Holt, Rinehart, and Winston. Philadelphia, 1991.

Tuinenga, P., *SPICE: A Guide to Circuit Simulation and Analysis Using PSpice*, 2nd. Ed., Prentice Hall, Englewood Cliffs, 1992.

PROBLEMS

4-1. An n-channel JFET is described by the following parameters:

$I_{DSS} = 4.5$ mA, $\qquad V_{PO} = -3.6$ V.

    a. If the JFET is in saturation, what gate-to-source voltage V_{GS} is necessary to achieve a drain current of 2.6 mA?

    b. What is the minimum V_{DS} that will satisfy the conditions stated in part a?

    c. If $V_{DS} = 2$ V, what gate-to-source voltage is necessary to achieve the same drain current?

4-2. A p-channel JFET is described by the following parameters:

$I_{DSS} = -4.0$ mA, $\qquad V_{PO} = +2.8$ V.

    a. If the JFET is in saturation, what source-to-gate voltage V_{SG} is necessary to achieve a drain current of -1.8 mA?

    b. What is the minimum V_{SD} that will satisfy the conditions stated in part a?

    c. If $V_{SD} = 1.5$ V, what gate-to-source voltage is necessary to achieve the same drain current?

 4-3. The parameters for a given JFET are

$\qquad I_{DSS} = 7.5$ mA, $\qquad V_{PO} = -4$ V.

The JFET is to be biased at

$I_D = 2$ mA, $\qquad V_{DS} = 6$ V,

with the circuit topology as shown.

    Determine the values of R_D and R_S to complete the design if $V_{DD} = 20$ V.

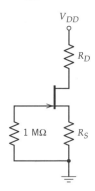

4-4. Given the n-channel JFET circuit shown. If the JFET is described by

$V_{PO} = -2.5$ V and $I_{DSS} = 4$ mA,

find I_D and V_{DS}.

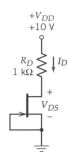

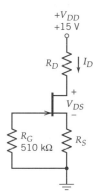 **4-5.** Complete the design of the n-channel JFET circuit shown for $I_D = 3$ mA and $V_{DS} = 5$ V. The JFET parameters are

$I_{DSS} = 7$ mA, $\qquad V_{PO} = -2.5$ V.

4-6. An n-channel, depletion-type MOSFET is described by the following parameters:

$I_{DSS} = 8.2$ mA, $\qquad V_{PO} = -3.1$ V.

    a. If the NMOSFET is in saturation, what gate-to-source voltage V_{GS} is necessary to achieve a drain current of 4.0 mA?

    b. What is the minimum V_{DS} that will satisfy the conditions stated in part a?

    c. If $V_{DS} = 2$ V, what gate-to-source voltage is necessary to achieve the same drain current?

    d. What is the output resistance of the NMOSFET at the conditions of part c?

4-7. An n-channel MOSFET has the following characteristics:

$V_{PO} = -3$ V, $\qquad V_A = 170$ V, $\qquad I_{DSS} = 8$ mA.

a. Determine the minimum drain–source voltage V_{DS} for the MOSFET to be in saturation.

b. Determine the output resistance of the MOSFET in the ohmic region when $V_{GS} = -1.5$ V.

c. Determine the output resistance of the MOSFET in the saturation region when $I_D = 2$ mA.

4-8. Assume an n-channel depletion-type MOSFET with the following parameters:

$$I_{DSS} = 10 \text{ mA}, \qquad V_{PO} = -5 \text{ V}.$$

Determine V_O and I_D for the circuit shown.

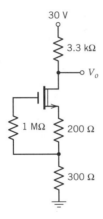

4-9. Find the drain current, drain–source voltage, and gate–source voltage for the given circuit. Assume $K = 0.25$ mA/V^2 and $V_T = 2.5$ V.

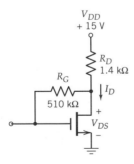

4-10. Determine the Q point for the circuit shown. Assume the p-channel MOSFET is described by

$$K = 0.3 \text{ mA/V}^2, \qquad V_T = -2.2 \text{ V}.$$

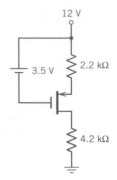

4-11. An n-channel, enhancement-type MOSFET is described by the following parameters:

$$K = 2.4 \text{ mA/V}^2, \qquad V_T = 1.2 \text{ V}.$$

a. If the NMOSFET is in saturation, what gate-to-source voltage V_{GS} is necessary to achieve a drain current of 4.0 mA?

b. What is the minimum V_{DS} that will satisfy the conditions stated in part a?

c. If $V_{DS} = -1.2$ V, what gate-to-source voltage is necessary to achieve the same drain current?

d. What is the output resistance of the NMOSFET at the conditions of part c?

4-12. For the n-channel MOSFET circuit shown, determine the drain current I_D and drain–source voltage V_{DS} using the following methods:

a. Analytical method: Use equations.

b. Load line analysis: Use SPICE to arrive at the transistor characteristic curves.

The MOSFET parameters are

$$V_T = 1.5 \text{ V}, \qquad V_A = 170 \text{ V},$$

$$K = 1.2 \text{ mA/V}^2.$$

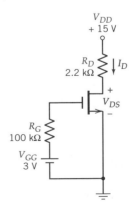

4-13. Design a circuit to bias an *n*-channel depletion MOSFET with $I_{DSS} = 8$ mA and $V_{PO} = -4$ V using the "bootstrapping" configuration shown. The design specifications require that $I_D = 4$ mA and *bias technique is used to preserve the high input resistance of the circuit.*

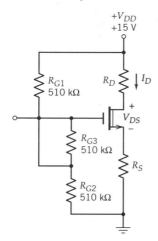

4-14. Complete the design of the *n*-channel depletion MOSFET circuit shown so that

$$I_D = 2 \text{ mA}, \qquad V_{DS} = 4 \text{ V}.$$

The MOSFET parameters are

$$I_{DSS} = 5 \text{ mA}, \qquad V_{PO} = -3 \text{ V}.$$

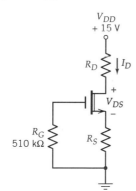

4-15. For the circuit shown, the *n*-channel JFET is described by

$$I_{DSS} = 8 \text{ mA}, \qquad V_{PO} = -6.5 \text{ V}.$$

    a. Complete the design of the circuit to achieve a *Q* point of

$$I_D = 6 \text{ mA},$$

by determining R_{G1} and R_{G2} for $R_{G1} \| R_{G2} = 54 \pm 5$ kΩ. Calculate V_{GS} and V_{DS}.

b. The JFET is replaced by a *p*-channel JFET with $I_{DSS} = -8$ mA and $V_{PO} = 4$ V. Draw the *p*-channel JFET circuit diagram so that the FET is biased in the saturation region. Find the *Q* point for the JFET (V_{DS}, I_D, and V_{GS}) using the resistor values found in part a.

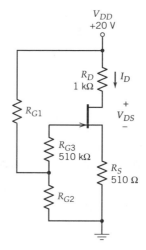

4-16. Complete the design of the *n*-channel enhancement MOSFET circuit shown so that

$$I_D = 2 \text{ mA},$$

The MOSFET parameters are

$$K = 1.3 \text{ mA/V}^2, \qquad V_T = 2 \text{ V}.$$

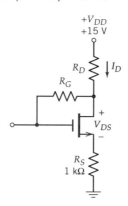

4-17. Use PSpice to generate the transistor characteristic curves for a *p*-channel depletion MOSFET with parameters

$$V_{PO} = 4 \text{ V}, \qquad I_{DSS} = -7 \text{ mA}$$

over the range $0 \le V_{SD} \le 15$ V.

Using the curves generated, determine the FET drain current and the V_{DS} in the circuit shown.

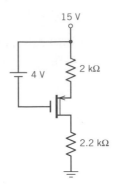

15 V

4 V   2 kΩ

2.2 kΩ

4-18. For the p-channel MOSFET circuit shown, determine the drain current I_D and drain–source voltage V_{DS} using the following methods:

a. Analytical method: Use equations.

b. Load line analysis: Use SPICE to arrive at the transistor characteristic curves.

The MOSFET parameters are

$V_{PO} = 2.5$ V, $V_A = 150$ V,

$I_{DSS} = -10$ mA.

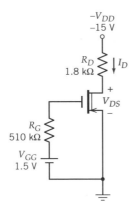

$-V_{DD}$
-15 V

R_D
1.8 kΩ I_D

$+$
V_{DS}
$-$

R_G
510 kΩ

V_{GG}
1.5 V

4-19. For the p-channel MOSFET circuit shown, determine the drain current I_D and drain–source voltage V_{DS} using the following methods:

a. Analytical method: Use equations.

b. Load line analysis: Use SPICE to arrive at the transistor characteristic curves.

The MOSFET specifications are

$V_{PO} = 3.5$ V, $V_A = 150$ V,

$I_{DSS} = -10$ mA.

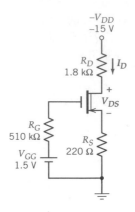

$-V_{DD}$
-15 V

R_D
1.8 kΩ I_D

$+$
V_{DS}
$-$

R_G
510 kΩ

R_S
220 Ω

V_{GG}
1.5 V

4-20. Plot the input and output characteristics of an n-channel enhancement MOSFET with $K = 1.2 \times 10^{-3}$ A/V^2 and $V_T = 3.5$ V. Design a circuit so that the MOSFET is biased at $V_{DS} = 5$ V and $I_D = 1$ mA with $V_{DD} = 10$ V.

4-21. Plot the input and output characteristics of an n-channel enhancement MOSFET with $K = 1.2 \times 10^{-3}$ A/V^2 and $V_T = 3.5$ V. Design a circuit so that the MOSFET is biased at $V_{DS} = 3.5$ V and $I_D = 3$ mA with $V_{DD} = 10$ V. Compare the region of FET operation with that of the previous problem.

4-22. Plot the input and output characteristics of a p-channel enhancement MOSFET with $K = 1.2 \times 10^{-3}$ A/V^2 and $V_T = -2$ V. Design a circuit so that the MOSFET is biased at $V_{SD} = 6$ V and $I_D = -1$ mA with $V_{DD} = -12$ V.

4-23. Plot the input and output characteristics of a p-channel enhancement MOSFET with $K = 1.2 \times 10^{-3}$ A/V^2 and $V_T = -2$ V. Design a circuit so that the MOSFET is biased at $V_{SD} = 3.5$ V and $I_D = -3$ mA with $V_{DD} = -12$ V. Compare the region of FET operation with that of the previous problem.

4-24. For the circuit shown, the MOSFET is described by

$I_{DSS} = 8$ mA, $I_D = 4$ mA, $V_{DS} = 8$ V,

$V_{PO} = -5$ V.

a. Find R_D, and R_{G1} and R_{G2} for $R_G = 1$ MΩ.

b. The MOSFET is replaced by one of the following parameters:

$I_{DSS} = 10$ mA, $V_{PO} = -6$ V.

Find the new Q point.

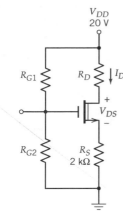

4-25. The output current of an FET current sources can be adjusted through the use of a resistor at the source of the FET as shown. The JFET parameters are given as

$I_{DSS} = 5$ mA, $V_{PO} = -2.5$ V.

For the given circuit:

    a. Determine the resistor value to obtain an output current I_D of 2 mA.

    b. Determine the resistor value to obtain an output current I_D of 3 mA.

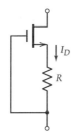

4-26. Current sources can be realized with p-channel as well as n-channel devices. The p-channel MOSFET parameters are given as

$I_{DSS} = -5$ mA, $V_{PO} = 2.0$ V.

For the given circuit:

    a. Determine the resistor value to obtain an output current I_O of 2 mA.

    b. Determine the resistor value to obtain an output current I_O of 3 mA.

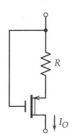

4-27. Given two identical n-channel JFETs with parameters

$I_{DSS} = 6$ mA   and $V_{PO} = -2.5$ V,

determine V_{DS1}, V_{DS2}, and I_{D2} for the circuit shown.

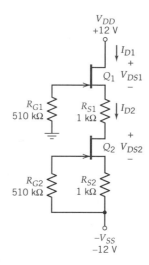

4-28. Find the output voltage V_O of the circuit shown for the following FET choices:

    a. Given two identical n-channel MOSFETs with parameters

$V_T = 2.5$ V, $K = 0.15$ mA/V^2.

    b. Given two different n-channel MOSFET with parameters

$V_{T1} = 1.5$ V, $V_{T2} = 3$ V, $K = 0.15$ mA/V^2.

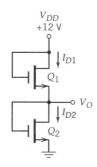

4-29. Given the voltage division circuit shown, the MOSFET is described by

$K = 1$ mA/V^2, $V_T = 2$V, $V_A = 100$ V.

What input voltage will result in an output voltage of 1.0 V?

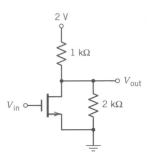

4-30. Given the voltage division circuit shown, the MOSFET is described by

$$I_{DSS} = 4 \text{ mA}, \qquad V_{PO} = -2 \text{ V}, \qquad V_A = 100 \text{ V}.$$

What input voltage will result in an output voltage of 1.0 V?

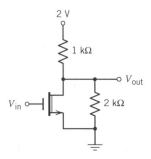

4-31. An n-channel enhancement MOS saturated load is driven with an n-channel enhancement MOS driver as shown. The input and output characteristics are shown in Figure 4.2-9 and 4.2-10. Show the load line and sketch the voltage transfer characteristic (V_O vs. V_I).

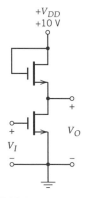

4-32. Use SPICE to determine the on and off resistance of the analog FET switch shown in Figure 4.5-1 as a function of the input voltage, v_s. The FET is described by the parameters

$$K = 3.0 \text{ mA/V}^2, \qquad V_T = 2.5 \text{ V},$$

$$V_A = 100 \text{ V}.$$

Assume the load resistance $R = 100 \ \Omega$, the input voltage exists over the range $0 \text{ V} < v_s < +10 \text{ V}$, and the control voltage v_c switches between 0 and $+10$ V.

4-33. Use SPICE to determine the on and off resistance the analog FET switch shown in Figure 4-5.3 as a function of the input voltage v_s. The FETs are described by the parameters:

$$K = 3.0 \text{ mA/V}^2, \qquad |V_T| = 2.5 \text{ V},$$

$$|V_A| = 100 \text{ V}.$$

Assume the load resisatnce $R = 200 \ \Omega$, the input voltage exists over the range $0 \text{ V} < v_s < +10 \text{ V}$, and the control voltage v_c switches between 0 and $+10$ V.

4-34. An n-channel MOSFET with the nominal characteristic parameters

$$V_{PO} = -2 \quad \text{and} \quad I_{DSS} = 8 \text{ mA}$$

is to be biased with quiescent conditions of

$$I_D = 3 \text{ mA} \quad \text{and} \quad V_{DS} = 4 \text{ V}$$

using resistors and a single 20-V power supply. The FET is subject to parameter variation:

$$-2.5 < V_{PO} < -1.5,$$

$$7 \text{ mA} < I_{DSS} < 10 \text{ mA}.$$

Determine the bias resistors necessary to establish quiescent conditions so that the drain current will not vary more than $\pm 2.5\%$ due to the FET parameter variation. Verify that your design meets specifications using PSpice.

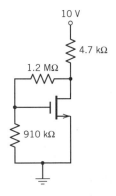

4-35. The n-channel MOSFET in the circuit shown is described by the parameters

$$V_T = 1.5 \text{ V} \quad \text{and} \quad K = 0.5 \text{ mA/V}^2.$$

Determine I_D and V_{DS} analytically. Verify your results using PSpice.

4-36. Design a bias circuit to achieve a quiescent condition of $I_D = 6$ mA and $V_{DS} = 6$ V using a single 20-V power supply and a FET with the following parameters:

$$V_{PO} = -2 \text{ V}, \qquad I_{DSS} = 12 \text{ mA}.$$

It is required that the drain of the FET be connected to the 20-V supply through a 2-kΩ resistor ($R_D = 2$ kΩ).

4-37. Design an FET constant-current source for the circuit configuration shown. Determine all resistance values so that

$$V_{CE} = 3 \text{ V} \quad \text{and} \quad I_C = 5 \text{ mA}.$$

The following components are available:

npn BJT	2N2222
Depletion NMOSFET	$I_{DSS} = 5$ mA, V_{PO} $= -3.5$ V

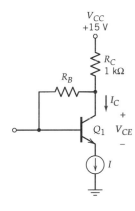

4-38. Design an inverting DC opamp amplifier with a voltage variable gain. The maximum required gain magnitude is 10. The following components are available:

n-JFET: $I_{DSS} = 8$ mA, $V_{PO} = -4$ V

Near-ideal opamp: μA741CN

Power supply: 0 to ±15 V

Standard value resistors

Show complete analysis of the design.

4-39. An *n*-channel MOSFET with the nominal characteristic parameters

$$V_T = 2 \text{ V} \quad \text{and} \quad K = 1.25 \text{ mA/V}^2$$

is required to be biased at the following quiescent conditions:

$$I_D = 2 \text{ mA} \quad \text{and} \quad V_{DS} = 3 \text{ V}.$$

The FET is subjected to the following parameter variations:

$$1 < V_T < 3 \text{ V},$$

$$1.00 \text{ mA/V}^2 < K \quad < 1.5 \text{ mA/V}^2.$$

The power supply voltage provided is +20 V.

Design a circuit to bias the FET so that the drain current does not vary more than ±10% due to the FET parameter variations. Verify that the design meets the stability specification using SPICE

4-40. A *p*-channel JFET with the nominal characteristic parameters

$$V_{PO} = 3 \text{ V} \quad \text{and} \quad I_{DSS} = -12 \text{ mA}$$

is required to be biased at the quiescent conditions

$$I_D = -2 \text{ mA} \quad \text{and} \quad V_{DS} - 5 \text{ V}.$$

The JFET is subjected to the following parameter variations:

$$2.5 < V_{PO} < 3.5 \text{ V},$$

$$-11 \text{ mA} < I_{DSS} < -13 \text{ mA}.$$

The power supply voltage provided is −24 V.

Design a circuit to bias the FET so that the drain current does not vary more than ±5% due to the FET parameter variations. Verify that the design meets the stability specification using SPICE

3,747,005
AUTOMATIC BIASED CONTROLLED AMPLIFIER
Ronald J. Freimark, Addison, and Ole K. Nilssen, Barrington
Hills, both of Ill., assignors to Motorola, Inc., Franklin, Ill.
Filed Feb. 1, 1971, Ser. No. 111,455
Int. Cl. H03f *3/04*

U.S. Cl. 330—22

5 Claims

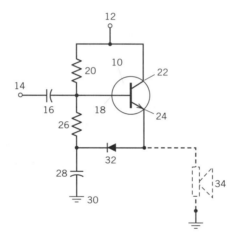

   An audio power amplifier comprising a transistor operated as an emitter follower which is quiescently biased so that the DC current through the output speaker is small. A feedback circuit connected between the output and the control electrode of the transistor compares the magnitude of the output peak voltage with the magnitude of the bias potential on the control electrode. When the peak output voltage exceeds the magnitude of the bias potential the feedback circuit increases the control bias potential to raise the operating point of the transistor. Where the magnitude of the peak voltage of the output does not exceed the magnitude of the bias potential, there is no feedback and the bias potential is maintained to keep the transistor at its quiescent operating point.

Linear Amplifiers: Small-Signal Analysis

A basic understanding of the operation of the four basic active elements has provided a good foundation for further study of electronic circuitry. Analog amplifiers constitute a major class of electronic circuitry and are a primary component in many applications. As the title of this section suggests, amplifiers typically operate within a region of incremental linearity. That is, a region where small variation in the input produces linearly amplified variation in the output. Analysis of such systems uses the principle of superposition: DC (or bias) conditions are separated from the AC (or variational) components of the input and output of an amplifier. The term *small-signal analysis* refers to the use of linear models. *Large-signal analysis* implies operation near the transition between operational regions of an active device: Such operation is typically non-linear and leads to distorted amplification.

The section begins with a review of two-port analysis. This review provides a basis for modeling transistors: They are most commonly modeled as two-port networks for small-signal analysis. At low frequencies the BJT is modeled by an *h*-parameter two-port and the FET as a modified hybrid-π two-port. Simple amplifiers are approached by observing the previously observed region of operation that appeared between the two logic states of an inverter. All single-transistor amplifier configurations are analyzed and performance characteristics compared.

Multiple-transistor amplifier circuitry is initially approached through cascading single-transistor amplifiers using capacitive coupling. Only after the basic concepts are mastered are the more complex circuits studied. Compound transistor configurations, such as the Darlington circuit, and direct-coupled amplifier stages are studied. As with single-transistor amplifiers, the previously observed linear region in ECL circuits leads to the study of emitter-coupled and source-coupled amplifiers. Common integrated circuit practices such as current source biasing and active loads are discussed.

Power amplifiers provide a good counterexample to the use of small-signal analysis. By necessity, the output of a power amplifier is not small and consequently may contain distorted components. Both harmonic and intermodulation distortion analysis techniques are introduced and compared. Amplifier conversion efficiency is discussed for class A, B, and AB power amplifiers. Thermal considerations are presented using simple heat transfer models and are related to power amplifier design criteria and limitations.

Feedback principles are introduced initially as a technique to stabilize amplifier gain, reduce distortion, and control impedance. The various configurations are introduced and analyzed through two-port network analysis techniques. Improvement in frequency response is mentioned as an additional benefit but discussion is left for the next section.

5

SINGLE-TRANSISTOR AMPLIFIERS

Circuits containing single BJTs and FETs are capable of providing amplification in six basic configurations: three configurations for each type of transistor. In the common-emitter and common-collector configurations for BJTs, the input signal is injected at the base of the transistor: in the common-base configuration it is injected at the emitter. Similarly, FET configurations have the input injected in the gate for common-source and common-drain configurations, and at the source for the common-gate configuration. These six single-transistor configurations are basic building blocks of amplifier design.

The focus of attention in this chapter is on the significant performance characteristics of each amplifier configuration: the voltage gain, current gain, input resistance, and output resistance. Each configuration has a unique set of performance characteristics that are dependent on the transistor characteristics as well as the configuration and value of surrounding circuit elements. This chapter primarily investigates discrete amplifiers where biasing is accomplished with voltage sources and resistances: Later chapters will highlight BJT biasing with current sources as developed in Section 3.6. The performance of transistor amplifiers with current source biasing can be directly derived from the results presented in this chapter. The frequency range of interest is the so-called midband region: that range of frequencies below which the transistor characteristics begin to change with increasing frequency and above which any capacitive elements (many circuits have *no* capacitors and therefore the frequency range begins at DC) have significant effect on the circuit performance.

Although amplifiers are usually comprised of more than one amplification stage, at least one of these six configurations is contained as an amplification stage in nearly every amplifier. The properties derived here are used in the analysis and design of multistage amplifiers presented in later chapters.

5.1 REVIEW OF TWO-PORT NETWORK BASICS

Electronic amplifiers are a subset of the system class commonly identified as *two-port networks*. In a two-port electronic network, signals are fed into a pair of terminals, amplified and/or modified by the system, and finally extracted at another pair of terminals. Each pair of terminals is identified as a port: Signals are fed into an *input port* and extracted from an *output port*. The modeling and analysis of transistor-based amplifiers as well as feedback systems[1] is greatly simplified through the use of two-port network principles.

---

[1] The interconnection of two-port networks as used in the analysis of feedback amplifiers will be discussed in Chapter 8.

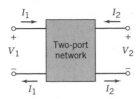

Figure 5.1-1
Two-port network sign conventions.

The items of interest in an electronic two-port network are the relationships between input port and output port voltages and currents. There are a few restrictions, upon which two-port analysis techniques are based, that must be identified:

- The network must be linear and time invariant.
- External connections may be made only to port terminals: No external connections can be made to any node internal to the port.
- All current entering one terminal of a port must exit the other terminal of that port.
- Sources and loads must be connected directly across the two terminals of a port.

Given the highly non-linear behavior of transistors, it may seem unusual to attempt to use two-port network analysis to describe transistor systems. It is possible, under small-signal conditions, however, to adequately model non-linear systems as *incrementally* linear. Electronic circuits previously discussed in this text have shown, along with non-linear operation, regions of linear operation. It is within these regions of linear operation that two-port analysis proves a particularly useful technique for the modeling of electronic systems.

As with all electronic systems, it is important to define sign conventions: Figure 5.1-1 is a representation of a two-port network with appropriate voltage and current polarity definitions. As is standard in electronic systems, the voltage and current at each port obey the passive sign convention. Typically, the input port is identified as port 1 and the output port as port 2.[2]

Three-terminal devices, such as transistors, can also be modeled using two-port techniques. One terminal is selected as a common terminal: that terminal is extended to both of the ports and becomes the reference (negative) terminal to each port.

There are six basic sets of equivalent descriptive parameters for every two-port network:

1. Impedance parameters (z parameters): port voltages in terms of port currents
2. Admittance parameters (y parameters): port currents in terms of port voltages
3. Hybrid parameters (h parameters): input voltage and output current in terms of input current and output voltage
4. Hybrid parameters (g parameters): input current and output voltage in terms of input voltage and output current

---

[2] The numbering of ports is not universal in acceptance. It does, however, allow for simple notation in the description of the parameters relating currents and voltages. In some electronic two-port descriptions of systems, the numbering system is replaced by a more descriptive identification of parameters. One common example of descriptive notation occurs in h-parameter modeling of BJTs as presented in Section 5.2.

5. Transmission parameters (*ABCD* parameters): input current and voltage in terms of output current and voltage

6. Transmission parameters (*𝒜ℬ𝒞𝒟* parameters): output current and voltage in terms of input current and voltage

The first four of these sets are of particular interest in the study of electronic feedback systems, set numbers three and four (hybrid parameters) are often used in the description of transistor properties, and the last two sets of parameters are particularly useful in the study of communication transmission systems. A short description of the first four two-port parameter set description follows.

Impedance Parameters (z Parameters)

The independent variables for this set of parameters are the port currents and the dependent variables are the port voltages: voltage as a function of current is an impedance. It is most common to write the equations in matrix form:

$$\begin{bmatrix} V_1 \\ V_2 \end{bmatrix} = \begin{bmatrix} z_{11} & z_{12} \\ z_{21} & z_{22} \end{bmatrix} \begin{bmatrix} I_1 \\ I_2 \end{bmatrix}. \tag{5.1-1}$$

The parameters $\{z_{ij}\}$ are called the impedance (or z) parameters of the network. For a linear, time-invariant network, the z parameters can be obtained by performing simple tests on the network:

$$z_{ij} = \left. \frac{V_i}{I_j} \right|_{I_{k \neq j} = 0}. \tag{5.1-2}$$

For a non-linear system that is operating in a region of linearity, a similar definition holds:

$$z_{ij} = \left. \frac{\partial V_i}{\partial I_j} \right|_{I_{k \neq j} = \text{const}}, \tag{5.1-3}$$

where the constant value of I_k is taken near the midpoint of the region of linearity. If a quiescent (zero-input) point exists, the constant value of I_k is chosen as its quiescent value.

EXAMPLE 5.1-1 Determine the z parameters for the given two-port network.

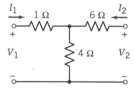

SOLUTION

The z parameters are defined for this linear system as open-circuit parameters: One of the currents is always set to zero. Zero current implies an open circuit in the appropriate path. Thus,

$$z_{11} = 1 + 4 = 5 \ \Omega \ \text{(zero current in 6-}\Omega\text{ resistor)},$$

$$z_{12} = 4 \ \Omega \qquad \text{(no voltage drop across 1-}\Omega\text{ resistor)},$$

$$z_{21} = 4 \ \Omega \qquad \text{(no voltage drop across 6-}\Omega\text{ resistor),}$$

$$z_{22} = 6 + 4 = 10 \ \Omega \quad \text{(zero current in 1-}\Omega\text{ resistor).}$$

Admittance Parameters (y Parameters)

Admittance parameters are defined with the independent variables are the port voltages and the dependent variables are the port currents: Current as a function of voltage carries the units of admittance. The y-parameter equations are

$$\begin{bmatrix} I_1 \\ I_2 \end{bmatrix} = \begin{bmatrix} y_{11} & y_{12} \\ y_{21} & y_{22} \end{bmatrix} \begin{bmatrix} V_1 \\ V_2 \end{bmatrix}. \tag{5.1-4}$$

The parameters can be determined by performing the tests

$$y_{ij} = \frac{I_i}{V_j} \bigg|_{V_{k \neq j} = 0} \tag{5.1-5}$$

or

$$y_{ij} = \frac{\partial I_i}{\partial V_j} \bigg|_{V_{k \neq j} = \text{const}}. \tag{5.1-6}$$

In a linear system the test for finding y parameters indicates that a voltage must be set to zero: Thus, the parameters are often called *short-circuit* admittance parameters.

EXAMPLE 5.1-2 Determine the y parameters for the network shown using standard phasor techniques, that is, find $Y(s)$.

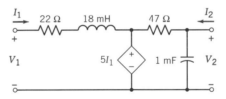

SOLUTION

It can be seen that this network is a linear system. Both two-port and phasor techniques are appropriate for such a system. The phasor equivalent impedance of the inductor and the capacitor are

$$Z_L = 0.018s, \qquad Z_C = \frac{1000}{s}$$

The y-parameter tests for a linear system are given by Equation 5.1-5:

$$y_{ij} = \frac{I_i}{V_j} \bigg|_{V_{k \neq j} = 0}.$$

In order to solve for y_{11} and y_{21}, the output terminals are short-circuited ($V_2 = 0$). A loop equation can then be written around the remaining left loop:

$$V_1 - 22I_1 - 0.018sI_1 - 5I_1 = 0,$$

which leads to

$$y_{11} = \left.\frac{I_1}{V_1}\right|_{V_2=0} = \frac{1}{27 + 0.018s} .$$

Since the capacitor has been shorted out by setting $V_2 = 0$, a loop equation can be written around the remaining right loop:

$$5I_1 + 47I_2 = 0 \Rightarrow I_2 = -0.1064I_1$$

and

$$y_{21} = \left.\frac{I_2}{V_1}\right|_{V_2=0} = \frac{-1}{253 + 0.1692s} .$$

The other two parameters, y_{12} and y_{22}, are obtained by shorting the input terminal ($V_1 = 0$). A loop around the left loop of this configuration yields

$$22I_1 + 0.018sI_1 + 5I_1 = 0 \Rightarrow I_1 = 0.$$

Obviously

$$y_{12} = \left.\frac{I_1}{V_2}\right|_{V_1=0} = 0.$$

With $I_1 = 0$, the current I_2 is the sum of the currents in the 47-Ω resistor and the capacitor (the dependent voltage source has zero value):

$$y_{22} = \left.\frac{I_2}{V_2}\right|_{V_1=0} = y_{47\Omega} + y_{1mF} = 0.02128 + 0.001s.$$

The y-parameter matrix is then given by

$$Y(s) = \begin{bmatrix} \dfrac{1}{27 + 0.018s} & 0 \\ \dfrac{-1}{253 + 0.1692s} & 0.02128 + 0.001s \end{bmatrix}.$$

Hybrid Parameters (h Parameters)

The independent variables are the input voltage and output current; the dependent variables are the input current and output voltage:

$$\begin{bmatrix} V_1 \\ I_2 \end{bmatrix} = \begin{bmatrix} h_{11} & h_{12} \\ h_{21} & h_{22} \end{bmatrix} \begin{bmatrix} I_1 \\ V_2 \end{bmatrix}. \tag{5.1-7}$$

Notice that these parameters, unlike the z and y parameters, do not all have the same dimensions. Each is different: h_{11} is the input port impedance; h_{12} is a dimensionless (input over output) voltage ratio; h_{21} is a dimensionless (output over input) current ratio; and h_{22} is output port admittance.

The parameters can be determined by performing the following tests:

$$h_{11} = \frac{V_1}{I_1}\bigg|_{V_2=0}, \qquad h_{12} = \frac{V_1}{V_2}\bigg|_{I_1=0},$$

(5.1-8)

$$h_{21} = \frac{I_2}{I_1}\bigg|_{V_2=0}, \qquad h_{22} = \frac{I_2}{V_2}\bigg|_{I_1=0},$$

or

$$h_{11} = \frac{\partial V_1}{\partial I_1}\bigg|_{V_2=\text{const}}, \qquad h_{12} = \frac{\partial V_1}{\partial V_2}\bigg|_{I_1=\text{const}},$$

(5.1-9)

$$h_{21} = \frac{\partial I_2}{\partial I_1}\bigg|_{V_2=\text{const}}, \qquad h_{22} = \frac{\partial I_2}{\partial V_2}\bigg|_{I_1=\text{const}}.$$

Hybrid Parameters (g Parameters)

The independent variables are the input current and output voltage; the dependent variables are the input voltage and output current.

$$\begin{bmatrix} I_1 \\ V_2 \end{bmatrix} = \begin{bmatrix} g_{11} & g_{12} \\ g_{21} & g_{22} \end{bmatrix} \begin{bmatrix} V_1 \\ I_2 \end{bmatrix}.$$

(5.1-10)

Notice that these parameters, like the h parameters, do not all have the same dimensions. Each is different: g_{11} is the input port admittance; g_{12} is a dimensionless (input over output) current ratio; g_{21} is a dimensionless (output over input) voltage ratio; and g_{22} is output port impedance.

The parameters can be determined by performing the following tests:

$$g_{11} = \frac{I_1}{V_1}\bigg|_{I_2=0}, \qquad g_{12} = \frac{I_1}{I_2}\bigg|_{V_1=0},$$

(5.1-11)

$$g_{21} = \frac{V_2}{V_1}\bigg|_{I_2=0}, \qquad g_{22} = \frac{V_2}{I_2}\bigg|_{V_1=0},$$

or

$$g_{11} = \frac{\partial I_1}{\partial V_1}\bigg|_{I_2=\text{const}}, \qquad g_{12} = \frac{\partial I_1}{\partial I_2}\bigg|_{V_1=\text{const}},$$

(5.1-12)

$$g_{21} = \frac{\partial V_2}{\partial V_1}\bigg|_{I_2=\text{const}}, \qquad g_{22} = \frac{\partial V_2}{\partial I_2}\bigg|_{V_1=\text{const}}.$$

TABLE 5.1-1 TWO-PORT NETWORK REPLACEMENT CIRCUIT ELEMENTS FOR NETWORK PARAMETERS

	Replacement Circuit Element	
Network Parameter	Same Port	Opposite Port
Impedance	Impedance	Current-controlled voltage source
Admittance	Admittance	Voltage-controlled current source
Current ratio	(Not applicable)	Current-controlled current source
Voltage ratio	(Not applicable)	Voltage-controlled voltage source

5.1.1 Circuit Representation of Two-Port Network

It is often important to create a simple equivalent circuit for a two-port network where the network parameters are known. While there are several techniques for the creation of these equivalent networks, the most useful in electronic applications involves Thévenin and Norton input and output realizations. In general, network parameters are replaced by simple circuit elements as shown in Table 5.1-1. Relationships between currents and voltages at the same port are treated as impedances or admittances, while relationships at different ports are treated as dependent sources.

EXAMPLE 5.1-3 A two-port network has been found to have the following g parameters:

$$g_{11} = 0.025 \text{ S}, \quad g_{12} = 47 \text{ mA/A},$$
$$g_{21} = 14 \text{ V/V}, \quad g_{22} = 270 \text{ }\Omega.$$

Determine an equivalent circuit representation of the two-port network.

SOLUTION

The matrix equation for the g-parameter representation of a two-port network can be written as two separate equations:

$$I_1 = g_{11}V_1 + g_{12}I_2, \quad V_2 = g_{21}V_1 + g_{22}I_2.$$

The equation for I_1 reveals that two currents must be added together to make the input port current. Referring to Table 5.1-1, one finds that the currents emanate from an admittance and a current-controlled current source: They must be connected as a Norton source. The equation for V_2 implies that two voltages must be added together. Table 5.1-1 shows the elements are an impedance and a voltage-controlled voltage

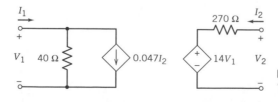

Figure 5.1-2
Two-port realization of Example 5.1-3.

source: They are connected in Thévenin fashion. The equivalent circuit representation is shown in Figure 5.1-2 (*Note:* For clarity, g_{11} is shown as a resistance rather than a conductance).

5.2 BJT Low-Frequency Models

In previous sections it has been shown that the Ebers–Moll model is an effective model in all regions of BJT operation. It has also been shown that for large-signal behavior, a simplified set of models can bring additional insight into the operation of an electronic circuit. These models are first-order approximations to BJT operation and exhibit some degree of error. When a BJT is operating in the forward-active region, it is appropriate to improve the modeling of its operation with a second-order approximation. As in most series approximations to behavior, second-order effects are a linearization, about the first-order (in this case, the quiescent) behavior, and are therefore added to the first-order behavior.

In the case of BJT operation, small-signal (often called AC) variations are linearized approximates to the operation of a transistor about its large-signal (often called DC or quiescent) behavior. Previous discussions of BJTs[3] have shown that the forward-active region (and, similarly, the inverse-active region) of BJT operation are approximately linear in nature. It is therefore quite reasonable to assume that small-signal approximations have great validity in the modeling of BJT operation. The small-signal approximations are made in the form of approximate models that take the form of two-port networks. Two-port network characterizations are valid for non-linear systems that have regions of linearity: The BJT operating in either of its active regions is such a system.

The first obvious problem in modeling a BJT as a two-port is that there are only three terminals rather than the four that seem to be necessary. This obstacle is overcome by assigning one BJT terminal *common* to both ports. Practice has made two possibilities for this common terminal standard: Either the emitter or the base is chosen to be the common terminal. Figure 5.2-1 shows an *npn* BJT as a two-port network with the emitter terminal chosen as common. In Chapter 3, it was shown that the major controlling port of a BJT is the base–emitter junction. Thus, the port containing the base–emitter junction is considered the input port, and the remaining port (which contains the collector terminal and the common terminal) is the output port. Since the terminal voltages and currents for BJTs have previously been identified with subscripts that are descriptive rather than the general numerical form of

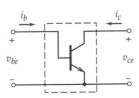

Figure 5.2-1
BJT as a two-port network (emitter as common terminal).

---

[3] See Section 3.5. In particular, discussions of the transfer relationship for the logic inverter and ECL OR gate showed a linear transition region between the two logic levels. These transition regions occurred when the BJT was operating in the forward-active region in transition between the cut-off and saturation regions.

Section 5.1, the descriptive subscripts will be continued in the BJT two-port representation.

The modeling process for transistor circuit performance is as follows:

1. Model the BJT with an appropriate DC model.
2. Determine the circuit quiescent (DC) conditions; verify BJT active region.
3. Determine the BJT AC model parameters from the quiescent conditions.
4. Create an AC equivalent circuit.
5. Determine the circuit AC performance by
    a. replacing the BJT by its AC model or
    b. using previously derived results for the circuit topology.
6. Add the results of the DC and AC analysis to obtain total circuit performance.

Techniques for the DC modeling of BJTs and the determination of quiescent conditions have been discussed thoroughly in Chapter 3: Discussions of AC modeling follow. The steps relating to circuit performance are discussed in later sections of this chapter. The choice of which set of two-port parameters will be most helpful in the modeling of a BJT is important. The good starting point is the Ebers–Moll model. The Ebers–Moll model for an *npn* BJT is shown in Figure 5.2-2.

In the forward-active region of operation, the base–emitter junction is forward-biased and the base–collector junction is reverse biased. Approximate models[4] for forward-biased and reverse-biased diodes lead to the base–collector junction diode taking on the appearance of a very large resistance, while the base–emitter junction diode can be linearized about the quiescent point to a much smaller resistance. The forward current source (described by α_F) remains significant while the reverse current source supplies very little current. This small-signal model is shown in Figure 5.2-3, with the two junction dynamic resistances defined as

$$r_{df} = \text{base–emitter diode dynamic forward resistance,}$$

$$r_{dr} = \text{base–collector diode dynamic reverse resistance.}$$

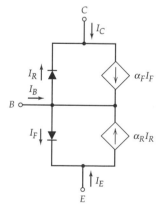

Figure 5.2-2
Ebers–Moll model of *npn* BJT.

---

[4] Models of a diode were derived in Chapter 2. The forward-biased model also includes a DC voltage source. The presence of that DC voltage source only affects quiescent behavior and not small-signal (AC) behavior. For that reason it is eliminated in these discussions.

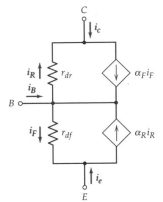

Figure 5.2-3

Forward-active small-signal model of a BJT.

The dynamic resistance of a junction was derived in Chapter 2 to be

$$r_d = \frac{\partial V}{\partial I} = \frac{\eta V_t}{I_s + I} = \frac{\eta V_t}{I_s} e^{-V/\eta V_t}. \tag{5.2-1}$$

The forward-biased value of this dynamic resistance is strongly dependent on the quiescent conditions, while the reverse-biased value is, in an active-region BJT, not. Thus r_{df} is strongly dependent on the quiescent conditions while r_{dr}, a much larger value, is not. While the derivative of this small-signal model was based on the Eber–Moll model of an *npn* BJT, the small-signal model itself is valid for *both npn and pnp* BJTs. Reversing the direction of the junctions only changes quiescent conditions, it does not change the expressions for dynamic resistance.

The analysis of electronic systems could be implemented using the model of Figure 5.2-3. However, two-port techniques simplify many aspects of analysis. The two-port realization of such a system would most effectively be accomplished by an *h*-parameter representation. This *h*-parameter realization is a common representation for small-signal, low-frequency BJT operation.[5] Figure 5.2-4 shows the *h*-parameter realization of a small-signal model of a BJT.

The governing equations[6] for the *h*-parameter model are

$$v_{be} = h_{ie}i_b + h_{re}v_{ce}, \tag{5.2-2a}$$

$$i_c = h_{fe}i_b + h_{oe}v_{ce}, \tag{5.2-2b}$$

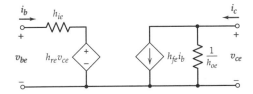

Figure 5.2-4

Common-emitter *h*-parameter model for BJT.

---

[5] Other possible models include the T model and the hybrid-π model. The *h*-parameter model is chosen as an introductory small-signal model for its simplicity. The hybrid-π model will be introduced in Chapter 10 where its added complexity is necessary to adequately describe frequency-dependent effects.

[6] Small letters with small subscripts indicate that these parameters are AC (or small-signal) parameters. They are a linearization of the transistor characteristics about a quiescent point and therefore depend on that quiescent point.

where descriptive subscripts have been chosen for the subscripts of the h-parameters. The second subscript defines which terminal is chosen as common (either the base or the emitter), and the first subscript identifies the function of the parameter:

h_{ie} = input impedance, emitter as common terminal

h_{re} = reverse voltage gain, emitter as common terminal

h_{fe} = forward current gain, emitter as common terminal

h_{oe} = output admittance, emitter as common terminal

The parameters[7] can be found as

$$h_{ie} = \left.\frac{\partial V_{BE}}{\partial I_b}\right|_{V_{CE}=V_{CEq}}, \qquad h_{re} = \left.\frac{\partial V_{BE}}{\partial V_{CE}}\right|_{I_B=I_{Bq}},$$

$$\text{(5.2-3)}$$

$$h_{fe} = \left.\frac{\partial I_C}{\partial I_B}\right|_{V_{CE}=V_{CEq}}, \qquad h_{oe} = \left.\frac{\partial I_C}{\partial V_{CE}}\right|_{I_B=I_{Bq}}.$$

Equations 5.2-3 can be revised using AC voltages and currents. Realizing that AC signals are variations about the quiescent point and that constant values have zero AC component, the h-parameters can also be defined as

$$h_{ie} = \left.\frac{v_{be}}{i_b}\right|_{v_{ce}=0}, \qquad h_{re} = \left.\frac{v_{be}}{v_{ce}}\right|_{i_b=0},$$

$$\text{(5.2-4)}$$

$$h_{fe} = \left.\frac{i_c}{i_b}\right|_{v_{ce}=0}, \qquad h_{oe} = \left.\frac{i_c}{v_{ce}}\right|_{i_b=0}.$$

All the signals in Equation 5.2-4 are AC signals. Setting an AC signal (i.e., i_b or v_{ce}) to zero holds the AC variation of that signal to zero, it does not change the quiescent conditions of the BJT.

Determination of the small-signal BJT h-parameters is accomplished by applying Equations 5.2-4 to the circuit of Figure 5.2-3.

5.2.1 Determination of h_{ie} and h_{fe}

The determination of these two parameters necessitates that the collector and emitter terminals be shorted, according to the requirements of Equation 5.2-4, in an AC sense.

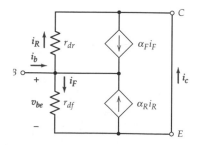

---

[7] The final subscript q indicates the quiescent value of the circuit parameter. Thus, I_{Bq} and V_{CEq} are the quiescent (DC) values of the base current and the collector–emitter voltage for the transistor.

A DC quiescent current flows the quiescent collector–emitter voltage is greater than $V_{CE(\text{sat})}$ and the BJT is in the forward-active region.

The small-signal ratios can be obtained by applying KCL at the base node:

$$i_b = i_F(1 - \alpha_F) + i_R(1 - \alpha_R), \tag{5.2-5}$$

where

$$i_F = \frac{v_{be}}{r_{df}}, \qquad i_R = \frac{v_{be}}{r_{dr}}. \tag{5.2-6}$$

The relationships of Equation 5.2-6 inserted into Equation 5.2-5 yield

$$i_b = \frac{v_{be}}{r_{df}}(1 - \alpha_F) + \frac{v_{be}}{r_{dr}}(1 - \alpha_R). \tag{5.2-7}$$

Equation 5.2-7 directly leads to the first of the h parameters. The relative size of the forward and reverse dynamic resistance of the junctions, $r_{dr} \gg r_{df}$, leads to a simplified expression for h_{ie}:

$$h_{ie} = \frac{v_{be}}{i_b} = \frac{r_{df}}{1 - \alpha_F} \left\| \frac{r_{dr}}{1 - \alpha_R} \approx \frac{r_{df}}{1 - \alpha_F} = (\beta_F + 1)r_{df}. \tag{5.2-8}$$

The value for h_{ie} is dependent on r_{df}, a quantity that strongly depends on the quiescent conditions: h_{ie} must also strongly depend on quiescent conditions. Equation 5.2-1 can be used to obtain an expression for h_{ie} in terms of the quiescent collector current and $\eta V_t \approx 26$ mV:

$$h_{ie} \approx (\beta + 1)\frac{\eta V_t}{I + I_S} \approx (\beta + 1)\frac{\eta V_t}{|I_C|}. \tag{5.2-9}$$

An additional equation must be written to find h_{fe}. At the collector node KCL gives

$$i_c = \alpha_F i_F - i_R = \alpha_F \frac{v_{be}}{r_{df}} - \frac{v_{be}}{r_{dr}}, \tag{5.2-10}$$

which, noting the relative dynamic resistance sizes, can be reduced to

$$\frac{v_{be}}{i_c} = \left(\frac{r_{df}}{\alpha_F} \left\| r_{dr}\right) \approx \frac{r_{df}}{\alpha_F}. \tag{5.2-11}$$

The required current ratio for h_{fe} is found using Equations 5.2-8 and 5.2-11:

$$h_{fe} = \frac{i_c}{i_b} \approx \frac{\alpha_F/r_{df}}{(1 - \alpha_F)/r_{df}} = \beta_F. \tag{5.2-12}$$

5.2.2 Determination of h_{re} and h_{oe}

The tests for these two h parameters (as seen in Equation 5.2-4) requires that the small-signal base current be set to zero. Again, remember that there is still a quiescent

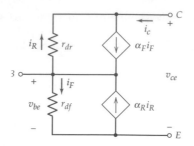

(DC) current flowing and the BJT is in the forward-active region. Kirchhoff's current law applied at the base node yields

$$i_F(1 - \alpha_F) + i_R(1 - \alpha_R) = 0, \tag{5.2-13}$$

where

$$i_F = \frac{v_{be}}{r_{df}}, \qquad i_R = \frac{v_{be} - v_{ce}}{r_{dr}}. \tag{5.2-14}$$

Then inserting Equation 5.2-14 into 5.2-13 yields

$$\frac{v_{be}}{r_{df}}(1 - \alpha_F) + \frac{v_{be} - v_{ce}}{r_{dr}}(1 - \alpha_R) = 0 \tag{5.1-15}$$

or

$$v_{be}\left(\frac{1 - \alpha_F}{r_{df}} + \frac{1 - \alpha_R}{r_{dr}}\right) = v_{ce}\left(\frac{1 - \alpha_R}{r_{dr}}\right). \tag{5.2-16}$$

The required voltage ratio can now be determined as

$$h_{re} = \frac{v_{be}}{v_{ce}} = \frac{r_{df}/(1 - \alpha_F)||r_{dr}/(1 - \alpha_R)}{r_{dr}/(1 - \alpha_R)}$$

$$\approx \frac{r_{df}}{r_{dr}}\frac{1 - \alpha_R}{1 - \alpha_F} = \frac{r_{df}}{r_{dr}}\frac{1 + \beta_F}{1 + \beta_R}. \tag{5.2-17}$$

The value for h_{re} is seen to be a *very* small quantity and can safely be assumed to be zero in the modeling of a BJT in either active region. In order to solve for h_{oe}, KCL is applied at the emitter node:

$$i_c = i_F - \alpha_R i_R = \frac{v_{be}}{r_{df}} - \alpha_F \frac{v_{be} - v_{ce}}{r_{dr}}$$

$$= v_{be}\left(\frac{1}{r_{df}} - \frac{\alpha_F}{r_{dr}}\right) + \alpha_F \frac{v_{ce}}{r_{dr}}. \tag{5.2-18}$$

Again using the property that $r_{df} \ll r_{dr}$, a simplification can be made:

$$i_c \approx \frac{v_{be}}{r_{df}} + \alpha_F \frac{v_{ce}}{r_{dr}}. \tag{5.2-19}$$

Equation 5.2-17 can be used to eliminate v_{be} from Equation 5.2-19:

$$i_c \approx \frac{v_{ce}}{r_{df}} \frac{r_{df}}{r_{dr}} \frac{1 - \alpha_R}{1 - \alpha_F} + \frac{v_{ce}}{r_{dr}} = \frac{v_{ce}}{r_{dr}} \left(\frac{1 + \beta_F}{1 + \beta_R} + \alpha_F \right). \tag{5.2-20}$$

The relationship for h_{oe} is obtained from Equation 5.2-20.

$$h_{oe} = \frac{i_c}{v_{ce}} \approx \frac{1}{r_{dr}} \left(\frac{1 + \beta_F}{1 + \beta_R} + \alpha_F \right). \tag{5.2-21}$$

This is a very small output conductance (the equivalent of a very large output resistance). The simplified expression of Equation 5.2-21 does not adequately model all the resistances contributing to output resistance for a real BJT. The actual output resistance is much smaller than predicted by the Ebers–Moll model. A more accurate predictor of the output resistance r_o is defined in terms of another quantity, the early voltage V_A[8]:

$$\frac{1}{h_{oe}} = r_o = \left| \frac{V_A}{I_C} \right|. \tag{5.2-22}$$

The early voltage for a BJT is usually determined experimentally by observing the slope of the I_C-versus-V_{CE} curves for a BJT in the forward-active region. It typically has a value of 100 V or more and is defined as

$$V_A = \frac{I_C}{\partial I_C / \partial V_{CE}}. \tag{5.2-23}$$

This output resistance is often large compared to the resistances connected to the collector–emitter port of the BJT and h_{oe} can, in that case, be assumed approximately

TABLE 5.2-1 BJT h-PARAMETER SUMMARY

Parameter	Value
h_{ee}	$\approx (\beta_F + 1) \dfrac{\eta V_t}{\vert I_C \vert}$
h_{fe}	$\approx \beta_F$
h_{re}	$\approx \dfrac{r_{df}}{r_{dr}} \dfrac{1 + \beta_F}{1 + \beta_R} \approx 0$
h_{oe}	$= \left\vert \dfrac{I_C}{V_A} \right\vert \approx 0$

---

[8] The early voltage is a parameter that describes the change in the width of the base region of a BJT that results from a change in base–collector junction voltage. A basic description of the effect can be found in most solid-state electronics texts but is beyond the scope of this text.

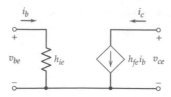

Figure 5.2-5
Simplified small-signal h-parameter model for BJT.

equal to zero. When the external resistances are not small compared to the output resistance, it is necessary to include h_{oe} in the model for the BJT.

The h-parameters for a BJT are summarized in Table 5.2-1. Only one parameter, h_{fe}, is essentially independent of quiescent conditions, while the other three parameters depend to varying degrees on the bias conditions, specifically on the quiescent collector current I_C.

The extremely small values of h_{re} and h_{oe} lead to a simplified h-parameter model of a BJT where these two parameters are assumed to be approximately zero, as shown in Figure 5.2-5. When large resistances are connected to the collector–emitter port of the BJT, it is important to reintroduce h_{oe} into the h-parameter model as a conductance shunting the dependent current source. In typical BJT applications, h_{re} is not significant.

EXAMPLE 5.2-1 Assume as given a silicon *npn* BJT with parameters

$$\beta_F = 150, \qquad V_A = 350 \text{ V}$$

operating in the given circuit at room temperature. Determine an appropriate small-signal h-parameter model for the BJT.

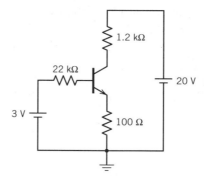

SOLUTION

The quiescent conditions must first be obtained: The KCL taken around the base–emitter loop yields

$$3 - 22{,}000 I_B - 0.7 - 100(151 I_B) = 0,$$

$$I_B = 62.0 \ \mu\text{A}, \qquad I_C = 150 I_B = 9.30 \text{ mA}.$$

Checking to see if the BJT is in the forward-active region by using KCL around the collector–emitter loop yields

$$V_{CE} = 20 - 1200 I_C - 100 \left(\frac{151}{150} I_C \right) = 7.90 \text{ V}.$$

This value is greater than $V_{CE(sat)}$, which implies that the BJT is in the forward-active region and the *h*-parameter model parameters itemized in Table 5.2-1 may be used:

$$h_{fe} \approx \beta_F = 150,$$

$$h_{ie} \approx (\beta_F + 1)\frac{\eta V_t}{|I_C|} = (151)\frac{26 \text{ mV}}{|9.30 \text{ mA}|} = 422 \text{ }\Omega,$$

$$h_{oe} = \left|\frac{I_C}{V_A}\right| = \left|\frac{930 \text{ mA}}{350 \text{ V}}\right| = 26.6 \text{ }\mu\text{S} \approx 0,$$

$$h_{re} \approx 0.$$

It is necessary to check to see if h_{oe} can be approximated as zero value. It is equivalent to a resistance r_o of 37.6 kΩ (the resistance is the inverse of h_{oe}), which is large with respect to the 1.2-kΩ and 100-Ω resistors connected to the collector–emitter port of the BJT: h_{oe} can be assumed in this application to be approximately zero in value.

The approximate small-signal model of this BJT operating in this circuit is therefore found to be as shown in the accompanying diagram.

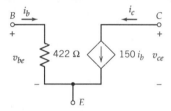

5.3 COMMON-EMITTER AMPLIFIERS

A description of the general performance of a common-emitter amplifier can be derived directly from the voltage transfer relationship of a BJT logic inverter. The BJT logic inverter, discussed in Section 3.5, has two distinct regions where the voltage output is relatively constant: when the BJT is in either the saturation or the cutoff region. Between these two regions lies a linear region where the variation in the output voltage about the quiescent point is directly proportional to the variation in the input voltage. This linear region proportionality constant for the logic inverter was seen in Example 3.5-1 to be negative and greater than 1 in magnitude: The input signal variations were amplified and inverted. The discussions of Section 3.5 were based on coarse, first-order approximations of the operation of a BJT in each of its regions of operation. This section develops the characteristics of this type of amplifier (and its close relatives) using the second-order *h*-parameter approximations developed earlier in this chapter.

Common-emitter amplifiers have the same general circuit topology as the logic inverter:

- The input signal enters the BJT at the base.
- The output signal exits the BJT at the collector.
- The emitter is connected to a constant voltage, often the ground (common) terminal, sometimes with an intervening resistor.

A simple common-emitter amplifier is shown in Figure 5.3-1. It is necessary that the quiescent point of the BJT be set wih the circuitry external to the transistor so that it is in the forward-active region. The values of the resistors, R_c and R_s, and the DC

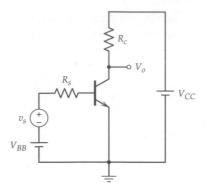

Figure 5.3-1
Simple common-emitter amplifier.

voltage sources, V_{CC} and V_{BB}, have therefore been chosen so that the BJT is in the forward-active region and the circuit will operate as an amplifier. The voltage source v_s is a small-signal AC source.

Once the circuit quiescent ($v_s = 0$) conditions have been calculated and it has been determined that the BJT is in the forward-active region of operation, the significant h-parameters can be calculated to form the small-signal model of the transistor:

$$h_{fe} = \beta_F, \qquad h_{oe} = \frac{I_C}{V_A} \approx 0,$$

$$h_{ie} = (\beta_F + 1) \left| \frac{\eta V_t}{I_C} \right|.$$

(5.3-1)

The small-signal (often called AC) circuit performance can now be calculated. Total circuit performance is the sum of quiescent and small-signal performance: The process is basically a superposition of the zero-frequency solution with the low-frequency AC solution.[9] In each case, the solutions are formed by setting the independent sources at all other frequencies to zero. Alternating currents circuit performance is obtained through analysis of a circuit obtained by setting the original circuit DC sources to zero and then replacing the BJT with its h-parameter two-port model. This circuit reconfiguration process, applied to Figure 5.3-1, is shown in Figure 5.3-2.

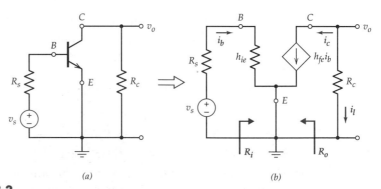

(a)   (b)

Figure 5.3-2
AC modeling of simple common-emitter amplifier: (a) small-signal circuit, DC source set to zero; (b) BJT replaced by h-parameter model.

---

[9] At high frequencies the BJT performance characteristics change. See Section III.

The small-signal performance characteristics[10] that are of interest in any amplifier are *current gain, input resistance, voltage gain,* and *output resistance.* For the simple common-emitter amplifier under consideration, these characteristics can be obtained from analysis of the circuit of Figure 5.3-2b. Definitions for these quantities often vary due to differing definition of the exact location of the point of measurement. Care must always be taken to *ensure that measurement points are clearly understood before any analysis begins.*

Current Gain
For this simple transistor amplifier, the current gain is defined as the ratio of load current to input current, that is,

$$A_I \equiv \frac{i_l}{i_b} = \frac{-i_c}{i_b}. \tag{5.3-2}$$

From the circuit of Figure 5.3-2b, it can easily be determined that the collector and base currents are related through the dependent current source by the constant h_{fe}. The current gain is dependent only on the BJT characteristics and independent of any other circuit element values. Its value is given by

$$A_I = -h_{fe}, \tag{5.3-3}$$

where the negative sign implies that the small-signal current input is inverted as well as amplified. If the resistor R_c is large, then the output resistance of the BJT must be considered in calculating the current gain. Including h_{oe} in the BJT h-parameter model divides the output current between the output resistance $r_o = 1/h_{oe}$ and the collector resistance:

$$A_I = -h_{fe} \frac{r_o}{r_o + R_c}. \tag{5.3-4}$$

Input Resistance
Input resistance is the measure of the AC input current as a function of the input voltage. If the AC input voltage v_b, is taken at the input terminal of the BJT (the base), the input resistance (shown in Figure 5.3-2b with the bold arrow and labeled as R_i) is given by

$$R_i \equiv \frac{v_b}{i_b} = h_{ie}. \tag{5.3-5}$$

The input resistance is unaffected by finite r_o.

Voltage Gain
The voltage gain is the ratio of output voltage to input voltage. If the input voltage is again taken to be the voltage at the input to the transistor, v_b, the voltage gain is defined as

$$A_V \equiv \frac{v_o}{v_b}. \tag{5.3-6}$$

---

[10]It should be remembered that the quantities discussed in this section and following sections are small-signal (or AC) quantities. The ratios considered here are the ratios of AC quantities not the ratios of total currents and/or voltages.

One of the best methods for obtaining this ratio is through the use of quantities already calculated or otherwise easily obtained:

$$A_V = \frac{v_o}{v_b} = \left(\frac{v_o}{i_l}\right)\left(\frac{i_l}{i_b}\right)\left(\frac{i_b}{v_b}\right). \tag{5.3-7}$$

The expressions for each quantity, when substituted into Equation 5.3-7, yield

$$A_V = (R_c)(A_I)\left(\frac{1}{R_i}\right) = \frac{-h_{fe}R_c}{h_{ie}}. \tag{5.3-8}$$

Again, the small-signal output voltage is an inverted, amplified replica of the input voltage. The voltage gain can be very large in magnitude. Equation 5.3-8 indicates it is limited only by the BJT parameter h_{fe} and the external resistance R_c. That conclusion is based, however, on the assumption that R_c is small with respect to r_o. If R_c approaches the magnitude of r_o, the gain becomes limited by the output resistance of the amplifier, and R_c must be replaced by $R_c \| r_o$ in Equation 3.5-8.

Often the voltage gain from the source to the load is of interest as well. This overall voltage gain can be defined as

$$A_{VS} \equiv \frac{v_o}{v_s}. \tag{5.3-9}$$

This ration can be directly derived from the simple voltage gain A_V with a voltage division using the amplifier input resistance and the source resistance:

$$A_{VS} = \frac{v_o}{v_s} = \left(\frac{v_o}{v_b}\right)\left(\frac{v_b}{v_s}\right) = A_V\left(\frac{R_i}{R_i + R_s}\right) \tag{5.3-10}$$

or

$$A_{VS} = \left(\frac{-h_{fe}R_c}{h_{ie}}\right)\left(\frac{h_{ie}}{h_{ie} + R_s}\right) = \frac{-h_{fe}R_c}{h_{ie} + R_s}. \tag{5.3-11}$$

Since A_{VS} is dependent on A_V, it is also limited in magnitude by finite r_o. To include the effect of finite r_o, R_c must be replaced by $R_c \| r_o$ in Equation 3.5-11.

Output Resistance

The output resistance is defined as the Thévenin resistance at the output of the amplifier (in this case the BJT collector) looking back into the amplifier. As in the case of voltage gain, the exact point of measurement is not always clear. Figure 5.3-2 indicates that R_o is measured without considering R_c: It is also possible to define an output resistance R_{ol} that includes the effects of collector resistance R_c. The AC equivalent circuit used to calculate the output resistance is shown in Figure 5.3-3a. The input independent source is set to zero and the Thévenin resistance is calculated looking into the output. It can easily be seen that $i_b = 0$ in this circuit. Kirchhoff's voltage law taken around the base loop gives

$$i_b(h_{ie} + R_s) = 0 \Rightarrow i_b = 0.$$

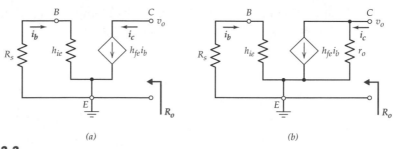

Figure 5.3-3
Circuit for calculation of output resistance: (*a*) small-signal output circuit, AC source set to zero; (*b*) more accurate BJT *h*-parameter model.

Zero base current implies that the dependent current source is also zero valued and the output resistance becomes

$$R_o \approx \infty. \tag{5.3-12}$$

While the output resistance is certainly large, a more revealing result could be obtained with a better model of the BJT. A more accurate *h*-parameter model of a BJT includes h_{oe} (as derived in Section 5.2). When this model is used to determine the output resistance of a simple common-emitter amplifier (see Fig. 5.3-3*b*), the output resistance is given by

$$R_o = r_o = \frac{1}{h_{oe}} = \left| \frac{V_A}{I_C} \right|. \tag{5.3-13}$$

The output resistance is therefore a large value that is dependent strongly on the quiescent conditions of the transistor.

As mentioned earlier, it is also possible to define output resistance of a common-emitter amplifier to include the collector resistor R_c:

$$R_{ol} = R_o \| R_c \approx R_c. \tag{5.3-14}$$

The common-emitter amplifier has been shown to have the following

- Moderate input resistance
- High output resistance
- High current gain
- Moderate to high voltage gain, depending on presence and size of any resistance connected to collector terminal

A summary of the common-emitter amplifier performance characteristics is found in Section 5.3.4.

EXAMPLE 5.3-1 Assume as given a silicon *npn* BJT with parameters

$$\beta_F = 150, \qquad V_A = 350 \text{ V}$$

operating in the given circuit at room temperature. Determine the amplifier small-signal performance characteristics.

SOLUTION

The modeling process for transistor circuit performance was described in Section 5.2 to contain the following steps:

1. Model the BJT with an appropriate DC model.
2. Determine the circuit quiescent (DC) conditions. Verify BJT active region.
3. Determine the BJT AC model parameters from the quiescent conditions.
4. Create an AC equivalent circuit.
5. Determine the circuit AC performance by
   a. replacing the BJT by its AC model or
   b. using previously derived results for the circuit topology.
6. Add the results of the DC and AC analysis to obtain total circuit performance.

Step 1: Model the BJT with an appropriate DC model. Hopefully, the BJT will be found to be in the forward-active region so that the circuit can operate as an amplifier. The DC forward-active model of the BJT is derived in Section 3.4 and is shown here. The appropriate transistor quantities that must be used in this case are $\beta_F = 150$ and $V_\gamma = 0.7$ V (silicon BJT).

Step 2: Determine the circuit quiescent (DC) conditions. Verify BJT active region. The BJT model of step 1 replaces the BJT as shown. Quiescent conditions are then calculated. Around the base–emitter loop, KCL yields

$$1.5 - 22{,}000(I_B) - 0.7 = 0,$$

22 V

1.2 kΩ

C V_o

$150I_B$

22 kΩ   0.7 V

1.5 V

B I_B E

so that

$$I_B = 36.36 \ \mu A,$$

$$I_C = 150I_B = 5.45 \ \text{mA}.$$

Both currents are positive: only V_{CE} must be checked to see if it is larger than $V_{CE(\text{sat})}$ = 0.2 V. Kirchhoff's current law applied to the collector—emitter loop yields

$$V_{CE} = 22 - 1200(I_C) = 15.46 \ \text{V}.$$

The transistor is in the forward-active region: The circuit will operate as an amplifier.

Step 3: Determine the BJT AC model parameters from the quiescent conditions. The relevant *h*-parameters are

$$h_{fe} \approx \beta_F = 150,$$

$$h_{ie} \approx (\beta_F + 1)\frac{\eta V_t}{|I_C|} = (151)\frac{26 \ \text{mV}}{|5.45 \ \text{mA}|} = 720 \ \Omega,$$

$$h_{oe} = \left|\frac{I_C}{V_A}\right| = \left|\frac{5.45 \ \text{mA}}{350 \ \text{V}}\right| = 15.6 \ \mu S \approx 0,$$

$$h_{re} = 0.$$

Step 4: Create an AC equivalent circuit.

The DC sources are set to zero and the output voltage becomes a small-signal quantity.

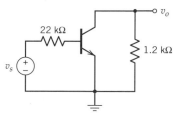

Step 5: Determine the circuit AC performance by (a) replacing the BJT by its AC model or (b) using previously derived results for the circuit topology. In this

case it appears easiest to use previously derived results. The input resistance at the base of the BJT is given by

$$R_i = h_{ie} = 720 \ \Omega.$$

If the input resistance is measured to the left of the 22-kΩ resistor, then

$$R_i' = 22 \ k\Omega + 720 \ \Omega = 22.7 \ k\Omega.$$

The current gain is given by

$$A_I = -h_{fe} = -150.$$

The voltage gain from the base of the BJT is given by

$$A_V = \frac{-h_{fe}R_c}{h_{ie}} = \frac{-150(1200)}{720} = -250.$$

The voltage gain from the source is given by

$$A_{VS} = \frac{-h_{fe}R_c}{h_{ie} + R_s} = \frac{-150(1200)}{720 + 22{,}000} = -7.92.$$

The output resistance is given by

$$R_o = \frac{1}{h_{oe}} = 64.17 \ k\Omega.$$

If the collector resistance is included in the output resistance,

$$R_{ol} = R_o \| R_c = \frac{(64{,}170)(1200)}{64{,}170 + 1200} = 1.18 \ k\Omega.$$

Step 6 is beyond the requirements of this problem, but all the data are present to add the quiescent solution to the small-signal solution for total circuit response.

5.3.1 Common-Emitter Amplifiers with Nonzero Emitter Resistance

Several of the characteristics of a common-emitter amplifier can be altered through the addition of a resistor connected between the emitter and ground as shown in Figure 5.3-4. It has previously been shown in Section 3.7 that the addition of such a resistor greatly improves bias stability due to BJT variations. The addition of an emitter resistor has the following effects on common-emitter amplifiers:

- Increased input resistance
- Increased output resistance
- Decreased voltage gain
- Unchanged current gain

Derivation of these effects follows the process described in Section 5.2.

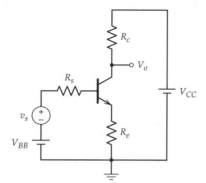

Figure 5.3-4
Common-emitter amplifier with emitter resistor.

In order to determine the AC performance of this amplifier, the quiescent conditions must first be obtained. If the BJT is operating in the forward-active region, an AC equivalent model of the circuit can be obtained (Figure 5.3-5a) and the BJT can be replaced by its h-parameter model[11] (based on the quiescent conditions) in the AC equivalent circuit (Figure 5.3-5b). Circuit performance parameters can now be obtained from this equivalent circuit.

Current Gain

The current gain is defined as the ratio of load current to input current, that is,

$$A_I \equiv \frac{i_l}{i_b} = \frac{-i_c}{i_b}. \tag{5.3-15}$$

The ratio of collector to base current remains unchanged from the simple common-emitter amplifier. The current gain is dependent only on the BJT characteristics and independent of any other circuit element values. Its value is given by

$$A_I = -h_{fe}. \tag{5.3-16}$$

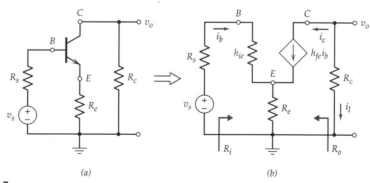

(a)   (b)

Figure 5.3-5
AC modeling of common-emitter amplifier with emitter resistor: (*a*) small-signal circuit, DC sources set to zero; (*b*) BJT replaced by h-parameter model.

---

[11]The initial calculations assume that h_{oe} is small and can be ignored. Calculations to determine the effect of nonzero h_{oe} follow in Section 5.3.2.

Input Resistance

The input resistance (shown in Figure 5.3-5*b*) is given by

$$R_i \equiv \frac{v_b}{i_b} = \frac{h_{ie}i_b + R_e(i_b + h_{fe}i_b)}{i_b} = h_{ie} + (h_{fe} + 1)R_e. \tag{5.3-17}$$

The addition of an emitter resistor has greatly increased the input resistance of the amplifier. From the input of the BJT, the emitter resistor appears to act as a resistor in series with h_{ie}, that is, $h_{fe} + 1$ times its true value.

Voltage Gain

The voltage gain is the ratio of output voltage to input voltage. If the input voltage is again taken to be the voltage at the input to the transistor v_b:

$$A_V \equiv \frac{v_o}{v_b}. \tag{5.3-18}$$

Using methods similar to those of the simple common-emitter amplifier to calculate the voltage gain yields

$$A_V = \frac{v_o}{v_b} = \left(\frac{v_o}{i_l}\right)\left(\frac{i_l}{i_b}\right)\left(\frac{i_b}{v_b}\right) \tag{5.3-19}$$

or

$$A_V = (R_c)(A_I)\left(\frac{1}{R_i}\right) = \frac{-h_{fe}R_c}{h_{ie} + (h_{fe} + 1)R_e}. \tag{5.3-20}$$

For large values of h_{fe}, the external resistors dominate the expression for voltage gain. A rough approximation that slightly overestimates the magnitude of the gain is

$$A_V \approx -\frac{R_c}{R_e}. \tag{5.3-21}$$

Often the voltage gain from the source to the load is of interest as well. This overall voltage gain can be defined as

$$A_{VS} \equiv \frac{v_o}{v_s}. \tag{5.3-22}$$

This ratio can be directly derived from the voltage gain A_V and a voltage division between the source resistance R_s and the amplifier input resistance R_i:

$$A_{VS} = \frac{v_o}{v_s} = \left(\frac{v_o}{v_b}\right)\left(\frac{v_b}{v_s}\right) = A_V\left(\frac{R_i}{R_i + R_s}\right) \tag{5.3-23}$$

or

$$A_{VS} = \left(\frac{-h_{fe}R_c}{R_i}\right)\left(\frac{R_i}{h_{ie} + (h_{fe} + 1)R_e + R_s}\right)$$

$$= \frac{-h_{fe}R_c}{h_{ie} + (h_{fe} + 1)R_e + R_s}$$

(5.3-24)

Output Resistance

The output resistance is defined as the Thévenin resistance at the output of the amplifier looking back into the amplifier. As in the case of the simple common-emitter amplifier, R_o is measured without considering R_c (Figure 5.3-5). Once again, the resistance looking into the collector of the BJT is very large, $R_o \approx \infty$.

EXAMPLE 5.3-2 Assume as given a silicon *npn* BJT with parameters

$$\beta_F = 150, \qquad V_A = 350 \text{ V}$$

operating in the given circuit at room temperature. Determine the following small-signal characteristics: voltage gain v_o/v_s, input resistance R_i, and output resistance R_o.

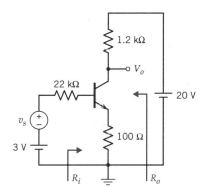

SOLUTION

The modeling process begins with solving for the quiescent circuit conditions. In this particular circuit, the identical transistor was placed in the same quiescent circuit in Example 5.2-1. There is no need to repeat identical operations to find the small-signal model of the BJT. The results of process steps 1–3 were

$$I_B = 62.0 \text{ } \mu\text{A}, \qquad I_C = 9.30 \text{ mA},$$

$$h_{fe} = 150, \qquad h_{ie} = 422 \text{ } \Omega, \qquad r_o = 37.6 \text{ k}\Omega.$$

The next step is to create an AC equivalent circuit and insert the BJT *h*-parameter model into the AC equivalent circuit or use previously derived results. The general topology of a common-emitter amplifier with an emitter resistor has previous results. The AC equivalent circuit is as shown.

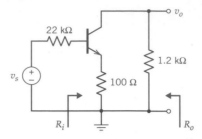

The voltage gain, as defined, is determined from Equation 5.3-24:

$$A_{VS} = \frac{-(150)(1200)}{422 + (151)(100) + 22{,}000} = -4.7972 \approx -4.80.$$

As shown, R_i is given by Equation 5.3-17:

$$R_i = 422 + (151)(100) = 15{,}552 \approx 15.5 \text{ k}\Omega.$$

The output resistance R_o is given by the parallel combination of the output resistance of the amplifier and the collector resistor R_c:

$$R_o \approx \infty \| 1.2 \text{ k}\Omega = 1.2 \text{ k}\Omega.$$

5.3.2 Effect of Nonzero h_{oe} on Common-Emitter Amplifiers with Emitter Resistor

As with simple common-emitter amplifiers, the effects of nonzero h_{oe} can only be seen when the collector resistance becomes large: when R_c approaches or exceeds r_o in magnitude. With simple common-emitter amplifiers, the most noticeable effect of nonzero h_{oe} is on the output resistance of the amplifier. Input resistance is unaffected, while the current and voltage gain are limited if the external collector resistance R_c is large. Nonzero h_{oe} in the presence of an emitter resistor effects all the common-emitter amplifier characteristics in a complex fashion.

Expressions for the input resistance, voltage gain, and current gain are obtained using the AC equivalent circuits of Figure 5.3-6. Figure 5.3-6b is obtained from the more traditional Figure 5.3-6a through a source transformation on the dependent current source and r_o. Symbolic manipulations of Kirchhoff's laws are facilitated through this well-known transformation.

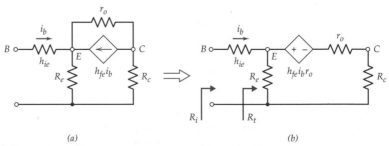

(a) (b)

Figure 5.3-6
AC equivalent circuits, nonzero h_{oe}.
(a) traditional representation
(b) after source transformation

Input Resistance

The input resistance can be determined by first finding R_t (as shown in Figure 5.3-6) and then adding the resistance h_{ie}:

$$R_i = h_{ie} + R_t. \tag{5.3-25}$$

In the usual Thévenin process, a voltage v is applied across R_e. The input current is then found to be

$$i_b = \frac{v}{R_e} + \frac{v - h_{fe}i_b r_o}{r_o + R_c}. \tag{5.3-26}$$

Collecting terms yields

$$i_b \frac{(1 + h_{fe})r_o + R_c}{r_o + R_c} = v\left(\frac{1}{R_e} + \frac{1}{r_o + R_c}\right), \tag{5.3-27}$$

which leads to the Thévenin resistance R_t:

$$R_t = \frac{v}{i_b} = R_e \frac{(1 + h_{fe})r_o + R_c}{r_o + R_c + R_e}. \tag{5.3-28}$$

The desired input resistance is found using Equation 5.3-25:

$$R_i = h_{ie} + R_t = h_{ie} + R_e \frac{(1 + h_{fe})r_o + R_c}{r_o + R_c + R_e}. \tag{5.3-29}$$

An interesting result of these calculations is that large values of the collector resistor R_c will *reduce* the value of the amplifier input resistance!

Voltage Gain

The voltage gain can be calculated as follows by continuing with the circuit of Figure 5.3-6 and many of the calculations used in determining the input resistance. The output voltage for this circuit is taken across the collector resistor R_c. It can be obtained through a simple voltage division from the voltage v (taken across R_e):

$$v_o = \frac{R_c}{R_c + r_o} (v - h_{fe}i_b r_o), \tag{5.3-30}$$

where, including the presence of h_{ie} in Figure 5.3-6,

$$v = v_s \frac{R_t}{h_{ie} + R_t}, \qquad i_b = \frac{v_s}{h_{ie} + R_t}. \tag{5.3-31}$$

Substitution of these two expressions into Equation 5.3-30 leads directly to the voltage gain:

$$A_V = \frac{v_o}{v_s} = -\frac{h_{fe}(R_c\|r_o)}{R_i} + \frac{R_c}{R_c + r_o}\frac{R_t}{R_i} \approx -\frac{h_{fe}(R_c\|r_o)}{R_i}. \tag{5.3-32}$$

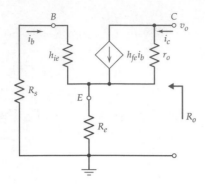

Figure 5.3-7

Circuit for calculation of output resistance.

For significantly large R_e,

$$R_e \gg \frac{\eta V_t}{|I_c|},$$

the numerator and denominator of this expression have approximately the same dependence on the relationship of R_c and r_o. The voltage gain is *approximately independent of r_o* and unchanged from the expression of Equation 5.3-20:

$$A_V \approx \frac{-h_{fe}R_c}{h_{ie} + (h_{fe} + 1)R_e}. \tag{5.3-33}$$

Current Gain

The current gain can be calculated using Equations 5.3-31 and the two expressions

$$i_o = \frac{v_o}{R_c}, \qquad v = i_b R_t \tag{5.3-34}$$

to obtain

$$i_o = \frac{v_o}{R_c} = \frac{1}{R_c + r_o} (i_b R_t - h_{fe} i_b r_o). \tag{5.3-35}$$

from which the current gain is calculated:

$$A_I = \frac{i_o}{i_b} = -\frac{h_{fe} r_o}{r_o + R_c} + \frac{R_t}{r_o + R_c} \approx -\frac{h_{fe} r_o}{r_o + R_c}. \tag{5.3-36}$$

Basically, the output current is divided between the collector resistor and r_o. It decreases as the collector resistance increases in magnitude.

Output Resistance

Calculations to determine output resistance including the effect of the BJT output resistance r_o are based on an AC equivalent circuit with a more complete *h*-parameter model of the BJT, as shown in Figure 5.3-7.

If a current i is applied to the collector terminal, the base current can be found through a current division:

$$i_b = \frac{-R_e i}{R_e + h_{ie} + R_s}.$$

(5.3-37)

The voltage that appears at the output terminals is then given by

$$v = [R_e || (R_s + h_{ie})]i + r_o(i - h_{fe}i_b)$$

(5.3-38)

or, including the relationship between i and i_b,

$$v = [R_e || (R_s + h_{ie})]i + \left(r_o + \frac{h_{fe}R_e r_o}{R_e + h_{ie} + R_s} \right)i.$$

(5.3-39)

Simple division leads to the expression for the output resistance:

$$R_o = \frac{v}{i} = [R_e || (R_s + h_{ie})] + r_o\left(1 + \frac{h_{fe}R_e}{R_e + h_{ie} + R_s} \right).$$

(5.3-40)

This is a *very* large resistance: The addition of an emitter resistor has significantly increased the output resistance over the simple common-emitter amplifier (the increase is by a factor on the order of $h_{fe} + 1$). The infinite output resistance approximation, $R_o \approx \infty$, is therefore valid over a greater range of load resistance R_c than in the simple common-emitter amplifier.

5.3.3 Coupling and Bypass Capacitors

Often the demands of properly biasing a BJT into the forward-active region and applying an AC signal to the input create conflicting circuit topology requirements. It is not always possible to put the AC input in series with a DC biasing voltage as has been previously described. In addition, the need for high voltage gain and good bias stability creates a design requirement conflict in the magnitude of the emitter resistor R_e. One *possible* design alternative that can resolve these conflicts involves the use of capacitors to couple the AC signal into the amplifier and/or to bypass the emitter resistor.

If an AC signal is imposed upon a capacitor, the complex impedance of the capacitor is given by

$$z_C = \frac{1}{j\omega C},$$

where C = capacitance of capacitor (F)
            ω = frequency of sinusoidal signal (rad/s)

At DC ($\omega = 0$), the capacitor has infinite impedance: A capacitor blocks DC signals. If either the frequency or the capacitance is large, the impedance approaches zero or becomes relatively small compared to adjacent circuit resistances. For purposes of demonstration, assume that a band of frequencies exist for which the impedance of all discrete capacitors in a circuit will be essentially zero. This band of frequencies is called the *midband frequency region*[12] of a circuit. If the AC input signals to a circuit

---

[12]Often call the *midband region*.

are within the midband frequency region of the circuit, each discrete capacitor will appear to be the equivalent of a short circuit. Each capacitor will, however, appear to be the equivalent of an open circuit to the bias (DC) circuitry.[13]

In the presence of coupling and/or bypass capacitors, the modeling of amplifier circuit performance is slightly altered. In determining quiescent conditions, each capacitor is replaced by an open circuit in addition to replacing the BJT with an appropriate model. The AC equivalent circuit is drawn with each capacitor replaced by a short.

EXAMPLE 5.3-3 A silicon BJT with parameters

$$\beta_F = 150, \qquad V_A = 350 \text{ V}$$

is placed in the amplifier shown. Assume the amplifier is operating in its midband frequency range, and determine the following circuit performance parameters: $A_V = v_o/v_s$, $A_I = i_{out}/i_{in}$, R_{in}, and R_{out}.

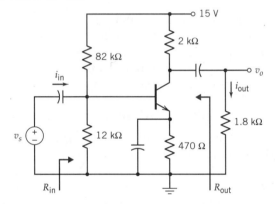

SOLUTION

The quiescent point of the BJT must be found first to ensure that it is in the forward-active region and to find its h-parameters. For DC operation, all capacitors appear as open circuits. The DC equivalent circuit, after replacing the base circuit with its Thévenin equivalent, is as shown:

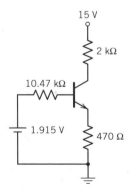

---

[13]The transition as frequency increases between an apparent open circuit and an apparent short circuit is beyond these early discussions. It is extensively covered in Section III, "Frequency Dependence."

The base and collector currents are

$$I_B = \frac{1.915 - 0.7}{10,470 + 151(470)} = 14.92 \ \mu A,$$

$$I_C = 150 I_B = 2.238 \ \text{mA}.$$

Here, V_{CE} must be checked to verify that the BJT is in the forward-active region.

$$V_{CE} = 15 - I_C(2000) - \frac{151}{150} I_C(470) = 9.47 \ \text{V} \geq 0.2 \ \text{V}.$$

The BJT h-parameters can now be determined:

$$h_{fe} = 150, \qquad h_{ie} = 151 \frac{\eta V_t}{|I_C|} = 1.755 \ \text{k}\Omega,$$

$$\frac{1}{h_{oe}} = \left| \frac{V_A}{I_C} \right| = 156.4 \ \text{k}\Omega.$$

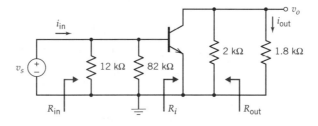

An AC equivalent circuit can now be determined as shown. All DC sources are set to zero and, since the circuit is in its midband frequency range, all capacitors are replaced by short circuits: the 470-W resistor is totally eliminated from the circuit. The AC equivalent circuit leads to the determination of amplifier performance characteristics.

$$A_V = -\frac{h_{fe} R_c}{h_{ie}} = -\frac{150(2000||1800)}{1755} = -81.0,$$

$$R_{\text{in}} = 82,000||12,000||R_i = 10,470||h_{ie} = 1.50 \ \text{k}\Omega,$$

$$R_{\text{out}} = 2000||r_o = 1980 \approx 2.0 \ \text{k}\Omega,$$

$$A_I = \frac{i_{\text{out}}}{i_{\text{in}}} = \left(\frac{i_{\text{out}}}{v_o} \right) \left(\frac{v_o}{v_s} \right) \left(\frac{v_s}{i_{\text{in}}} \right) = \left(\frac{1}{1800} \right)(-81.0)(1500) = -67.5.$$

5.3.4 Summary of Common-Emitter Amplifier Properties

A summary of common-emitter (CE) amplifier performance characteristics as derived in this section is given in Table 5.3-1.

TABLE 5.3-1 COMMON-EMITTER AMPLIFIER CHARACTERISTICS[a]

Parameter	CE	CE + R_e
A_I	$-h_{fe}$	$-h_{fe}$
R_i	h_{ie}	$h_{ie} + (h_{fe} + 1)R_e$
A_V	$\dfrac{-h_{fe}R_c}{h_{ie}}$	$\dfrac{-h_{fe}R_c}{h_{ie} + (h_{fe} + 1)R_e} \approx -\dfrac{R_c}{R_e}$
R_o	$\dfrac{1}{h_{oe}} \approx \infty$	$\approx \infty$

[a]These characteristics are based on approximations assuming that the BJT output resistance r_o is much larger than the sum of the total resistances connected to the collector and emitter terminals of the transistor. Large resistances necessitate the use of the expressions derived in Section 5.3.3.

5.4 COMMON-COLLECTOR AMPLIFIERS

Common-collector amplifiers have the following general circuit topology:

- The input signal enters the BJT at the base.
- The output signal exits the BJT at the emitter.
- The collector is connected to a constant voltage, often the ground (common) terminal, sometimes with an intervening resistor.

A simple common-collector amplifier is shown in Figure 5.4-1. The collector resistor R_c is unnecessary in many applications; it is shown here for generality. The quiescent point of the BJT must be set with the circuitry external to the transistor so that it is in the forward-active region. The values of the resistors, R_c and R_s, and the DC voltage sources, V_{CC} and V_{BB}, have therefore been chosen so that the BJT is in the forward-active region and the circuit will operate as an amplifier.

Once the circuit quiescent conditions have been calculated and it has been determined that the BJT is in the forward-active region of operation, the significant h parameters are calculated to form the small-signal model of the transistor:

$$h_{fe} = \beta_F, \qquad h_{oe} = \left|\frac{I_C}{V_A}\right| \approx 0,$$

$$h_{ie} = (\beta_F + 1)\left|\frac{\eta V_t}{I_C}\right|. \tag{5.4-1}$$

The small-signal circuit performance can now be calculated. Total circuit performance is the sum of quiescent and small-signal performance. The process of AC modeling of the circuit and replacing the BJT with an appropriate AC model, applied to Figure 5.4-1, is shown in Figure 5.4-2.

The small-signal performance can be obtained from analysis of the circuit of Figure 5.4-2b. Definitions for these quantities often vary due to differing definition

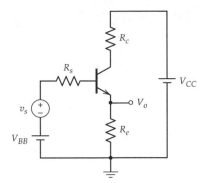

Figure 5.4-1
Typical common-collector amplifier.

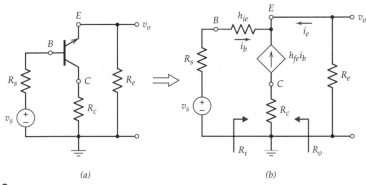

(a)   (b)

Figure 5.4-2
AC modeling of common-collector amplifier: (*a*) small-signal circuit, DC sources set to zero; (*b*) BJT replaced by *h*-parameter model.

of the exact location of the point of measurement. Care must always be taken to *ensure that measurement points are clearly understood before any analysis begins.* Many similarities exist between the circuit topologies of a common collector amplifier and a common-emitter amplifier with an emitter resistor: Those similarities will be utilized in calculating common-collector amplifier performance characteristics.

Current Gain
For this simple transistor amplifier, the current gain is defined as the ratio of load current to input current, that is

$$A_I \equiv \frac{i_l}{i_b} = \frac{-i_e}{i_b}. \tag{5.4-2}$$

From the circuit of Figure 5.4-2*b*, it can be determined that the emitter and base currents are related through the dependent current source by the constant $h_{fe} + 1$. The current gain is dependent only on the BJT characteristics and independent of any other circuit element values. Its value is given by

$$A_I = h_{fe} + 1. \tag{5.4-3}$$

Input Resistance
The input resistance (shown in Figure 5.4-2*b*) is given by

$$R_i \equiv \frac{v_b}{i_b} = \frac{h_{ie}i_b + R_e(i_b + h_{fe}i_b)}{i_b} = h_{ie} + (h_{fe} + 1)R_e. \tag{5.4-4}$$

This result is identical to that for a common-emitter amplifier with an emitter resistor. The input resistance to a common-collector amplifier is large for typical values of the load resistance R_e.

Voltage Gain

The voltage gain is the ratio of output voltage to input voltage. If the input voltage is again taken to be the voltage at the input to the transistor v_b:

$$A_V \equiv \frac{v_o}{v_b}. \tag{5.4-5}$$

Using previously introduced methods to calculate the voltage gain yields

$$A_V = \frac{v_o}{v_b} = \left(\frac{v_o}{i_l}\right)\left(\frac{i_l}{i_b}\right)\left(\frac{i_b}{v_b}\right), \tag{5.4-6}$$

or, replacing each term with its equivalent expression,

$$A_V = (R_e)(A_l)\left(\frac{1}{R_i}\right) = \frac{(h_{fe} + 1)R_e}{h_{ie} + (h_{fe} + 1)R_e}. \tag{5.4-7}$$

Equation 5.4-7 is somewhat less than unity. A form of the voltage gain expression that shows this approximation is

$$A_V = 1 - \frac{h_{ie}}{h_{ie} + (h_{fe} + 1)R_e} = 1 - \frac{h_{ie}}{R_i} \approx 1. \tag{5.4-8}$$

Often the voltage gain from the source to the load is of interest as well. This overall voltage gain can be defined as

$$A_{VS} \equiv \frac{v_o}{v_s}. \tag{5.4-9}$$

This ratio can be directly derived from the voltage gain A_V and a voltage division between the source resistance R_s and the amplifier input resistance R_i:

$$A_{VS} = \frac{v_o}{v_s} = \left(\frac{v_o}{v_b}\right)\left(\frac{v_b}{v_s}\right) = A_V\left(\frac{R_i}{R_i + R_s}\right). \tag{5.4-10}$$

Appropriate substitutions lead to

$$A_{VS} = 1 - \frac{h_{ie} + R_s}{R_i + R_s}. \tag{5.4-11}$$

Output Resistance

The output resistance is defined as the Thévenin resistance at the output of the amplifier looking back into the amplifier. The circuit of Figure 5.4-3 defines the necessary topology and circuit variables for output resistance calculations.

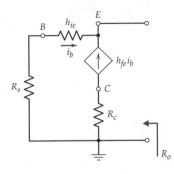

Figure 5.4-3
Common-collector output
resistance AC equivalent circuit.

If a voltage v is applied to the output terminals, the base current is found to be

$$i_b = \frac{-v}{R_s + h_{ie}}. \tag{5.4-12}$$

The total current flowing into the emitter of the BJT is given by

$$i = -i_b - h_{fe}i_b, \tag{5.4-13}$$

from which the output resistance is calculated:

$$R_o = \frac{v}{i} = \frac{R_s + h_{ie}}{h_{fe} + 1}. \tag{5.4-14}$$

The output resistance for a common-collector amplifier is typically small.
The common-collector amplifier has been shown to have the following:

- High input resistance
- Low output resistance
- High current gain
- Low voltage gain

A summary of common-collector performance characteristics, as derived in this section, is given in Table 5.4-1.

TABLE 5.4-1 COMMON-COLLECTOR AMPLIFIER CHARACTERISTICS[a]

A_I	$h_{fe} + 1$
R_i	$h_{ie} + (h_{fe} + 1)R_e$
A_V	$1 - \dfrac{h_{ie}}{R_i} \approx 1$
R_o	$\dfrac{R_s + h_{ie}}{h_{fe} + 1}$

[a]These characteristics are based on approximations assuming that the BJT output resistance r_o is much larger than the sum of the total resistance connected to the collector and emitter terminals of the transistor. Large resistances necessitate the use of the expressions derived in Section 5.4.1.

EXAMPLE 5.4-1 A silicon BJT with parameters

$$\beta_F = 100, \qquad V_A = 250 \text{ V}$$

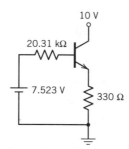

is placed in the amplifier shown. Assuming the amplifier is operating in its midband frequency range, determine the following circuit performance parameters: $A_V = v_o/v_s$, $A_I = i_{out}/i_{in}$, R_{in}, and R_{out}.

SOLUTION

The quiescent point of the BJT must be found first to ensure that it is in the forward-active region and to find its h-parameters. For DC operation, all capacitors appear as open circuits. The DC equivalent circuit, after replacing the base circuit with its Thévenin equivalent, is as shown.

The base and collector currents are

$$I_B = \frac{7.523 - 0.7}{20,310 + 101(330)} = 127.2 \ \mu\text{A},$$

$$I_C = 100 I_B = 12.72 \text{ mA}.$$

Here, V_{CE} must be checked to verify that the BJT is in the forward-active region:

$$V_{CE} = 10 - \frac{151}{150} I_C(330) = 5.76 \text{ V} \geq 0.2 \text{ V}.$$

The BJT h-parameters can now be determined:

$$h_{fe} = 100, \qquad h_{ie} = 101 \frac{\eta V_t}{|I_C|} \approx 210 \ \Omega, \qquad \frac{1}{h_{oe}} = \left| \frac{V_A}{I_C} \right| = 19.65 \text{ k}\Omega.$$

An AC equivalent circuit can now be determined as shown. All DC sources are set to zero and, since the circuit is in its midband frequency range, all capacitors are replaced by short circuits. The AC equivalent circuit leads to the determination of amplifier performance characteristics. Care must be taken concerning parameter definitions:

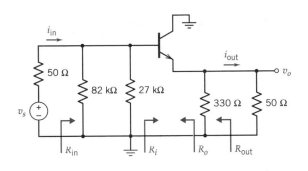

$$R_i = h_{ie} + (h_{fe} + 1)R_e$$

$$= 210 + (101)(330\|50) = 4.596 \text{ k}\Omega,$$

$$R_{in} = 82 \text{ k}\Omega\|27 \text{ k}\Omega|R_i = 3.75 \text{ k}\Omega,$$

$$A_v = \frac{v_o}{v_s} = \frac{v_o}{v_b}\frac{v_b}{v_s} = \left(1 - \frac{h_{ie}}{R_i}\right)\frac{R_{in}}{R_{in} + 50} = \left(1 - \frac{210}{4596}\right)\frac{3750}{3800} = 0.942,$$

$$R_v = \frac{R_s + h_{ie}}{h_{fe} + 1} = \frac{50\|82,000\|27,000 + 210}{101} = 2.573 \ \Omega,$$

$$R_{out} = R_o\|330 = 2.55 \ \Omega,$$

$$A_I = \frac{i_{out}}{i_{in}} = \left(\frac{i_b}{i_{in}}\right)\left(\frac{i_o}{i_b}\right)\left(\frac{i_{out}}{i_o}\right)$$

$$= \left(\frac{82,000\|27,000}{82,000\|27,000 + 3750}\right)(101)\left(\frac{330}{330 + 50}\right) = 74.0.$$

5.4.1 Effect of Nonzero h_{oe} on Common-Collector Amplifiers

The basic circuit topology of a common-collector amplifier is essentially the same as a common-emitter amplifier with an emitter resistor. The only significant difference is the location of the output. Many of the calculations necessary to obtain amplifier performance characteristics can be drawn from those performed in Section 5.3.2. Figure 5.4-4 serves as a reference for amplifier performance calculations.

Current Gain
The current gain is defined as the ratio of load to base current:

$$A_I = \frac{-i_e}{i_b}. \tag{5.4-15}$$

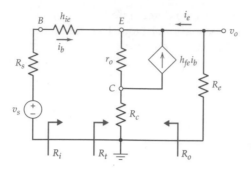

Figure 5.4-4
AC equivalent circuit, common-collector amplifier.

In the circuit diagram of Figure 5.4-4 it can be seen that

$$i_b = \frac{v_o}{R_t}, \qquad -i_e = \frac{v_o}{R_e}. \tag{5.4-16}$$

The quantity R_t is measured here in the identical fashion as with a common-emitter amplifier with an emitter resistor. Using the results obtained in Section 5.3 (Equation 5.3-28), the current gain becomes

$$A_I = \frac{R_t}{R_e} = \frac{(h_{fe} + 1)R_e[r_o/(r_o + R_c + R_e)]}{R_e}, \tag{5.4-17}$$

which can be simplified to become

$$A_I = \frac{(h_{fe} + 1)r_o + R_c}{r_o + R_c + R_e}. \tag{5.4-18}$$

Current gain is reduced by the presence of nonzero h_{oe}.

Input Resistance
The input resistance of a common-collector amplifier is identical to the input resistance of a common-emitter amplifier with an emitter resistor:

$$R_i = h_{ie} + (h_{fe} + 1)R_e \frac{r_o}{r_o + R_c + R_e}. \tag{5.4-19}$$

As in the common-emitter case, the input resistance is reduced.

Voltage Gain
The voltage gain can be obtained as a voltage division between h_{ie} and R_t:

$$A_V = \frac{v_a}{v_b} = \frac{R_t}{h_{ie} + R_t} = 1 - \frac{h_{ie}}{R_i}, \tag{5.4-20}$$

which is the same as Equation 5.4-8 except for the definition of R_i. Here, R_i is given by Equation 5.4-19: The slight decrease in R_i due nonzero h_{oe} will also very slightly reduce the gain.

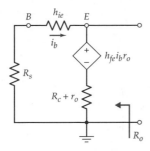

Figure 5.4-5
AC equivalent circuit for output resistance calculation.

Output Resistance

The output resistance is best calculated by setting the input source to zero and performing a source transformation on the dependent current source as shown in Figure 5.4-5. In order to simplify the diagram, the BJT output resistance r_o has been combined with the collector resistor R_c into a single resistance $R_c + r_o$.

If a voltage v is applied across the output terminals, the total current flowing into the terminal is given by

$$i = -i_b + \frac{v - h_{fe}i_b r_o}{r_o + R_c},$$

(5.4-21)

where

$$i_b = -\frac{v}{h_{ie} + R_s}.$$

(5.4-22)

Substitution of Equation 5.4-22 into Equation 5.4-21 leads to

$$R_o = \frac{v}{i} \approx \frac{h_{ie} + R_s}{1 + (1 + h_{fe}r_o)/(r_o + R_c)}$$

$$\approx \frac{h_{ie} + R_s}{1 + h_{fe}[r_o/(r_o + R_c)]}$$

(5.4-23)

Nonzero h_{oe} will slightly increase the output resistance.

5.5 Common-Base Amplifiers

Common-base amplifiers have the following general circuit topology:

- The input signal enters the BJT at the emitter.
- The output signal exits the BJT at the collector.
- The base is connected to a constant voltage, often the ground (common) terminal, sometimes with an intervening resistor.

A simple common-base amplifier is shown in Figure 5.5-1. The quiescent point of the BJT must be set with the circuitry external to the transistor so that it is in the forward-active region. The values of the resistors R_c, the base resistors R_{b1} and R_{b2},

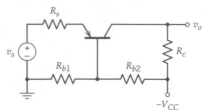

Figure 5.5-1
Typical common-base amplifier.

and the DC voltage sources V_{CC} and V_{BB} have therefore been chosen so that the BJT is in the forward-active region and the circuit will operate as an amplifier.

The circuit quiescent conditions must be determined to ensure that the BJT is in the forward-active region of operation. The significant h-parameters are then calculated to form the small-signal model of the transistor:

$$h_{fe} = \beta_F, \qquad h_{oe} = \left|\frac{I_C}{V_A}\right| \approx 0,$$

$$h_{ie} = (\beta_F + 1)\left|\frac{\eta V_t}{I_C}\right|. \tag{5.5-1}$$

The small-signal circuit performance can now be calculated. Total circuit performance is, as usual, the sum of quiescent and small-signal performance. The process of AC modeling of the circuit and replacing the BJT with an appropriate AC model, applied to Figure 5.5-1, is shown in Figure 5.5-2. In order to simplify calculations, the base resistors have been combined as a parallel combination:

$$R_b = R_{b1} \| R_{b2}. \tag{5.5-2}$$

The small-signal performance can be obtained from analysis of the circuit of Figure 5.5-2b. Definitions for these quantities often vary due to differing definition of the exact location of the point of measurement. Care must always be taken to *ensure that measurement points are clearly understood before any analysis begins.* A few similarities exist between the circuit topologies of a common-base amplifier and the other amplifier types previously discussed: Those similarities will be utilized in calculating performance characteristics.

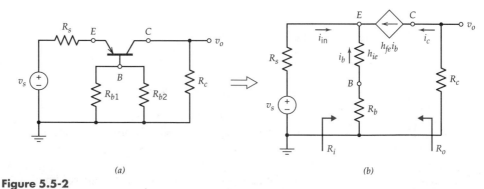

(a)   (b)

Figure 5.5-2
AC modeling of common-base amplifier: (a) small-signal circuit, DC sources set to zero; (b) BJT replaced by h-parameter model.

Current Gain

For this simple transistor amplifier, the current gain is defined as the ratio of load current to input current, that is,

$$A_I \equiv \frac{i_l}{i_{in}} = \frac{-i_c}{i_e}. \tag{5.5-3}$$

This current ratio for a BJT in the forward-active region is well known:

$$A_I = \frac{h_{fe}}{h_{fe} + 1}. \tag{5.5-4}$$

This gain is very nearly unity.

Input Resistance

The input resistance is defined as

$$R_i = \frac{v_e}{i_{in}} = \frac{v_e}{i_b} \frac{i_b}{i_{in}} = (R_b + h_{ie}) \frac{1}{h_{fe} + 1}. \tag{5.5-5}$$

Common-base amplifiers have low input resistance.

Voltage Gain

The voltage gain is the ratio of output voltage to input voltage. If the input voltage is again taken to be the voltage at the input to the transistor v_e:

$$A_V \equiv \frac{v_o}{v_e}. \tag{5.5-6}$$

Using previously introduced methods to calculate the voltage gain yields

$$A_V = \frac{v_o}{v_e} = \left(\frac{v_o}{i_l}\right)\left(\frac{i_l}{i_{in}}\right)\left(\frac{i_{in}}{v_e}\right), \tag{5.5-7a}$$

or, after replacing each term with its value,

$$A_V = R_c A_I \frac{1}{R_i} = \frac{h_{fe}R_c}{h_{ie} + R_b}. \tag{5.5-7b}$$

The voltage gain can be large and is noninverting.

Output Resistance

The output resistance is defined as the Thévenin resistance at the output of the amplifier looking back into the amplifier. The circuit topology for this operation is the same as for the common-emitter amplifier with an emitter resistor (with the names of the resistors changed). The results are the same:

$$R_o \approx \infty, \tag{5.5-8}$$

a very large value.

TABLE 5.5-1 SUMMARY OF COMMON-BASE AMPLIFIER PERFORMANCE CHARACTERISTICS[a]

A_I	$\dfrac{h_{fe}}{h_{fe} + 1}$
R_i	$\dfrac{h_{ie} + R_b}{h_{fe} + 1}$
A_V	$\dfrac{h_{fe}R_c}{h_{ie} + R_b}$
R_o	$\approx \infty$

[a]These characteristics are based on approximations assuming that the BJT output resistance r_o is much larger than the sum of the total resistance connected to the collector and emitter terminals of the transistor. Large resistances necessitate the use of the expressions derived in Section 5.5.1.

The common-base amplifier has been shown to have the following:

- Low input resistance
- High output resistance
- Low gain
- Moderate to high voltage gain

A summary of common-base performance characteristics, as derived in this section, is given in Table 5.5-1. Increased voltage gain and decreased input resistance can be obtained with a bypass capacitor from the BJT base terminal to ground, thereby making $R_b = 0$ in the AC sense.

EXAMPLE 5.5-1 A silicon BJT with parameters

$$\beta_F = 120, \qquad V_A = 200 \text{ V}$$

is placed in the amplifier shown. Determine the following performance parameters: $A_V = v_o/v_s$, R_{in}, and R_{out}.

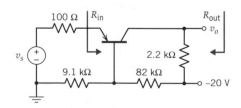

SOLUTION

The DC equivalent circuit is as shown (a Thévenin equivalent of the base biasing circuit was made). The quiescent conditions can be calculated beginning with the base current. Around the base–emitter loop, KVL yields

$$121I_B(100) - 0.7 + I_B(8191) + 1.998 = 0$$

or

$$I_B = -63.96 \text{ μA}, \qquad I_C = \beta_F I_B = -7.675 \text{ mA}.$$

The collector–emitter voltage is

$$V_{CE} = -20 - (121I_B)(100) - (I_C)(2200) = -2.341 \text{ V}.$$

The constraints for a *pnp* BJT in the forward-active region are met.

The *h* parameters for the *pnp* BJT are obtained in the same manner as *npn* BJT parameters. At this quiescent point they are given by

$$h_{fe} = 120, \qquad h_{ie} = 121\frac{\eta V_t}{|I_C|} \approx 410 \text{ Ω}, \qquad \frac{1}{h_{oe}} = \left|\frac{V_A}{I_C}\right| = 26.06 \text{ kΩ}.$$

The AC equivalent circuit is as shown. The circuit performance parameters can be calculated as

$$R_{in} = \frac{h_{ie} + R_b}{h_{fe} + 1} = \frac{410 + 8191}{121} = 71.1 \text{ Ω},$$

$$R_{out} = 2.2 \text{ kΩ},$$

$$A_V = \frac{v_o}{v_s} = \left(\frac{v_o}{v_e}\right)\left(\frac{v_e}{v_s}\right) = \left(\frac{h_{fe}R_c}{h_{ie} + R_b}\right)\left(\frac{R_{in}}{R_{in} + 100}\right)$$

$$= \left(\frac{(120)(2200)}{410 + 8191}\right)\left(\frac{71.1}{71.1 + 100}\right) = 12.76.$$

5.5.1 Effect of Nonzero h_{oe} on Common-Base Amplifiers

Nonzero h_{oe} is modeled by altering the circuit of Figure 5.5-2b to include the output resistance r_o of the BJT. After a source transformation is performed on the dependent current source, the AC equivalent circuit becomes the circuit shown in Figure 5.5-3.

Input Resistance

If a voltage v_e is applied to the emitter of the amplifier (assuming the source and source resistor are removed), the input current is given by

$$i_{in} = \frac{v_e}{h_{ie} + R_b} + \frac{v_e - h_{fe}i_b r_o}{r_o + R_c},\tag{5.5-9}$$

where

$$i_b = \frac{-v_e}{h_{ie} + R_b}.\tag{5.5-10}$$

Substituting Equation 5.5-10 into 5.5-9 yields

$$i_{in} = \frac{v_e}{h_{ie} + R_b} + \frac{v_e}{r_o + R_c}\left(1 + \frac{h_{fe}r_o}{h_{ie} + R_b}\right).\tag{5.5-11}$$

The input resistance can now be determined as

$$R_i = \frac{v_e}{i_{in}} = \frac{h_{ie} + R_b}{1 + (h_{ie} + R_b + h_{fe}r_o)/(r_o + R_c)}.\tag{5.5-12}$$

In typical common-base amplifiers, R_b is not a large resistance. It is therefore reasonable to assume that

$$h_{ie} + R_b \ll h_{fe}r_o.\tag{5.5-13}$$

Under that assumption, the input resistance expression reduces to

$$R_i \approx \frac{h_{ie} + R_b}{1 + h_{fe}[r_o/(r_o + R_c)]}.\tag{5.5-14}$$

The presence of nonzero h_{oe} increases the input resistance.

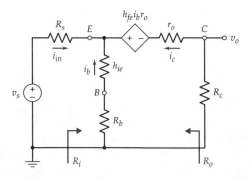

Figure 5.5-3
Common-base amplifier AC equivalent circuit including r_o.

Current Gain

The current gain is defined as the output current divided by the input current:

$$A_I = \frac{-i_c}{i_{in}}.$$

(5.5-15)

The output current was expressed as a component of Equation 5.5-11:

$$-i_c = \frac{v_e}{r_o + R_c}\left(1 + \frac{h_{fe}r_o}{h_{ie} + R_b}\right).$$

(5.5-16)

Algebraic manipulations on Equations 5.5-11 and 5.5-16 lead to

$$A_I = \frac{h_{fe}r_o + h_{ie} + R_b}{(h_{fe} + 1)r_o + h_{ie} + R_b + R_c}.$$

(5.5-17)

The current gain remains essentially unchanged by the presence of nonzero h_{oe}. Small external resistances will very slightly increase the gain while extremely large R_c will decrease the gain slightly.

Voltage Gain

The output voltage can be determined through a voltage division between r_o and R_c in Figure 5.5-3:

$$v_o = (v_e - h_{fe}i_b r_o)\frac{R_c}{R_c + r_o}.$$

(5.5-18)

The emitter voltage v_e and the base current i_b are related by h_{ie} and R_b through Equation 5.5-10. Substitution of this expression yields

$$v_o = v_e\left(1 + \frac{h_{fe}r_o}{h_{ie} + R_b}\right)\frac{R_c}{R_c + r_o}.$$

(5.5-19)

The voltage gain is easily obtained:

$$A_V = \frac{v_o}{v_e} = \left(1 + \frac{h_{fe}r_o}{h_{ie} + R_b}\right)\frac{R_c}{R_c + r_o} \approx \frac{h_{fe}R_c}{h_{ie} + R_b}\frac{r_o}{R_c + r_o}.$$

(5.5-20)

The voltage gain is reduced by the presence of nonzero h_{oe}.

Output Resistance

The output resistance is the same as that seen for a common-emitter amplifier with an emitter resistor (Equation 5.3-40). The external resistors have changed name so that the expression becomes

$$R_o = [R_s||(R_b + h_{ie})] + r_o\left(1 + \frac{h_{fe}R_s}{R_s + h_{ie} + R_b}\right).$$

(5.5-21)

This is a very large resistance.

5.6 COMPARISON OF BJT AMPLIFIER TYPES

Single-BJT amplifiers have been shown to fall into three general categories: common-emitter (both with and without an emitter resistor), common-collector, and common-base. The performance characteristics for each type of amplifier are summarized in Table 5.6-1.

The common-emitter configuration appears the most useful of the three types: It provides both significant current and voltage gain (in each case with an inversion). The common-emitter amplifier input and output impedances are high. In fact, this configuration is the most versatile of the three types and will often form the major gain portion of multiple-transistor amplifiers.

The common-collector and common-base configurations amplify either current or voltage, but not both. Neither inverts either the voltage or current signal. These configurations find greatest utility as impedance-matching (or buffer) stages of multistage amplifiers. Common-collector amplifier stages are capable of easily driving low-impedance loads. Common-base stages can impedance match a very low impedance source.

Quantitative expressions for the amplifier performance characteristics are found in Table 5.6-2. These expressions are based on the often realistic assumption concerning the relative size of the external resistors (connected to the collector and emit-

TABLE 5.6-1 QUALITATIVE COMPARISON OF BJT AMPLIFIER CONFIGURATIONS

Parameter	CE	CE + R_e	CC	CB
A_I	High	High	High	Approximately unity
R_i	Medium	High	High	Low
A_V	High	Medium	Approximately unity	Medium to high
R_o	High	Very high	Low	Very high

TABLE 5.6-2 SUMMARY OF BJT AMPLIFIER PERFORMANCE CHARACTERISTICS

Parameter	CE	CE + R_e	CC	CB
A_I	$-h_{fe}$	$-h_{fe}$	$h_{fe} + 1$	$\dfrac{h_{fe}}{h_{fe} + 1}$
R_i	h_{ie}	$h_{ie} + (h_{fe} + 1)\boldsymbol{R_e}$	$h_{ie} + (h_{fe} + 1)\boldsymbol{R_e}$	$\dfrac{h_{ie} + \boldsymbol{R_b}}{h_{fe} + 1}$
A_V	$\dfrac{-h_{fe}\boldsymbol{R_c}}{h_{ie}}$	$\dfrac{-h_{fe}\boldsymbol{R_c}}{h_{ie} + (h_{fe} + 1)\boldsymbol{R_e}} \approx -\dfrac{\boldsymbol{R_c}}{\boldsymbol{R_e}}$	$1 - \dfrac{h_{ie}}{R_i} \approx 1$	$\dfrac{h_{fe}\boldsymbol{R_c}}{h_{ie} + \boldsymbol{R_b}}$
R_o	$\dfrac{1}{h_{oe}} \approx \infty$	$\approx \infty$	$\dfrac{R_s + h_{ie}}{h_{fe} + 1}$	$\approx \infty$

NOTES: All measurements are made at the input port and/or the output port of the BJT. The bold italic resistances are the total Thévenin resistances connected to that BJT port. Expressions assume that $h_{oe}(\boldsymbol{R_e} + \boldsymbol{R_c}) \ll 1$.

ter terminals) and the BJT output resistance: $h_{oe}(R_e + R_c) \ll 1$. High-resistance loads require the more exact expressions for amplifier performance, which do not depend on that assumption. These expressions can be found in previous subsections concerning the effects of nonzero h_{oe}.

5.7 FET LOW-FREQUENCY MODELS

The FET saturation region with small-signal input variations about the DC quiescent operating point (Q point) is considered approximately linear in nature. Therefore like the BJT, small-signal approximations, in the form of approximate circuit networks, are commonly used to model FET operation in the saturation region. The model is used to approximate all FET types.

Figure 5.7-1a shows a "generic" FET biased in the saturation region by a gate–source voltage V_{GSQ}, with a drain–source voltage V_{DSQ} and drain current I_{DQ}. The effect of small changes of the bias voltages on I_D can be found by independently varying V_{GSQ} and V_{DSQ}, as shown in Figures 5.7-1b and c, respectively.

The change in I_D with respect to small variations in V_{GS} and V_{DS} are expressed as conductance parameters:

1. *Small-signal transconductance or mutual conductance g_m,* often referred to simply as "transconductance." The mathematical relation for determining transconductance corresponding to the measurement in Figure 5.7-1b is

$$g_m = \left.\frac{\partial I_D}{\partial V_{GS}}\right|_{V_{DS}=\text{const}} \approx \left.\frac{\Delta I_D}{\Delta V_{GS}}\right|_{V_{DS}=\text{const}}. \tag{5.7-1}$$

2. *Small-signal drain conductance or output conductance g_d,* corresponding to the measurement in Figure 5.7-1c, is defined as

$$g_d = r_d^{-1} = \left.\frac{\partial I_D}{\partial V_{DS}}\right|_{V_{GS}=\text{const}} \approx \left.\frac{\Delta I_D}{\Delta V_{DS}}\right|_{V_{GS}=\text{const}}, \tag{5.7-2}$$

where r_d is the drain resistance or output resistance in Ohms.

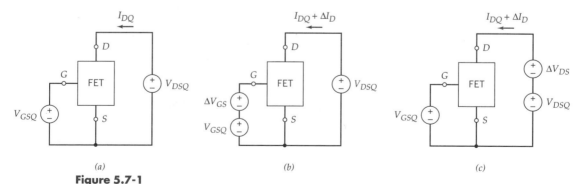

Figure 5.7-1
(a) Generic FET biased in saturation region by gate–source voltage V_{GSQ}, with a drain–source voltage V_{DSQ} and drain current I_{DQ}. (b) Method for determining gate small-signal transconductance with small variations in V_{GS} causing small changes in I_D. (c) Method for determining drain small-signal conductance with small variations in V_{DS} causing small changes in I_D.

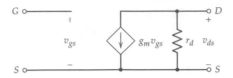

Figure 5.7-2
Low-frequency small-signal model of FET.

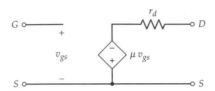

Figure 5.7-3
Equivalent low-frequency model of FET with voltage-controlled voltage source.

Both g_m and g_d have units of conductance and are expressed in Siemens [S].

Consider the general case where both V_{DS} and V_{GS} changed simultaneously. The corresponding total change in the drain current is

$$\Delta I_D = \left(\left. \frac{\partial I_D}{\partial V_{GS}} \right|_{V_{DS}=\text{const}} \right) \Delta V_{GS} + \left(\left. \frac{\partial I_D}{\partial V_{DS}} \right|_{V_{GS}=\text{const}} \right) \Delta V_{DS}. \tag{5.7-3}$$

If ΔI_D, ΔV_{DS}, and ΔV_{GS} are time-varying quantities about the Q point, they are AC signals. Equation 5.7-3 can then be rewritten as

$$i_d = \left(\left. \frac{\partial I_D}{\partial v_{GS}} \right|_Q \right) v_{gs} + \left(\left. \frac{\partial I_D}{\partial v_{DS}} \right|_Q \right) v_{ds}. \tag{5.7-4}$$

Substituting the small-signal conductances into Equation 5.7-4 yields

$$i_d = g_m v_{gs} + g_d v_{ds}. \tag{5.7-5}$$

Recall from Chapter 4 that the gate current is assumed to be zero. Therefore,

$$\Delta I_G = 0. \tag{5.7-6}$$

Equations 5.7-5 and 5.7-6 relate small-signal quantities that can are represented by the small-signal equivalent circuit of Figure 5.7-2.

The voltage-controlled current source in the low-frequency small-signal FET model in Figure 5.7-2 may be changed to a voltage-controlled voltage source by source transformation. The source-transformed FET model is shown in Figure 5.7-3. The gain or amplification factor μ is defined as

$$\mu = \frac{g_m}{g_d} = g_m r_d. \tag{5.7-7}$$

The FET models shown in Figures 5.7-2 and 5.7-3 are valid for all types of FETs biased in the saturation region. The actual values of the elements in the model vary for different FETs.

The value of $r_d = g_d^{-1}$ can be obtained from the output characteristic curve of the

FET at the desired Q point. From Equation 5.7-2, g_d is the slope of the line defined by I_D and V_{DS} at the desired Q point defined by a constant V_{GSQ},

$$g_d = \left.\frac{\Delta i_D}{\Delta v_{DS}}\right|_{V_{GSQ}} = r_d^{-1}. \tag{5.7-8}$$

Typical values for $r_d = g_d^{-1}$ range from 10 to 100 kΩ. Here, r_d can also be determined from I_D and the Early voltage V_A using the relationship

$$r_d = \frac{V_A}{I_D}. \tag{5.7-9}$$

The small-signal value for g_m for the JFET and depletion MOSFET is found by taking the derivative of I_D from Table 4.4-1 with respect to V_{GS},

$$
\begin{aligned}
g_m &= \frac{\partial I_D}{\partial V_{GS}} = \frac{\partial[I_{DSS}(1 - V_{GS}/V_{PO})^2]}{\partial V_{GS}} \\
&= \frac{-2I_{DSS}(1 - V_{GS}/V_{PO})}{V_{PO}} = 2\frac{I_{DSS}}{V_{PO}^2}(V_{GS} - V_{PO}).
\end{aligned}
\tag{5.7-10}
$$

But by definition, $I_D = I_{DSS}(1 - V_{GS}/V_{PO})^2$ or

$$I_{DSS} = \frac{I_D}{(1 - V_{GS}/V_{PO})^2}. \tag{5.7-11}$$

Substituting Equation 5.7-11 into 5.7-10 and simplifying yields an alternate expression for the small-signal transconductance of depletion-type FETs:

$$g_m = \frac{2I_D}{V_{GS} - V_{PO}} = 2\sqrt{I_D\left(\frac{I_{DSS}}{V_{PO}^2}\right)}. \tag{5.7-12}$$

For the enhancement MOSFET, g_m is found by taking the derivative of I_D from Table 4.4-1 with respect to V_{GS},

$$
\begin{aligned}
g_m &= \frac{\partial I_D}{\partial V_{GS}} = \frac{\partial[K(V_{GS} - V_T)^2]}{\partial V_{GS}} \\
&= \frac{2I_D}{V_{GS} - V_T} \\
&= 2K(V_{GS} - V_T) \\
&= 2\sqrt{I_D K}.
\end{aligned}
\tag{5.7-13}
$$

Typical values for g_m range from 10^{-4} to 10^{-2} S.

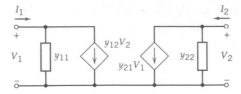

Figure 5.7-4
Generic y-parameter two-port network.

An alternate method for arriving at the small-signal FET model is through the use of two-port analysis. The y-parameter two-port parameters are used since the independent variables are port voltages (V_{GS} and V_{DS}) and the dependent variable is the drain current I_D. As in the BJT, one terminal of the FET is assigned as a common terminal to both ports. In practice, the FET source is assigned as the common terminal.

The generic y-parameter two-port network is shown in Figure 5.7-4. Recall the y-parameter equations from Equations 5.1-4 and 5.1-6:

$$\begin{bmatrix} I_1 \\ I_2 \end{bmatrix} = \begin{bmatrix} y_{11} & y_{12} \\ y_{21} & y_{22} \end{bmatrix} \begin{bmatrix} V_1 \\ V_2 \end{bmatrix}$$

and

$$y_{ij} = \left. \frac{\partial I_i}{\partial V_j} \right|_{V_{k \neq j} = \text{const}} .$$

For the FET, v_{gs} and v_{ds} are represented as V_1 and V_2 in the two-port network, respectively. The currents I_1 and I_2 are I_G and I_D, respectively. However, the gate current in the FET is approximately zero. Therefore, $I_1 = 0$, which implies that the dependent current source $y_{12}V_2 = 0$, and the two-port network exhibits an infinite input impedance ($y_{11} = 0$). The remaining two parameters of the y-parameter two-port network are the transadmittance y_{21} and the output admittance y_{22}. The real components of the transadmittance and output admittance were found in Equations 5.7-1 and 5.7-2:

$$|y_{21}| = g_m, \qquad (5.7\text{-}14)$$

$$|y_{22}| = g_d. \qquad (5.7\text{-}15)$$

Often y parameters with descriptive subscripts are used to specify parameters in manufacturers' data sheets. The second subscript defines which terminal is chosen as common (the source), and the first subscript identifies the function of the parameter:

$y_{fs} = |y_{21}| = g_m =$ input admittance: source as common terminal (5.7-14)
$y_{os} = |y_{22}| = g_d =$ output admittance: source as common terminal (5.7-15)

The small-signal parameters for the FET are summarized in Table 5.7-1. The FET parameters are dependent on quiescent conditions.

TABLE 5.7-1 FET SMALL-SIGNAL PARAMETERS

Small-Signal Parameter	Value
JFET and depletion MOSFET: $y_{fs} = \|y_{21}\| = g_m$	$2\dfrac{I_{DSS}}{V_{PO}^2}(V_{GS} - V_{PO}) = \dfrac{2I_D}{V_{GS} - V_{PO}} = 2\sqrt{I_D\dfrac{I_{DSS}}{V_{PO}^2}}$
JFET and depletion MOSFET: $y_{os} = \|y_{22}\| = g_d$	$\left.\dfrac{\Delta i_D}{\Delta V_{DS}}\right\|_{V_{GSQ}} = \dfrac{I_D}{V_A} = r_d^{-1}$
Enhancement MOSFET: $y_{fs} = \|y_{21}\| = g_m$	$2K(V_{GS} - V_T) = \dfrac{2I_D}{V_{GS} - V_T} = 2\sqrt{I_D K}$
Enhancement MOSFET: $y_{os} = \|y_{22}\| = g_d$	$\left.\dfrac{\Delta i_D}{\Delta V_{DS}}\right\|_{V_{GSQ}} = \dfrac{I_D}{V_A} = r_d^{-1}$

EXAMPLE 5.7-1 Given a depletion NMOSFET with parameters

$$I_{DSS} = 5 \text{ mA}, \qquad V_{PO} = -2 \text{ V}$$

operating in the given circuit, determine the small-signal transconductance for the FET.

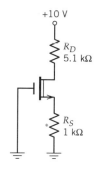

+10 V

R_D
5.1 kΩ

R_S
1 kΩ

SOLUTION

The quiescent conditions must first be obtained.
Note that $V_G = 0$. Therefore,

$$V_{GS} = V_G - V_S = -V_S = -I_D(1000).$$

Solvings for V_{DS},

$$V_{DS} = 10 - I_D(1000 + 5100)$$
$$= 10 - I_D(6100).$$

Assume that the FET is in saturation. That is,

$$V_{DS} > V_{GS} - V_{PO}.$$

Then,

$$I_D = I_{DSS}\left(1 - \frac{V_{GS}}{V_{PO}}\right)^2$$

$$= I_{DSS}\left(1 + \frac{I_D \cdot 10^3}{V_{PO}}\right)^2$$

$$= 5 \times 10^{-3}(1 - 500I_D)^2$$

$$= 5 \times 10^{-3}(1 - 1000I_D + 250 \times 10^3 I_D^2)$$

$$= 5 \times 10^{-3}(1 - 1000I_D + 250 \times 10^3 I_D^2).$$

Use the quadratic equation to solve for I_D. The two roots of the second-order equation are

$$I_D = 1.07 \text{ mA} \quad \text{or} \quad 3.725 \text{ mA}.$$

For $I_D = 1.07$ mA,

$$V_{GS} = -I_D(1000) = -1.07 \times 10^{-3}(1000) = -1.07 \text{ V},$$

$$V_{DS} = 10 - I_D(6100)$$

$$= 10 - 1.07 \times 10^{-3}(6.1 \times 10^3)$$

$$= 3.47 \text{ V}.$$

Similarly, for $I_D = 3.725$ mA,

$$V_{GS} = -I_D(1000) = -3.725 \times 10^{-3}(1000) = -3.725 \text{ V},$$

$$V_{DS} = 10 - I_D(6100)$$

$$= 10 - 3.725 \times 10^{-3}(6.1 \times 10^3)$$

$$= -12.7 \text{ V}. \quad \Leftarrow \quad \text{This solution is clearly not valid.}$$

Therefore, the drain current $I_D = 1.07$ mA.
  Confirm that the FET is in saturation:

$$V_{DS} > V_{GS} - V_{PO} = -1.07 + 2 = 0.93 \text{ V}$$

and

$$3.47 \text{ V} > 0.93 \text{ V}. \quad \Leftarrow \quad \text{The FET is in saturation.}$$

The transconductance of the FET is

$$g_m = \frac{-2I_D}{V_{PO} - V_{GS}} = \frac{-2(1.07 \times 10^{-3})}{-2 - (-1.07)} = 2.3 \text{ mS}.$$

5.8 COMMON–SOURCE AMPLIFIERS

Between the ohmic and cutoff regions of the FET characteristics lies a linear region where the variation in the output voltage about the quiescent point is directly proportional to the variation in the input voltage. This section develops the characteristics of the common–source amplifier using the small-signal parameters developed in Section 5.7.

Common–source amplifiers have the following general circuit topology:

- The input signal enters the FET at the gate.
- The output signal exits the FET at the drain.
- The source terminal is connected to a constant voltage, often the ground (common) terminal, sometimes with an intervening resistor.

A simple common–source amplifier is shown in Figure 5.8-1. Although Figure 5.8-1 shows a common source enhancement NMOSFET amplifier, the small-signal analysis of the amplifier is valid for all types of FETs. As in BJT amplifiers, the quiescent point of the FET must be set with circuitry external to the transistor so that it is in the FET saturation region. The values of the resistors R_D and R_G and the DC voltage sources V_{GG} and V_{DD} have therefore been chosen so that the FET is in the saturation region and the circuit will operate as an amplifier. Since the gate current is zero, $V_{GG} = V_G$, and the gate-to-source voltage V_{GS} must have a value that places the operation of the FET in the saturation region. The voltage source v_s is a small-signal AC source with source resistance R_S.

Once the quiescent conditions ($v_s = 0$) have been calculated, and it has been determined that the FET is in the saturation region, the significant small-signal conductances can be calculated by referring to Table 5.7-1.

For the enhancement NMOSFET,

$$g_m = 2K(V_{GS} - V_T) = 2\sqrt{I_D K},$$

$$g_d = \frac{\Delta i_D}{\Delta v_{DS}}\bigg|_{V_{GSQ}} \quad \text{(from characteristic curves).}$$

(5.8-1)

The small-signal circuit performance can now be calculated. Total circuit performance is the sum of the quiescent and small-signal performance. The process of AC modeling of the circuit and replacing the FET with an appropriate AC model, applied to Figure 5.8-1, is shown in Figure 5.8-2.

The small-signal performance can be obtained from analysis of the circuit of Figure 5.8-2b. The small-signal characteristics that are of interest are input resistance,

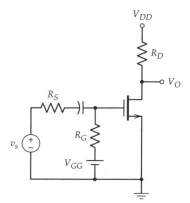

Figure 5.8-1
Simple common–source amplifier.

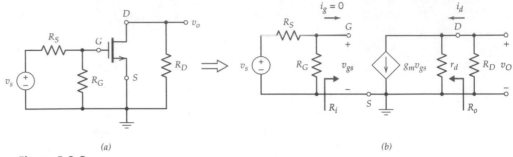

(a)　　　　　　　　　　　　　　　(b)

Figure 5.8-2
AC modeling of common–source amplifier: (a) small-signal circuit, DC sources set to zero; (b) FET replaced by small-signal model.

voltage gain, and output resistance. Definitions of these quantities may vary due to differing definitions of the exact location of the point of measurement.

Input Resistance
The input resistance (shown in Figure 5.8-2b) is given by

$$R_i \equiv \frac{v_{gs}}{i_g} = \infty. \tag{5.8-2}$$

Because the gate current is zero, the input impedance is infinite.

Voltage Gain
The voltage gain is the ratio of output voltage to input voltage. If the input voltage is taken to be the voltage input to the FET, v_g,

$$A_V \equiv \frac{v_o}{v_g}. \tag{5.8-3}$$

Using previously introduced methods to calculate the voltage gain yields

$$A_V = \frac{v_o}{v_g} = \frac{v_o}{v_{gs}} = \left(\frac{v_o}{i_1}\right)\left(\frac{i_1}{v_{gs}}\right), \tag{5.8-4}$$

where $i_1 \equiv$ the current through the parallel combination of the load resistor R_D and r_d.

But

$$i_1 = -g_m v_{gs}. \tag{5.8-5}$$

Therefore, the voltage gain is

$$A_V = (R_D \| r_d)\left(\frac{-g_m v_{gs}}{v_{gs}}\right) = -g_m(R_D \| r_d). \tag{5.8-6}$$

The voltage gain from the source is of interest as well. This overall voltage gain can be defined as

$$A_{VS} \equiv \frac{v_o}{v_s}.$$ (5.8-7)

This ratio can be directly derived from the voltage gain A_V and a voltage division between the source resistance R_S and the bias resistance R_G (the amplifier input resistance R_i is infinite),

$$A_{VS} = \frac{v_o}{v_s} = \left(\frac{v_o}{v_{gs}}\right)\left(\frac{v_{gs}}{v_s}\right) = A_V\left(\frac{R_G||R_i}{(R_G||R_i) + R_s}\right)$$

$$= A_V\left(\frac{R_G}{R_G + R_S}\right).$$ (5.8-8)

The appropriate substitutions lead to

$$A_{VS} = -g_m(r_d||R_D)\left(\frac{R_G}{R_G + R_S}\right).$$ (5.8-9)

Output Resistance

The output resistance is defined as the Thévenin resistance at the output of the amplifier looking back into the amplifier. The circuit of Figure 5.8-3 defines the necessary topology and circuit variables for output resistance calculations.

Since $i_g = 0$, the dependent source is also zero valued and the output resistance is the parallel combination of r_d and the resistance of the zero-valued dependent source,

$$R_O = r_d||\infty = r_d = \left|\frac{V_A}{I_D}\right|,$$ (5.8-10)

where V_A is the early voltage of the FET and I_D is the quiescent drain current.

The output resistance is dependent on the quiescent conditions of the FET. It is also possible to define the output resistance of a common–source amplifier to include the drain resistor R_D,

$$R_{OI} = R_O||R_D = r_d||R_D.$$ (5.8-11)

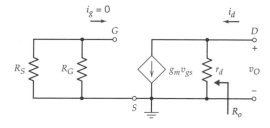

Figure 5.8-3
Circuit for calculating output resistance of common–source amplifier.

EXAMPLE 5.8-1   Assume as given an n-JFET with parameters

$$V_{PO} = -5 \text{ V}, \qquad I_{DDS} = 6 \text{ mA},$$
$$I_S = 1 \text{ pA}, \qquad \text{Early voltage} = V_A = 100 \text{ V}$$

operating in the given circuit at room temperature, determine the amplifier small-signal performance characteristics in its midband frequency range.

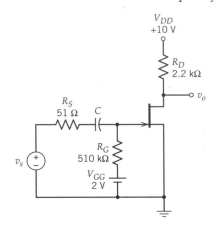

SOLUTION

The modeling process for FET circuit performance contains the following steps:

1.   Determine the quiescent (DC) conditions; verify FET in saturation region.
2.   Determine the FET AC model parameters from the quiescent conditions.
3.   Create the AC equivalent circuit.
4.   Determine the AC performance by replacing the FET by its AC model.
5.   Add the results of the DC and AC analysis to obtain total circuit performance.

Step 1:   Determine the circuit quiescent (DC) conditions. Verify that the FET is in the saturation region. Using the equations in Table 4.4-1 for the n-JFET,

$$V_{GS} = V_G - V_S = -2 - 0 = -2 \text{ V},$$
$$I_D = I_{DSS}\left(1 - \frac{V_{GS}}{V_{PO}}\right)^2 = 6 \times 10^{-3}\left(1 - \frac{-2}{-5}\right)^2 = 2.16 \text{ mA}.$$

So the drain–source voltage is

$$V_{DS} = V_{DD} - I_D R_D = 10 - (2.16 \times 10^{-3})(2200) = 5.25 \text{ V}.$$

The condition for saturation is

$$V_{DS} \geq V_{GS} - V_{PO} = -2 - (-5) = 3 \text{V},$$
$$5.25 \text{ V} > 3 \text{ V}.$$

The circuit is confirmed to be in saturation.

Step 2: Determine the FET AC model parameters from the quiescent conditions. The relevant parameters are

$$r_d = \frac{V_A}{I_D} = \frac{100}{2.16 \times 10^{-3}} = 46.3 \text{ k}\Omega,$$

$$g_m = \frac{-2I_{DSS}(1 - V_{GS}/V_{PO})}{V_{PO}} = \frac{-2(6 \times 10^{-3})[1 - (-2/-5)]}{-5} = 1.44 \text{ mS}.$$

Step 3: Create an AC equivalent circuit as shown. The DC sources are set to zero and the output voltage becomes a small-signal quantity.

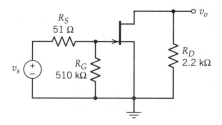

Step 4: Determine the AC performance by replacing the FET by its AC model. The small-signal circuit model of the amplifier is as shown.

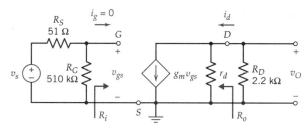

The input resistance to the gate of the FET is

$$R_i = \infty.$$

If the input resistance is measured to the left of R_s, then

$$R_{iS} = R_S + (R_G \| R_i) = 51 + 510{,}000 \approx 510 \text{ k}\Omega.$$

The voltage gain from the gate of the FET is

$$A_V = -g_m(r_d \| R_D) = -1.44 \times 10^{-3}(46{,}300 \| 2200) = -3.$$

The voltage gain from the source is given by

$$A_{VS} = -g_m(r_d \| R_D) \frac{R_G}{R_G + R_S}$$

$$= -1.44 \times 10^{-3}(46{,}300 \| 2200) \frac{510{,}000}{510{,}000 + 51} \approx A_V = -3.$$

The output resistance is

$$R_O = r_d || \infty = r_d = \left| \frac{V_A}{I_D} \right| = \left| \frac{100}{2.16 \times 10^3} \right| = 46.3 \text{ k}\Omega.$$

If the drain resistance is included in the output resistance,

$$R_{OI} = R_O || R_D = r_d || R_D = 46{,}300 || 2200 = 2.1 \text{ k}\Omega.$$

Step 6 is beyond the requirements of this problem, but all the data is present to add the quiescent solution to the small-signal solution for total circuit response.

5.8.1 Common–Source Amplifiers with Nonzero Source Resistance

Several characteristics of the common–source amplifier (Figure 5.8-4) can be altered through the addition of a resistor connected between the source and common (ground). The source resistor has the following effects on common source amplifiers:

- Increased output resistance
- Decreased voltage gain

Derivation of these effects follows the process described in Section 5.7.

In order to determine the AC performance of this amplifier, the quiescent conditions must first be obtained. If the FET is operating in the saturation region, an AC equivalent model of the circuit can be obtained (Figure 5.8-5a) and the FET can be replaced by its small-signal model (based on the quiescent conditions) in the AC equivalent circuit (Figure 5.8-5b). Circuit performance can then be obtained from this equivalent circuit.

Input Resistance

The input resistance (shown in Figure 5.8-5b) is the same as the input resistance for a common–source amplifier without a source resistor,

$$R_i = \infty. \tag{5.8-12}$$

The addition of a source resistor has not changed the input resistance of the amplifier.

Voltage Gain

The voltage gain is the ratio of the output voltage to input voltage. If the input voltage is again taken to be the voltage at the input to the FET, v_g,

$$A_V \equiv \frac{v_o}{v_g}. \tag{5.8-13}$$

Although the voltage gain can be found using the small-signal model of Figure 5.8-5b, the alternate source-transformed form of the FET small-signal model of Figure 5.7-3 may be used. The small-signal model of the common–source amplifier with source resistor using the small-signal model of the FET with voltage-controlled voltage source is shown in Figure 5.8-6.

Using the methods similar to those of the simple common–source amplifier to calculate the voltage gain yields

$$A_V = \frac{v_o}{v_g} = \left(\frac{v_o}{i_1} \right) \left(\frac{i_1}{v_{gs}} \right) \left(\frac{v_{gs}}{v_g} \right). \tag{5.8-14a}$$

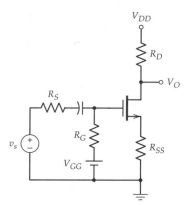

Figure 5.8-4
Common–source amplifier with emitter resistor.

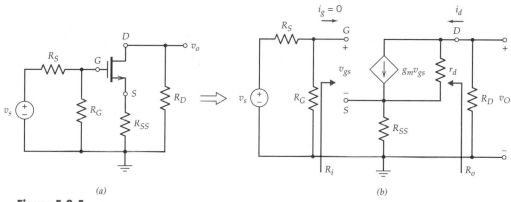

(a)   (b)

Figure 5.8-5
AC modeling of common–source amplifier with source resistor: (a) small-signal circuit, DC sources set to zero; (b) FET replaced by small-signal model.

Equation 5.8-14a is rewritten as

$$A_v = R_D\left(\frac{-g_m v_{gs} r_d/(R_{SS} + r_d + R_D)}{v_{gs}}\right)\left(\frac{v_{gs}}{v_g}\right)$$

$$= -\frac{g_m r_d R_D}{R_{SS} + r_d + R_D}\frac{v_{gs}}{v_g}.$$

(5.8-14b)

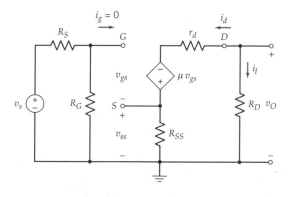

Figure 5.8-6
Alternate small-signal model of common–source amplifier with source resistor using small-signal model of FET with voltage-controlled voltage source.

By voltage division, the source voltage v_{ss} is

$$v_{ss} = \frac{g_m v_{gs} r_d R_{SS}}{R_{SS} + r_d + R_D}. \tag{5.8-15}$$

Using Equation 5.8-15, the gate–source voltage is

$$v_{gs} = v_g - v_{ss} = v_g - \frac{g_m v_{gs} r_d R_{SS}}{R_{SS} + r_d + R_D}. \tag{5.8-16}$$

From Equation 5.8-16, the expression for the gate voltage v_g as a function of the gate–source voltage v_{gs} is

$$v_g = v_{gs}\left(1 + \frac{g_m r_d R_{SS}}{R_{SS} + r_d + R_D}\right). \tag{5.8-17}$$

Substituting Equation 5.8-17 into 5.8-14 results in the expression for the voltage gain for the common-source amplifier with FET source terminal resistor,

$$A_V = -\frac{g_m r_d R_D}{R_{SS} + r_d + R_D} \frac{v_{gs}}{v_{gs}[1 + g_m r_d R_{SS}/(R_{SS} + r_d + R_D)]}$$

$$= -\frac{g_m r_d R_D}{R_{SS} + r_d + R_D} \frac{1}{[r_d + R_D + (1 + g_m r_d)R_{SS}]/(R_{SS} + r_d + R_D)}$$

$$= -\frac{g_m r_d R_D}{r_d + R_D + (1 + g_m r_d)R_{SS}}.$$

The voltage gain from the input voltage source is defined as

$$A_{VS} \equiv \frac{v_o}{v_s}. \tag{5.8-19}$$

This ratio can be directly derived from the voltage gain following the derivation for the common-source amplifier *without* source resistor:

$$A_{VS} = \frac{v_o}{v_s} = \left(\frac{v_o}{v_g}\right)\left(\frac{v_g}{v_s}\right) = A_V\left(\frac{v_g}{v_s}\right). \tag{5.8-20}$$

Using voltage division to determine v_g as a function of v_s,

$$A_{VS} = A_V \frac{v_g}{v_s} = A_V \frac{v_s R_G/(R_G + R_S)}{v_s}$$

$$= -\frac{g_m r_d R_D}{r_d + R_D + (1 + g_m r_d)R_{SS}} \frac{R_G}{R_G + R_S}. \tag{5.8-21}$$

Output Resistance

The output resistance is defined as the Thévenin resistance at the output of the amplifier looking back into the amplifier. As in the case of the simple common–source amplifier, R_o is measured without considering R_D.

Calculations to determine the output resistance of the common–source amplifier with source resistor are based on the small-signal model of Figure 5.8-5*b* and is similar to the calculations performed previously for the common–source amplifier without source resistor. The circuit of Figure 5.8-7 defines the necessary topology and circuit variables for output resistance calculations.

The Thévenin voltage and current at the output are v_t and i_t, respectively, where the output resistance R_o is the Thévenin resistance defined as

$$R_o = \frac{v_t}{i_t}. \tag{5.8-22}$$

Solving for the Thévenin current as a function of the Thévenin voltage,

$$i_t = \frac{v_t + g_m v_{gs} r_d}{r_d + R_{SS}} - \frac{v_t + g_m r_d(v_g - v_{ss})}{r_d + R_{SS}}. \tag{5.8-23}$$

Since R_S and the gate bias resistor R_G are grounded (the independent voltage source is set to zero), $v_g = 0$. Therefore, Equation 5.8-23 is rewritten as

$$i_t = \frac{v_t - g_m r_d v_{ss}}{r_d + R_{SS}}. \tag{5.8-24}$$

The voltage at the source of the FET, v_{ss}, is simply

$$v_{ss} = i_t R_{SS}. \tag{5.8-25}$$

Substituting Equation 5.8-25 into 5.8-24 yields

$$i_t = \frac{v_t - g_m r_d(i_t R_{SS})}{r_d + R_{SS}}. \tag{5.8-26}$$

Rearranging Equation 5.8-26 and solving for the Thévenin voltage v_t in terms of the Thévenin current i_t,

$$v_t = i_t[r_d + (1 + g_m r_d)R_{SS}]. \tag{5.8-27}$$

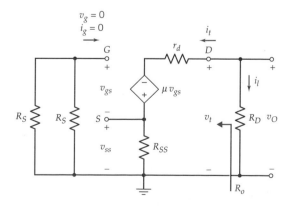

Figure 5.8-7
Circuit for calculating output resistnce of common–source amplifier with source resistor.

Applying the definition of the Thévenin resistance shown in Equations 5.8-22 to 5.8-27 yields the output resistance

$$R_o = \frac{v_t}{i_t} = [r_d + (1 + g_m r_d)R_{SS}]. \tag{5.8-28}$$

EXAMPLE 5.8-2 An enhancement NMOSFET with parameters

$$K = 2.96 \text{ mA/V}^2, \qquad V_T = 2\text{V}, \qquad r_d = 30 \text{ k}\Omega, \qquad I_D = 5 \text{ mA}$$

is placed in the amplifier shown. Assume that the amplifier is operating in its mid-band frequency range, and determine the following circuit parameters: $A_{VS} = v_o/v_s$, R_{in}, and R_{out}.

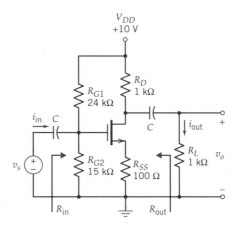

SOLUTION

The quiescent point of the FET must be found first to ensure that it is in the saturation region. For DC operation, all capacitors appear as open circuits. By voltage division, the gate voltage is

$$V_G = \frac{V_{DD}R_{G2}}{R_{G1} + R_{G2}} = \frac{10(15,000)}{24,000 + 15,000} \approx 3.8 \text{ V}.$$

The voltage at the source terminal of the FET is

$$V_S = I_D R_{SS} = (5 \times 10^{-3})(100) = 0.5 \text{ V}.$$

Therefore, the gate–source voltage is

$$V_{GS} = V_G - V_S = 3.8 - 0.5 = 3.3 \text{ V}.$$

The drain–source voltage is

$$V_{DS} = V_{DD} - I_D(R_D + R_{SS}) = 10 - 5 \times 10^{-3}(1100) = 4.5 \text{ V}.$$

For the FET to be in saturation, $V_{DS} \geq V_{GS} - V_T$. For the circuit in this example

$$4.5 \text{ V} > 3.3 \text{ V} - 2 \text{ V}$$

$$> 1.3 \text{ V} \quad \Rightarrow \quad \text{FET is in saturation.}$$

The small-signal transconductance is

$$g_m = 2\sqrt{I_D K} = 2\sqrt{(5 \times 10^{-3})(2.96 \times 10^{-3})} \approx 7.7 \text{ mS.}$$

The AC equivalent circuit shown can be found by setting all DC sources to zero and, since the circuit is in its midband frequency range, all capacitors are replaced by short circuits.

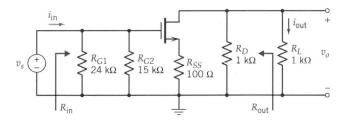

The AC equivalent circuit leads to the given small-signal equivalent circuit for the amplifier:

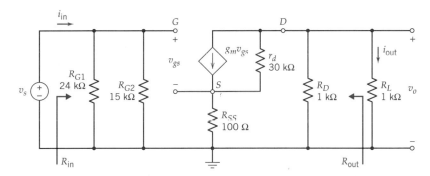

$$A_{VS} = A_V$$

$$= -\frac{g_m r_d (R_D \| R_L)}{R_{SS} + r_d + R_D \| R_L + g_m r_d R_{SS}}$$

$$= -\frac{7.7 \times 10^{-3}(30{,}000)(500)}{100 + 30{,}000 + 500 + 7.7 \times 10^{-3}(30{,}000)(100)} \approx -2.2,$$

$$R_{in} = 24{,}000 \| 15{,}000 = 9.23 \text{ k}\Omega,$$

$$R_{out} = R_D \| R_o = R_D \| (r_d + R_{SS} + g_m r_d R_{SS})$$

$$= 1\text{k} \| [30{,}000 + 100 + 7.7 \times 10^{-3}(30{,}000)(100)] \approx 980 \ \Omega.$$

5.8.2 Summary of Common–Source Amplifier Properties

A summary of common–source (CS) amplifier performance characteristics, as derived in this section, is given in Table 5.8-1.

TABLE 5.8-1 COMMON–SOURCE AMPLIFIER CHARACTERISTICS

Parameter	CS	CS + R_{SS}
R_i	∞	∞
A_V	$-g_m(R_D \parallel r_d)$	$\dfrac{g_m r_d R_D}{[r_d + R_D + (1 + g_m r_d)R_{SS}]}$
R_o	r_d	$r_d + (1 + g_m r_d)R_{SS}$

5.9 COMMON–DRAIN AMPLIFIERS

Common–drain amplifiers have the following general circuit topology:

- The input signal enters the FET at the gate.
- The output signal exits the FET at the source.
- The drain terminal is connected to a constant voltage, often the ground (common) terminal, sometimes with an intervening resistor.

A simple common–drain amplifier is shown in Figure 5.9-1. Although Figure 5.9-1 shows a common–drain n-JFET amplifier, the small-signal analysis of the amplifier is valid for all types of FETs. As in BJT amplifiers, the quiescent point of the FET must be set with circuitry external to the transistor ensuring linear operation, which for the FET is the saturation region. The values of the resistors R_D and R_G and the DC voltage sources V_{GG} and V_{DD} have therefore been chosen so that the FET is in the saturation region and the circuit will operate as an amplifier. Since the gate current is zero, $V_{GG} = V_G$, and the gate to source voltage V_{GS} must have a value that places the operation of the FET in the saturation region. The voltage source v_s is a small-signal AC source with source resistance R_s.

Once the quiescent conditions ($v_s = 0$) have been calculated, and it has been determined that the FET is in the saturation region, the significant small-signal conductances can be calculated by referring to Table 5.7-1.

For the n-JFET,

$$g_m = \frac{2I_D}{V_{GS} - V_{PO}},$$

$$g_d = \frac{\Delta i_D}{\Delta v_{DS}}\bigg|_{V_{GSQ}} = \frac{|I_{DQ}|}{|V_A|}$$

from the characteristic curves, where V_A is the early voltage.

The small-signal circuit performance can now be calculated. Total circuit performance is the sum of the quiescent and small-signal performance. The process of AC modeling of the circuit and replacing the FET with an appropriate AC model, applied to Figure 5.9-1, is shown in Figure 5.9-2.

The small-signal performance can be obtained from analysis of the circuit of Figure 5.9-2b. The small-signal characteristics that are of interest are input resistance,

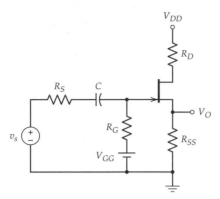

Figure 5.9-1
Simple common–drain amplifier.

voltage gain, and output resistance. Definitions of these quantities may vary due to differing definition of the exact location of the point of measurement.

Input Resistance

The input resistance (shown in Figure 5.9-2b) is given by

$$R_i \equiv \frac{v_{gs}}{i_g} = \infty. \tag{5.9-2}$$

Because the gate current is zero, the input impedance is infinite.

Voltage Gain

The voltage gain is the ratio of the input voltage to input voltage. If the input voltage is again taken to be the voltage at the input to the FET, v_g,

$$A_V \equiv \frac{v_o}{v_g}. \tag{5.9-3}$$

Although the voltage gain can be found using the small-signal model of Figure 5.9-2b, the alternate source-transformed form of the FET small-signal model of Figure

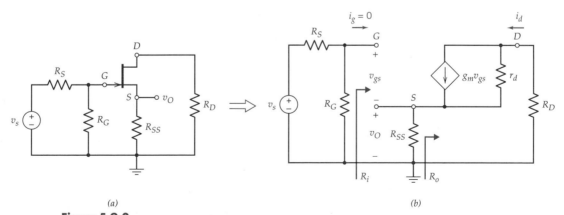

(a)   (b)

Figure 5.9-2
AC modeling of common–drain amplifier: (a) small-signal circuit, DC sources set to zero; (b) FET replaced by small-signal model.

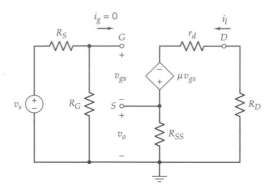

Figure 5.9-3
Alternate small-signal model of common–drain amplifier using small-signal model of FET with voltage-controlled voltage source.

5.9-3 may be used. The small-signal model of the common–source amplifier with source resistor using the small-signal model of the FET with voltage-controlled voltage source is shown in Figure 5.9-3.

Using the methods similar to those of the simple common–source amplifier with source resistor to calculate the voltage gain yields

$$A_V = \frac{v_o}{v_g} = \left(\frac{v_o}{i_1}\right)\left(\frac{i_1}{v_{gs}}\right)\left(\frac{v_{gs}}{v_g}\right). \tag{5.9-4}$$

Equation 5.9-4 is rewritten as

$$A_v = (R_{SS}) \left(\frac{g_m v_{gs} r_d / (R_{SS} + r_d + R_D)}{v_{gs}}\right)\left(\frac{v_{gs}}{v_g}\right)$$

$$= \frac{g_m r_d R_{SS}}{R_{SS} + r_d + R_D} \frac{v_{gs}}{v_g}. \tag{5.9-5}$$

By voltage division, the output voltage v_o is

$$v_o = \frac{g_m v_{gs} r_d R_{SS}}{R_{SS} + r_d + R_D}. \tag{5.9-6}$$

Using Equation 5.9-6, the gate–source voltage is

$$v_{gs} = v_g - v_o = v_g - \frac{g_m v_{gs} r_d R_{SS}}{R_{SS} + r_d + R_D}. \tag{5.9-7}$$

From Equation 5.9-7, the expression for the gate voltage v_g as a function of the gate–source voltage v_{gs} is

$$v_g = v_{gs}\left(1 + \frac{g_m r_d R_{SS}}{R_{SS} + r_d + R_D}\right). \tag{5.9-8}$$

Substituting Equation 5.9-8 into 5.9-5 results in the expression for the voltage gain for the common–drain amplifier,

$$A_V = \frac{g_m r_d R_{SS}}{R_{SS} + r_d + R_D} \frac{v_{gs}}{v_{gs}(1 + g_m r_d R_{SS}/(R_{SS} + r_d + R_D))}$$

$$= \frac{g_m r_d R_{SS}}{R_{SS} + r_d + R_D} \frac{1}{[r_d + R_D + R_{SS}(1 + g_m r_d)]/(R_{SS} + r_d + R_D)}$$

$$= \frac{g_m r_d R_{SS}}{r_d + R_D + R_{SS}(1 + g_m r_d)}. \tag{5.9-9}$$

For $g_m r_d R_{SS} \gg R_D + r_d$, $A_V \approx 1$. The common–drain configuration is therefore called the source follower, since the output follows the input voltage. The common–drain amplifier is the FET counterpart to the BJT common–collector amplifier.

The voltage gain from the input voltage source is defined as

$$A_{VS} \equiv \frac{v_o}{v_s}. \tag{5.9-10}$$

This ratio can be directly derived from the voltage gain following the derivation for the common-source amplifier:

$$A_{VS} = \frac{v_o}{v_s} = \left(\frac{v_o}{v_g}\right)\left(\frac{v_g}{v_s}\right) = A_V \frac{v_g}{v_s}. \tag{5.9-11}$$

Using voltage division to determine v_g as a function of v_s,

$$A_{VS} = A_V \frac{v_g}{v_s} = A_V \frac{v_s R_G/(R_G + R_S)}{v_s}$$

$$= \frac{g_m r_d R_{SS}}{r_d + R_D + R_{SS}(1 + g_m r_d)} \frac{R_G}{R_G + R_S} \tag{5.9-12}$$

$$\approx \frac{R_G}{R_G + R_S}, \qquad g_m r_d R_{SS} \gg R_D + r_d.$$

Output Resistance

The output resistance is defined as the Thévenin resistance at the output of the amplifier looking back into the amplifier. Calculations to determine the output resistance of the common–drain amplifier are based on the small-signal model of Figure 5.9-3 and is similar to the calculations performed previously for the common–source amplifier with source resistor. The circuit of Figure 5.9-4 defines the necessary topology and circuit variables for output resistance calculations.

The Thévenin voltage and current at the output are v_t and i_t, respectively, where the output resistance R_o is the Thévenin resistance, defined as

$$R_o = \frac{v_t}{i_t}. \tag{5.9-13}$$

Solving for the Thévenin current as a function of the Thévenin voltage,

$$i_t = \frac{v_t - g_m v_{gs} r_d}{r_d + R_D} = \frac{v_t - g_m r_d(v_g - v_t)}{r_d + R_D}. \tag{5.9-14}$$

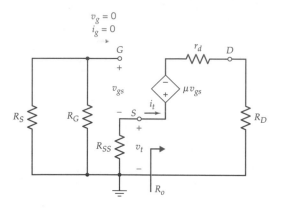

Figure 5.9-4
Circuit for calculating output resistance of common–drain amplifier.

Since R_S and the gate bias resistor R_G are grounded (the independent voltage source is set to zero), $v_g = 0$. Therefore, Equation 5.9-14 is rewritten as

$$i_t = \frac{v_t + g_m r_d v_t}{r_d + R_D}.$$
(5.9-15)

Rearranging Equation 5.9-15 and solving for the Thévenin voltage v_t in terms of the Thévenin current i_t,

$$i_t = \frac{v_t(1 + g_m r_d)}{r_d + R_D}.$$
(5.9-16)

Applying the definition of the Thévenin resistance shown in Equation 5.9-13 to 5.9-16 yields the output resistance

$$R_o = \frac{v_t}{i_t} = \frac{r_d + R_D}{1 + g_m r_d}.$$
(5.9-17)

The output resistance of the common–source amplifier is typically small.

EXAMPLE 5.9-1 A *p*-JFET with parameters

$$V_{PO} = 3 \text{ V}, \qquad I_{DSS} = -13.8 \text{ mA}, \qquad I_S = 1 \text{ pA}, \qquad r_d = 30 \text{ k}\Omega$$

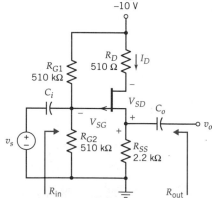

operating in the given circuit at room temperature is placed in the amplifier shown. Assume that the amplifier is operating in its midband frequency range, and determine the following circuit parameters: $A_{VS} = v_o/v_s$, R_{in}, and R_{out}.

SOLUTION

The quiescent point of the FET must be found first to ensure that it is in the saturation region. For DC operation, all capacitors appear as open circuits. By voltage division, the gate voltage is

$$V_G = \frac{V_{DD}R_{G2}}{R_{G1} + R_{G2}} = \frac{-10(510,000)}{510,000 + 510,000} \approx -5 \text{ V}.$$

The drain current I_D is

$$I_D = I_{DSS}\left(1 - \frac{V_{GS}}{V_{PO}}\right)^2$$

$$= I_{DSS}\left(1 - \frac{-V_{SG}}{V_{PO}}\right)^2$$

$$= -13.8 \times 10^{-3}\left(1 - \frac{-V_{SG}}{3}\right)^2.$$

But

$$V_{SG} = V_S - V_G = I_D R_{SS} - V_G.$$

Substituting into the equation for I_D in terms of I_{DSS},

$$I_D = -13.8 \times 10^{-3}\left(1 - \frac{-(I_D R_{SS} - V_G)}{3}\right)^2$$

$$= -13.8 \times 10^{-3}\left(1 - \frac{-(2200I_D + 5)}{3}\right)^2.$$

By rearranging the equation, the drain current can be found by solving the second-order polynomial equation

$$0 = I_D^2 + 7.41 \times 10^{-3}I_D + 13.2 \times 10^{-6}.$$

Using the quadratic equation, the solutions to the drain current are I_D of approximately -3.0 and -4.4 mA. To determine which of the two solutions for I_D are valid, test the values in the KVL equation for the source–drain loop,

$$0 = I_D(R_{SS} + R_D) - V_{SD} - V_{DD},$$

or

$$V_{SD} = I_D(R_{SS} + R_D) - V_{DD},$$

where $V_{DD} = -10$ V.

For $I_D = -4.4$ mA, the resultant $V_{SD} = -1.92$ V. Clearly, this result indicates that since $V_{SD} < 0$, the FET is not in saturation. Additionally, since

$$V_{SG} = V_S - V_G = I_D R_{SS} - V_G$$
$$= (-4.4 \times 10^{-3})(2200) - (-5) = -4.68 \text{ V}.$$

and $V_{SG} < -V_{PO}$, the FET is in cutoff.

For $I_D = -3.0$ mA, the resultant $V_{SD} = 1.87$ V. Additionally, since

$$V_{SG} = V_S - V_G = I_D R_{SS} - V_G$$
$$= (-3.0 \times 10^{-3})(2200) - (-5) = -1.6 \text{ V}$$

and

$$V_{SD} \geq_. V_{SG} + V_{PO}$$

or

$$1.87 \geq -1.6 + 3 = 1.4 \text{ V},$$

the FET is in saturation. Therefore, the valid solution for the drain current is

$$I_D = -3.0 \text{ mA}.$$

The small-signal transconductance is

$$g_m = \frac{2I_D}{V_{GS} - V_{PO}} = \frac{2(-3 \times 10^{-3})}{-(-1.6) - 3} \approx 4.3 \text{ mS}.$$

The AC equivalent circuit shown can be found by setting all DC sources to zero and, since the circuit is in its midband frequency range, all capacitors are replaced by short circuits.

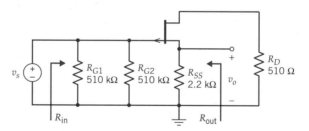

The small-signal model for *p*- and *n*-channel devices are identical and are based on the *Q* points determined by the external circuitry. The AC equivalent circuit leads

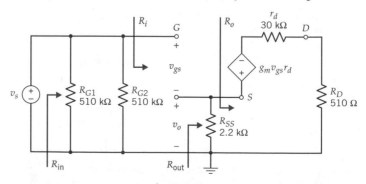

to the small-signal equivalent circuit shown for the *p*-channel JFET common drain amplifier:

The voltage gain of the amplifier is

$$A_{VS} = A_V \frac{v_g}{v_s} = A_V \frac{v_s R_G / (R_G + R_S)}{v_s}$$

$$= \frac{g_m r_d R_{SS}}{r_d + R_D + R_{SS}(1 + g_m r_d)} \frac{R_G}{R_G + R_S}.$$

But since $R_S = 0$, the gain of the amplifier is

$$A_{VS} = A_V = \frac{g_m r_d R_{SS}}{R_{SS} + r_d + R_D + g_m r_d R_{SS}}$$

$$= \frac{(4.3 \times 10^{-3})(30{,}000)(2.2\text{k})}{30\text{k} + 510 + (2200)[1 + (4.3 \times 10^{-3})(30{,}000)]} = 0.896 \approx 0.90.$$

Here, R_{in} is given as

$$R_{in} = R_i \| (R_{G1} \| R_{G2}) = \infty \| (510{,}000 \| 510{,}000) = 255 \text{ k}\Omega.$$

and R_{out} as

$$R_{out} = R_o \| R_{SS}$$

$$= \frac{r_d + R_D}{1 + g_m r_d} \| R_{SS}$$

$$= \frac{30{,}000 + 510}{1 + (4.3 \times 10^{-3})(30{,}000)} \| 2200 = 235 \| 2200 \approx 212 \ \Omega.$$

A summary of common–drain (CD) amplifier performance characteristics, as derived in this section, is given in Table 5.9-1.

TABLE 5.9-1 COMMON–DRAIN AMPLIFIER CHARACTERISTICS

Parameter	CD
R_i	∞
A_V	$\dfrac{g_m r_d R_{SS}}{r_d + R_D + R_{SS}(1 + g_m r_d)} \approx 1, \ g_m r_d R_{SS} \gg R_D + r_d$
R_o	$\dfrac{r_d + R_D}{1 + g_m r_d}$

▬▬▬▬▬ 5.10 COMMON–GATE AMPLIFIERS

Common–gate amplifiers have the following general circuit topology:

- The input signal enters the FET at the source.
- The output signal exits the FET at the drain.
- The gate terminate is connected to a constant voltage, often the ground (common) terminal, sometimes with an intervening resistor.

A simple common–gate amplifier is shown in Figure 5.10-1. Although Figure 5.10-1 shows a common–gate depletion NMOSFET amplifier, the small-signal analysis of the amplifier is valid for all types of FETs. As in BJT amplifiers, the quiescent point of the FET must be set with circuitry external to the transistor so that it is in the FET saturation region. The values of the resistors R_D and R_G and the DC voltage sources V_{GG} and V_{DD} have therefore been chosen so that the FET is in the saturation region and the circuit will operate as an amplifier. Since the gate current is zero, $V_{GG} = V_G$ and the gate to source voltage V_{GS} must have a value that places the operation of the FET in the saturation region. The voltage source v_s is a small-signal AC source with source resistance R_S.

Once the quiescent conditions ($v_s = 0$) have been calculated, and it has been determined that the FET is in the saturation region, the significant small-signal conductances can be calculated by referring to Table 5.7-1.

For the depletion NMOSFET,

$$g_m = \frac{2I_D}{V_{GS} - V_{PO}}, \qquad g_d = \left.\frac{\Delta i_D}{\Delta V_{DS}}\right|_{V_{GSQ}} = \frac{|I_{DQ}|}{|V_A|} \qquad (5.10\text{-}1a)$$

from the characteristic curves, where V_A is the early voltage.

The small-signal circuit performance can now be calculated. Total circuit performance is the sum of the quiescent and small-signal performance. The process of AC modeling of the circuit and replacing the FET with an appropriate AC model, applied to Figure 5.10-1, is shown in Figure 5.10-2.

To simplify the analysis of the circuit, the source-transformed version of the FET model of Figure 5.10-3 is used to obtain the small-signal performance of the amplifier. The small-signal characteristics that are of interest are input resistance, voltage gain, and output resistance. Definitions of these quantities may vary due to differing definitions of the exact location of the point of measurement.

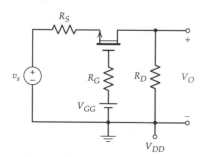

Figure 5.10-1
Simple common–gate amplifier.

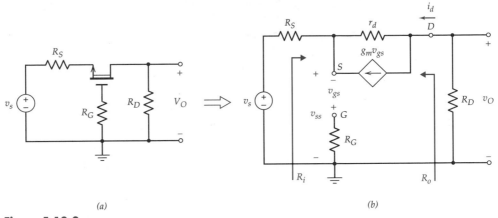

Figure 5.10-2.
AC modeling of common–gate amplifier: (*a*) small-signal circuit, DC sources set to zero; (*b*) FET replaced by small-signal model.

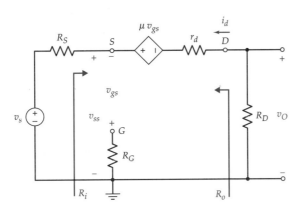

Figure 5.10-3.
Alternate small-signal model of common–gate amplifier using small-signal model of FET with voltage-controlled voltage source.

Input Resistance

The input resistance is defined as the Thévenin resistance at the input of the amplifier looking back into the amplifier. Calculations to determine the input resistance of the common–gate amplifier are based on the small-signal model of Figure 5.10-4 and is similar to the calculations performed previously for the common–source amplifier with source resistor. The circuit of Figure 5.10-4 defines the necessary topology and circuit variables for output resistance calculations.

The Thévenin voltage and current at the input are v_t and i_t, respectively, where the input resistance R_i is the Thévenin resistance, defined as

$$R_i = \frac{v_t}{i_t}. \tag{5.10-2}$$

Solving for the Thévenin current as a function of the Thévenin voltage,

$$i_t = \frac{v_t - g_m v_{gs} r_d}{r_d + R_D} = \frac{v_t - g_m r_d(v_g - v_t)}{r_d + R_D}. \tag{5.10-3}$$

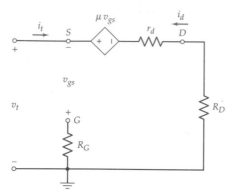

Figure 5.10-4.
Circuit for calculating input resistance of common–gate amplifier.

Since the gate bias resistor R_G is grounded (the independent voltage source is set to zero), $v_g = 0$. Therefore, Equation 5.9-14 is rewritten as

$$i_t = \frac{v_t + g_m r_d v_t}{r_d + R_D}. \tag{5.10-4}$$

Rearranging Equation 5.10-4 and solving for the Thévenin voltage v_t in terms of the Thévenin current i_t,

$$i_t = \frac{v_t(1 + g_m r_d)}{r_d + R_D}. \tag{5.10-5}$$

Applying the definition of the Thévenin resistance shown in Equations 5.10-2–5.10-5 yields the input resistance

$$R_i = \frac{v_t}{i_t} = \frac{r_d + R_D}{1 + g_m r_d}. \tag{5.10-6}$$

The input resistance of the common–gate amplifier is typically small and is identical to the output resistance of the common–drain amplifier.

Voltage Gain

The voltage gain is the ratio of the output voltage to input voltage. If the input voltage is taken to be the voltage at the input to the FET, v_{ss},

$$A_V \equiv \frac{v_o}{v_{ss}}. \tag{5.10-7}$$

Although the voltage gain can be found using the small-signal model of Figure 5.10-2b, the alternate source-transformed form of the FET small-signal model of Figure 5.10-3 may be used.

Using the methods discussed for other FET amplifier configurations to calculate the voltage gain yields

$$A_V = \frac{v_o}{v_{ss}} = \left(\frac{v_o}{-i_d}\right)\left(\frac{-i_d}{v_{ss}}\right). \tag{5.10-8}$$

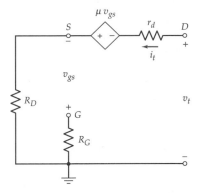

Figure 5.10-5.
Circuit for calculating output resistance of common–gate amplifier.

Equation 5.10-8 is rewritten as

$$A_v = R_D \frac{(v_{ss} - g_m v_{gs} r_d)/(r_d + R_D)}{v_{ss}}.$$ (5.10-9)

The gate–source voltage is

$$v_{gs} = v_g - v_{ss} = 0 - v_s = -v_{ss}.$$ (5.10-10)

Substituting Equation 5.10-10 into 5.10-9 results in the expression for the voltage gain for the common–gate amplifier,

$$A_v = \frac{(1 + g_m r_d)R_D}{r_d + R_D}.$$ (5.10-11)

Output Resistance

The output resistance is defined as the Thévenin resistance at the output of the amplifier looking back into the amplifier. Calculations to determine the output resistance of the common gate amplifier are based on the small-signal model of Figure 5.10-3. The circuit of Figure 5.10-5 defines the necessary topology and circuit variables for output resistance calculations.

The Thévenin voltage and current at the output are v_t and i_t, respectively, where the output resistance R_o is the Thévenin resistance, defined as

$$R_o = \frac{v_t}{i_t}.$$ (5.10-12)

Solving for the Thévenin current as a function of the Thévenin voltage,

$$i_t = \frac{v_t + g_m v_{gs} r_d}{r_d + R_S} = \frac{v_t + g_m r_d(v_g - i_t R_S)}{r_d + R_S}.$$ (5.10-13)

Since the gate bias resistor R_G is grounded (the independent voltage source is set to zero), $v_g = 0$. Therefore, Equation 5.10-13 is rewritten as

$$i_t = \frac{v_t - i_t g_m r_d R_S}{r_d + R_S}.$$

(5.10-14)

Rearranging Equation 5.10-14 and solving for the Thévenin voltage v_t in terms of the Thévenin current i_t,

$$i_t = \frac{v_t}{r_d + R_S + g_m r_d R_S}.$$

(5.10-15)

Applying the definition of the Thévenin resistance shown in Equations 5.10-12 to 5.10-15 yields the output resistance

$$R_o = r_d + (1 + g_m r_d)R_S.$$

(5.10-16)

The output resistance of the common–gate amplifier is typically large.

EXAMPLE 5.10-1 An enhancement NMOSFET with parameters

$$K = 2.96 \text{ mA/V}^2, \qquad V_T = 2 \text{ V}, \qquad r_d = 30 \text{ k}\Omega, \qquad I_D = 5 \text{ mA}$$

is placed in the amplifier shown. Assume that the amplifier is operating in its mid-band frequency range, and determing the following parameters: $A_{VS} = v_O/v_s$, R_{in}, and R_{out}.

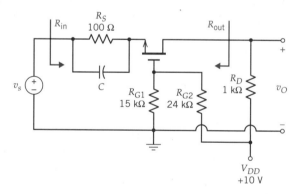

SOLUTION

The quiescent point of the FET must be was found in Example 5.8-2:

$$V_{GS} = 3.3 \text{ V}, \qquad V_{DS} = 4.5 \text{ V}$$

and the FET is in saturation.

The small-signal transconductance is

$$g_m = 2\sqrt{I_D K} = 2\sqrt{(5 \times 10^{-3})(2.96 \times 10^{-3})} \approx 7.7 \text{ mS}.$$

Since the circuit shown si in its midband frequency range, all capacitors are replaced by short circuits.

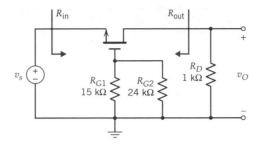

The AC equivalent circuit leads to the small-signal equivalent circuit shown for the common–gate amplifier.

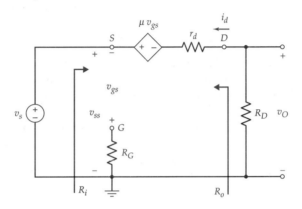

The voltage gain of this amplifier is

$$A_{VS} = A_V = \frac{1 + g_m r_d R_D}{r_d + R_D} = \frac{1 + (7.7 \times 10^{-3})(30{,}000)(1000)}{30{,}000 + 1000} \approx 7.5.$$

Here, R_{in} is given as

$$R_{in} = R_i = \frac{r_d + R_D}{1 + g_m r_d}$$

$$= \frac{30{,}000 + 1000}{1 + (7.7 \times 10^{-3})(30{,}000)} \approx 134 \ \Omega.$$

and R_{out} as

$$R_{out} = R_o \| R_D$$
$$= [r_d + R_S(1 + g_m r_d)] \| R_D.$$

But $R_S = 0$, so

$$R_{out} = r_d \| R_D = 30{,}000 \| 1000 = 968 \ \Omega.$$

A summary of common–gate (CG) amplifier performance characteristics, derived in this section, is given in Table 5.10-1.

TABLE 5.10-1 COMMON–GATE AMPLIFIER CHARACTERISTICS

Parameter	CG
R_i	$\dfrac{(r_d + R_D)}{(1 + g_m r_d)}$
A_V	$\dfrac{(1 + g_m r_d)R_D}{(r_d + R_D)}$
R_o	$r_d + (1 + g_m r_d)R_S$

5.11 COMPARISON OF FET AMPLIFIER TYPES

Single FET amplifiers have been shown to fall into three general categories: common–source (both with and without a source resistor), common–drain, and common–gate. The performance characteristics for each type of amplifier are summarized in Table 5.11-1.

The common–source configuration appears the most useful of the three types: It provides significant voltage gain with inversion. The common–source amplifier input and output resistances are high. In fact, this configuration is the most versatile of the three types and will often form the major voltage gain stage portion of multiple-transistor amplifiers.

TABLE 5.11-1 QUALITATIVE COMPARISON OF FET AMPLIFIER CONFIGURATIONS

Parameter	CS	CS + R_S	CD	CG
R_i	Very high	Very high	Very high	Low
A_V	High	Medium	Approximately unity	Medium to high
R_o	High	Very high	Low	High

TABLE 5.11-2 SUMMARY OF FET AMPLIFIER PERFORMANCE CHARACTERISTICS

Parameter	CS	CS + R_S	CD	CG
R_i	∞	∞	∞	$\dfrac{r_d + R_D}{1 + g_m r_d}$
A_V	$-g_m(R_D \| r_d)$	$\dfrac{-g_m r_d R_D}{r_d + R_D + R_{SS}(1 + g_m r_d)}$	$\dfrac{g_m r_d R_{SS}}{r_d + R_D + R_{SS}(1 + g_m r_d)} \approx 1,$ $g_m r_d R_{SS} \gg R_D + r_d$	$\dfrac{(1 + g_m r_d)R_D}{r_d + R_D}$
R_o	r_d	$r_d + R_{SS}(1 + g_m r_d)$	$\dfrac{r_d + R_D}{1 + g_m r_d}$	$r_d + R_{SS}(1 + g_m r_d)$

NOTES: All measurements are made at the input and/or output port of the FET.

The common–drain configuration is, for all practical purposs, a unity gain, non-inverting buffer amplifier for impedance matching between electronic circuit stages. Common–drain amplifiers are capable of easily driving low-resistance loads. Common–gate amplifiers can act as an impedance-matching stage from low- to high-resistance electronic circuit stages.

Quantitative expressions for the amplifier performance are found in Table 5.11-2.

5.12 BIASING TO ACHIEVE MAXIMUM SYMMETRICAL SWING

Amplification can be restricted by the size and positioning of the available undistorted output swing. Amplifiers consisting strictly of a transistor and resistors can be biased so that the transistor quiescent point lies in the middle of the linear output voltage range provided by the power supply. This location of the Q point allows symmetrical excursion of signal voltages about a central value: Distortion (amplifier saturation) will occur for equal magnitude excursions in the positive and negative directions.

Amplifiers that have capacitively coupled loads or resistors that are shunted by capacitors in the output voltage–current relationship do not have the entire range of the power supply for output voltage swing. Placing the Q point in the middle of the power supply rails will not provide equal output voltage swing and will greatly limit the utility of the amplifier. It is therefore important to be able to easily choose a Q point that will provide the maximum symmetrical swing for the amplifier given the various design constraints.

A technique that analytically determines the Q point for maximum symmetrical swing is based on transistor output load lines. The output curve for a transistor (either BJT or FET) is typically given as a current versus voltage curve. On this curve DC and AC load lines can be drawn as shown in Figure 5.12-1.

The two load lines always intersect at the quiescent point: If there is zero AC, the output must be on the Q point of the DC load line. It is also true that the magnitude of the midband AC load line slope for all practical amplifier circuits is greater than that of the DC load line. The increased magnitude AC load line slope decreases the available oscillatory swing along the abscissa to a value often significantly less than the power supply limits. The maximum supply limit $\{S\}$ is typically the magnitude of the power supply: The minimum supply limit $\{C\}$ is determined by the edge of the linear region of the transistor (for a BJT it is approximately $V_{CE(\text{sat})}$).

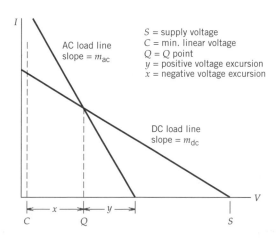

S = supply voltage
C = min. linear voltage
Q = Q point
y = positive voltage excursion
x = negative voltage excursion

AC load line slope = m_{ac}

DC load line slope = m_{dc}

Figure 5.12-1
Typical amplifier AC and DC load lines.

For maximum symmetrical swing it is necessary to choose a Q point that allows equal oscillatory space in both the positive and negative directions. A choice either to the right or left of this optimum value will decrease the symmetrical output voltage swing capability of the amplifier. In Figure 5.12-1, the Q point is chosen so that

$$x = y.$$

Simple geometry applied to the figure yields the expressions

$$y = \frac{m_{dc}}{m_{ac}}[S - Q], \qquad x = Q - C. \tag{5.12-1}$$

solving for the Q point yields

$$\frac{m_{dc}}{m_{ac}}[S - Q] = Q - C, \tag{5.12-2}$$

$$Q\left(1 + \frac{m_{dc}}{m_{ac}}\right) = \frac{m_{dc}}{m_{ac}}S + C. \tag{5.12-3}$$

The quiescent point for maximum symmetrical swing is then determined to be

$$Q = \frac{(m_{dc}/m_{ac})S + C}{1 + m_{dc}/m_{ac}}. \tag{5.12-4}$$

This optimum Q point can easily be determined from only the ratio of the slopes of the AC and DC load lines and the limits on output voltage (the supply voltage and the minimum voltage edge of the linear region of the transistor).

EXAMPLE 5.12-1 Design a resistor bias network that will achieve maximum symmetrical swing for the BJT output configuration shown. The silicon BJT parameters are given as

$$\beta_F = 150, \qquad V_A = 350 \text{ V}.$$

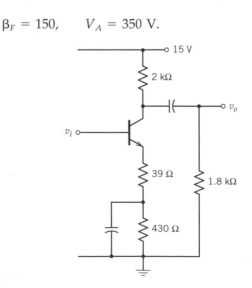

SOLUTION

The output V–I relationship for BJTs is a plot of I_C versus V_{CE}. The load lines must therefore express that relationship.

The expression for the DC load line is

$$V_{CE} = 15 - I_C\left(2000 + \frac{151}{150}\, 469\right)$$

from which is determined

$$m_{dc} = \frac{1}{2000 + (151/150)469} = 0.4045 \times 10^{-3}.$$

The voltage limits on V_{CE} are given by

$$S = 15 \text{ V}, \qquad C = V_{CE(sat)} = 0.2 \text{ V}.$$

The AC load line slope is given by

$$m_{ac} = \frac{1}{(151/150)39 + 2000\|1800} = 1.0136 \times 10^{-3}$$

The Q point can then be calculated as

$$Q = V_{CEQ} = \frac{(0.4045/1.0136)15 + 0.2}{1 + (0.4045/1.0136)} = 4.42 \text{ V}.$$

With a quiescent collector current

$$I_{CQ} = 4.279 \text{ mA}.$$

The collector–emitter voltage will vary from approximately 0.2 to 8.64 V with a symmetrical excursion of 4.22 V about a quiescent value of 4.42 V.

The most obvious resistor bias scheme is the self-bias circuit. The base of the BJT must be at

$$V_{BQ} = -I_{EQ}(469) + 0.7 = 2.72 \text{ V}.$$

Without any restrictions on the input resistance of the amplifier, a wide variety of resistor pair values will achieve proper biasing. A common rule of thumb for bias stability is the current through the base bias resistors should be at least 10 times the BJT base current. Using that rule, the maximum resistance that can be connected between the base and ground is

$$R_{b2(max)} = \frac{2.72}{10(4.279 \times 10^{-3}/150)} = 9.53 \text{ k}\Omega.$$

The closest 5% standard resistor value is 9.1 kΩ. Thus choose $R_{b2} = 9.1$ kΩ. The value of the resistor between the power supply and the BJT base is given by

$$R_{b1} = \frac{15 - 2.72}{(4.279 \times 10^{-3}/150) + (2.72/9100)} = 37.5 \text{ k}\Omega.$$

The nearest 5% standard resistor value to this optimum value is 36 kΩ. It will provide proper operation within acceptable production standards. The final circuit topology is as shown.

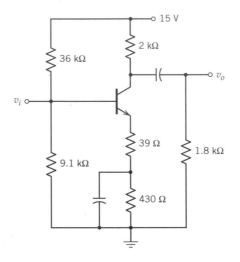

A PSpice simulation of the maximum symmetrical swing properties of this circuit follows. The rounding of resistor values has introduced a slight error in both the Q point and the swing. Distortion is the cause of dissimilar limiting of the signal at the two extremes.

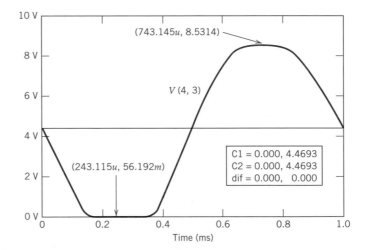

5.13 CONCLUDING REMARKS

The small-signal model of the BJT was developed in this chapter using two-port network analysis and the Ebers–Moll transistor model. Three configurations of BJT amplifiers were described: common–emitter, common–collector, and common–base. The current gain, input resistance, output resistance, and voltage gain of each of the configurations were found.

The FET small-signal model was also developed. Three configurations of FET amplifiers were described: common–source, common–drain, and common–gate. The input resistance, voltage gain, and output resistance were found.

The modeling process for transistor circuit performance contain the following steps:

1. Determine the quiescent (DC) conditions. Verify that the BJT is in the forward-active region or the FET is in the saturation region.
2. Determine the transistor small-signal parameters from the quiescent conditions.
3. Create the AC equivalent circuit.
4. Determine the AC performance by replacing the transistor by its small-signal model.
5. Add the results of the DC and AC analysis to obtain total circuit performance.

Restrictions on amplification was shown to be dependent on transistor quiescent conditions. A technique that analytically determines the Q point for maximum symmetrical swing based on transistor load lines was introduced.

SUMMARY DESIGN EXAMPLE

Whenever the physical distance between a signal source and the load is more than a very small fraction of a wavelength, the wires connecting the source and load appear to act as a transmission line as shown. The voltage and current waveforms traveling on a transmission have amplitudes related by the *characteristic impedance* of the transmission line Z_o. This characteristic impedance is a function of the geometry of the transmission line and the dielectric material surrounding the line.

The interfaces between the source, transmission line, and the load must be carefully controlled in order to avoid signal reflections: Signal reflections produce an "echo" effect. In order to ensure that there are no signal reflections at the source end or the load end of the transmission line, the output resistance of the source and the input resistance of the load must be equal to the characteristic impedance of the transmission line.

Design an amplifier that will accept an input signal from a matched 50 Ω transmission line system. The amplifier will have a voltage gain of ± 10.

Solution

In order to have no reflections at the load, the input resistance of the amplifier must be 50 Ω. The only amplifier topologies that have small input resistance while maintaining a voltage gain are the common-base and common-gate configurations. Of these two, the common-base configuration is much more common: The common-base configuration is chosen for this design. Assume typical transistor parameters: $\beta_F = 150$, $V_A = 160$.

The basic topology of the common-base receiving circuit is shown below. Note that the source is an AC source with an output resistance matched to the characteristic impedance of the transmission line.

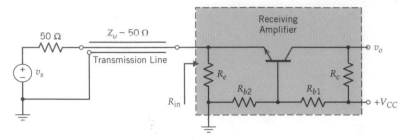

The expression for the input resistance and voltage gain of a common-base amplifier are found in Table 5.5-1:

$$R_i = \frac{h_{ie} + R_b}{h_{fe} + 1} \qquad A_V = \frac{h_{fe}R_c}{h_{ie} + R_b}.$$

The design requirement for a 50-Ω input resistance implies that

$$R_{\text{in}} = R_e \| R_i = R_e \left\| \frac{h_{ie} + R_b}{h_{fe} + 1} \right.,$$

$$50 = R_e \left\| \frac{h_{ie} + R_b}{h_{fe} + 1} \right. = R_e \left\| \frac{(\beta_F + 1)(\eta V_t/I_c) + R_b}{h_{fe} + 1} \right. = R_e \left\| \left(\frac{\eta V_t}{I_C} + \frac{R_b}{h_{fe} +} \right) \right.$$

The choice of R_c is somewhat arbitrary: 50 Ω is the lowest limit with large values limiting the symmetrical swing of the amplifier. Choose R_e = **100 Ω**. The quiescent collector current can also be somewhat arbitrarily chosen (the input resistance requirement and the previous choice of R_c place a lower limit of 0.26 mA on I_c). Choose I_c = 2 mA. This choice constrains R_b:

$$50 = 100 \left\| \left(\frac{26 \text{ mV}}{2 \text{ mA}} + \frac{R_b}{151} \right) \right. \Rightarrow R_b = 13.137 \text{ k}\Omega.$$

With this bias condition the h parameters of the BJT are given by

$$h_{fe} = \beta_F = 150, \qquad h_{ie} = (\beta_F + 1) \frac{\eta V_t}{I_c} = 151(13) = 1.963 \text{ k}\Omega,$$

$$h_{oe}^{-1} = \frac{V_A}{I_c} = \frac{160 \text{ V}}{2 \text{ mA}} = 80 \text{ k}\Omega.$$

The design requirement for a gain of 10 yields the value of the collector resistor:

$$10 = \frac{h_{fe}R_c}{h_{ie} + R_b} = \frac{150R_c}{15.1 \text{ k}\Omega} \Rightarrow R_c = 1.007 \text{ k}\Omega$$

$$\approx \textbf{1.01 k}\Omega \quad \text{(standard value)}.$$

For DC conditions, the transmission line acts as a short circuit. Thus, the effective emitter resistor is the parallel combination of the source output resistance and R_e:

$$R_{e(\text{eff})} = 50 \| R_e = 33.33 \ \Omega.$$

In order to achieve maximum output swing the power supply must be chosen so that the maximum collector current is double the quiescent collector current:

$$V_{CC} > 4 \text{ mA}(R_{e(\text{eff})} + R_c) = 4.173 \text{ V}.$$

Choose $V_{CC} = $ **5 V**.

The only unknown component values are the base resistors R_{b1} and R_{b2}. These are determined in the usual fashion:

$$\frac{R_{b2}}{R_{b1} + R_{b2}} 5 = V_{bb} = \frac{(\beta_F + 1)I_c}{\beta_F} R_{e(eff)} + 0.7 + \frac{R_b I_c}{\beta_F + 1} = 0.94023 \text{ V},$$

$$\frac{R_{b1} R_{b2}}{R_{b1} + R_{b2}} = R_b = 13.137 \text{ k}\Omega.$$

Simultaneous solution of these two equations leads to

$$R_{b1} = 69.86 \text{ k}\Omega \approx \textbf{69.8 k}\Omega, \qquad R_{b2} = 16.18 \text{ k}\Omega \approx \textbf{16.2 k}\Omega.$$

COMPONENT SUMMARY:

For a common-base amplifier, the BJT parameters are $\beta_F = 150, V_A = 160$ V:

$$R_{b1} = 69.8 \text{ k}\Omega, \qquad R_{b2} = 16.2 \text{ k}\Omega, \qquad R_c = 1.01 \text{ k}\Omega,$$
$$R_e = 100 \ \Omega, \qquad V_{CC} = 5 \text{ V}.$$

REFERENCES

Colclaser, R. A. and Diehl-Nagle, S., *Materials and Devices for Electrical Engineers and Physicists,* McGraw-Hill Book Company, New York, 1985.

Ghausi, M. S., *Electronic Devices and Circuits: Discrete and Integrated,* Holt, Rinehart and Winston, New York, 1985.

Gray, P. R., and Meyer, R. G., *Analysis and Design of Analog Integrated Circuits,* 3rd. Ed., John Wiley & Sons, Inc., New York, 1993.

Millman, J., *Microelectronics, Digital and Analog Circuits and Systems,* McGraw-Hill Book Company, New York, 1979.

Millman, J. and Halkias, C. C., *Integrated Electronics: Analog and Digital Circuits and Systems,* McGraw-Hill Book Company, New York, 1972.

Sedra, A. S. and Smith, K. C., *Microelectronic Circuits,* 3rd. Ed., Holt, Rinehart, and Winston. Philadelphia, 1991.

PROBLEMS

5-1. Determine the h parameters for the given two-port network.

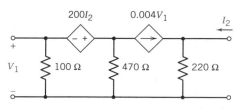

5-2. A two port network has the following y parameters:

$$y_{11} = 0.03 \text{ S}, \qquad y_{12} = 0.00001 \text{ S},$$
$$y_{21} = 0.05 \text{ S}, \qquad y_{22} = 0.02 \text{ S}.$$

Determine the equivalent circuit representation of the two-port network.

5-3. The h-parameter representation for a BJT in a particular common-emitter application is

$$h_{ie} = 1.2 \text{ k}\Omega, \qquad h_{re} = 0.1 \text{ mV/V},$$
$$h_{fe} = 150, \qquad h_{oe} = 10 \ \mu\text{S}.$$

Determine the equivalent BJT g parameters for this application.

5-4. A silicon BJT is described by $\beta_F = 160$ and $V_A = 120$ V. Determine h_{ie}, h_{fe}, and h_{oe} for the following quiescent conditions:

a. $I_c = 2$ mA

b. $I_c = 0.5$ mA

5-5. Determine the significant BJT h parameters when operating in the given circuit.

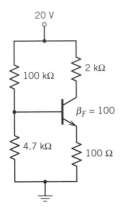

20 V

100 kΩ

2 kΩ

$\beta_F = 100$

4.7 kΩ

100 Ω

5-6. A transistor with $\beta_F = 200$ is used in the circuit shown. Assume the capacitors are infinite.

a. Determine the quiescent conditions for the transistor.

b. Determine the BJT h parameters.

c. Determine the AC circuit parameters $A_V = v_o/v_i$, R_i as shown, and R_o as shown.

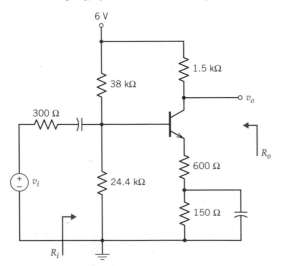

6 V

38 kΩ

1.5 kΩ

v_o

300 Ω

v_i

24.4 kΩ

600 Ω

R_o

150 Ω

R_i

5-7. Complete the given design by finding the value for R_E to establish the given qui-

escent conditions. Draw the midband small-signal equivalent circuit for the circuit below. Determine the appropriate h parameters for the model and find midband input resistance R_{in}, output resistance R_{out}, and voltage gain of the circuit.

Assume that $V_{BE} = 0.7$ V and $\beta_F = 200$.

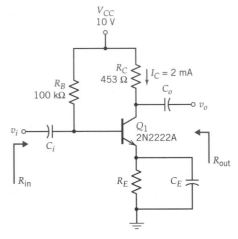

V_{CC}
10 V

R_B
100 kΩ

R_C
453 Ω

$I_C = 2$ mA

C_o

v_o

Q_1
2N2222A

v_i

C_i

R_{in}

R_E

C_E

R_{out}

5-8. For the common-collector circuit shown, assume a silicon BJT with $\beta_F = 75$.

a. Determine the value of R_B so that $I_C = 7$ mA.

b. Determine the amplifier performance parameters $A_I = i_o/i_i$, $A_V = v_o/v_i$, R_i as shown, and R_o as shown.

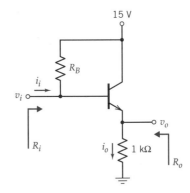

15 V

R_B

i_i

v_i

v_o

R_i

i_o

1 kΩ

R_o

5-9. Complete the given design by finding the value for R_B so that the quiescent $I_C = 2$ mA. Determine the quiescent voltages and currents. Draw the midband small-signal equivalent circuit for the circuit shown and determine the midband input resistance R_{in}, output resistance R_{out}, and voltage gain of the circuit.

Assume that $V_{BEQ} = 0.7$ V and $\beta_F = 180$.

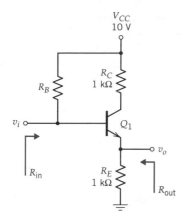

5-10. Draw the midband small-signal equivalent circuit for the circuit shown. Find the expressions for the midband input resistance R_{in}, output resistance R_{out}, and voltage gain of the circuit.

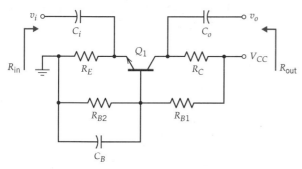

5-11. Complete the design of the given circuit for $R_{in} = 1.5 \text{ k}\Omega$ and determine the quiescent currents and voltages. Draw the midband small-signal equivalent circuit. Determine the appropriate h parameters for the small-signal model and find the midband voltage gain, input resistance R_{in}, and output resistance R_{out}. Assume that

$$V_{BEQ} = 0.7 \text{ V}, \qquad \beta_F = 150, \qquad I_C = -2 \text{ mA}.$$

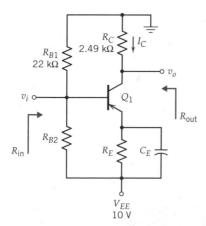

5-12. Complete the design of the circuit shown and draw the midband small-signal equivalent circuit. Determine the quiescent currents and voltages. Draw the small-signal equivalent circuit. Determine the appropriate h parameters for the small-signal model and find the midband voltage gain, input resistance R_{in}, and output resistance R_{out}.

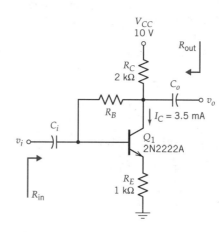

5-13. For the circuit shown:

a. Determine the quiescent currents and voltages.

b. Determine the overall midband voltage gain A_{vs}, R_{in}, and R_{out}.

c. Determine the maximum symmetrical swing and corresponding maximum peak-to-peak input signal.

Assume the transistor parameters $\beta_F = 220$ and $V_{BE} = 0.7 \text{ V}$. All capacitors are sufficiently large to have small reactances.

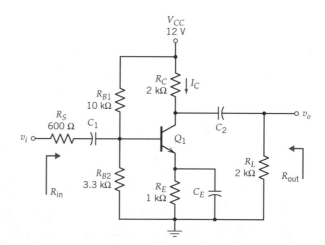

5-14. For the circuit shown:

a. Determine the quiescent currents and voltages.

b. Determine the overall midband voltage gain A_{vs} and the input and output resistances R_{in} and R_{out}.

c. Determine the maximum symmetrical output voltage swing and corresponding maximum peak-to-peak input signal.

Assume the following transistor parameters in SPICE: BF = 180, VA = 100, IS = 6.8 F.

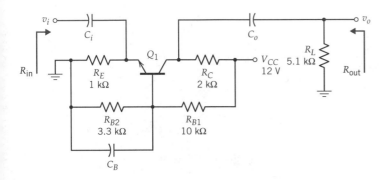

5-15. For the amplifier circuit shown:

a. Complete the design of the common-base BJT amplifier shown below for a bias stability of 1% change in I_C for a 10% change in β_F. The *pnp* transistor parameters are

$$V_A = 150 \text{ V}, \qquad \beta_F = 120.$$

b. Determine the midband voltage gain of the amplifier.

c. Find the input resistance R_{in} and output resistance R_{out}.

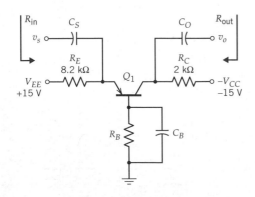

5-16. Design a noninverting amplifier with a gain of $10 < A_v < 25$ with an input resistance of 50 Ω ± 20% and an output resistance of 600 Ω ± 20%. Use a 2N3906 *pnp* BJT with the charactertistic curves found in the Appendix. Determine the maximum undistorted output signal for the design and the corresponding input signal. Available power supply voltage is ±12 V.

5-17. An amplifier with low input resistance and moderate voltage gain is to be developed from the self-bias circuit shown.

a. Indicate the proper input and output terminals to meet the design goals. Insert appropriate bypass capacitors to improve the performance characteristics.

b. Determine the voltage gain and input resistance of the design.

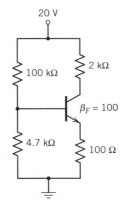

5-18. Design an impedance buffer/transformer using a 2N2222 *npn* BJT. The specifications for the buffer/transformer are

Input resistance: 12 kΩ ± 20%

Output resistance: 51 Ω ± 20%

Available power supply voltages: ±12 V

Determine the voltage and current gains of the circuit.

5-19. The amplifier shown was *supposed* to have been designed for a voltage gain:

$$A_{vs} = -33 \pm 20\%.$$

However, its output is $v_o = 0$ for all input voltages v_i. Something is wrong with this amplifier. The transistor parameters are

$$\beta_F = 180, \qquad V_{BE} = 0.7 \text{ V}.$$

All capacitors are sufficiently large to have small reactances.

a. Why doesn't the amplifier work properly?

b. Redesign the bias network for the new quiescent conditions by solving for I_C from h_{ie} using the small-signal model with $A_{vs} = v_o/v_s = -33$. There is no specification on R_{in}.

c. Determine A_{vs}, R_{in}, and R_{out}.

d. Determine the maximum symmetrical output voltage swing and corresponding maximum peak-to-peak input signal.

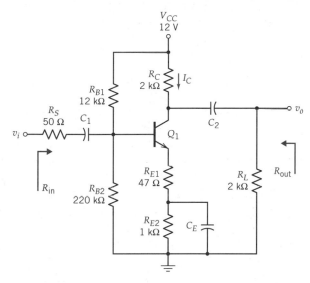

5-20. Table 5.7-1 itemizes several alternate expressions for the determination of the mutual transconductance g_m for FETs. Show the equivalence of the various forms for each type of FET.

5-21. Assume a JFET with parameters

$$V_{PO} = -4 \text{ V}, \qquad I_{DSS} = 10 \text{ mA}, \qquad V_A = 120 \text{ V}$$

operating in the circuit shown.

a. Determine the FET small-signal parameters.

b. Determine the amplifier small-signal performance characteristics in its midband frequency range.

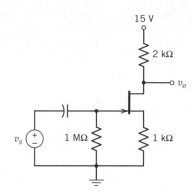

5-22. Find the midband voltage gain for the circuit shown. The n-channel JFET parameters are

$$I_{DSS} = 7 \text{ mA}, \qquad V_{PO} = -4 \text{ V}, \qquad V_A = 120 \text{ V}.$$

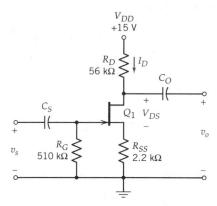

5-23. Complete the given design of the n-channel depletion MOSFET circuit to achieve a Q point: $I_D = 1 \text{ mA}$, $V_{DS} = 4 \text{ V}$.

Draw the midband small-signal model and find midband input resistance R_{in}, output resistance R_{out}, and voltage gain of the amplifier. The MOSFET parameters are

$$V_A = 100 \text{ V}, \qquad I_{DSS} = 8 \text{ mA}, \qquad V_{PO} = -2\text{V}.$$

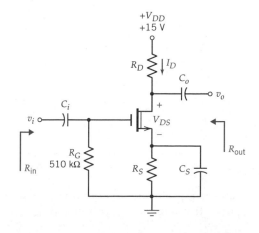

5-24. The MOSFET amplifier circuit shown is to be designed for a midband voltage gain of 8. The circuit is to be biased for the following quiescent condition:

$$I_D = 5 \text{ mA}, \qquad V_{DS} = 5 \text{ V}.$$

The transistor parameters are

$$I_{DSS} = 8 \text{ mA}, \qquad V_{PO} = -3 \text{ V}, \qquad V_A = 60 \text{ V}.$$

Complete the design.

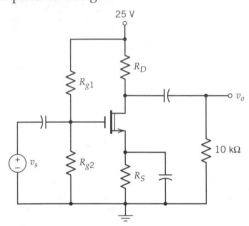

5-25. Complete the design of the MOSFET circuit shown so that $I_D = 4$ mA and $V_{DS} = 5$ V.

The MOSFET parameters are

$$I_{DSS} = 5 \text{ mA}, \quad V_{PO} = -3 \text{ V}, \qquad V_A = 100 \text{ V}.$$

Find the voltage gain $A_{vs} = v_o/v_s$, R_{in} and R_{out}. Include the SPICE netlist and DC midband voltage gain R_{in} and R_{out} analyses of the amplifier. All capacitors are sufficiently large to have small reactances.

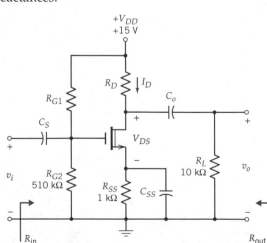

5-26. Complete the design of the amplifier shown to establish operation at $V_{GS} = V_{DS}$.

$= 2V_T$ with equal values of R_S and R_D. The input resistance R_{in} is 730 kΩ ± 5%. The n-channel enhancement MOSFET parameters are

$$V_T = 1.5 \text{ V}, \quad K = 0.5 \text{ mA/V2}, \quad V_A = 100 \text{ V}.$$

a. Find I_D and V_{DS}.

b. Find R_S, R_D, R_{G1}, and R_{G2}.

c. Determine the midband voltage gain of the amplifier.

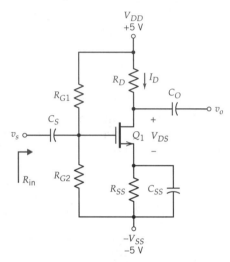

5-27. Complete the given design of the common–gate amplifier so that $I_D = -1$ mA. Determine the quiescent voltages and currents, the voltage gain $A_{vs} = v_o/v_s$, R_{in}, and R_{out}. The FET parameters of interest are

$$I_{DSS} = -6 \text{ mA}, \qquad V_{PO} = 2.5 \text{ V},$$
$$V_A = 100 \text{ V}.$$

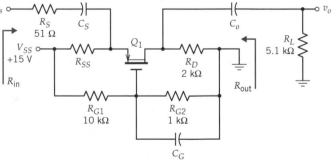

5-28. Design a noninverting amplifier with midband voltage gain $A_{vs} \geq 2$ and output resistance, $R_{out} = 1$ kΩ ± 20%, input resistance, $R_{in} = 100$ Ω ± 25 Ω using the 2N5461 p-channel JFET. The 2N5461 specification sheet is found in the Appendix. Use the typical values of the parameters: That is, use the average of the maximum and minimum values, unless only maximum or minimum values are given. The gate–source cutoff voltage $V_{GS(off)}$ specification is equivalent to FET pinchoff voltage V_{PO}, and the output conductance $R_e\{y_{os}\}$ specification is equivalent to g_d.

a. What is the quiescent condition of the JFET?

b. Confirm the design using SPICE.

5-29. The amplifier shown has, as its output, $v_o = 0$ for all input voltages v_i. Something is wrong with this amplifier. The 2N5485 transistor parameters are given on the specification sheet found in the Appendix. Use the typical values of the parameters: That is, use the average of the maximum and minimum values, unless only maximum or minimum values are given. The gate–source cutoff voltage $V_{GS(off)}$ specification is equivalent to FET pinchoff voltage V_{PO} and the output conductance $R_e(V_{os})$ specification is equivalent to g_d. All capacitors are sufficiently large to have small reactances.

a. Why doesn't the amplifier work properly?

b. Redesign the bias network for the new quiescent conditions of $I_D = 1$ mA and $V_{DS} = 4$ V.

c. Determine A_{vs}, R_{in}, and R_{out}. Include the SPICE netlist and DC, midband voltage gain, R_{in}, and R_{out} analyses of the amplifier.

d. Determine the maximum symmetrical output voltage swing and corresponding maximum peak-to-peak input signal.

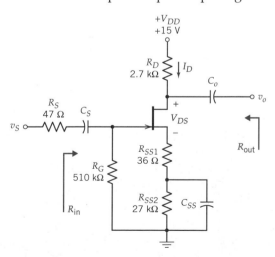

5-30. What is the midband voltage gain and output resistance of a common-drain amplifier when the output is taken from the source of the redesigned circuit in the previous problem? What are the new quiescent conditions, midband voltage gain, input resistance, and output resistance of the common-drain amplifier if $R_{SS1} = 2.7$ kΩ, $R_{SS2} = 0$ Ω, and C_{SS} removed?

5-31. Design a circuit to bias an n-channel depletion MOSFET described by the following $V_A = 100$ V, $I_{DSS} = 9$ mA, and $V_{PO} = -3$ V using the configuration shown. The design specifications require that $I_D = 2$ mA and $V_{DS} = 3$ V. The bootstrapping bias technique is used to preserve the high input resistance of the circuit. Draw the midband small-signal model and find midband intput resistance R_{in}, output resistance R_{out}, and voltage gain of the amplifier.

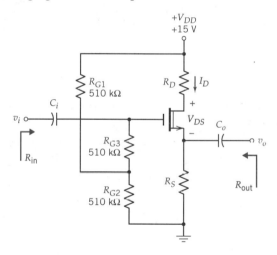

5-32. Assume an n-channel depletion-type MOSFET with parameters

$$I_{DSS} = 9 \text{ mA}, \qquad V_{PO} = -4.5 \text{ V},$$
$$V_A = -140 \text{ V}.$$

Determine the midband AC voltage gain

$$A_v = \frac{v_o}{v_i}.$$

Note: This circuit is *not* one of the topology types for which simple AC parameter expressions have previously been determined. The gain must be derived from the AC transistor model inserted into the given circuit.

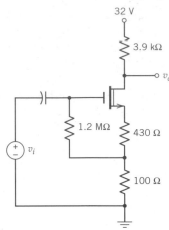

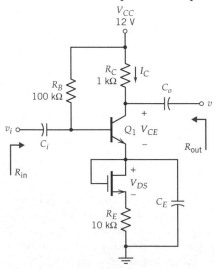

5-33. The common-emitter amplifier with constant-current source biasing shown does not meet the required design specifications. The specifications are, for Q_1,

$$I_C = 2 \text{ mA},$$

$V_{CE} = 3 \text{ V}$ for power dissipation of 6 mW for BJT,

$$\beta_F = 220, \quad V_{BE} = 0.7 \text{ V},$$

and for Q_2,

$$V_T = 2 \text{ V}, \quad K = 500 \text{ μA/V}^2.$$

All capacitors are sufficiently large to have small reactances.

   a. Why does the circuit not meet specifications.?

   b. Redesign the amplifier to meet the specifications.

   c. Find the voltage gain and input and output resistances.

   d. Estimate the total power dissipated.

5-34. The common-collector buffer amplifier (using Q_1) shown does not meet the required design specifications. The FET Q_2 forms a constant-current source. The specifications are

$$I_C = -2 \text{ mA}$$

Q_1 maximum DC power dissipation: 15 mW
$$(P_D = I_C V_{CE} \le 15 \text{ mW})$$

The transistor parameters are

$$Q_1: \beta_F = 120, V_{BE} = -0.7 \text{ V}$$
$$Q_2: K = 500 \text{ μA/V}^2, V_T = -2 \text{ V}$$

All capacitors are sufficiently large to have small reactances.

   a. Why does the circuit shown not meet the specifications?

   b. Redesign the bias network to fulfill the design specifications. Find the new quiescent currents and voltages $V_{DSQ2}, I_{DQ2}, V_{GSQ2}, I_{CQ1}, V_{CEQ1}$, and I_{BQ1}. (*Hint:* Design for $I_C = -2$ mA and $V_{EC} < 7.5$ V and alter R_D and R_B.)

   c. What is the total (DC) power dissipation?

   d. Determine the new overall midband voltage gain $A_{vs} = v_o/v_{ss}$, R_{in}, and R_{out}.

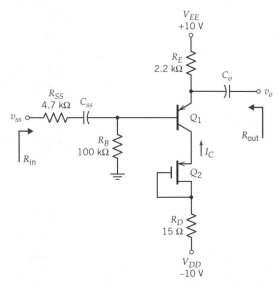

5-35. In the given circuit determine the proper connection for v_o, resistor values, and a power supply V_{DD} such that the following design goals are satisfied: $R_{in} = 50$ kΩ and $R_{out} = 10$ kΩ, as shown on the diagram, and

$$|A_v| = \left| \frac{v_o}{v_i} \right| = 7.0$$

with quiescent drain current $I_D = 2$ mA. Assume an n-channel JFET with

$$I_{DSS} = 5 \text{ mA}, \quad V_{PO} = -4 \text{ V}, \quad V_A = -100 \text{ V}.$$

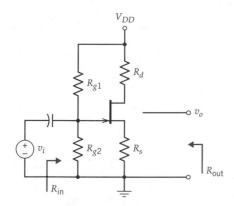

5-36. For the circuit shown, the BJT has parameters

$$\beta_F = 150, \quad V_A = 250 \text{ V}.$$

a. Determine values of the unknown resistors such that maximum symmetrical swing of the output, v_o, will occur. It is required that the input resistance $R_{in} \approx 24$ kΩ.

b. Using the values found above, determine the voltage gain

$$A_v = \frac{v_o}{v_i}$$

and the output resistance R_{out}.

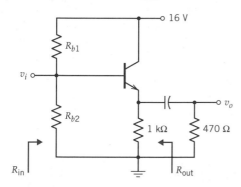

5-37. Complete the design of the circuit shown for maximum symmetrical collector current swing. Let $\beta_F = 180$ and $V_\gamma = 0.7$ V, use the rule-of-thumb relationship between R_E and R_B for stable operation (1% change in I_C for 10% change in β_F). Determine the quiescent conditions.

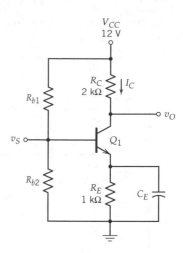

5-38. Design an inverting amplifier with a gain of $35 < |A_v| < 45$ with an input resistance of 12 kΩ ± 20% and an output resistance of 3.5 kΩ ± 20%. Use a 2N2222 BJT and design for maximum symmetrical swing. Available power supply voltage is 15 V. How will the performance of the amplifier be altered if an emitter degenerative resistor (an emitter resistor that is not bypassed with a capacitor) of 51 Ω is used in the circuit?

5-39. Complete the design of the circuit shown for maximum symmetrical swing in the midband frequency range. The circuit is required to be have a bias stability of 1% change in I_C for a 10% change in β_F. The transistor has a $\beta_F = 180$ and $V_A = 200$ V.

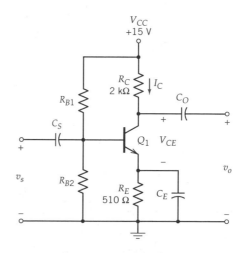

5-40. Complete the design of the circuit shown for maximum symmetrical swing in the midband frequency range. The circuit is required to be have a bias stability of 1% change in

I_C for a 10% change in β_F. The *pnp* BJT has a β_F = 120 and V_A = 120 V.

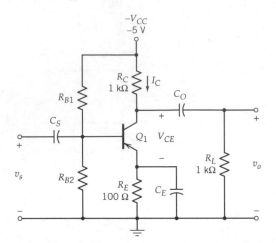

5-41. Designing for maximum symmetrical swing using FETs is challenging due to the dependence of the minimum linear voltage on quiescent conditions. For a FET this voltage occurs at the transition between the ohmic and sat-

uration regions. For the circuit application shown, complete the design so that maximum symmetrical swing is achieved. Explain your methodology for determining the Q point. The FET is described by

$$K = 1 \text{ mA/V}^2, \qquad V_T = 1.2 \text{ V}, \qquad V_A = 110 \text{ V}.$$

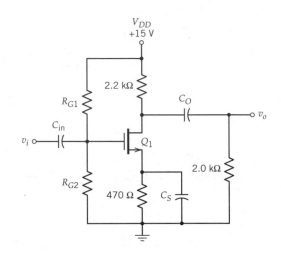

CHAPTER 6

MULTIPLE-TRANSISTOR AMPLIFIERS

Previous discussions of transistor amplifiers have shown that single-transistor amplifiers have a wide range of possible design characteristics described by the principal performance characteristics: voltage and current gain and input and output resistance. There are, however, many circumstances when the desired overall performance characteristics for an amplifier cannot be met by a single-transistor amplifier. That is, the required combination of amplification, input resistance, and output resistance may be beyond the capability of a single-transistor amplifier. In these circumstances it is necessary to employ amplifiers that use more than a single transistor.

The obvious approach to changing amplifier performance characteristics is to cascade (connect the output of a given stage directly into the input of the following stage) single-transistor amplifiers. In this manner, additional gain or the modification of the input–resistance or the output resistance can be easily accomplished. This simple approach to multiple-transistor amplifiers is discussed in Section 6.1. A major drawback to this approach is the relatively high number of electronic components necessary to accomplish a design. In addition, if the simple stages are coupled with capacitors (a common practice to decouple the DC bias conditions of each stage), the low-frequency performance of the amplifier can be severely degraded.

Several transistor cascades are commonly packaged as a single unit. Most common among these cascade types are the Darlington configurations. These cascade configurations have the advantage of simple external bias circuitry, high performance, and DC coupling of stages. Analysis of these types is discussed in Section 6.2.

Differential amplifiers are another type of high-performance multiple-transistor amplifier. They commonly are based on the high gain of emitter-coupled and source-coupled transistor amplifiers. Operational amplifiers are a common form of differential amplifier that typically have emitter-coupled or source-coupled amplifiers as input stages.

In addition, the practice of biasing transistor amplifiers with transistor current sources will be explored in this chapter. This bias technique is particularly suited to integrated amplifiers contained on a single semiconductor chip where transistors are more practical and economical than resistors.

Active (as opposed to resistive) loads will be explored as a means of increasing gain and modifying output resistance without the use of extremely large resistors and power supplies.

6.1 MULTISTAGE AMPLIFIERS USING SIMPLE STAGES CASCADED

Among the many multistage connections possible in large amplifiers, the cascade connection is the simplest. In a cascade connection, the output voltage and current of an amplifier stage are passed directly to the input of the next amplifier stage. A

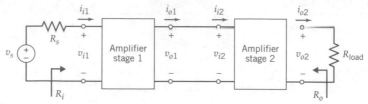

Figure 6.1-1
Amplifier consisting of two cascaded stages.

two-stage cascade-connected amplifier is represented by Figure 6.1-1. In this amplifier, the outputs v_{o1} and i_{o1} of the first amplifier stage become the inputs v_{i2} and i_{i2} to the second stage without any modification. For the discussions in this section, it is assumed that each of the amplifier stages consists of one of the simple single-transistor amplifiers described in Chapter 5.[1]

The advantage of cascade-connected amplifiers becomes apparent in the analysis procedure. The overall voltage gain, for example, of the amplifier of Figure 6.1-1 is given by the ratio of the load voltage to the source voltage:

$$A_V = \frac{v_{o2}}{v_s}.$$

(6.1-1)

A simple progression through the stages of the amplifier expands this expression to more familiar expressions:

$$
\begin{aligned}
A_V = \frac{v_{o2}}{v_s} &= \left(\frac{v_{o2}}{v_{i2}}\right)\left(\frac{v_{i2}}{v_{o1}}\right)\left(\frac{v_{o1}}{v_{i1}}\right)\left(\frac{v_{i1}}{v_s}\right) \\
&= (A_{V2})(1)(A_{V1})\left(\frac{R_i}{R_i + R_s}\right),
\end{aligned}
$$

(6.1-2)

where A_{V1} = voltage gain of first amplifier stage

A_{V2} = voltage gain of second amplifier stage

The total voltage gain of a cascade-connected amplifier can be expressed as a product of the gains of the individual stages and simple voltage divisions. The beauty of cascade-connected amplifiers comes in the second of the multiplicative terms of Equation 6.1-2: It is an expression of identity and consequently can be eliminated from consideration. The expressions for the gains of the individual stages were developed in Chapter 5 and may be used directly in calculating the overall gain. It is, however, important to note that *each stage presents a load to the previous stage*: Its input resistance is part of the total load that is apparent to a previous stage. In a similar fashion the current gain of a cascade-connected amplifier is the product of the current gains of the individual stages and simple current divisions. Again the individual stages interact, and the expressions for current gain must include the effects of that interaction. Total input resistance R_i or total output resistance R_o may also be modified by the interaction of individual stages.

---

[1] More complicated individual stages may be cascade connected. The restriction of individual stages to simple single-transistor ones is for demonstration purposes only: later sections will include more complicated stages that are cascade connected.

EXAMPLE 6.1-1 The two-stage cascade-connected amplifier shown is composed of two identical amplifier stages. The two silicon BJTs have characteristic parameters

$$\beta_F = 150, \qquad V_A = 350.$$

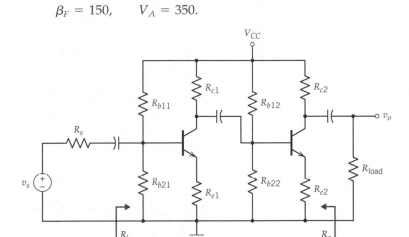

The circuit resistors are given as

$$R_{b11} = R_{b12} = 82 \text{ k}\Omega, \qquad R_{c1} = R_{c2} = 2.2 \text{ k}\Omega,$$

$$R_{b21} = R_{b22} = 12 \text{ k}\Omega, \qquad R_{e1} = R_{e2} = 430 \text{ }\Omega,$$

$$R_s = 100 \text{ }\Omega, \qquad R_{\text{load}} = 2.7 \text{ k}\Omega$$

The power supply voltage is

$$V_{CC} = 15 \text{ V}.$$

Determine the AC voltage and current gain (from the source to the load) and the input and output resistances (as shown) for this amplifier.

SOLUTION

The determination of the performance of multistage amplifiers follows the same basic steps that were derived in Chapter 5:

1. Model the transistors with an appropriate DC model (circuit or analytic).
2. Determine the circuit quiescent (DC) conditions; verify active region for BJTs or saturation region for FETs.
3. Determine the transistor AC model parameters from the quiescent conditions.
4. Create an AC equivalent circuit.
5. Determine the AC performance of each amplifier stage by replacing the transistors by their respective AC models or using previously derived results for the circuit topology.
6. Combine the stage performance quantities to obtain total amplifier performance.
7. Add the results of the DC and AC analysis to obtain total circuit performance.

Only the fifth step has been modified to reflect the new condition that several stages may comprise the total amplifier.

Beginning with the amplifier stage containing Q_1, the DC equivalent circuit, after replacing the base circuit with its Thévenin equivalent, is as shown.

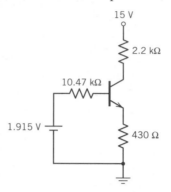

The base and collector currents are calculated to be

$$I_B = \frac{1.915 - 0.7}{10{,}470 + 151(430)} = 16.11 \ \mu A,$$

$$I_C = 150 I_B = 2.417 \ \text{mA}.$$

The voltage V_{CE} must be checked to verify that the BJT is in the forward-active region:

$$V_{CE} = 15 - I_C(2200) - \frac{151}{150} I_C(430) = 8.64 \ \text{V} \geq 0.2 \ \text{V}.$$

After verification, the BJT h-parameters can be determined:

$$h_{fe} = 150, \qquad h_{ie} = 151 \frac{\eta V_t}{|I_C|} = 1.624 \ \text{k}\Omega,$$

$$h_{oe} = \left| \frac{I_C}{V_A} \right| = \frac{1}{144.8 \ \text{k}\Omega}.$$

In this amplifier the two stages are identically biased. Consequently the BJT quiescent conditions in each amplifier stage and the resultant BJT h-parameters are the same. When the stages are *not* identical, the bias conditions and transistor parameters must be obtained for each stage.

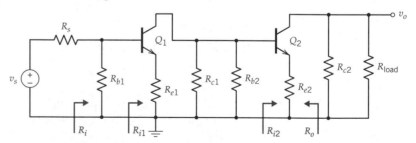

The AC equivalent circuit for the two-stage amplifier is as shown, where

$$R_{b1} = R_{b11} \| R_{b21} = 10.47 \ \text{k}\Omega, \qquad R_{b2} = R_{b12} \| R_{b22} = 10.47 \ \text{k}\Omega.$$

This AC model depicts a cascade-connected amplifier consisting of two common-emitter (CE) with emitter–resistor (CE + R_e) amplifier stages. The performance characteristics for simple BJT amplifier stages is given in Table 5.6-2; the characteristics will be reproduced here.[2] The input resistance of a CE + R_e amplifier stage is

$$R_i = h_{ie} + (h_{fe} + 1)R_e.$$

Since the stages are identical ($R_{e1} = R_{e2}$),

$$R_{i1} = R_{i2} = 1624 + (151)430 = 66.55 \text{ k}\Omega$$

The total amplifier input resistance is the parallel combination of the first-stage bias resistors and input resistance:

$$R_i = R_{b1}\|R_{i1} = 10{,}470\|66{,}550 = 9.0467 \text{ k}\Omega \approx \textbf{9.05 k}\Omega.$$

The total amplifier output resistance is the output resistance of the second stage:

$$R_o = R_{o2} \approx R_{c2} = \textbf{2.2 k}\Omega.$$

The voltage gain for a CE + R_e amplifier is

$$A_V = \frac{-h_{fe}R_c}{h_{ie} + (h_{fe} + 1)R_e} = \frac{-h_{fe}R_c}{R_i}.$$

The only quantity yet undetermined in this gain expression is R_c, the total resistance connected to the collector terminal of the amplifier stage BJT. For the first amplifier stage, R_c is the parallel combination of R_{b2}, R_{c1}, and the input resistance to the second stage, R_{i2}. For the second stage, R_c is the parallel combination of R_{c2} and R_{load}. Notice that even though the amplifier stages are identical, the different loading of each stage produces a variation in some of the stage performance characteristics. The voltage gain for each amplifier stage is calculated to be

$$A_{V1} = \frac{-150(2.2\|10.47\|66.55)}{66.55} = -3.9887,$$

$$A_{V2} = \frac{-150(2.2\|2.7)}{66.55} = -2.7323.$$

The overall voltage gain is given by the product of the individual stage gains and a voltage division at the input, as originally derived in Equation 6.1-2:

$$A_V = \frac{v_o}{v_s} = (A_{V2})(A_{V1})\left(\frac{R_i}{R_i + R_s}\right)$$

$$= (-3.9887)(-2.7323)\left(\frac{9.05}{9.05 + 0.1}\right) \approx \textbf{10.8}.$$

---

[2] The BJT parameter $1/h_{oe}$ is sufficiently large in comparison to the other circuit resistances that it has been ignored in all calculations.

The current gain can be derived from the voltage gain and the various resistances:

$$A_I = \frac{i_{load}}{i_{source}} = \left(\frac{i_{load}}{v_o}\right)\left(\frac{v_o}{v_s}\right)\left(\frac{v_s}{i_{source}}\right) = \left(\frac{1}{R_{load}}\right)(A_V)(R_i + R_s) \approx \mathbf{36.5}.$$

All required amplifier performance characteristics have been calculated: The problem statement in this example does not require that AC and DC performance be added.

While the capacitive coupling of amplifier stages has its primary advantage in the decoupling of the individual stage quiescent conditions, there are disadvantages that must be considered. The addition of capacitors may seriously degrade the low-frequency response of an amplifier. In addition, the decoupling of quiescent conditions necessitates the individual biasing of each transistor into the forward-active (for BJTs) or saturation (for FETs) regions. Individual biasing can significantly increase the number of bias elements (in discrete amplifiers these elements are usually resistors), which will increase the size, cost, and power consumption of an amplifier. It is therefore advantageous, whenever possible, to directly couple amplifier stages. Capacitive coupling of the input source and the load is often unavoidable due to the DC offsets often necessary in these simple amplifier stages.

An example of a two-stage cascade-connected amplifier with directly coupled stages is shown in Figure 6.1-2. The bias for the second transistor stage (the BJT for this amplifier) is dependent on the quiescent conditions of the first transistor stage. The choice to use dependent biasing rather than capactively coupled independent stages has eliminated, in this case, two bias resistors and a capacitor from the design.

Under certain circumstances, it might be possible to eliminate other elements from this amplifier. The purpose of R_g is to ensure that the quiescent voltage at the gate of Q_1 is at zero potential. If the designer is absolutely sure the input source v_s has no DC bias, the resistor R_g and its associated input capacitor could also be eliminated from this design.

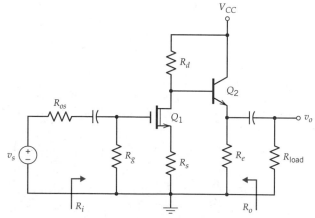

Figure 6.1-2
Cascade-connected amplifier with direct coupling of stages.

EXAMPLE 6.1-2 The two-stage cascade-connected amplifier shown in Figure 6.1-2 is composed of two directly connected simple stages. The transistors have characteristic parameters:

$$\beta_F = 150, \qquad V_A = 350 \text{ V} \quad \text{(silicon BJT)},$$

$$V_{PO} = -3.5 \text{ V}, \qquad I_{DSS} = 10 \text{ mA}, \qquad V_A = 250 \text{ V} \quad \text{(FET)}.$$

The circuit resistors are given as

$$R_g = 1 \text{ M}\Omega, \qquad R_e \quad = 2.7 \text{ k}\Omega,$$

$$R_d = 1.5 \text{ k}\Omega, \qquad R_{os} = 100 \text{ }\Omega,$$

$$R_s = 100 \text{ }\Omega, \qquad R_{\text{load}} = 2.2 \text{ k}\Omega. \quad .$$

The power supply voltage is

$$V_{CC} = 15 \text{ V}.$$

Determine the AC voltage and current gains (from the source to the load) and the input and output resistances (as shown) for this amplifier.

SOLUTION

The quiescent conditions for the FET are determined with the usual expressions relating I_D and V_{GS} in the saturation region:

$$I_D = I_{DSS}\left(1 - \frac{V_{GS}}{V_{PO}}\right)^2 = -\frac{V_{GS}}{R_s}$$

$$= 10 \text{ mA}\left(1 - \frac{V_{GS}}{-3.5}\right)^2 = -\frac{V_{GS}}{130}.$$

Solving this equation yields the FET quiescent conditions:

$$V_{GS} = -0.7833 \text{ V}, \qquad I_D = 6.025 \text{ mA}.$$

The BJT quiescent conditions depend on the DC circuit shown. The base and collector currents are found by writing a loop equation around the base–emitter loop:

$$15 - 1500(6.025 \times 10^{-3} + I_B) - 0.7 - (R_e)151I_B = 0.$$

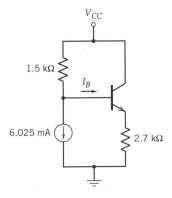

The BJT base and collector currents are found to be

$$I_B = 12.86 \ \mu A, \qquad I_C = 1.93 \ mA.$$

Checking to see if the transistors are in the appropriate regions yields

$$V_{CE} = 15 - I_C(2700) = 9.79 \ V \geq 0.2,$$

$$V_{DS} = 15 - 1500(I_D - I_B) - 100I_D = 5.3 \ V \geq V_{GS} - V_{PO} = 2.71.$$

The transistor AC parameters then are calculated to be

$$h_{fe} = \beta_F = 150, \qquad h_{ie} = (\beta_F + 1)\frac{\eta V_t}{|I_C|} = 2.04 \ k\Omega,$$

$$g_m = \frac{2I_D}{V_{GS} - V_{PO}} = 4.425 \ mA/V, \qquad r_d = \left|\frac{V_A}{I_D}\right| = 41.49 \ k\Omega,$$

$$\mu = g_m r_d = 183.6.$$

The AC model for the circuit is composed of a common-source amplifier stage followed by a common-collector stage. The performance parameters for these simple stages were derived in Chapter 4.

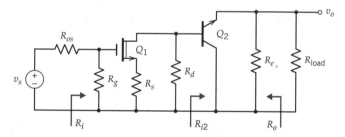

The total input resistance R_i is the parallel combination of R_g and the input resistance of the common-source stage:

$$R_i = R_g \| \infty = R_g = \mathbf{1 \ M\Omega}.$$

The input resistance of the common-collector stage is given by

$$R_{i2} = h_{ie} + (h_{fe} + 1)(R_e \| R_{\text{load}}) = 185.1 \ k\Omega.$$

The gain of the common-source stage is

$$A_{V1} = \frac{-\mu(R_d \| R_{i2})}{r_d + (R_d \| R_{i2}) + (\mu + 1)R_s} = -4.447.$$

The gain of the common-collector stage is

$$A_{V2} = 1 - \frac{h_{ie}}{R_{i2}} = 0.9890.$$

The overall gain is given by

$$A_V = A_{V1}A_{V2} \frac{R_i}{R_i + R_{os}} = -4.40.$$

The current gain can be derived from the voltage gain and the various resistances:

$$A_I = \frac{i_{load}}{i_{source}} = \left(\frac{i_{load}}{v_o}\right)\left(\frac{v_o}{v_s}\right)\left(\frac{v_s}{i_{source}}\right) = \left(\frac{1}{R_{load}}\right)(A_V)(R_i + R_s) \approx 2000.$$

The output resistance calculation is the most difficult to perform. The output resistance of the common-collector stage depends on the output resistance of the common-source stage:

$$R_{o2} = \frac{(R_{o1}\|R_d) + h_{ie}}{h_{fe} + 1}$$

The output resistance of the common-source stage is

$$R_{o1} = r_d + (\mu + 1)R_s = 59.95 \text{ k}\Omega.$$

Therefore,

$$R_{o2} = 23.2 \ \Omega,$$

and the total output resistance is given by

$$R_o = R_{o2}\|R_e = 23.0 \ \Omega.$$

6.1.1 Design Choices for Transistor Configuration in Cascade-Connected Amplifier

The derivations and calculations of voltage and current gain and input and output resistance in previous examples apply to any configuration of cascade-connected amplifier. As each stage is added, the gain expression is increased by only one additional term. The individual stages may be any of the three BJT or three FET configurations and any number of these stages may be connected in any order. There are, however, better engineering design choices than others.

The basic reasons for choosing a multiple-stage amplifier over a single-stage amplifier include one or more of the following performance characteristics:

- Increased amplification
- Input impedance modification
- Output impedance modification

Common-emitter and common-source amplifier stages are ideal for increased amplification purposes. Each type exhibits significant voltage gain *and* significant current gain. In addition, when cascade connected, the relatively high input resistance of these stage configurations does not significantly load previous amplification stages (voltage gain is a function of the load). Common-base and common-gate amplifiers exhibit good voltage gain but no current gain. The relatively low input resistance of these stages also produces a gain-reducing load to previous stages. Consequently, these stages are most useful as low-input-resistance first stages. Common-collector and common-drain amplifiers have good current gain but no voltage gain. Their relatively high input resistance allows them to follow an amplification stage without seriously decreasing the voltage gain. These two configurations are most useful for low-output-resistance final stages. In summary, the design choices for a cascade-connected amplifier are usually based on the following principles:

- First stage should *not* be common-collector or common-drain:
  - **a.** Low-input-resistance cases: common-base or common-gate
  - **b.** Other cases: common-emitter or common-source
- Intermediate stages should be common-emitter or common-source.
- Final stage should *not* be common-base or common-gate:
  - **a.** Low-output-resistance cases: common-collector or common-drain
  - **b.** Other cases: common-emitter or common-source

Special-purpose amplifiers may violate these design principles, as seen in Section 6.2.

6.2 DARLINGTON AND OTHER SIMILAR CONFIGURATIONS

In addition to cascade-connected single-transistor amplifiers, there exist several two-transistor configurations that are commonly analyzed and often packaged as signal stages. These particular combinations are unusual in that they seem to violate the configuration principles of good amplifier design as itemized in the last section. The configurations do, however, produce amplifiers with specific, significantly enhanced properties.

Most common among these configurations are the two Darlington BJT configurations and the cascode configuration. Mixed BJT and FET combinations are currently not as common as the single-type configurations. The BiFET (BiCMOS) Darlington seems most promising of these newer technology combinations.

6.2.1 Darlington Configurations

The Darlington two-transistor amplifier configurations consist of a pair of BJTs connected so that the emitter of the input transistor couples directly, in an AC sense, into the base of the output transistor. A DC biasing current source is also commonly present at this transistor interconnection node. The output is taken at either the collector or the emitter of the second transistor. Thus, in a Darlington-connected BJT pair, the first transistor stage is a common-collector while the second stage can be either common-collector or common-emitter. Figure 6.2-1 shows the two basic Darlington configurations.[3]

The Darlington configurations incorporate an additional transistor so that overall circuit performance is altered significantly:

- Current gain is increased.
- Input resistance is increased.

---

[3] Other slight variations occur. These variations are due to a variety of possible connections for the voltage and current biasing sources, V_{CC} and I_{bias}.

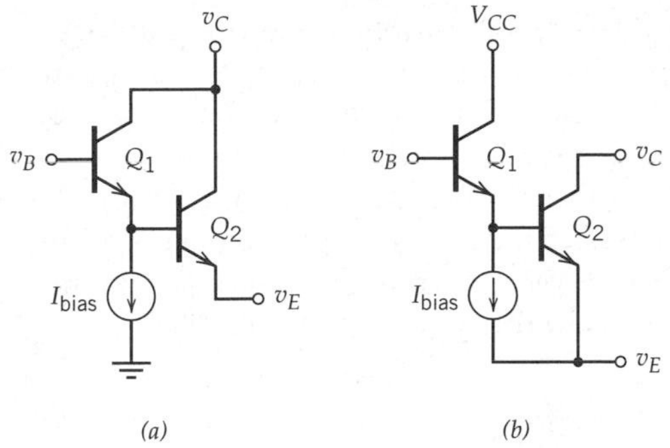

Figure 6.2-1
Darlington configurations: (*a*) dual common-collector pair; (*b*) CC–CE pair.

These increases in current gain and input resistance are by a factor of approximately β_F. The exact change in circuit performance requires careful analysis.

It is common practice to analyze these two-transistor pairs as a single subcircuit. This subcircuit takes on many of the characteristic properties of a single BJT, modified by the presence of two transistors. The subcircuit nodes that correspond to the so-called composite transistor are labeled in Figure 6.2-2 as B^C, C^C, and E^C.

The fourth terminal in Darlington subcircuits exists solely for DC biasing. Commercially packaged Darlingtons allow for the fourth terminal in several different ways:

- Providing a separate external connection (a four-terminal package)
- Internally providing for the bias current or voltage source with additional circuitry within the package (a three-terminal package)
- In the case of the bias current source, occasionally eliminating it entirely (a three-terminal package). This case is equivalent to setting the bias current source I_{bias} to zero.

Dual Common-Collector Darlington Configuration
The dual common-collector Darlington configuration (Figure 6.2-2) is the most common of the two Darlington configurations; a wide variety of commercial packages of this configuration of transistors is available. In this configuration, the base of the composite transistor is the base of the first BJT (Q_1), the two BJT collectors are connected together to form the composite collector, and the emitter of the second BJT (Q_2) forms the composite emitter. As one might expect, the composite transistor can operate in all four regions that are usually associated with BJTs: the cutoff, saturation,

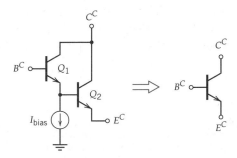

Figure 6.2-2
Modeling dual common-collector Darlington pair as a composite transistor.

forward-active, and inverse-active regions. Simplified models, and their equivalent circuits, of composite transistor operation in each region are of the same form as those developed for BJTs in Section 3.4. However, the composite transistor will have characteristic parameters in each of these regions that vary from typical BJT values.

In the *cutoff region*, the composite transistor operates as three terminals with open circuits between. In order to turn the subcircuit on, both BJT base–emitter junctions must become forward biased. Since the composite base–emitter consists of two BJT base–emitter junctions, the typical turn-on base–emitter voltage for this composite transistor[4] is approximately double that of a single BJT:

$$V_{BE(\text{on})} \approx 1.2 \text{ V}. \tag{6.2-1}$$

The *saturation region* is modeled by two voltage sources: one modeling the base–emitter terminal pair and the other modeling the collector–emitter terminal pair. The composite base–emitter is modeled by two ON BJTs. Thus, in the saturation region,

$$V_{BE(\text{sat})} \approx 1.6 \text{ V}. \tag{6.2-2}$$

The transition from the saturation region into the forward-active region occurs when the collector–emitter voltage of Q_1 becomes too large. At this collector–emitter voltage, Q_2 is in the forward-active region ($V_{CE} > 0.2$ V). Thus the composite collector–emitter voltage is the sum of the saturation collector–emitter voltage of Q_1 and the forward-active base–emitter voltage of Q_2:

$$V_{CE(\text{sat})} \approx 0.9 \text{ V}. \tag{6.2-3}$$

The *forward-active region* is modeled by the ratio of collector current to base current and the base–emitter voltage. The current ratio can be derived by observing that the composite collector current is the sum of the individual collector currents:

$$I_C = I_{C1} + I_{C2}. \tag{6.2-4}$$

Then

$$\begin{aligned} I_C &= \beta_{F1}I_{B1} + \beta_{F2}I_{B2} \\ &= \beta_{F1}I_{B1} + \beta_{F2}[(\beta_{F1} + 1)I_{B1} - I_{\text{bias}}]. \end{aligned} \tag{6.2-5}$$

For large values of β_F, this expression is commonly approximated as

$$I_C \approx \beta_{F1}\beta_{F2}I_{B1} - \beta_{F2}I_{\text{bias}}. \tag{6.2-6}$$

Incremental changes in the composite collector current are proportional to the *product* of the BJT current gains. Consequently, it can be very large. The composite base–emitter voltage is the sum of the forward-active base–emitter voltages for the two BJTs:

$$V_\gamma \approx 1.4 \text{ V}. \tag{6.2-7}$$

---

[4] For simplicity of discussion, all values given in this section will be typical for configurations using silicon *npn* transistors. Darlington pairs constructed with *pnp* transistors will have similar values with appropriate sign differences. These sign differences are discussed in Chapter 3.

The emitter current is calculated to be

$$-I_E = (\beta_{F1} + 1)(\beta_{F2} + 1)I_B - (\beta_{F2} + 1)I_{\text{bias}}. \tag{6.2-8}$$

Due to the addition, in the composite transistor, of a fourth terminal with nonzero current, the current flowing out the emitter terminal is, in general, not equal to the sum of the base and emitter currents.

The *inverse-active region* is similar to the forward-active region. Darlington circuits rarely enter into this region since the multiple-transistor design brings no benefits over single transistors operating in this region.

Small-Signal h-Parameters When a Darlington pair in the forward-active region is used in an amplifier, it is appropriate to model the composite transistor with an h-parameter model. The composite h-parameters can be obtained from the h-parameter models of the individual BJTs, as is shown in Figure 6.2-3.

The collector current is the sum of the individual collector currents:

$$i_c = i_{c1} + i_{c2}. \tag{6.2-9}$$

Then

$$
\begin{aligned}
i_c &= h_{fe1}i_{b1} + h_{fe2}i_{b2} \\
&= h_{fe1}i_{b1} + h_{fe2}(h_{fe1} + 1)i_{b1} \\
&= (h_{fe2} + 1)(h_{fe1} + 1)i_{b1} - i_{b1}.
\end{aligned}
\tag{6.2-10}
$$

For large values of the individual BJT current gain, this expression is commonly approximated as

$$i_c \approx h_{fe1}h_{fe2}i_{b1}$$

or

$$h_{fe} \approx h_{fe1}h_{fe2} = \beta_{F1}\beta_{F2}. \tag{6.2-11}$$

The composite current gain is the product of the current gains. In general, this is a very large quantity. The input resistance of the composite circuit is given as

$$h_{ie} = h_{ie1} + (h_{fe1} + 1)h_{ie2}. \tag{6.2-12}$$

The input resistance has been multiplied significantly and is also a very large quantity. It should be noted that the composite transistor input resistance parameter h_{ie}

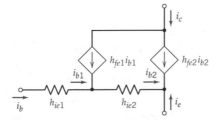

Figure 6.2-3
The AC model for dual common-collector Darlington pair.

cannot be obtained from the composite collector current in the same manner as with a BJT: The value obtained by that process will be much too small. However, similar operations can be performed. The exact value of h_{ie} depends on the collector current I_C as well an on I_{bias}, but a good approximation can be made that is independent of I_{bias}. Two common bias schemes are as follows:

- $I_{\text{bias}} = 0$
- I_{bias} set so that the two BJT collector currents are identical

In the first case, the two collector currents are approximately[5] different by a factor of β_{F2}:

$$I_{C1} \approx \frac{I_C}{\beta_{F2} + 1}, \qquad I_{C2} \approx \frac{\beta_{F2} I_C}{\beta_{F2} + 1}. \tag{6.2-13}$$

The individual BJT parameters can be calculated from these collector currents using Equation 5.2-9:

$$h_{ie1} = (h_{fe1} + 1)\frac{\eta V_t}{|I_{C1}|} = (\beta_{F1} + 1)(\beta_{F2} + 1)\frac{\eta V_t}{|I_C|}, \tag{6.2-14}$$

$$h_{ie2} = (h_{fe2} + 1)\frac{\eta V_t}{|I_{C2}|} = \frac{(\beta_{F2} + 1)^2}{\beta_{F2}}\frac{\eta V_t}{|I_C|}. \tag{6.2-15}$$

These two values are combined with Equation 6.2-11 to find h_{ie}:

$$h_{ie} = \left[(\beta_{F1} + 1)(\beta_{F2} + 1) + \frac{(h_{fe1} + 1)(\beta_{F2} + 1)^2}{\beta_{F2}} \right] \frac{\eta V_t}{|I_C|}, \tag{6.2-16}$$

which, for large current gains, can be approximated as

$$h_{ie} \approx 2(\beta_{F1} + 1)(\beta_{F2} + 1)\frac{\eta V_t}{|I_C|} \approx 2\beta_{F1}\beta_{F2}\frac{\eta V_t}{|I_C|}. \tag{6.2-17}$$

In the second case, $I_{C1} = I_{C2} = 0.5I_C$. This equality of current leads to

$$h_{ie} = (\beta_{F1} + 1)\frac{2\eta V_t}{|I_C|} + (\beta_{F1} + 1)(\beta_{F2} + 1)\frac{2\eta V_t}{|I_C|}, \tag{6.2-18}$$

which, for large current gains, can be approximated as

$$h_{ie} = 2(\beta_{F1} + 1)(\beta_{F2} + 2)\frac{\eta V_t}{|I_C|} \approx 2\beta_{F1}\beta_{F2}\frac{\eta V_t}{|I_C|}. \tag{6.2-19}$$

Notice that the approximate expressions contained in Equations 6.2-17 and 6.2-19 are identical: They are a good approximation for h_{ie} in either bias scheme.

---

[5] There is an additional term of $1 + 1/h_{fe1}$ in the ratio. For reasonably large current gains this term is approximately unity. This slight variation has been ignored to simplify the derivation without significantly changing the results.

Consideration of the output admittance parameter h_{oe} does not significantly alter the expressions for h_{fe} or h_{ie}. The composite output admittance parameter is dependent on the output admittance of both transistors:

$$h_{oe} \approx h_{oe2} + (1 + h_{fe2})h_{oe1}. \qquad (6.2\text{-}20)$$

One must remember that each output admittance parameter depends on the collector current in each transistor, not the total collector current. If the bias current I_{bias} is zero,

$$h_{oe} \approx 2\left|\frac{I_C}{V_{A2}}\right|. \qquad (6.2\text{-}21)$$

If the I_{bias} is set so that the two transistor currents are equal, the output admittance is greatly increased:

$$h_{oe} \doteq (h_{fe2} + 2)\left|\frac{I_C}{V_A}\right|. \qquad (6.2\text{-}22)$$

Each of these two expressions has the effect of reducing the output resistance of the composite transistor as compared to a single transistor. The reduction factor lies between 2 and $h_{fe2} + 2$ depending on the value of I_{bias}.

EXAMPLE 6.2-1 The given circuit is constructed with silicon BJTs with parameters

$$\beta_F = 100, \qquad V_A = 400.$$

Determine the two voltage gains

$$A_{V1} = \frac{v_{o1}}{v_s}, \qquad A_{V2} = \frac{v_{o2}}{v_s},$$

the current gains (defined as the ratio of current in the output resistors to the current in the source), and the indicated AC resistances.

Solution

The given circuit employs a dual common collector Darlington pair with the bias current source set to zero value. The procedure for finding the required AC circuit parameters follows the same guidelines as those presented in Chapter 5.

The quiescent conditions for the composite transistor are obtained by inserting a DC model of the transistor into the circuit. The two significant composite transistor parameters are given by Equations 6.2-5 and 6.2-7:

$$V_\gamma = 1.4 \text{ V}, \qquad \beta_F = 10,200.$$

After the base circuit is replaced by its Thévenin equivalent, the circuit shown here must be analyzed to find the quiescent conditions:

$$I_B = \frac{1.714 - 1.4}{13,714 + 10,201(20)} = 1.443 \ \mu\text{A},$$

$$I_C = 10,200 I_B = 14.723 \text{ mA},$$

$$I_E = (101)^2 I_B = -14.724 \text{ mA}.$$

The collector–emitter voltage must be checked to make sure the composite transistor is in the forward-active region:

$$V_{CE} = 20 - 1000(I_C) - 20(-I_E) = 4.98 \text{ V} > 0.9 \text{ V}.$$

The composite transistor is in the forward-active region. The AC parameters can now be determined:

$$h_{ie} \approx 2\beta_{F1}\beta_{F2} \frac{\eta V_t}{|I_C|} = 35.32 \text{ k}\Omega, \qquad h_{fe} \approx \beta_{F1}\beta_{F2} = 10,000.$$

The amplifier circuit is, depending on which output is taken, of the form of a common-emitter (with an emitter resistor) or a common-collector amplifier. The amplifier performance summary of Section 5.6 is used to determine the performance of this amplifier. The input resistance at the base of the composite transistor is given by

$$R_i = h_{ie} + h_{fe}(20) = \mathbf{235 \text{ k}\Omega}.$$

The input resistance to the total amplifier is

$$R_{in} = R_i \| 15 \text{ k}\Omega \| 160 \text{ k}\Omega = \textbf{12.96 k}\boldsymbol{\Omega}.$$

The composite common-emitter gain A_{V1} is given as

$$A_{V1} = \frac{12.96 \text{ k}\Omega}{12.96 \text{ k}\Omega + 50 \text{ }\Omega} \left(-\frac{10,000(1 \text{ k}\Omega)}{235 \text{ k}\Omega} \right) = \textbf{−42.5}.$$

The corresponding current gain is easily derived from previously obtained quantities:

$$A_{I1} = \frac{i_{o1}}{i_s} = \left(\frac{i_{o1}}{v_o}\right)\left(\frac{v_o}{v_s}\right)\left(\frac{v_s}{i_s}\right) = \frac{1}{R_{\text{load}}} A_{V1} R_{in}$$

$$= \frac{1}{1 \text{ k}\Omega} (-42.5)(12.96 \text{ k}\Omega) = \textbf{−551}.$$

The composite common-collector gain A_{V2} is given as

$$A_{V2} = \frac{12.96 \text{ k}\Omega}{12.96 \text{ k}\Omega + 50 \text{ }\Omega} \left(1 - \frac{35.32 \text{ k}\Omega}{235 \text{ k}\Omega} \right) = \textbf{0.850}.$$

The corresponding current gain is given by

$$A_{I2} = \frac{i_{o2}}{i_s} = \left(\frac{i_{o2}}{v_o}\right)\left(\frac{v_o}{v_s}\right)\left(\frac{v_s}{i_s}\right) = \left(\frac{1}{R_{\text{load}}}\right)(A_{V2})(R_{in})$$

$$= \left(\frac{1}{20 \text{ }\Omega}\right)(0.850)(12.96 \text{ k}\Omega) = \textbf{551}.$$

As expected, the two current gains are very nearly the same in magnitude.

The output resistance of the common-emitter (with an emitter resistor) amplifier is approximately the load resistor:

$$R_{o1} \approx \textbf{1.0 k}\boldsymbol{\Omega}.$$

If h_{oe} is considered to be nonzero, the output resistance can be calculated using Equations 6.2-21 and 5.3-40:

$$h_{oe} \approx 2 \left| \frac{I_C}{V_{A2}} \right| = 2 \frac{14.723 \text{ mA}}{400 \text{ V}} = 73.6 \text{ }\mu S \Rightarrow r_o = 13.6 \text{ k}\Omega,$$

$$R_o = \frac{v}{i} = [R_e \| (R_s + h_{ie})] + r_o\left(1 + \frac{h_{fe}R_e}{R_e + h_{ie} + R_s}\right) = 90.4 \text{ k}\Omega,$$

$$R_{o1} = 1.0 \text{ k}\Omega \| 90.4 \text{ k}\Omega = 990 \text{ }\Omega \approx \textbf{1.0 k}\boldsymbol{\Omega}.$$

The common-collector output resistance is given by

$$R_{o2} = \frac{(50 \text{ }\Omega \| 13.714 \text{ k}\Omega) + 35.32 \text{ k}\Omega}{10,000 + 1} = \textbf{3.54 }\boldsymbol{\Omega}.$$

Common-Collector–Common-Emitter Darlington Pair

The common-collector–common-emitter (CC–CE) Darlington pair is not as common as the dual common-collector pair. The electrical properties of the two types are very similar, except the CC–CE pair always requires at least four terminal connections. If the CC–CE pair is to be commercially packaged, it either requires the inclusion of an internal DC current source I_{bias} or necessitates that the source be zero valued (replaced by an open).

The analysis procedure of this configuration is similar to that of the dual common-collector Darlington. Figure 6.2-4 identifies the correspondence between the circuit terminals and those of the composite transistor equivalent.

In the cutoff region, the composite transistor operates as three terminals with open circuits between. In order to turn the subcircuit on, both BJT base–emitter junctions must become forward biased. Since the composite base–emitter consists of two BJT base–emitter junctions, the typical turn-on base–emitter voltage for this composite transistor is approximately double that of a single BJT:

$$V_{BE(on)} \approx 1.2 \text{ V.}$$

The saturation region is modeled by two voltage sources: one modeling the base–emitter terminal pair and the other modeling the collector–emitter terminal pair. The composite base–emitter is modeled by two ON BJTs. Thus, in the saturation region,

$$V_{BE(sat)} \approx 1.6 \text{ V.} \tag{6.2-23a}$$

The composite collector–emitter voltage is the saturation collector–emitter voltage of Q_2:

$$V_{CE(sat)} \approx 0.2 \text{ V.} \tag{6.2-23b}$$

The forward-active region is modeled by the ratio of collector current to base current and the base-emitter voltage. The current ratio can be derived by observing that the composite collector current is the same as the collector current of Q_2:

$$I_C = I_{C2}. \tag{6.2-24}$$

Then

$$I_C = \beta_{F2}I_{B2}$$
$$= \beta_{F2}[(\beta_{F1} + 1)I_{B1} - I_{bias}]. \tag{6.2-25}$$

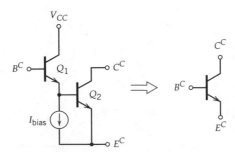

Figure 6.2-4
Modeling CC–CE Darlington pair as composite transistor.

For large values of β_F, this expression is the same as the dual common-collector Darlington:

$$I_C \approx \beta_{F1}\beta_{F2}I_{B1} - \beta_{F2}I_{\text{bias}}. \tag{6.2-26}$$

Incremental changes in the composite collector current are proportional to the product of the BJT current gains; consequently it can be very large. The composite base–emitter voltage is the sum of the forward-active base–emitter voltages for the two BJTs:

$$V_\gamma \approx 1.4 \text{ V}. \tag{6.2-27}$$

The emitter current is calculated to be

$$
\begin{aligned}
-I_E &= (\beta_{F2} + 1)I_{B2} + I_{\text{bias}} \\
&= (\beta_{F1} + 1)(\beta_{F2} + 1)I_B - (\beta_{F2})I_{\text{bias}}.
\end{aligned}
\tag{6.2-28}
$$

Due to the addition, in the composite transistor, of a fourth terminal with nonzero current, the current flowing out of the emitter terminal is, in general, not equal to the sum of the base and emitter currents.

The inverse-active region is similar to the forward-active region. Darlington circuits rarely enter into this region since the multiple transistor design brings no benefits over single transistors operating in this region.

Small-Signal h-Parameters   When a Darlington pair is in the forward-active region, the composite h-parameters can be obtained from the h-parameter models of the individual BJTs as is shown in Figure 6.2-5.

The collector current is the collector current of the Q_2:

$$i_c = i_{c2}. \tag{6.2-29}$$

Then

$$
\begin{aligned}
i_c &= h_{fe2}i_{b2} \\
&= h_{fe2}(h_{fe1} + 1)i_{b1}.
\end{aligned}
\tag{6.2-30}
$$

For large values of the individual BJT current gain, this expression is approximated as

$$i_c \approx h_{fe1}h_{fe2}i_{b1} \quad \text{or} \quad h_{fe} \approx h_{fe1}h_{fe2}. \tag{6.2-31}$$

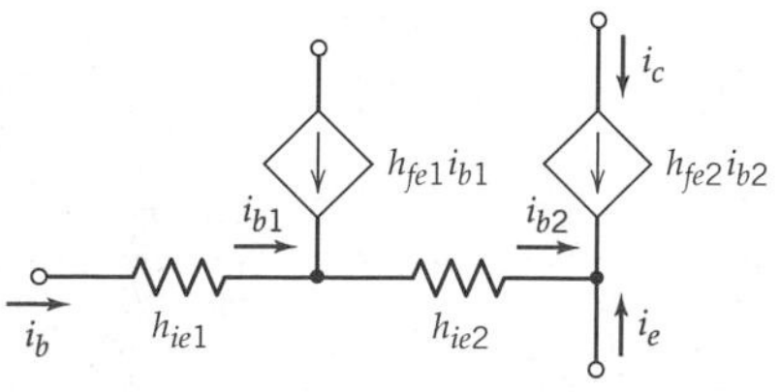

Figure 6.2-5
The AC model for CC–CE Darlington pair.

The composite current gain is the product of the current gains. In general, this is a very large quantity. The input resistance of the composite circuit is given as

$$h_{ie} = h_{ie1} + (h_{fe1} + 1)h_{ie2}. \tag{6.2-32}$$

The input resistance has been multiplied significantly and is also a very large quantity. It should be noted that the composite transistor input resistance parameter h_{ie} cannot be obtained from the composite collector current in the same manner as with a BJT: The value obtained by that process will be much too small. However, similar operations can be performed. The exact value of h_{ie} depends on the collector current I_C as well an on I_{bias}, but a good approximation can be made that is independent of I_{bias}. Two common bias schemes are as follows:

- $I_{\text{bias}} = 0$
- I_{bias} set so that the two BJT collector currents are identical

In the first case, the two collector currents differ[6] by a factor of β_{F2}:

$$I_{C1} \approx \frac{I_C}{\beta_{F2}}, \qquad I_{C2} = I_C. \tag{6.2-33}$$

The individual BJT parameters can be calculated from these collector currents using Equation 5.2-9:

$$h_{ie1} = (h_{fe1} + 1) \frac{\eta V_t}{|I_{C1}|} = (\beta_{F1} + 1)(\beta_{F2}) \frac{\eta V_t}{|I_C|}, \tag{6.2-34}$$

$$h_{ie2} = (h_{fe2} + 1) \frac{\eta V_t}{|I_{C2}|} = (\beta_{F2} + 1) \frac{\eta V_t}{|I_C|}. \tag{6.2-35}$$

These two values are combined with Equation 6.2-32 to find h_{ie}:

$$h_{ie} = [(\beta_{F1} + 1)\beta_{F2} + (h_{fe1} + 1)(\beta_{F2} + 1)] \frac{\eta V_t}{|I_C|}, \tag{6.2-36}$$

which, for large current gains, can be approximated as

$$h_{ie} \approx 2\beta_{F1}\beta_{F2} \frac{\eta V_t}{|I_C|}. \tag{6.2-37}$$

In the second case, $I_{C1} = I_{C2} = I_C$. This equality of current leads to

$$h_{ie} = (\beta_{F1} + 1) \frac{\eta V_t}{|I_{C1}|} + (\beta_{F1} + 1)(\beta_{F2} + 1) \frac{\eta V_t}{|I_C|} \tag{6.2-38}$$

---

[6] As with the dual common-collector Darlington, there is an additional term of $1 + 1/h_{fe1}$ in the ratio. For reasonably large current gains this term is approximately unity. This slight variation has been ignored to simplify the derivation without significantly changing the results.

which, for large current gains, can be approximated as

$$h_{ie} = (\beta_{F1} + 1)(\beta_{F2} + 2) \frac{\eta V_t}{|I_C|} \approx \beta_{F1}\beta_{F2} \frac{\eta V_t}{|I_C|} \qquad (6.2\text{-}39)$$

Notice that the approximate expressions contained in Equations 6.2-37 and 6.2-39 differ by a factor of 2: They present a good approximation for the range of h_{ie}. Consideration of the output admittance parameter h_{oe} does not significantly alter the expressions for h_{fe} or h_{ie}. The composite output admittance parameter is that of Q_2:

$$h_{oe} = h_{oe2} = \left| \frac{I_C}{V_{A2}} \right|. \qquad (6.2\text{-}40)$$

6.2.2 Cascode Configuration

The cascode configuration consists of a pair of BJTs connected in a common-emitter–common-base configuration as shown in Figure 6.2-6. The principal attribute of the cascode configuration of interest here is extremely high output impedance. Other attributes that will be investigated in later chapters include very good (compared to common-emitter) high-frequency performance and very little reverse transmission. The property of small reverse transmission (signals applied to the output appearing at the input) aids in the design of high-frequency tuned amplifiers.

The large-signal operating conditions are characterized by the regional parameters

$$
\begin{aligned}
V_{BE(on)} &= 0.6 \text{ V}, & V_{\gamma} &= 0.8 \text{ V}, \\
V_{BE(sat)} &= 0.8 \text{ V}, & V_{CE(sat)} &= 0.4 \text{ V},
\end{aligned}
\qquad (6.2\text{-}41)
$$

with the additional qualifiers on the above that

$$V_C \geq V_{bias} - 0.6, \qquad V_E \leq V_{bias} - 1.0. \qquad (6.2\text{-}42)$$

Small-Signal h-Parameters

When the BJTs of a cascode are in the forward-active region, two of the composite h parameters can be obtained directly from Table 5.6-2:

$$h_{ie} = h_{ie1} \approx (\beta_{F1} + 1) \frac{\eta V_t}{|I_C|}, \qquad (6.2\text{-}43)$$

$$h_{fe} = h_{fe1} \frac{h_{fe2}}{h_{fe2} + 1} \approx h_{fe1}. \qquad (6.2\text{-}44)$$

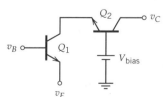

Figure 6.2-6
Cascode configuration.

The output admittance parameter h_{oe} is calculated by calculating the output admittance of the small-signal equivalent circuit with the base current set to zero (Figure 6.2-7).

The test for h_{oe} requires that there be zero base current: This condition implies that the output impedance of the common-emitter BJT is in parallel with the input impedance of the common-base BJT. Application of a unity current to the output terminals yields

$$i_{b2} = \frac{-r_{o1}}{r_{o1} + h_{ie2}}, \qquad V_1 = \frac{r_{o1}h_{ie2}}{r_{o1} + h_{ie2}}.$$

The resultant voltage at the output terminals is determined by finding the current through h_{oe2} and using KVL:

$$I_2 = 1 - h_{fe2}i_{b2} = 1 + h_{fe2}\frac{r_{o1}}{r_{o1} + h_{ie2}}.$$

Then,

$$V = V_1 + I_2 r_{o2}$$

or

$$V = \frac{r_{o1}h_{ie2}}{r_{o1} + h_{ie2}} + \left(1 + h_{fe2}\frac{r_{o1}}{r_{o1} + h_{ie2}}\right)r_{o2}.$$

The output admittance is the ratio of the current I to the voltage V:

$$h_{oe} = \frac{1}{r_{o1}h_{ie2}/(r_{o1} + h_{ie2}) + [1 + h_{fe2}r_{o1}/(r_{o1} + h_{ie2})]r_{o2}} \tag{6.2-45}$$

$$\approx \frac{h_{oe2}}{1 + h_{fe2}}.$$

The output admittance is an extremely small quantity. It can be calculated from the quiescent conditions to be

$$h_{oe} \approx \frac{1}{h_{fe2}}\left|\frac{I_C}{V_{A2}}\right|. \tag{6.2-46}$$

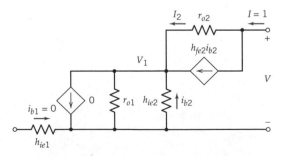

Figure 6.2-7
Equivalent circuit for calculation of cascode h_{oe}.

In that the output impedance is the reciprocal of output admittance, the output impedance of a cascode configuration has been shown to be larger than that of a single common-emitter stage by a factor of $1 + h_{fe}$.

An FET variation of the cascode configuration is often found in analog design. The FET cascode consists of a common-source–common-gate connection. As one might expect, the only major change over a single common-source stage is increased output impedance. Since the amplification of a common-source stage is highly dependent on the output impedance r_d of the FET, the FET cascode will have greatly improved voltage gain. One drawback is that maximum symmetrical swing may be degraded with the FET cascode.

6.2.3 BiFET Darlington Configuration

No useful all-FET Darlington circuits exist: The potential for increased input impedance and current gain have little significance in FET circuitry. There is, however, a useful Darlington circuit connection that combines the properties of FETs and BJTs. In this configuration the input BJT of a Darlington pair is replaced by an FET (Figure 6.2-8). The connection retains the near-infinite input impedance of the FET and, with the addition of the BJT, attains a larger effective composite transconductance g_m and a lower output resistance r_d. Common technologies currently in use that allow BJTs and FETs to be produced on the same substrate include the BiCMOS and BiFET technologies.[7]

Previous analysis of similar circuitry suggests that the formation of a composite FET is appropriate. Quiescent analysis of the BiCMOS Darlington leads to the following relationships:

$$I_D^C = K^C(V_{GS}^C - V_T^C)^2 - \beta_F I_{\text{bias}} \tag{6.2-47}$$

where

$$I_D^C = \text{composite drain current}$$

$$V_{GS}^C = \text{composite gate–source voltage}$$

$$V_T^C = \text{composite threshold voltage}, = V_T + V_\gamma$$

$$K_C = \text{composite transconductance factor}, = (1 + \beta_F)K$$

Similarly, a small-signal analysis of the composite transistor can be performed in a similar manner as with previous configurations. Each transistor is replaced by its

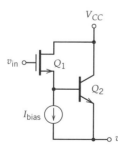

Figure 6.2-8
BiCMOS Darlington configuration. (As with BJT Darlington circuits, exact connection of biasing current source may vary in particular applications. Effect is only in quiescent conditions and individual small-signal parameters of each transistor: general performance *relationships* are unchanged.)

---

[7] BiCMOS implies CMOS and BJT transistors on the same substrate. Although BiFET is a more general term, it usually implies JFETs and BJTs.

appropriate small-signal model, a voltage source is applied to the composite transistor gate–source terminals, and the short-circuit current gain g_m is calculated:

$$g_m^C = \frac{(1 + h_{fe})g_m}{1 + h_{ie}(g_m + 1/r_d)}. \tag{6.2-48}$$

The Thévenin output resistance r_d is calculated by setting the composite transistor gate–source voltage to zero and applying a current to the drain–source terminals. The results of such operations leads to

$$r_d^C = \frac{r_d + (\mu + 1)h_{ie}}{h_{fe} + 1}. \tag{6.2-49}$$

As expected, the addition of a common-collector stage to the FET did not significantly change the voltage gain of the composite transistor ($\mu = g_m r_d$).

6.3 EMITTER-COUPLED AND SOURCE-COUPLED PAIRS

Emitter-coupled and source-coupled pairs of transistors are widely used to form the basic element of differential amplifiers with the most common application being the input stage of an OpAmp. The output of such amplifier stages is proportional to the difference in the input voltages and has a high common-mode rejection ratio (CMRR). A high CMRR allows the input to be differential or single (one input at a fixed voltage). The output can be similarly differential or single (as in the OpAmp).

Another useful property of emitter-coupled and source-coupled amplifier stages is that they can be cascaded without interstage coupling capacitors: Any common-mode DC offset that may be present has a negligible effect on the performance of any stage. Input impedance is typically high for FET pairs and moderate for BJT pairs. Output impedance is moderate for all types.

6.3.1 Emitter-Coupled Pairs

The basic topology for an emitter-coupled pair is shown in Figure 6.3-1. In order to form a good differential amplifier, the circuit must be a symmetric network: The

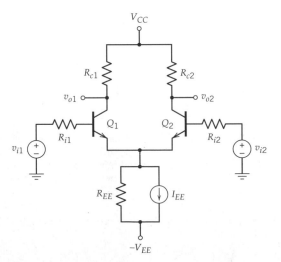

Figure 6.3-1
Typical emitter-coupled pair circuit diagram.

transistors must have similar properties and corresponding collector resistances should have identical values ($R_{c1} = R_{c2} = R_c$). These collector resistances are often discrete resistors; in integrated circuit applications, they are more often active loads.[8] In integrated circuit realizations of an emitter-coupled pair, the biasing network is typically in the form of a transistor current source; in other realizations it can be as simple as a single resistor. For generality, the biasing network is shown in its most general form as a Norton equivalent source.

The topology of this amplifier is very similar to that of the input section of an emitter-coupled logic (ECL) gate. Previous explorations of that logic gate family[9] found that a linear region of amplification exists in the transition region between the two logic levels. This region extends over a very small range of input voltage differences: The difference between the voltages at the two BJT base terminals cannot exceed a few tenths of a volt before one of the transistors enters the cutoff region and the amplifier saturates. Still, as has been seen in the case of the OpAmp, it is not necessary for the input voltage difference to be large in order to create a highly useful device.

DC Analysis of Emitter-Coupled Pair

If the circuit is truly symmetric ($R_{i1} = R_{i2} = R_i$), it can be easily shown that the quiescent BJT collector currents are equal and have value:

$$I_C = \beta_F \frac{V_{EE} + I_{EE}R_{EE} - V_{BE(on)}}{2(\beta_F + 1)R_{EE} + R_i}. \tag{6.3-1}$$

For values of R_{EE} larger than the output resistance R_i of the sources, a good approximation[10] for the collector current is

$$I_C \approx \frac{\alpha_E}{2}\left(I_{EE} + \frac{V_{EE} - V_{BE(on)}}{R_{EE}}\right).$$

From this derived value of I_C and the BJT characteristic parameters, the usual small-signal BJT h-parameters can be derived:[11]

$$h_{fe} = \beta_F, \qquad h_{ie} = (\beta_F + 1)\frac{\eta V_T}{I_C}, \qquad h_{oe} = \frac{I_C}{V_A}.$$

Since the BJT quiescent conditions are identical, each of the two BJTs has these same h-parameters.

AC Analysis of Emitter-Coupled Pair

The most significant performance parameter for a differential amplifier is the differential-mode gain. Also of significance is the CMRR: the ratio of differential-mode gain to common-mode gain.[12] The best path to find these performance parameters is through AC analysis and the use of superposition. The AC equivalent circuit for zero v_{i2} is shown in Figure 6.3-2.

---

[8] Active loads are discussed in Section 6.5
[9] Emitter-coupled logic gates are discussed in Section 3.5.
[10] This approximation is particularly useful if R_{i1} and R_{i2} are not equal but are each smaller than R_{EE}.
[11] The BJT must be in the forward-active region; it is important to check V_{CE} to make sure before continuing.
[12] For a more complete discussion of CMMR, refer to Section 1.4.

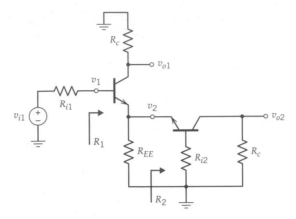

Figure 6.3-2

The AC equivalent circuit for emitter-coupled pair (single input source).

The gain path to v_{o1} is a common-emitter (with an emitter resistor) amplifier stage, while the gain path to v_{o2} is a two-stage cascaded amplifier: a common-collector stage followed by a common-base stage. Table 5.6-2 provides small-signal amplifier performance parameters to guide the AC analysis of this circuit.[13] Two derived input resistances aid in calculations. The resistance R_2 is the input resistance of a common-base amplifier:

$$R_2 = \frac{h_{ie} + R_{i2}}{h_{fe} + 1}. \tag{6.3-2}$$

Here, R_1 is the input resistance of a common-collector amplifier:

$$R_1 = h_{ie} + (h_{fe} + 1)(R_{EE}\|R_2). \tag{6.3-3}$$

Typically the biasing current source output resistance R_{EE} is much greater than the input resistance to the common-base stage, R_2 (a very small value). This relationship leads to a common approximation, $R_{EE}\|R_2 \approx R_2$, and

$$R_1 \approx h_{ie} + (h_{fe} + 1)\frac{h_{ie} + R_{i2}}{h_{fe} + 1} = 2h_{ie} + R_{i2}. \tag{6.3-4}$$

This approximation for R_1 will be used in early discussions and will be tested for validity in Example 6.3-2.

The voltage gain to each of the output terminals can now be easily determined. The voltage gain to the v_{o1} terminal is simply the product of a common-emitter amplifier gain and a voltage division:

$$
\begin{aligned}
A_{V1} &= \frac{v_{o1}}{v_{i1}} \\
&= \left(\frac{v_{o1}}{v_1}\right)\left(\frac{v_1}{v_{i1}}\right) \\
&= \left(\frac{-h_{fe}R_c}{R_1}\right)\left(\frac{R_1}{R_1 + R_{i1}}\right) \\
&\approx \frac{-h_{fe}R_c}{2h_{ie} + R_{i1} + R_{i2}}.
\end{aligned}
\tag{6.3-5}
$$

---

[13]Table 5.6-2 assumes that the resistors R_c are small compared to $1/h_{oe}$. This may not always be the case. Section 6.4 discussed the effect of large load resistances (in the form of active loads) on amplifier performance.

The voltage gain to the v_{o2} terminal is the product of a common-base amplifier gain, a common-collector amplifier gain, and a voltage division:

$$A_{V2} = \frac{v_{o2}}{v_{i1}}$$

$$= \left(\frac{v_{o2}}{v_2}\right)\left(\frac{v_2}{v_1}\right)\left(\frac{v_1}{v_{i1}}\right)$$

$$= \left(\frac{h_{fe}R_c}{h_{ie} + R_{i2}}\right)\left(\frac{R_1 - h_{ie}}{R_1}\right)\left(\frac{R_1}{R_1 + R_{i1}}\right)$$

Rearranging terms gives the expression

$$A_{V2} = \left(\frac{h_{fe}R_c}{R_1 + R_{i1}}\right)\left(\frac{R_1 - h_{ie}}{h_{ie} - R_{i2}}\right) = -A_{V1}\frac{R_1 - h_{ie}}{h_{ie} + R_{i2}}. \tag{6.3-6}$$

Cancellation and substitution of the approximate expression for R_1 leads to a final gain expression:

$$A_{V2} \approx -A_{V1}\frac{2h_{ie} + R_{i2} - h_{ie}}{h_{ie} + R_{i2}} = -A_{V1}. \tag{6.3-7}$$

The two gain expressions given in Equations 6.3-5 and 6.3-7 are equal in magnitude:

$$A_{V2} = -A_{V1} = A, \tag{6.3-8}$$

where

$$A \approx \frac{h_{fe}R_c}{2h_{ie} + R_{i1} + R_{i2}}. \tag{6.3-9}$$

Application of symmetry to this circuit leads to similar expressions for the voltage gains to each output from input v_{i2} when v_{i1} is set to zero value:

$$\frac{v_{o1}}{v_{i2}} = A, \qquad \frac{v_{o2}}{v_{i2}} = -A. \tag{6.3-10}$$

The total voltage transfer relationships can then be expressed as a superposition of the results of the derivations:

$$v_{o1} = -Av_{i1} + Av_{i2}, \qquad v_{o2} = Av_{i1} - Av_{i2}. \tag{6.3-11}$$

The differential output v_{od} depends only on the differential input:[14]

$$v_{od} = v_{o1} - v_{o2}$$

$$= (-Av_{i1} + Av_{i2}) - (Av_{i1} - Av_{i2}) \tag{6.3-12}$$

$$= -2A(v_{i1} - v_{i2}),$$

---

[14]This expression is exactly twice that derived in Chapter 1. The difference is due to the differential nature of the output of this amplifier compared to the single-sided output of an OpAmp. The difference also applies to the common-mode gain; hence the CMRR of the amplifier is unchanged.

or

$$v_{od} = \frac{-h_{fe}R_c(v_{i1} - v_{i2})}{h_{ie} + \frac{1}{2}(R_{i1} + R_{i2})}. \tag{6.3-13}$$

The common output v_{oc} is approximately zero:

$$
\begin{aligned}
v_{oc} &= \tfrac{1}{2}(v_{o1} + v_{o2}) \\
&= \tfrac{1}{2}[(-Av_{i1} + Av_{i2}) + (Av_{i1} - Av_{i2})] \approx 0.
\end{aligned} \tag{6.3-14}
$$

The output resistance of each gain path is extremely large looking into the BJT collectors. The output resistance for the total amplifier is approximately equal to R_c.

It appears that an emitter-coupled BJT amplifier pair forms a very good differential amplifier with the following properties:

Gain: $A_{VD} = (v_{o1} - v_{o2})/(v_{i1} - v_{i2}) = -h_{fe}R_c/[h_{ie} + \frac{1}{2}(R_{i1} + R_{i2})]$
Input resistance: $R_i \approx 2h_{ie} + R_{i2}$
Output resistance: $R_o = R_c$

EXAMPLE 6.3-1 Determine the differential voltage gain for the given circuit. Assume identical BJTs with

$$\beta_F = 120.$$

SOLUTION

The output resistance of the current source is much larger than either of the output resistances of the sources; thus

$$I_C \approx \frac{\alpha_F}{2}\left(I_{EE} + \frac{V_{EE} - V_{BE\text{(on)}}}{R_{EE}}\right)$$

$$= 0.4959\left(6 \times 10^{-3} + \frac{14.3}{39{,}000}\right) = 3.157 \text{ mA}.$$

A check of V_{CE} must be made:

$$V_{CE} = V_C - V_E$$
$$= (15 - I_C R_c) - (0 - I_B R_i - V_{BE(\text{on})})$$
$$= 4.58 - (-0.702)$$
$$= 5.28 \text{ V}.$$

The BJT is in the forward-active region. The significant BJT parameters are then

$$h_{fe} = 120, \qquad h_{ie} = 996.5 \ \Omega.$$

The differential gain is given by

$$A_{\text{VD}} = \frac{v_{o1} - v_{o2}}{v_{i1} - v_{i2}} = \frac{-h_{fe} R_c}{h_{ie} + \frac{1}{2}(R_{i1} + R_{i2})} = \frac{-120(3300)}{996.5 + 75} = -370.$$

EXAMPLE 6.3-2 Use exact expressions for currents, resistances, and gains in the circuit of Example 6.3-1 to determine the single and differential voltage gains and the CMRR for the circuit.

SOLUTION

The exact collector current expression is

$$I_C = \beta_F \frac{V_{EE} + I_{EE} R_{EE} - V_{BE(\text{on})}}{2(\beta_F + 1)R_{EE} + R_i}$$
$$= 120 \frac{15 + 234 - 0.7}{2(121)39{,}000 + 75} = 3.157 \text{ mA}.$$

There is no significant change from the approximate value of the previous example: The BJTs are in the forward-active region. The BJT h parameters are

$$h_{fe} = 120, \qquad h_{ie} = 996.5 \ \Omega.$$

The resistance R_2 is given by

$$R_2 = \frac{h_{ie} + R_{i2}}{h_{fe} + 1} = \frac{996.5 + 75}{121} = 8.86 \ \Omega$$

and the resistance R_1 is given by

$$R_1 = h_{ie} + (h_{fe} + 1)(R_{EE} \| R_2)$$
$$= 996.5 + (121)(39{,}000 \| 8.86) = 2.068 \text{ k}\Omega.$$

The load resistance for the common-collector stage is given by

$$R_E = R_{EE} \| R_2 = 8.8535 \ \Omega,$$

which leads to the gain expressions

$$A_{V1} = \frac{-120(3300)}{2068} \frac{2068}{2068 + 75} = -184.807,$$

$$A_{V2} = -A_{V1} \frac{2068 - 996.5}{996.5 + 75} = 184.768.$$

The total transfer relationships can then be written as

$$v_{o1} = -184.807v_{i1} + 184.768v_{i2},$$

$$v_{o2} = 184.768v_{i1} - 184.807v_{i2}.$$

The differential gain is given by

$$v_{od} = v_{o1} - v_{o2} = -369.575(v_{i1} - v_{i2}),$$

and the common-mode gain is

$$v_{oc} = \frac{v_{o1} + v_{o2}}{2} = -0.039234\left(\frac{v_{i1} + v_{i2}}{2}\right),$$

which yields

$$\text{CMRR} = 20 \log_{10}\left|\frac{-369.575}{-0.039234}\right| = 79.5 \text{ dB}.$$

6.3.2 Source-Coupled Pairs

Source-coupled amplifier pairs are the FET equivalent of emitter-coupled BJT amplifiers. The basic topology is shown in Figure 6.3-3.[15] As in the emitter-coupled pair, it is important that the two FETs have similar properties. Similarly the drain resistances R_d are to have similar values and may be composed of simple discrete resistors or active loads. The bias network is almost always a current source but is shown here as its Norton equivalent.

DC Analysis of Source-Coupled Pair

DC analysis of FET circuits is inherently more complicated than BJT circuits due to the highly non-linear large-signal behavior of FETs. Thus a closed-form expression for the quiescent FET drain current is not easily derived. If, however, the circuit is truly symmetric, the DC analysis can be accomplished through the use of a "half-circuit," as shown in Figure 6.3-4.

Half-circuit analysis uses the property that the drain currents for the two FETs are identical; thus the effect of the Norton source on *one* of the FETs appears to be modified. This modification changes the apparent Norton source: The value of the Norton current is halved and the Norton resistance appears to be doubled. The drain

---

[15]Here the FETs are shown as depletion-mode *n*-channel FETs for simplicity. Any other type of FET can be used in a source-coupled pair. Many commercially available OpAmps use source-coupled JFETs or MOSFETs for differential amplifier input stages.

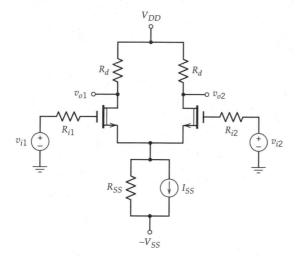

Figure 6.3-3
Typical source-coupled pair circuit diagram.

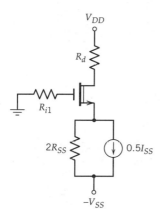

Figure 6.3-4
Half-circuit equivalent for quiescent condition calculation.

current and gate–source voltage for this single FET quiescent equivalent circuit can be obtained using the half-circuit quiescent equivalent and the techniques outlined in Chapter 4. These derived quiescent conditions and the FET characteristic parameters lead directly to the small-signal FET parameters:

$$g_m = \frac{2I_D}{V_{GS} - V_{PO}} \quad \text{or} \quad g_m = 2K(V_{GS} - V_T),$$

$$r_d = \left| \frac{V_A}{I_D} \right|.$$

(6.3-15)

AC Analysis of Source-Coupled Pair

As in the emitter-coupled pair, a good method of AC analysis is through the use of superposition. The AC equivalent circuit for zero v_{i2} is shown in Figure 6.3-5.

The gain path to v_{o1} is a common-source (with a source resistor) amplifier stage, while the gain path to v_{o2} is a two-stage cascaded amplifier, a common-drain stage followed by a common-gate stage. Table 5.11-2 provides small-signal amplifier performance parameters to guide the AC analysis of this circuit. Two derived input resistances aid in calculations. The resistance R_2 is the input resistance of a common-gate amplifier:

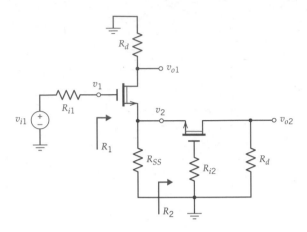

Figure 6.3-5
The AC equivalent circuit for source-coupled pair (single input source).

$$R_2 = \frac{r_d + R_d}{1 + g_m r_d}. \tag{6.3-16}$$

Here, R_1 is the input resistance of a common-drain amplifier:

$$R_1 \approx \infty.$$

Typically the biasing current source output resistance R_{SS} is much greater than the input resistance to the common-gate stage, R_2 (a very small value). This relationship leads to a common approximation, $R_S = R_{SS}\|R_2 \approx R_2$.

The voltage gain to each of the output terminals can now be easily determined. The voltage gain to the v_{o1} terminal is simply the product of a common-source amplifier gain and a voltage division:

$$
\begin{aligned}
A_{V1} &= \frac{v_{o1}}{v_{i1}} \\[2mm]
&= \left(\frac{v_{o1}}{v_1}\right)\left(\frac{v_1}{v_{i1}}\right) \\[2mm]
&= \left(\frac{-g_m r_d R_d}{r_d + R_d + (1 + g_m r_d)R_S}\right)\left(\frac{1}{1}\right) \\[2mm]
&\approx \frac{-g_m r_d R_d}{2(r_d + R_d)}.
\end{aligned}
\tag{6.3-17}
$$

The voltage gain to the v_{o2} terminal is the product of a common-base amplifier gain, a common-collector amplifier gain, and a voltage division:

$$
\begin{aligned}
A_{V2} &= \frac{v_{o2}}{v_{i1}} \\[2mm]
&= \left(\frac{v_{o2}}{v_2}\right)\left(\frac{v_2}{v_1}\right)\left(\frac{v_1}{v_{i1}}\right) \\[2mm]
&= \left(\frac{(1 + g_m r_d)R_d}{r_d + R_d}\right)\left(\frac{g_m r_d R_S}{r_d + R_d + (1 + g_m r_d)R_S}\right)\left(\frac{1}{1}\right).
\end{aligned}
$$

Rearranging terms leads to

$$A_{V2} = \frac{g_m r_d R_d}{r_d + R_d + (1 + g_m r_d)R_S} \frac{(1 + g_m r_d)R_S}{r_d + R_d}$$

$$= -A_{V1} \frac{(1 + g_m r_d)R_S}{r_d + R_d}, \tag{6.3-18}$$

or, when the approximate value of R_S is inserted,

$$A_{V2} \approx -A_{V1} \frac{(1 + g_m r_d)(r_d + R_d)/(1 + g_m r_d)}{r_d + R_d} = -A_{V1}. \tag{6.3-19}$$

Once again, the two gain expressions (given in Equations 6.3-17 and 6.3-19) are equal in magnitude[16]:

$$A_{V2} = -A_{V1} = A,$$

where

$$A \approx \frac{g_m r_d R_d}{2(r_d + R_d)}. \tag{6.3-20}$$

Application of symmetry to this circuit leads to similar expressions for the voltage gains to each output from input v_{i2} when v_{i1} is set to zero value:

$$\frac{v_{o1}}{v_{i2}} = A, \qquad \frac{v_{o2}}{v_{i2}} = -A. \tag{6.3-21}$$

The total output voltage expressions can then be expressed as a superposition of the results of the derivations:

$$v_{o1} = -Av_{i1} + Av_{i2}, \qquad v_{o2} = Av_{i1} - Av_{i2}. \tag{6.3-22}$$

The differential output v_{od} depends only on the differential input:

$$v_{od} = v_{o1} - v_{o2} \tag{6.3-23}$$

$$= (-Av_{i1} + Av_{i2}) - (Av_{i1} - Av_{i2})$$

$$= -2A(v_{i1} - v_{i2}),$$

$$v_{od} = \frac{-g_m r_d R_d}{r_d + R_d}(v_{i1} - v_{i2}). \tag{6.3-24}$$

The common output v_{oc} is approximately zero:

$$v_{oc} = \tfrac{1}{2}(v_{o1} + v_{o2}) \tag{6.3-25}$$

$$= \tfrac{1}{2}[(-Av_{i1} + Av_{i2}) + (Av_{i1} - Av_{i2})] \approx 0.$$

---

[16]The equivalence is true only for the approximation $R_S = R_{SS}\|R_2 \approx R_2$. The gain expressions of Equations 6.3-17 and 6.3-18 are exact.

If it is assumed that R_{SS} is large with respect to R_2, the output resistance at the output terminal of the FET is given by

$$R_o \approx 2r_d + R_d. \tag{6.3-26}$$

The total output resistance (including effects of the load resistance R_d) is given by

$$R_o \approx R_d\left[1 - \frac{R_d}{2(r_d + R_d)}\right]. \tag{6.3-27}$$

A source-coupled FET amplifier pair also forms a very good differential amplifier with the following properties:

Gain: $A_{VD} = (v_{o1} - v_{o2})/(v_{i1} - v_{i2}) = -g_m r_d R_d/(r_d + R_d)$
Input resistance: $R_i \approx \infty$
Output resistance: $R_o \approx R_d[1 - R_d/2(r_d + R_d)]$

EXAMPLE 6.3-3 The source-coupled amplifier of Figure 6.3-3 is constructed with two identical JFETs with properties

$$V_{PO} = -4\ \text{V}, \qquad I_{DSS} = 8\ \text{mA}, \qquad V_A = -160\ \text{V}$$

and circuit parameters

$$V_{DD} = V_{SS} = +15\ \text{V}, \qquad I_{SS} = 6\ \text{mA},$$
$$R_{SS} = 39\ \text{k}\Omega, \qquad R_d = 3.3\ \text{k}\Omega.$$

Determine the circuit performance parameters and CMRR.

SOLUTION

The DC analysis uses the half-circuit analysis technique to obtain values of I_D and V_{GS}. The two significant equations are

$$I_D = 8\ \text{mA}\left(1 + \frac{V_{GS}}{4}\right)^2, \qquad I_D = \frac{15 - V_{GS}}{78,000} + 3\ \text{mA}.$$

The solutions to the above are

$$I_D = 3.2111 \text{ mA}, \qquad V_{GS} = -1.4658 \text{ V}.$$

The term V_{DS} must be checked:

$$V_{DS} = 15 - (3.2111 \times 10^{-3})(3300) - 1.4658$$
$$= 2.938 \geq -1.4658 + 4 = 2.534 \text{ V}.$$

The JFET is in the saturation region. The AC small-signal parameters can now be determined:

$$g_m = \frac{2(3.2111 \times 10^{-3})}{2.5342} = 2.534 \text{ mA/V},$$

$$r_d = \left| \frac{-160}{3.2111 \times 10^{-3}} \right| = 49.83 \text{ k}\Omega.$$

The AC analysis can now proceed:

$$A_{VD} = \frac{-g_m r_d R_d}{r_d + R_d} = \frac{-2.534 \times 10^{-3}(49,830)(3300)}{(49,830 + 3300)} = \mathbf{-7.843,}$$

$$R_o \approx R_d \left(1 - \frac{R_d}{2(r_d + R_d)} \right) = 3300 \left(1 - \frac{3300}{2(49,830 + 3300)} \right)$$
$$= \mathbf{3.197 \text{ k}\Omega.}$$

The CMRR calculations need the more exact expressions for resistances and gain. The resistance R_2 is given by

$$R_2 = \frac{r_d + R_d}{1 + g_m r_d} = \frac{49,830 + 3300}{1 + 126.27} = 417.45 \text{ }\Omega,$$

and the total resistance seen by the source of the FETs is

$$R_S = R_{SS} \| R_2 = 39,000 \| 417.45 = 413.03 \text{ }\Omega.$$

The two voltage gain expressions (using exact values of R_S) are given by Equation 6.3-17,

$$A_{V1} = \frac{-g_m r_d R_d}{r_d + R_d + (1 + g_m r_d) R_S}$$

$$= \frac{-126.27(3300)}{49,830 + 3300 + (1 + 126.27)413.03} = -3.94238,$$

and Equation 6.3-18,

$$A_{V2} = -A_{V1} \frac{(1 + g_m r_d)R_S}{r_d + R_d}$$

$$= -3.94238 \frac{(1 + 126.27)413.03}{49,830 + 3300} = 3.90064.$$

The differential-mode gain is given by

$$A_{DM} = A_{V1} - A_{V2} = -7.843,$$

and the common-mode gain is given by

$$A_{CM} = A_{V1} + A_{V2} = -0.04175$$

for

$$CMRR = 20 \log_{10} \left| \frac{-7.84302}{-0.0417465} \right| = 45.5 \text{ dB}.$$

6.3.3 Variations on the Theme

In an effort to improve the various performance characteristics of emitter-coupled and source-coupled amplifiers, designers often change the circuit topology while keeping its basic characteristics. Two common techniques for altering circuit performance are as follows:

- Including a resistor between the emitter or source terminal and the common terminal
- Using a compound transistor (e.g., Darlington) instead of a single transistor

Additional resistors have the effect of increasing the range of input voltages over which the amplifier is linear and increasing the input resistance of emitter-coupled amplifiers. Compound transistors can have a variety of effects. Figure 6.3-6 shows a

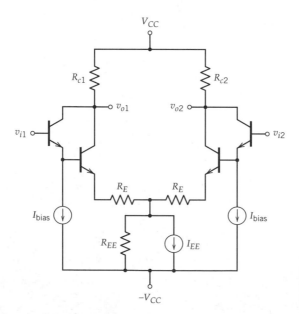

Figure 6.3-6
Simplified input stage schematic: LM318 OpAmp.

simplified schematic of the input stage of an OpAmp that uses compound Darlington transistors and additional emitter resistors to alter performance characteristics.

While the circuit of Figure 6.3-6 shows a multiple BJT differential amplifier, the compound transistors are not limited to BJTs. BiFET and BiCMOS technologies allow for the mixing of transistor types in the differential pair as long as the circuit remains symmetric. Analysis of such amplifiers follows the same basic techniques demonstrated in this section. The compound transistor parameters rather than single-transistor parameters are used and appropriate resistances (R_1 and/or R_2 in the previous derivations) are modified to reflect any additional emitter or source resistors.

6.3.4 Summary

The basic performance characteristics of basic emitter-coupled and source-coupled amplifiers are summarized in Table 6.3-1.

TABLE 6.3-1 SUMMARY OF DIFFERENTIAL PAIR AMPLIFIER PROPERTIES

Parameter	Emitter Coupled	Source Coupled
Differential voltage gain	$\dfrac{-h_{fe}R_c}{h_{ie} + \frac{1}{2}(R_{i1} + R_{i2})}$	$\dfrac{-g_m r_d R_d}{r_d + R_d}$
Input resistance	$2h_{ie} + R_i$	$\approx \infty$
Output resistance[a] At output terminal of transistor	$\approx \infty$	$\approx 2r_d + R_d$
Includes effect of load resistance R_c or R_d	$\approx R_c$	$\approx R_d\left(1 - \dfrac{R_d}{2(r_d + R_d)}\right)$

6.4 Transistor Current Sources

Current sources may be used to bias the transistor circuit in the forward-active region for a small-signal amplifier. The biasing schemes explored thus far apply a voltage across the base–emitter junction of the BJT or the gate–source of the FET. Current source biasing may be used in lieu of a base or gate resistor bias network to set the quiescent collector or drain current and is the preferred method for biasing differential amplifiers. Bias networks in integrated circuits most often depend on current sources.

6.4.1 Simple Bipolar Current Source

A simple BJT current source that delivers an approximately constant collector current I_c is shown in Figure 6.4-1. For this current source to be operational, the collector voltage of the BJT must be sufficiently more positive than the emitter voltage such that the transistor operates in the forward-active region. The constant current is supplied by the collector of the transistor.

The transistor collector current can be obtained by applying the node voltage method at V_B, that is, by summing the currents at the base of Q,

$$0 = I_1 + I_2 + I_B. \tag{6.4-1}$$

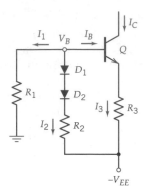

Figure 6.4-1
Simple BJT current source.

Using KVL around the base–emitter loop yields an expression for V_B,

$$V_B = I_3R_3 + V_{BE} - V_{EE}. \qquad (6.4\text{-}2)$$

The current I_1 can be expressed in terms of I_3 by substituting Equation 6.4-2 for the base voltage,

$$I_1 = \frac{V_B}{R_1} = \frac{I_3R_3 + V_{BE} - V_{EE}}{R_1}. \qquad (6.4\text{-}3)$$

The current I_2 is found in a similar manner,

$$I_2 = \frac{V_B - 2V_d + V_{EE}}{R_2} = \frac{I_3R_3 + V_{BE} - 2V_d}{R_2}, \qquad (6.4\text{-}4)$$

where V_d is the voltage drop across each of the diodes D_1 and D_2.
    The base current is simply

$$I_B = \frac{I_3}{\beta_F + 1}. \qquad (6.4\text{-}5)$$

Substitution of Equations 6.4-3 to 6.4-5 into Equation 6.4-1 yields an expression in terms of the emitter current I_3,

$$0 = \frac{I_3R_3 + V_{BE} - V_{EE}}{R_1} + \frac{I_3R_3 + V_{BE} - 2V_d}{R_2} + \frac{I_3}{\beta_F + 1}. \qquad (6.4\text{-}6)$$

Solving for I_3 in Equation 6.4-6 yields

$$I_3 = \frac{R_2(V_{EE} - V_{BE}) + R_1(2V_d - V_{BE})}{R_2R_3 + R_1R_3 + R_1R_2/(\beta_F + 1)}. \qquad (6.4\text{-}7)$$

The collector current is found to be

$$\begin{aligned}
I_C &= \frac{\beta_F}{\beta_F + 1} \frac{R_2(V_{EE} - V_{BE}) + R_1(2V_d - V_{BE})}{R_2R_3 + R_1R_3 + R_1R_2/(\beta_F + 1)} \\
&= \beta_F \frac{R_2(V_{EE} - V_{BE}) + R_1(2V_d - V_{BE})}{(\beta_F + 1)(R_2R_3 + R_1R_3) + R_1R_2}.
\end{aligned} \qquad (6.4\text{-}8)$$

An additional benefit of using diodes in this and similar biasing networks is the reduction of collector current dependence on the base–emitter temperature. The reduction of collector current dependence on the BJT junction temperature increases the stability of the operating point.

6.4.2 Current Mirror

The simple current source in Figure 6.4-1 is useful but requires two diodes, one transistor, and three resistors to configure. A two-transistor current source, which uses fewer components (in particular, resistors), is shown in Figure 6.4-2. This configuration is commonly called a current mirror.

The constant current is supplied from the collector of Q_2. The base–collector voltage of Q_1 is equal to zero, ensuring that the BJT is operating in the forward-active region. This connection is referred to as a diode-connected BJT. If the transistors are identical, both Q_1 and Q_2 have the same base–emitter voltages; therefore $I_{B1} = I_{B2}$ and $I_{C1} = I_{C2}$ since $\beta_{F1} = \beta_{F2} = \beta_F$. Applying KCL at the collector node of Q_1,

$$I_{\text{ref}} = I_{C1} + I_{B1} + I_{B2} = I_{C1} + \frac{2I_{C1}}{\beta_F} = I_{C1}\left(1 + \frac{2}{\beta_F}\right). \tag{6.4-9}$$

Therefore, the collector current of Q_2 is

$$I_{C2} = I_{C1} = \frac{I_{\text{ref}}}{1 + 2/\beta_F}. \tag{6.4-10}$$

Solving for I_{ref} in terms of the applied voltage V_{CC} and resistance R,

$$I_{\text{ref}} = \frac{V_{CC} - V_\gamma}{R} \tag{6.4-11}$$

since $V_{CE1} = V_\gamma$.

Substituting Equation 6.4-11 into Equation 6.4-10 and solving for I_{C1},

$$I_{C1} = \frac{V_{CC} - V_\gamma}{(1 + 2/\beta_F)R}. \tag{6.4-12}$$

So the Q_2 collector current (constant current) is

$$I_{C2} = I_{C1} = \frac{V_{CC} - V_\gamma}{(1 + 2/\beta_F)R}. \tag{6.4-13}$$

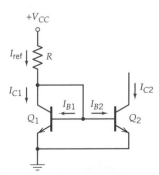

Figure 6.4-2
Simple BJT current mirror.

The load current I_{C2} is also dependent on V_{CE2}. For a constant base current, the collector current will increase slowly as the collector–emitter voltage increases. In the current mirror of Figure 6.4-2, $V_{CE2} \geq V_{CE1}$ since $V_{CE1} = V_{BE1}$. The value of V_{CE2} depends on the bias voltage at the collector of Q_2. Typically, $V_{CE2} \gg V_{CE1}$. Depending on the quiescent conditions of Q_2, I_{C2} may be up to 10% larger than I_{C1}.

Stability is an important parameter of constant-current sources. Variations in parameter values will cause an undesirable fluctuation in I_{C2}. The stability factors defined in Section 3.7 are

$$S_I = \frac{\partial I_{C2}}{\partial I_{CO}}, \qquad S_V = \frac{\partial I_{C2}}{\partial V_{BE}}, \qquad S_\beta = \frac{\partial I_{C2}}{\partial \beta_F}. \tag{6.4-14}$$

Using Equation 6.4-13,

$$S_I = \frac{\partial}{\partial I_{CO}}\left[\frac{V_{CC} - V_\gamma}{(1 + 2/\beta_F)R} + (\beta_F + 1)I_{CO}\right] = (\beta_F + 1), \tag{6.4-15}$$

$$S_V = \frac{\partial}{\partial V_{BE}} \frac{V_{CC} - V_{BE}}{(1 + 2/\beta_F)R} = -\frac{1}{(1 + 2/\beta_F)R}, \tag{6.4-16}$$

$$S_\beta = \frac{\partial}{\partial \beta_F} \frac{V_{CC} - V_\gamma}{(1 + 2/\beta_F)R} = \frac{2(V_{CC} - V_\gamma)}{R(\beta_F + 2)^2}. \tag{6.4-17}$$

By increasing R, increased stability against variations in base–emitter voltage and quiescent point shifts due to varying β can be obtained.

The advantage of the current mirror in Figure 6.4-2 is the reduction in the number of resistors that makes the circuit more applicable to integrated circuits.

6.4.3 Current Mirror with Additional Stability

A third transistor can be added to gain more stability in the quiescent point. Using three identical transistors, Q_3 in a common–collector configuration at the base of Q_1 and Q_2 is used to provide additional stability. The three-transistor current source is shown in Figure 6.4-3.

Since the base–emitter voltages of Q_1 and Q_2 are identical, $I_{B1} = I_{B2}$ and $I_{C1} = I_{C2}$ with $\beta_{F1} = \beta_{F2} = \beta_{F3} = \beta_F$. The emitter current of Q_3 is

$$-I_{E3} = I_{B1} + I_{B2} = \frac{I_{C1}}{\beta_F} + \frac{I_{C2}}{\beta_F} = \frac{2I_{C2}}{\beta_F}. \tag{6.4-18}$$

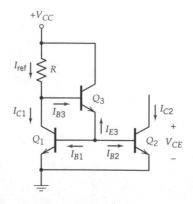

Figure 6.4-3
Three-transistor current mirror.

The base current of Q_3 is therefore

$$I_{B3} = \frac{-I_{E3}}{\beta_F + 1} = \frac{2I_{C2}}{\beta_F(\beta_F + 1)}. \qquad (6.4\text{-}19)$$

Using Equation 6.4-19 and the fact that $I_{C1} = I_{C2}$, the reference current can be found,

$$I_{\text{ref}} = I_{C1} + I_{B3} = I_{C2} + \frac{2I_{C2}}{\beta_F(\beta_F + 1)}. \qquad (6.4\text{-}20)$$

The reference current is found by applying KVL from I_{ref} through the base–emitter junctions of Q_3 and Q_1,

$$I_{\text{ref}} = \frac{V_{CC} - V_{BE1} - V_{BE3}}{R} = \frac{V_{CC} - 2V_\gamma}{R}. \qquad (6.4\text{-}21)$$

Rearranging Equation 6.4-20 to solve for I_{C2} using the expression for I_{ref} in Equation 6.4-21,

$$\begin{aligned}
I_{C2} &= \frac{I_{\text{ref}}}{1 + 2/\{\beta_F(\beta_F + 1)\}} \\
&= \frac{V_{CC} - 2V_\gamma}{R[1 + 2/\{\beta_F(\beta_F + 1)\}]}.
\end{aligned} \qquad (6.4\text{-}22)$$

The current gain stability factor is found by differentiating Equation 6.4-22 with respect to β_F,

$$\begin{aligned}
S_\beta &= \frac{\partial}{\partial \beta_F} \frac{V_{CC} - 2V_\gamma}{R[1 + 2/\{\beta_F(\beta_F + 1)\}]} \\
&= \frac{2(2\beta_F + 1)(V_{CC} - 2V_\gamma)}{R[\beta_F(\beta_F + 1) + 2]^2} = \frac{2I_{\text{ref}}(2\beta_F + 1)}{[\beta_F(\beta_F + 1) + 2]^2}.
\end{aligned} \qquad (6.4\text{-}23)$$

Recall that in the two-transistor current mirror in Section 6.4.2, the β stability factor was approximately proportional to β_F^{-2}. For the three-transistor current mirror, the β stability factor is approximately proportional to β_F^{-3}. Therefore, the three-transistor current mirror has greater stability than the two-transistor current mirror by a factor of β_F. This indicates that the collector current is tightly bound to the desired value caused by operating temperature swings or through parameter changes as a result of interchanging individual transistors with slightly different characteristics, which is commonly encountered when mass producing BJT circuits.

6.4.4 Wilson Current Source

The Wilson current source is used when low sensitivity to transistor base currents is desired and is shown in Figure 6.4-4. The transistors are again assumed to be identical for this analysis.

Applying KCL to the collector of Q_3 yields an expression for the emitter current of Q_2,

$$-I_{E2} = I_{B1} + I_{B3} + I_{C3} = \frac{I_{C1}}{\beta_F} + I_{C3}\left(1 + \frac{1}{\beta_F}\right). \qquad (6.4\text{-}24)$$

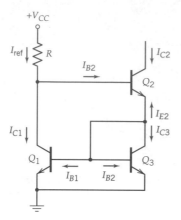

Figure 6.4-4
Wilson current mirror.

However, since Q_1 and Q_3 are identical and have identical values of base–emitter voltage, the collector currents are also equal ($I_{B1} = I_{B3}$ and $I_{C1} = I_{C3}$ with $\beta_{F1} = \beta_{F2} = \beta_{F3} = \beta_F$). Therefore, Equation 6.4-24 can be put in terms of only I_{C3},

$$-I_{E2} = I_{C3}\left(1 + \frac{2}{\beta_F}\right). \tag{6.4-25}$$

From Equation 6.4-25 the collector current of Q_2 can be determined,

$$I_{C2} = -I_{E2}\frac{\beta_F}{\beta_F + 1} = I_{C3}\left(1 + \frac{2}{\beta_F}\right)\frac{\beta_F}{\beta_F + 1}. \tag{6.4-26}$$

Applying KCL at the base of Q_2,

$$I_{\text{ref}} = I_{C1} + I_{B2} = I_{C1} + \frac{I_{C2}}{\beta_F}. \tag{6.4-27}$$

But since the transistors are identical, $I_{C1} = I_{C3}$; so, rearranging Equation 6.4-27 yields

$$I_{C3} = I_{\text{ref}} - \frac{I_{C2}}{\beta_F}. \tag{6.4-28}$$

Substituting Equation 6.4-28 into Equation 6.4-26 yields

$$I_{C2} = \left(I_{\text{ref}} - \frac{I_{C2}}{\beta_F}\right)\left(1 + \frac{2}{\beta_F}\right)\frac{\beta_F}{\beta_F + 1}. \tag{6.4-29}$$

Rearranging Equation 6.4-29 yields the Q_2 collector current in terms of the reference current,

$$I_{C2} = I_{\text{ref}}\frac{\beta_F^2 + 2\beta_F}{\beta_F^2 + 2\beta_F + 2} = I_{\text{ref}}\left(1 - \frac{2}{\beta_F^2 + 2\beta_F + 2}\right). \tag{6.4-30}$$

From Equation 6.4-30 it is apparent that the output current differs from the reference current by a small amount corresponding to approximately $2/\beta_F^2$.

6.4.5 Widlar Current Source

In the two- and three-transistor current mirrors, the load current I_{C2} is approximately equal to the reference current for large β_F. To supply a load current of milliamperes with power supply voltages on the order of 10 V, the reference resistor must be several thousands of ohms. This poses little problem for the designer. However, if the load current is small (on the order of 5 μA), the required reference resistor approaches or exceeds 1 MΩ. In OpAmp integrated circuits, the low-input-current requirement mandates a low bias current for the input emitter-coupled differential amplifier. A resistor of this magnitude is prohibitive in terms of the area that it would require on the chip. It is also difficult to fabricate accurate resistance values on integrated circuit chips.

The load and reference currents can be made to be substantially different by forcing the base–emitter voltages of Q_1 and Q_2 to be unequal in the two-transistor current mirror of Figure 6.4-2. One way to achieve this goal of substantially different load and reference currents is the addition of an emitter resistor to Q_2, as is done in the Widlar current source shown in Figure 6.4-5.

Applying KVL to the base–emitter loop of Q_1 and Q_2,

$$0 = V_{BE1} - V_{BE2} + I_{E2}R_2. \tag{6.4-31}$$

The collector currents of Q_1 and Q_2 can be related to the base–emitter voltages by using the Ebers–Moll equation 3.1-3b:

$$I_C = -I_{CS}(e^{V_{BC}/\eta V_T} - 1) + \alpha_F I_{ES}(e^{V_{BE}/\eta V_T} - 1). \tag{6.4-32}$$

Equation 3.1-8 gives the saturation current values:

$$\alpha_F I_{ES} = I_S. \tag{6.4-33}$$

For a transistor operating in the forward-active region, $V_{BC} \leq 0$. Therefore, Equation 6.4-32 simplifies to

$$I_C \approx I_S(e^{V_{BE}/\eta V_T} - 1). \tag{6.4-34}$$

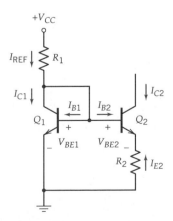

Figure 6.4-5
Widlar current source.

So the base–emitter voltage of a transistor is

$$V_{BE} = \eta V_T \ln\left(\frac{I_C}{I_S} + 1\right). \tag{6.4-35}$$

Using Equation 6.4-35, the loop equation 6.4-31 is given as

$$
\begin{aligned}
0 &= \eta_1 V_T \ln\left(\frac{I_{C1}}{I_{S1}} + 1\right) - \eta_2 V_T \ln\left(\frac{I_{C2}}{I_{S2}} + 1\right) + I_{R2} R_2 \\
&= \eta_1 V_T \ln\left(\frac{I_{C1}}{I_{S1}} + 1\right) - \eta_2 V_T \ln\left(\frac{I_{C2}}{I_{S2}} + 1\right) - \frac{\beta_F + 1}{\beta_F} I_{C2} R_2.
\end{aligned} \tag{6.4-36}
$$

If the transistors are identical, $I_{S1} = I_{S2}$ and $\eta_1 = \eta_2$. Then Equation 6.4-36 may be simplified as

$$\eta V_T \ln \frac{I_{C1} + I_S}{I_{C2} + I_S} = \frac{\beta_F + 1}{\beta_F} I_{C2} R_2. \tag{6.4-37}$$

Since the saturation current is very small compared to the collector current for BJTs in the forward-active region,

$$\eta V_T \ln \frac{I_{C1}}{I_{C2}} = \frac{\beta_F + 1}{\beta_F} I_{C2} R_2. \tag{6.4-38}$$

This is a transcendental function that must be solved by iterative techniques for a given R_2 and I_{C1} (or I_{ref}) to find I_{C2}.

EXAMPLE 6.4-1 For the Widlar current source shown in Figure 6.4-5, the required load current is I_{C2} = 5 μA, I_{C1} = 1 mA, V_{CC} = 5 V, V_{BE} = 0.7 V, V_T = 0.026 V, β_F = 200, and η = 1. Find the two resistance values and the reference current.

SOLUTION
From Equation 6.4-38,

$$0.026 \ln \frac{10^{-3}}{5 \times 10^{-6}} = \frac{201}{200} (5 \times 10^{-6}) R_2,$$

yielding the value

$$R_2 = 27.5 \text{ k}\Omega \approx \textbf{27 k}\Omega.$$

Using KCL at the collector node of Q_1,

$$I_{\text{ref}} = I_{c1} + I_{B1} + I_{B2}$$

$$= I_{C1}\left(1 + \frac{1}{\beta_F}\right) + \frac{I_{C2}}{\beta_F}$$

$$= 10^{-3}\left(1 + \frac{1}{200}\right) + \frac{5 \times 10^{-6}}{200} = \textbf{1.005 mA.}$$

The reference resistor is found by applying KVL to the collector–base–emitter loop of Q_1,

$$R_1 = \frac{V_{CC} - V_{BE}}{I_{\text{ref}}} = \frac{5 - 0.7}{1.005 \times 10^{-3}} = 4.28 \text{ k}\Omega \approx \textbf{4.3 k}\boldsymbol{\Omega}.$$

An example of current source biasing of an input emitter-coupled differential amplifier for an OpAmp (μA741) is shown in Figure 6.4-6. It is a Widlar current source consisting of Q_{12}, Q_{11}, Q_{10}, R_4, and R_5 with the load current taken off of Q_{10}. Analysis of this circuit begins with the reference current I_{ref}:

$$I_{\text{ref}} = \frac{V_{CC} - (-V_{EE}) - V_{BE12} - V_{BE11}}{R_5}$$

$$= \frac{30 - 1.4}{39,000} = 0.73 \text{ mA,}$$

where $V_{BE12} = V_{BE12} = V_\gamma - 0.7$ V.

The output current is found by the transcendental function

$$\eta V_T \ln \frac{I_{C11}}{I_{C10}} = \frac{\beta_F + 1}{\beta_F} I_{C10} R_4 \approx I_{C10} R_4.$$

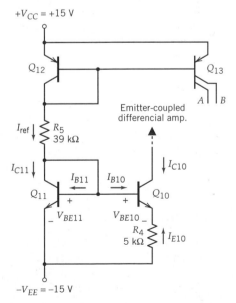

+V_{CC} = +15 V

Q_{12}

Q_{13}

A | B

Emitter-coupled differencial amp.

I_{ref} | R_5 39 kΩ

I_{C11}

I_{C10}

I_{B11} | I_{B10}

Q_{11} + | + Q_{10}

V_{BE11} | V_{BE10}

R_4 5 kΩ | I_{E10}

$-V_{EE}$ = -15 V

Figure 6.4-6
Widlar current source used to bias input stage of μA741 OpAmp.

```
****************************************************************************
Widlar circuit
****    CIRCUIT DESCRIPTION
****************************************************************************
VCC     1   0    15V
Q12     1   2    2     QPNP
R5      2   3    39K
Q11     3   3    4     QNPN
VEE     4   0    −15V
VTEST   0   5    0V
Q10     5   3    6     QNPN
R4      6   4    5K
.MODEL QPNP  PNP(IS=1p BF=200 BR=6)
.MODEL QNPN  NPN(IS=1p BF=200 BR=6)
.DC VCC 0 15 15
.PRINT DC I(R5)I(VTEST)
.OPTIONS NOPAGE
.END

****    BJT MODEL PARAMETERS
          QPNP            QNPN
          PNP             NPN
     IS   1.000000E−12  1.000000E−12
     BF   200             200
     NF   1               1
     BR   6               6
     NR   1               1

****    DC TRANSFER CURVES     TEMPERATURE = 27.000 DEG C
VCC     I(R5)    I(VTEST)
0.000E+00 3.586E−04 1.598E−05
1.500E+01 7.422E−04 1.887E−05     NOTE: These are the results desired.
****************************************************************************
```

Figure 6.4-7
SPICE simulation of OpAmp Widlar current source.

The load current can be found by using a package like MathCAD in the solver mode to evaluate the transcendental relationship. An alternate way to find the load current is through SPICE simulations. The SPICE output file for the Widlar current source in Figure 6.4-6 is shown in Figure 6.4-7. The transistors were assumed to have the following simple parameters:

$$IS = 1 \text{ pA}, \qquad BF = 200, \qquad BR = 6.$$

To obtain the load current, a zero-value voltage source (VTEST) was placed between the collector of Q_{10} and ground. For convenience, a DC sweep was performed to be able to print out the current through R_5 and the load current. Here, Q_{13} is ignored without error for this analysis.

The result of the SPICE simulation shows that $I_{ref} = 0.74$ mA and $I_{C10} = 19$ μA.

6.4.6 Simple MOSFET Current Mirror

In MOS integrated circuits, the FETs are biased in their saturation regions through the use of a MOSFET constant-current source. A simple enhancement-mode NMOSFET current mirror is shown in Figure 6.4-8. The FETs M_1 and M_2 are con-

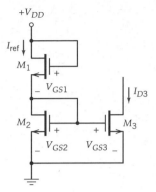

Figure 6.4-8
Simple enhancement NMOSFET current mirror.

nected as saturated devices. The reference current I_{ref} and the load current I_{D3} are determined by using the expression for the drain current,

$$I_{ref} = K_1(V_{GS1} - V_{T1})^2 = K_2(V_{GS2} - V_{T2})^2, \qquad (6.4\text{-}39)$$

$$I_{D3} = K_3(V_{GS3} - V_{T3})^2. \qquad (6.4\text{-}40)$$

The NMOSFET M_3 is assumed to be in saturation.

Applying KVL, the gate–source voltage of M_1 is found:

$$V_{GS1} = V_{DD} - V_{GS2}. \qquad (6.4\text{-}41)$$

Substituting Equation 6.4-41 into Equation 6.4-39 yields

$$K_1(V_{DD} - V_{GS2} - V_{T1})^2 = K_2(V_{GS2} - V_{T2})^2. \qquad (6.4\text{-}42)$$

Rearranging the equation yields

$$\left[1 + \left(\frac{K_1}{K_2}\right)^{1/2}\right]V_{GS2} = \left(\frac{K_1}{K_2}\right)^{1/2}(V_{DD} - V_{T1}) + V_{T2}. \qquad (6.4\text{-}43)$$

Solving for V_{GS2},

$$V_{GS2} = \frac{(K_1/K_2)^{1/2}(V_{DD} - V_{T1}) + V_{T2}}{1 + (K_1/K_2)^{1/2}}. \qquad (6.4\text{-}44)$$

But $V_{GS3} = V_{GS2}$, so the load current I_{D3} is found by substituting Equation 6.4-44 into Equation 6.4-40.

EXAMPLE 6.4-2 Given an enhancement-mode NMOSFET in the circuit of Figure 6.4-8, find the ratio of the transconductance of the FETs, M_1 and M_2, to achieve a current of

$$I_{D3} = 25 \ \mu A.$$

The other FET parameters are

$$V_{DD} = 10 \text{ V}, \qquad K_3 = 0.1 \text{ mA/V}^2,$$
$$V_{T1} = V_{T2} = V_{T3} = 1 \text{ V}, \qquad I_{\text{ref}} = 100 \text{ μA}.$$

SOLUTION

Solve for $V_{GS3} = V_{GS2}$ using Equation 6.4-40:

$$V_{GS2} = V_{GS3} = \sqrt{\frac{I_{D3}}{K_3}} + V_{T3} = \sqrt{\frac{25 \times 10^{-6}}{1 \times 10^{-4}}} + 1 = 1.5 \text{ V}.$$

Rearranging Equation 6.4-35 yields the required transconductance ratio of M_1 and M_2,

$$\frac{K_1}{K_2} = \left(\frac{V_{GS2} - V_{T2}}{V_{DD} - V_{GS2} - V_{T1}}\right)^2$$
$$= \left(\frac{1.5 - 1}{15 - 1.5 - 1}\right)^2 = \textbf{1.6}.$$

6.5 ACTIVE LOADS

In integrated circuit amplifiers, transistor current source configurations are often used as active loads. Since small-signal gain in amplifiers is directly proportional to the load resistance, large loads are desired to achieve large gain. However, increasing the load requires that a large power supply voltage must be used in order to keep to transistor in the proper region of operation. The quiescent point will also be altered. Both of these consequences to the use of a large resistor for a load are undesirable.

Active loads using current source configurations are commonly used to provide the high load required for increasing small-signal gain. The small-signal output resistance of the current source configuration is used as the load to the amplifying circuit. A Norton and Thévenin equivalent model of a transistor current source is shown in Figure 6.5-1.

For the Norton equivalent model, the analysis and design of current sources, discussed in Section 6.4 using the Ebers–Moll model, is used to determine I_O. The output resistance R_O is found by determining the performing small-signal analysis on the current source. Since a transistor is used in place of a resistor, the load is

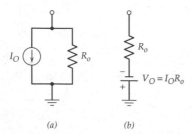

(a)   (b)

Figure 6.5-1

(a) Norton and (b) Thévenin equivalent models of constant-current sources.

dependent on both the inherent transistor characteristics and the quiescent point of the circuit. In particular, the early voltage V_A is of great importance when determining the active load of a small-signal transistor amplifier.

6.5.1 Common Emitter Amplifier with Active Load

A common-emitter amplifier with an active load is shown in Figure 6.5-2. The active load consists of a simple *pnp* current mirror. The *pnp* current mirror provides the load for the *npn* BJT common-emitter amplifier. The current mirror also determines the range of bias currents over which the common-emitter will be in the forward-active region.

The current mirror transistors Q_1 and Q_2 are assumed to be identical. For Q_2 to be in the forward-active region, its collector current and the collector current of the common-emitter BJT, Q_3, must be greater than $I_{C,\text{lower}}$, where $I_{C,\text{lower}}$ is defined by

$$I_{C,\text{lower}} = \frac{V_{CC} - V_\gamma}{(1 + 2/\beta_{FQ1})R} \approx I_{\text{ref}}. \tag{6.5-1}$$

This current effectively clips and distorts the output at I_{c3}. From the Ebers–Moll equations and Equation 6.4-35, the base–emitter bias voltage of Q_3 required to establish this lower current limit of operation is

$$V_{BE3(\text{lower})} = \eta V_t \ln\left(\frac{I_{C3,\text{lower}}}{I_{S1}} + 1\right). \tag{6.5-2}$$

Therefore, the input voltage V_i must exceed $V_{BE3,\text{lower}}$ for operation of Q_3 in the forward-active region.

The characteristic curve of the active load transistor Q_1 superimposed on the set of output characteristic curves for Q_3 is shown in Figure 6.5-3. The *pnp* BJT, Q_1, acts as a load line of resistance $1/h_{oe1}$ on the collector of the common-emitter-configured *npn* transistor Q_3. Since the collector current is small, the load is very large. If a resistive load was used instead of an active load, the circuit would require a very large power supply voltage to establish an end of the resistive load line. The active load line formed by Q_1 transitions from the cutoff to the forward-active region at $I_{C1} = I_{C,\text{lower}}$. The upper limit of amplifier operation for Q_3 is established when the active

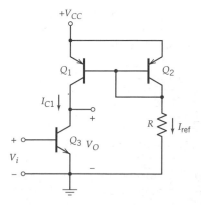

Figure 6.5-2
Common emitter amplifier with *pnp* current mirror active load.

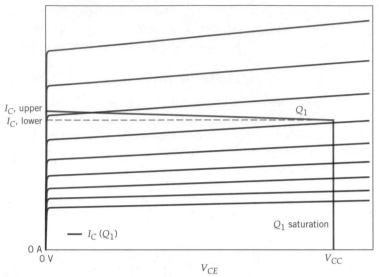

Figure 6.5-3
Characteristic curve of active load device Q_1 superimposed on output characteristic curves for Q_3.

load line crosses the point where Q_3 saturates. This point establishes the upper limit, $I_{C,\text{upper}}$, on Q_3 and is approximately

$$I_{C,\text{upper}} = I_{C,\text{lower}} + \frac{V_{CC} - V_{CE3(\text{sat})}}{1/h_{oe1}}. \tag{6.5-3}$$

The corresponding base–emitter voltage of Q_3 is

$$V_{BE3(\text{upper})} = \eta V_t \ln\left(\frac{I_{C,\text{upper}}}{I_{S1}} + 1\right). \tag{6.5-4}$$

A plot of the Q_3 input voltage versus the collector voltage and collector current of Q_3 is shown in Figure 6.5-4. The Q_3 input voltage versus the collector voltage in Figure 6.5-4 is a transfer characteristic of the common-emitter amplifier.

Small-signal analysis is used to determine the common-emitter load. The AC and small-signal models of the common emitter circuit in Figure 6.5-2 are shown in Figure 6.5-5.

To find the Thévenin resistance R_{TH} of the *pnp* current mirror, the resistance looking into the nodes C_3, C_1, and E_1 is found. For the common-emitter circuit in Figure 6.5-2, the load (Thévenin) resistance is $R_O = h_{oe1}^{-1}$. From Equation 5.2-22, the output resistance is

$$R_O = \frac{1}{h_{oe1}} = \frac{|V_{A1}|}{I_{C1}}, \tag{6.5-5}$$

where V_{A1} is the early voltage of Q_1. Therefore, R_O is very large for small values of collector current.

The simplified small-signal model of the common-emitter amplifier with active load in Figure 6.5-5 is shown in Figure 6.5-6.

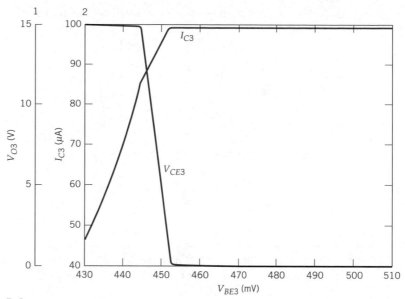

Figure 6.5-4
The Q_3 collector current and voltage as function of input (base–emitter) voltage.

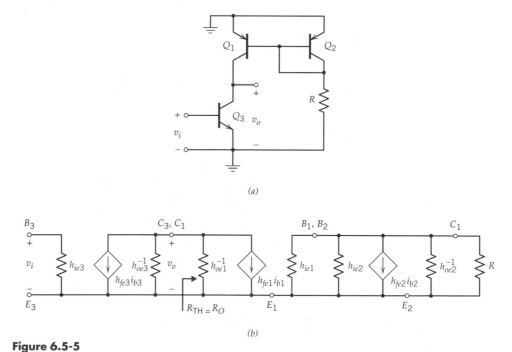

Figure 6.5-5
(*a*) AC and (*b*) small-signal model of common emitter amplifier with active load in Figure 6.5-2.

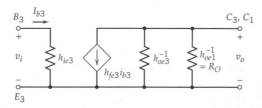

Figure 6.5-6
Simplified small-signal model of common emitter active load amplifier in Figure 6.5-5.

The small-signal output voltage of the amplifier is

$$v_o = -h_{fe3}i_{b3}(h_{oe3}^{-1}\|R_O).$$ (6.5-6)

But $i_{b3} = v_i/h_{ie3}$, so the gain of the amplifier is

$$A_V = \frac{v_o}{v_i} = -\frac{h_{fe3}}{h_{ie3}}(h_{oe3}^{-1}\|R_O),$$ (6.5-7)

where, from Equation 5.2-9,

$$h_{ie3} \approx (\beta_{F3} + 1)\frac{\eta V_t}{|I_{C3}|}.$$ (6.5-8)

EXAMPLE 6.5-1 Design a common-emitter amplifier with an active load in the configuration shown in Figure 6.5-2. The lower current limit is required to be 100 μA. The following transistors are available for use:

> 2N2222 *npn*: IS = 14.4 × 10^{-15}, BF = 255, BR = 6, VA = 75
> 2N2907 *pnp*: IS = 650.6 × 10^{-18}, BF = 232, BR = 4, VA = 116

The power supply is +15 V.
Find the input resistance, small-signal gain, and other component values.

Solution 1

Design the common-emitter amplifier using the configuration in Figure 6.5-2. Find the resistor value R for the lower collector current limit of 100 μA. Rearranging Equation 6.5-1,

$$R = \frac{V_{CC} - V_{\gamma 1}}{(1 + 2/\beta_{FQ1})I_{C,\text{lower}}} = \frac{15 - 0.7}{(1 - 2/232)10^{-4}}$$

$$= \mathbf{143\ k\Omega} \quad (1\% \text{ standard value}).$$

The upper limit of the collector current is found by first determining $1/h_{oe1}$:

$$R_O = \frac{1}{h_{oe1}} \approx \frac{V_{A1}}{I_{C,\text{lower}}} = \frac{116}{10^{-4}} = 1.16\ M\Omega.$$

The upper current limit is therefore approximately

$$I_{C,\text{upper}} \approx I_{C,\text{lower}} + \frac{V_{CC} - V_{CE3(\text{sat})}}{R_o}$$

$$= 10^{-4} + \frac{15 - 0.2}{1.16 \times 10^6} = 113\ \mu A.$$

For maximum output swing, select a bias point halfway between $I_{C,\text{upper}}$ and $I_{C,\text{lower}}$,

$$I_{C,\text{bias}} = 106.5\ \mu A.$$

The base–emitter bias voltage for Q_3 is then

$$V_{BE3(\text{bias})} = \eta V_t \ln\left(\frac{I_{C,\text{bias}}}{I_{S3}} + 1\right)$$

$$= 0.026 \ln\left(\frac{10^{-4}}{14.3 \times 10^{-15}} + 1\right) = 591 \text{ mV},$$

where $\eta = 1$.

The input resistance R_i is given as

$$R_i = h_{ie3} \approx (\beta_{F3} + 1)\frac{\eta V_t}{|I_{C3}|} = (256)\frac{0.026}{106.5 \times 10^{-6}} = \textbf{62.5 k}\boldsymbol{\Omega}.$$

The small-signal gain is then

$$A_V = \frac{v_o}{v_i} = -\frac{h_{fe3}}{h_{ie3}}(h_{oe3}^{-1}\|R_o)$$

$$= -\frac{255}{62{,}500}(1.16 \times 10^6 \ \| \ 1.16 \times 10^6) = \textbf{-2370}.$$

SOLUTION SPICE

The circuit in Figure 6.5-2 is simulated using SPICE with the value of R determined above for a reference current of 100 μA.

```
******************************PSpice*****************************************
Amplifier with Active Load
**** CIRCUIT DESCRIPTION
****************************************************************************

.OPTION NOPAGE
VCC    1   0   15V
Q1     3   2   1   Q2N2907
Q2     2   2   1   Q2N2907
Rref   2   0   143K
Q3     3   4   0   Q2N2222
VB     4   0   585mV
.model Q2N2907 PNP(IS=650.6E-18 BF=232 BR=4 VA=116)
.model Q2N2222 NPN(IS=14.3E-15 BF=255 BR=6 VA=75)
.TF  V(3)  VB
.DC LIN VB  550m   650m   0.1m
.PROBE
.END
**** BJT MODEL PARAMETERS
        Q2N2907              Q2N2222
        PNP                  NPN
        IS 650.600000E-18  14.300000E-15
        BF 232               255
        NF  1                1
        VAF 116              75
        BR  4                6
        NR  1                1

****   SMALL-SIGNAL BIAS SOLUTION   TEMPERATURE = 27.000 DEG C
NODE  VOLTAGE  NODE  VOLTAGE  NODE  VOLTAGE  NODE  VOLTAGE
( 1)  15.0000  ( 2)  14.3340  ( 3)   8.1330  ( 4)   .5850
```

```
VOLTAGE SOURCE CURRENTS
NAME      CURRENT

VCC       -2.049E-04
VB        -3.730E-07

TOTAL POWER DISSIPATION 3.07E-03 WATTS
****   SMALL-SIGNAL CHARACTERISTICS
V(3)/VB = -1.904E+03
INPUT RESISTANCE AT VB = 6.934E+04
OUTPUT RESISTANCE AT V(3) = 4.706E+05
JOB CONCLUDED
TOTAL JOB TIME    29.71
```

The results of the SPICE simulation are as follows:

> Small-signal gain: -1900
> Input resistance: 69.3 kΩ
> Output resistance: 471 kΩ

These results are in favorable agreement with the analytical results.

___

6.5.2 Common Source Amplifier with Active Load

An enhancement NMOSFET common source amplifier (M_1) with a PMOSFET current mirror active load is shown in Figure 6.5-7. FETs M_2 and M_3 form a current mirror where the reference current is determined by the resistor R. For operation in the saturation region,

$$I_{ref} = I_{D3} = K_3(V_{GS3} - V_{T3})^2, \tag{6.5-9}$$

where

$$V_{GS3} = -V_{SG3} = -(V_{DD} - I_{ref}R). \tag{6.5-10}$$

By substituting Equation 6.5-10 into Equation 6.5-9, the reference current is found as a function of R.

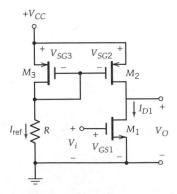

Figure 6.5-7
Common source amplifier with active load.

The transistor M_1 is in cutoff when $V_{GS} < V_T$. Under this condition, no current flows through M_2 and, because it is a current mirror, M_3. As V_{GS} increases above V_T, M_1 enters saturation and forces M_2 into the ohmic region and lowers V_O. Further increase in the input voltage forces M_2 into saturation, decreasing the output voltage. Finally, M_1 enters the ohmic region while M_2 and M_3 remain in the saturation region.

The transfer function of the active load common–source amplifier of Figure 6.5-7 is shown in Figure 6.5-8.

The small-signal model of the amplifier is used to determine the gain of the amplifier. The AC model of Figure 6.5-7 is shown in Figure 6.5-9.

From the AC model in Figure 6.5-9, the small-signal model can be derived and is shown in Figure 6.5-10. The load resistor consists only of the drain resistance of M_2.

The drain resistance is found by applying Equation 5.8-10,

$$r_d = \frac{|V_A|}{|I_D|}.$$

(6.5-11)

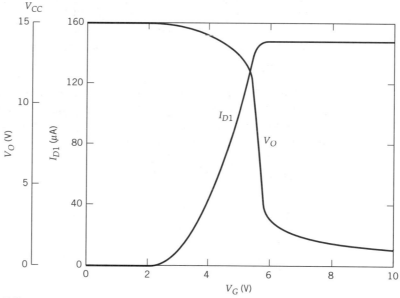

Figure 6.5-8
Transfer characteristic of active load common source amplifier in Figure 6.5-7.

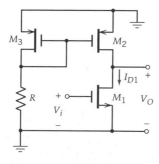

Figure 6.5-9
The AC model of common–source amplifier with active load in Figure 6.5-7.

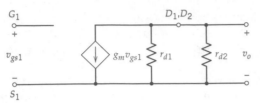

Figure 6.5-10
Small-signal model of common–source amplifier with active load.

The small-signal transconductance is defined in Equation 5.7-13 as

$$g_m = 2K(V_{GS} - V_T) = 2\sqrt{I_D K}. \qquad (6.5\text{-}12)$$

The gain of the amplifier is

$$A_V = -g_{m1}(r_{d1}\|r_{d2}). \qquad (6.5\text{-}13)$$

6.5.3 Emitter-Coupled Differential Amplifier with Active Load

An emitter-coupled differential amplifier with an active load current mirror is shown in Figure 6.5-11. It is assumed that Q_3 and Q_4 and Q_1 and Q_2 are matched pairs. In this circuit, the collector current is controlled by the Q_1 side of the emitter-coupled pair. It is also a single-ended output amplifier, which eliminates the common-mode problem and has higher CMRR than a single-output differential amplifier with resistive loads.

The AC model of the emitter-coupled single-ended differential amplifier with an active load of Figure 6.5-11 is shown in Figure 6.5-12. The gain for the circuit in Figure 6.5-12 is performed by using a single input at v_{i1} and grounding v_{i2}.

Since Q_3 and Q_4 form a current mirror, the collector current $i_{c1} = h_{fe1}i_{b1}$ through transistor Q_1 is equal to the collector current through Q_2. Since the output resistance of the common base transistor, Q_2, is much greater than that of the common emitter transistor, Q_4, the output voltage is then

$$v_o = \frac{h_{fe1}i_{b1}}{h_{oe4}}. \qquad (6.5\text{-}14)$$

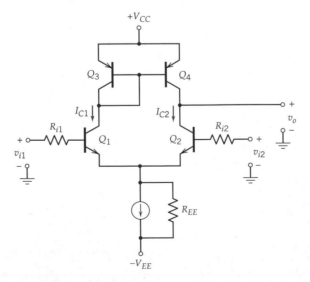

Figure 6.5-11
Emitter-coupled single-ended output differential amplifier with active load current mirror.

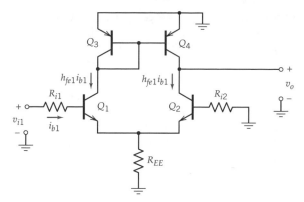

Figure 6.5-12
The AC model of emitter-coupled differential amplifier in Figure 6.5-11.

The emitter resistance of Q_1 must be found to determine the base current i_{b1}. The emitter of Q_1 is connected to R_{EE} and the emitter of Q_2. Here, Q_2 appears as a common-base configuration with an input resistance

$$R_{E2} \approx \frac{h_{ie2} + R_{i2}}{h_{fe2} + 1} \, \Big\| \, R_{EE} \approx \frac{h_{ie2} + R_{i2}}{h_{fe2} + 1} \tag{6.5-15}$$

for a large h_{oe4}^{-1} and R_{EE}.

The base current for Q_1 is found by using Equation 6.5-15,

$$i_{b1} = \frac{v_{i1}}{(R_{i1} + h_{ie1}) + \{(h_{fe1} + 1)[(h_{ie2} + R_{i2})/h_{fe2} + 1]\}}. \tag{6.5-16}$$

By substituting Equation 6.5-16 into Equation 6.5-14, the output voltage is

$$v_o = \frac{h_{fe1} v_{i1}}{h_{oe4}(2R_{i1} + 2h_{ie1})} = \frac{h_{fe1} v_{i1}}{2h_{oe4}(R_{i1} + h_{ie1})}. \tag{6.5-17}$$

Therefore, the gain of the amplifier is

$$A_v = \frac{v_o}{v_{i1}} = \frac{h_{fe1}}{2h_{oe4}(R_{i1} + h_{ie1})}. \tag{6.5-18}$$

This result is exactly one-half of that predicted by Equation 6.3-13, where the load resistance is given by

$$R_c = \frac{1}{h_{oe4}}.$$

The factor of 2 is the result of the single-ended rather than differential output.

6.6 CONCLUDING REMARKS

The range of transistor amplifiers available for use was expanded in this chapter to include the use of multiple-transistor applications. This extension allows the designer to create amplifiers that have a combination of amplification, input resistance, and output resistance that is not within the capabilities of single-transistor amplifiers. Cascaded-stage designs, consisting of several simple stages in series, are the dominant form of multiple-transistor amplifiers. Closely related to cascaded-stage designs are the Darlington and similar grouped-transistor amplification stages.

Other common multiple-transistor amplification stages are based on differential inputs rather than inputs solely referenced to a common terminal. Emitter- and source-coupled transistor pairs form the basis for differential amplifiers. Among the many needs for differential inputs are many OpAmp applications.

While the addition of additional amplification stages complicates the process, the modeling and analysis techniques for multiple-transistor amplifiers follow the same basic procedure previously described:

1. Determine the quiescent (DC) conditions; verify all transistors are in the proper operating region.
2. Determine the small-signal parameters for each transistor from the quiescent conditions.
3. Create an AC equivalent circuit.
4. Determine the AC performance for the circuit.
5. Add the results of the DC and AC analysis to obtain total circuit performance.

Although each step of the procedure may be more complicated, simple steps taken together provide the desired results.

Integrated circuit amplifier designs are an area where the use of multiple transistors prevails. In discrete applications resistors are cost effective and reliable; in integrated circuit applications they create a multitude of design obstacles. The use of multiple-transistor current sources as amplifier bias networks solves many of the problems. Current sources can also provide high-resistance active loads for many of the single- and multiple-transistor amplifier types. Active loads provide high gain without the use of large resistors or high-voltage sources.

While the interconnection of many transistors in simple amplifiers can satisfy many design goals, it does not always provide the optimum design. The sensitivity of these designs to transistor parameter variation is, on occasion, a concern. High-power applications also put constraints on amplifier designs. These design concerns and several solutions are explored in Chapters 7 and 8.

SUMMARY DESIGN EXAMPLE

A particular electronic application requires the amplification of the difference of two voltages by a factor of 200 ± 20. Each input source has an output resistance of 50 Ω. The output of the amplifier must also be differential with each output having an output resistance of 2.7 kΩ. Available voltages sources are at ground potential and ± 10 V. It is expected that this design may eventually be realized in a bipolar integrated circuit application where the individual BJTs are described by

$$\beta_F = 200, \qquad V_A = 160.$$

Solution

The design requirements suggest an emitter-coupled differential amplifier. The possibility of eventual integrated circuit realization implies that the differential pair should be biased with a BJT current source. The basic topology of the circuit is therefore as shown (the load and source resistors are specified above).

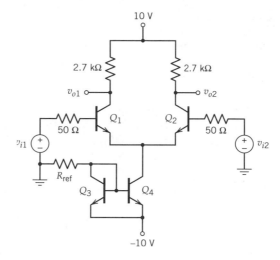

The differential gain of this circuit is given by

$$A_{VD} = \frac{-h_{fe}R_c}{h_{ie} + \frac{1}{2}(R_{i1} + R_{i2})}.$$

This gain requirement restricts the differential pair BJT parameter h_{ie}:

$$h_{ie} = \frac{h_{fe}R_c}{A_{VD}} - \frac{1}{2}(R_{i1} + R_{i2}) = 2650 \ \Omega.$$

Such a value can only be achieved with a specific collector current in each of the differential pair BJTs:

$$h_{ie} = (\beta_F + 1)\frac{\eta V_t}{I_C} \Rightarrow I_C = 1.972 \text{ mA}.$$

The bias current provided by the current source must be the sum of the differential pair emitter currents:

$$I_{bias} = 2\left(\frac{\beta_F + 1}{\beta_F}\right)I_C = 3.964 \text{ mA}.$$

This bias current can be directly related to the collector currents of the two current source BJTs:

$$I_{bias} \approx I_{C4} + (10 - V_\gamma)h_{oe4} = I_{C4}\left(1 + \frac{9.3}{V_A}\right) \Rightarrow I_{C4} = 3.746 \text{ mA}.$$

The current source bias resistor value can then be determined:

$$R_{ref} = \frac{10 - V_\gamma}{3.746(1 + \frac{2}{200})} = 2.458 \text{ k}\Omega \approx \textbf{2.46 k}\boldsymbol{\Omega}.$$

SPICE simulation shows the differential gain of this amplifier to be ≈ 193, which is within specifications.

REFERENCES

Small-Signal Transistors, FET's and Diodes Device Data Book, Motorola, Inc., Phoenix, 1991

Bipolar Power Transistor Data Book, 7th. Ed., Motorola, Inc., Phoenix, 1992

Ghausi, M. S., *Electronic Devices and Circuits: Discrete and Integrated,* Holt, Rinehart and Winston, New York, 1985.

Gray, P. R., and Meyer, R. G., *Analysis and Design of Analog Integrated Circuits,* 3rd. Ed., John Wiley & Sons, Inc., New York, 1993.

Horowitz, P., and Hill, W., *The Art of Electronics,* 2nd. Ed. Cambridge University Press, Cambridge, 1992.

Millman, J., *Microelectronics, Digital and Analog Circuits and Systems,* McGraw-Hill Book Company, New York, 1979.

Sedra, A. S. and Smith, K. C., *Microelectronic Circuits,* 3rd. Ed., Holt, Rinehart, and Winston. Philadelphia, 1991.

Tuinenga, P., *SPICE: A Guide to Circuit Simulation and Analysis Using PSpice,* 2nd. Ed., Prentice Hall, Englewood Cliffs, 1992.

PROBLEMS

6-1. The amplifier shown uses transistors with the following characteristics:

$$\beta_{F1} = 220, \qquad \beta_{F2} = 180,$$

$$V_{A1} = V_{A2} = 200 \text{ V}.$$

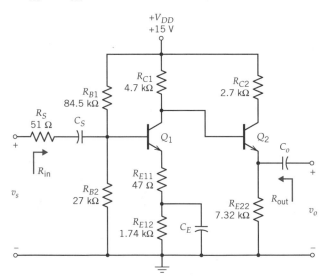

a. Determine the quiescent currents and voltages of each transistor.

b. Find the overall voltage gain A_{vs}.

c. Find the input resistance R_{in}.

d. Find the output resistance R_{out}.

6-2. Determine the voltage gain

$$A_V = \frac{v_o}{v_s},$$

current gain

$$A_I = \frac{i_o}{i_s},$$

input resistance R_{in}, and output resistance R_{out} for the given two-stage cascade amplifier. The silicon BJTs are identical and are described by $\beta_F = 150$.

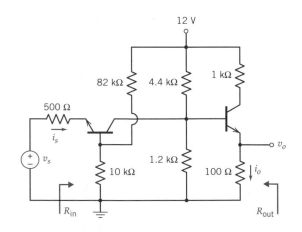

6-3. The amplifier shown uses identical transistors with the following characteristics:

$$\beta_F = 180, \qquad V_A = 200 \text{ V}.$$

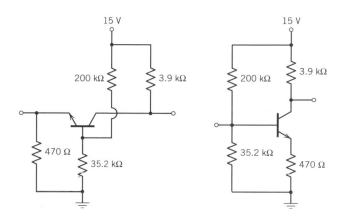

a. Find the quiescent point of the transistors

b. Find the midband voltage gain of the amplifier.

c. Confirm the results using SPICE.

6-4. It is usually not a good design choice to cascade two simple common-collector amplifier stages. Investigate this cascade by performing the following design:

a. Using a silicon BJT with $\beta_F = 150$, design a simple one-transistor common-collector amplifier to meet the following design goals:

- Emitter resistor: $R_e = 1 \text{ k}\Omega$
- Power supply: $V_{CC} = 10 \text{ V}$
- Q point: $V_{CEQ} = 5 \text{ V}$
- Stage input resistance: $R_{in} = 20 \text{ k}\Omega$ (includes bias resistors)

b. Determine the voltage gain, current gain, and output resistance of this simple stage.

c. Capacitively cascade two stages of the design from part a; compare performance characteristics (including input resistance) of this two-stage amplifier with the single-stage amplifier of part b. Comment on results.

6-5. In most multistage amplifier designs, common-base amplifier stages are best used as input stages to an amplifier cascade: They have little value when used as intermediate or final stages. Investigate this simple design principle by comparing the voltage and current gain of two-stage, capacitively-coupled cascades consisting of the two given amplifiers in the two configurations: CB–CE and CE–CB. Assume silicon BJTs with $\beta_F = 100$.

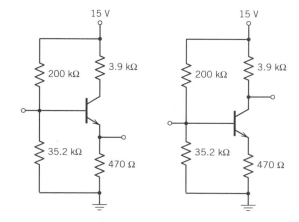

6-6. In most multistage amplifier designs, common-collector amplifier stages are best used as output stages to an amplifier cascade: They have little value when used as input or intermediate stages. Investigate this simple design principle by comparing the voltage and current gain of two-stage, capacitively coupled cascades consisting of the two given amplifiers in the two configurations: CC–CE and CE–CC. Assume silicon BJTs with $\beta_F = 100$.

6-7. In most multistage amplifier designs, common-gate amplifier stages are best used as input stages to an amplifier cascade: They have little value when used as intermediate or final stages. Investigate this simple design principle by comparing the voltage and current gain of two-stage, capacitively coupled cascades consisting of the two given amplifiers in the two configurations: CG–CS and CS–CG. Assume FET parameters:

$$I_{DSS} = 4 \text{ mA}, \qquad V_{PO} = -2 \text{ V},$$

$$V_A = 150 \text{ V}.$$

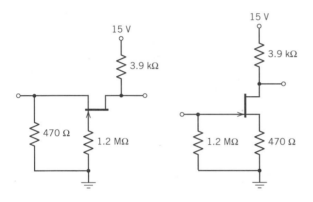

6-8. The amplifier shown uses identical transistors with the following characteristics:

$$I_{DSS} = 10 \text{ mA}, \qquad V_{PO} = -2 \text{ V},$$

$$V_A = 100 \text{ V}.$$

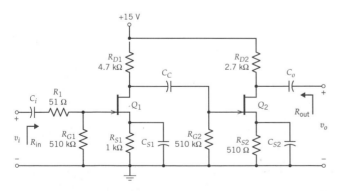

a. Find the overall voltage gain A_{vs}.
b. Find the input resistance R_{in}.
c. Find the output resistance R_{out}.

 6-9. For the circuit shown, the transistor parameters are

$$Q_1: \qquad \beta_F = 200, \qquad V_\gamma = 0.7 \text{ V},$$

$$V_A = 250 \text{ V},$$

$$Q_2: \qquad I_{DSS} = 10 \text{ mA}, \qquad V_{PO} = -5 \text{ V},$$

$$V_A = 250 \text{ V}.$$

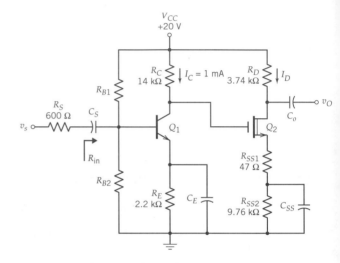

a. Find R_{B1} and R_{B2} for $I_C = 1$ mA. Use the "rule-of-thumb" relationship between R_E and R_B for stable operation (1% change in I_C for 10% change in β_F).
b. Find the quiescent condition of the two transistors.
c. Find the voltage gain of the amplifier.
d. Find the input resistance R_{in} of the amplifier.

 6-10. For the given Darlington circuit:

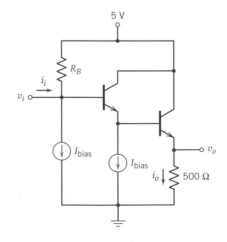

a. Determine the values of the two identical current sources I_{bias} and the bias resistor R_B so that the quiescent BJT collector currents are the same value and the quiescent output voltage is 2.5 V. The silicon BJTs are identical with

$$\beta_F = 200, \qquad V_A = 350 \text{ V}.$$

b. Determine the small-signal h parameters for the composite Darlington transistor.

c. Find the current gain of the circuit:

$$A_I = \frac{i_o}{i_s}.$$

6-11. Many Darlington silicon power transistor pairs provide built-in base–emitter shunt resistors as an aid to biasing. One such Darlington pair is the 2N6387, for which a typical circuit diagram is shown. For DC analysis purposes these resistors act as current sources of value

$$I_{bias} = \frac{V_\gamma}{R_{shunt}}.$$

In AC analysis the resistors slightly decrease the effective value of the two dominant h parameters, h_{ie} and h_{fe}, in each BJT. For this Darlington pair assume each individual BJT in the typical circuit diagram is described by $\beta_F = 60$.

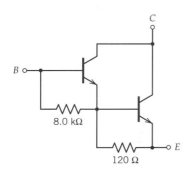

a. Design a common-collector amplifier using the following parts to achieve a quiescent collector current for the Darlington pair of 3 A:

- DC power supply: $V_{CC} = 48$ V
- Load resistor: $R_L = 8\ \Omega$ (at emitter)
- Single-bias resistor: $R_B =$ any value (use only one resistor)

b. Determine the effective small-signal h parameters of each of the Darlington transistors (find the h parameters for each BJT with its base–emitter junction shunted) and the Darlington pair as a whole.

c. Determine the performance characteristics of the amplifier, that is, find A_V, A_I, and R_{in}.

6-12. The transistors in the circuit shown have parameters

$$I_{DSS} = 2 \text{ mA}, \qquad V_{PO} = -2, \qquad V_A = 100,$$
$$\beta_F = 150, \qquad V_A = 200.$$

The quiescent conditions have been found to be

$$|I_{DQ}| = 2 \text{ mA}, \qquad V_{DSQ} = 14 \text{ V},$$
$$|I_{CQ}| = 1.515 \text{ mA}, \qquad V_{CEQ} = 6.21 \text{ V}.$$

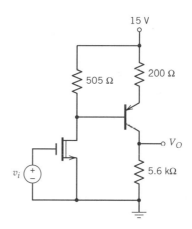

a. Determine the small-signal parameters for the two transistors.

b. Determine the voltage gain of the circuit.

6-13. For the amplifier shown, the transistor characteristics are

$$I_{DSS} = 8 \text{ mA}, \qquad V_{PO} = -5 \text{ V}, \qquad V_A = 100 \text{ V}$$

for the JFET and

$$\beta_F = 200, \qquad V_A = 100 \text{ V}$$

for the BJT.

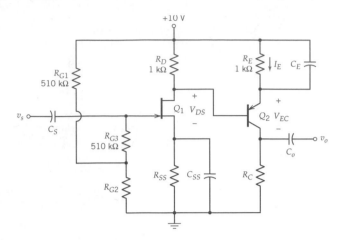

a. Complete the design of the amplifier so that $I_E = 3$ mA and $V_{DS} = V_{EC} = 5$ V. Find the quiescent point of all of the transistors.

b. Find the midband voltage gain of the amplifier.

c. Confirm the results using SPICE.

 6-14. Complete the design of the amplifier shown. The transistor parameters are

$$I_{DSS} = -8 \text{ mA}, \qquad V_{PO} = 3.5\text{V}, \qquad V_A = 100 \text{ V}$$

for the JFET and

$$\beta_F = 120, \qquad V_A = 150 \text{ V}$$

for the BJT.

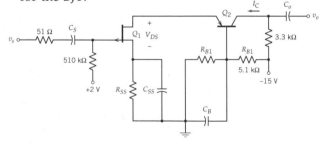

a. Find R_{SS} and R_{B2} so that

$$I_C = -1 \text{ mA}, \qquad V_{DS} = -3.9 \text{ V}.$$

Determine the quiescent points of the transistors.

b. Determine the midband voltage gain of the amplifier

c. Confirm the results using SPICE.

 6-15. For the BiFET Darlington configuration shown, the bias voltage V_g is adjusted so that the quiescent output voltage is 2.5 V. The transistor parameters are

$$\beta_F = 200, \qquad\qquad V_A = 150 \text{ V},$$
$$K = 1 \text{ mA/V}^2, \qquad V_T = 2 \text{ V}.$$

a. Determine V_g.

b. Find the voltage gain of the circuit:

$$A_V = \frac{v_o}{v_i}.$$

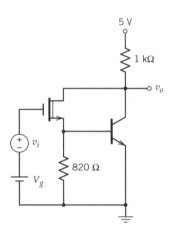

6-16. Determine the quiescent currents and voltages and the midband voltage gain of the Darlington common-base amplifier shown. Identical transistors are used with parameters

$$\beta_F = 180, \qquad V_A = 200 \text{ V}.$$

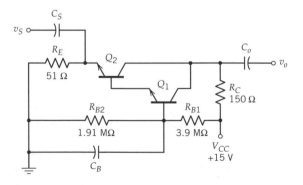

 6-17. For the circuit shown, the transistors parameters are

$$\beta_F = 180, \qquad V_\gamma = 0.7 \text{ V}, \qquad V_A = 250 \text{ V}.$$

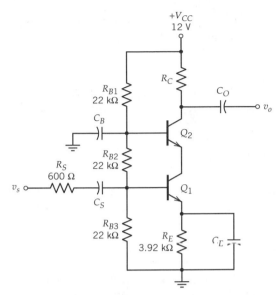

a. Complete the design by finding R_C.

b. Find the quiescent condition of the two transistors.

c. Find the voltage gain of the amplifier.

6-18. Determine the quiescent currents and voltages and the midband voltage gain of the FET/bipolar cascode amplifier shown. The transistor parameters are

Q_1: $V_{PO} = -4$ V, $I_{DSS} = 5$ mA,

      $V_A = 200$ V,

Q_2: $\beta_F = 220$, $V_A = 250$ V.

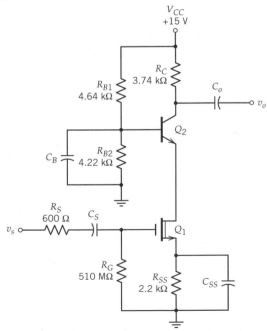

6-19. The differential amplifier shown utilizes a simple emitter resistor to establish quiescent conditions. Assume identical BJTs with

$\beta_F = 120$.

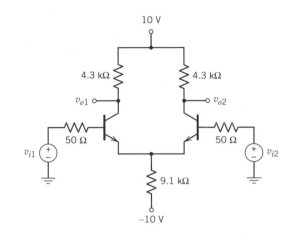

a. Determine the differential-mode gain.

b. Determine the common-mode rejection ratio (CMRR).

6-20. Complete the design of the differential amplifier shown so that CMRR = 43 dB. Assume identical transistors with $\beta_F = 120$.

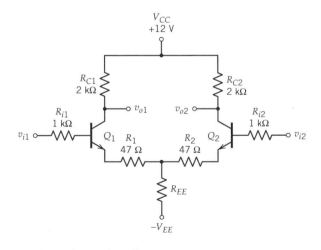

a. Find R_{EE} and V_{EE}.

b. Confirm the results using SPICE.

6-21. For the differential amplifier shown, find i_L in terms of the common- and differential-mode input signals. Assume identical transistors with $\beta_F = 120$.

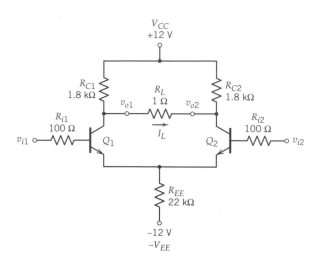

6-22. The transistors in the differential amplifier shown are identical and have the following characteristics:

$$I_{DSS} = 10 \text{ mA}, \qquad V_{PO} = -4.5 \text{ V},$$

$$V_A = 100 \text{ V}.$$

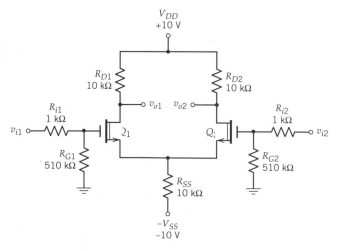

a. Find the CMRR.

b. Determine the output voltage: $v_{o2} - v_{o1}$.

c. Find the output resistance looking into the drain of Q_2.

d. Use SPICE to confirm the output voltage found in part b and the output resistance found in part c.

6-23. Design a 100-μA Widlar current source utilizing two identical silicon BJTs with parameters

$$\beta_F = 150 \quad \text{and} \quad V_A = 250 \text{ V},$$

two resistors, and a 10-V battery. In order to extend battery life, total power consumption must be less than 10 mW.

6-24. It is often necessary in integrated circuit applications to design a current source with multiple outputs. The general topology of a simple current mirror with three outputs is shown.

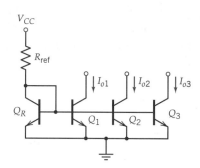

a. Determine the magnitude of output currents as a function of the circuit parameters.

b. Generalize the results of part a for circuits with n outputs.

c. Design a four-output current source using silicon BJTs with $\beta_F = 120$ and a voltage source of 10 V DC to have identical output currents of 300 μA.

6-25. Complete the design of the amplifier shown so that the quiescent value of the output voltage v_o has value +1 V. The transistors are identical and have the characteristics

$$\beta_F = 200, \qquad V_A = 200 \text{ V}.$$

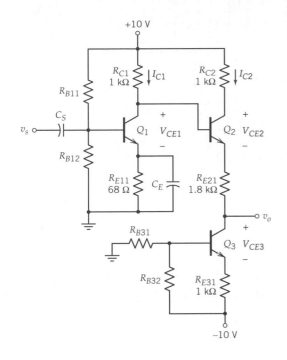

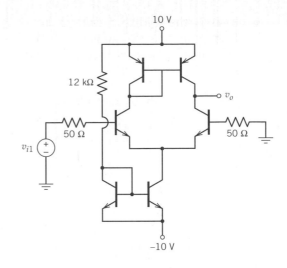

a. Find the quiescent point of the transistors

b. Find the midband voltage gain of the amplifier.

c. Confirm the results using SPICE.

6-26. Design a simple enhancement-mode NMOSFET current mirror for a load current of 50 μA. Use identical transistors with the parameters $K = 0.5$ mA/V^2 and $V_T = 2$ V. Confirm the analytical design using SPICE.

6-27. Design a simple depletion-mode NMOSFET current mirror for a load current of 30 μA. Use identical transistors with the parameters $I_{DSS} = 10$ mA and $V_{PO} = -4$ V. Confirm the analytical design using SPICE.

6-28. The emitter-coupled amplifier shown utilizes a current source to establish quiescent conditions and another to provide a high-resistance load. Assume BJTs with

$$\beta_F = 120, \qquad V_A = 250 \text{ V}.$$

a. Determine the voltage gain and the input resistance of the circuit.

b. Verify analytic results using SPICE.

6-29. The differential amplifier shown uses identical BJTs. The BJT parameters are

$$\beta_F = 200, \qquad V_A = 250 \text{ V}.$$

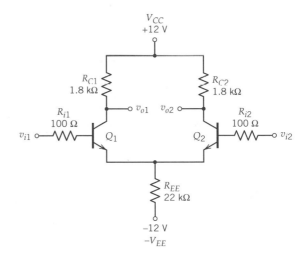

a. Determine the midband differential- and common-mode voltage gains.

b. Design a Wilson current mirror using three identical BJTs to replace R_{EE} using the ±12-V power rails. The BJTs for the current mirror are identical to the BJTs for the differential amplifier.

c. What is the midband differential- and common-mode voltage gains of the differential amplifier biased by the Wilson current mirror?

(F) **6-30.** Using appropriate analysis and complete explanations:

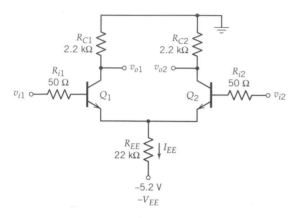

a. Find the quiescent condition for the differential amplifier shown. The transistors are identical with parameters $\beta_F = 200$, $V_\gamma = 0.7$ V, and $V_A = 250$ V.

b. How large can R_{C1} and R_{C2} become in this design if the bias current I_{EE} is to remain constant at the value found in part a and still remain a viable amplifier?

c. Design a simple two-transistor current mirror to replace R_{EE}.

d. What is the impact on the amplifier CMRR if R_{EE} is replaced by a simple current mirror in the differential amplifier design?

e. What are the design advantages to replacing R_{C1} and R_{C2} with an active load designed from a simple two-transistor *pnp* current mirror?

6-31. In order to boost the input resistance of a differential amplifier, resistors are often added between the emitters of the emitter-coupled BJT pair. For the circuit shown calculate the differential voltage gain, input resistance, and CMRR. Compare appropriate results with a circuit without these emitter resistors (Examples 6.3-1 and 6.3-2). Assume identical BJTs with

$\beta_F = 120$.

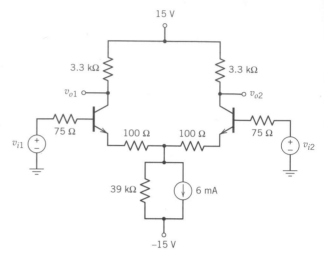

6-32. The Darlington differential amplifier shown is designed using identical transistors with

$$\beta_F = 120, \qquad V_A = 200 \text{ V}.$$

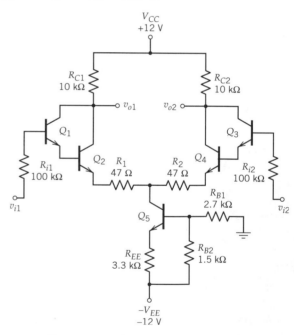

a. Determine the quiescent operating conditions.

b. Find the CMRR.

c. Determine the output voltage $v_{o2} - v_{o1}$.

(☀) **6-33.** Design a common emitter amplifier with an active load. The lower current limit specification is 50 μA. The following transistors are available:

- 2N2222A *npn*: IS = 14.4f, BF = 255, VA = 175
- 2N2907 *pnp*: IS = 650E-18, BF = 232, VA = 116

The power supply is +24 V.

    a. Find the input resistance of the amplifier.

    b. Find the small-signal midband gain.

    c. Confirm the analytical results with SPICE.

6-34. Design a common source amplifier with an active load using enhancement NMOSFETs so that the gain of the amplifier is -25 ± 5. The following transistors are available:

```
.model PMOSFET PMOS(VTO=2 KP=20E-6
LAMBDA=0.01)
.model NMOSFET NMOS(VTO=2 KP=206E-6
LAMBDA=0.01)
```

Confirm the analytical results with SPICE.

6-35. A differential *npn* BJT amplifier is biased by a 150-μA emitter bias source. The collector load resistors are mismatched by 5%. Find the offset voltage required at the input so that the differential output voltage is zero.

6-36. A differential amplifier is designed with two *npn* BJT transistors: one with $\beta_F = 120$ and the other with $\beta_F = 150$. If all other transistor parameters and external components are matched, determine the resulting input offset voltage.

6-37. Find the current gain stability factor S_β for the Wilson current source using matched BJTs.

6-38. Find the current gain stability factor S_β for the Widlar current source using matched BJTs.

6-39. A BJT Widlar current source is designed for a load current of 100 μA using identical transistors with $\beta_F = 150$ and $V_A = 150$ V. If the fabrication process guarantees a $\beta_F = 150 \pm 10$, what is the resulting maximum deviation in load current from the original design?

6-40. A BJT Wilson current source is designed for a load current of 100 μA using identical transistors with $\beta_F = 150$ and $V_A = 150$ V. If the fabrication process guarantees a $\beta_F = 150 \pm 10$, what is the resulting maximum deviation in load current from the original design? Compare with the results of the previous problem and comment on the benefits of each design.

6-41. Design a BJT differential amplifier that uses a Widlar current source that amplifies a differential input signal of 0.25 V to provide a differential output signal of 5 V with differential input resistance of greater than 50 kΩ. Limit the signal amplitude across each base–emitter junction to 5 mV to ensure linear operation. The BJTs are matched and have the following characteristics:

$$\beta_F = 200, \qquad V_A = 170 \text{ V}.$$

Confirm the analytical design with SPICE.

6-42. Consider the BJT differential amplifier in Example 6.3-1. Determine the largest input common-mode signal that can be applied to the amplifier with the BJTs remaining in the linear region of operation.

CHAPTER 7

POWER AMPLIFIERS AND OUTPUT STAGES

In general, analog amplifiers discussed in the previous chapters are described as small-signal circuits whose purpose is to increase the amplitude of the input signal or act as an impedance buffer between other amplifier stages. In reality, these amplifiers accept signals over a broad range of amplitudes. This chapter describes amplifiers (or amplification stages) that are capable of delivering high power levels at the output.

Power amplifiers must be capable of delivering a specific amount of power to a load, such as a stereo amplifier to a pair of speakers or a radio frequency amplifier to a broadcasting antenna. They are also used as output stages in integrated circuits. Because high output powers may be involved, the efficiency of the amplifier to convert a low-power signal to high power becomes increasingly important. Inefficiency causes unwanted increases in transistor operating temperature and may lead to accelerated device failure. Because power amplifier stages may deliver high output power to low-impedance loads, they are significantly different from the low-power small-signal amplifiers.

The large-signal nature of power amplifiers warrants special design consideration that may not be significant for small-signal amplifiers. Therefore, small-signal approximations and models either are not appropriate or must be used with care. In large-signal operation, signals transverse the extremes of the forward-active region of the transistor causing distortion at the output. That is, when a pure sinusoid is introduced to a power amplifier, the output may no longer be a pure sinusoid but a signal composed of the original sinusoid (at some amplitude determined by the amplification) and its Fourier (harmonic) components. The relative amplitude of the fundamental component decreases with respect to the Fourier components as distortion increases.

Distortion is defined in one of several ways depending on the particular application of the circuit. Measures of distortion include the following:

- Total harmonic distortion (THD), commonly used for audio circuits
- Intermodulation distortion (IMD) using the two-tone ratio method for radio and microwave frequency circuits
- Second- and third-order intercept points, also used for radio and microwave circuits
- Composite second-order (CSO) and composite third-order beat ratio (CTBR), used for cable television transmission circuits

Description of the THD and IMD methods will be discussed, and the relationship between the two measurements will be presented.

The design and analysis concepts introduced in this chapter are as follows:

- Analysis of distortion
- Design of large-signal (power) amplifiers with particular interest in reduction of distortion
- Thermal consideration that must be taken into account when designing power amplifiers

7.1 POWER AMPLIFIER CLASSIFICATION

In previous chapters, amplifiers were classified in terms of their circuit configurations (common-emitter, common-collector, common-base, common-drain, common-source, and common-gate). Another classification scheme is used when classifying power amplifiers. This scheme labels circuits according to the portion of the period of the output waveform during which the transistors conduct. The measure of designation is the conduction angle of each transistor in the circuit, assuming a sinusoidal input.

Regardless of the classification of the power amplifier, it must be capable of handling large-signal amplitudes where the current and voltage swings may be a significantly large fraction of the bias values. In such cases, small-signal analysis and models may not be appropriate and large-signal (DC) transfer characteristics must be used. The limitation on small-signal operation is discussed in this section.

7.1.1 Classification Scheme

The amplifier classifications by conduction angle for sinusoidal inputs are shown in Table 7.1-1. The output (collector or drain) current through a transistor for class A, B, AB, and C power amplifiers are shown in Figure 7.1-1. In Figure 7.1-1a, the current flows through the transistor over the whole period: This is classified as class A operation. In class B operation, shown in Figure 7.1-1b, each transistor only conducts over half the period of the input sinusoidal waveform. Class AB operation, shown in Figure 7.1-1c, illustrates current flow through each transistor for greater than a half cycle but less than the full cycle of the input sinusoid. Figure 7.1-1d shows class C operation where each transistor conducts over less than half cycle of the input sinusoid. Class D operation is not shown since the conduction angle could vary with time over the entire cycle.

Initially, it may seem as if class A operation is the only configuration that will yield low distortion signals. Class B and class AB power amplifiers assure signal continuity by making use of arrangements of two transistors that allow each tran-

TABLE 7.1-1 AMPLIFIER CLASSIFICATIONS

Amplifier Class	Individual Transistor Conduction Angle
A	360°
B	180°
AB	180–360°
C	Fixed drive, $<180°$
D	Switched operation, conduction angle may vary with time from 0° to 360° or may be fixed

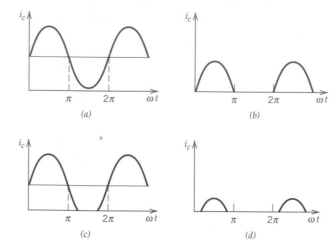

Figure 7.1-1

Transistor current for various amplifier classes: (*a*) class A amplifier, full-period current flow; (*b*) class B amplifier, half-period current flow; (*c*) class AB amplifier, greater-than-half-period current flow; (*d*) class C amplifier, less-than-half-period current flow.

sistor to share portions of the input signal conduction angle. Class C amplifiers provide single-frequency sinusoidal output by driving resonant circuits over a small portion of the cycle; the continuity of the sinusoidal output is assured by the tuned circuit.[1] In class D operation, there are two inputs, the input signal and a sampling square wave or pulse-width-modulated signal. In essence, the sampling signal forces the transistors to either turn on or off over the interval of the sampling wave, yielding a sampled version of the input signal. This sampled signal is then filtered to yield the desired waveform.

Classes A, B, and AB are studied in some detail in Sections 7.2, 7.3, and 7.4, respectively. Class C and D amplifiers are not discussed since their analysis is beyond the scope of this book.

7.1.2 Limits on Distortionless Small-Signal Operation

To investigate the limitations of analyses using "small-signal" models, the limiting mechanisms must be understood. Graphical method load line analysis demonstrates

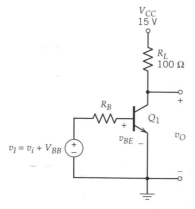

Figure 7.1-2

Common-emitter power amplifier.

---

[1] Class C amplifiers are used for narrow-band signal applications. The focus of this section is on broadband amplifiers.

the large-signal nature of power amplifiers. The power amplifier is assumed to operate with large current and voltage swings that may be a significant fraction of its quiescent current and voltage.

As an example, consider a simple common-emitter power amplifier as shown in Figure 7.1-2.[2] The input and output characteristics are shown in Figure 7.1-3.

The common emitter power amplifier shown in Figure 7.1-3 has the following

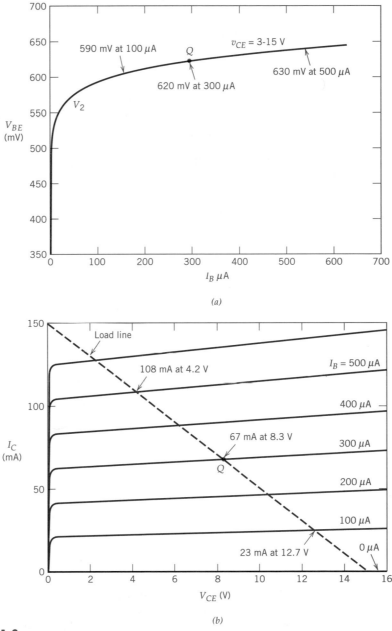

(a)

(b)

Figure 7.1-3
Characteristics of common-emitter power amplifier of Figure 7.1-2: (*a*) input Characteristics, (*b*) output characteristics.

---

[2] This analysis is for demonstration purposes only. Analysis of other BJT amplifier configurations and FET amplifier configurations can be performed in a similar fashion.

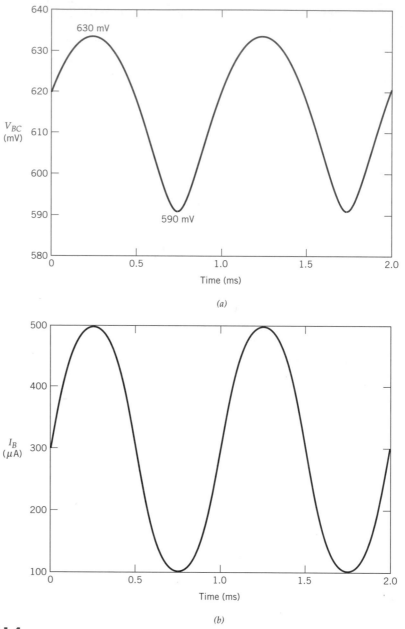

Figure 7.1-4

Voltage and current waveforms for power amplifier of Figure 7.1-2: (*a*) base–emitter voltage; (*b*) base current; (*c*) collector current; (*d*) collector-emitter voltage. Input base current is a pure sinusoid. Other waveforms are distorted.

quiescent conditions: I_{BQ} = 300 mA, V_{BEQ} = 0.62 V, I_{CQ} = 67 mA, and V_{CEQ} = 8.3 V. For a large swing on the base current of ±200 μA, the base–emitter voltage swings from 0.59 to 0.63 V as shown in the input characteristics. The result is an asymmetric base–emitter voltage excursion about the operating point of −0.03 and +0.01 V. The output characteristics shows a corresponding swing in collector currents of 23–108 mA; resulting in an asymmetric current excursion about the operation point of −44 and +41 mA.

The voltage and current waveforms can be obtained from the dynamic operating

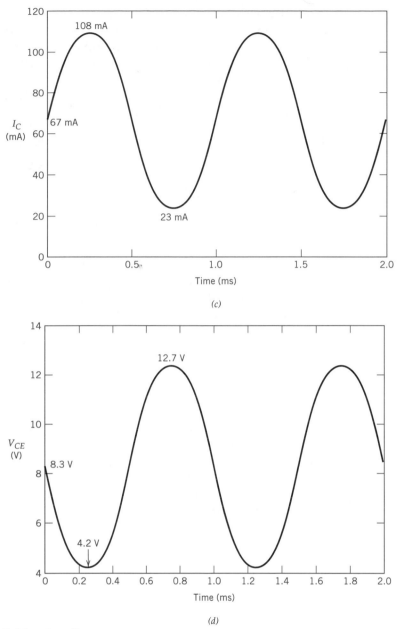

Figure 7.1-4 (continued)

curves of Figure 7.1-3 and is shown in Figure 7.1-4. For a pure sinusoidal input base current, the base–emitter voltage, collector current, and collector–emitter voltage waveforms are shown to be suffering from varying degrees of distortion. From the output characteristics, the collector–emitter voltage waveform has a quiescent value of 8.3 V and has a voltage swing from 4.2 to 12.7 V. The output excursions about the operating point are −4.1 and +4.4 V, which is clearly asymmetrical.

For small-signal operation, the input signal is small, which yields negligible distortion of the output waveform. In power amplifier, however, the output signals are large and potentially distorted. Therefore, simple linear small-signal analysis cannot necessarily be used effectively. Instead, large-signal transfer characteristics are best used to determine the operating characteristics of the power amplifier.

The limit of distortionless small-signal analysis is difficult to define since it is dependent on amplifier distortion specifications. However, some sense of the limitation of small-signal analysis can be found by determining the relationship between the base–emitter voltage swing and the output current.

Assume that the base–emitter voltage is

$$v_{BE} = v_{be} + V_{BEQ}, \qquad (7.1\text{-}1)$$

where
$$V_{BEQ} = \text{quiescent base–emitter voltage (DC)}$$
$$v_{be} = \text{AC component of base–emitter voltage,}$$
$$= V_1 \cos \omega t$$

From the Ebers–Moll equations 3.1-3, the emitter current is

$$I_E = -I_{ES}(e^{v_{BE}/\eta V_t} - 1) + \alpha_R I_{CS}(e^{V_{BC}/\eta V_t} - 1), \qquad (7.1\text{-}2)$$

where the base–emitter voltage is defined by Equation 7.1-1.

For a strongly forward-biased transistor, $V_{BE} \gg \eta V_t$ and $V_{BC} \ll -\eta V_t$. Equation 7.1-2 can be simplified to

$$I_E \approx -I_{ES} e^{v_{BE}/\eta V_t} = -I_{ES} e^{V_{BE}/\eta V_t} e^{V_1 \cos \omega t/\eta V_t}$$
$$= -I_{ES} e^{V_{BE}/\eta V_t} e^{x \cos \omega t}, \qquad (7.1\text{-}3)$$

where $x = V_1/\eta V_t$ to normalize the drive voltage.

It is apparent from Equation 7.1-3 that the emitter current and, in turn, the collector current of the transistor in Figure 7.1-2 are proportional to the normalized ratio of $e^{x \cos \omega t}/e^x$ for any fixed value of x. Figure 7.1-5 shows the normalized emitter current (proportional to the collector current) as a function of time over two cycles of the variation in the base–emitter voltage. The plot clearly shows that by the time $x = 10$, the emitter current is clearly distorted. Small-signal analysis can still be used under these circumstances but will not identify the distortion contained in the output signal.

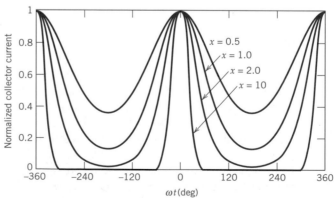

Figure 7.1-5
Normalized emitter currents vs. angle for varying base–emitter voltages.

7.2 **CLASS A POWER AMPLIFIERS**

7.2.1 **Common Collector**

Consider a common collector power amplifier driven by a voltage V_i, shown in Figure 7.2-1. The linear BJT models of Figure 3.4-1 for the cutoff, forward-active, and saturation regions are used to find the large-signal transfer function for the common collector power amplifier.

The power amplifier in Figure 7.2-1 is redrawn using the equivalent models for the cutoff, forward-active, and saturation regions in Figure 7.2-2.

The transistor Q_1 is cut off when $V_{BE} < V_{BE(\text{on})}$. No current flows through Q_1, effectively creating an open circuit at the emitter, base, and collector nodes as shown in Figure 7.2-2a. Therefore, $V_O = 0$ until the transistor turns on with $V_i > V_{BE(\text{on})}$. Recall that a typical value of $V_{BE(\text{on})}$ for silicon transistors is 0.6 V.

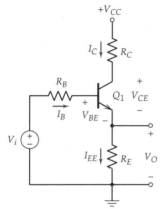

Figure 7.2-1
Common-collector power amplifier.

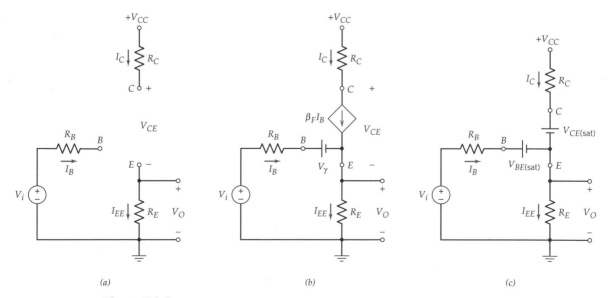

(a)   *(b)*   *(c)*

Figure 7.2-2
Equivalent models for (*a*) cutoff, (*b*) forward-active, and (*c*) saturation regions for common-emitter power amplifier in Figure 7.2-1.

When Q_1 turns on, the transistor operates in the forward-active region and the amplifier is modeled as shown in Figure 7.2-2b. The output voltage is

$$V_O = I_{EE}R_E. \tag{7.2-1}$$

where $I_{EE} = -I_E$.

The base–emitter loop equation is written to find I_{EE} in terms of V_i,

$$
\begin{aligned}
0 &= V_i - I_B R_B - V_\gamma - I_{EE}R_E \\
&= V_i - \frac{I_{EE}}{\beta_F + 1}R_B - V_\gamma - I_{EE}R_E.
\end{aligned} \tag{7.2-2}
$$

Solving for I_{EE} yields

$$I_{EE} = \frac{(\beta_F + 1)(V_i - V_\gamma)}{R_B + (\beta_F + 1)R_E}. \tag{7.2-3}$$

The transfer function is found by substituting Equation 7.2-3 into Equation 7.2-1:

$$V_O = I_{EE}R_E = \frac{(\beta_F + 1)(V_i - V_\gamma)}{R_B + (\beta_F + 1)R_E}R_E, \tag{7.2-4}$$

where the slope of the transfer function in the forward-active region is $(\beta_F + 1)R_E/[R_B + (\beta_F + 1)R_E]$.

In saturation, the amplifier is modeled as shown in Figure 7.2-2c. The typical collector–emitter voltage is $V_{CE(\text{sat})} = 0.2$ V and the base–emitter voltage is $V_{BE(\text{sat})} = 0.8$ V. It is useful to redraw the circuit in Figure 7.2-2c to analyze this case, as shown in Figure 7.2-3.

The transfer function for the saturation model of the common-collector amplifier is easily found by applying the superposition principle to Figure 7.2-3,

$$
\begin{aligned}
V_O &= \frac{V_i(R_E \parallel R_C)}{R_B + (R_E \parallel R_C)} - \frac{V_{BE(\text{sat})}(R_E \parallel R_C)}{R_B + (R_E \parallel R_C)} \\
&\quad - \frac{V_{CE(\text{sat})}(R_B \parallel R_E)}{R_C + (R_B \parallel R_E)} + \frac{V_{CC}(R_B \parallel R_E)}{R_C + (R_B \parallel R_E)} \\
&= \frac{(V_i - V_{BE(\text{sat})})(R_E \parallel R_C)}{R_B + (R_E \parallel R_C)} + \frac{(V_{CC} - V_{CE(\text{sat})})R_B \parallel R_E)}{R_C + (R_B \parallel R_E)}.
\end{aligned} \tag{7.2-5}
$$

The large-signal gain V_o/V_i is proportional to $(R_E \parallel R_C/[R_B + (R_E \parallel R_C)])$.

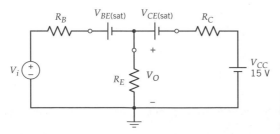

Figure 7.2-3
Redrawn saturation model of power amplifier in Figure 7.2-2c.

The boundary between the forward-active and cutoff regions can be found by solving the collector–emitter loop equation using the forward-active model of the circuit:

$$V_{CE(sat)} = V_{CC} - I_C R_C - I_{EE} R_E. \tag{7.2-6}$$

But

$$I_C = \frac{\beta_F}{\beta_F + 1} I_{EE}, \qquad I_{EE} = (\beta_F + 1)I_B, \text{ so,}$$

$$V_{CE(sat)} = V_{CC} - I_{EE}\left(\frac{\beta_F}{\beta_F + 1} R_C + R_E\right). \tag{7.2-7}$$

The value for V_i at the boundary between the forward-active and saturation regions is found by substituting Equation 7.2-3 into Equation 7.2-7 and rearranging the equation,

$$V_{i(f\text{-}a/sat)} = \frac{(V_{CC} - V_{CE(sat)})[R_B + (\beta_F + 1)R_E]}{(\beta_F + 1)R_E + \beta_F R_C} + V_{BE(sat)}. \tag{7.2-8}$$

To find the output voltage at the forward-active and saturation region boundary, substitute Equation 7.2-8 into Equation 7.2-4 to yield

$$V_{O(f\text{-}a/sat)} = \frac{(\beta_F + 1)(V_{CC} - V_{CE(sat)})}{(\beta_F + 1)R_E + \beta_F R_C} R_E. \tag{7.2-9}$$

A typical transfer characteristic of the circuit in Figure 7.2-1 is shown in Figure 7.2-4.

The transfer characteristic in Figure 7.2-4 indicates that when the input voltage exceeds $V_{i(f\text{-}a/sat)}$, then the output voltage of the common-collector power amplifier will begin to "clip" and distort the output. The output will also clip when the input voltage is less than V_{BE}.

If the common-collector amplifier is operating in the "linear" region of the large-signal transfer characteristic, it may be assumed that there is little nonlinear distortion. In this case, the power and efficiency calculations are straightforward.

Recall that because power amplifiers may be required to deliver large currents and voltages at the output, the conversion efficiency of the circuit is of interest. That

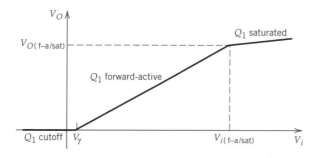

Figure 7.2-4
Transfer characteristic of common-collector power amplifier in Figure 7.2-1.

is, higher efficiency power amplifiers will produce higher output power signals at the same amount of applied DC power than lower efficiency power amplifiers.

The DC power that the amplifier requires from the power supply is

$$P_{DC} = V_{CC}I_{CQ}. \tag{7.2-10}$$

The AC power output of the common-collector amplifier is found by multiplying the RMS values of the load current and voltage in terms of the peak values,

$$P_{AC} = \frac{I_{EP}}{\sqrt{2}} \frac{V_{EP}}{\sqrt{2}} = \frac{I_{EP}^2 R_E}{2} = \frac{V_{EP}^2}{2R_E} \tag{7.2-11}$$

where I_{EP} = peak emitter current

V_{EP} = peak excursion from quiescent voltage

and $V_{EP} = V_{O(\text{f-a/sat})}$ and $I_{EP} = V_{O(\text{f-a/sat})/R_E}$ referred to Figures 7.2-1 and 7.2-4.

The rms values of the peak emitter current and voltage are found from their maximum and minimum values,

$$\frac{I_{EP}}{\sqrt{2}} = \frac{1}{\sqrt{2}} \frac{I_{E,\max} - I_{E,\min}}{2}, \tag{7.2-12}$$

$$\frac{V_{EP}}{\sqrt{2}} = \frac{1}{\sqrt{2}} \frac{V_{E,\max} - V_{E,\min}}{2}. \tag{7.2-13}$$

The peak values of the emitter current and voltage are shown in Figure 7.2-5.

The AC power in terms of the peak current and voltage values are

$$P_{AC} = \tfrac{1}{2}(I_{EP}V_{EP}) = \tfrac{1}{8}(I_{E,\max} - I_{E,\min})(V_{E,\max} - V_{E,\min}). \tag{7.2-14}$$

The conversion efficiency of the applied DC power to the total AC power is defined as

$$\eta = \frac{\text{AC power delivered to load}}{\text{DC power supplied}} \times 100\%$$

$$= \frac{P_{AC}}{P_{DC}} \times 100\%. \tag{7.2-15}$$

Substituting Equations 7.2-14 and 7.2-10 into 7.2-15 yields an expression for the conversion efficiency of the common-collector power amplifier in terms of the maximum and minimum values of the output current and voltage,

$$\eta = \frac{(I_{E,\max} - I_{E,\min})(V_{E,\max} - V_{E,\min})}{8V_{CC}I_{CQ}} \times 100\%. \tag{7.2-16}$$

For *maximum* attainable ideal efficiency for the common-collector power amplifier configuration, the output current and voltage is set to its extremes. That is,

$$I_{E,\min} = 0, \qquad I_{E,\max} = 2I_{EQ},$$

$$V_{E,\min} = 0, \qquad V_{E,\max} = V_{CC} - V_{CE(\text{sat})}$$

for $V_{CC} - V_C = 0$.

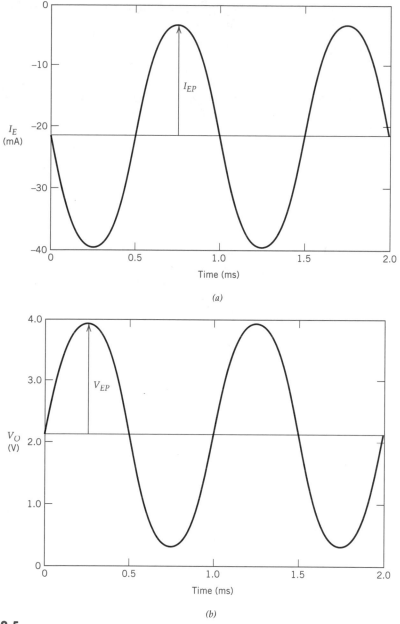

Figure 7.2-5
Peak values of (*a*) emitter current and (*b*) emitter (load) voltage derived from current and voltage waveforms.

The maximum attainable efficiency is, for large β,

$$\eta_{max} = \frac{P_{AC}}{P_{DC}} \times 100\%$$

$$= \frac{I_{EP}V_{EP}}{2(V_{CC}I_{CQ})} \times 100\% \qquad (7.2\text{-}17)$$

$$= \frac{(2I_{EQ}/2)(V_{CC} - V_{CE(sat)})/2}{2(V_{CC}I_{CQ})} \times 100\% \approx 25\%,$$

where $V_{CC} \gg V_{CE(sat)}$. Naturally, the power amplifier shown in Figure 7.2-1 will never be able to reach an efficiency of 25% for finite values of R_C, which lowers the peak output current and voltage.

EXAMPLE 7.2-1   The common-collector power amplifier shown in Figure 7.2-1 has the following circuit element and transistor parameter values:

$$V_{CC} = 15 \text{ V}, \qquad R_C = 100 \text{ } \Omega, \qquad R_E = 100 \text{ } \Omega,$$

$$R_B = 10 \text{ k}\Omega, \qquad \beta_F = 160.$$

Determine the efficiency of the power amplifier assuming no additional losses due to thermal effects.

SOLUTION

From Equation 7.2-15 the efficiency is

$$\eta = \frac{P_{AC}}{P_{DC}} \times 100\%.$$

But from Equation 7.2-11,

$$P_{AC} = \frac{V_{EP}^2}{2R_E} = \frac{1}{2R_E}\left(\frac{V_{O(f\text{-a}/sat)} - 0}{2}\right)^2 = \frac{V_{O(f\text{-a}/sat)}^2}{8R_E}.$$

The power delivered by the power supply is related to the peak output voltage by

$$P_{DC} = V_{CC}I_{CQ} = V_{CC}\left(\frac{\beta_F}{\beta_F + 1} \frac{V_{O(f\text{-a}/sat)}}{2R_E}\right).$$

The efficiency is then

$$\eta = \frac{V_{O(f\text{-a}/sat)}^2/8R_E}{V_{CC}\{[\beta_F/(\beta_F + 1)](V_{O(f\text{-a}/sat)}/2R_E)\}} = \frac{\beta_F + 1}{\beta_F} \frac{V_{O(f\text{-a}/sat)}}{4V_{CC}}.$$

Substituting Equation 7.2-9 for the maximum output voltage into the expression for efficiency above yields

$$\eta = \frac{\beta_F + 1}{\beta_F} \frac{\{(\beta_F + 1)(V_{CC} - V_{CE(sat)})/[\beta_F + 1]R_E + \beta_F R_C]\}R_E}{4V_{CC}}$$

$$= \left[\frac{161}{160} \frac{161(15 - 0.2)(100)/(161)(100) + (160)(100)}{(4)(15)}\right] \times 100\%$$

$$= 12.5\%.$$

As expected, the efficiency for the common-collector power amplifier of Figure 7.2-1 with the circuit values of this example is less than the ideal maximum efficiency of 25%.

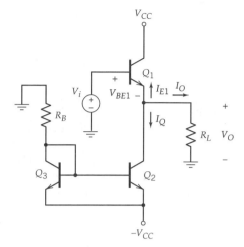

Figure 7.2-6

Common-collector power amplifier biased by current mirror (formed by Q_2, Q_3, and R_B).

Integrated circuits also use power output stages for their low distortion and low output resistance characteristic. Consider the common-collector circuit biased by a current mirror shown in Figure 7.2-6. Like its discrete counterpart, when this circuit is to be used as the output amplifier stage in an integrated circuit, it must be capable of handling large-signal amplitudes where the current and voltages swings may be a significantly large fraction of the bias values. In such cases, small-signal analysis and models may not be appropriate and large signal (DC) transfer characteristics must be used.

In the circuit of Figure 7.2-6, Q_2, Q_3, and R_B forms a current mirror to bias the common-collector circuit formed by Q_1. The large-signal transfer characteristic is

$$V_O = V_i - V_{BE1}, \tag{7.2-18}$$

where $V_O = I_O R_L$. Using the simplified Ebers–Moll emitter current expression in Equation 3.1-3 when Q_1 is forward biased,

$$I_{E1} \approx -I_{ES}(e^{V_{BE1}/\eta V_t} - 1) \approx -I_{ES}e^{V_{BE1}/\eta V_t}. \tag{7.2-19}$$

But

$$-I_{E1} = I_Q + \frac{V_O}{R_L} \tag{7.2-20}$$

when Q_2 is in the forward-active region.

Solving for V_{BE1} by substituting Equation 7.2-20 into Equation 7.2-19 yields

$$V_{BE1} = \eta V_t \ln \frac{I_Q + V_O/R_L}{I_{ES}}. \tag{7.2-21}$$

By substituting Equation 7.2-21 into Equation 7.2-18, a non-linear equation for the transfer function is found under the assumption that the load resistor R_L is small when compared to the output resistance of the transistors,

$$V_O = V_i - \eta V_t \ln \frac{I_Q + V_O/R_L}{I_{ES}}, \tag{7.2-22}$$

or

$$V_i = \eta V_t \ln \frac{I_Q + V_O/R_L}{I_{ES}} + V_O. \tag{7.2-23}$$

Equation 7.2-23 is a transfer function for the common-collector power amplifier, and its transfer characteristics are plotted in Figure 7.2-7 for $\eta = 1$.

Referring to the transfer characteristic shown in Figure 7.2-7, consider the case when the load resistor is large (designated $R_{L \text{ large}}$). When the load resistor is large, the natural logarithm in Equation 7.2-23 remains relatively constant with changing V_O. Physically, this means that for large load resistors, the current in the load is small and that $I_{E1} \approx I_Q$. This also implies that V_{BE1} is approximately constant. Therefore, when both Q_1 and Q_2 are in the forward-active region, the transfer characteristic for $R_{L \text{ large}}$ is nearly a straight line offset by the quiescent base–emitter voltage V_{BEQ1} on the V_i axis.

As V_i is made a large positive input, the Q_1 collector–base junction is forward biased and saturates the transistor so that the output is

$$V_{O(\text{max})} = V_{CC} - V_{CE1(\text{sat})}, \tag{7.2-24}$$

where the input voltage is

$$V_{i(\text{max})} = V_{CC} - V_{CE1(\text{sat})} + V_{BE1}. \tag{7.2-25}$$

When V_i is made large and negative, Q_2 saturates and the output voltage is

$$V_{O(\text{min})} = -V_{CC} + V_{CE2(\text{sat})}, \tag{7.2-26}$$

where the input voltage is

$$V_{i(\text{max})} = -V_{CC} + V_{CE2(\text{sat})} + V_{BE1}. \tag{7.2-27}$$

For a large and negative input voltage with small values of the load resistance, $R_{L \text{ small}}$, the natural logarithm in Equation 7.2-23 approaches negative infinity at

$$V_O = -I_Q R_{L \text{ small}}. \tag{7.2-28}$$

In this case, the load current is equal to I_Q. Therefore, no current flows in Q_1, sending it into cutoff. Then V_O no longer increases with V_i.

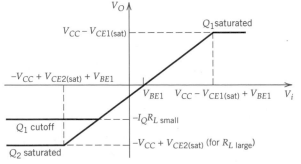

Figure 7.2-7

Transfer characteristic of common-collector power amplifier of Figure 7.2-6.

For a large positive input voltage with $R_{L\text{ small}}$, the transfer characteristic is similar to the case for $R_{L\text{ large}}$.

The maximum efficiency of the current source biased common-collector output stage is also 25%.

7.2.2 Common-Emitter

Common-emitter power amplifier stages are frequently used as output stage drivers of multitransistor amplifiers. However, common-emitter output stages are not often used in integrated circuit design because of the superior characteristics of the common-collector stages (low output resistance and low distortion).

Consider a common-collector power amplifier driven by a voltage V_i, shown in Figure 7.2-7. As in the common-collector power amplifier analysis, the linear BJT models of Figure 3.4-1 for the cutoff, forward-active, and saturation regions are used to find the large-signal transfer function for the common-collector power amplifier.

The common-emitter power amplifier in Figure 7.2-8 is analyzed in the same way as the common-collector configuration using the equivalent models for the cutoff, forward-active, and saturation regions.

Transistor Q_1 does not conduct until the base–emitter voltage is $V_{BE\text{(on)}} \approx 0.6$ V. Since Q_1 is in cutoff, the V_i must exceed $V_{BE\text{(on)}}$ for the transistor to enter the forward-active region. The output voltage at this point in the transfer characteristic is $V_O = V_{CC}$.

In the forward-active region, the output voltage is

$$V_O = V_{CC} - \frac{R_C(V_1 - V_\gamma)}{R_B/\beta_F + [\beta_F/(\beta_F + 1)]R_E}. \tag{7.2-29}$$

In the saturation region, the output voltage of the common-emitter is

$$V_O = \frac{(V_i - V_{BE\text{(sat)}})(R_E \parallel R_C)}{R_B + (R_E \parallel R_C)} \\ + \frac{V_{CE\text{(sat)}}R_C + V_{CC}(R_E \parallel R_B)}{R_C + (R_E \parallel R_B)}. \tag{7.2-30}$$

The transfer characteristic is shown in Figure 7.2-9.

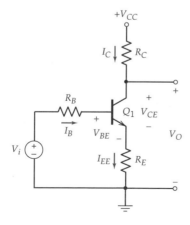

Figure 7.2-8
Common-emitter power amplifier.

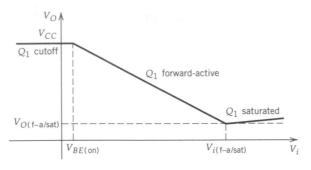

Figure 7.2-9

Transfer characteristic of common-emitter power amplifier in Figure 7.2-8.

The AC power output of the common-emitter amplifier is found by multiplying RMS values of the load current and voltage in terms of the peak values,

$$P_{AC} = \frac{I_{CP}}{\sqrt{2}} \frac{V_{CP}}{\sqrt{2}} = \frac{I_{CP}^2 R_C}{2} = \frac{V_{CP}^2}{2R_C}, \quad (7.2\text{-}31)$$

where

$$I_{CP} = \text{peak collector current}$$

$$V_{CP} = \text{peak voltage from collector to ground}$$

and $V_{CP} = V_{CC}$ and $I_{CP} = (V_{CC} - V_{O(f\text{-}a/sat)})/R_C$; refer to Figure 7.2-9.

The RMS values of the peak emitter current and voltage are found from their maximum and minimum values,

$$\frac{I_{CP}}{\sqrt{2}} = \frac{1}{\sqrt{2}} \frac{I_{C,max} - I_{C,min}}{2}, \quad (7.2\text{-}32)$$

$$\frac{V_{CP}}{\sqrt{2}} = \frac{1}{\sqrt{2}} \frac{V_{C,max} - V_{C,min}}{2}. \quad (7.2\text{-}33)$$

The AC power in terms of the peak current and voltage values are

$$P_{AC} = \frac{I_{CP}V_{CP}}{2} = \frac{(I_{C,max} - I_{C,min})(V_{C,max} - V_{C,min})}{8}. \quad (7.2\text{-}34)$$

Substituting Equations 7.2-34 into Equation 7.2-15 yields an expression for the conversion efficiency of the common-emitter power amplifier in terms of the maximum and minimum values of the output current and voltage,

$$\eta = \frac{(I_{C,max} - I_{C,min})(V_{C,max} - V_{C,min})}{8V_{CC}I_{CQ}} \times 100\%. \quad (7.2\text{-}35)$$

For *maximum* attainable ideal efficiency for the common-emitter power amplifier configuration, the output current and voltage is set to its extremes. That is,

$$I_{C,min} = 0, \qquad I_{C,max} = 2I_{CQ},$$

$$V_{C,min} = V_{CE(sat)}, \qquad V_{C,max} = V_{CC}$$

for $V_E = 0$.

The maximum attainable efficiency is, for large β,

$$
\begin{aligned}
\eta_{max} &= \frac{P_{AC}}{P_{DC}} \times 100\% \\
&= \frac{I_{CP}V_{CP}}{2(V_{CC}I_{CQ})} \times 100\% \\
&= \frac{(2I_{CQ}/2)[(V_{CC} - V_{CE(sat)})]}{2(V_{CC}I_{CQ})} \times 100\% \approx 25\%,
\end{aligned}
\tag{7.2-36}
$$

where $V_{CC} \gg V_{CE(sat)}$.

Naturally, the power amplifier shown in Figure 7.2-8 will never be able to reach an efficiency of 25% for finite values of R_E, which lowers the peak output current and voltage.

A common-emitter integrated circuit power output stage is shown in Figure 7.2-10. The transistor Q_2 is part of a constant current source that establishes the Q point for transistor Q_1. The large-signal characteristic is derived by first inspecting the load resistance node,

$$
I_O = I_Q - I_{C1}.
\tag{7.2-37}
$$

The output voltage V_O is given as

$$
V_O = I_O R_L.
\tag{7.2-38}
$$

The collector current I_{C1} for Q_1 assuming operation in the forward-active region is found using the Ebers–Moll equation 3.1-3,

$$
I_{C1} \approx \alpha_F I_{ES} e^{V_{BE1}/\eta V_t} = \frac{\beta_F}{\beta_F + 1} I_{ES} e^{V_{BE1}/\eta V_t}.
\tag{7.2-39}
$$

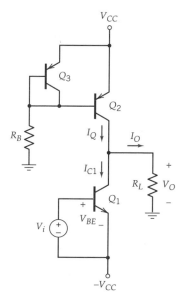

Figure 7.2-10
Common-emitter power amplifier biased by current mirror (formed by Q_2, Q_3, and R_B).

But

$$I_{C1} = I_Q - \frac{V_O}{R_L} \tag{7.2-40}$$

when Q_2 is in the forward-active region.

Substituting Equation 7.2-40 into Equation 7.2-39 and knowing that $V_{BE1} = V_i$ yields

$$\frac{\beta_F}{\beta_F + 1} I_{ES} e^{V_{BE1}/\eta V_t} = I_Q - \frac{V_O}{R_L}. \tag{7.2-41}$$

The output voltage is

$$V_O = -R_L \left(\frac{\beta_F}{\beta_F + 1} I_{ES} e^{V_{BE1}/\eta V_t} - I_Q \right). \tag{7.2-42}$$

If R_L is small, much of I_Q flows through the load when V_i is reduced. With V_i small or negative, the first term in the parentheses in Equation 7.2-42 becomes negligible. Therefore, for $R_{L\ \text{small}}$, the output voltage is

$$V_O = I_Q R_{L\ \text{small}}. \tag{7.2-43a}$$

If R_L is large, V_O increases with decreasing V_i until Q_2 saturates. Therefore for $R_{L\ \text{large}}$,

$$V_O = V_{CC} - V_{CE2(\text{sat})}. \tag{7.2-43b}$$

For either small or large load resistance, as V_i increases and become more positive, I_Q increases and V_O becomes negative until Q_1 saturates.

Equation 7.2-42 is a non-linear equation for the transfer function of the common-emitter amplifier biased by a current source in Figure 7.2-10 and is plotted in Figure 7.2-11 for $\eta = 1$.

From Equation 7.2-42, the transfer function is an exponential. Therefore, the transfer characteristics shows a curvature between the extreme output voltages. This causes distortion of the output signal that is significantly more pronounced than that of the common-collector amplifier of Figure 7.2-6. As in the common-collector power amplifier in Figure 7.2-6, the upper limit of the transfer function depends on the size of R_L.

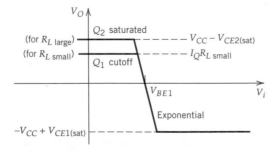

Figure 7.2-11
Transfer characteristic of common-emitter power amplifier of Figure 7.2-10.

The maximum efficiency of the current source biased common-emitter output stage is also 25%.

7.2.3 Transformer-Coupled Class A Power Amplifier

If a load is connected directly to the transistor as in the common-emitter power amplifier in Figure 7.2-8, the quiescent current must pass through the load. The load increases the required DC power and reduces efficiency since the quiescent current through the load does not contribute to the AC signal component. To circumvent this problem, the load is commonly transformer coupled to the transistor as shown in Figure 7.2-12.

The transformer is used as an impedance-matching element. To transfer a significant amount of power to a low-impedance load, such as a the voice coil of a loudspeaker, which is typically between 4 and 15 Ω, it is necessary to use an output matching transformer.

For an ideal transformer, the AC voltage–current relations are

$$v_1 = \frac{n_1}{n_2} v_2, \qquad i_1 = \frac{n_2}{n_1} i_2, \tag{7.2-44}$$

where v_1 and v_2 are the voltages across the primary and secondary windings, respectively, and i_1 and i_2 are the currents through the primary and secondary windings, respectively. The terms n_1 and n_2 are the number of turns in the primary and secondary, respectively, of the transformer.

Using Equation 7.2-44, the effective input resistance looking into the primary winding is

$$R_L' = \frac{v_1}{i_1} = \left(\frac{n_1}{n_2}\right)^2 R_L. \tag{7.2-45}$$

The DC and AC load lines for the circuit in Figure 7.2-12 are shown on the output transistor characteristics in Figure 7.2-13.

The DC load line is nearly vertical due to the very small primary resistance in

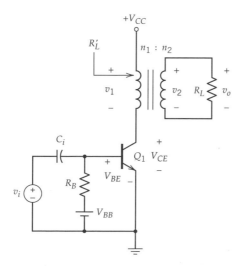

Figure 7.2-12
Class A power amplifier with transformer-coupled output.

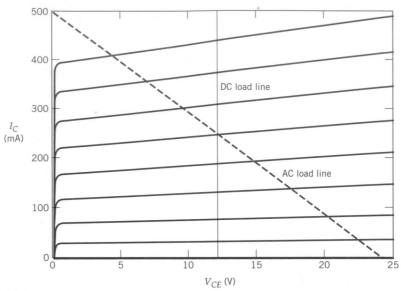

Figure 7.2-13
Output characteristic of transformer-coupled class A power amplifier shown in Figure 7.2-12.

the transformer. If, however, a emitter resistor is present, the DC load line will have a slope of $\dfrac{1}{R_E}$. The maximum peak-to-peak out voltage is $2V_{CC}$, and the maximum peak-to-peak output current is $2I_C$. Therefore, the maximum peak-to-peak AC power is

$$P_{AC} = \tfrac{1}{2}V_P I_P = \tfrac{1}{2}V_{CC}I_C. \tag{7.2-46}$$

The maximum possible efficiency for a transformer-coupled class A power amplifier is

$$\eta_{\max} = \frac{P_{AC}}{P_{DC}} \times 100\% = \frac{V_{CC}I_C}{2V_{CC}I_C} \times 100\% = 50\%, \tag{7.2-47}$$

which is double the maximum efficiency of class A amplifiers that directly drives a load resistor. Naturally, the efficiency of real transformer-coupled class A power amplifiers will be lower than 50% due to additional losses (e.g., transformer losses).

EXAMPLE 7.2-2 Using graphical techniques, design a transformer-coupled class A transistor amplifying stage to meet the following requirements:

Load resistance	2 kΩ
Output transformer efficiency	67% (it is a *bad* transformer)
Power output	1 W (maximum, sinusoidal)
Temperature	27°C
Supply voltage	12 V
Frequency	400 Hz

A 2N6474 *npn* transistor is available with the following SPICE characteristic parameters:

$$IS = 2.45 \text{ pA}, \quad BF = 208, \quad BR = 13,$$

$$XTI = 3, \quad XTB = 1.5, \quad VAF = 100.$$

SOLUTION

From the specifications, the class A amplifier may be assumed to enter the saturation and cutoff regions of the transistor characteristics for power delivery in excess of 1 W of full-load power (P_{FL}).

Since the transformer efficiency is 67%, the amplifier stage must supply

$$P_{AC} = \frac{P_{FL}}{\eta_{xfmr}} = \frac{1.0}{0.67} = 1.5 \text{ W}.$$

Assume that the amplifier is configured identically to Figure 7.2-12.

The peak-to-peak collector current can be found by applying the expression for the AC power delivered and by considering the saturation region of the transistor:

$$P_{AC} = 1.5 = \frac{V_P}{\sqrt{2}} \frac{I_P}{\sqrt{2}} \frac{V_{CC} - \frac{1}{2}V_{CE(sat)}}{\sqrt{2}} \frac{2I_C}{2\sqrt{2}} = \frac{I_C(V_{CC} - \frac{1}{2}V_{CE(sat)})}{2}.$$

Rearranging the equation above, solve for the maximum (peak-to-peak) collector current,

$$I_{C,max} = 2I_C = \frac{4P_{AC}}{V_{CC} - \frac{1}{2}V_{CE(sat)}} = \frac{4(1.5)}{12 - 0.1} \approx 0.5 \text{ A}.$$

The peak-to-peak output voltage is

$$V_{C,max,p\text{-}p} = 2V_{CC} - V_{CE(sat)}.$$

Therefore, the load seen by the transistor is

$$R'_L = \frac{V_{C,max,p\text{-}p}}{I_{C,max}} = \frac{2(12) - 0.2}{0.5} \approx 48 \text{ }\Omega.$$

The quiescent point is

$$V_{CE} = 12 \text{ V}, \quad I_C = 0.25 \text{ A},$$

resulting in a standby dissipation of $V_{CE}I_C = 3$ W. The load line analysis is as shown. The quiescent base current is approximately $I_B = 3$ mA. Also drawn on the load line graph is the 3-W constant power dissipation hyperbola. A rough calculation may be made to determine the base bias resistor R_{BB}. Assuming that $V_{BB} = V_{CC}$,

$$R_{BB} = \frac{V_{CC} - V_{BE}}{I_B} = \frac{12 - 0.7}{3 \times 10^{-3}} \approx 3.9 \text{ k}\Omega.$$

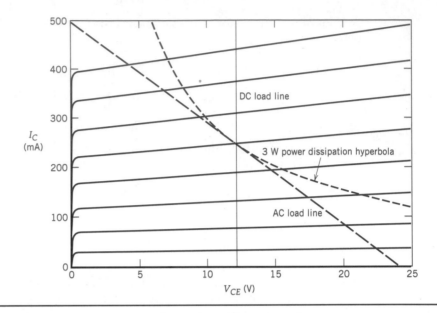

7.3 CLASS B POWER AMPLIFIERS

Class B amplifiers are widely used because of the low standby DC current requirements, unlike class A amplifiers that require an operating point characterized by a large collector current. The reduction in the standby or DC current requirement in class B operation is achieved by biasing each of the two transistors at cutoff. This in turn reduces the quiescent collector power dissipation in the transistors.

Since class B power amplifier operation has transistors conducting in only one half of the period of the input signal, two transistors are necessary in a *push–pull* arrangement to add the two halves of the cycle for the reconstruction of the entire amplified sinusoid, shown in Figure 7.3-1. The transformers load coupling shown in the Figure 7.2-1 is widely used. The collector supply is fed into the center tap of the output transformer primary. The base bias current is fed into the center tap of the input transformer secondary.

A graphical construction for determining the output waveforms of a single class B transistor stage (half of the push–pull stage) is shown in Figure 7.3-2.

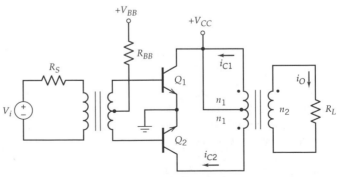

Figure 7.3-1
Class B common-emitter push–pull power amplifier.

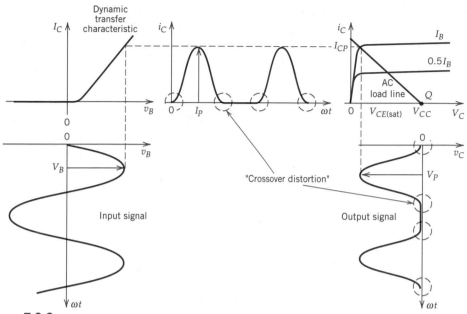

Figure 7.3-2
Waveforms of single class B transistor stage (half of a push–pull stage) constructed with graphical methods.

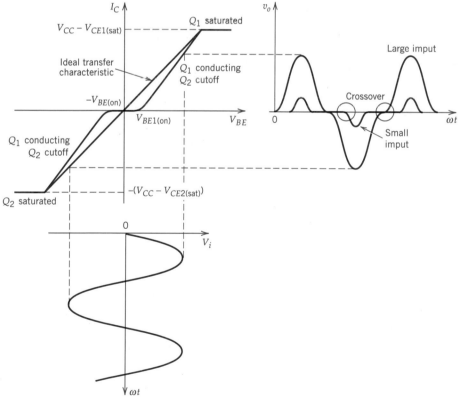

Figure 7.3-3
Output waveforms of class B amplifier for varying input signals.

A disadvantage of using class B operation is the inherently non-linear nature of the input characteristic of BJTs. This leads to distortion near the zero crossings of the output current i_O. The input characteristics and the resulting output in Figure 7.3-2 clearly shows the load current distortion. This distortion is referred to as the *crossover distortion*. Crossover distortion is best understood by studying Figure 7.3-3. The non-linear input–output characteristic of the push–pull configuration is shown. The exponential nature of the curve indicates high input resistance at low signal levels. Little base current flows until the base–emitter voltage exceeds V_γ. The resulting collector current will also be small until the input voltage is sufficiently high since the I_C is essentially proportional to the base current. The resulting output from a push–pull class B configuration is a sinusoid that exhibits crossover distortion.

It is customary to treat only one transistor when analyzing the push–pull configuration since each is operating at identical currents and voltages into an identical load, under the assumption that the transistors have identical characteristics. The output characteristic with load line of one of the transistors in the class B amplifier of Figure 7.3-1 is shown in Figure 7.3-4.

The DC load line per transistor is one-half of the resistance of the output transformer primary. If high-quality output transformers are used, the resistance of the primary can be assumed to be negligible. Therefore, the DC load line has an infinite negative slope. The AC load line has a slope of $-1/R_L'$ where

$$R_L' = \left(\frac{n_1}{n_2}\right)^2 R_L. \tag{7.3-1}$$

Here, R_L' is the AC load per transistor, n_1 is one-half the total number of primary turns on the output transformer, n_2 is the number of secondary turns on the output transformer, and R_L is the actual load.

To analyze the class B push–pull configuration, it is assumed that each transistor is operating at identical levels driving idential loads. The DC load per stage consists of one-half of the total primary DC resistance, which, for good transformers, can be considered negligible. Each collector circuit has only one-half of the total primary turns or n_1 turns (refer to Figure 7.3-1) for an AC resistance of R_L' per transistor or a total AC primary resistance of $4R_L'$.

In Figure 7.3-4, the load line indicates an operating point at $I_C = 0$ and $V_{CE} = V_{CC}$. A load line joins this point with $I_C = I_{CP}$ and $V_{CE} = 0.2$ V. Then the AC resistance is

$$R_L' = \frac{V_{CC} - V_{CE(sat)}}{I_{CP}} \approx \frac{V_{CC}}{I_{CP}} \quad \text{for } V_{CC} \gg V_{CE(sat)}. \tag{7.3-2}$$

The power delivered per transistor to the load is

$$P_{AC,\text{per transistor}} = \frac{1}{2} \frac{V_{CC} - V_{CE(sat)}}{\sqrt{2}} \frac{I_{CP}}{\sqrt{2}} \approx \frac{V_{CC} I_{CP}}{4}. \tag{7.3-3}$$

The factor of $\frac{1}{2}$ is used because each transistor passes a half-wave signal. For the push–pull pair,

$$P_{AC} = \frac{V_{CC} - V_{CE(sat)}}{\sqrt{2}} \frac{I_{CP}}{\sqrt{2}} \approx \frac{V_{CC} I_{CP}}{2}. \tag{7.3-4}$$

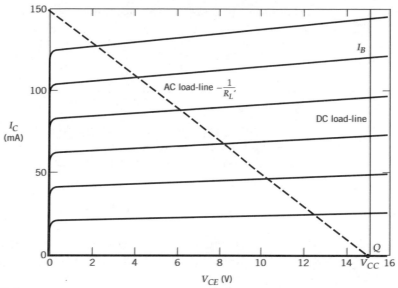

Figure 7.3-4

Output characteristic with load line of one of the transistors in class B amplifier of Figure 7.3-1.

The DC power delivered to the class B push–pull amplifier is negligible during standby operation. When a sinusoid of peak amplitude I_{CP} is produced by the amplifier, each transistor conducts during one-half of the period of the sinusoid,

$$I_{DC,Q1} = I_{DC,Q2} = \frac{1}{T} \int_0^{T/2} I_{CP} \sin \omega t \, dt. \tag{7.3-5}$$

Solving Equation 7.3-5 yields

$$\begin{aligned} I_{DC,Q1} &= I_{DC,Q2} \\ &= \int_0^{T/2} I_{CP} \sin \omega t \, dt \\ &= \frac{\omega}{2\pi} \int_0^{\pi/\omega} I_{CP} \sin \omega t \, dt \\ &= \frac{I_{CP}}{\pi}. \end{aligned} \tag{7.3-6}$$

Therefore, the total current delivered by the power supply is

$$I_{DC} = I_{DC,Q1} + I_{DC,Q2} = \frac{2I_{CP}}{\pi}. \tag{7.3-7}$$

Then the total power delivered by the power supply is

$$P_{DC} = V_{CC}I_{DC} = \frac{2V_{CC}I_{CP}}{\pi}. \tag{7.3-8}$$

The maximum efficiency of the class B push–pull power amplifier is found to be

$$\eta_{max} = \frac{P_{AC}}{P_{DC}} \times 100\% = \frac{V_{CC}I_{CP}/2}{2V_{CC}I_{CP}/\pi} \times 100\%$$

$$= \frac{\pi}{4} \times 100\% \approx 78\%.$$

(7.3-9)

For class B power amplifiers, η is linearly dependent on signal strength, while in class A amplifier, it is dependent on the square of the signal strength.

Should a class B power amplifier be driven to a fraction of its total allowable swing,

$$\Delta V_C = kV_{C,max}, \qquad \Delta I_C = kI_{C,max}.$$

(7.3-10)

where k is a fraction of the total allowable swing, and ΔV_C and ΔI_C are the actual voltage and current swings, respectively. Therefore, the actual full-load power is

$$P_{FL,actual} = \tfrac{1}{4}[(kV_{C,max})(kI_{C,max})].$$

(7.3-11)

EXAMPLE 7.3-1 Design a 10-W servo amplifier to meet the following specifications:

Rated load power	10 W
Load	500 Ω (a small instrument motor)
Overload capacity	10%
Input resistance	50 kΩ minimum
Output transformer efficiency	80%
Carrier frequency	400 Hz
Power supplies	28 V

Two matched MJE15028 *npn* transistors are available.

SOLUTION

The circuit arrangement of Figure 7.3-1 will be used.
   The full-load power is $P_{FL} = 10$ W. To supply P_{FL} to the load,

$$\text{Transformer primary power} = \frac{P_{FL}}{\eta_{xfmr}} = \frac{10}{0.8} = 12.5 \text{ W}.$$

For a 10% overload capacity, the power amplifier must be capable of handling

$$(110\%)(12.5) = 13.7 \text{ W}.$$

Therefore, each transistor in the class B push–pull amplifier must supply 13.7 W/2 = 6.85 W.
   From the load line analysis shown,

$$V_{C,max} = V_{CC} - V_{CE(sat)} \approx V_{CC} = 28 \text{ V}.$$

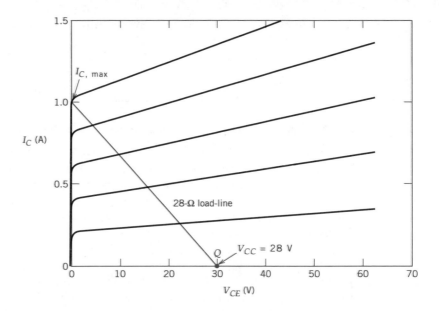

For each transistor under overload conditions,

$$P_{AC \text{ per transistor}} = \tfrac{1}{4}(V_{C,max}I_{C,max}).$$

Solving for $I_{C,max}$,

$$I_{C,max} = \frac{4P_{AC \text{ per transistor}}}{V_{C,max}} = \frac{4(6.85)}{28} \approx 1.0 \text{ A}.$$

Therefore, the AC load that each transistor must drive is

$$R'_L = \frac{V_{C,max}}{I_{C,max}} = \frac{28}{1} = 28 \ \Omega$$

(which is the negative reciprocal of the slope of the load line), and the total primary AC load resistance is

$$R'_L = 2^2(28) = 112 \ \Omega.$$

The output transformer turns ratio is then

$$\frac{n_1}{n_2} = \sqrt{\frac{R'_L}{R_L}} = \sqrt{\frac{28}{500}} = 0.24$$

The full-load power for each transistor is

$$P_{FL, \text{ per transistor}} = \frac{P_{FL}}{2} = \frac{12.5}{2} = 6.25 \text{ W}.$$

The actual swings for the rated full load are found using the laws of similar triangles applied to the load line graph,

$$P_{FL,\text{actual}} = \tfrac{1}{4}[(\Delta I_C)(\Delta V_C)] = \tfrac{1}{4}[(kI_{C,max})(kV_{C,max}].$$

This yields $k = 0.89$. Therefore, the maximum output swings are

$$\Delta I_C = k I_{C,\max} \approx 0.9 \text{ A}, \qquad \Delta V_C = K V_{C,\max} \approx 25 \text{ V},$$

and the actual full-load power is $(0.9)(25) = 5.6$ W per transistor, or 11.2 W total.

7.3.1 Complementary Class B (Push–Pull) Output Stage

A simplified version of a complementary class B amplifier arrangement that is often used as an integrated circuit output stage is shown in Figure 7.3-5. This arrangement uses complementary transistors (one *pnp* and one *npn*) in common-collector configuration. By using complementary transistors, the need for input and output transformers is eliminated. The transfer characteristic is identical to that of the transformer-coupled class B power amplifier.

The principle of operation of the simplified complementary class B output stage is as follows. For $|V_i| < V_{BE(on)}$, both Q_1 and Q_2 are cut off. When $V_i > V_{BE(on)}$, Q_1 turns on and enters the forward saturation region. Here, Q_2 remains in cutoff. The upper limit on V_o occurs when Q_1 saturates at $V_i = V_{CC} - V_{CE1(sat)}$. For negative values of V_i, the same situation holds true when the roles of Q_1 and Q_2 are reversed. The lower limit on V_o occurs when Q_2 saturates at $V_i = -(V_{CC} - V_{CE2(sat)})$.

The CMOS power buffers in the inverter configuration, as discussed in Chapter 4, are commonly used to interface CMOS logic to saturated logic (e.g., TTL). This practice is common in instances where a conventional CMOS gate may not be able to supply the required input power, while at the same time provide sufficiently low output voltage for the TTL gates.

It is common to use feedback to reduce or eliminate crossover distortion.[3] One arrangement employs feedback with a unity gain OpAmp configuration as shown in Figure 7.3-6.

Define the open-loop gain of the OpAmp as A, which is very large ($\approx 200,000$). Then the output voltage from the OpAmp is

$$V_A = A(V_2 - V_1). \tag{7.3-12}$$

From Equation 7.3-12, when $V_A = V_{BE(on)} = 0.6$ V,

$$\frac{V_A}{A} = \frac{0.6}{A} = V_2 - V_1. \tag{7.3-13}$$

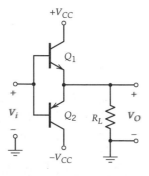

Figure 7.3-5
Simplified complementary class B output stage.

---

[3] Chapter 8 provides a complete discussion of the use of the reduction of distortion due to feedback.

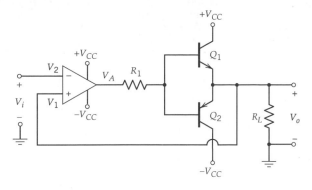

Figure 7.3-6
Use of feedback to eliminate crossover distortion in class B amplifiers.

Therefore, when $V_1 = 0$,

$$V_2 = \frac{0.6}{A} = \frac{0.6}{200 \times 10^3} \ll 0.6 \text{ V.} \tag{7.3-14}$$

Equation 7.3-14 indicates that for a very small input at $V_2 = V_i$, the output voltage from the push–pull stage is zero. Therefore, the crossover region in the transfer characteristic has been reduced from $\pm V_{BE(\text{on})}$ to less than ± 5 μV. So essentially, there is no crossover distortion to speak of when using the configuration in Figure 7.3-6.

7.4 CLASS AB POWER AMPLIFIERS

In Section 7.3, class B power amplifiers were shown to have a maximum efficiency of approximately 78.5%. Unfortunately, feedback was required to eliminate the nonlinear transfer characteristic caused by crossover distortion.

An alternate method to feedback for reducing or nearly eliminating crossover distortion is to bias each transistor in the push–pull configuration so that they are barely in the forward-active region with $V_i = 0$. One possible configuration to achieve this bias condition to eliminate crossover distortion is shown in Figure 7.4-1. This class of amplifier is called a class AB power amplifier/output stage since the transistors are always biased as in class A operation but is biased at a fraction of the peak load current as in class B operation.

In Figure 7.4-1, the resistors R_1 and R_2 are adjusted so that transistors Q_1 and Q_2 are just barely forwarded biased, or $V_{BQ1} = V_{BQ2} = V_{BE(\text{on})}$ for matched transistors. The operation of the circuit in Figure 7.4-1 is as follows: Both Q_1 and Q_2 are

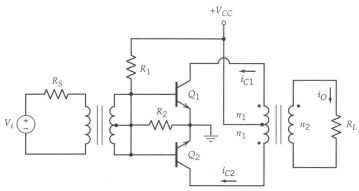

Figure 7.4-1
Class AB power amplifier configuration.

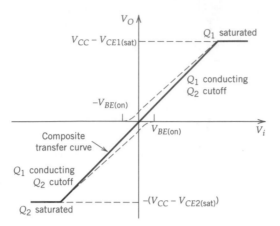

Figure 7.4-2
Transfer characteristic of class AB power
amplifier shown in Figure 7.4-1.

biased so that the base–emitter voltage of each transistor is $V_{BE(on)}$ for $V_i = 0$. For
an input to Q_1 of less than V_i, the transistor is cut off. For $V_i \geq V_{BE(on)}$, Q_1 turns
on and in the forward-active region. The output voltage due to Q_1 saturates at
$V_{CC} - V_{CE1(sat)}$. Conversely for Q_2, if $V_i > V_{BE(on)}$ the transistor is cut off; Q_2 turns
on when $V_i \leq V_{BE(on)}$. The output voltage due to Q_2 saturates at $-(V_{CC} - V_{CE2(sat)})$.
The transfer characteristic is shown in Figure 7.4-2. The characteristic curve for each
transistor is shown as well as the resulting composite transfer characteristic. The
composite transfer characteristic is found by adding the transfer characteristics of
the two transistors. The resulting composite transfer characteristic shows no crossover distortion.

An integrated circuit implementation of a class AB output stage is shown in
Figure 7.4-3a. The class AB output stage shown is used in the μA741 OpAmp. A
simplified version of the μA741 output stage is shown in Figure 7.4-3b.

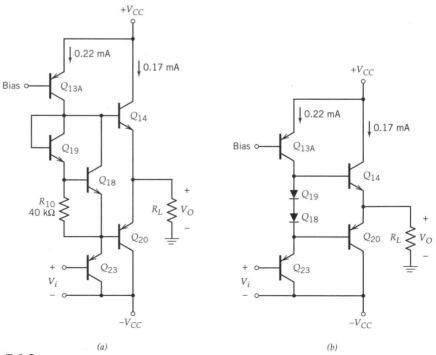

Figure 7.4-3
μA741 OpAmp output stage: (a) actual schematic; (b) simplified schematic.

In analyzing Figure 7.4-3b, Q_{13A} acts as a constant current source of 0.22 mA. Here, Q_{23} is the input transistor to the class AB output stage. With $V_i = 0$, the output complementary transistor pair Q_{14} and Q_{20} are biased at a collector current of about 0.17 mA by the diodes Q_{19} and Q_{18}. The current through Q_{19} and Q_{18} produces a voltage

$$V_{BE19} + V_{BE18} = V_{BE14} + V_{EB20}. \tag{7.4-1}$$

As V_i goes negative, the base of Q_{20} and V_O follows (since Q_{23} and Q_{20} are in common-collector configurations), with Q_{20} drawing current from R_L. When V_i is $-V_{CC}$, the output voltage is limited to

$$V_{O(neg)} = -V_{CC} - V_{BE23} - V_{BE20}. \tag{7.4-2}$$

The negative voltage limit is about 0.7 V + 0.7 V = 1.4 V more positive than the negative rail voltage ($-V_{CC}$).

As V_i increases in the positive direction from $V_i = 0$, the output voltage and the base of Q_{20} follows with Q_{14} delivering the current to the load. When V_i is $+V_{CC}$, the output voltage is limited by Q_{13A} saturating. The positive limit of the output voltage is

$$V_{O(pos)} = V_{CC} - V_{CE13A(sat)} - V_{BE14}. \tag{7.4-3}$$

The positive voltage limit is approximately 0.7 V + 0.2 V = 0.9 V less than the positive voltage rail (V_{CC}).

The conversion efficiency of a Class AB amplifier is somewhat less that that of a Class B amplifier. This reduction in efficiency is due to an additional term in the DC power due to the quiescent current necessary to achieve minimal crossover distortion:

$$P_{DC} = \frac{2V_{CC}I_{CP}}{\pi} + V_{CC}I_{Bias} \tag{7.4-4}$$

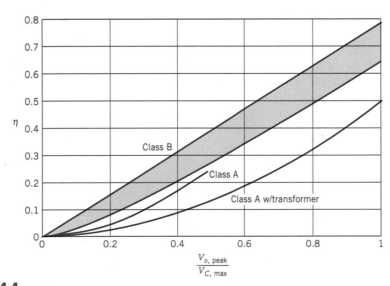

Figure 7.4-4
Typical conversion efficiency of well-designed power amplifiers as a function of output signal amplitude.

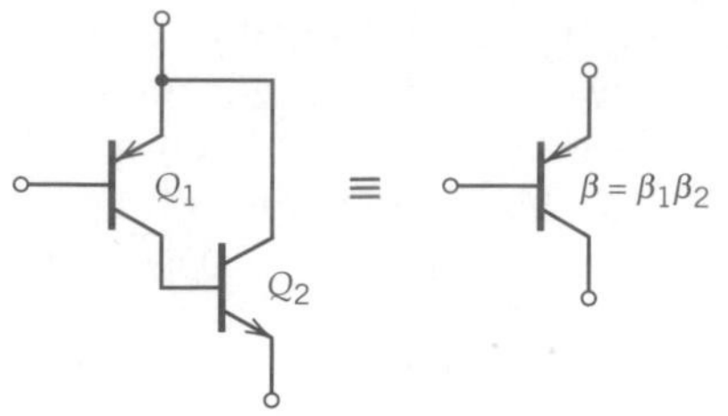

Figure 7.4-5
Compound *pnp* BJT.

Therefore, the maximum efficiency for a Class AB amplifier is given by:

$$\eta_{max} = \frac{\dfrac{V_{CC}I_{CP}}{2}}{\dfrac{2V_{CC}I_{CP}}{\pi} + V_{CC}I_{\text{Bias}}} \times 100\% = \frac{\pi I_{CP}}{4I_{CP} + 2\pi I_{\text{Bias}}} \times 100\%. \quad (7.4\text{-}5)$$

It is interesting to compare the efficiency of well-designed amplifiers as a function of output signal strength (Figure 7.4-4). The output power of all amplifiers is proportional to the square of output signal strength. Class A amplifiers are characterized by constant DC power: this results in a parabolic efficiency curve. Class B amplifiers are characterized by DC power that is proportional to signal strength: a linear efficiency curve results. Class AB amplifiers lie between the two other classes with Class B as an upper limit.

In order to increase the gain of output stages in integrated circuits while reducing base currents, compound transistors in dual common-collector Darlington pairs are commonly used (see Section 6.2.1). Compound *pnp* Darlington configurations are also used. Since good-quality *pnp* BJTs are difficult to fabricate in integrated circuits, an alternate compound configuration shown in Figure 7.4-5 is often used. This compound transistor has a current gain of $\beta \approx \beta_1\beta_2$.

7.5 DISTORTION

When the output signal waveform of an amplifier differs in general shape from the input signal waveform, the output is said to be distorted. In particular, if a single-frequency input to an amplifier results in an output composed of the input frequency and other frequencies, the amplifier has distorted the signal. The creation of additional frequencies is typically the result of non-linear distortion.

Earlier, it was shown that large input signals to amplifiers caused the amplifier under test to yield distorted output signals. In this section, a systematic description of distortion is developed. With this description, two common industry definitions of distortion are developed and related to each other.

An incrementally linear power amplifier[4] has a transfer characteristic described by

$$v_O = V_{\text{DC}} + a_1 v_i, \quad (7.5\text{-}1)$$

where $v_O \equiv$ output voltage

   $v_i \equiv$ input voltage (AC and DC components possible)

   $a_1 \equiv$ voltage gain

   $V_{\text{DC}} \equiv$ DC offset voltage at output

---

[4] Incremental linearity was previously discussed in Chapter 5.

The principles of superposition applies to incrementally linear systems in a unique manner. That is, if

$$v_{O1} = V_{DC} + a_1 v_{i1}, \qquad v_{O2} = V_{DC} + a_1 v_{i2}, \tag{7.5-2}$$

then the total output of a incrementally linear system is,

$$v_O = V_{DC} + a_1(v_{i1} + v_{i2}). \tag{7.5-3}$$

In electronic amplifiers, the DC component, V_{DC}, consists of the bias or quiescent point of the circuit.

Superposition no longer applies when the transfer characteristic is non-linear. Figure 7.5-1 shows a transfer characteristic of a power amplifier.

The region in Figure 7.5-1 that is not clipped is described by the power series,

$$v_O = V_{DC} + a_1 v_i + a_2 v_i^2 + a_3 v_i^3 + \dots. \tag{7.5-4}$$

For a sinusoidal input $v_i = X \cos \omega t$, the power series in Equation 7.5-4 is

$$v_O = V_{DC} + a_1 X \cos \omega t + a_2 (X \cos \omega t)^2$$
$$+ a_3 (X \cos \omega t)^3 + \dots. \tag{7.5-5}$$

Using trigonometric identities, Equation 7.5-5 is rewritten as

$$v_O = (V_{DC} + \tfrac{1}{2} a_2 X^2) + (a_1 X + \tfrac{1}{4} 3 a_3 X^3)\cos \omega t$$
$$+ \tfrac{1}{2} a_2 X^2 \cos 2\omega t + \tfrac{1}{4} a_3 X^3 \cos 3\omega t + \dots, \tag{7.5-6}$$

where the fourth and higher harmonics may be ignored with negligible error.[5] Inspection of Equation 7.5-6 shows that harmonics are generated by the non-linear transfer characteristic. Harmonics are sinusoidal terms that are multiples of the fundamental frequency ω, which is the same frequency as the input signal. For linear power amplifiers, the presence of harmonics is undesirable and are the direct result of non-linear distortion.

Additionally, there is a shift in the DC or quiescent point by the addition of a term containing the constant for the squared term in the power series. The fundamental ($\cos \omega t$) term also has increased beyond the amplification constant associated with the fundamental by a term proportional to the constant for the cubed term in the power series.

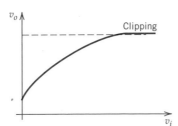

Figure 7.5-1
Non-linear transfer characteristic of power amplifier.

---

[5] This approximation is valid in typical amplifiers where distortion is relatively small. In cases of extremely large distortion, higher-order terms must be considered.

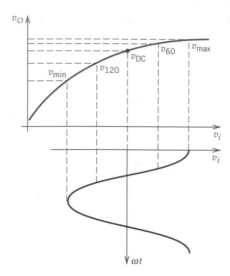

Figure 7.5-2

Graphical determination of distortion content in output voltage.

Equation 7.5-6 can be represented by a Fourier series,

$$v_O = V_{DC} + A_0 + A_1 \cos \omega t + A_2 \cos 2\omega t$$
$$+ A_3 \cos 3\omega t + \dots, \quad (7.5\text{-}7)$$

where A_0, A_1, A_2, and A_3 are the Fourier coefficients. One method that may be used to determine the Fourier coefficients is a method due to Espley.[6] A pure sinusoidal signal voltage is applied to the input of the power amplifier. The input and output waveforms are sampled at several time intervals to obtain the Fourier coefficients of Equation 7.5-7. To evaluate the four coefficients in Equation 7.5-7, the values of the output at the four input voltages are determined as shown in Figure 7.5-2. For this example, the samples are taken at $\omega t = 0°$, 60°, 120°, 180°, with the corresponding output voltages designated as V_{max}, V_{60}, V_{120}, V_{min}. The value at $\omega t = 90°$ is V_{DC} (the quiescent point).

As noted earlier, the DC or quiescent point of the output has shifted away from the average value of the signal.

The values of the Fourier coefficients A_0, A_1, A_2, and A_3 in Equation 7.5-7 can theoretically be determined by substituting the values of the angle (ωt) into Equation 7.5-7 and solving the four simultaneous equations. That is,

$$\begin{aligned}
V_{max} &= V_{DC} + A_0 + A_1 + A_2 + A_3 & (\omega t = 0°), \\
V_{60} &= V_{DC} + A_0 + \tfrac{1}{2}A_1 - \tfrac{1}{2}A_2 - A_3 & (\omega t = 60°), \\
V_{120} &= V_{DC} + A_0 - \tfrac{1}{2}A_1 - \tfrac{1}{2}A_2 + A_3 & (\omega t = 120°), \\
V_{min} &= V_{DC} + A_0 - A_1 + A_2 - A_3 & (\omega t = 180°).
\end{aligned} \quad (7.5\text{-}8)$$

The solution to the four simultaneous equations in Equation 7.5-8 yields the expressions for the Fourier coefficients in Equation 7.5-7,

---

[6] D. C. Espley, "The Calculation of Harmonic Production in Thermionic Valves with Resistive Loads," *Proceedings of the IRE*, Vol. 21, 1933, pp. 1439–1446.

$$A_0 = \tfrac{1}{6}(V_{\max} + V_{\min}) + \tfrac{1}{3}(V_{60} + V_{120}) - V_{DC},$$

$$A_1 = \tfrac{1}{3}(V_{\max} - V_{\min}) + \tfrac{1}{3}(V_{60} - V_{120}),$$

$$A_2 = \tfrac{1}{3}(V_{\max} + V_{\min}) - \tfrac{1}{3}(V_{60} + V_{120}), \qquad (7.5\text{-}9)$$

$$A_3 = \tfrac{1}{6}(V_{\max} - V_{\min}) - \tfrac{1}{3}(V_{60} - V_{120}).$$

The harmonic distortion is defined as

$$D_2 \equiv \frac{|A_2|}{|A_1|}, \qquad D_3 \equiv \frac{|A_3|}{|A_1|} \qquad (7.5\text{-}10)$$

and is commonly given as a percentage or in decibels where D_2 is the second-harmonic distortion and D_3 is the third-harmonic distortion. In terms of decibels, the second and third harmonic distortions are

$$D_2 \text{ (dB)} \equiv 20 \log \frac{|A_2|}{|A_1|}, \qquad D_3 \text{ (dB)} \equiv 20 \log \frac{|A_3|}{|A_1|} \qquad (7.5\text{-}11)$$

and are negative numbers indicating that the voltage amplitude of the harmonic components is less than the fundamental. The THD is commonly given as a percentage and is expressed as the ratio of the RMS values of all the harmonic terms to the effective value of the fundamental, and is used extensively in audio amplifier specifications,

$$\text{THD} = \frac{\sqrt{A_2^2 + A_3^2 + \cdots}}{A_1} \times 100\% = \sqrt{D_2^2 + D_3^2 + \cdots} \times 100\%. \qquad (7.5\text{-}12)$$

If the distortion is not negligible, the total power delivered at the output load R_l is

$$P_O = (A_1^2 + A_2^2 + A_3^2 + \cdots) \frac{R_L}{2}$$

$$= \left[1 + \left(A_1 \frac{\text{THD}}{100} \right)^2 \right] P_1, \qquad (7.5\text{-}13)$$

where the fundamental power is

$$P_1 = \tfrac{1}{2} A_1^2 R_L. \qquad (7.5\text{-}14)$$

As an example, state-of-the-art audio power amplifiers that incorporate negative feedback to compensate for non-linearlity typically have THD of less than 0.003% at low frequencies and low output power levels.[7] In cable television applications, the radio frequency (RF) power amplifiers also require very low distortion levels to deliver acceptable picture quality.

Because the distortion levels are so low, the method of Espley using the direct measure of the transfer characteristic yields inaccurate distortion data due to noise and other spurious factors. Instead, it is common practice to test amplifier non-linearity directly by using a pure sinusoidal input. At the output, the harmonics are measured directly using a spectrum analyzer, which displays the RMS magnitude (or power) of the frequency components of the output signal. The fundamental and harmonics appear on the CRT of the spectrum analyzer as spikes. Measurement of

---

[7] Negative feedback is discussed in Chapters 8 and 11.

the peak of the harmonic spikes relative to the fundamental spike yields the amplifier THD. Other analyzers (e.g., audio analyzers) perform this operation automatically and display the THD on its front panel.

Another method, commonly used in RF and microwave electronics, for finding the amount of distortion of an amplifier is the two-tone ratio (TTR) method for determining harmonic and intermodulation distortion products. In the TTR method, two signals with identical amplitudes that are separated by some frequency are combined and used as the input to the amplifier. The output frequency components are then analyzed for distortion.

Consider the non-linear transfer function of Equation 7.5-4, rewritten here for convenience:

$$v_O = V_{DC} + a_1 v_i + a_2 v_i^2 + a_3 v_i^3 + \dots.$$

Two sinusoids are added so that the input signal is

$$v_i = X_1 \cos \omega_1 t + X_2 \cos \omega_2 t. \tag{7.5-15}$$

Substituting Equation 7.5-15 into Equation 7.5-4 yields

$$v_O = V_{DC} + a_1(X_1 \cos \omega_1 t + X_2 \cos \omega_2 t) + a_2(X_1 \cos \omega_1 t + X_2 \cos \omega_2 t)^2 + a_3(X_1 \cos \omega_1 t + X_2 \cos \omega_2 t)^3 + \dots. \tag{7.5-16}$$

By using trigonometric identities, Equation 7.5-16 is put in a form similar to Equation 7.5-6 to yield the magnitude and frequency components of the output signal,

$$
\begin{aligned}
v_O = &[V_{DC} + \tfrac{1}{2}a_2(X_1 + X_2)] && \text{(DC term)} \\
&+[a_1 X_1 + \tfrac{3}{2}a_3(X_1 X_2^2 + \tfrac{1}{2}X_1^3)]\cos \omega_1 t && \\
&+[a_1 X_2 + \tfrac{3}{2}a_3(X_2 X_1^2 + \tfrac{1}{2}X_2^3)]\cos \omega_2 t && \text{(fundamental)} \\
&+ \tfrac{1}{2}a_2(X_1^2 \cos 2\omega_1 t + X_2^2 \cos 2\omega_2 t) && \text{(second harmonics)} \\
&+a_2 X_1 X_2[\cos(\omega_1 - \omega_2)t + \cos(\omega_1 + \omega_2)t] && \text{(second-order IMD products)} \\
&+ \tfrac{1}{4}a_3(X_1^3 \cos 3\omega_1 t + X_2^3 \cos 3\omega_2 t) && \text{(third harmonics)} \\
&+ \tfrac{3}{4}a_3\{X_1^2 X_2[\cos(2\omega_1 - \omega_2)t + \cos(2\omega_1 + \omega_2)t]\} && \\
&+ \tfrac{3}{4}a_3\{X_1 X_2^2[\cos(2\omega_2 - \omega_1)t + \cos(2\omega_2 + \omega_1)t]\} && \text{,(third-order IMD products)}
\end{aligned}
$$

$$\tag{7.5-17}$$

In the TTR method, new sum and difference frequencies of the two input frequencies are created and are called intermodulation distortion (IMD) products or sometimes "beat" frequencies. The new sum and difference frequencies are not harmonically related to either of the two fundamental input frequencies.

There are several reasons for using the TTR method over single-frequency harmonic measurement methods. They include the following:

- Difficulty exists in generating a pure sinusoid. All real signal generators have some harmonic content.

- In single-octave[8] systems (those systems that only operate in one octave) with multiple sinusoids closely spaced in frequency within the single octave, third-order IMD products are of interest.

---

[8] An octave consists of a range of frequencies spanned by a factor of 2.

The relationship between the results of the TTR results and the harmonic coefficients for the Fourier series in Equation 7.5-7 can be found. The combination of the equal-amplitude TTR input signals are adjusted to equal the total input power of a single sinusoid for the harmonic measurement. This relationship implies that the RMS power of the two combined sinusoids for the TTR is equal to the RMS value of the single input for the harmonic measurement,

$$X^2 = \left(\frac{X_1}{\sqrt{2}}\right)^2 + \left(\frac{X_2}{\sqrt{2}}\right)^2, \tag{7.5-18}$$

where X_1 and X_2 are the TTR inputs and X is the input for the single-frequency harmonic measurement.

For equal-amplitude TTR inputs, $X_1 = X_2$, so

$$X_1 = X_2 = \frac{X}{\sqrt{2}}. \tag{7.5-19}$$

Using Equation 7.5-19, the following relationships are found:

$$\frac{\text{Fundamental two tone}}{\text{Fundamental single tone}} = \frac{1}{\sqrt{2}},$$

$$\frac{\text{Second-harmonic two tone}}{\text{Second-harmonic single tone}} = \frac{1}{2},$$

$$\frac{\text{Second IMD}}{\text{Second-harmonic single tone}} = 1, \tag{7.5-20}$$

$$\frac{\text{Third-harmonic two tone}}{\text{Third-harmonic single tone}} = \frac{1}{2\sqrt{2}},$$

$$\frac{\text{Third IMD}}{\text{Third-harmonic single tone}} = \frac{3}{2\sqrt{2}}.$$

Therefore, given TTR measurement results, the Fourier coefficients in Equation 7.5-7 and the constants in the power series in Equation 7.5-4 can be found by applying Equation 7.5-20.

In Figure 7.5-1, as the output voltage starts to clip, another form of distortion takes place. This type of distortion is called *gain compression* and is due to a gradual decrease in the voltage gain of the amplifier, which, in simple terms, reduces the gain coefficient to the fundamental signal in the Fourier series in Equation 7.5-4. This has the effect of increasing the ratio of the non-linear distortion products to the fundamental.

The push–pull arrangement of the class B power amplifier (shown again in Figure 7.5-3) not only increases efficiency but, despite the effects of crossover distortion, has unique distortion canceling properties.

Consider the base current in Q_1:

$$i_{B1} = I_{B(\text{max})}\cos \omega t. \tag{7.5-21}$$

The resulting Q_1 collector current can be expressed as a Fourier series:

$$i_{C1} = I_{CQ1} + A_0 + A_1\cos \omega t + A_2\cos 2\omega t$$
$$+ A_3\cos 3\omega t + \dots. \tag{7.5-22}$$

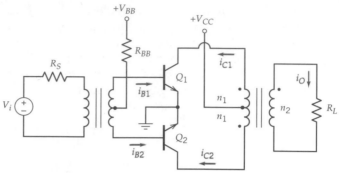

Figure 7.5-3
Class B push–pull arrangement.

The corresponding base current in Q_2 is

$$i_{B2} = -i_{B1} = I_{B(max)} \cos (\omega t + \pi). \qquad (7.5\text{-}23)$$

Equation 7.5-23 implies that

$$i_{B2}(\omega t) = i_{B1}(\omega t + \pi). \qquad (7.5\text{-}24)$$

There, the Q_2 collector current is

$$i_{C2} = I_{CQ2} + A_0 + A_1\cos(\omega t + \pi) + A_2\cos 2(\omega t + \pi) \\ + A_3\cos 3(\omega t + \pi) + \dots. \qquad (7.5\text{-}25)$$

Simplifying Equation 7.5-5 yields

$$i_{C2} = I_{CQ2} + A_0 - A_1\cos \omega t + A_2\cos 2\omega t \\ - A_3\cos 3\omega t + \dots. \qquad (7.5\text{-}26)$$

Since the collector currents of Q_1 and Q_2 are in opposite directions through the output transformer primary, the total output current is proportional to the difference in collector currents:

$$i_{xfmr} = k(i_{C1} - i_{C2}) \\ = 2k(A_1\cos \omega t + A_3\cos 3\omega t + \dots), \qquad (7.5\text{-}27)$$

where k is a proportionality factor.

Equation 7.5-7 shows that there are only odd harmonics in the output signal when matched transistors are used. Even harmonics have been canceled out.

Because no even harmonics are present in the output of a push–pull amplifier, such a circuit will give more output per active device for a given amount of THD.

SPICE performs Fourier analysis on any waveform in a transient analysis using the Fourier command,

```
.FOUR <frequency value> <output variable>
```

The **.FOUR** command computes the amplitude and phase of any waveform with respect to frequency. The Fourier components may then be plotted.

SPICE can also calculated the intermodulation distortion components of a waveform using the command

```
.DISTO <RLname> <number of points> <fstart> <fstop> <reference
power>
```

 PSpice does not support the **.DISTO** command.

7.6 THERMAL CONSIDERATIONS

The removal of heat from the BJT collector–base junction or the FET channel warrants considerable attention in power transistors due to the high power delivered to the load. For power BJTs,[9] metallic heat sinks are often necessary to remove the heat to the surrounding air. When possible, the metallic portion of the transistor case should make contact with the heat sink. However, it is common practice to use a mica washer as electrical insulation between the case and the heat sink. The washer is used to electrically isolate the heat sink from the case because in some power BJT packages, the collector is directly attached to the case.

A specification for the maximum allowable collector junction temperature is found in the manufacturer's transistor specifications. This temperature is usually 150°C for silicon devices. Exceeding this temperature may cause irreparable damage to the transistor. The operating junction temperature T_j is dependent on the ambient temperature T_a (typically 25°C), the thermal resistance θ_T of the heat transfer path from the junction to the surrounding, and the power dissipated P_D. This relationship is expressed as

$$T_j = T_a + \theta_T P_D \qquad (°C). \tag{7.6-1}$$

The total thermal resistance θ_T has the units of degrees Celsius per watt.

The power dissipated by the transistor is almost entirely at the collector junction. Therefore, the power dissipated is

$$P_D = V_{CE} I_C. \tag{7.6-2}$$

The maximum power dissipation is described by the maximum dissipation hyperbola and is

$$P_{D(max)} = (V_{CE} I_C)_{max} = \frac{T_{j(max)} - T_a}{\theta_T}. \tag{7.6-3}$$

The total thermal resistance is found by solving the heat transfer network analogous to an electrical network shown in Figure 7.6-1.

The current source represents the heat generated (or power dissipated by the BJT), and the three resistances represent the thermal resistances of the junction to

---

[9] The discussion that follows is equally applicable to FETs.

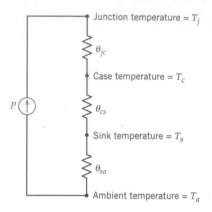

Figure 7.6-1
Heat transfer model.

case (θ_{jc}), case to heat sink (θ_{cs}), and heat sink to the ambient environment (θ_{sa}). The total thermal resistance is

$$\theta_T = \theta_{jc} + \theta_{cs} + \theta_{sa} = \theta_{ja}. \tag{7.6-4}$$

For power transistors, the values of θ_{jc} is typically about 10°C/W. If the transistor case and the heat sink are insulated by a mica washer, θ_{cs} has values of around 0.5°C/W. The primary mode of heat removal from a power transistor is conduction by the heat sink and convection from the heat sink to the ambient surroundings. By blowing air over the heat sink, more heat may be removed.

Casual inspection of Equation 7.6-4 indicates that by removing the heat sink in the thermal path, the limitation on the power handling capacity is eliminated. However, by doing so, the case to ambient convection and radiation has significantly higher thermal resistance thereby reducing the ability of the device to dissipate heat: Power handling capacity is actually dimished for this case.

Not shown in Figure 7.6-1 are associated thermal capacitances in parallel with each resistance. The capacitance is used to model transient thermal behavior. The thermal time constant (resistance times capacitance) is long for the external components of the thermal model but short for the transistor itself.

Thermal runaway can result if the rate of increase in the power dissipation with junction temperature exceeds the ability of the heat transfer network to remove the heat. The dissipated heat energy removed by the networks is found by differentiating Equation 7.6-1 with respect to T_j. The necessary condition for thermal stability is

$$\frac{\partial V_{CE} I_C}{\partial T_j} < \frac{1}{\theta_T}. \tag{7.6-5}$$

For a constant collector–emitter voltage, the stability condition is

$$V_{CE} \frac{\partial I_C}{\partial T_j} < \frac{1}{\theta_T}. \tag{7.6-6}$$

Variations in the current amplification factor and in the leakage current of the transistor causes the change in collector current with temperature, and is ultimately dependent on the bias circuitry. With leakage current suspected of being the chief contributor to the collector current excursions,

$$V_{CE} \frac{\partial I_C}{\partial I_{CO}} \left(\frac{\partial I_{CO}}{\partial T_j} \right)_{max} < \frac{1}{\theta_T}. \tag{7.6-7}$$

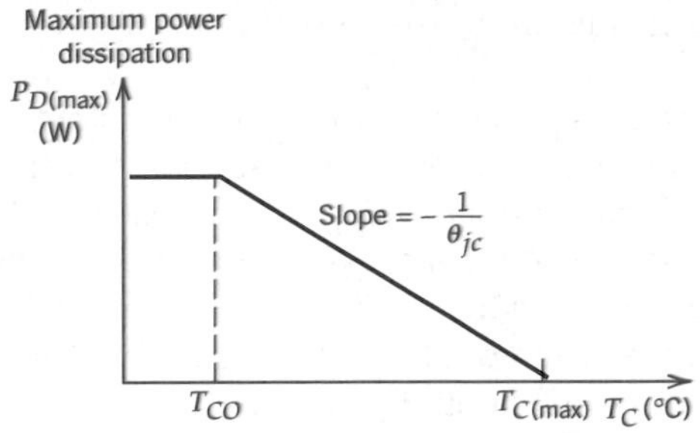

Figure 7.6-2
Transistor power dissipation derating curve.

From Section 3.7, the stability factor is

$$S_I = \frac{\partial I_C}{\partial I_{CO}}. \tag{7.6-8}$$

For a thermally stable network, Equation 7.6-7 is rewritten as

$$\left(\frac{\partial I_{CO}}{\partial T_j}\right)_{max} < \frac{1}{\theta_T V_{CE} S_I}. \tag{7.6-9}$$

Despite using stable bias arrangements and effective heat transfer configurations, the transistor case temperature cannot be held at ambient temperature. Therefore, manufacturers provide a power–temperature derating curve (shown in Figure 7.6-2).

In Figure 7.6-2, T_{CO} is the temperature where the derating begins. The maximum safe power dissipation, as specified by the manufacturer, is achieved when $T_C = T_{C(max)}$. The curve in Figure 7.6-2 can also be used to find the power dissipation as a function of the junction temperature T_j.

EXAMPLE 7.6-1 For a case temperature of 120°C for a 50 W power transistor rated at 35°C, find the maximum power dissipated. The slope of the derating curve is 0.55 W/°C.

SOLUTION

$$P_{D(max)} = 50 \text{ W} \quad \text{at } 35°C.$$

Using the equation of a line, the power dissipated at 120°C can be found:

$$(120°C - 35°C)(0.55 \text{ W/°C}) = 46.75 \text{ W}.$$

Therefore, the device power dissipation at 120°C is

$$P_D = (50 \text{ W} - 46.75 \text{ W}) = 3.25 \text{ W} \quad \text{at } 120°C.$$

7.7 CONCLUDING REMARKS

Different classes of amplifiers designed for the transfer of high power to loads were analyzed. The analysis of the amplifiers required the use of large-signal analysis methods. Because of the large-signal excursions, the output waveforms experience distortion.

The most commonly used power amplifier/output stage classifications are classes A, B, and AB. In class A operation, the transistor conducts over the whole period (360°) of the input signal. In class B operation, each of the two transistors conducts over half the period (180°). In class AB operation, each transistor conducts over a time greater than half the period.

Since high powers may be delivered by these circuits, the conversion of DC power to signal power must be efficient. The theoretical maximum efficiencies for the different classes of amplifiers are 25% for class A, 50% for class A output transformer coupled amplifiers, 78.5% for class B, and less than 78.5% for class AB operation.

Distortion analysis was presented with a comparison between the total harmonic distortion and two-tone ratio methods of measurement.

Since power amplifiers deliver high power to their loads, the transistor may dissipate large amounts of power in the form of heat. Therefore, a heat transfer model was analyzed and related to the stability factor.

SUMMARY DESIGN EXAMPLE: PUBLIC ADDRESS (PA) SYSTEM AMPLIFIER

In most audio electronic systems, the object is to drive one or more speakers. Typically, these speakers can be modeled as 8 Ω devices. Therefore, the output stage of audio electronic systems require the ability to provide maximum power transfer to a very low resistance device. One common method to drive low-resistance devices is through the use of an output transformer. The purpose of the output transformer is to provide impedance matching between the active devices and the speaker.

The output stage of a PA system will require the application of an electronic signal to an 8Ω speaker. Although PA systems generally do not require high audio fidelity, a low-distortion system is desired since some music may be amplified on the system. For a low- to moderate-power PA system, the rated (or full) load power can be in the order of 25 W, with an overload capacity of 10%. A single-rail, +24 V power supply is commonly used. The output resistance of the circuit that drives the output stage is 600 Ω.

Design an output power amplifier stage to fulfill the requirements given.

SOLUTION:

Several alternate output power stage topologies can be used in the design: class A, class B, and class AB. A transformer-coupled class AB power amplifier topology, as

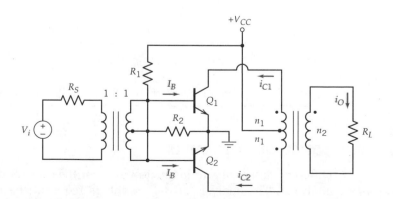

shown, is chosen because of its low-distortion characteristic and since power dissipation is not specified. Assuming 80% output transformer efficiency, the transformer primary power is

$$\text{Transformer primary power} = \frac{P_{FL}}{\eta_{xfmr}} = \frac{25}{0.8} = 31.25 \text{ W}.$$

For a 10% overload capacity, the power amplifier must be capable of handling

$$(110\%)(31.25) = 34.4 \text{ W}.$$

Therefore, each transistor in the class AB push–pull amplifier must supply 34.4 W/2 = 17.2 W.

The maximum collector voltage for each transistor is

$$V_{C,\max} = V_{CC} - V_{CE(\text{sat})} \approx V_{CC} = 24 \text{ V}.$$

For each transistor under overload conditions,

$$P_{AC \text{ per transistor}} = \tfrac{1}{4} V_{C,\max} I_{C,\max}.$$

Solving for $I_{C,\max}$,

$$I_{C,\max} = \frac{4 P_{AC \text{ per transistor}}}{V_{C,\max}} = \frac{4(17.2)}{24} \approx 2.87 \text{ A}.$$

Therefore, the AC load that each transistor must drive is

$$R_L' = \frac{V_{C,\max}}{I_{C,\max}} = \frac{24}{2.87} = 8.4 \ \Omega,$$

and the total primary AC load resistance is

$$R_L' = 2^2(8.4) = 33.6 \ \Omega.$$

The output transformer turns ratio is then

$$\frac{n_1}{n_2} = \sqrt{\frac{R_L'}{R_L}} = \sqrt{\frac{8.36}{8}} \approx 1.$$

The full-load power for each transistor is

$$P_{FL, \text{ per transistor}} = \tfrac{1}{2} P_{FL} = \tfrac{1}{2}(31.25) = 5.6 \text{ W}.$$

Assuming identical transistors and a base–emitter turn on voltage of $V_{BE(\text{on})} = 0.6$ V,

$$V_{BE(\text{on})} = 0.6 \approx \frac{V_{CC} R_2}{R_1 + R_2} = \frac{24(600)}{R_1},$$

where $R_1 \parallel R_2 = 600 \ \Omega$.

Then $R_1 = 24 \text{ k}\Omega$ and $R_2 = 620 \ \Omega$.

REFERENCES

Carson, R. S., *Radio Concepts: Analog*, Wiley, New York, 1990.

Clarke, K. K., and Hess, D. T., *Communication Circuits: Analysis and Design*, Addison-Wesley, Reading, MA, 1971.

Colclaser, R. A., Neaman, D. A., and Hawkins, C. F., *Electronic Circuit Analysis: Basic Principles*, Wiley, New York, 1984.

Espley, D. C., "The Calculation of Harmonic Production in Thermionic Valves with Resistive Loads," *Proc. IRE*, Vol. 21, 1933, pp. 1439–1446.

Ghausi, M. S., *Electronic Devices and Circuits: Discrete and Integrated*, Holt, Rinehart, and Winston, New York, 1985.

Gray, P. R., and Meyer, R. G., *Analysis and Design of Analog Integrated Circuits*, 3rd ed., Wiley, New York, 1993.

Millman, J., and Halkias, C. C., *Integrated Electronics: Analog and Digital Circuits and Systems*, McGraw-Hill, New York, 1972.

Sedra, A. S., and Smith, K. C., *Microelectronic Circuits*, 3rd ed., Holt, Rinehart, and Winston, Philadelphia, 1991.

PROBLEMS

7-1. The design goals of the class A amplifier shown include maximum sinusoidal power delivered to the load of at least 40 mW and minimum possible power supply current. Complete the design (including a realization of the current source) using a selection of resistors and BJTs with characteristics

$$\beta_F = 150, \qquad V_A = 200.$$

7-2. The design goals of the class A amplifier shown include maximum sinusoidal power delivered to the load of at least 0.6 W and minimum possible power supply current. Complete the design (including a realization of the current source) using a selection of resistors and BJTs with characteristics

$$\beta_F = 100, \qquad V_A = 200.$$

7-3. A power amplifier to drive a load, $R_L = 220\ \Omega$, with the basic topology shown is under design. The silicion power BJT is described by

$$\beta_F = 80, \qquad V_A = 100\ \text{V}.$$

The pertinent design goals are maximum symmetricl output voltage swing and $R_{in} \approx 12\ \text{k}\Omega$.

a. Complete the design by determining the proper bias resistors.

b. Compute the maximum conversion efficiency of the design.

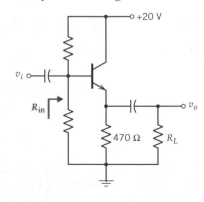

7-4. The amplifier of problem 7-3 is to be redesigned for a new load, $R_L = 82\ \Omega$. The other design goals remain the same.

   a. Complete the design by determining the proper bias resistors.

   b. Computer the maximum conversion efficiency of the design.

7-5. The common-drain class A amplifier shown uses devices for which

$$V_{PO} = -2\ V, \quad I_{DSS} = 5\ mA, \quad V_A = 120\ V.$$

For linear operation (the FETs must be within the saturation region), what is the range of output voltages obtained with $R_L = \infty$? What is the range for $R_L = 100\ \Omega$? Verify the analytic resulting using SPICE.

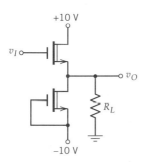

+10 V

v_I

v_O

R_L

−10 V

7-6. The BiFET common-emitter class A amplifier shown uses devices described by

$$\beta_F = 150 \quad (BJT),$$

$$V_{PO} = -2\ V, \quad I_{DSS} = 7.5\ mA \quad (FET).$$

   a. For linear operation, what is the maximum symmetrical range of output voltages possible?

   b. Determine the conversion efficiency when the circuit is operating with maximum symmetrical output

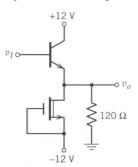

+12 V

v_I

v_o

120 Ω

−12 V

7-7. Redesign the class A amplifier of problem 7-5 by replacing the FET current with a BJT current source. For the redesigned amplifier:

   a. What is the maximum symmetrical range of output voltages possible maintaining linear operation?

   b. What is the conversion efficiency when the circuit is operating with maximum symmetrical output?

7-8. The circuit shown uses BJTs described by

$$\beta_F = 120, \quad V_A = 160\ V.$$

   a. For linear operation, what is the maximum **symmetrical** range of output voltages possible?

   b. Determine the conversion efficiency when the circuit is operating with maximum symmetrical output

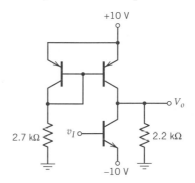

+10 V

V_o

2.7 kΩ v_I 2.2 kΩ

−10 V

7-9. Complete the design of the class A transformer-coupled power amplifier shown to drive an 8-Ω speaker. The required input resistance of the amplifier is $R_{in} = 600\ \Omega$. The power amplifier is required to deliver 10 W of power (AC) to the 8-Ω speaker load. Design the biasing network so that a 1% change in I_C corresponds to a 10% change in β_F.

   a. Show all circuit values and the transformer turns ratio.

   b. Specify minimum transistor power dissipation rating.

   c. Confirm the operation of the design using SPICE. The transformer coupling efficiency is 0.999 and the BJT characteristics are $\beta_F = 70, I_S = 0.3\ pA$, and $V_A = 75\ V$.

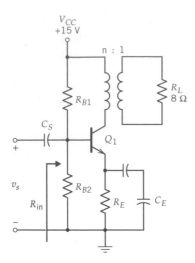

7-10. Given a class A BJT power amplifier with a load resistance of 4 Ω and power transistor rating of

$$P_{C,\text{max}} = 10 \text{ W}, \qquad V_{CE(\text{sat})} = 0.2 \text{ V},$$
$$V_{CE(\text{max})} = 60 \text{ V}.$$

   a. Determine the maximum attainable voltage swing at the output, the maximum power dissipated by the load, and the efficiency when transformer coupling is not used.

   b. Repeat part a when a transformer coupling is used with a transformer turns ratio of 2.

7-11. For the complementary push–pull amplifier shown, determine the peak-to-peak voltage of the largest possible undistorted sinusoidal output. The silicon transistor is described by $\beta_F = 75$.

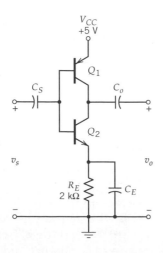

7-12. For the class B amplifier shown, find the following:

   a. Maximum undistorted peak output voltage

   b. Maximum DC power consumed

   c. Output AC power

   d. Maximum amplifier efficiency η

Assume that the transistors are matched with $\beta_F = 100, \qquad I_S = 0.03 \text{ pA}, \qquad V_{\gamma D1} = 0.7 \text{ V}.$

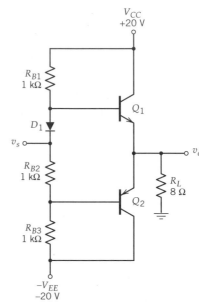

7-13. The push–pull power amplifier shown uses identical transistors with $\beta_F = 75$. Find the maximum positive and negative values of V_o for:

   a. $R_L = 10 \text{ k}\Omega$

   b. $R_L = 1 \text{ k}\Omega$

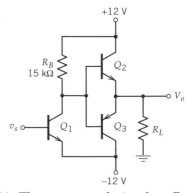

7-14. The common-drain class B amplifier shown uses devices for which

$$V_T = 1 \text{ V}, \qquad K = 200 \text{ μA/V}^2.$$

For a sinusoidal input:

a. At what input voltages do the respective FETs enter the saturation region?

b. What is the maximum peak output voltage level?

c. What is the maximum conversion efficiency of this circuit?

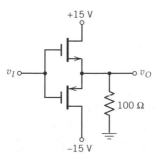

7-15. Complete the design of the class B push–pull amplifier shown, which uses matched BJTs with $\beta_F = 75$, $I_S = 0.03$ pA, and $V_A = 120$ V. Assume 99% transformer efficiency, rated load power of 12 W, and an overload capacity of 10%.

a. Determine the input transformer ratio n_p for maximum power transfer.

b. Determine the output transformer ratio $n_1 : n_2$.

c. Find the maximum values of i_{C1}, i_{C2}, and i_o.

d. Find the power delivered to the load per transistor, $P_{AC \text{ per transistor}}$, collector power dissipation P_C, and DC power dissipation P_{DC}.

d. Calculate the amplifier efficiency η.

e. Plot the transfer characteristic of the power amplifier using SPICE.

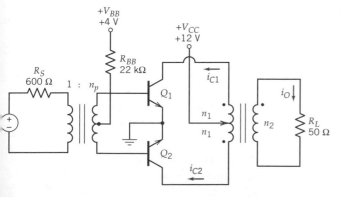

7-16. The class AB amplifier shown uses two 10 kΩ resistors to establish quiescent current through two diodes bridging the base–emitter junctions of a matched BJT pair. This quiescent current increases the DC power drawn from the power supplies and therefore decreases the efficiency of the amplifier. The transistors have characteristics

$$\beta_F = 150, \qquad V_A = 200.$$

a. Determine the maximum conversion efficiency of this amplifier.

b. What is the conversion efficiency if the output is reduced to one-half the maximum possible undistorted amplitude?

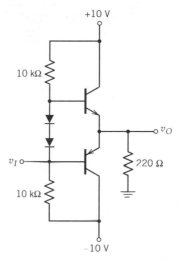

7-17. A class AB amplifier with the general topology shown is under design. The power transistors being used are silicion Darlington pairs (shown in schematic form). A Zener diode is being used to eliminate crossover distortion and a current source provides biasing.

a. What should the Zener voltage V_z be to eliminate crossover distortion?

b. What is the maximum sinusoidal amplitude of the output voltage?

c. The typical DC current gain of the Darlington pairs is 2500, and the Zener diode requires a minimum diode current of 2 mA to ensure regulation. What is the maximum value of the resistor R necessary to keep the Zener diode in regulation while achieving maximum output power?

d. Determine the maximum efficiency of this circuit with the above determined diode, transistor, and resistor parameters.

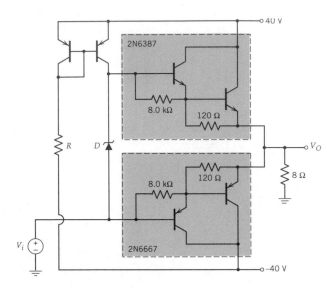

7-18. A class AB amplifier with the general topology shown is under design. The transistors being used are silicon Darlington pairs (shown in schematic form). A Zener diode is being used to eliminate crossover distortion. This Zener diode requires a minimum diode current of 500 μA to ensure regulation. The individual BJTs in the Darlington pairs are silicon with $\beta_F = 60$.

a. What should the diode Zener voltage V_z be in order to eliminate crossover distortion?

b. If the design goals include delivering 55 W of signal power to the load, what is the minimum undistorted output voltage swing?

c. For purposes of simple analysis, the resistors shunting the BJT base–emitter junctions can be considered to act as current sources of value

$$I_{bias} = \frac{V_\gamma}{R_{shunt}}.$$

Determine the maximum value of the resistor labeled R to meet the design goal (55 W of signal power delivered to the load).

d. Determine the maximum efficiency of this circuit using the design value of the resistor determined in part c and the Zener voltage determined in part a.

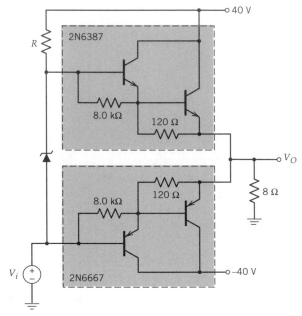

7-19. Complete the design of the class AB power amplifier shown, which uses matched BJTs with $\beta_F = 50$, $I_S = 0.3$ pA, and $V_A = 150$ V. Assume 99% transformer efficiency, rated load power of 15 W, and an overload capacity of 15%.

a. Determine the input transformer ratio n_p for maximum power transfer.

b. Determine the output transformer ratio $n_1 : n_2$.

c. Find the maximum values of i_{C1}, i_{C2}, and i_o.

d. Find the power delivered to the load per transistor $P_{AC\ per\ transistor}$, collector power dissipation P_C, and DC power dissipation P_{DC}.

e. Calculate the amplifier efficiency η.

f. Plot the transfer characteristic of the power amplifier using SPICE.

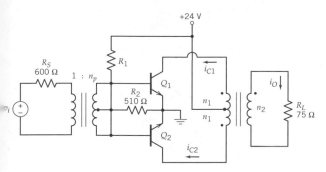

(F) **7-20.** Complete the design of a direct-coupled output push–pull amplifier, shown to achieve a maximum output to an 8-Ω load. The circuit uses matched transistors with the following specifications:

$$P_{C,max} = 8 \text{ W}, \quad i_{C,max} = 1.5 \text{ A}, \quad \beta_F = 60$$

$$V_A = 120 \text{ V}, \quad I_S = 0.05 \text{ pA}, \quad V_{CEmax} = 55 \text{ V}.$$

a. Determine the maximum input voltage for an undistorted output signal.

b. Plot the transfer characteristic of the power amplifier using SPICE.

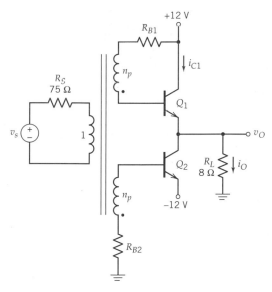

7-21. Design a class B power amplifier stage to deliver an average power of 75 W into a 12-Ω load. The power supply must be 5 V greater than the peak sinusoidal output signal.

a. Draw the schematic of the power amplifier.

b. Determine the required power supply voltage, peak power supply current, total

DC power consumer by the circuit, and efficiency of the amplifier.

c. Determine the maximum possible power dissipation per transistor for a sinusoidal input signal.

7-22. Design a class AB direct-coupled output power amplifier using matched BJTs with $\beta_F = 60$, $I_S = 0.05$ pA, and $V_A = 120$ V. The amplifier is required to provide a rated load power of 15 W, and an overload capacity of 15%. Assume 99% transformer efficiency and a power supply of ±24 V.

(F) **7-23.** Complete the design of the class AB power amplifier shown for $R_{B1} \parallel R_{B2} = R_{B3} \parallel R_{B4} = 12$ kΩ. The transistors are matched and have the parameters $\beta_F = 60$, $I_S = 0.033$ pA, and $V_A = 120$ V.

a. Design the bias network to eliminate crossover distortion (adjust to $V_{BE(on)} = 0.6$ V).

b. What is the maximum undistorted power delivered to the load?

c. Determine the DC power dissipation P_{DC} and the efficiency η of the amplifier for maximum output current.

d. Plot the transfer characteristic of the power amplifier using SPICE.

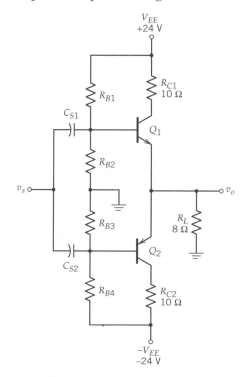

7-24. A direct-coupled class AB power amplifier is shown using matching transistors with parameters $\beta_F = 75$, $I_S = 0.033$ pA, and $V_A = 100$ V. Resistors R_{E1} and R_{E2} are included to guard against the possibility of transistor thermal runaway.

    a. For $R_L = \infty$, find the quiescent current through each transistor and calculate v_o.

    b. For $R_L = \infty$, find the collector current through each transistor when $v_s = +5$ V and calculate v_o.

    c. For $R_L = 75$ Ω, find the collector current through each transistor when $v_s = +5$ V and calculate v_o.

    d. Plot the transfer characteristic of the power amplifier using SPICE for $R_L = 75$ Ω.

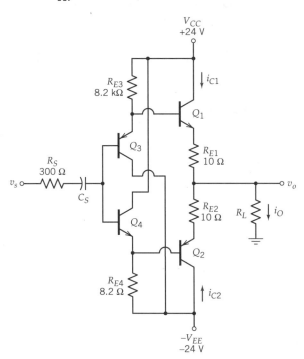

7-25. A class B amplifier has the transfer characteristic shown. Determine the second- and third-harmonic distortion (in decibels) and the total harmonic distortion for the following amplitude input sinusoids:

    a. 14 cos(ωt) volts

    b. 8 cos(ωt) volts

The scale is 10 V per major division.

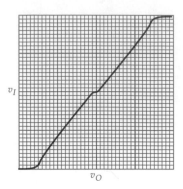

7-26. A class A amplifier has the transfer characteristic shown (the quiescent point for the amplifier is the central point of the diagram). Determine the second- and third-harmonic distortion (in decibels) and the total harmonic distortion for the following amplitude input sinusoids:

    a. 0.1 cos(ωt) volts

    b. 0.2 cos(ωt) volts

The vertical scale is 5 V per major division and the horizontal scale is 0.2 V per major division.

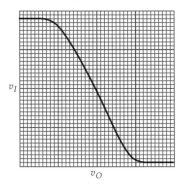

7-27. The input to an amplifier is a pure, undistorted sinusoid. One cycle of the resultant output waveform is shown. The vertical scale is 1 V per division and the horizontal scale is 50 μs per division.

    a. Determine the amplifier second- and third-harmonic distortion by sampling the waveform at five (5) appropriately positioned data points (remember these points are not uniformly spaced in time) and performing appropriate calculations.

    b. What is the total harmonic distortion of the amplifier with this input?

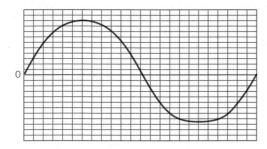

7-28. The input to an amplifier is a pure, undistorted sinusoid. One cycle of the resultant output waveform is shown. The vertical scale is 0.5 V per division and the horizontal scale is 20 ms per division.

a. Determine the amplifier second- and third-harmonic distortion by sampling the waveform at five (5) appropriately positioned data points (remember these points are not uniformly spaced in time) and performing appropriate calculations.

b. What is the total harmonic distortion of the amplifier with this input?

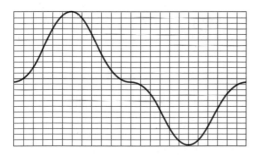

7-29. Amplifiers may distort the input signal in rather unusual ways. For instance, a class A, common-emitter amplifier with an emitter-resistor responds to an overly large input as shown. Determine the second, third, and total harmonic distortion of the output waveform shown.

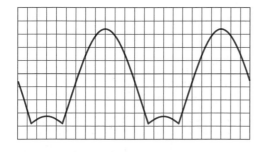

7-30. A class A, common-emitter amplifier with an emitter resistor responds to an overly large amplitude input signal by "reflecting" a portion of the signal (see output waveform shown in the previous problem). This reflection occurs while the BJT is in the saturation region of operation. Investigate this phenomenon by using simple BJT regional models to analyze the output voltage waveform of circuit shown for the following amplitude input sinusoids:

a. $v_i(t) = 0.5 \sin(\omega t)$
b. $v_i(t) = 1.5 \sin(\omega t)$
c. $v_i(t) = 3.0 \sin(\omega t)$

The silicon BJT is described by $\beta_F = 100$.

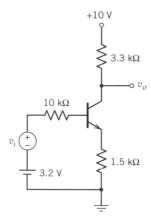

7-31. Perform the distortion investigation of the circuit shown in the previous problem using SPICE. Comment on results.

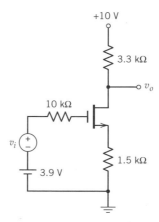

7-32. It seems reasonable to assume that the reflection distortion found in common-emitter amplifiers with an emitter resistor might also be present in common-source amplifiers with a source resistor. Investigate the distortion present in the FET circuit shown using SPICE and compare results to those found in problems 7-27 or 7-28. Comment on the similarities and/or differences in the output of the two circuits. Explain what characteristics in FET and BJT performance cause these similarities and/or differences. The FET in the given circuit is described by

$$V_T = 1.0 \text{ V}, \qquad K = 5 \text{ mA/V}^2, \qquad V_A = 160 \text{ V}.$$

7-33. An amplifier is characterized using the two-tone measurement method with a total average input power of 100 μW. The resulting intermodulation results were $IMD_2 = -35$ dB and $IMD_3 = -40$ dB. The amplifier is not experiencing gain compression.

    a. Find the second- and third-harmonic ratios (to the fundamentals).

    b. Determine the Fourier coefficients of the output signal.

    c. Use SPICE or other computer software package to plot the transfer function of the output compared to the ideal undistorted transfer characteristic.

    d. Maintaining a 5-dB difference in IMD_2 and IMD_3, determine the distortion level where there is a 5% difference between the transfer function of the amplifier and the ideal.

7-34. Consider the class B push–pull amplifier shown in Figure 7.5-3. The amplifier distortion characteristics are given as $IMD_2 = -30$ dB and $IMD_3 = -40$ dB for the input signal power used.

    a. If the base current in Q_1 is $i_{B1} = I_{B1}(\cos \omega_o t + 0.02 \cos 2\omega_o t)$, find the corresponding Q_2 collector current and the total output current at the load.

    b. If the base current in Q_1 is $i_{B1} = I_{B1}(\cos \omega_o t + 0.02 \cos 3\omega_o t)$, find the corresponding Q_2 collector current and the total output current at the load.

    c. Use SPICE to simulate parts a and b. The transistors are matched with parameters $\beta_F = 75$, $I_S = 0.033$ pA, and $V_A = 100$ V.

7-35. Data from a transistor specification sheet shows that the maximum junction temper-

ature is 150°C and the maximum allowable dissipation at any temperature is 15 W. The transistor should be derated above 25°C ambient. Assume a heat sink is used with $\theta_{cs} = 0.7$°C/W and $\theta_{sa} = 1.5$°C/W. Let $P_D = 7$ W at a 40°C ambient. Find:

    a. Junction temperature

    b. Case temperature

    c. Heat sink temperature.

7-36. A thermal equivalent circuit has $\theta_{sa} = 0.2$°C/mW, $\theta_{cs} = 0.1$°C/mW, and $\theta_{jc} = 0.1$° C/mW for a particular transistor.

    a. If the thermal system (transistor and heat sink) dissipates 0.1 W into an ambient environment at 30°C, find the junction temperature T_j, case temperature T_c, and heat sink temperature T_s.

    b. If $T_j = 150$°C with $T_A = 50$°C, find T_C and P_D.

7-37. A typical transistor may have thermal data as follows: "Total device dissipation at $T_A = 25$°C is 1 W; derate above 25°C, 10 mW/°C." What is the safe power dissipation at an ambient temperature of:

    a. 75°C

    b. 0°C

    c. If $\theta_{jc} = 0.03$°C/mW, what is θ_{ca} (case-to-ambient thermal resistance)?

7-38. Given a BJT with parameters

$$\beta_F = 75, \qquad I_S = 0.033 \text{ pA},$$
$$V_A = 100 \text{ V}$$

and a power supply voltage of 15 V, assume that the BJT SPICE parameters XTB = 1.5 and XTI = 3, yielding $V_{BE} = 0.7$ V at 25°C and decreasing linearly by 1.3 mV/°C.

    a. Design a stable biasing circuit for $I_C = 5$ mA ± 0.05 mA and $V_{CE} = 5$ V ± 1 V for 55°C ≤ T ≤ 125°C. Assume that the BJT SPICE parameters XTB = 1.5 and XTI = 3, yielding $V_{BE} = 0.7$ V at 25°C and decreasing linearly by 1.3 mV/°C.

    b. Find the junction temperature at ambient temperatures of −55 and 125°C.

7-39. A transistor is specified for a maximum allowable case temperature of 150°C. The maximum allowable dissipation is 150 mW, and the transistor should be derated above $T_C = 25$°C. If $\theta_{cs} = 0.1$°C/mW and $\theta_{sa} = 0.4$°C/mW, find the temperature of the heat sink when:

a. 150 mW is being dissipated and $T_C = 25$°C

b. 150 mW is being dissipated and $T_C = 0$°C

c. As much power as possible is being dissipated and $T_C = 100$°C

7-40. The given power derating curve of an FET shows maximum power dissipation as a function of ambient temperature. If $\theta_{ca} = 2$°C/ mW, find:

a. θ_{ja}

b. θ_{jc}

c. Maximum safe value of P_D if $T_A = 100$°C

d. How much power may be dissipated if $T_C = 100$°C

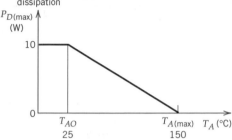

CHAPTER 8

FEEDBACK AMPLIFIER PRINCIPLES

Feedback is the process of combining a portion of the output of a system with the system input to achieve modified performance characteristics. Negative feedback is especially important in amplifier design as it produces several significant benefits. The primary benefits are as follows:

- The gain of the amplifier is stabilized against variation in the characteristic parameters of the active devices due to voltage or current supply changes, temperature changes, or device degradation with age. Similarly, amplifier performance is stabilized within a group of amplifiers that have, by necessity, active devices with different characteristic parameters.
- The input and output impedances of the amplifier can be selectively increased or decreased.
- Non-linear signal distortion is reduced.
- The midband frequency range is increased. Discussion of this aspect is delayed until Chapter 11.

It is a rare occurrence when benefits come without a price. In the case of negative feedback, these benefits are accompanied by two primary drawbacks:

- The gain of the circuit is reduced. In order to regain the losses due to feedback, additional amplification stages must be included in the system design. This adds complexity, size, weight, and cost to the final design.
- There is a possibility for oscillation to occur. Oscillation will destroy the basic gain properties of the amplifier. Discussion of this aspect will also be delayed until Chapter 11.

In this chapter the benefits of negative feedback are considered. Basic definitions are followed by a general discussion of the properties of a feedback system. Amplifiers are divided into four categories of feedback topology and the specific properties of each topological type are derived. While the emphasis of discussions must focus on circuit analysis techniques, a clear understanding of feedback in general, and effects of circuit topology in particular, is a necessity for good feedback amplifier design.

The summary design example explores the common design practice of modifying existing circuitry to meet new, but similar, performance specifications. Minimal alteration of the existing design is explored as a possible, discretionary design criterion.

8.1 BASIC FEEDBACK CONCEPTS

The basic topology of a feedback amplifier is shown in Figure 8.1-1. This figure shows a feedback system in its most general form: Each signal symbolized with $\{X_{(\ )}\}$ can take the form of either a voltage or a current and travels only in the direction of the indicated arrows. The triangular symbol is a linear amplifier of gain A, as has been described in several of the preceding chapters. The rectangle indicates a feedback network that samples the output signal, scales it by a factor f and passes it forward to the input of the system. The circular symbol is a summing (or mixing) junction that subtracts the feedback signal X_f from the inputs. Subtraction of the two inputs at the summing junction is a key factor in negative feedback systems.

The system can be mathematically modeled in the following fashion. The output of the amplifier X_o is related to its input signal X_δ by a linear amplification factor (gain) A, often called the *forward* or *open-loop gain*:

$$X_o = A(X_\delta). \tag{8.1-1}$$

Since the quantities X_o and X_δ can be either voltage or current signals, the forward gain A can be a voltage gain, a current gain, a transconductance, or a transresistance.[1] The feedback signal X_f (a fraction of the output signal X_o) is then subtracted from the input signal X_i to form the difference signal X_δ:

$$X_\delta = (X_i - X_f) = (X_i - fX_o), \tag{8.1-2}$$

where f is the *feedback ratio* defining the relationship between X_f and X_o:

$$X_f = fX_o. \tag{8.1-3}$$

The feedback ratio f can also be a ratio of voltage, currents, transconductance, or transresistance. In order to have stable negative feedback, it is necessary that the mathematical sign of f be the same as that of A.[2] Thus the product Af, called the *loop gain*, is a positive, dimensionless quantity. The input–output relationship for the overall system is derived from Equations 8.1-1 and 8.1-2:

$$X_o = \frac{A}{1 + Af} X_i = A_f X_i. \tag{8.1-4}$$

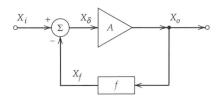

Figure 8.1-1
Basic negative-feedback topology.

---

[1] Gain quantities are ratios of an output quantity to an input quantity. Transresistance implies the ratio of an output voltage to an input current: Transconductance is the ratio of an output current to an input voltage.

[2] This mathematical sign identity is necessary in the midband frequency region: Chapter 11 explores frequency-dependent phase shifts in the gain and feedback quantities and their consequences.

The overall gain of the system including the effects of feedback is then written as

$$A_f = \frac{X_o}{X_i} = \frac{A}{1 + Af} \tag{8.1-5}$$

Notice that A_f does not need to be either a voltage or current gain: It will have the same dimensions as the forward gain A. Equation 8.1-5 has special significance in the study of feedback systems and is called the *basic feedback* equation. The denominator of the basic feedback equation is identified as the *return difference D*, also referred to as the amount of feedback:

$$D = 1 + Af. \tag{8.1-6}$$

The return difference, for negative feedback systems, has magnitude larger than unity (in the midband frequency region) and is often specified in decibels:

$$D_{dB} = 20 \log_{10}|1 + Af|. \tag{8.1-7}$$

The return difference quantifies the degradation in gain due to the addition of feedback to the system. It also plays a significant role in quantifying changes in input and output impedance and frequency bandwidth.

The derivation of the basic feedback equation is based on two basic idealized assumptions:

- The reverse transmission through the amplifier is zero (applying a signal at the output produces no signal at the input).
- The forward transmission (left to right in Figure 8.1-1) through the feedback network is zero.

While these assumptions are impossible to meet in practice, a reasonable approximation is obtained with the following realistic requirements:

- The reverse transmission through the amplifier is negligible compared to the reverse transmission through the feedback network
- The forward transmission through the feedback network is negligible compared to the forward transmission through the amplifier.

In most feedback amplifiers the amplifier is an active device with significant forward gain and near-zero reverse gain; the feedback network is almost always a passive network. Thus, in the forward direction, the large active gain will exceed the passive attenuation significantly. Similarly, in the reverse direction, the gain of the feedback network, albeit typically small, is significantly greater than the near-zero reverse gain of the amplifier. In almost every electronic application, the requirements stated for the use of the basic feedback equation are easily met by the typical feedback amplifier.

Some of the drawbacks and benefits of feedback systems can be investigated on a simple level by looking at the properties of the basic feedback equation (Equation 8.1-5):[3]

---

[3] The change in the input and output impedances cannot be investigated at this level: It is necessary to specify the nature (voltage or current) or the input and output quantities. Discussion of these properties of feedback systems is found in Section 8.3.

1. *The gain of the circuit is reduced.* It has been already shown that the overall gain is the gain of the simple amplifier without feedback divided the return difference, which is larger in magnitude than 1.

2. *There is a possibility for oscillation to occur.* It will be shown in Chapters 10 and 11 that the gain decreases and the phase of the gain of an amplifier changes as frequency increases. This change in phase may cause the loop gain Af to change sign from positive to negative. If this change in sign occurs at the same frequency that the magnitude of the loop gain approaches unity, the return difference approaches zero at that frequency. Division by zero indicates an instability: That instability is realized as an oscillation.

3. *The gain of the amplifier is stabilized against variation in the characteristic parameters of the active devices.* It has been shown in previous chapters that the gain A of an amplifier is highly dependent on the parameters of the active devices. These parameters are highly dependent on temperature, bias conditions, and manufacturing tolerances. It is therefore desirable to design amplifiers that are reasonably insensitive to the variation of the device parameters.

The relationship between the differential change in gain due to device parameter variation with and without feedback is obtained by differentiating Equation 8.1-5:

$$dA_f = \frac{1}{(1 + Af)^2} dA, \tag{8.1-8}$$

which is more typically expressed as

$$\left| \frac{dA_f}{A_f} \right| = \left| \frac{1}{1 + Af} \right| \left| \frac{dA}{A} \right|. \tag{8.1-9}$$

Stable negative feedback amplifiers require that the return difference have magnitude greater than unity:

$$(1 + Af) > 1; \tag{8.1-10}$$

thus the variation of the overall amplifier gain A_f is reduced by a factor of the return ratio.

EXAMPLE 8.1-1 A feedback amplifier is constructed with an amplifier that is subject to a 3% variation in gain as its fundamental forward-gain element. It is desired that the feedback amplifier have no more than 0.1% variation in its overall gain due to the variation in this element. Determine the necessary return difference to achieve this design goal.

SOLUTION

Equation 8.1-9 is the significant relationship

$$\left| \frac{dA_f}{A_f} \right| = \left| \frac{1}{1 + Af} \right| \left| \frac{dA}{A} \right|.$$

The significant properties are

$$0.001 \geq \left| \frac{1}{1 + Af} \right| 0.03 \Rightarrow D = 1 + Af \geq 30.$$

The minimum necessary return ratio is 30, more often identified as its decibel equivalent,

$$D_{db} = 20 \log_{10} D = 29.54 \text{ dB}.$$

Equation 8.1-9 is useful for small changes in amplification due to parameter variation. If the changes are large, the mathematical process must involve differences rather than differentials:

$$\Delta A_f = A_{2f} - A_{1f} = \frac{A_2}{1 + A_2 f} - \frac{A_1}{1 + A_1 f}. \tag{8.1-11}$$

In order to put this into the same format as Equation 8.1-9, it is necessary to divide both sides of the equation by A_{1f}.

$$\left| \frac{\Delta A_f}{A_{1f}} \right| = \left| \frac{A_2}{1 + A_2 f} \left(\frac{1 + A_1 f}{A_1} \right) - 1 \right| = \left| \frac{A_2 - A_1}{1 + A_2 f} \left(\frac{1}{A_1} \right) \right| \tag{8.1-12}$$

or

$$\left| \frac{\Delta A_f}{A_{1f}} \right| = \left| \frac{1}{1 + A_2 f} \right| \left| \frac{\Delta A}{A_1} \right| = \left| \frac{1}{1 + (A_1 + \Delta A) f} \right| \left| \frac{\Delta A}{A_1} \right|. \tag{8.1-13}$$

EXAMPLE 8.1-2 A feedback amplifier is constructed with an amplifier with nominal gain $A = 100$ that is subject to a 30% variation in gain as its fundamental forward-gain element. It is desired that the feedback amplifier have no more than 1% variation in its overall gain due to the variation in this element. Determine the necessary return difference to achieve this design goal.

SOLUTION

This problem statement is a duplicate of Example 8.1-1 with the variation increased by a factor of 10. Use of Equation 8.1-9 will produce the same results ($D = 30$). The large variation, however, requires the use of Equation 8.1-13:

$$|0.01| \geq \left| \frac{1}{1 + (A_1 + \Delta A) f} \right| |0.3| \Rightarrow 1 + (A_1 + \Delta A) f \geq 30,$$

where $A_1 = 100$ and $\Delta A = \pm 0.3 A_1 = \pm 30$. Solving the above inequality using both values of ΔA leads to two values of the feedback ratio f.

$$f \geq 0.2290 \quad \text{or} \quad f \geq 0.4225.$$

Good design practice indicates that *both* inequalities should be satisfied. Thus

$$D = (1 + A_1 f) = [1 + 100(0.4225)] = 43.25$$

and

$$D_{dB} = 20 \log_{10} D = 32.72 \text{ dB}.$$

This result is about 3 dB more than predicted using the methods in Example 8.1-1.

4. *Non-linear signal distortion is reduced.* Stabilization of gain with parameter variation suggests that the gain will be stabilized with respect to other gain-changing effects. One such effect is non-linear distortion: This type of distortion is a variation of the gain with respect to input signal amplitude. A simple example of non-linear distortion is shown in Figure 8.1-2. Here the transfer characteristic of a simple amplifier is approximated by two regions, each of which is characterized by different amplification, A_1 and A_2. To this transfer characteristic, a small amount of feedback is applied so that $A_1 f = 1$, and the resultant feedback transfer characteristic is also shown. As can be easily seen, the overall feedback transfer characteristic also consists of two regions with overall amplification A_{1f} and A_{2f}. In this demonstration, the amplification ratios are

$$\frac{A_1}{A_2} = 3, \quad \& \quad \frac{A_{1f}}{A_{2f}} = 1.5$$

Feedback has significantly improved the linearity of the system and reduced non-linear distortion. Larger amounts of feedback (increasing the feedback ratio f) will continue to improve the linearity. For this example, increasing the feedback ratio by a factor of 5 will result in a ratio of overall gain in the two regions of 1.067 (as compared to 1.5 above).

It should be noted that the saturation level of an amplifier is not significantly altered by the introduction of negative feedback. Since the incremental gain in saturation is essentially zero, the feedback difference is also zero. No significant change to the input occurs and the output remains saturated.

Another possible viewpoint on gain stabilization comes from another form of the basic feedback equation:

$$A_f = \frac{A}{1 + Af} = \frac{1}{f}\left(1 - \frac{1}{1 + Af}\right) \approx \frac{1}{f}. \tag{8.1-14}$$

For large return difference ($D = 1 + Af$) the overall gain with feedback is dominated by the feedback ratio f.

5. *The midband frequency range is increased.* While this property is discussed extensively in Chapter 11, it can be considered a special case of the reduction in gain variation. As frequencies increase, the performance parameters of an amplifier de-

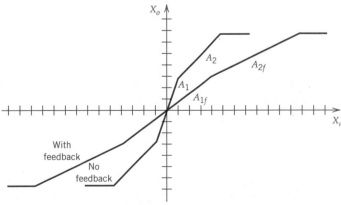

Figure 8.1-2
Effect of feedback on an amplifier transfer characteristic.

grade. Similarly, coupling and bypass capacitors will degrade low-frequency performance. Feedback reduces the effects of these frequency-dependent degradations and thereby increases the frequency band over which the amplifier has stable gain.

8.2 FEEDBACK AMPLIFIER TOPOLOGIES

As has been seen in the previous section, general discussions provide great insight into many of the properties of feedback systems. In order to consider the design of electronic feedback amplifiers, it is necessary, however, to specify the details of the feedback sampling and mixing processes and the circuits necessary to accomplish these operations. The sampling and mixing processes have a profound effect on the input impedance, the output impedance, and the definition of the forward-gain quantity that undergoes *quantified* change due to the application of feedback.[4] This section will analyze the various idealized feedback configurations: Section 8.3 will look at practical feedback configurations.

As has been previously stated, both the mixing and the sampling process for a feedback amplifier can utilize either voltages or currents. Voltage mixing (subtraction) implies a series connection of voltages at the input of the amplifier: Current mixing implies a shunt connection. Voltage sampling implies a shunt connection of the sampling probes across the output voltage: Current sampling implies a series connection so that the output current flows into the sampling network. Either type of mixing can be combined with either type of sampling. Thus, a feedback amplifier may have one of four possible combinations of the mixing and sampling processes. These four combinations are commonly identified by a compound term: (mixing topology)–(sampling topology). The four types are as follows:

- Shunt–shunt feedback (current mixing and voltage sampling)
- Shunt–series feedback (current mixing and current sampling)
- Series–shunt feedback (voltage mixing and voltage sampling)
- Series–series feedback (voltage mixing and current sampling)

The four basic feedback amplifier topologies are shown schematically in Figure 8.2-1. A source and a load resistance have been attached to model complete operation. In each diagram the input, feedback, and output quantities are shown properly as voltages or currents. Forward gain A must be defined as the ratio of the output sampled quantity divided by the input quantity that undergoes mixing. As such it is a transresistance, current gain, voltage gain, or transconductance. The feedback network, as described by the feedback ratio f, must sample the output quantity and present a quantity to the mixer that is of the same type (current or voltage) as the input quantity. As such it is a transconductance, current gain, voltage gain, or transresistance. Table 8.2-1 lists the appropriate quantities mixed at the input, the output sampled quantity, the forward gain, and the feedback ratio for each of the four feedback amplifier topologies. It is important to remember that the product, Af, must be dimensionless and, in the midband region of operation, positive.

In the previous section, all drawbacks and benefits of feedback were discussed *except* the modification of input and output impedance. The specific definitions of

---

[4] For any gain topology, all forward-gain quantities may undergo change: The gain quantity listed in Table 8.2-1 is the single quantity that is altered as described by the basic feedback equation (Equation 8.1-5).

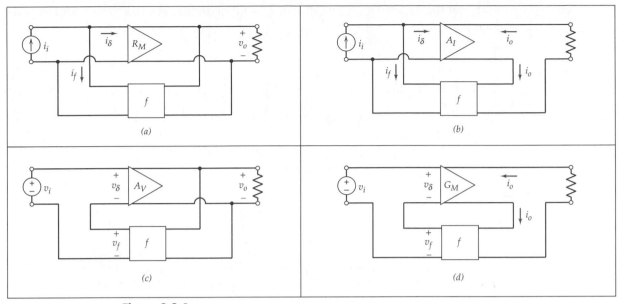

Figure 8.2-1
Feedback amplifier topologies: (*a*) shunt–shunt feedback; (*b*) shunt–series feedback; (*c*) series–shunt feedback; (*d*) series–series feedback.

TABLE 8.2-1 FEEDBACK AMPLIFIER TOPOLOGY PARAMETERS

Parameter	Shunt–Shunt	Shunt–Series	Series–Shunt	Series–Series
Input quantity, X_i	Current i_s	Current i_s	Voltage v_s	Voltage v_s
Output quantity, X_o	Voltage v_o	Current i_o	Voltage v_o	Current i_o
Forward gain, A	Transresistance R_M	Current gain A_I	Voltage gain A_V	Transconductance G_M
Feedback ratio, f	i_f/v_o	i_f/i_o	v_f/v_o	v_f/i_o

the four feedback amplifier topologies allow for that discussion to begin here. The *mixing process alters the input impedance* of a negative-feedback amplifier. Shunt mixing decreases the input resistance and series mixing increases the input impedance:

1. *Shunt mixing decreases the input resistance.* For the *shunt–shunt* feedback amplifier (Figure 8.2-2), the voltage across its input terminals (arbitrarily identified as v) and the input current i_i are related by the feedback amplifier input resistance R_{if}:

$$v = i_i R_{if}. \tag{8.2-1}$$

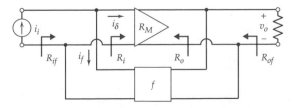

Figure 8.2-2
Input and output resistance for shunt–shunt feedback.

Similarly the forward gain amplifier has input quantities related by its input impedance R_i:

$$v = i_\delta R_i. \tag{8.2-2}$$

The two input currents i_i and i_δ are related through the forward-gain and the feedback ratio:

$$i_\delta = i_i - i_f = i_i - i_\delta(R_M f) \Rightarrow i_i = i_\delta(1 + R_M f). \tag{8.2-3}$$

Therefore, combining Equations 8.2-1 and 8.2-3 yields

$$R_{if} = \frac{v}{i_i} = \frac{v}{i_\delta(1 + R_M f)} = \frac{R_i}{1 + R_M f}. \tag{8.2-4}$$

The input resistance to feedback amplifier is the input resistance of the forward-gain amplifier reduced by a factor of the return difference. *Shunt–series* feedback amplifier input resistance is similarly derived (replacing R_M by A_I. The same basic reduction in input resistance occurs:

$$R_{if} = \frac{R_i}{1 + A_I f}. \tag{8.2-5}$$

2. *Series mixing increases input resistance.* For the *series–series* feedback amplifier of Figure 8.2-3, the voltage across its input terminals, v_i, and the input current (arbitrarily identified as i) are related by the feedback amplifier input resistance R_{if}:

$$v_i = i R_{if}. \tag{8.2-6}$$

Similarly, the forward-gain amplifier has input quantities related by its input impedance R_i:

$$v_\delta = i R_i. \tag{8.2-7}$$

The two input voltages v_i and v_δ are related through the forward-gain and the feedback ratio:

$$v_\delta = v_i - v_f = v_i - v_\delta(G_M f) \Rightarrow v_i = v_\delta(1 + G_M f). \tag{8.2-8}$$

Therefore, combining Equations 8.2-6 and 8.2-8 yields

$$R_{if} = \frac{v_i}{i} = \frac{v_\delta(1 + G_M f)}{i} = R_i(1 + G_M f). \tag{8.2-9}$$

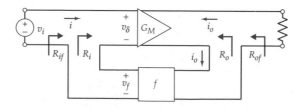

Figure 8.2-3
Input and output resistance for series–series feedback.

The input resistance to feedback amplifier is the input resistance of the forward-gain amplifier increased by a factor of the return difference. *Series–shunt* feedback amplifier input resistance is similarly derived (replacing G_M by A_V). The same basic reduction in input resistance occurs:

$$R_{if} = R_i(1 + A_V f). \tag{8.2-10}$$

It should be noted that resistors shunting the input, such as biasing resistors, often do not fit within the topological standards of series mixing. Thus, they must be considered separate from the feedback amplifier in order to properly model feedback amplifier characteristics using the techniques outlined in this, and following, chapters. Examples of such resistors will be found in the next section of this chapter.

The sampling process alters the output impedance of the feedback amplifier. Here shunt sampling decreases the output resistance and series sampling increases the output resistance.

1. *Shunt sampling decreases the output resistance.* For the *shunt–shunt* feedback amplifier of Figure 8.2-2, the output resistance is measured by applying a voltage source of value v to the output terminals with the input i_i set to zero value. A simplified schematic representation of that measurement is shown in Figure 8.2-4. In this figure, the forward-gain amplifier has been shown with its appropriate gain parameter, R_M, and output resistance R_o.

The output resistance of the feedback system is the ratio

$$R_{of} = \frac{v}{i}. \tag{8.2-11}$$

The current i is calculated from Ohm's law at the output of the amplifier:

$$i = \frac{v - R_M i_d}{R_o}. \tag{8.2-12}$$

In the case where the input current has been set to zero,

$$i_\delta = -i_f = -fv \tag{8.2-13}$$

Combining Equations 8.2-12 and 8.2-13 yields

$$i = \frac{v - R_M(-fv)}{R_o} \Rightarrow R_{of} = \frac{v}{i} = \frac{R_o}{1 + R_M f} \tag{8.2-14}$$

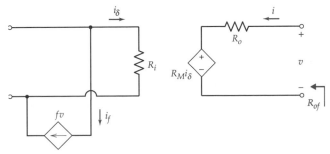

Figure 8.2-4
Schematic representation of shunt–shunt feedback for output resistance calculations.

The output resistance of the feedback amplifier is the output resistance of the forward-gain amplifier decreased by a factor of the return difference. *Series–shunt* feedback amplifier output resistance is similarly derived (replacing R_M by A_V). The same basic reduction in input resistance occurs:

$$R_{of} = \frac{R_o}{1 + A_V f}.$$
(8.2-15)

Resistors that shunt the output terminals, such as a load resistor, are considered as part of the feedback amplifier. The forward-gain parameter (R_M or A_V) must be calculated in a consistent fashion with the consideration of these elements.

2. *Series sampling increases the output resistance.* For the *series–series* feedback amplifier of Figure 8.2-3, the output resistance is measured by applying a current source of value i to the output terminals with the input v_i set to zero value. A simplified schematic representation of that measurement is shown in Figure 8.2-5. In this figure, the forward-gain amplifier has been shown with its appropriate gain parameter A_V and output resistance R_o.

The output resistance of the feedback system is the ratio

$$R_{of} = \frac{v}{i}.$$
(8.2-16)

The voltage v is given by

$$v = (i - G_M v_\delta) R_o.$$
(8.2-17)

Since the input voltage v_i has been set to zero value,

$$v_\delta = -v_f = -fi.$$
(8.2-18)

Combining Equations 8.2-17 and 8.2-18 yields

$$v = [i - G_M(-fi)]R_o = i(1 + G_M f)R_o.$$
(8.2-19)

The output resistance is then given by

$$R_{of} = \frac{v}{i} = (1 + G_M f)R_o.$$
(8.2-20)

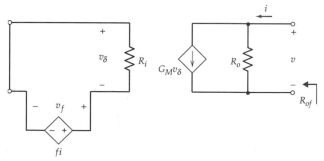

Figure 8.2-5
Schematic representation of series–series feedback for output resistance calculations.

TABLE 8.2-2 SUMMARY OF FEEDBACK AMPLIFIER RESISTANCE PARAMETERS

Parameter	Shunt–Shunt	Shunt–Series	Series–Shunt	Series–Series
Input resistance	$R_{if} = \dfrac{R_i}{1 + R_M f}$	$R_{if} = \dfrac{R_i}{1 + A_I f}$	$R_{if} = R_i(1 + A_V f)$	$R_{if} = R_i(1 + G_M f)$
Output resistance	$R_{of} = \dfrac{R_o}{1 + R_M f}$	$R_{of} = R_o(1 + A_I f)$	$R_{of} = \dfrac{R_o}{1 + A_V f}$	$R_{of} = R_o(1 + G_M f)$

NOTES: Series mixing feedback amplifiers that have resistors shunting the input require special care in the application of these formulas. Also, feedback amplifiers with series sampling and loads that shunt the output require similar care. See Section 8.3 for examples of proper calculations.

The output resistance of the feedback amplifier is the output resistance of the forward-gain ampliifer increased by a factor of the return difference. *Shunt–series* feedback amplifier output resistance is similarly derived (replacing G_M by A_I). The same basic increase in input resistance occurs:

$$R_{of} = (1 + A_I f)R_o. \tag{8.2-21}$$

A summary of the effect of feedback topology on input and output resistance is found in Table 8.2-2. It should be noted that resistances shunting the output, such as load resistances, do not fit within the topological standards of series sampling. Thus, they must be considered separate from the feedback amplifier in order to properly model feedback amplifier characteristics using the techniques outlined in this, and following, chapters. The forward-gain parameters A_I and G_M must be calculated excluding these resistances. Examples of such resistances will be found in the next section of this chapter.

8.3 PRACTICAL FEEDBACK CONFIGURATIONS

Previous discussions of feedback and feedback configurations have been limited to idealized systems and amplifiers. The four idealized feedback schematic diagrams of Figure 8.2-1 identify the forward-gain amplifier and the feedback network as two-port networks with a very specific property: Each is a device with one-way gain. Realistic electronic feedback amplifiers can only approximate that idealized behavior. In addition, in practical feedback amplifiers there is always some interaction between the forward-gain amplifier and the feedback network. This interaction most often takes the form of input and output resistive loading of the forward-gain amplifier. The division of the practical feedback amplifier into its forward-gain amplifier and feedback network is also not always obvious. These apparent obstacles to using idealized feedback analysis can be resolved through the use of two-port network relationships in the derivation of practical feedback amplifier properties. Once amplifier gain and impedance relationships have been derived, the utility of the two-port representations becomes minimal and is typically discarded.

Feedback topology is determined through careful observation of the interconnection of the feedback network and forward-gain amplifier. Shunt mixing occurs at the input terminal of the amplifier. Thus *shunt mixing* is identified by a connection of feedback network and the forward-gain amplifier at the input terminal of the first active device within the amplifier:

- At the base of a BJT for a common-emitter or common-collector first stage
- At the emitter of a BJT for a common-base first stage
- At the gate of an FET for a common-source or common-drain first stage
- At the source of an FET for a common-gate first stage

Series mixing occurs in a loop that contains the input terminal of the forward-gain amplifier and the controlling port of the first active device. The controlling port of a BJT in the forward-active region is the base–emitter junction: An FET in the saturation region is controlled by the voltage across the gate–source input port. *Series mixing* is characterizied by a circuit element that is *both* connected to the output *and* in series with the input voltage and the input port of the first active device.

Identification of the sampling is derived from direct observation of the connection of the output of the basic forward amplifier and the feedback network. Shunt sampling is typically characterized by a direct connection of the feedback network to the output node: Series sampling implies a series connection of the amplifier output, the feedback network, and the load. Two tests performed at the feedback amplifier output can aid in the determination of sampling toplogy:

- If the feedback quantity vanishes for a short-circuit load, the output voltage must be the sampled quantity. Thus zero feedback for a short-circuit load implies *shunt sampling*.
- If the feedback quantity vanishes for an open-circuit load, the output current must be the sampled quantity. Thus zero feedback for an open-circuit load implies *series sampling*.

After the topological type has been identified, each amplifier must be transformed into a form that allows for the use of the idealized feedback formulations. This transformation includes modeling the amplifier and the feedback network with a particular two-port representation that facilitates combination of elements. Once the transformations are accomplished, the amplifier performance parameters are easily obtained using the methods previously outlined in this chapter. The particular operations necessary to transform each of the four feedback amplifier topological types require separate discussion. The examples in this section demonstrate the typical interconnection of a basic forward amplifier and a feedback network for each feedback topology. While BJT amplifiers form the examples of this section, the general approach applies to all feedback amplifiers regardless of active device type.

8.3.1 Shunt–Shunt Feedback

Figure 8.3-1 shows the small-signal model of a typical shunt–shunt feedback amplifier. In this representation, the forward-gain amplifier and the feedback network have been replaced by their equivalent y-parameter two-port network representations so that parallel parameters can be easily combined. A resistive load has been applied to the output port; and, since shunt–shunt feedback amplifiers are transresistance amplifiers, a Norton equivalent source has been shown as the input. It should also be noted that the forward-gain parameter of each two-port network, y_{21}, is the transadmittance.

The basic feedback equation for a transresistance amplifier takes the form:

$$R_{Mf} = \frac{R_M}{1 + R_M f}.$$

(8.3-1)

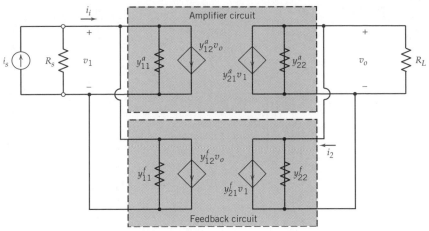

Figure 8.3-1
Two-port realization of shunt–shunt feedback amplifier.

The application of the basic feedback equation to this circuit in its current form is not immediately clear. It is necessary to transform the feedback amplifier circuit into a form that allows for easy application of the basic feedback equation, Equation 8.3-1. Such a transformation must meet the previously stated feedback requirements:

- The forward-gain amplifier is to be a forward transmission system only; its reverse transmission must be negligible.
- The feedback network is to be a reverse transmission system that presents a feedback current, dependent on the output voltage, to the amplifier input port.

While a mathematically rigorous derivation of the transformation is possible, greater insight to the process comes with a heuristic approach.

The two-port y-parameter representation, in conjunction with the shunt–shunt connection, is used to describe the two main elements of this feedback amplifier so that all the input port elements of both two-port networks are in parallel. Similarly, all output port elements are in parallel. It is well known that circuit elements in parallel may be rearranged and, so long as they remain in parallel, the circuit continues to function in an identical fashion. Hence, it is possible, for analysis purposes only, to conceptually move elements from one section of the circuit into another (from the feedback circuit to the amplifier circuit or the reverse). The necessary conceptual changes made for the transformation are:

- The source resistance, the load resistance, and all input and output admittances y_{11} and y_{22} are placed in the modified amplifier circuit.[5]
- All forward transadmittances y_{21} (represented by current sources dependent on the input voltage v_1) are placed in the modified amplifier circuit.
- All reverse transadmittances y_{12} (represented by current sources dependent on the output voltage v_o) are placed in the modified feedback circuit.

---

[5] Inclusion of the source and load resistance in the amplifier seems, at first, counterproductive. It is necessary, however, to include these resistances so that the use of the feedback properties produces correct results for input and output resistance (after appropriate transformations).

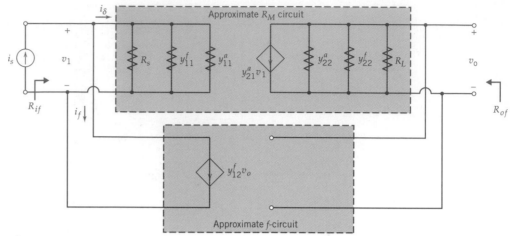

Figure 8.3-2
Redistributed shunt–shunt realization.

The dependent current sources can be easily combined:

$$y_{12}^t = y_{12}^a + y_{12}^f, \tag{8.3-2}$$

and

$$y_{21}^t = y_{21}^a + y_{21}^f. \tag{8.3-3}$$

In virtually every practical feedback amplifier the reverse transadmittance of the forward-gain amplifier is much smaller than that of the feedback network ($y_{12}^a \ll y_{12}^f$) and the forward transadmittance of the feedback network is much smaller than that of the forward-gain amplifier ($y_{21}^f \ll y_{21}^a$). Thus approximate simplifications of the amplifier representation can be made:

$$y_{12}^t = y_{12}^a + y_{12}^f \approx y_{12}^f, \tag{8.3-4}$$

$$y_{21}^t = y_{21}^f + y_{21}^a \approx y_{21}^a. \tag{8.3-5}$$

The shunt–shunt feedback amplifier circuit of Figure 8.3-1 is, with these changes and approximations, thereby transformed into the circuit shown in Figure 8.3-2.

This transformed circuit is composed of two simple elements:

- The original amplifier with its input shunted by the source resistance and the feedback network short-circuit input admittance y_{11}^f and its output shunted by the load resistance and the feedback network short-circuit output admittance y_{22}^f
- A feedback network composed solely of the feedback network reverse transadmittance y_{12}^f

It is also important to notice that the input resistance R_{if} of this circuit includes the source resistance R_s: as such it is not the same as the input resistance of the true amplifier, R_{in}. The input resistance of the true amplifier can be obtained as

$$R_{in} = \left(\frac{1}{R_{if}} - \frac{1}{R_s} \right)^{-1}. \tag{8.3-6}$$

Similarly the output resistance R_{of} of this circuit includes the load resistance R_L: Similar operations may be necessary to obtain the true output resistance of the amplifier.

The y-parameters of the feedback network can be obtained as outlined in Chapter 5:

$$y_{11}^f = \left.\frac{i_f}{v_1}\right|_{v_o=0}, \qquad y_{22}^f = \left.\frac{i_2}{v_0}\right|_{v_1=0}, \qquad y_{12}^f = \left.\frac{i_f}{v_0}\right|_{v_1=0}, \qquad (8.3\text{-}7)$$

where i_2 is the current entering the output port of the feedback network (see Figure 8.3-1). With the determination of these two-port parameters, the circuit has been transformed into a form that is compatible with all previous discussions. The forward-gain parameter (in this case G_M) of the loaded basic amplifier must be calculated, while the feedback ratio has been determined from the two-port analysis of the feedback network:

$$f = y_{12}^f. \qquad (8.3\text{-}8)$$

In the case of totally resistive feedback networks, the shunting resistances can be found in a simple fashion:

- Resistance $r_{\text{in}} = (y_{11}^f)^{-1}$ is found by setting the output voltage to zero value, $v_o = 0$, and determining the resistance from the input port of the feedback network to ground.
- Resistance $r_{\text{out}} = (y_{22}^f)^{-1}$ is found by setting the input voltage to zero value, $v_i = 0$, and determining the resistance from the output port of the feedback network to ground.

The feedback ratio f is simply the ratio of the feedback current i_f to the output voltage when the input port of the feedback network, v_i, is set to zero value.

All idealized feedback methods can be applied to this transformed amplifier and all previously derived feedback results are valid.

EXAMPLE 8.3-1 Determine the small-signal midband voltage gain

$$A_{vf} = \frac{v_o}{v_i}$$

and the indicated input and output resistances for the feedback amplifier shown. The silicon BJT is described by $\beta_F = 150$.

SOLUTION

This amplifier has typical shunt–shunt topology with a simple resistor as the feedback network. This resistor is directly connected to the base of the input of a common-emitter amplifier: This connection signifies shunt mixing. Similarly the direct connection to the output node signifies shunt sampling (if the load is replaced by a short circuit, the feedback goes to zero). As outlined above, analysis of feedback amplifiers follows a distinct procedure:

DC analysis: As in all amplifier designs, the DC quiescent conditions must be determined so that the BJT *h*-parameters can be determined. The primary equation of interest for this amplifier is a loop equation passing through the DC source and the BJT base–emitter junction:

$$16 - 3900(151 I_B) - 3300 I_B - 0.7 - 27(151 I_B) = 0.$$

This equation leads to the quiescent conditions

$$I_B = 25.66 \ \mu\text{A}, \qquad I_C = 3.849 \ \text{mA}, \qquad V_{CE} = 0.785 \ \text{V}.$$

The significant BJT *h*-parameters are determined from these conditions to be

$$h_{fe} = 150, \qquad h_{ie} = 1.020 \ \text{k}\Omega.$$

Partition the circuit into its functional modules: The AC equivalent amplifier circuit must be partitioned into its basic functional modules so that analysis can proceed. In addition a Norton equivalent of the source must be made so that the source resistance shunts the input. The functional modules necessary are the basic forward amplifier, the feedback network, and the source. The load resistance and the output resistance of the source must be included in the basic forward amplifier module.

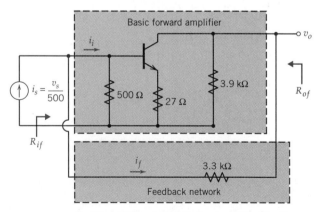

Partitioned AC equivalent feedback amplifier circuit

Load the basic forward amplifier: The input and output of the basic forward amplifier must be loaded with the short-circuit input and output admittances of the feedback network. That process is accomplished by shunting the input with a duplicate of the feedback network whose output has been shorted and shunting the output with a duplicate of the feedback network whose input has been shorted. The

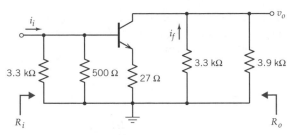

Loaded basic forward amplifier

results of such operations are as shown. Notice that the feedback network resistances (in this case the 3.3-kΩ resistor) may appear in this loaded circuit more than once. After identifying the feedback current i_f on the output portion of the circuit, the feedback quantity f can also be determined from this circuit from the duplicate of the feedback network loading the output:

$$f = \left.\frac{i_f}{v_o}\right|_{v_i=0} = \frac{-1}{3300}.$$

Determine the performance of the loaded forward amplifier: The determination of the AC performance of the loaded forward amplifier follows the procedures that have been developed in previous chapters. The input resistance is given by

$$R_i = 3300 \parallel 500 \parallel [h_{ie} + (1 + h_{fe})27]$$

$$= 434.2 \parallel [1020 + (1 + 150)27] = 400.1 \ \Omega.$$

The output resistance is

$$R_o = 3900 \parallel 3300 = 1.788 \ \text{k}\Omega.$$

The amplifier forward transresistance is given by

$$R_M = \frac{v_o}{i_i} = \left(\frac{v_o}{v_i}\right)\left(\frac{v_i}{i_i}\right) = (A_V)(R_i)$$

$$= \left(\frac{-150(3900 \parallel 3300)}{1020 + (1 + 150)27}\right)(400.12) = -21.05 \ \text{k}\Omega$$

Apply the feedback relationships to obtain total circuit performance: All of the necessary quantities have been determined for the use of the feedback relationships.

$$D = 1 + Af = 1 + R_M f = 1 - 21{,}050\left(\frac{-1}{3300}\right) = 7.3782,$$

$$R_{if} = \frac{R_i}{D} = \frac{400.12}{7.3782} = 54.23, \qquad R_{of} = \frac{R_o}{D} = \frac{1.7875}{7.3782} = 242.3,$$

$$R_{Mf} = \frac{R_M}{D} = \frac{-21{,}050}{7.3782} = -2.8525 \ \text{k}\Omega.$$

The feedback circuit performance parameters are obtained through simple relationships. The voltage gain is obtained with the current–voltage relationship of a Norton–Thévenin transformation:

$$A_{Vf} = \frac{v_o}{v_s}\Big|_f = \left(\frac{v_o}{i_s}\Big|_f\right)\left(\frac{i_s}{v_s}\right) = R_{Mf}\left(\frac{1}{500}\right)$$

$$= \frac{-2852.7}{500} = -5.705 \approx -5.70.$$

The output resistance is as shown and does not require modification:

$$R_{out} = R_{of} \approx 242 \; \Omega.$$

The input resistance must undergo the transformation of Equation 8.3-6:

$$R_{in} = \left(\frac{1}{R_{if}} - \frac{1}{500}\right)^{-1} = \left(\frac{1}{54.23} - \frac{1}{500}\right)^{-1} \approx 60.8 \; \Omega.$$

It is helpful to reiterate the procedure for finding the performance parameters of a feedback amplifier. In order to find the feedback amplifier performance parameters, the following procedure is followed:

- Perform a DC analysis to obtain active device performance parameters.
- Draw an AC equivalent circuit of the entire circuit.
- Partition the circuit into its functional modules.
- Load the basic forward amplifier.
- Determine the performance of the loaded forward amplifier.
- Apply the feedback relationships to obtain total circuit performance.
- If necessary, transform these performance parameters into equivalent performance parameters as specified.

This procedure can be followed with all four feedback topologies in the same manner as has been shown in Example 8.3-1. The specifics of loading the basic forward amplifier and the application of the feedback formulas will vary according to the topology.

8.3.2 Shunt–Series Feedback

Figure 8.3-3 shows the small-signal model of a typical shunt–series feedback amplifier. In this representation, the forward-gain amplifier and the feedback network have been replaced by their equivalent g-parameter two-port network representations so that input and output parameters can be easily combined. A resistive load has been applied to the output port; and, since shunt–series feedback amplifiers are current amplifiers, a Norton equivalent source has been shown as the input. The forward-gain parameter of each two-port, g_{21}, is the voltage gain.

The basic feedback equation for a current amplifier takes the form:

$$A_{If} = \frac{A_I}{1 + A_I f}. \tag{8.3-9}$$

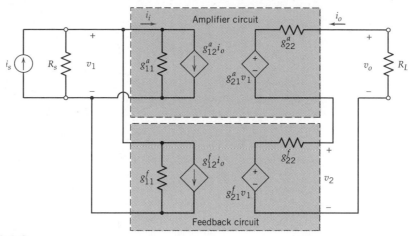

Figure 8.3-3
Two-port realization of shunt–series feedback amplifier.

Once again, it is necessary to transform the feedback amplifier circuit into a form that allows for easy application of the basic feedback equation, Equation 8.3-9. Such a transformation must meet the previously stated feedback requirements:

- The forward-gain amplifier is to be a forward transmission system only—its reverse transmission must be negligible.

- The feedback network is to be a reverse transmission system that presents a feedback current, dependent on the output voltage, to the amplifier input port.

The two-port g-parameter representation, in conjunction with the shunt–series connection, is used to describe the two main elements of this feedback amplifier so that all the input port elements of both two-port networks are in parallel. In contrast, all output port elements are in series. For analysis purposes only, elements are conceptually moved from one section of the circuit into another (from the feedback circuit to the amplifier circuit or the reverse). The necessary conceptual changes made for the transformation are as follows:

- The source resistance, all input admittance g_{11}, and output impedances g_{22} are placed in the modified amplifier circuit. The load resistance is kept separate.

- All forward voltage gains g_{21} (represented by voltage sources dependent on the input voltage v_1) are placed in the modified amplifier circuit.

- All reverse current gains g_{12} (represented by current sources dependent on the output current i_o) are placed in the modified feedback circuit.

The dependent current sources can be easily combined:

$$g_{12}^t = g_{12}^a + g_{12}^f, \tag{8.3-10}$$

and

$$g_{21}^t = g_{21}^a + g_{21}^f. \tag{8.3-11}$$

In virtually every practical feedback amplifier the reverse current gain of the forward-gain amplifier is much smaller than that of the feedback network ($g_{12}^a \ll g_{12}^f$)

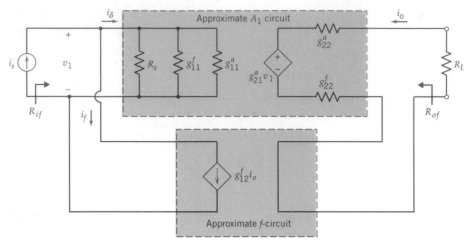

Figure 8.3-4
Redistributed shunt–series realization.

and the forward voltage gain of the feedback network is much smaller than that of the forward-gain amplifier ($g^f_{21} \ll g^a_{21}$). Thus approximate simplifications of the amplifier representation can be made:

$$g^t_{12} = g^a_{12} + g^f_{12} \approx g^f_{12}, \tag{8.3-12}$$

$$g^t_{21} = g^a_{21} + g^f_{21} \approx g^a_{21}. \tag{8.3-13}$$

The shunt–series feedback amplifier circuit of Figure 8.3-3 is, with these changes and approximations, thereby transformed into the circuit shown in Figure 8.3-4.

This transformed circuit is composed of two simple elements:

- The original amplifier with its input shunted by the source resistance and the feedback network open-circuit input admittance g^f_{11} and its output in series with the feedback network short-circuit output admittance g^f_{22} and the load resistance
- A feedback network composed solely of the feedback network reverse transadmittance g^f_{12}

As in the other case of shunt mixing, it is important to notice that the input resistance R_{if} of this circuit includes the source resistance R_s: as such it is not the same as the input resistance of the true amplifier R_{in}. The input resistance of the true amplifier can be obtained as

$$R_{in} = \left(\frac{1}{R_{if}} - \frac{1}{R_s} \right)^{-1}. \tag{8.3-14}$$

The output resistance R_{of} of this circuit does not include the load resistance R_L.

The g-parameters of the feedback network can be obtained as outlined in Chapter 5:

$$g^f_{11} = \left. \frac{i_f}{v_1} \right|_{i_o=0}, \qquad g^f_{22} = \left. \frac{v_2}{i_o} \right|_{v_1=0}, \qquad g^f_{12} = \left. \frac{i_f}{i_o} \right|_{v_1=0}, \tag{8.3-15}$$

where v_2 is the voltage across the output port of the feedback network (see Figure 8.3-3). With the determination of these two-port parameters, the circuit has been transformed into a form that is compatible with all previous discussions. The forward-gain parameter (in this case A_I) of the loaded basic amplifier must be calculated, while the feedback ratio has been determined from the two-port analysis of the feedback network:

$$f = g_{12}^f. \tag{8.3-16}$$

In the case of totally resistive feedback networks, the loading resistances can be found in a simple fashion:

- Resistance $r_{in} = (g_{11}^f)^{-1}$ is found by setting the output current to zero value, $i_o = 0$, and determining the resistance from the input port of the feedback network to ground. Resistance r_{in} shunts the input.
- Resistance $r_{out} = g_{22}^f$ is found by setting the input voltage to zero value, $v_i = 0$, and determining the resistance from the output port of the feedback network to ground. Resistance r_{out} is in series with the output.

The feedback ratio f is simply the ratio of the feedback current i_f to the output current when the input port of the feedback network v_i is set to zero value.

EXAMPLE 8.3-2 The circuit shown is a shunt–series feedback amplifier. Determine the midband voltage gain

$$A_v = \frac{v_o}{v_s}$$

and the indicated input and output resistances. The silicion BJTs are described by

$$\beta_F = 150.$$

SOLUTION

As in Example 8.3-1, the direct connection of the feedback network to the base of the input BJT signifies shunt mixing. At the output, however, the feedback is connected to the emitter of the output BJT, whereas the output is at the collector. Simple shorting of the output voltage has no effect on the feedback, while opening the collector of the output BJT eliminates all feedback: The sampling topology must be series sampling.

DC analysis: As in all amplifier designs, the DC quiescent conditions must be determined so that the BJT *h*-parameters can be determined. After setting all capacitors to open circuits, the usual analysis techniques lead to

$$I_{c1} = 1.136 \text{ mA} \Rightarrow h_{ie1} = 3.46 \text{ k}\Omega,$$

$$I_{c2} = 7.494 \text{ mA} \Rightarrow h_{ie2} = 524 \text{ }\Omega.$$

Partition the circuit into its functional modules: The AC equivalent amplifier circuit must be partitioned into its basic functional modules so that analysis can proceed. In addition a Norton equivalent of the source must be made so that the source resistance shunts the input. The functional modules necessary are the basic forward amplifier, the feedback network, the source, and the load. The output resistance of the source must be included in the basic forward amplifier module.

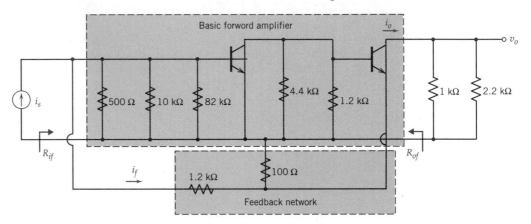

Partitioned AC equivalent feedback amplifier circuit

Replacing the source with its Norton equivalent leads to the relationship

$$i_s = \frac{v_s}{500}.$$

Load the basic forward amplifier: The basic forward amplifier must have its input and output ports loaded. The input port is shunted with the feedback network whose output port has been open circuited (shown as a series connection of 1.2 kΩ and

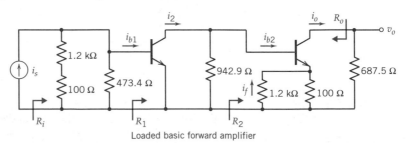

Loaded basic forward amplifier

100Ω). The output must have the feedback network, with its input port shorted to ground, placed in series with the output current (here at the emitter of the output BJT). This basic loading is in the accompanying figure. Resistors in parallel (from the partitioned circuit) have been combined in order to simplify later calculations.

In addition, the feedback ratio f is easily determined from the output loading circuit. The feedback current is identified in the output circuit (feedback current is subtracted from the input, thus it flows toward the output port) and a ratio is formed:

$$f = \frac{i_f}{i_o} = \left(\frac{i_f}{i_{e2}}\right)\left(\frac{i_{e2}}{i_o}\right) = \left(\frac{100}{100 + 1200}\right)\left(\frac{151}{150}\right) = 0.07744.$$

Determine the performance of the loaded forward amplifier: The input resistance of each amplifier stage is calculated as follows:

$$R_1 = h_{ie1} = 3.46 \text{ k}\Omega,$$

$$R_2 = h_{ie2} + 151(1200 \parallel 100) = 14.46 \text{ k}\Omega.$$

The input resistance of the basic loaded amplifier is given by

$$R_i = 1300 \parallel 473.4 \parallel R_1 = 315.4 \ \Omega.$$

The current gain can be calculated as

$$A_I = \left(\frac{i_o}{i_s}\right) = \left(\frac{i_o}{i_{b2}}\right)\left(\frac{i_{b2}}{i_2}\right)\left(\frac{i_2}{i_{b1}}\right)\left(\frac{i_{b1}}{i_s}\right),$$

$$= (150)\left(\frac{942.9}{942.9 + 14{,}460}\right)(150)\left(\frac{1300 \parallel 473.4}{1300 \parallel 473.4 + 3460}\right) = 125.6.$$

The output resistance is approximately infinite.

Apply the feedback relationships to obtain total circuit performance: With the calculation of the input and output resistance, the current gain, and the feedback ratio, it becomes possible to use the feedback relationships to find circuit performance including the effects of both the loading and feedback:

$$D = 1 + A_I f = 1 + 125.6(0.07744) = 10.72,$$

$$A_{If} = \frac{A_I}{D} = \frac{125.6}{10.72} = 11.71,$$

$$R_{if} = \frac{R_i}{D} = \frac{315.4}{10.72} = 29.42, \qquad R_{of} \approx \infty.$$

The defined performance characteristics are altered forms of the above relationships:

$$A_{Vf} = \left.\frac{v_o}{v_s}\right|_f = \left(\frac{v_o}{i_o}\right)\left(\left.\frac{i_o}{i_s}\right|_f\right)\left(\frac{i_s}{v_s}\right) = (1000 \parallel 2200)(11.71)\left(\frac{1}{500}\right) = 16.1,$$

$$R_{in} = \left(\frac{1}{R_{if}} - \frac{1}{R_s}\right)^{-1} = 31.3 \ \Omega,$$

$$R_{out} = R_{of} \parallel 1000 = 1 \text{ k}\Omega.$$

8.3.3 Series–Shunt Feedback

Figure 8.3-5 shows the small-signal model of a typical series–shunt feedback amplifier. In this representation, the forward-gain amplifier and the feedback network have been replaced by their equivalent h-parameter two-port network representations. A resistive load has been applied to the output port; and, since series–shunt feedback amplifiers are voltage amplifiers, a Thévenin equivalent source has been shown as the input. The forward-gain parameter of each two-port, h_{21}, is the current gain.

The basic feedback equation for a voltage amplifier takes the form

$$A_{Vf} = \frac{A_V}{1 + A_V f}. \tag{8.3-17}$$

Again a transformation of the circuit is necessary. The two-port h-parameter representation, in conjunction with the series–shunt connection, is used to describe the two main elements of this feedback amplifier so that all the input port elements of both two-port networks are in series. All output port elements are in parallel. For analysis purposes only, elements are conceptually moved from one section of the circuit into another.

- The load resistance, all input impedances h_{11}, and all output admittances h_{22} are placed in the modified amplifier circuit.
- All forward current gains h_{21} (represented by current sources dependent on the input current i_i) are placed in the modifed amplifier circuit.
- All reverse voltage gains h_{12} (represented by voltage sources dependent on the output voltage v_o) are placed in the modified feedback circuit.

The usual combinations and approximations hold:

$$h_{12}^t = h_{12}^a + h_{12}^f \approx h_{12}^f, \tag{8.3-18a}$$

$$h_{21}^t = h_{21}^a + h_{21}^f \approx h_{21}^a. \tag{8.3-18b}$$

The series–shunt feedback amplifier circuit of Figure 8.3-5 is, with these changes and approximations, thereby transformed into the circuit shown in Figure 8.3-6.

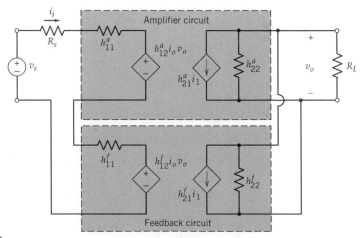

Figure 8.3-5
Two-port realization of series–shunt feedback amplifier.

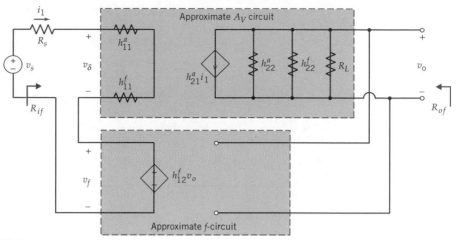

Figure 8.3-6
Redistributed series–shunt realization.

This transformed circuit is composed of two simple elements:

- The original amplifier with its output shunted by the load resistance and the feedback network open-circuit output admittance h_{22}^f and its input in series with the feedback network short-circuit input admittance h_{11}^f and the source resistance
- A feedback network composed solely of the feedback network reverse voltage gain h_{12}^f.

It is important to notice that the output resistance R_{of} of this circuit includes the load resistance R_L: as such, it may not be the same as the output resistance of the true amplifier, R_{out}. If the load resistance is not included in R_{out}, the correct expression is

$$R_{\text{out}} = \left(\frac{1}{R_{of}} - \frac{1}{R_L} \right)^{-1}. \tag{8.3-19}$$

The input resistance R_{if} of this circuit does not include the source resistance R_s or any other resistances in series with the input of the basic forward amplifier.

The h-parameters of the feedback network can be obtained as outlined in Chapter 5:

$$h_{11}^f = \left.\frac{v_1}{i_f}\right|_{v_o=0}, \qquad h_{22}^f = \left.\frac{i_2}{v_0}\right|_{i_1=0}, \qquad h_{12}^f = \left.\frac{v_f}{v_0}\right|_{i_1=0}, \tag{8.3-20}$$

where i_1 is the current entering the input port of the feedback network (see Figure 8.3-6). With the determination of these two-port parameters, the circuit has been transformed into a form that is compatible with all previous discussions. The forward-gain parameter (in this case A_V) of the loaded basic amplifier must be calculated, while the feedback ratio has been determined from the two-port analysis of the feedback network:

$$f = h_{12}^f. \tag{8.3-21}$$

In the case of totally resistive feedback networks, the loading resistances can be found in a simple fashion:

- Resistance $r_{in} = h_{11}^f$ is found by setting the output voltage to zero value, $v_o = 0$, and determining the resistance from the input port of the feedback network to ground. Resistance r_{in} is in series with the input.
- Resistance $r_{out} = (h_{22}^f)^{-1}$ is found by setting the input current to zero value, $i_1 = 0$, and determining the resistance from the output port of the feedback network to ground. Resistance r_{out} shunts the output.

The feedback ratio f is simply the ratio of the feedback voltage v_f to the output voltage when the input current of the feedback network i_1 is set to zero.

EXAMPLE 8.3-3 Determining the midband voltage gain

$$A_v = \frac{v_o}{v_s}$$

and the indicated input and output resistances for the circuit shown. The silicon BJTs are described by

$$\beta_F = 150.$$

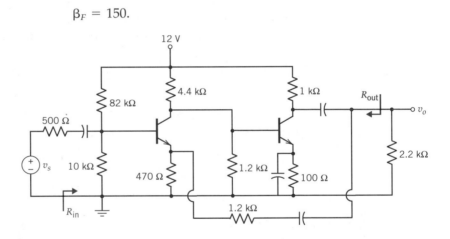

SOLUTION

In this circuit the feedback network is connected at the emitter of the input BJT. Feedback in the form of a voltage will appear here in series with a Thévenin source and the base–emitter junction of the BJT: This is series mixing. The direct connection of the feedback network to the output node implies shunt mixing (replacing the load with a short eliminates all feedback).

DC analysis: The DC portion of this circuit is identical to that of Example 8.3-2. The only changes have been in the feedback (which is capacitively coupled) and the resistors bypassed by capacitors. Thus the DC quiescent conditions of the BJTs remain unchanged. The h-parameters are also the same as in Example 8.3-2:

$$h_{ie1} = 3.46 \text{ k}\Omega, \qquad h_{ie2} = 524 \text{ }\Omega.$$

Partition the circuit into its functional modules: The AC equivalent amplifier circuit must be partitioned into its basic functional modules so that analysis can proceed. The functional modules necessary are the basic forward amplifier, the feedback net-

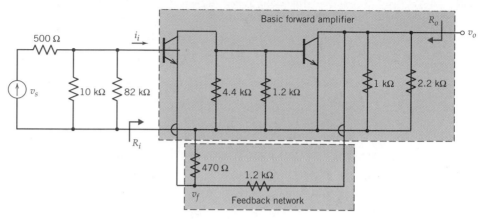

Partitioned AC equivalent feedback amplifier circuit

work, and the source. Of particular interest is that the bias resistors at the input of the first BJT must be partitioned within the source. This partitioning is due to the requirement of series mixing that items at the input must be in series. The load resistance is included in the basic forward amplifier module.

Load the basic forward amplifier: The basic forward amplifier must have its input and output ports loaded. The output port is shunted with the feedback network whose input port has been open circuited (shown as a series connection of 1.2 kΩ and 470 Ω). The input must have the feedback network, with its output port shorted to ground, placed in series with the input active device (shown as the parallel connection 1.2 kΩ and 470 Ω in series with the emitter of the BJT). This basic loading is shown in the accompanying figure. Resistors in parallel (from the partitioned circuit) have been combined in order to simplify later calculations.

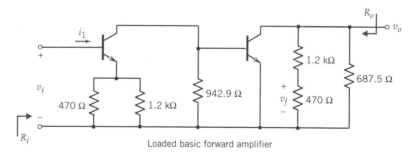

Loaded basic forward amplifier

In addition, the feedback ratio f is easily determined from the output loading circuit. The feedback voltage is identified on the output loading circuit (feedback voltage is subtracted from the input, thus its positive pole is toward the output port) and a ratio is formed:

$$f = \frac{v_f}{v_o} = \frac{470}{470 + 1200} = 0.2814.$$

Determine the performance of the loaded forward amplifier: The input and output resistance of the loaded amplifier are calculated to be

$$R_i = h_{ie1} + 151(470 \parallel 1200) = 54.46 \text{ k}\Omega,$$
$$R_o = 687.5 \parallel 1670 = 487.0 \text{ } \Omega.$$

The voltage gain is given by

$$A_V = \left(\frac{v_o}{v_i}\right) = \left(\frac{v_o}{v_1}\right)\left(\frac{v_1}{v_i}\right)$$

$$= \left(\frac{-150(487.0)}{524}\right)\left(\frac{-150(942.9 \parallel 524)}{54,460}\right) = 129.4$$

Apply the feedback relationships to obtain total circuit performance: All necessary components are now available to use the feedback formulas:

$$D = 1 + A_V f = 37.40, \qquad A_{Vf} = \frac{A_V}{D} = 3.46,$$

$$R_{if} = R_i D = 2.037 \text{ M}\Omega, \qquad R_{of} = \frac{R_o}{D} = 13.02 \ \Omega.$$

In order to match the performance parameter definitions of the stated problem, there must be a few alterations:

$$R_{\text{in}} = R_{if} \parallel 10,000 \parallel 82,000 = 8.87 \text{ k}\Omega,$$

$$R_{\text{out}} = \left(\frac{1}{R_{of}} - \frac{1}{2200}\right)^{-1} = 13.1 \ \Omega,$$

$$A_V = \frac{v_o}{v_s} = \left(\frac{v_o}{v_i}\right)\left(\frac{v_i}{v_s}\right) = A_{Vf} \frac{R_{\text{in}}}{R_{\text{in}} + 500} = 3.27.$$

8.3.4 Series–Series Feedback

Figure 8.3-7 shows the small-signal model of a typical series–series feedback amplifier. In this representation, the forward-gain amplifier and the feedback network have been replaced by their equivalent z-parameter two-port network representations. A

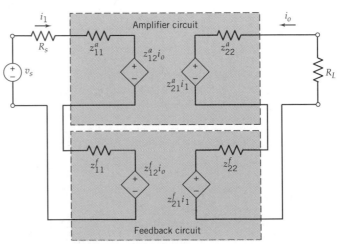

Figure 8.3-7
Two-port realization of series–series feedback amplifier.

resistive load has been applied to the output port; and, since series–series feedback amplifiers are transconductance amplifiers, a Thévenin equivalent source has been shown as the input. The forward-gain parameter of each two port, z_{21}, is the transimpedance.

The basic feedback equation for a transconductance amplifier takes the form

$$G_{mf} = \frac{G_M}{1 + G_M f}. \tag{8.3-22}$$

Again a transformation of the circuit is necessary. The two-port z-parameter representation, in conjunction with the series–series connection, is used to describe the two main elements of this feedback amplifier so that all the input port elements of both two-port networks are in series. All output port elements are also in series. For analysis purposes only, elements are conceptually moved from one section of the circuit into another:

- The input and output impedances z_{11} and z_{22} are placed in the modified amplifier circuit.
- All forward transresistance z_{21} (represented by voltage sources dependent on the input current i_1) are placed in the modified amplifier circuit.
- All reverse transresistance z_{12} (represented by voltage sources dependent on the output current i_o) are placed in the modified feedback circuit.

The usual combinations and approximations hold:

$$z_{12}^t = z_{12}^a + z_{12}^f \approx z_{12}^f, \tag{8.3-23}$$

$$z_{21}^t = z_{21}^a + z_{21}^f \approx z_{21}^a. \tag{8.3-24}$$

The series–series feedback amplifier circuit of Figure 8.3-7 is, with these changes and approximations, thereby transformed into the circuit shown in Figure 8.3-8.

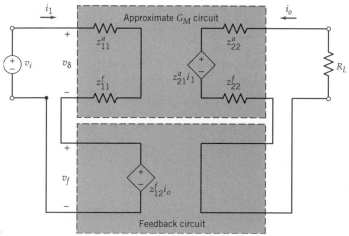

Figure 8.3-8
Redistributed series–series realization.

This transformed circuit is composed of two simple elements:

- The original amplifier with its input in series with the feedback network short-circuit input impedance z_{11}^f and the source resistance and its output in series with the feedback network short-circuit output impedance z_{22}^f and the load
- A feedback network composed solely of the feedback network reverse voltage gain z_{12}^f

The z-parameters of the feedback network can be obtained as outlined in Chapter 5:

$$z_{11}^f = \left.\frac{v_1}{i_1}\right|_{i_o=0}, \qquad z_{22}^f = \left.\frac{v_0}{i_o}\right|_{i_1=0}, \qquad z_{12}^f = \left.\frac{v_f}{i_o}\right|_{i_1=0}, \qquad (8.3\text{-}25)$$

where i_1 is the current entering the input port of the feedback network (see Figure 8.3-7). With the determination of these two-port parameters, the circuit has been transformed into a form that is compatible with all previous discussions. The forward-gain parameter (in this case G_M) of the loaded basic amplifier must be calculated, while the feedback ratio is determined from the two-port analysis of the feedback network:

$$f = z_{12}^f. \qquad (8.3\text{-}26)$$

In the case of totally resistive feedback networks, the loading resistances can be found in a simple fashion:

- Resistance $r_{in} = z_{11}^f$ is found by setting the output current to zero value, $i_o = 0$, and determining the resistance from the input port of the feedback network to ground. Resistance r_{in} is in series with the input.
- Resistance $r_{out} = z_{22}^f$ is found by setting the input current to zero value, $i_1 = 0$, and determining the resistance from the ouput port of the feedback network to ground. Resistance r_{out} is in series with the output.

The feedback ratio f is simply the ratio of the feedback voltage v_f to the output voltage when the input current of the feedback network i_1 is set to zero.

EXAMPLE 8.3-4   Determine the following midband performance parameters for the following circuit shown: the voltage gain

$$A_v = \frac{v_o}{v_s}$$

and the indicated input and output resistances. The transistors are silicion with parameters

$$\beta_F = 150.$$

SOLUTION

In this circuit, the feedback network is connected to the base of the input BJT, but the input is at the emitter. Again, feedback will appear in series with the input and

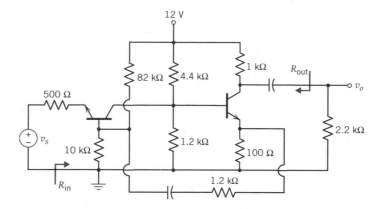

the base–emitter junction of the BJT: This is series mixing. The connection of the feedback network at the output characterizes series sampling (opening *both* load resistors at the collector of the output BJT eliminated all feedback).

DC analysis: As in all amplifier designs, the DC quiescent conditions must be determined so that the BJT *h*-parameters can be determined. After setting all capacitors to open circuits, the usual analysis techniques lead to

$$I_{c1} = 1.074 \text{ mA} \Rightarrow h_{ie1} = 3.66 \text{ k}\Omega,$$

$$I_{c2} = 8.030 \text{ mA} \Rightarrow h_{ie2} = 489 \text{ }\Omega.$$

Partition the circuit into its functional modules: The amplifier is partitioned into its two functional modules. Series mixing implies the need for a Thévenin source. Since there is series mixing at both input and output port, the source and load resistances are excluded from the basic forward amplifier.

The AC model of the partitioned amplifier is shown with all parallel resistances combined.

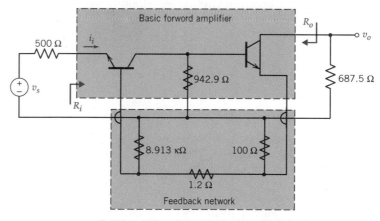

Partitioned AC equivalent feedback amplifier circuit

Load the basic forward amplifier: The input loading is achieved by setting the output current i_o to zero. This opens the output port of the feedback network and forms the input loading circuit connected to the base of the input BJT. The output loading is achieved by setting the input current i_i to zero. This opens the input port of the feedback network and forms the output loading circuit connected to the emitter of the output BJT.

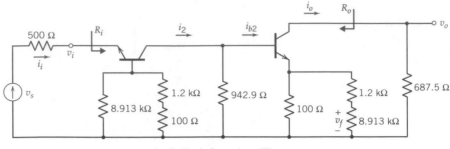

Loaded basic forward amplifier

The feedback quantity v_f is then drawn on the output loading circuit (positive end toward the load). The feedback ratio is determined to be

$$f = \frac{v_f}{i_o} = \frac{-151}{150} \frac{100}{100 + 1200 + 8913} (8913) = -87.85.$$

Determine the performance of the loaded forward amplifier: The following relationships can be determined for the loaded basic forward amplifier.

The input resistance to the total circuit is given by the input resistance of a common base amplifier:

$$R_i = \frac{h_{ie1} + R_b}{h_{fe} + 1} = \frac{3660 + (8913 \,\|\, 1300)}{151} = 31.72.$$

The input resistance to the common emitter stage is given by

$$R_{i1} = h_{ie2} + 151[100 \,\|\, (1200 + 8913)] = 15{,}441.$$

With these two relationships, the forward transconductance can be calculated:

$$G_M = \frac{i_o}{v_i} = \left(\frac{i_o}{i_{b2}}\right)\left(\frac{i_{b2}}{i_2}\right)\left(\frac{i_2}{i_i}\right)\left(\frac{i_i}{v_i}\right)$$

$$= (-150)\left(\frac{942.9}{942.9 + R_{i2}}\right)\left(\frac{150}{151}\right)\left(\frac{1}{R_i}\right) = -0.2703.$$

The output resistance of this amplifier is approximately infinite.

Apply the feedback relationships to obtain total circuit performance: With the forward gain, feedback ratio, input resistance, and output resistance calculated for the basic loaded forward amplifier, the effects of feedback can now be easily calculated:

$$D = 1 + G_M f = 1 + (-0.2703)(-87.85) = 24.75,$$

$$R_{if} = R_i D = 785, \qquad R_{of} = R_o D \approx \infty,$$

$$G_{Mf} = \frac{G_M}{D} = -0.010923.$$

These results are then transformed into the proper performance parameters as defined in the statement of the problem:

$$A_{Vf} = \frac{v_o}{v_s}\Big|_f = \left(\frac{v_o}{i_o}\right)\left(\frac{i_o}{v_i}\Big|_f\right)\left(\frac{v_i}{v_s}\Big|_f\right) = (1000 \parallel 2200)(G_{Mf})\left(\frac{R_{if}}{R_{if}+500}\right)$$

$$= -4.59,$$

$$R_{\text{in}} = R_{if} = 785\ \Omega, \qquad R_{\text{out}} = R_{of} \parallel 1000 = 1\ \text{k}\Omega.$$

8.4 CONCLUDING REMARKS

The fundamental advantages of the use of feedback in electronic amplifiers have been described in this chapter. Reduction in non-linear distortion, variation in amplifier performance parameters, and control of input and out impedance are all significant benefits available through the use of feedback.

The analysis of feedback amplifiers can be a complex process complicated by the

TABLE 8.4-1 FEEDBACK AMPLIFIER ANALYSIS

Characteristic	Topology			
	Shunt–Shunt	**Shunt–Series**	**Series–Shunt**	**Series–Series**
Input, X_i	Current i_s	Current i_s	Voltage v_s	Voltage v_s
Output, X_o	Voltage v_o	Current i_o	Voltage v_o	Current i_o
Signal source	Norton	Norton	Thévenin	Thévenin
Input circuit	Include shunting resistances; set $v_o = 0$	Include shunting resistances; set $i_o = 0$	Exclude all shunt resistances; set $v_o = 0$	Exclude all shunt resistances; set $i_o = 0$
Output circuit	Include shunting resistances; set $v_i = 0$	Exclude all shunt resistances; set $v_i = 0$	Include shunting resistances; set $i_i = 0$	Exclude all shunt resistances; set $i_i = 0$
Feedback ratio, f	i_f/v_o	i_f/i_o	v_f/v_o	v_f/i_o
Forward gain, A	Transresistance R_M	Current gain A_I	Voltage gain A_V	Transconductance G_M
Input resistance	$R_{if} = \dfrac{R_i}{(1+R_Mf)}$	$R_{if} = \dfrac{R_i}{(1+A_If)}$	$R_{if} = R_i(1+A_Vf)$	$R_{if} = R_i(1+G_Mf)$
Output resistance	$R_{of} = \dfrac{R_o}{(1+R_Mf)}$	$R_{of} = R_o(1+A_If)$	$R_{of} = \dfrac{R_o}{(1+A_Vf)}$	$R_{of} = R_o(1+G_Mf)$

NOTES: *Input/output circuit*: These procedures give the basic forward amplifier without feedback but including the effects of loading due to the feedback network.
Resistance: The resistance modified at *shunted* ports will include all shunting resistances that were included in the basic forward amplifier. The resistance modified at *series* ports will only include the resistances included in the basic forward amplifier. The true amplifier input and output impedances must be modified to reflect the point of measurement desired.

interaction of the basic forward amplifier and the feedback network. It is, however, possible to simplify this analysis process through the use of a few basic feedback relationships and a systematic analysis method. This analysis method consists of a seven-step process:

- Identify the feedback topology.
- Perform a DC analysis to obtain active device performance parameters.
- Partition the AC equivalent circuit into its functional modules.
- Load the basic forward amplifier.
- Determine the performance of the loaded forward amplifier.
- Apply the feedback relationships to obtain total circuit performance.
- If necessary, transform these performance parameters into equivalent performance parameters as specified.

Vital to the analysis method are the partitioning of the circuit into its functional modules and loading the basic forward amplifier. Table 8.4-1 serves as an aid in that process, itemizing the process variation due to feedback topology.

Another significant benefit of feedback, the increase in the midband frequency range, will be extensively discussed in Chapter 11.

SUMMARY DESIGN EXAMPLE

The design process often involves the modification of an existing design to meet new, but similar, performance specifications. As an example, the feedback amplifier shown is constructed with enhancement mode FETs with the following properties:

$$K = 1 \text{ mA/V}, \qquad V_T = 1.5 \text{ V}, \qquad V_A = 160 \text{ V}.$$

This amplifier is known to have high input resistance ($\approx 260 \text{ k}\Omega$), moderate output resistance ($\approx 300 \ \Omega$), and a midband voltage gain

$$A = \frac{v_o}{v_s} \approx 18.3.$$

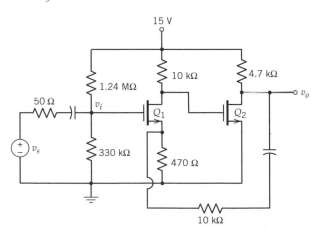

The quiescent conditions for the transistors are known to be

$$I_{D1} = 1.19 \text{ mA}, \qquad I_{D2} = 2.47 \text{ mA}.$$

It is desired to design an amplifier with similar input and output resistance and a midband voltage gain of 15. Redesign the amplifier to meet the new design goals.

Solution

If at all possible, a redesign process should minimize the number of changes. For the given amplifier, the FET AC parameters will remain the same if the quiescent drain currents are not changed. Thus, it seems advantageous to retain the same Q point. This proposed redesign constraint limits changes to the feedback resistor (10 kΩ) and the load resistor (4.7 kΩ). While variation in either will alter the maximum symmetrical output swing, voltage gain is more strongly dependent on the feedback resistor: Less variation in quiescent output voltage will result in changing the feedback resistor. It is therefore decided to alter only the feedback resistor to accomplish the new design goals.

This amplifier is a series–shunt feedback amplifier: The gain parameter that is stabilized is the voltage gain from the input port to the output port. In order to meet the new design goals the gain with feedback must be (the input resistance of this amplifier simply is the parallel combination of the bias resistors and does not change with variation in feedback):

$$A_{vf} = A \frac{R_{\text{in}}}{R_{\text{in}} + 50} = 15.0029.$$

The expression for the *loaded gain* of the amplifier *without feedback* is given by

$$A_v = \frac{-g_{m1}r_{d1}(10{,}000)}{r_{d1} + 10{,}000 + g_{m1}r_{d1}(470 \| R_f)} \left(-g_{m2}[r_{d2} \| 4700 \| (470 + R_f)] \right).$$

The feedback quantity is given by

$$f = \frac{v_f}{v_o} = \frac{470}{470 + R_f}.$$

If the quiescent condition does not change with the redesign, the FET parameters can easily be found to be

$$g_m = 2\sqrt{I_D K} \Rightarrow g_{m1} = 2.18 \text{ mA/V}, \qquad g_{m2} = 3.14 \text{ mA/V},$$

$$r_d = \frac{V_A}{I_D} \Rightarrow r_{d1} = 134 \text{ k}\Omega, \qquad r_{d2} = 6.48 \text{ k}\Omega.$$

Thus, for the original design the gain is given by

$$A_v = (-10.6)(-9.7) = 102.8, \qquad f = 0.04489, \qquad A_{vf} = 18.31.$$

Two alternatives exist for finding the value of R_f so that the design goals are met:

- A mathematical search of the exact gain expression
- Reasonable approximations of the gain expression

In this case, neither provides a significant advantage in the trade-off between difficulty and accuracy of the result. A mathematical search provides $R_f = 7.882$ kΩ. One reasonable set of approximations are shown below.

In the expression for A_v, the parallel combination of R_f with 470 Ω is unlikely to change significantly with reasonable variation in R_f. A good approximation to the unloaded gain is therefore

$$A_v = (-10.6)\{-g_{m2}[r_{d2} \parallel 4700 \parallel (470 + R_f)]\}$$
$$= 0.03328[4382 \parallel (470 + R_f)].$$

The gain with feedback can then be written as

$$A_{vf} = \frac{0.03328\{4382(470 + R_f)/[4382 + (470 + R_f)]\}}{1 + 0.03328\{4382 (470 + R_f)/[4382 + (470 + R_f)]\}[470/(470 + R_f)]}$$

Significant cancellation occurs:

$$A_{vf} = \frac{0.03328[4382(470 + R_f)]}{4382 + (470 + R_f) + 0.03328(4382)(470)}.$$

This expression can easily be solved for R_f so that the gain $A_{vf} = 15.0029$:

$$R_f = \frac{A_{vf}[4382 + 0.03328(4382)(470)]}{0.03328(4382) - A_{vf}} - 470 = 7.892 \text{ kΩ}.$$

The variation between this approximate solution and the exact solution is 0.127%. The closest standard value resistor is 7.87 kΩ.

REDESIGN SUMMARY:

The original circuit remains essentially the same with the exception that the feedback resistor is reduced in value from 10 to 7.87 kΩ.

REFERENCES

Ghausi, M. S., *Electronic Devices and Circuits: Discrete and Integrated*, Holt, Rinehart, and Winston, New York, 1985.

Gray, P. R., and Meyer, R. G., *Analysis and Design of Analog Integrated Circuits*, 3rd ed., Wiley, New York, 1993.

Hurst, P. J., "A Comparison of Two Approaches to Feedback Circuit Analysis," *IEEE Trans. Ed.*, Vol. 35, No. 3, 1992.

Millman, J., *Microelectronics, Digital and Analog Circuits and Systems*, McGraw-Hill, New York, 1979. ·

Nilsson, J. W., *Electric Circuits*, 3rd ed., Addison-Wesley, New York, 1990.

Rosenstark, S., *Feedback Amplifier Principles*, Macmillan, New York, 1986.

Sedra, A. S., and Smith, K. C., *Microelectronic Circuits*, 3rd ed., Holt, Rinehart, and Winston, New York, 1991.

PROBLEMS

8-1. The voltage gain of an amplifier is subject to a variation given by

$$A = 238 \pm 4 \text{ V/V}.$$

It is desired to use this amplifier as the fundamental forward-gain element in a feedback amplifier. The feedback amplifier must have no more than 0.15% variation in its voltage gain.

    a. Determine the necessary feedback ratio to achieve this variation while achieving maximum gain.

    b. What is the gain of the feedback amplifier?

8-2. An amplifier with an open-loop voltage gain of $A_V = 1000 \pm 100$ is given. The amplifier must be altered so that the voltage gain varies by no more than $\pm 0.5\%$.

    a. Find the feedback factor f to achieve this variation while achieving maximum gain.

    b. Find the gain of the feedback amplifier.

8-3. A feedback amplifier must be designed to have a closed-loop voltage gain of 120 utilizing a feedback network with a feedback factor $f = 0.02$. Determine the range of values of the gain A_V for which the closed-loop gain A_{Vf} is

    a. 120 ± 1

    b. 120 ± 0.2

8-4. Design a feedback amplifier that has a nominal gain of 200 with no more than 1% variation in overall gain due to the variation of individual elements. The individual elements available are amplifier stages having a gain of 2000 with 25% variation. Other constraints require that feedback be applied only to each individual stage and that such stages be connected in cascade.

8-5. Design a feedback amplifier that has a nominal gain of 200 with no more than 1% variation in overall gain due to the variation of individual elements. The individual elements available are amplifier stages having a gain of 2000 with 25% variation. Other constraints require that stages be connected in cascade and that

feedback be applied only to the total cascade-connected amplifier.

8-6. The gain of an amplifier is described by the relationship

$$A(v) = 10 \left| \arctan\left(\frac{v}{4}\right) \right| + 5$$

for input voltages ranging from -10 to $+10$ V. Feedback is applied to the circuit so that the feedback ratio $f = 0.1$

    a. Plot the voltage transfer relationship of the original amplifier.

    b. On the same graph, plot the voltage transfer relationship of the feedback amplifier.

    c. Quantitatively, compare the linearity of the feedback amplifier to that of the amplifier without feedback. Comment on results.

8-7. The gain of an amplifier is described by the relationship:

$$A(v) = 10 \left| \arctan\left(\frac{v}{4}\right) \right| + 5.$$

    a. Determine the input v necessary to produce an output of value 50.

    b. If a sinusoid of the amplitude, v (determined in part a), is the input to this amplifier, what is the total harmonic distortion of the output?

    c. Feedback is applied to the circuit so that the feedback ratio $f = 0.1$. What value of the input to the new amplifier will produce an output of 50?

    d. Repeat part b with an input of amplitude, v, as determined in part c. Compare results.

8-8. The gain of an amplifier is described by the relationship

$$A(v) = \tfrac{1}{10}v^2 + \tfrac{1}{2}v + 4$$

for input voltages ranging from -5 to $+5$ V. Feedback is applied to the circuit so that the feedback ratio $f = 0.6$

    a. Plot the voltage transfer relationship of the original amplifier.

    b. On the same graph, plot the voltage

transfer relationship of the feedback amplifier.

c. Quantitatively, compare the linearity of the feedback amplifier to that of the amplifier without feedback. Comment on results.

8-9. The gain of an amplifier is described by the relationship

$A(v) = \frac{1}{10}v^2 + \frac{1}{2}v + 4.$

a. Determine the input v necessary to produce an output of value 20.

b. If a sinusoid of the amplitude, v (determined in part a), is the input to this amplifier, what is the total harmonic distortion of the output?

c. Feedback is applied to the circuit so that the feedback ratio $f = 0.6$. What value of the input to the new amplifier will produce an output of 20?

d. Repeat part b with an input of amplitude, v, as determined in part c. Compare results.

8-10. Despite the prominent crossover distortion inherent in class B amplifiers, several OpAmp types use a class B output stage. A simple model for such an OpAmp is shown. The application of feedback in the form of resistors connecting the input and output OpAmp terminals significantly reduces the crossover distortion so that the OpAmp can be used as an effective circuit element that is essentially linear.

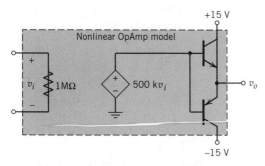

a. Determine the amplitude of an input sinusoidal voltage v_i that will produce a 1-V amplitude output for the circuit shown. Determine the total harmonic distortion of the output using SPICE (either with a **.FOUR** or a **.DISTO** statement). Assume the OpAmp is loaded

with a 470-Ω resistor and the BJTs are silicon with $\beta_F = 75$.

b. Design an OpAmp inverting amplifier with a voltage gain of 10 using the given OpAmp model and appropriate resistors. Using SPICE, apply a 0.1-V sinusoid to this amplifier (loaded with a 470-Ω resistor) and determine the total harmonic distortion of the output. Compare results to those obtained in part a

8-11. It is not intuitively obvious as to whether local or global feedback is most beneficial in a multistage amplifier. The two amplifiers shown are examples of each for a two-stage amplifier. The two feedback ratios f_1 and f_2 are chosen so that the total gain A_f of each circuit is identical. The total gain includes all feedback effects and is defined as

$$A_f = \frac{v_o}{v_i}.$$

Investigate which feedback amplifier configuration has greater resistance in the total gain A_f to variation in the gain of the individual stages A.

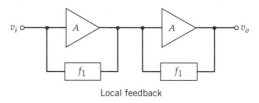

Local feedback

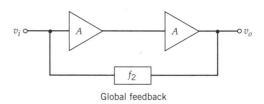

Global feedback

8-12. Complex amplifiers may incorporate both local and global feedback. The amplifier shown is an example of such a complex amplifier. Derive the transfer characteristic for this amplifier.

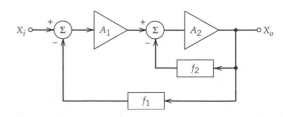

8-13. The BJT in the shunt–shunt feedback amplifier of Example 8.3-1 is replaced by a BJT described by

$\beta_F = 120$.

    a. Determine the change in the small-signal midband gain of the amplifier due to this substitution.

    b. Verify the gain with SPICE simulation.

    c. Verify that the gain variation is within that predicted by the gain stabilization formulas.

8-14. The shunt–shunt feedback amplifier of Example 8.3-1 is to be redesigned to have a small-signal midband gain of

$$A_{vf} = \frac{v_o}{v_i} = -6.0 \pm 0.05.$$

This change is to be accomplished by alteration of the value of the feedback resistor *only*.

    a. Determine the new resistance value that will meet the gain specifications.

    b. Determine the change in the input resistance due to the change.

    c. Verify results using SPICE.

8-15. For the single-stage amplifier shown, the FET parameters are

$K = 1 \text{ mA/V}^2, \quad V_T = 1.5 \text{ V}, \quad V_A = 160 \text{ V}.$

The FET is replaced by another described by

$K = 1.2 \text{ mA/V}^2, \quad V_T = 1.6 \text{ V}, \quad V_A = 150 \text{ V}.$

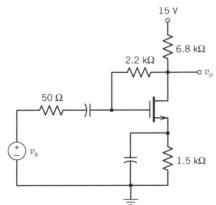

    a. Determine the change in the small-signal midband gain of the amplifier due to this substitution:

$$A = \frac{v_o}{v_s}.$$

    b. Verify that the gain variation is within that predicted by the gain stabilization formulas.

8-16. The design goal for the given feedback amplifier is a voltage gain:

$$A = \frac{v_o}{v_s} = -5.$$

Redesign the amplifier to achieve that voltage gain goal. The FET is described by

$K = 1.2 \text{ mA/V}^2, \quad V_T = 1.6 \text{ V},$
$V_A = 150 \text{ V}.$

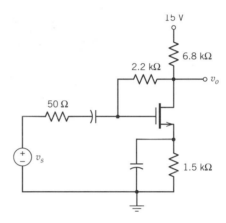

8-17. For the circuit shown, assume the transistors are biased into the forward-active region.

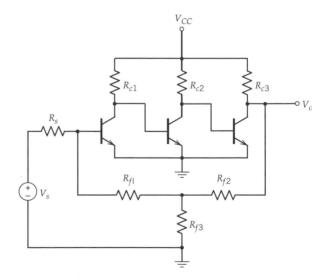

    a. Identify the topology of the feedback.

    b. Draw a circuit diagram of the loaded basic forward amplifier.

c. Write the expression for the feedback ratio f.

d. Write the expression for the midband voltage gain in terms of the circuit and BJT parameters.

8-18. For the single-stage amplifier shown, determine the value of R_f to achieve an overall midband voltage gain of magnitude 8. The FET parameters are

$$K = 1 \text{ mA/V}^2, \qquad V_T = 1.5 \text{ V}, \qquad V_A = 160 \text{ V}.$$

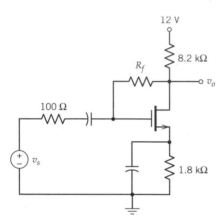

8-19. Complete the design of the circuit shown to achieve a closed-loop gain:

$$A_{Vf} = -10.$$

The JFET characteristic parameters are

$$V_{\text{PO}} = -3.5 \text{ V}, \qquad I_{DSS} = 8 \text{ mA},$$
$$V_A = 120 \text{ V}.$$

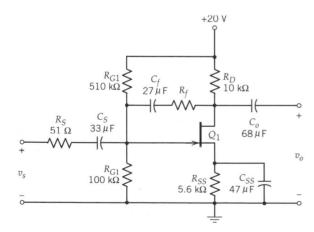

8-20. Complete the design of the circuit shown for a closed-loop gain $A_{Vf} = -10$.

The JFET characteristic parameters are

$$V_{\text{PO}} = -4 \text{ V}, \qquad I_{DSS} = 8 \text{ mA}, \qquad V_A = 120 \text{ V}.$$

The JFET is to be biased such that

$$V_{GS} = -1 \text{ V}, \qquad V_{DS} = 7 \text{ V}.$$

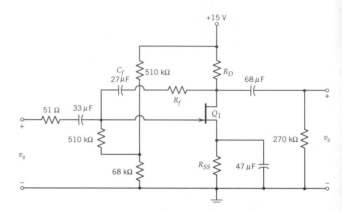

8-21. The performance parameters of the given OpAmp circuit can be determined using feedback techniques. The characteristics of the OpAmp are

Input resistance	1 MΩ
Output resistance	75 Ω
Voltage gain	500 kV/V

a. Identify the feedback topology.

b. Draw a circuit diagram for the loaded basic forward amplifier.

c. Determine the voltage gain for the total circuit using feedback techniques. Compare results to those determined using the techniques described in Chapter 1.

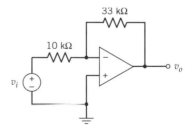

8-22. Use the principles of feedback analysis to find the closed-loop voltage gain, the input resistance R'_{if} and output resistance R'_{of}. The OpAmp open-loop gain is $A_V = 10\text{k V/V}$, input resistance of the OpAmp is 1 MΩ, and the output resistance of the OpAmp is 75 Ω.

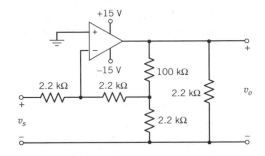

8-23. Determine the closed-loop gain, input resistance, and output resistance of the amplifier shown. The transistors have the following characteristics:

$$\beta_F = 200, \qquad V_A = 120 \text{ V} \quad (npn),$$

$$\beta_F = 200, \qquad V_A = 100 \text{ V} \quad (pnp).$$

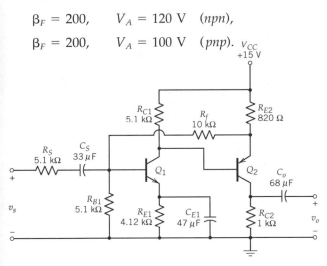

8-24. Determine the closed-loop gain, input resistance, and output resistance of the amplifier shown. The transistors are identical and have the following characteristic parameters:

$$\beta_F = 200, \qquad V_A = 120 \text{ V}.$$

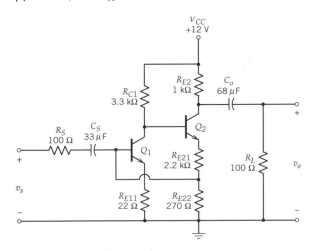

8-25. Determine the closed-loop gain, input resistance (not including the source resistor R_S), and output resistance of the amplifier shown. The transistors are identical and have the following characteristics:

$$\beta_F = 180, \qquad V_A = 200 \text{ V}.$$

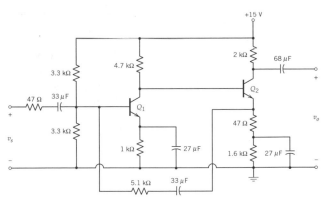

8-26. Determine the closed-loop gain, input resistance (not including the source resistor R_S), and output resistance of the amplifier shown. The transistors are identical and have the following characteristic parameters:

$$\beta_F = 200, \qquad V_A = 120 \text{ V}.$$

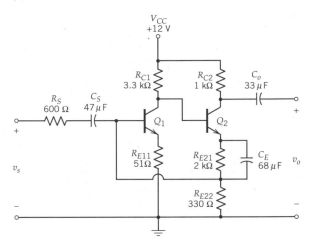

8-27. The shunt–series feedback amplifier of Example 8.3-2 is to be redesigned to have a voltage gain

$$A_v = \frac{v_o}{v_s} = 20 \pm 0.1.$$

The change is to be accomplished by alteration of the feedback (1.2-kΩ) resistor only. Complete this redesign and determine the effect of the change on input and output resistance of the amplifier. Use SPICE to verify the redesigned amplifier performance.

 8-28. In an attempt to create a circuit with a midband voltage gain

$$A_v = \frac{v_o}{v_s}$$

of 20, the circuit shown was created. The p-channel MOSFET transistors have parameters

$$V_{PO} = 2 \text{ V}, \qquad I_{DSS} = -5 \text{ mA}, \qquad V_A = 200 \text{ V}.$$

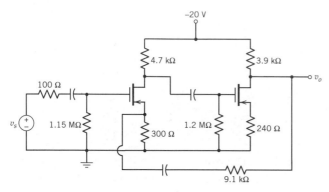

a. Identify the circuit topology.

b. Determine the voltage gain of the circuit without modification.

c. Rework the design so that the correct voltage gain (20) is attained by changing the feedback (9.1-kΩ) resistor only.

8-29. Determine the closed-loop gain, input resistance, and output resistance of the amplifier shown. The NMOSFETs are identical with the following characteristic parameters:

$$K = 1.25 \text{ mA/V}^2, \qquad V_T = 2\text{V},$$
$$V_A = 120 \text{ V}.$$

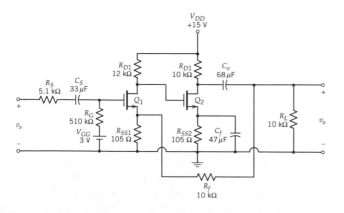

 8-30. The BJT cascode amplifier shown uses identical transistors with the parameters:

$$\beta_F = 200, \qquad V_A = 200 \text{ V}.$$

a. Complete the design of the amplifier for $I_E = -2$ mA, and $V_{CE1} = V_{CE2} = 5$ V. Find the quiescent point of all of the transistors.

b. Find the closed-loop gain, input resistance, and output resistance of the amplifier.

c. Confirm the closed-loop gain using SPICE.

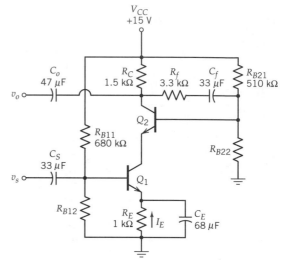

 8-31. The amplifier shown is to be used as the basis for a series–shunt feedback amplifier with a voltage gain of 12. The JFETs are described by

$$V_{PO} = -2 \text{ V}, \qquad I_{DSS} = 4 \text{ mA}, \qquad V_A = 200 \text{ V}.$$

a. Complete the design by inserting a single feedback resistor–capacitor network.

b. Verify the design using SPICE.

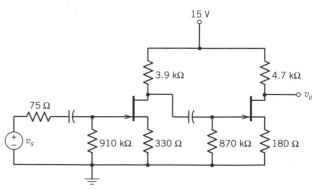

8-32. For the amplifier shown, the transistor characteristics are

$$I_{DSS} = 8 \text{ mA}, \quad V_A = 100 \text{ V}, \quad V_{PO} = -5 \text{ V}$$

for the JFET and

$$\beta_F = 200, \quad V_A = 100 \text{ V}$$

for the BJT.

a. Complete the design of the amplifier for

$$I_D = 1 \text{ mA}, \quad I_E = 3 \text{ mA},$$
$$V_{EC} = V_{DS} = 5 \text{ V}.$$

Find the quiescent point of all of the transistors.

b. Find the closed-loop gain, input resistance, and output resistance of the amplifier.

c. Confirm the closed-loop gain using SPICE.

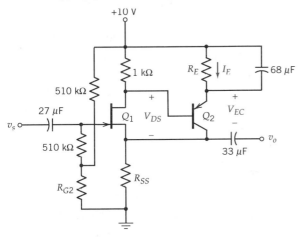

8-33. For the differential amplifier shown, find the closed-loop gain, input resistance, and output resistance of the amplifier. Assume identical transistors with $\beta_F = 120$. Use SPICE to confirm closed-loop gain of the amplifier.

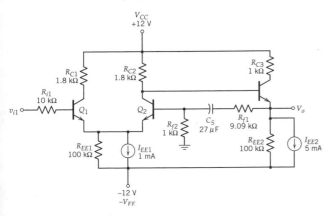

8-34. The transistors in the differential amplifier shown are identical and have the following characteristics:

$$I_{DSS} = 10 \text{ mA}, \quad V_A = 100 \text{ V},$$
$$V_{PO} = -4.5 \text{ V}.$$

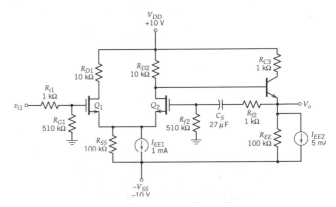

a. Find the closed-loop gain, input resistance, and output resistance of the amplifier.

b. Use SPICE to confirm closed-loop gain of the amplifier.

8-35. The series–shunt feedback amplifier of Example 8.3-3 is to be redesigned to have larger voltage gain. This change is to be accomplished by alteration of the feedback (1.2-kΩ) resistor *only*. If the requirement on the midband input resistance R_{in} is reduced to

$$R_{in} \geq 8.80 \text{ k}\Omega$$

from the current value (8.87 kΩ), what is the maximum midband voltage gain that can be achieved?

8-36. The performance parameters of the given OpAmp circuit can be determined using feedback techniques. The characteristics of the OpAmp are

Input resistance	2 MΩ
Output resistance	50 Ω
Voltage gain	800 kV/V

a. Identify the feedback topology.

b. Draw a circuit diagram for the loaded basic forward amplifier.

c. Determine the voltage gain for the total

circuit using feedback techniques. Compare results to those determined using the techniques described in Chapter 1.

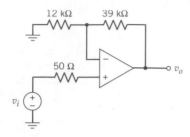

8-37. Many circuits incorporate both local and global feedback: The simple OpAmp circuit shown is an example of one such circuit. For simplicity, assume the ideal OpAmp expressions derived in Chapter 1 can be used to describe the local feedback characteristics of the amplifier.

    a. Identify the global feedback topology.

    b. Use feedback techniques to determine the overall voltage gain and input resistance of the circuit.

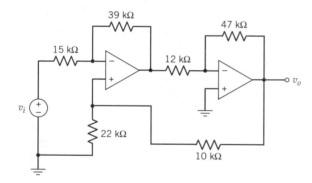

   8-38. The characteristics of the amplifier shown include high input resistance with moderate voltage gain and output resistance. It is desired to reduce the output resistance of the amplifier through the use of feedback without significantly altering the input resistance. If the transistors in the circuit have parameters

$$I_{DSS} = 2.5 \text{ mA}, \quad V_{PO} = -2.5, \quad V_A = 100\text{V},$$

$$\beta_F = 150, \quad V_A = 200 \text{ V},$$

the quiescent conditions have been found to be

$$|I_{DQ}| = 2.10 \text{ mA}, \quad V_{DSQ} = 14.7 \text{ V},$$

$$|I_{CQ}| = 1.95 \text{ mA}, \quad V_{CEQ} = -6.49 \text{ V}.$$

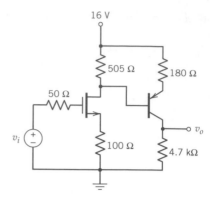

    a. What feedback topology will lower the output resistance while maintaining high input resistance?

    b. Design a feedback network that will meet the following design goals:

- Transistor quiescent conditions are not changed.
- Voltage gain is no less than 10.
- Output resistance is minimal.

    c. Determine the output resistance of the amplifier designed in part b.

   8-39. For the feedback amplifier shown, the silicon transistors are described by

$$\beta_F = 120.$$

Assume v_i has zero DC component.

    a. Identify the feedback topology.

    b. Determine the quiescent conditions.

    c. Determine the voltage gain

$$A_V = \frac{v_o}{v_i}.$$

    d. Determine the input resistance R_{in}.

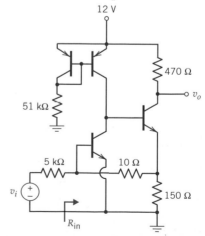

8-40. For the feedback amplifier shown, the silicon transistors are described by

$$\beta_F = 100.$$

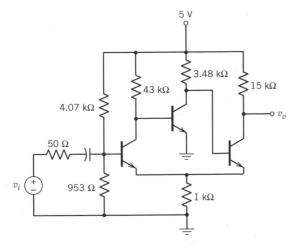

a. Identify the feedback topology.

b. Determine the quiescent conditions.

c. Determine the voltage gain

$$A_V = \frac{v_o}{v_i}.$$

d. Verify the results of parts b and c using SPICE.

3,562,660
OPERATIONAL AMPLIFIER
Robert A. Pease, Wilmington, Mass., assignor to Teledyne, Inc.,
Los Angeles, Calif., a corporation of Delaware
Filed Dec. 26, 1967, Ser. No. 693,603
Int. Cl. H03f *3/04, 3/68*

U.S. Cl. 330—30

4 Claims

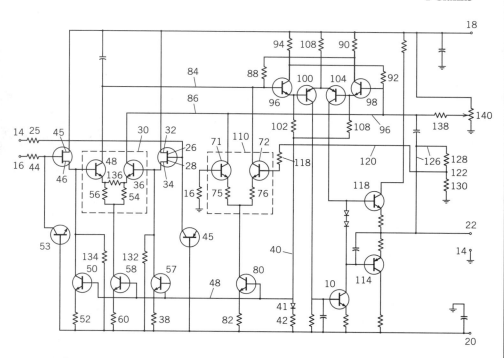

A differential operational amplifier having two input amplifier stages with independent inputs and paralleled outputs, the inputs of one stage being connectable to a low level, high impedance source, the input or inputs of the other stage being connectable to negative feedback resistors for adjusting the closed loop gain of the amplifier. The respective outputs of the stages are applied in parallel to a differential-to-single-ended amplifier stage. By making each input amplifier stage with high common mode rejection ability, common mode errors are eliminated at the front end of the amplifier. By matching the characteristics of these stages, overall gain is adjustable without introducing common mode errors or affecting the signal source.

SECTION III

Frequency-Dependent Behavior

In the previous sections, all electronic circuit operation was considered to be either at near-zero frequency or in the midband region of operation. The *midband* range of frequencies, vitally important to amplifier discussions, is characterized by two basic simplifying assumptions:

- The midband frequencies are large enough so that discrete circuit capacitors appear to have negligible impedance with respect to the resistances in the circuit.
- The midband frequencies are small enough so that the active elements (transistors, OpAmps, etc.) appear to have frequency-invariant properties.

It is the purpose of this section to explore the variation of circuit behavior over the entire range of frequencies. In addition to exploring frequency dependence, the time-domain equivalence of these effects is explored.

A review of the characteristics of ideal filters and frequency response plots leads into the design of active filters. Active Butterworth, Chebyshev, and elliptic filter design is discussed in this section using OpAmps as the active circuit elements. Discussion of passive filters and filters with other active elements is saved until the chapter in Section IV on communication circuits. The frequency response limitations of OpAmps provide an introduction to limitations common in other devices.

Transistor amplifier frequency dependence is first discussed through the effects of coupling and bypass capacitors. Modified models for the diode, BJT, and FET are introduced to model the high-frequency limitations of common devices and the result of these limitations on the frequency response of amplifier circuits. The effect of feedback on frequency response is initially presented as a special case of stabilization: here stabilization against variation in element value change due to frequency. Feedback effects on pole migration is emphasized. Compensation against possible instabilities or oscillations is explored extensively.

The discussion of linear oscillators is based on a foundation derived from feedback amplifier principles. Linear oscillators are shown to be amplifiers driven into a region of oscillatory instability characterized by the Barkhausen criterion. Colpitts, Hartley, and Wein-Bridge topologies form the basis for primary oscillator discussion. Stabilization of the frequency of oscillation through the use of a crystal is explored. Nonlinear oscillators (or waveform generators) are explored through discussions of multivibrator circuits. Other nonlinear circuits, such as pulse generators and Schmitt triggers, are derived from multivibrator characteristics. An introduction to arbitrary waveform generation concludes the section.

CHAPTER 9

ACTIVE FILTERS

An electronic filter is defined as a device that separates, passes, or suppresses an electronic signal (or group of signals) from a mixture of signals. Most common among all possible filters are those that separate signals according to the signal frequency content. This type of filter is found in virtually every common electronic system. As an example, consider a radio receiver. The antenna system for a radio receiving system receives a wide range of frequencies associated with many distinct radio stations. A radio receiver attached to that antenna system can be tuned to receive any particular station: Its input filter section passes only those frequencies associated with that station and blocks all other frequencies. Of interest in this chapter are the characterization and design of such frequency-selective filters.

Electronic frequency-selective filters are divided into two major groups based on the circuit elements making up the filter: passive filters and active filters. Passive electronic filters are commonly constructed using the basic lumped-parameter passive building blocks familiar to all electrical engineers: resistors, capacitors, and inductors. These filters are of great value particularly in high-power and high-frequency applications.[1] In low-frequency applications, the high-value inductances required in passive filter designs present several problems. Large value inductors are physically large, heavy, non-linear, and usually have a relatively large loss factor. In addition, the magnetic fields they generate are a source of electromagnetic interference. In many designs one or more of these features is undesirable. It is therefore often necessary to design filters that contain only resistors and capacitors. The addition of another common building block, the operational amplifier, allows for the design of these lower frequency filters as efficient and cost-effective devices. The inclusion of an active device in the design has led to calling these filters *active filters*.

The approach to active filter design and characterization developed in this chapter begins with a review of frequency response characterization in the form of Bode plots. This form of graphical representation of filter response leads to a presentation of the four basic ideal types of frequency-selective filters: high-pass, low-pass, band-pass, and band-stop. Since ideal filter response is not achievable in the real world, several common, realistic filter response approximations are discussed and compared. The design process necessary to realize the four basic filter types using OpAmps, resistors, and capacitors is developed and explored.

Resistor size, weight, and power dissipation can also present problems to the filter designer. In integrated circuits, one low-frequency solution to the problem is switched-capacitor filters. These filters are closely related to standard active filters

---

[1] A discussion of the uses of passive filters in communication circuits can be found in Chapter 15. Of particular interest is the discussion of trade-offs between active and passive filters.

where the resistors are replaced by rapidly switched capacitors. A discussion of this variation of active filters is found in Section 9.7.

Although OpAmps are highly useful electronic building blocks, they do have limitations. Most significant in filter design is the degradation in the voltage gain with frequency. The chapter ends with a discussion of the limitations of active filters designed with the use of OpAmps.

9.1 BODE PLOTS

Characterization of electronic systems in the frequency domain is directly related to time-domain system response. The response of a linear time-invariant system $x_o(t)$ to an arbitrary input $x_i(t)$ can be characterized, in the time domain, by its impulse response $h(t)$:

$$x_o(t) = \int_0^t h(t - \tau)x_i(\tau) \, d\tau. \tag{9.1-1}$$

While the convolution integral of Equation 9.1-1 has great utility in many situations,[2] a frequency-domain characterization of the system is more typical. The frequency-domain response of a system can be obtained by taking the Fourier transform of the output quantity $x_o(t)$:

$$X_o(\omega) = \mathscr{F}\{x_o(t)\} = \int_{\infty}^{\infty} x_o(t)e^{-j\omega t}dt, \tag{9.1-2}$$

which reduces to

$$X_o(\omega) = H(\omega)X_i(\omega). \tag{9.1-3}$$

The spectrum of the output quantity $X_o(\omega)$ is the product of the input quantity spectrum $X_i(\omega)$ and the transfer function of the system, $H(\omega)$. The impulse response and the transfer function of a system are related by the Fourier transform:

$$H(\omega) = \mathscr{F}\{h(t)\}. \tag{9.1-4}$$

This frequency-domain transfer function is characterized by its two parts:

- $|H(\omega)|$—the magnitude response
- $\angle H(\omega)$—the phase response

One of the most useful representations of these two quantities is a pair of plots. The typical format for these plots is the magnitude (on either a linear or a decibel scale) or the phase (in either degrees or radians) as the ordinate and frequency (on a logarithmic scale) as the abscissa. While many computer simulation programs exist for the efficient creation of such plots, much insight into the functioning of a filter (or any electronic circuit) and the dependence of the responses to circuit parameter vari-

---

[2] Time-domain characteristics of electronic circuits and the relationship of these characteristics to the frequency-domain characteristics are discussed in Chapters 10–13.

ation can be gained by the manual creation of simple straight-line approximations to these magnitude and phase plots. Plots of the magnitude of the transfer function, in decibels, and the phase of the transfer function are called *Bode plots*.[3] The two plots together form a *Bode diagram*.

Mathematical Derivations

The response of a linear system to sinusoidal inputs can be described as the ratio of two polynomials in frequency ω:

$$H(\omega) = K_o \frac{Z_m(\omega)}{P_n(\omega)}, \tag{9.1-5}$$

where $Z_m(\omega)$ is a mth-order polynomial whose roots identify the m zeroes of the response. The roots of $P_n(\omega)$ identify the n poles. For all real electronic systems, the response must not increase as frequency becomes infinite; thus the number of poles must be equal to or larger than the number of zeroes, $n \geq m$. Equation 9.1-5 can be expanded to be of the form:

$$H(\omega) = K_o \frac{1 + (j\omega)a_1 + (j\omega)^2 a_2 + \cdots + (j\omega)^m a_m}{1 + (j\omega)b_1 + (j\omega)^2 b_2 + \cdots + (j\omega)^n b_n}. \tag{9.1-6}$$

Each of the two polynomials $P_n(\omega)$ and $Z_m(\omega)$ can be written as a factored product of multiples of four types of simple function, hereafter identified as *factors*, $\{F(\omega)\}$:

1. K_o—a constant
2. $j\omega$—a root at the origin
3. $1 + j\omega/\omega_o$—a simple root at $\omega = \omega_o$
4. $1 + 2\zeta(j\omega/\omega_o) + (j\omega/\omega_o)^2$—a complex conjugate root pair

Each of these four simple functions has a straight-line approximate Bode representation. The use of decibels (a logarithmic function) as a vertical scale for the magnitude plot and a linear scale for phase converts the product of the factors into the *sum* of the factor magnitudes (in decibels) and the *sum* of the factor phases. Therefore the total system response plot becomes the sum of the plots of the individual simple factors.

9.1.1 Bode Plots of Factors

The Bode straight-line approximate representation of each of the four simple factors is unique. The representations for a constant and a root at zero frequency are exact: The other two representations are less accurate. A discussion of each Bode representation follows: A summary can be found in Table 9.1-1.

A Constant

The most simple of the factors is a constant:

$$F(\omega) = K_o. \tag{9.1-7}$$

---

[3] Named for H. W. Bode. One early reference is his book *Network Analysis and Feedback Design*, Van Nostrand, New York, 1945.

TABLE 9.1-1 BODE FACTOR MAGNITUDE AND PHASE PLOTS

Factor[a]	Bode Magnitude Plot	Bode Phase Plot
K_o	0 dB, ω, $-20 \log (K_o)$	0°, ω
$j\omega$	0 dB, 1, ω, -20 dB/decade	0°, ω, $-90°$
$1 + \dfrac{j\omega}{\omega_o}$	0 dB, ω_o, ω, -20 dB/decade	0°, $0.1\omega_o$, ω_o, $10\omega_o$, ω, $-45°$/decade, $-90°$
$1 + 2\zeta\,\dfrac{j\omega}{\omega_o} + \left(\dfrac{j\omega}{\omega_o}\right)^2$	0 dB, ω_o, ω, -40 dB/decade	0°, $10^{-\zeta}\omega_o$, ω_o, $10^{\zeta}\omega_o$, ω, $\dfrac{-90°}{\zeta}$/decade, $-180°$

Here the magnitude is constant at $20\ log|K_o|$. The phase plot is also constant at $\theta = 0°$ or $\theta = 180°$ depending on the mathematical sign of K_o.

A Pole or Zero at Zero Frequency
A pole or zero at the origin also has a simple Bode representation. The factor is of the form

$$F(\omega) = j\omega. \tag{9.1-8}$$

The magnitude plot is a straight line on a logarithmic frequency scale:

$$|F(\omega)|_{\text{dB}} = 20\ log|j\omega| = 20\ log(\omega). \tag{9.1-9}$$

If the factor is in the numerator (indicating a zero), this straight line has a slope of +20 dB/decade[4]: Denominator factors (poles) have a slope of −20 dB/decade. In each case the line passes through the point (0 dB, 1 rad/s).

The factor phase is constant at $\theta = 90°$ (or π radians). Consequently the phase plot for zeroes will be at +90° and at −90° for poles.

A Simple Pole or Zero

Single poles and zeroes not at the origin have more complex plots. The factor form for a simple pole is

$$F(\omega) = 1 + \frac{j\omega}{\omega_o}, \tag{9.1-10}$$

where ω_o is the pole (zero) frequency. The magnitude of the factor is given by

$$|F(\omega)|_{dB} = 20 \log \sqrt{1 + \left(\frac{\omega}{\omega_o}\right)^2} = 10 \log \left[1 + \left(\frac{\omega}{\omega_o}\right)^2\right]. \tag{9.1-11}$$

The phase is given by

$$\angle F(\omega) = \tan^{-1}\left(\frac{\omega}{\omega_o}\right). \tag{9.1-12}$$

These two components of the simple factor are shown in Figure 9.1-1 as if the factor were in the denominator (representing a pole). Also shown in the figure are the Bode straight-line approximate plots.

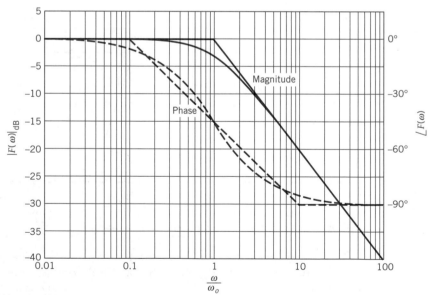

Figure 9.1-1
Simple pole bode diagram.

---

[4] A decade is a change in frequency by a factor of 10. A slope of 20 dB/decade is also identified in some sources as 6 dB/octave, where an octave is a frequency change by a factor of 2.

The Bode plot of the magnitude is an asymptotic approximation. If $\omega \ll \omega_o$, $F(\omega) \approx 1$ and the magnitude plot is constant at 0 dB. If $\omega \gg \omega_o$,

$$|F(\omega)|_{dB} \approx 20 \log \sqrt{\left(\frac{\omega}{\omega_o}\right)^2} = 20 \log(\omega) - 20 \log(\omega_o). \tag{9.1-13}$$

This approximation is a straight line with slope of 20 dB/decade that intersects the 0-dB line at $\omega = \omega_o$. The approximation has its greatest error ($\sim$3 dB) at $\omega = \omega_o$.

As can be seen in Figure 9.1-1, the factor phase never exceeds 90° and essentially all phase change takes place within ± 1 decade of ω_o. Beyond one decade the phase is approximated by a constant:

$$\omega < 0.1\omega_o, \qquad \angle F(\omega) = 0°,$$
$$\omega > 10\omega_o, \qquad \angle F(\omega) = 90°.$$

Within ± 1 decade of ω_o, the phase can be approximated by a straight line of slope 45°/decade:

$$0.1\omega_o < \omega < 10\omega_o, \qquad \angle F(\omega) = 45°\left[log\left(\frac{10\omega}{\omega_o}\right)\right].$$

This straight-line Bode approximation has a maximum error of $\sim$5.7°.

Complex-Conjugate Pole or Zero Pairs

A complex-conjugate root pair presents a more complex relationship. The appropriate form of the factor is given by

$$F(\omega) = 1 + 2\zeta\frac{j\omega}{\omega_o} + \left(\frac{j\omega}{\omega_o}\right)^2, \tag{9.1-14}$$

where ω_o is identified as the resonant frequency and ζ is the damping factor.[5] A plot of the magnitude and phase response for a complex pair of poles with the damping factor as a parameter ($0.2 \leq \zeta \leq 0.9$ in increments of 0.1) is shown in Figure 9.1-2.

The variation of the plot with damping coefficient makes straight-line approximations near the resonant frequency questionable. Still, at about a third of a decade or larger from the resonant frequency, reasonable approximations of the magnitude plot can be made. Once again, if $\omega \ll \omega_o$, $F(\omega) \approx 1$ and the approximate magnitude plot is constant at 0 dB. If $\omega \gg \omega_o$,

$$|F(\omega)|_{dB} \approx 20 \log\left|\left(\frac{j\omega}{\omega_o}\right)^2\right| = 40 \log \omega - 40 \log \omega_o. \tag{9.1-15}$$

In this region the magnitude has a slope of 40 dB/decade with an intercept of the 0-dB axis at $\omega = \omega_o$. Near the resonant frequency, there can be a large difference between the approximate Bode magnitude plot and the true magnitude plot. Depending on the application, corrections to the curve may be necessary. These corrections can usually be accomplished with the addition of only a few data points and judicial interpolation. While calculation of a variety of points is possible, two signif-

---

[5] The damping factor lies in the range $0 \leq \zeta < 1$ for complex-conjugate root pairs.

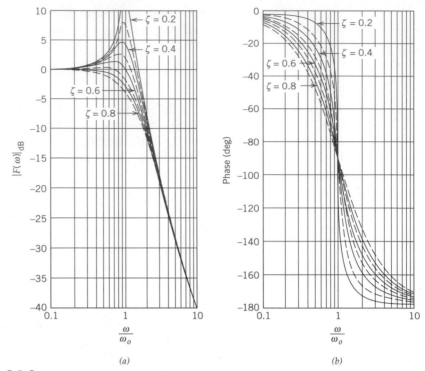

Figure 9.1-2
Complex-conjugate pair pole frequency response: (*a*) magnitude; (*b*) phase.

icant data points are the peak magnitude of the "bump" in the curve (if one exists) and the magnitude at the resonant frequency. For damping coefficients less than $1/\sqrt{2}$, a peak in the magnitude response occurs at

$$\omega_{peak} = \omega_o\sqrt{1 - 2\zeta^2}, \tag{9.1-16}$$

which has value (on a linear scale)

$$|F(\omega_{peak})| = \frac{1}{2\zeta\sqrt{1 - \zeta^2}}. \tag{9.1-17}$$

The magnitude at the resonant frequency is given by

$$|F(\omega_o)| = \frac{1}{2\zeta}. \tag{9.1-18}$$

The phase plot for a complex root pair is also complicated by the damping coefficient. At frequencies much smaller than the resonant frequency, the phase is near constant at 0°: At frequencies much larger than ω_o, the phase is about 180°. Near ω_o the Bode approximate curve must be a straight line. There are two reasonable approximations: a ζ-independent approximation and an approximation that depends on ζ. Either approximation is valid in most applications.[6]

---

[6] The authors prefer the ζ-*dependent* approximation since it has less phase error and reduces to the more simple case for $\zeta = 1$. The ζ-*independent* approximation has wide usage due to its more easily constructed format.

The most simple Bode approximation to the phase plot ignores the variation of the true phase plots with the damping coefficient ζ. In this ζ-independent approximation, the Bode phase plot consists of three straight lines. More than one decade from the resonant frequency, the approximate phase is constant:

$$\omega < 0.1\omega_o, \qquad \angle F(\omega) = 0°,$$
$$\omega > 10\omega_o, \qquad \angle F(\omega) = 180°.$$

Within ± 1 decade of ω_o, the phase can be approximated by a straight line of slope $90°/\text{decade}$:

$$0.1\omega_o < \omega < 10\omega_o, \qquad \angle F(\omega) = 90°\left[\log\left(\frac{10\omega}{\omega_o}\right) \right]$$

The ζ-dependent Bode phase approximation follows the variation in phase with damping coefficient and keeps the three-straight-lines approach, changing the frequency ranges and the slope near the resonant frequency. The constant phase regions are a function of ζ and are described as

$$\omega < \omega_o/10^\zeta, \qquad \angle F(\omega) = 0°,$$
$$\omega > 10^\zeta\omega_o, \qquad \angle F(\omega) = 180°.$$

The phase between the two constant phase regions is approximated by a straight line of slope $(90/\zeta)°/\text{decade}$ passing through the point $\omega_o, 90°$:

$$\omega_o/10^\zeta < \omega < 10^\zeta\omega_o, \qquad \angle F(\omega) = \frac{90°}{\zeta}\left[\log\left(\frac{\omega}{\omega_o}\right) \right] + 90°.$$

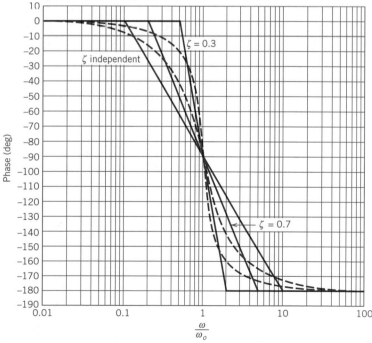

Figure 9.1-3
The ζ-dependent Bode phase approximation.

Figure 9.1-3 shows that ζ-dependent Bode phase approximation for two values of ζ along with the actual phase plots and the ζ-independent Bode phase approximation.

The Bode approximate plots for each of the four factors are summarized in Table 9.1-1 as if the factors are in the denominator. Factors in the numerator will have *positive* slopes and mathematical signs but otherwise be similar in shape. Total Bode plots are the sum of the plots of the factors.

EXAMPLE 9.1-1 Draw the Bode diagram for the following transfer function:

$$H(\omega) = \frac{0.2j\omega}{1 + j\omega/62.5 + (j\omega)^2/2500}$$

SOLUTION

This transfer function is the product of three factors:

- A constant
- A zero at the origin
- A complex-conjugate pair of poles

The Bode plots will be the sum of the individual factor Bode plots. The magnitude plot for the constant is a horizontal line at

$$20 \log 0.2 = -14 \text{ dB}.$$

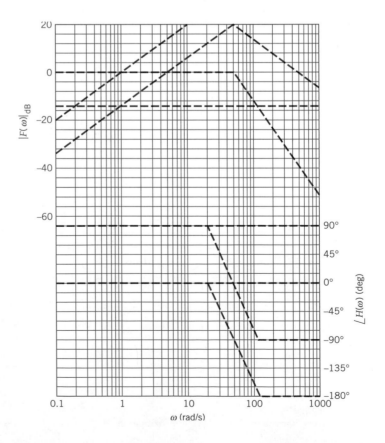

The magnitude plot for the zero at the origin has slope of +20 dB/decade and passes through (0 dB, 0). The complex pair of poles has a resonant frequency given by

$$\omega_o^2 = 2500 \Rightarrow \omega_o = 50$$

and a damping coefficient of

$$\frac{2\zeta}{\omega_o} = \frac{1}{62.5} \Rightarrow \zeta = 0.4.$$

Since the complex-conjugate pair lies in the denominator, the slopes are negative. The magnitude plot is as shown.

The phase plot for the constant and the zero at the origin are simple horizontal lines at $0°$ and $90°$. Each is in the numerator: The phases are positive. For the complex-conjugate pole phase plot, the range of ω where the phase changes must be calculated:

$$10^\zeta = 10^{0.4} = 2.51.$$

Thus the phase change takes place between

$$\frac{\omega_o}{2.51} < \omega < 2.51\omega_o \Rightarrow 19.9 < \omega < 126$$

and since the factor is in the denominator, the phase change has a slope of

$$\frac{-90}{\zeta} = -225°/\text{decade}.$$

9.2 FILTER CHARACTERISTICS

Filters are classified according to the function that they perform. Ideal frequency selective filters have three fundamental design goals:

- To have no attenuation of input signals over a range of frequencies, called the *passband*
- Over another, distinct range of frequencies, called the *stopband*, to have complete attenuation of input signals
- To have no distortion of signals whose frequency content lies within the passband

The relative frequency location of the passband and the stopband give rise to the four most common types of filters: low-pass, high-pass, band-pass, and band-stop. The magnitude of the ideal frequency response characteristics of these four filter types is depicted in Figure 9.2-1.

An ideal low-pass transfer characteristic is shown in Figure 9.2-1a with the passband extending from zero frequency to a cutoff frequency, $\omega = \omega_c$, and the stopband extending from ω_c to infinite frequency. The filter gain in the passband is unity and zero in the stopband. The ideal high-pass filter of Figure 9.2-1b is the inverse of the low-pass filter: The stopband extends from zero frequency to $\omega = \omega_c$ and the passband from ω_c to infinite frequency. A band-pass filter (Figure 9.2-1c) has a single

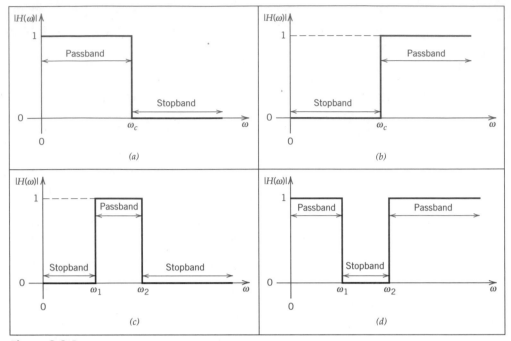

Figure 9.2-1
Filter idealized characteristics: (*a*) low pass; (*b*) high pass; (*c*) band pass; (*d*) band stop.

passband extending in the frequency range $\omega_1 < \omega < \omega_2$, while all other frequencies are stopped. Band-stop filters (Figure 9.2-1*d*) have a single stopband extending in the frequency range $\omega_1 < \omega < \omega_2$, while all other frequencies are passed. Band-stop filters are often called band elimination filters or notch filters.

The requirement for constant magnitude of the frequency response in the passband is a consequence of the need to pass signals whose frequency content lies within the passband without distortion. In order that a signal pass through a filter without distortion, the output of the filter must be an amplified duplicate of the input signal possibly delayed by a time lag t_d[7]:

$$x_o(t) = K_o x_i(t - t_d). \tag{9.2-1}$$

The frequency-domain equivalent of Equation 9.2-1 is obtained through the use of the Fourier transform:

$$X_o(\omega) = \mathcal{F}\{x_o(t)\} = \int_{-\infty}^{\infty} x_o(t)e^{-j\omega t}\, dt. \tag{9.2-2}$$

Here $X_o(\omega)$ is the spectrum of $x_o(t)$. Replacing $x_o(t)$ by its definition (Equation 9.2-1) yields the frequency-domain relationship between the input signal spectrum and the undistorted output signal spectrum:

$$X_o(\omega) = K_o e^{-j\omega t_d} X_i(\omega). \tag{9.2-3}$$

---

[7] The time lag for many low-frequency filters can be approximated as $t_d \approx 0$. As frequencies become large, the physical size of an electronic filter and the internal capacitance of its elements can make time lag a significant property of the filter.

The transfer characteristic of the nondistorting filter, $H(\omega)$, is interpreted from Equation 9.2-3 to be the multiplier of the input signal spectrum:

$$H(\omega) = K_o e^{-j\omega t_d}. \tag{9.2-4}$$

The properties of a nondistorting filter are apparent. It is necessary that, over the range of frequencies contained in the input signal, the magnitude of the filter transfer function be constant with respect to frequency and the phase shift induced by the filter be linear with respect to frequency:

$$|H(\omega)| = K_o, \qquad \angle H(\omega) = -\omega t_d. \tag{9.2-5}$$

Notice that for zero time delay ($t_d = 0$), the phase shift induced by the filter is also zero.[8]

Real filters can only approximate the ideal transfer relationships depicted in Figure 9.2-1. The abrupt transition between the passband and the stopband of the various types of ideal filters is, unfortunately, impossible to achieve with a finite number of real circuit elements. In addition, zero gain in the stopband and no variation in gain in the passband are, at best, difficult to achieve. It is necessary to relax the design goals for these filters in order to be able to design and fabricate practical, real filters. A realistic set of *practical* design goals for frequency-selective filters follows:

- In the passband, to have relatively small attenuation (or a gain) of input signals over a range of frequencies. The variation in gain over the passband is specified by a maximum variation value γ_{max}.
- Over the stopband, to have relatively large attenuation of input signals specified by a minimum attenuation value γ_{min}.
- To have a transition region lying between passband and stopband specified by a passband cutoff frequency ω_c and a stopband edge ω_s.
- To have minimal distortion of signals whose frequency content lies within the passband.

Figure 9.2-2 is a graphical representation of these design goals applied to a low-pass filter. Here the separation of the passband and the stopband by a transition band is

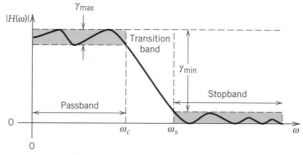

Figure 9.2-2
Realistic low-pass filter design goals and one possible solution.

---

[8] Often a constant phase shift of $\pm 180°$ ($\pm \pi$ radians) is also acceptable: Constant $\pm 180°$ phase shift is a simple inversion of the signal when passing through the filter.

apparent. The ratio of the frequencies that specify the transition band, ω_c / ω_s, is called the selectivity factor and indicates the "sharpness" of the filter. The shaded areas represent tolerances in the passband and stopband gains: One of many possible filter responses that meet the design goals is presented as an arbitrary example.

As the tolerances for a filter are reduced (γ_{max} reduced, γ_{min} increased, and ω_c / ω_s increased), the filter approaches ideal characteristics. While this goal may seem worthwhile in all cases, reducing tolerances increases the complexity of the circuitry involved to accomplish the more stringent design goals. Good design involves a balance between constraints and complexity.

There are several families of filters that meet the practical design goals of realistic filters in an efficient manner. Each family meets the goals in a unique manner. Three of the families that are common and will be discussed at length in this chapter are as follows:

- Butterworth filters
- Chebyshev filters
- Elliptic filters

Butterworth filters are smooth filters. There is no ripple in the response in either the passband or the stopband and the transition between bands is monotonic. The edge of the passband is the frequency at which the magnitude of the filter transfer function. $|H(\omega)|$ drops by γ_{max}. The edge of the stopband occurs at the frequency at which $|H(\omega)|$ drops by γ_{min} below the nominal passband gain. Chebyshev filters have ripple in either the passband or in the stopband: Once out of the region of ripple they are monotonic. Chebyshev filters can achieve a smaller transition region as compared to an equivalent-complexity Butterworth filter. Elliptic filters achieve an even smaller transition region by allowing an equal amount of ripple in both the passband and in the stopband.

All real filters introduce some phase shift. In good filters this shift is reasonably linear with frequency throughout most of the passband: It can be interpreted as a time delay between the input and the output. As the order of the filter increases, the delay typically increases as well. Near the junction between the passband and the transition region the phase shifts become increasingly nonlinear: Frequencies near the edge of the passband suffer from phase distortion, as well as magnitude distortion. Discussion of filter design criteria in the following sections will be based primarily on the magnitude of the filter frequency response.

9.3 BUTTERWORTH FILTERS

Butterworth filters, often called maximally flat filters, are arguably the most common type of electronic filter. They meet the realistic goals of filters presented in the last section in a unique manner: They are *smooth* filters that transition monotonically from the passband to the stopband. The polynomials that characterize the Butterworth response are functions of only two parameters: the order of the filter, n, and the 3-dB frequency ω_o. The response of an nth order low-pass Butterworth filter is an all-pole response characterized by the nth Butterworth polynomial $B_n(\omega)$:

$$A_V(\omega) = \frac{A_{Vo}}{B_n(\omega)}. \tag{9.3-1}$$

The magnitude of the nth Butterworth polynomial applied to low-pass filters is given by

$$|B_n(\omega)| = \sqrt{1 + \left(\frac{\omega}{\omega_o}\right)^{2n}} \qquad (9.3\text{-}2)$$

There are several interesting properties of the Butterworth polynomials. At $\omega = \omega_o$ the magnitude of every Butterworth polynomial is

$$|B_n(\omega_o)| = \sqrt{1 + (1)^{2n}} = \sqrt{2}. \qquad (9.3\text{-}3)$$

Thus ω_o is the frequency at which the output signal of the filter (a voltage or current) is reduced by a factor of $\sqrt{2}$ or, equivalently, the output power is reduced to one-half its passband value. The half-power frequency, here ω_o, is also commonly called the 3-dB frequency:

$$20 \log (\sqrt{2}) = 3.01030 \text{ dB} \approx 3 \text{ dB}. \qquad (9.3\text{-}4)$$

Since all Butterworth filters have a response that is reduced by 3 dB at the resonant frequency, it is often common to specify Butterworth filters with $\gamma_{\max} = 3$ dB.

The slope of the filter magnitude response at $\omega = 0$ is zero, and at high frequencies, $\omega \gg \omega_o$, is $-20n$ decibels per decade. Thus the order of the filter will determine the magnitude of the response in the stopband. The magnitude frequency response plot for the first six orders of Butterworth low-pass filter is presented in Figure 9.3-1.

Butterworth polynomials are generated in order to achieve a smooth filter. In Section 9.1 it was demonstrated that the response of a system is determined only by the roots of the transfer characteristic equation. For a Butterworth filter, the roots of

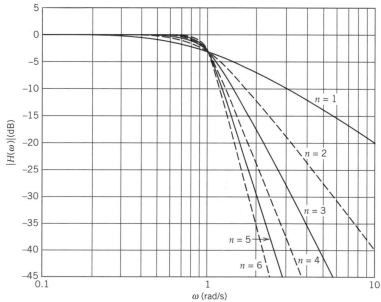

Figure 9.3-1
Butterworth filter frequency response.

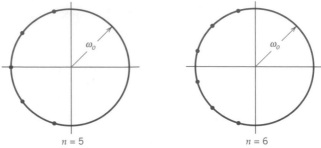

Figure 9.3-2
Butterworth polynomial pole locations.

an nth order polynomial are chosen so that the first $2n - 1$ derivatives of the magnitude of the response are zero at $\omega = 0$. These roots can easily be derived to be

$$-r_{k,n} = \omega_o e^{j(2k+n-1)\pi/2n}, \qquad k = 1, 2, \ldots, n, \tag{9.3-5}$$

where $r_{k,n}$ is the kth root of the nth order Butterworth polynomial. These roots lie in the complex plane on a circle of radius ω_o and are separated by an angle of by π/n. Odd-order Butterworth polynomials have one real root at $-\omega_o$ and $(n - 1)/2$ complex conjugate pairs. Even-order polynomials have $n/2$ complex conjugate pairs and no real roots. Figure 9.3-2 shows the location of the roots of the fifth and sixth Butterworth polynomials.

A Butterworth polynomial can therefore be characterized by ω_o and the damping coefficient associated with each of the complex conjugate pairs. These damping coefficients can be calculated simply as

$$\zeta_{i,n} = \left| \cos\left(\frac{(2i + n - 1)\pi}{2n}\right) \right|, \qquad i = 1, 2, \ldots, n/2. \tag{9.3-6}$$

where $\zeta_{i,n}$ is the ith damping coefficient of the nth order Butterworth polynomial. While computation of the damping coefficients is quite simple, repetitive use of the

TABLE 9.3-1 BUTTERWORTH FILTER DAMPING COEFFICIENTS ζ FOR FIRST ELEVEN POLYNOMIALS

1					
2	0.7071				
3	0.5000				
4	0.3827	0.9239			
5	0.3090	0.8090			
6	0.2588	0.7071	0.9660		
7	0.2225	0.6235	0.9010		
8	0.1951	0.5556	0.8315	0.9808	
9	0.1736	0.5000	0.7660	0.9397	
10	0.1564	0.4540	0.7071	0.8910	0.9877
11	0.1423	0.4154	0.6549	0.8413	0.9595

same quantities makes a table of the first several order coefficients useful: Table 9.3-1 lists the damping coefficients for the first 11 Butterworth polynomials. It is important to remember that *odd-order Butterworth polynomials also have a real root.*

EXAMPLE 9.3-1 A low-pass Butterworth filter is to be designed to meet the following design criteria:

Passband

Nominal gain $A_{Vo} = 1$

$\gamma_{max} = 3$ dB at frequency $f_{3dB} = 3$ kHz

Stopband

$f_s = 20$ kHz

$\gamma_{min} = 40$ dB minimum

Determine the order of filter necessary to achieve the design goals and the filter transfer function $H(\omega)$.

SOLUTION

The resonant frequency of the filter response is

$$\omega_o = 2\pi f_{3dB} = 18.85 \times 10^3.$$

The order of the filter can be determined with Equation 9.3-2:

$$|B_n(\omega)| = \sqrt{1 + \left(\frac{\omega}{\omega_o}\right)^{2n}} \Rightarrow 10^{40/20} = \sqrt{1 + \left(\frac{2\pi f_s}{2\pi f_o}\right)^{2n}},$$

which can be rearranged to solve for n:

$$n = \frac{\log(10^{40/10} - 1)}{2\log(f_s/f_o)} = 2.427.$$

Since fractional order filters are not realistic, a *third-order filter is necessary* to accomplish the design goals. A third-order filter is characterized by a real pole and a single complex conjugate pair with damping coefficient (from Table 9.3-1), $\zeta = 0.5$.
The filter transfer function is given by ($\omega_o = 18.85 \times 10^3$)

$$H(\omega) = \frac{1}{(1 + j\omega/\omega_o)[1 + j\omega/\omega_o + (j\omega/\omega_o)^2]}$$

9.3.1 Alternate Definitions

The definition of the Butterworth polynomials as given in Equation 9.3-2 is extremely useful for the particular (and common) case where the edge of the passband is defined by the 3-dB frequency. There are, however, many cases where the edge of the passband must be defined by a variation in the gain, γ_{max}. In those situations a slight variation in definition is necessary. The Butterworth polynomials can be defined with a third degree of freedom:

$$|B_n(\omega)| = \sqrt{1 + \epsilon^2 \left(\frac{\omega}{\omega_c}\right)^{2n}}. \tag{9.3-7}$$

Here ω_c is the edge frequency of the passband. At $\omega = \omega_c$, the magnitude of all Butterworth polynomials is

$$|B_n(\omega)| = \gamma_{max} = \sqrt{1 + \epsilon^2}. \tag{9.3-8}$$

Notice that $\epsilon = 1$ relates to the standard definition of Butterworth polynomials given previously. The order of the filter is determined from Equation 9.3-7 using the attenuation necessary in the stopband. Once n and ϵ are known, the resonant frequency of the filter can be determined:

$$\omega_o = \frac{\omega_c}{\sqrt[n]{\epsilon}} \tag{9.3-9}$$

EXAMPLE 9.3-2 A low-pass Butterworth filter is to be designed to meet the following design criteria:

Passband

Nominal gain $A_{Vo} = 1$

Passband edge $f_c = 3$ kHz

$\gamma_{max} = 1$ dB maximum

Stopband

$f_s = 15$ kHz

$\gamma_{min} = 40$ dB minimum

Determine the Butterworth design parameters.

SOLUTION

A 1-dB variation in gain in the passband sets the value of ϵ (Equation 9.3-8):

$$\sqrt{1 + \epsilon^2} = 10^{1/20} \Rightarrow \epsilon = 0.50885.$$

The order of the filter can now be obtained by solving Equation 9.3-7:

$$|B_n(\omega)| = \sqrt{1 + \epsilon^2 \left(\frac{\omega}{\omega_c}\right)^{2n}} \Rightarrow 10^{40/20} = \sqrt{1 + 0.50885^2 \left(\frac{2\pi f_s}{2\pi f_c}\right)^{2n}}.$$

This has a solution $n = 3.28$: A fourth-order filter is necessary.
   The resonant frequency of the filter is given by Equation 9.3-9:

$$\omega_o = \frac{\omega_c}{\sqrt[n]{\epsilon}} = \frac{2\pi(3000)}{\sqrt[4]{0.50885}} = 22.318 \times 10^3 \text{ rad/s}, \qquad f_o = 3.552 \text{ kHz}.$$

The design goals of this filter can be met by a fourth-order Butterworth filter with resonant frequency 3.552 kHz. The fourth-order Butterworth polynomial has two pairs of complex conjugate poles with damping coefficients 0.3827 and 0.9329.

9.3.2 High-Pass Butterworth Characterization

High-pass Butterworth filters are characterized by Butterworth polynomials with the quantities ω and ω_o interchanged. This interchange has the effect of adding two zero-frequency zeroes to the transfer function. The magnitude of the Butterworth polynomials applied to high-pass filters is given by

$$|B_n(\omega)| = \sqrt{1 + \left(\frac{\omega_o}{\omega}\right)^{2n}} = \sqrt{1 + \epsilon^2 \left(\frac{\omega_c}{\omega}\right)^{2n}}. \qquad (9.3\text{-}10)$$

With the interchange of variable, all design processes are similar. The only exception is in the determination of the resonant frequency when the passband edge frequency is used. The high-pass relationship is

$$\omega_o = \omega_c \sqrt[n]{\epsilon} \qquad (9.3\text{-}11)$$

EXAMPLE 9.3-3 A high-pass Butterworth filter is to be designed to meet the following design criteria:

Passband

Nominal gain $A_{Vo} = 1$
Passband edge $f_c = 3$ kHz
$\gamma_{\max} = 0.5$ dB

Stopband

$f_s = 1$ kHz
$\gamma_{\min} = 50$ dB minimum

Determine the Butterworth design parameters.

SOLUTION

A 0.5-dB variation in gain in the passband sets the value of ϵ (Equation 9.3-8):

$$\sqrt{1 + \epsilon^2} = 10^{0.5/20} \Rightarrow \epsilon = 0.3493.$$

The order of the filter can now be obtained by solving Equation 9.3-10 modified to allow for $\epsilon \neq 1$:

$$|B_n(\omega)| = \sqrt{1 + \epsilon^2 \left(\frac{\omega_c}{\omega}\right)^{2n}} \Rightarrow 10^{50/20} = \sqrt{1 + 0.3493^2 \left(\frac{2\pi f_c}{2\pi f_s}\right)^{2n}}.$$

This has a solution $n = 6.197$: A seventh-order filter is necessary.

The resonant frequency of the filter is given by Equation 9.3-11:

$$\omega_o = \omega_c \sqrt[n]{\epsilon} = [2\pi(3000)]\sqrt[7]{0.3493}$$

$$= 16.220 \times 10^3 \text{ rad/s}, \qquad f_o = 2.581 \text{ kHz}.$$

The design goals of this filter can be met by a seventh-order Butterworth filter with resonant frequency 2.581 kHz. The seventh-order Butterworth polynomial has one real pole and three pairs of complex conjugate poles with damping coefficients 0.2225, 0.6235, and 0.9010.

9.4 OPAMP REALIZATIONS OF BUTTERWORTH FILTERS

One of the major problems in designing passive high-order filters is the interaction of cascaded filter stages. Each passive stage presents a load to preceding and following stages that can vary the design parameters of the filter. While this is not an insurmountable problem, elimination of stage interaction can make the design process significantly more simple. While limited to the maximum range of frequencies at which they can operate, OpAmp circuits provide the required isolation of stages in many electronic applications.

Individual OpAmp circuits can be placed in cascade without interaction of the individual stages: Each stage does not typically present a gain-changing load to either preceding or following stages. Thus the voltage transfer function of a cascade connection of several OpAmp stages is the product of the individual transfer functions. Mathematically one can represent the total voltage transfer function, $A_{VT}(\omega)$, of a series of N individual stages in terms of the gain of the individual stages:

$$A_{VT}(\omega) = \prod_{i=1}^{N} A_{Vi}(\omega). \qquad (9.4\text{-}1)$$

The voltage transfer relationship represented by Equation 9.4-1 is particularly useful when using OpAmp circuits to realize the transfer function of a filter. As with the Butterworth polynomials, filter transfer functions are specified as a product of first- and second-order polynomials in $j\omega$. If simple OpAmp circuits are constructed to have the transfer characteristic of these first- and second-order polynomials, then a high-order filter can be realized with a cascade of the simple OpAmp circuits. The OpAmp circuits must meet the following design criteria:

- The resonant frequency ω_o must be variable in both first-and second-order stages.
- The damping coefficient ζ of second-order stages must be variable.

In the case of Butterworth low-pass and high-pass filters, ω_o is the 3-dB frequency and ζ is the damping coefficient (tabulated in Table 9.3-1). Fortunately circuits that meet these design criteria are readily available to provide all types of frequency selective filters. The specifics of each type are sufficiently different so that discussion of low-pass, high-pass, band-pass, and band-stop filters must be separated.

9.4.1 Low-Pass OpAmp Filters

Figure 9.4-1 presents the schematic representation of two stages with appropriate first- and second-order low-pass voltage transfer relationships.

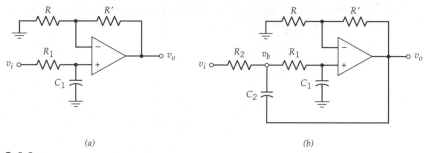

(a)   (b)

Figure 9.4-1
Low-pass section realizations: (a) first order; (b) second order.

First-Order Low-Pass Stage

The response of the first-order OpAmp filter stage (Figure 9.4-1a) is obtained by noting that the circuit is a noninverting amplifier in series with a low-pass passive RC filter. The transfer characteristic is therefore given by

$$\frac{v_o}{v_i} = \frac{1 + R'/R}{1 + j\omega R_1 C_1}. \tag{9.4-2}$$

This circuit meets the design goal of adjustable resonant frequency. The resonant frequency ω_o of this circuit is determined by the input RC time constant:

$$\omega_o = \frac{1}{R_1 C_1} \tag{9.4-3}$$

In addition the low-frequency gain of the circuit is adjustable through the elements R' and R:

$$A_{Vo} = 1 + R'/R. \tag{9.4-4}$$

Second-Order Low-Pass Stage

The second-order stage (Figure 9.4-1b) was developed by R. P. Sallen and E. L. Key and is therefore known as the Sallen and Key circuit. It includes, as its core element, a first-order stage whose transfer characteristic has been derived to be

$$\frac{v_o}{v_b} = \frac{1 + R'/R}{1 + j\omega R_1 C_1}. \tag{9.4-5}$$

The additional elements present necessitate an additional relationship so that the total transfer characteristic can be determined. If the currents are summed at the node identified with v_b, the resultant equation is

$$\frac{v_i - v_b}{R_2} + (v_o - v_b)(-j\omega C_2) - \frac{v_b}{R_1 + 1/(j\omega C_1)} = 0 \tag{9.4-6}$$

Equations 9.4-5 and 9.4-6 are combined to determine the transfer characteristic:

$$\frac{v_o}{v_i} = \frac{1 + R'/R}{1 + j\omega[(R_1 + R_2)C_1 - (R'/R)R_2 C_2] + (j\omega)^2 (R_1 R_2 C_1 C_2)}. \tag{9.4-7}$$

This is immediately recognizable as a second-order low-pass characteristic with the following parameters:

$$A_{Vo} = 1 + \frac{R'}{R} \tag{9.4-8}$$

$$\omega_o = \frac{1}{\sqrt{R_1 R_2 C_1 C_2}}, \tag{9.4-9}$$

$$\frac{2\zeta}{\omega_o} = (R_1 + R_2)C_1 - \frac{R'}{R} R_2 C_2. \tag{9.4-10}$$

This is the best of all possible worlds: There are only three constraints on the design and five quantities to vary. While an infinite variety of possible solutions for any specific design exists, there are two specific cases that are of particular interest and that commonly occur in electronic design. These cases are characterized by (1) a need for unity gain in the passband or (2) uniform time constants in the filter section.

Unity-Gain Design

Often it is important that a filter have unity gain in the passband. This restriction leads to the following constraints:

$$A_{Vo} = 1 \Rightarrow \frac{R'}{R} = 0. \tag{9.4-11}$$

In order to achieve the gain requirements, a short circuit is connected between the output and the inverting input of the OpAmp and resistor R is omitted ($R = \infty$). The transfer relationship reduces to

$$\frac{v_o}{v_i} = \frac{1}{1 + j\omega[(R_1 + R_2)C_1] + (j\omega)^2(R_1 R_2 C_1 C_2)}. \tag{9.4-12}$$

The pertinent design parameters are then

$$\omega_o = \frac{1}{\sqrt{R_1 R_2 C_1 C_2}}, \tag{9.4-13}$$

$$\frac{2\zeta}{\omega_o} = (R_1 + R_2)C_1. \tag{9.4-14}$$

EXAMPLE 9.4-1 Design a fourth-order Butterworth low-pass filter with unity gain in the passband that has a 3-dB frequency at 1 kHz.

SOLUTION

A fourth-order filter is constructed from two second-order stages. Each stage will have its resonant frequency at

$$\omega_o = 2\pi f_{3dB} = 6.2832 \text{ krad/s}.$$

The damping coefficients of the two Butterworth stages are obtained from the fourth-order row of Table 9.3-1:

$$\zeta_1 = 0.3827, \qquad \zeta_2 = 0.9239.$$

While there appears to only be two constraints on the values chosen for the elements, realistic solutions are of vital importance. Resistance values must be large with respect to the output resistance of the OpAmp and small with respect to OpAmp input resistance. Limited commercial capacitor availability suggests that they be chosen first. A directed approach usually produces appropriate results, but iteration may be necessary.

Stage with $\zeta_1 = 0.3827$
In working with the constraints of Equations 9.4-13 and 9.4-14, a good start is the balance between C_1 and the sum of R_1 and R_2:

Choose $C_1 = \textbf{0.010 μF}$.

The two constraint equations become

$$R_1 + R_2 = 12.182 \text{ k}\Omega, \qquad R_1 R_2 C_2 = 2.5330.$$

Here the resultant sum of resistor values is such that the resistors will probably fall into the acceptable range. Attention must now be placed on C_2. Whenever the sum and the product of two variables are known, real, nonzero solutions exist only if

$$R_1 R_2 < \tfrac{1}{4}[(R_1 + R_2)^2].$$

Here, C_2 is chosen so that criterion is met reasonably.

Choose $C_2 = \textbf{0.082 μF} \Rightarrow R_1 R_2 = 30.89 \times 10^6$.

This leads to the resistor values[9]

$$R_1 = 8582.4 \approx \textbf{8.56 k}\Omega, \qquad R_2 = 3599.3 \approx \textbf{3.61 k}\Omega.$$

Stage with $\zeta_1 = 0.9239$
The same process is followed as for the stage with $\zeta_1 = 0.3827$. In order to reduce complexity, it is convenient to try the same capacitor values:

Choose $C_1 = \textbf{0.010 μF} \Rightarrow R_1 + R_2 = 29.409 \text{ k}\Omega$ and $R_1 R_2 C_2 = 2.5330$.
Choose $C_2 = \textbf{0.082 μF} \Rightarrow R_1 R_2 = 30.89 \times 10^6$.

This leads to the resistor values

$$R_1 = 1091 \approx \textbf{1.09 k}\Omega, \qquad R_2 = 28{,}318 \approx \textbf{28.4 k}\Omega.$$

The completed design takes the final form shown. While it is necessary to place these stages in cascade, the order of the stages is not significant.

---

[9] Resistor values are rounded to standard values.

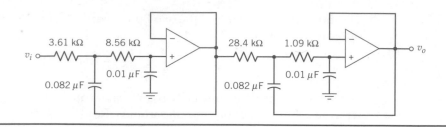

Uniform Time Constant Designs

Another interesting special case is when the value of the low-frequency gain is not significant. In that case, it is convenient to set the capacitors and associated resistors to specific values:

$$C_1 = C_2 = C_c, \qquad R_1 = R_2 = R_c. \tag{9.4-15}$$

The pertinent filter parameters are then reduced to

$$\omega_o = \frac{1}{R_c C_c}, \tag{9.4-16}$$

$$2\zeta = 2 - \frac{R'}{R}. \tag{9.4-17}$$

This specific choice of resistor and capacitor values separates the two main parameters of the filter stage. The resonant frequency ω_o is controlled only by R_c and C_c: The damping coefficient is controlled only by the gain of the individual stage.

EXAMPLE 9.4-2 Design a fifth-order Butterworth low-pass filter with a 3-dB frequency at 1 kHz.

SOLUTION

A fifth-order filter is comprised of a single first-order stage and two second-order stages. The lack of a requirement on the passband gain suggests a uniform time constant filter would be appropriate. Table 9.3-1 yields the two damping coefficients necessary for a fifth-order filter:

$$\zeta_1 = 0.3090, \qquad \zeta_2 = 0.8090.$$

The filter resonant frequency is set to the 3-dB frequency:

$$\omega_o = 2\pi f_{3dB} = 6.2832 \text{ krad/s}.$$

Any reasonable pair of resistor and capacitor values that satisfy Equation 9.4-16 are appropriate. One pair that utilizes standard value components is

$$R_c = 5.9 \text{ k}\Omega, \qquad C_c = 0.027 \text{ μF}.$$

The gain, as established by the ratio of resistors R and R', of each second-order stage is obtained from Equation 9.4-17:

$$\frac{R'}{R} = 2 - 2\zeta = 1.382, 0.382.$$

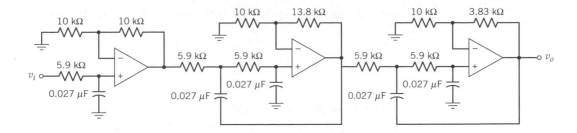

If R is set to 10 kΩ in each stage, the two values for R' are (standard values)

$$R' \approx 13.8 \text{ k}\Omega, \qquad R' = 3.83 \text{ k}\Omega.$$

The gain of the first-order stage is not significant to the performance of the filter as specified. A gain of two allows the use of identical resistors for R and R'. The final design of the specified low-pass filter is shown in schematic form.

9.4.2 High-Pass OpAmp Filters

The two low-pass circuits of Figure 9.4-1 can be converted into high-pass filter sections simply by interchanging the position of the numbered capacitors with the numbered resistors. Interchanging these elements retains the same number of transfer function poles and adds zero-frequency zeroes. Figure 9.4-2 presents the schematic representation of these two high-pass stages.

First-Order High-Pass Stage

The first-order OpAmp filter stage (Figure 9.4-2a) is a noninverting amplifier in series with a high-pass passive RC filter. The transfer characteristic is therefore given by

$$\frac{v_o}{v_i} = \frac{(j\omega R_1 C_1)(1 + R'/R)}{1 + j\omega R_1 C_1}. \tag{9.4-18}$$

This circuit meets the design goal of adjustable resonant frequency. The resonant frequency ω_o of this circuit is determined by the input RC time constant:

$$\omega_o = \frac{1}{R_1 C_1}. \tag{9.4-19}$$

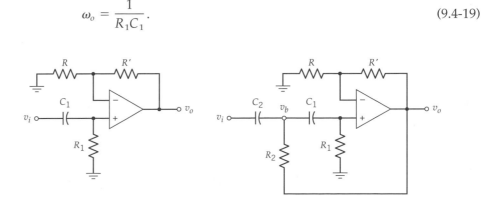

(a)   (b)

Figure 9.4-2
High-pass section realizations: (a) first order; (b) second order.

In addition the high-frequency gain of the circuit is adjustable through the elements R' and R:

$$A_{Vo} = 1 + \frac{R'}{R}. \tag{9.4-20}$$

Second-Order High-Pass Stage

The second-order high-pass stage (Figure 9.4-2b) transfer function is obtained in much the same fashion as the low-pass stage. The transfer characteristic is

$$\frac{v_o}{v_i} = \frac{(j\omega)^2(R_1R_2C_1C_2)(1 + R'/R)}{1 + j\omega[R_2(C_1 + C_2) - (R'/R)R_1C_1] + (j\omega)^2(R_1R_2C_1C_2)}. \tag{9.4-21}$$

This is immediately recognizable as a second-order high-pass characteristic with the following parameters:

$$A_{Vo} = 1 + \frac{R'}{R} \quad , \tag{9.4-22}$$

$$\omega_o = \frac{1}{\sqrt{R_1R_2C_1C_2}}, \tag{9.4-23}$$

$$\frac{2\zeta}{\omega_o} = R_2(C_1 + C_2) - \frac{R'}{R}R_1C_1. \tag{9.4-24}$$

The passband gain and the resonant frequency are the same as for the low-pass case. The *damping coefficient expression is different.* The same two specific design cases are of particular interest (1) unity gain in the passband and (2) uniform time constants in the filter section.

Unity-Gain Designs

Often it is important that a filter have unity gain in the passband. This restriction leads to the following constraints:

$$A_{Vo} = 1 \Rightarrow \frac{R'}{R} = 0. \tag{9.4-25}$$

In order to achieve the gain requirements, a short circuit is connected between the output and the inverting input of the OpAmp and resistor R is omitted ($R = \infty$). The transfer relationship reduces to

$$\frac{v_o}{v_i} = \frac{(j\omega)^2(R_1R_2C_1C_2)}{1 + j\omega[R_2(C_1 + C_2)] + (j\omega)^2(R_1R_2C_1C_2)}. \tag{9.4-26}$$

The pertinent parameters are then

$$\omega_o = \frac{1}{\sqrt{R_1R_2C_1C_2}}, \tag{9.4-27}$$

$$\frac{2\zeta}{\omega_o} = R_2(C_1 + C_2). \tag{9.4-28}$$

Uniform Time Constant Designs

Another interesting special case is when the value of the high-frequency gain is not significant. In that case, it is convenient to set the capacitors and associated resistors to specific values:

$$C_1 = C_2 = C_c, \qquad R_1 = R_2 = R_c.$$

The pertinent filter parameters are then reduced to

$$\omega_o = \frac{1}{R_c C_c}, \tag{9.4-29}$$

$$2\zeta = 2 - \frac{R'}{R}. \tag{9.4-30}$$

Notice that these parameters are unchanged from the low-pass case. This specific choice of resistor and capacitor values separates the two main parameters of the filter stage. The resonant frequency ω_o is controlled only by R_c and C_c: The damping coefficient is controlled only by the gain of the individual stage.

EXAMPLE 9.4-3 Design a third-order Butterworth high-pass filter with unity gain in the passband and a 3-dB frequency of 2 kHz.

SOLUTION

Third-order filters are comprised of a first-order stage followed by a second-order stage with damping coefficient $\zeta = 0.5$.

The resonant frequency is given by

$$\omega_o = 2\pi f_{3\text{dB}} = 12.5664 \times 10^3 \text{ rad/s}.$$

The first-order stage can be realized with element values (rounded to standard values):

$$R_{1(\text{first})} = 7.96 \text{ k}\Omega, \qquad C_{1(\text{first})} = 0.01 \text{ } \mu\text{F}.$$

The second-order stage requires a bit more effort. For simplicity, identical capacitors with the same value as the first-order stage capacitor will be utilized:

$$C_{1(\text{second})} = C_{2(\text{second})} = 0.01 \text{ } \mu\text{F}.$$

Equation 9.4-28 then reduces to

$$R_{2(\text{second})} = 3.97 \text{ k}\Omega.$$

Equation 9.4-27 yields the other resistor value:

$$R_{1(\text{second})} R_{2(\text{second})} = 63.326 \times 10^6 \Rightarrow R_{2(\text{second})} = 16.0 \text{ k}\Omega.$$

The final design is as shown.

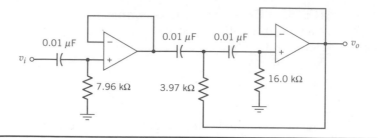

9.4.3 Band-Pass and Band-Stop OpAmp Filters

In many cases band-pass and band-stop filters can be achieved with the series or parallel connection of high-pass and low-pass filters. These filters are particularly useful if the passband or stopband is relatively large.

The series connection of a low-pass filter with a high-pass filter will produce a band-pass filter if there exists a region of common passband. The band-pass filter will extend from the low edge of the high-pass filter passband to the high edge of the low-pass filter passband. This concept is pictured in Figure 9.4-3

Similarly, the parallel connection of a low-pass filter and a high-pass filter will produce a band-stop filter if there is a region of common stopband. The stopband will extend from the low edge of the low-pass filter stopband to the high edge of the high-pass filter stopband. This concept is pictured in Figure 9.4-4

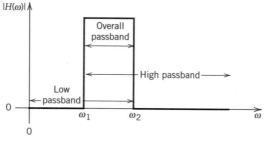

Figure 9.4-3
Cascaded band-pass filter characteristic.

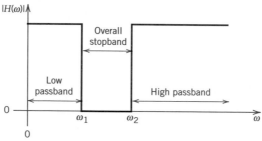

Figure 9.4-4
Parallel band-stop filter characteristic.

EXAMPLE 9.4-4 Design a Butterworth band-pass filter using a minimum number of OpAmps to meet the following design goals:

Midband Region

Low 3-dB frequency: 500 Hz

High 3-dB frequency: 2.35 kHz

Voltage gain: any value (A_{Vo})

Stopband Region

$$f \le 150 \text{ Hz}: |A_V|_{dB} \le |A_{Vo}|_{dB} - 50 \text{ dB}$$
$$f \ge 10 \text{ kHz}: |A_V|_{dB} \le |A_{Vo}|_{dB} - 50 \text{ dB}$$

Verify the design meets the design goals using SPICE.

SOLUTION

This filter can be constructed with the series connection of a low-pass filter and a high-pass filter. Minimum number of OpAmps implies that the order of the filter should be kept as small as possible while still meeting specifications. The absence of a midband gain requirement suggests (but does not require) uniform time constant designs.

Low-Pass Section

The significant radian frequencies of interest are the 3-dB frequency and the edge of the stopband frequency:

$$\omega_{o(\text{lp})} = 2\pi(2350) = 14.77 \times 10^3 \text{ rad/s,}$$

$$\omega_s = 2\pi(10{,}000) = 62.83 \times 10^3 \text{ rad/s.}$$

Fifty decibels of attenuation translates into the magnitude of the Butterworth polynomial at the edge of the stopband:

$$|B_n(\omega)| = 10^{50/20} = 316.23.$$

The order of this filter is found by solving Equation 9.3-2:

$$|B_n(\omega)| = \sqrt{1 + \left(\frac{\omega_s}{\omega_o}\right)^{2n}} \Rightarrow 316.2 = \sqrt{1 + (4.255)^{2n}},$$

$$n = \frac{\log(316.2^2 - 1)}{2\log(4.255)} = 3.975.$$

A fourth-order low-pass filter is necessary. The necessary damping coefficients are found in Table 9.3-1:

$$\zeta_1 = 0.3827, \quad \zeta_2 = 0.9239.$$

Uniform time constant design controls the damping coefficient with the amplifier gain:

$$\frac{R'}{R} = 2 - 2\zeta = 1.2346, 0.1522.$$

If R is chosen arbitrarily as 10 kΩ, then the two feedback resistors are

$$R'_{\text{lp1}} = 12.3 \text{ k}\Omega, \quad R'_{\text{lp2}} = 1.52 \text{ k}\Omega.$$

The resonant frequency of the filter is chosen so that

$$\omega_o = \frac{1}{R_{c(\text{lp})}C_{c(\text{lp})}}.$$

Two standard value components that will satisfy this relationship are

$$R_{c(lp)} = 1.74 \text{ k}\Omega, \qquad C_{c(lp)} = 0.039 \text{ } \mu\text{F}.$$

High-Pass Section
The significant frequencies of interest are

$$\omega_{o(hp)} = 2\pi(500) = 3.142 \times 10^3 \text{ rad/s},$$
$$\omega_s = 2\pi(150) = 942.5 \text{ rad/s}.$$

It is found that a fifth-order high-pass filter is needed:

$$n = \frac{\log(316.2^2 - 1)}{2 \log(3.333)} = 4.781.$$

The damping coefficients are

$$\zeta_1 = 0.3090, \qquad \zeta_2 = 0.8090,$$

which lead to resistor values

$$R = 10 \text{ k}\Omega, \qquad R'_{hp1} = 13.8 \text{ k}\Omega \quad \text{and} \quad R'_{hp2} = 3.82 \text{ k}\Omega.$$

The other components are chosen to have the proper 3-dB frequency:

$$R_{c(hp)} = 8.16 \text{ k}\Omega, \qquad C_{c(hp)} = 0.039 \text{ } \mu\text{F}.$$

The final design is a series connection of the two sections as shown.

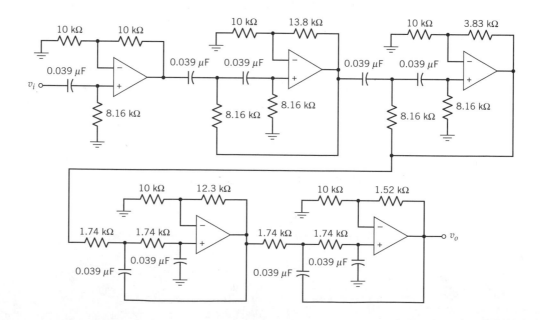

PSpice Verification

Many versions of SPICE do not allow full OpAmp modeling of a circuit this large. In order to alleviate this problem, a simple model of the OpAmp is produced using a **.SUBCKT** statement. The SPICE input file is given by:

```
A Butterworth Band-Pass Filter
VI 1 0 AC 1
*the simple model of an OpAmp
.SUBCKT OA 1 2 3
RIN 1 2 10MEG
E1 3 0 1 2 100MEG
.ENDS
*the fifth order high pass filter section
CH1 2 0.039U
RH1 2 0 8.16K
R1 3 0 10K
RP1 3 4 10K
XH1 2 3 4 OA
CH2 4 5 0.039U
CH3 5 6 0.039U
RH2 5 8 8.16K
RH3 6 0 8.16K
R2 7 0 10K
RP2 7 8 13.8K
XH2 6 7 8 OA
CH4 8 9 0.039U
CH5 9 10 0.039U
RH4 9 12 8.16K
RH5 10 0 8.16K
R3 11 0 10K
RP3 11 12 3.83K
X3 10 11 12 OA
*the fourth order low pass section
RL1 12 13 1.74K
RL2 13 14 1.74K
CL1 13 16 0.039U
CL2 14 0 0.039U
R4 15 0 10K
RP4 15 16 12.3K
X4 14 15 16 OA
RL3 16 17 1.74K
RL4 17 18 1.74K
CL3 17 20 0.039U
CL4 18 0 0.039U
R5 19 0 10K
RP5 19 20 1.52K
X5 18 19 20 OA
*control—filteroutput at node 20
.AC DEC 50 10 100K
.PROBE
.END
```

The response plot is also shown. Notice each of the predicted design goals are very close to theory. The midband gain is approximately the product of the passband gains of each stage:

$$A_{Vo} = 20 \ log[(2)(2.38)(1.383)(2.23)(1.152)]$$

$$= 20 \ log \ 16.91 = 24.56 \ dB.$$

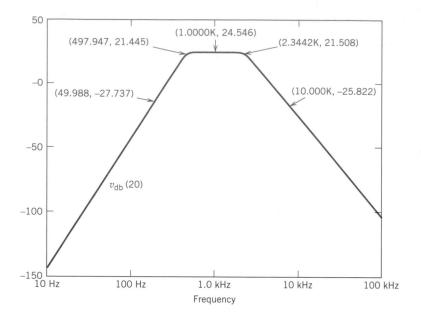

Each 3-dB frequency is exact to within a fraction of a hertz and the stopbands begin at the correct frequencies. Notice the asymmetry of the response plot due to the different order filters in high-pass and low-pass sections.

9.5 OTHER FILTER TYPES

Two other filter families that are commonly used in electronic filter design are the Chebyshev and elliptic designs. Both filter types allow for a small amount of ripple in the passband, stopband, or, in the case of the elliptic filter, both. The disadvantage of the ripple in the filter response is counteracted by a steeper transition band for a given filter order. That is, for a low-pass filter of a given order, the Chebyshev filter has a lower stopband frequency than the Butterworth filter, and the elliptic filter has a lower stopband frequency than the filter with a Chebyshev response.

9.5.1 Chebyshev Filter

When filter design specifications allow for a small amount of ripple in either the passband or the stopband, an all-pole filter called the Chebyshev filter can be used.

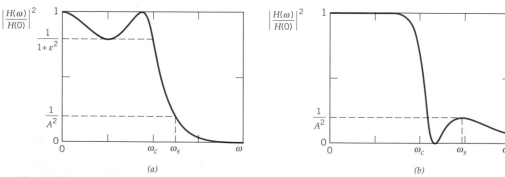

Figure 9.5-1
Chebyshev low-pass filter response: (a) type 1; (b) type 2.

Two types of Chebyshev filters can be specified as shown in Figure 9.5-1: one containing ripple in the passband, classified as type 1, and the other with the ripple in the stopband, called type 2.

In most electronic applications, the Chebyshev type 1 response, with ripple in the passband, is the more common of the two types. The Chebyshev type 1 filter transfer function is[10]

$$\left|\frac{H(\omega)}{H_o}\right|^2 = \frac{1}{1 + \epsilon^2 C_n^2(\omega/\omega_c)},\qquad(9.5\text{-}1)$$

where $H(\omega)/H_o$ is the normalized magnitude of the filter transfer function, the $C_n(\omega/\omega_c)$ are the Chebyshev polynomials defined from a recursive formula given in Table 9.5-1, and ϵ is a parameter related to the ripple defined in Equation 9.3-8: $\gamma_{\max} = \sqrt{1 + \epsilon^2}$.

From Table 9.5-1, it is evident that, for $x = 0$, the Chebyshev polynomial $C_n(0)$ is 1 when n is even and zero when n is odd. The resulting response is

$$\left|\frac{H(0)}{H_o}\right|^2 = \begin{cases} \dfrac{1}{1 + \epsilon^2}, & n \text{ even,} \\ 1 & n \text{ odd.} \end{cases}\qquad(9.5\text{-}2)$$

The two general shapes of the Chebyshev type 1 low-pass filter for n odd and even are shown in Figure 9.5-2. The squared magnitude frequency response oscillates between 1 and $1/(1 + \epsilon^2)$ within the passband and has a value of $1/(1 + \epsilon^2)$ at the cutoff frequency ω_c. The response is monotonic outside the passband. The stopband edge is defined by ω_s corresponding to a magnitude of A^{-2}.

The roots of $1 + \epsilon^2 C_n^2(\omega/\omega_c)$ in the denominator of the Chebyshev type 1 transfer function (the poles) lie on an ellipse on the complex plane, as shown in Figure 9.5-3.

TABLE 9.5-1 CHEBYSHEV POLYNOMIALS

n	$C_n(x)$, where $x = \omega/\omega_c$
0	1
1	x
2	$2x^2 - 1$
3	$4x^3 - 3x$
4	$8x^4 - 8x^2 + 1$
5	$16x^5 - 20x^3 + 5x$
6	$32x^6 - 48x^4 + 18x^2 - 1$
7	$64x^7 - 112x^5 + 56x^3 - 7x$
8	$128x^8 - 256x^6 + 160x^4 - 32x^2 + 1$
9	$256x^9 - 1280x^7 + 432x^5 - 120x^3 + 9x$

---

[10]The Chebyshev type 2 filter transfer function is

$$\left|\frac{H(\omega)}{H_o}\right|^2 = \frac{\epsilon^2 C_n^2(\omega_c/\omega)}{1 + \epsilon^2 C_n^2(\omega_c/\omega)}.$$

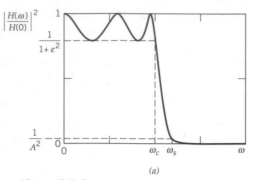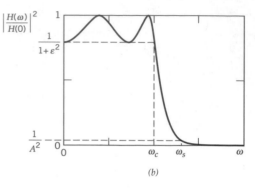

(a)   (b)

Figure 9.5-2
Frequency Response of (a) odd-order Chebyshev type 1 Filter and (b) even-order Chebyshev type 1
Filter.

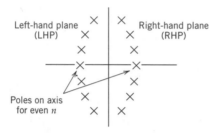

Figure 9.5-3
Poles of low-pass Chebyshev type 1 filter.

For simplicity, let $\omega_r = \omega/\omega_c$. The amount of ripple in the response is related to the eccentricity of the ellipse.

Using the left-hand plane poles only, the transfer function can be written as

$$\frac{H(\omega_r)}{H_o} = \frac{K}{\displaystyle\prod_{k=1}^{n}[(j\omega_r - s_k)]} = \frac{K}{V_n(j\omega_r)}, \tag{9.5-3}$$

where s_k are the pole locations, $V_n(j\omega_r) = (j\omega_r)^n + b_{n-1}(j\omega_r)^{n-1} + \cdots + b_1(j\omega_r) + b_o$, and K is a normalizing constant that makes

$$\frac{H(0)}{H_o} = \begin{cases} \dfrac{1}{\sqrt{1+\epsilon^2}}, & n \text{ even,} \\ 1, & n \text{ odd.} \end{cases}$$

The normalizing constant is therefore

$$K = \begin{cases} V_n(0) = b_0, & n \text{ odd,} \\ \dfrac{V_n(0)}{\sqrt{1+\epsilon^2}}, & n \text{ even.} \end{cases}$$

Table 9.5-2 provides the normalized polynomials for Chebyshev type 1 filters for orders $n = 1, \ldots, 9$ and ϵ corresponding to 0.5, 1, 2, and 3-dB ripples in the passband.

TABLE 9.5-2 NORMALIZED CHEBYSHEV POLYNOMIALS

n	Polynomial

0.5-dB Ripple ($\epsilon = 0.3493$)

1. $j\omega + 2.863$
2. $(j\omega)^2 + 1.426(j\omega) + 1.516$
3. $(j\omega + 0.626)[(j\omega + 0.313)^2 + 1.022^2]$
4. $[(j\omega + 0.175)^2 + 1.016^2][(j\omega + 0.423)^2 + 0.421^2]$
5. $(j\omega + 0.362)[(j\omega + 0.112)^2 + 1.012^2][(j\omega + 0.293)^2 + 0.625^2]$
6. $[(j\omega + 0.078)^2 + 1.009^2][(j\omega + 0.212)^2 + 0.738^2][(j\omega + 0.290)^2 + 0.270^2]$
7. $(j\omega + 0.256)[(j\omega + 0.057)^2 + 1.006^2][(j\omega + 0.160)^2 + 0.807^2][(j\omega + 0.231)^2 + 0.448^2]$
8. $[(j\omega + 0.044)^2 + 1.005^2][(j\omega + 0.124)^2 + 0.852^2][(j\omega + 0.186)^2 + 0.569^2][(j\omega + 0.219)^2 + 0.200^2]$
9. $(j\omega + 0.198)[(j\omega + 0.034)^2 + 1.004^2][j\omega + 0.099)^2 + 0.883^2][(j\omega + 0.152)^2 + 0.655^2]$
   $[(j\omega + 0.186)^2 + 0.349^2]$

1.0-dB Ripple ($\epsilon = 0.5089$)

1. $j\omega + 1.962$
2. $(j\omega)^2 + 1.098j\omega + 1.103$
3. $(j\omega + 0.494)[(j\omega + 0.247)^2 + 0.966^2]$
4. $[(j\omega + 0.140)^2 + 0.983^2][(j\omega + 0.337)^2 + 0.407^2]$
5. $j\omega + 0.289)[(j\omega + 0.090)^2 + 0.990^2][(j\omega + 0.234)^2 + 0.612^2]$
6. $[(j\omega + 0.062)^2 + 0.993^2][(j\omega + 0.170)^2 + 0.727^2][(j\omega + 0.232)^2 + 0.266^2]$
7. $(j\omega + 0.205)[(j\omega + 0.046)^2 + 0.995^2][(j\omega + 0.128)^2 + 0.798^2][(j\omega + 0.185)^2 + 0.443^2]$
8. $[(j\omega + 0.035)^2 + 0.997^2][(j\omega + 0.100)^2 + 0.845^2][(j\omega + 0.149)^2 + 0.564^2][(j\omega + 0.176)^2 + 0.198^2]$
9. $(j\omega + 0.159)[(j\omega + 0.028)^2 + 0.997^2][(j\omega + 0.080)^2 + 0.877^2][(j\omega + 0.122)^2 + 0.651^2]$
   $[(j\omega + 0.150)^2 + 0.346^2]$

2.0-dB Ripple ($\epsilon = 0.7648$)

1. $j\omega + 1.308$
2. $(j\omega)^2 + 0.804j\omega + 0.637$
3. $(j\omega + 0.369)[(j\omega + 0.184)^2 + 0.923^2]$
4. $[(j\omega + 0.105)^2 + 0.958^2][(j\omega + 0.253)^2 + 0.397^2]$
5. $(j\omega + 0.218)[(j\omega + 0.068)^2 + 0.974^2][(j\omega + 0.177)^2 + 0.602^2]$
6. $[(j\omega + 0.047)^2 + 0.982^2][(j\omega + 0.128)^2 + 0.719^2][(j\omega + 0.175)^2 + 0.263^2]$
7. $(j\omega + 0.155)[(j\omega + 0.035)^2 + 0.987^2][(j\omega + 0.097)^2 + 0.791^2][(j\omega + 0.140)^2 + 0.440^2]$
8. $[(j\omega + 0.027)^2 + 0.990^2][(j\omega + 0.075)^2 + 0.839^2][(j\omega + 0.113)^2 + 0.561^2[[(j\omega + 0.133)^2 + 0.197^2]$
9. $(j\omega + 0.121)[(j\omega + 0.021)^2 + 0.992^2][(j\omega + 0.060)^2 + 0.872^2][(j\omega + 0.092)^2 + 0.647^2]$
   $[(j\omega + 0.113)^2 + 0.345^2]$

3.0-dB Ripple ($\epsilon = 0.9953$)

1. $j\omega + 1.002$
2. $(j\omega)^2 + 0.645j\omega + 0.708$
3. $(j\omega + 0.299)[(j\omega + 0.149)^2 + 0.904^2]$
4. $[(j\omega + 0.085)^2 + 0.947^2][(j\omega + 0.206)^2 + 0.392^2]$
5. $(j\omega + 0.178)[(j\omega + 0.055)^2 + 0.966^2][j\omega + 0.144)^2 + 0.597^2]$
6. $[(j\omega + 0.038)^2 + 0.976^2][(j\omega + 0.104)^2 + 0.715^2][(j\omega + 0.143)^2 + 0.262^2]$
7. $(j\omega + 0.127)[(j\omega + 0.028)^2 + 0.983^2][(j\omega + 0.079)^2 + 0.788^2][(j\omega + 0.114)^2 + 0.437^2]$
8. $[(j\omega + 0.022)^2 + 0.987^2][(j\omega + 0.062)^2 + 0.837^2][(j\omega + 0.092)^2 + 0.559^2][(j\omega + 0.109)^2 + 0.196^2]$
9. $(j\omega + 0.098)[(j\omega + 0.017)^2 + 0.990^2][(j\omega + 0.049)^2 + 0.870^2][(j\omega + 0.075)^2 + 0.646^2]$
   $[(j\omega + 0.092)^2 + 0.344^2]$

The required order of a Chebyshev type 1 low-pass filter depends on the following factors:

- Cutoff frequency
- Stopband frequency
- Stopband attenuation
- Passband ripple

Design specifications usually provide the required minimum attenuation desired at a certain stopband frequency instead of an exact attenuation. The required order of the filter, n, is found by applying the following equation:

$$n = \frac{\log(g + \sqrt{g^2 - 1})}{\log(\omega_r + \sqrt{\omega_r^2 - 1})}, \tag{9.5-4}$$

where

$$A = \left| \frac{H_o}{H(\omega_r)} \right|, \tag{9.5-5}$$

$$g = \sqrt{\frac{A^2 - 1}{\epsilon^2}}. \tag{9.5-6}$$

EXAMPLE 9.5-1 Find the transfer function for a low-pass filter with the following specifications:

a) Cutoff frequency $f_c = 1000$ Hz
b) Acceptable passband ripple of 2 dB
c) Stopband attenuation ≥ 20 dB beyond 1300 Hz

SOLUTION

The specifications imply the following relationships:

$$20 \log \left| \frac{H(2\pi \times 1000)}{H_o} \right| = 20 \log \left[\frac{1}{\sqrt{1 + \epsilon^2}} \right] = -2 \text{ dB},$$

$$20 \log \left| \frac{H(2\pi \times 1300)}{H_o} \right| = 20 \log \left[\frac{1}{\sqrt{A^2}} \right] = -20 \text{ dB}.$$

The solutions to the two relationships above are $\epsilon = 0.7648$ and $A = 10$. The relative stopband frequency is $1300/1000 = 1.3$.

The required order of the Chebyshev type 1 low-pass filter is found by using Equations 9.5-4 to 9.5-6,

$$g = \sqrt{\frac{A^2 - 1}{\epsilon^2}} = 13.01,$$

$$n = \left\lceil \frac{\log(13.01 + \sqrt{13.01^2 - 1})}{\log(1.3 + \sqrt{1.3 - 1})} \right\rceil = \lceil 4.3 \rceil = 5,$$

where the partial brackets indicate the operation to round up to the next integer.

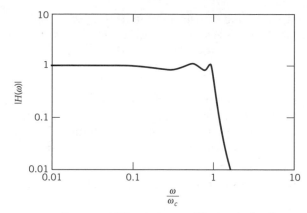

Response of fifth-order low-pass filter transfer function

Using Table 9.5-2 for $n = 5$ and 2-dB ripple, the transfer function is

$$\frac{H(\omega/\omega_c)}{H_o}$$

$$= \frac{1}{\left\{\left(j\dfrac{\omega}{\omega_c} + 0.218\right)\left[\left(j\dfrac{\omega}{\omega_c} + 0.068\right)^2 + 0.974^2\right]\left[\left(j\dfrac{\omega}{\omega_c} + 0.177\right)^2 + 0.602^2\right]\right\}}$$

$$= \frac{1}{\left[\left(\dfrac{j\omega}{0.218\omega_c} + 1\right)\left(\dfrac{(j\omega)^2}{0.953\omega_c^2} + \dfrac{0.136(j\omega)}{0.953\omega_c} + 1\right)\left(\dfrac{(j\omega)^2}{0.394\omega_c^2} + \dfrac{0.354(j\omega)}{0.394\omega_c} + 1\right)\right]}.$$

High-Pass Characterization

High-pass Chebyshev filters are characterized by using the analog-to-analog transformation. The Chebyshev low-pass to high-pass transformation, like the Butterworth low-pass to high-pass analog-to-analog transformation, has the effect of adding two zero-frequency zeroes to the transfer function. The generic procedure for the

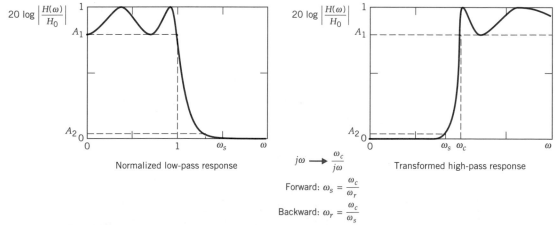

Figure 9.5-4
Analog-to-Analog Transformation: Low-Pass to High-Pass

analog-to-analog transformation from low- to high-pass shown in Figure 9.5-4 is not restricted to Chebyshev filters, but may be used for other filter types.

The general procedure to find the transfer function of a high-pass filter is as follows:

1. Perform a backward transformation using the high-pass filter specifications for the cutoff frequency ω_c and the stopband frequency ω_s, which is associated with some level of attenuation. The backward transformation of frequencies yields the normalized low-pass ratio of stopband to cutoff frequency ω_r.

2. Use the normalized low-pass ratio of stopband to cutoff frequency ω_r to find the order of the filter.

3. Use Table 9.5-2 to find the transfer function of the normalized low-pass filter.

4. Perform a forward transformation on the transfer function by replacing all $j\omega$ with $\omega_c/(j\omega)$.

EXAMPLE 9.5-2 Find the transfer function for a high-pass filter with the following specifications:

(a) Cutoff frequency $f_c = 1000$ Hz

(b) Acceptable passband ripple of 2 dB

(c) Stopband attenuation ≥ 20 dB below 500 Hz

SOLUTION

Following the high-pass to low-pass transformation procedure, a backward transformation is performed that yields

$$\omega_r = \frac{\omega_c}{\omega_s} = \frac{2\pi(1000)}{2\pi(500)} = 2.$$

The specifications imply the following relationships:

$$20 \log\left|\frac{H(2\pi \times 1000)}{H_o}\right| = 20 \log\left(\frac{1}{\sqrt{1 + \epsilon^2}}\right) = -2 \text{ dB},$$

$$20 \log\left|\frac{H(2\pi \times 500)}{H_o}\right| = 20 \log\left(\frac{1}{\sqrt{A^2}}\right) = -20 \text{ dB}.$$

The solutions to the two relationships above are $\epsilon = 0.7648$ and $A = 10$.

The required order of the Chebyshev type 1 low-pass filter is found by using Equations 9.5-4 to 9.5-6,

$$g = \sqrt{\frac{A^2 - 1}{\epsilon^2}} = \sqrt{\frac{10^2 - 1}{0.7648^2}} = 13.01,$$

$$n = \left\lceil \frac{\log(13.01 + \sqrt{13.01^2 - 1})}{\log(2 + \sqrt{2 - 1})} \right\rceil = \lceil 2.96 \rceil = 3.$$

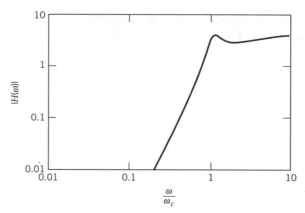

Response of third-order high-pass filter transfer function

Using Table 9.5-2 for n values of 3 and 2-dB ripple, the transfer function is

$$\frac{H(\omega/\omega_c)}{H_o} = \frac{1}{\left\{\left(j\frac{\omega}{\omega_c} + 0.299\right)\left[\left(j\frac{\omega}{\omega_c} + 0.149\right)^2 + 0.904^2\right]\right\}}$$

$$= \frac{1}{\left\{\left(j\frac{\omega}{\omega_c} + 0.299\right)\left[\left(j\frac{\omega}{\omega_c}\right)^2 + 0.298\left(j\frac{\omega}{\omega_c}\right) + 0.839\right]\right\}}.$$

Perform a forward transformation to yield the high-pass filter transfer function:

$$\frac{H(\omega)}{H_o}\bigg|_{j\omega\to\omega_c/j\omega}$$

$$= \frac{1}{[(\omega_c/(j\omega)) + 0.299][(\omega_c/(j\omega))^2 + 0.298(\omega_c/(j\omega)) + 0.839]}.$$

The normalized high-pass filter function is

$$\frac{H(\omega/\omega_c)}{H_o}\bigg|_{j\omega\to\omega_c/j\omega}$$

$$= \frac{(j\omega/\omega_c)^3}{[1 + 0.299(j\omega/\omega_c)][1 + 0.298(j\omega/\omega_c) + 0.839(j\omega/\omega_c)^2]}.$$

9.5.2 OpAmp Realization of Chebyshev Filters

Since the Chebyshev type 1 low-pass filter is an all-pole filter, the OpAmp circuit implementation is identical to the Butterworth filter circuit configurations. Both the unity gain and Sallen and Key configurations can be used to implement Chebyshev type 1 filters.

Low-Pass OpAmp Filters

Although the OpAmp circuit configurations for the Chebyshev type 1 and the Butterworth low-pass filters are identical, the calculations for the components of the circuit components differ. The calculations for the first-order low-pass stages of Chebyshev type 1 and Butterworth filters is identical. However, the calculation of the component values for the Chebyshev type 1 transfer function of second-order low-pass stages requires some modification from the Butterworth calculations.

These modifications for calculating second-order stages of Chebyshev type 1 low-pass filters are as follows: If

$$H\left(\frac{\omega}{\omega_c}\right) = \prod_{i=1}^{n} \frac{1}{(j\omega)^2/(a_i\omega_c^2) + b_i(j\omega)/(a_i\omega_c) + 1} \, , \tag{9.5-7}$$

then the "relative" resonant frequency for the two second-order stages are

$$\omega_{oi} = \sqrt{a_i}\,\omega_c. \tag{9.5-8}$$

The damping coefficient is

$$2\zeta_i = \frac{b_i}{\sqrt{a_i}}. \tag{9.5-9}$$

Then Equation 9.4-9 is replaced by

$$\omega_{oi} = \frac{1}{\sqrt{R_1 R_2 C_1 C_2}}. \tag{9.5-10}$$

Using the uniform time constant form of the Sallen and Key second-order stage where $R_1 = R_2$ and $C_1 = C_2$, the gain can be expressed as

$$A_{Vi} = 3 - 2\zeta_i. \tag{9.5-11}$$

EXAMPLE 9.5-3   Design a low-pass filter with the following specifications:

(a) Cutoff frequency $f_c = 1000$ Hz
(b) Acceptable passband ripple of 3 dB
(c) Stopband attenuation ≥ 20 dB above 1400 Hz

Implement the filter transfer function with the Sallen and Key configuration.

SOLUTION

The specifications imply the following relationships:

$$20 \log \left| \frac{H(2\pi \times 1000)}{H_o} \right| = 20 \log\left(\frac{1}{\sqrt{1 + \epsilon^2}}\right) = -3 \text{ dB,}$$

$$20 \log \left| \frac{H(2\pi \times 1400)}{H_o} \right| = 20 \log\left(\frac{1}{\sqrt{A^2}}\right) = -20 \text{ dB.}$$

The solutions to the two relationships above are $\epsilon = 1$ and $A = 10$.

The required order of the Chebyshev type 1 low-pass filter is found by using Equations 9.5-4 to 9.5-6,

$$g = \sqrt{\frac{A^2 - 1}{\epsilon^2}} = \sqrt{\frac{10^2 - 1}{1^2}} = 9.95,$$

$$n = \left\lceil \frac{\log(9.95 + \sqrt{9.95^2 - 1})}{\log(1.4 + \sqrt{1.4^2 - 1})} \right\rceil = \lceil 3.45 \rceil = 4.$$

Using Table 9.5-2 for n values of 4 and 3-dB ripple, the transfer function is

$$\frac{H(\omega/\omega_c)}{H_o}$$

$$= \frac{1}{\left\{\left[\left(j\frac{\omega}{\omega_c} + 0.085\right)^2 + 0.947^2\right]\left[\left(j\frac{\omega}{\omega_c} + 0.206\right)^2 + 0.392^2\right]\right\}}$$

$$= \frac{1}{\left\{\left[\left(j\frac{\omega}{\omega_c}\right)^2 + 0.170\left(j\frac{\omega}{\omega_c}\right) + 0.904\right]\left[\left(j\frac{\omega}{\omega_c}\right)^2 + 0.412\left(j\frac{\omega}{\omega_c}\right) + 0.196\right]\right\}}$$

$$= \frac{5.644}{[(j\omega)^2/(0.904\omega_c^2) + 0.170j\omega/(0.904\omega_c) + 1][(j\omega)^2/(0.196\omega_c^2) + 0.412j\omega/(0.196\omega_c) + 1]}.$$

The relative resonant frequencies for the two second-order stages are

$$\omega_{o1} = \sqrt{0.904}\omega_c = (0.951)(2\pi)(1000) = 5.98 \text{ krad/s,}$$

$$\omega_{o2} = \sqrt{0.196}\omega_c = (0.443)(2\pi)(1000) = 2.78 \text{ krad/s.}$$

The damping coefficients are

$$2\zeta_1 = \frac{0.170}{\sqrt{0.904}} = 0.179,$$

$$2\zeta_2 = \frac{0.412}{\sqrt{0.196}} = 0.931.$$

Let the capacitors be equal and $C = 0.1 \text{ }\mu\text{F}$. Let the frequency controlling resistances within each stage be identical. Then, the frequency controlling resistors for the two stages are

$$R_1 = \frac{1}{\omega_{o1}C} = \frac{1}{(5980)(0.1 \times 10^{-6})} = 3.6 \text{ k}\Omega,$$

$$R_2 = \frac{1}{\omega_{o2}C} = \frac{1}{(2780)(0.1 \times 10^{-6})} \approx 1.6 \text{ k}\Omega \quad \text{(standard value).}$$

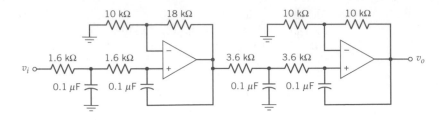

The gain for the two stages are

$$A_{v1} = 3 - (2\xi_1) = 2.82, \qquad A_{v2} = 3 - (2\xi_1) = 2.07.$$

If $R'_{gain1} = R'_{gain2} = 10$ kΩ, then $R_{gain1} \approx 18$ kΩ and $R_{gain2} \approx 10$ kΩ. The SPICE simulation is as follows:

```
*Chebyshev low-pass filter: 4th order with 3 dB ripple, cutoff = 1KHz
V1    1    0    AC    IV
*First Sallen and Key stage
R11   1    12   1.6K
R12   12   13   1.6K
C11   12   14   0.1u
C12   13   0    0.1u
R13   14   15   18K
R14   15   0    10K
X1    15   13   14       OpAmp
*Second Sallen and Key Stage
R21   14   22   3.6K
R22   22   23   3.6K
C21   22   24   0.1u
C22   23   0    0.1u
R23   24   25   10K
R24   25   0    10K
X2    25   23   24       OpAmp
*
*Subcircuit for Ideal OpAmp
.SUBCKT OpAmp 101 102 103
*              |   |   |
*   inverting input |   |
*      non-inverting input |
*                  output
*
   Rin     101   102    2MEG
   Rout    104   103    75
   Eopamp 104    0      102    101    200K
.ENDS
*
.AC   DEC   50   1   10K
.PROBE
.END
```

The SPICE frequency response is as follows:

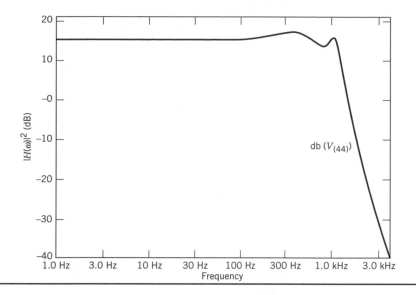

High-Pass OpAmp Filters

Unlike Butterworth filters, the Chebyshev high-pass OpAmp filter is not arrived at by simple interchanges of the frequency controlling resistances and capacitances. As shown in Section 9.5-1, the transfer function of the Chebyshev high-pass filter is found by performing analog-to-analog transformations. The high-pass transfer function is then implemented by a circuit using the same techniques outlined for the Chebyshev low-pass filter.

EXAMPLE 9.5-4 Design a high-pass filter with the following specifications:

    (a) Cutoff frequency f_c = 1000 Hz

    (b) Acceptable passband ripple of 3 dB

    (c) Stopband attenuation $\geq$20 dB below 600 Hz

Implement the filter transfer function with the Sallen and Key configuration.

SOLUTION

The specifications imply the following relationships:

$$20 \log \left| \frac{H(2\pi \times 1000)}{H_o} \right| = 20 \log \left[\frac{1}{\sqrt{1 + \epsilon^2}} \right] = -3 \text{ dB},$$

$$20 \log \left| \frac{H(2\pi \times 600)}{H_o} \right| = 20 \log \left[\frac{1}{\sqrt{A^2}} \right] = -20 \text{ dB}.$$

The solutions to the two relationships above are $\epsilon = 1$ and $A = 10$.

An analog-to-analog transformation is performed to determine the transformed normalized low-pass characteristics. The backward transformation from high to low-pass yields

$$\omega_r = \frac{\omega_c}{\omega_s} = \frac{2\pi(1000)}{2\pi(600)} = 1.67.$$

The required order of the normalized Chebyshev type 1 low-pass filter is found by using Equations 9.5-4 to 9.5-6,

$$g = \sqrt{\frac{A^2 - 1}{\epsilon^2}} = \sqrt{\frac{10^2 - 1}{1^2}} = 9.95,$$

$$n = \left\lceil \frac{\log(9.95 + \sqrt{9.95^2 - 1})}{\log(1.67 + \sqrt{1.67 - 1})} \right\rceil = \lceil 3.27 \rceil = 4.$$

Using Table 9.5-2 for $n = 4$ and 3-dB ripple, the transfer function is

$$\frac{H(\omega)}{H_o} = \frac{1}{[(j\omega + 0.085)^2 + 0.947^2][(j\omega + 0.206)^2 + 0.392^2]}$$

$$= \frac{1}{[(j\omega)^2 + 0.170(j\omega) + 0.904][(j\omega)^2 + 0.412(j\omega) + 0.196]}$$

Another analog-to-analog transformation is performed to yield the normalized high-pass transfer function by replacing $j\omega$ with $\omega_c/j\omega$:

$$\frac{H(\omega)}{H_o}$$

$$= \frac{1}{[(\omega_c/(j\omega))^2 + 0.170(\omega_c/(j\omega)) + 0.904][(\omega_c/(j\omega))^2 + 0.412(\omega_c/(j\omega)) + 0.196]}$$

$$= \frac{(j\omega/\omega_c)^2}{[1 + 0.170(j\omega/\omega_c) + 0.904(j\omega/\omega_c)^2][1 + 0.412(j\omega/\omega_c) + 0.196(j\omega/\omega_c)^2]}$$

$$= \frac{(j\omega/\omega_c)^2}{\left[1 + \dfrac{0.188}{1/0.904}\left(\dfrac{j\omega}{\omega_c}\right) + \dfrac{1}{1/0.904}\left(\dfrac{j\omega}{\omega_c}\right)^2\right]\left[1 + \dfrac{2.112}{1/0.196}\left(\dfrac{j\omega}{\omega_c}\right) + \dfrac{1}{1/0.196}\left(\dfrac{j\omega}{\omega_c}\right)^2\right]}.$$

The relative resonant frequencies for the two second-order stages are

$$\omega_{o1} = \sqrt{\frac{1}{0.904}}\,\omega_c = (1.052)(2\pi)(1000) = 6.61 \text{ krad/s},$$

$$\omega_{o2} = \sqrt{\frac{1}{0.196}}\,\omega_c = (2.259)(2\pi)(1000) = 14.18 \text{ krad/s}.$$

The damping coefficients are

$$2\zeta_1 = \frac{0.188}{\sqrt{1/0.904}} = 0.179,$$

$$2\zeta_2 = \frac{2.112}{\sqrt{1/0.196}} = 0.935.$$

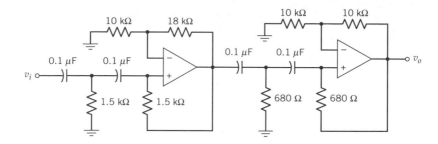

Note that the damping coefficients are identical (except for round-off error) to that of the fourth-order low-pass example in Example 9.5-3.

Let the capacitors be equal and $C = 0.1\ \mu F$. Let the frequency controlling resistances within each stage be identical. Then, the frequency controlling resistors for the two stages are

$$R_1 = \frac{1}{\omega_{o1}C} = \frac{1}{6610(0.1 \times 10^{-6})} \approx 1.5\ \text{k}\Omega,$$

$$R_2 = \frac{1}{\omega_{o2}C} = \frac{1}{14{,}180(0.1 \times 10^{-6})} \approx 680\ \Omega \quad \text{(standard value)}.$$

The gain for the two stages are

$$A_{v1} = 3 - (2\xi_1) = 2.82, \qquad A_{v2} = 3 - (2\xi_1) = 2.07.$$

If $R'_{\text{gain1}} = R'_{\text{gain2}} = 10\ \text{k}\Omega$, then $R_{\text{gain1}} \approx 18\ \text{k}\Omega$ and $R_{\text{gain2}} \approx 10\ \text{k}\Omega$: The gain resistors are identical to the fourth-order low-pass example in Example 9.5-3.

```
SPICE simulation:
High-pass filter: 4th order Chebyshev
*
*Chebyshev high-pass filter: 4th order with 3 dB ripple, cutoff = 1KHz
V1    1   0    AC     IV
C11   1   12   0.1u
C12   12  13   0.1u
R11   12  14   1.5K
R12   13  0    1.5K
R13   14  15   18K
R14   15  0    10K
X1    15  13   14     OpAmp
C21   14  22   0.1u
C22   22  23   0.1u
R21   22  24   680
R22   23  0    680
R23   24  25   10K
R24   25  0    10K
X2    25  23   24     OpAmp
*
*Subcircuit for Ideal OpAmp
.SUBCKT OpAmp 101  102  103
*              |   |   |
*    inverting input |  |
*        non-inverting input |
```

```
*                     output
*
  Rin    101   102   2MEG
  Rout   104   103   75
  Eopamp 104   0     102   101   200K
.ENDS
*
.AC   DEC   50   1   100K
.PROBE
.END
```

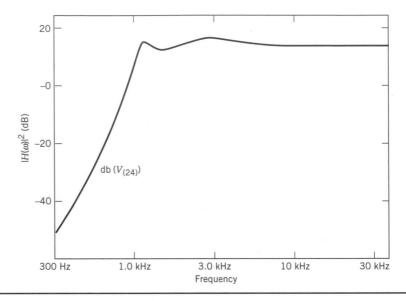

9.5-3 Elliptic Filter

By allowing ripple in both the passband and stopband, the elliptic filter achieves a steep transition region. The advantage of the elliptic filter lies in the fact that for given passband ripple, the transition band is steeper than the Chebyshev filter for a given order filter. The filter is normally designed so that the peak of the stopband ripple is well within the specified stopband attenuation. Since the mathematical development of the elliptic filter is quite complex, a detailed mathematical discussion is not presented.

The squared magnitude frequency response of a normalized low-pass elliptic filter of order n is defined by

$$|H(\omega)|^2 = \frac{1}{1 + \epsilon^2 R_n^2(\omega)}, \tag{9.5-12}$$

where $R_n^2(\omega)$ is the Chebyshev rational function determined from the specified ripple characteristics. Figure 9.5-5 shows typical plots of elliptic low-pass filters.

In elliptic filters, it is convenient to define the parameter ω_r as the sharpness of transition,

$$\omega_r = \frac{\omega_s}{\omega_c}. \tag{9.5-13}$$

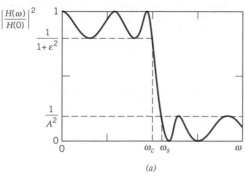 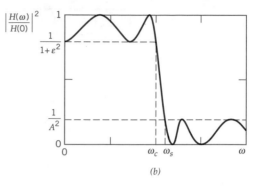

(a)   (b)

Figure 9.5-5
Elliptic low-pass filter: (*a*) odd order; (*b*) even order.

A large ω_r indicates a large transition region, while a small ω_r is indicative of a narrower transition region.

The general form of the transfer function for normalized elliptic low-pass filter for odd- and even-order filters are

$$
H(\omega = \begin{cases} \dfrac{H_o}{j\omega + s_0} \displaystyle\prod_{i=1}^{(n-1)/2} \dfrac{(j\omega)^2 + A_{0i}}{(j\omega)^2 + B_{1i}(j\omega) + B_{0i}}, & n \text{ odd,} \qquad (9.5\text{-}14) \\[3em] H_o \displaystyle\prod_{i=1}^{n/2} \dfrac{(j\omega)^2 + A_{0i}}{(j\omega)^2 + B_{1i}(j\omega) + B_{0i}}, & n \text{ even.} \qquad (9.5\text{-}15) \end{cases}
$$

where
n = order of filter
H_o = normalizing magnitude scaling factor
s_0 = single-pole location
A_{0i} = coefficients of numerator quadratic terms
B_{1i}, B_{0i} = coefficients of denominator quadratic terms
(determines pole locations)

The stopband and passband gains are found using the procedure shown for Chebyshev filters. The normalizing frequency used to find the actual from the normalized transfer function is

$$
\omega_n = \sqrt{\omega_c \omega_s}. \qquad (9.5\text{-}16)
$$

Because of the complexity of the mathematical form of the elliptic filter, simple tables similar to Butterworth and Chebyshev filters are not possible. Table 9.5-3 is a table for normalized elliptic filters for stopband gains of -20, -40, and -60 dB and passband ripples of 0.5, 1, 2, and 3 dB. Using the allowable ripple, stopband attenuation, and transition sharpness, ω_r, the order of the elliptic filter is selected from the table.

Convenient identification of the required order of the elliptic filter for a given stopband attenuation, relative to the peak response, is found in graphical form in Figure 9.5-6. For example, for a stopband attenuation (including ripple in the passband) of 60 dB and $\omega_r = 4.2$, a third-order elliptic filter is required.

TABLE 9.5-3 COEFFICIENTS FOR ELLIPTIC FILTER TRANSFER FUNCTIONS

N	i	A_{0i}	B_{0i}	B_{1i}	H_o/S_o	Ω_r
			Passband Ripple = 0.5 dB, Stopband Gain = −20 dB			
2	1	5.33789	0.566660	0.809390	0.100220	2.76267
3	1	1.75640	0.808321	0.359160	0.306214	1.42189
					0.667292	
4	1	4.38105	0.611195	0.931959	0.100219	1.13188
	2	1.21841	0.927132	0.136543		
5	1	1.65076	0.827787	0.412816	0.303895	1.04465
	2	1.07211	0.973640	0.049395	0.667292	
	1	4.36790	0.611899	0.933855	0.100218	1.01553
6	2	1.19243	0.934830	0.156221		
	3	1.02486	0.990620	0.017576		
	1	1.64918	0.828092	0.413652	0.303861	1.00545
7	2	1.06401	0.976479	0.056384	0.667292	
	3	1.00870	0.996681	0.006219		
8	1	4.36811	0.611846	0.933864	0.100192	1.00192
	2	1.19207	0.934928	0.156548		
	3	1.02213	0.991634	0.020051		
	4	1.00306	0.998827	0.002197		
9	1	1.64927	0.828047	0.413695	0.303786	1.00068
	2	1.06390	0.976512	0.056505	0.667292	
	3	1.00775	0.997041	0.007093		
	4	1.00108	0.999586	0.000775		
			Passband Ripple = 0.5 dB; Stopband Gain = −40 dB			
2	1	16.91940	0.179222	0.486687	0.100001×10^{-1}	8.48925
3	1	3.55131	0.422730	0.352622	0.476460×10^{-1}	2.71147
					0.417693	
4	1	9.15630	0.274972	0.710817	0.100001×10^{-1}	1.62842
	2	1.84784	0.651784	0.213089		
5	1	2.77710	0.502279	0.488797	0.450049×10^{-1}	1.27264
	2	1.35383	0.808559	0.117164	0.417693	
6	1	8.84271	0.281790	0.725407	0.100001×10^{-1}	1.12697
	2	1.64666	0.706755	0.288016		
	3	1.16198	0.899904	0.061260		
7	1	2.73666	0.507485	0.497501	0.448511×10^{-1}	1.06110
	2	1.27960	0.841245	0.155925	0.417693	
	3	1.07732	0.949005	0.031209		

TABLE 9.5-3 (continued)

N	i	A_{0i}	B_{0i}	B_{1i}	H_o/S_o	Ω_r
8	1	8.82410	0.282208	0.726296	0.100001×10^{-1}	1.02987
	2	1.63523	0.710223	0.292713		
	3	1.13002	0.917716	0.080817		
	4	1.03765	0.974360	0.015692		
9	1	2.73426	0.507796	0.498030	0.448397×10^{-1}	1.01471
	2	1.27524	0.843263	0.158322	0.417693	
	3	1.06253	0.958272	0.040984		
	4	1.01851	0.987192	0.007839		

Passband Ripple = 0.5 dB, Stopband Gain = −60 dB

N	i	A_{0i}	B_{0i}	B_{1i}	H_o/S_o	Ω_r
2	1	53.50591	0.056674	0.275431	0.999962×10^{-3}	26.76230
3	1	7.54288	0.201322	0.258673	0.709340×10^{-2} 0.307274	5.67937
4	1	17.20286	0.143036	0.529149	0.999995×10^{-2}	2.68325
	2	3.11024	0.396229	0.197979		
5	1	4.56247	0.304405	0.433499	0.607279×10^{-2} 0.307274	1.77664
	2	1.93110	0.581652	0.137383		
6	1	15.26368	0.157911	0.565141	0.999998×10^{-3}	1.40138
	2	2.38314	0.493638	0.315921		
	3	1.47091	0.727066	0.089646		
7	1	4.30526	0.319374	0.457276	0.596437×10^{-2} 0.307274	1.21984
	2	1.66031	0.659035	0.212758		
	3	1.25499	0.828773	0.056241		
8	1	15.04762	0.159784	0.569572	0.999987×10^{-3}	1.12427
	2	2.30879	0.506752	0.331430		
	3	1.34713	0.781898	0.136095		
	4	1.14323	0.895182	0.034429		
9	1	4.27529	0.321225	0.460193	0.595147×10^{-2} 0.307274	1.07151
	2	1.63040	0.669021	0.222400		
	3	1.19187	0.864916	0.084292		
	4	1.08212	0.936789	0.020762		

Passband Ripple = 1 dB, Stopband Gain = −20 dB

N	i	A_{0i}	B_{0i}	B_{1i}	H_o/S_o	Ω_r
2	1	4.42342	0.497233	0.676727	0.100185	2.32474
3	1	1.58565	0.790229	0.282927	0.281080 0.565168	1.30797
4	1	3.81475	0.536633	0.768217	0.100185	1.09029
	2	1.15956	0.926578	0.099029		

TABLE 9.5-3 (continued)

N	i	A_{0i}	B_{0i}	B_{1i}	H_o/S_o	Ω_r
5	1	1.51852	0.808049	0.318242	0.279829	1.02826
	2	1.04886	0.975703	0.032771	0.565168	
6	1	3.80873	0.537071	0.769217	0.100184	1.00902
	2	1.14367	0.933194	0.110761		
	3	1.01550	0.992101	0.010654		
7	1	1.51782	0.808242	0.318627	0.279814	1.00290
	2	1.04421	0.977941	0.036573	0.565168	
	3	1.00497	0.997446	0.003444		
8	1	3.80866	0.537076	0.769228	0.100185	1.00093
	2	1.14350	0.933266	0.110888		
	3	1.01405	0.992834	0.011881		
	4	1.00160	0.999176	0.001111		
9	1	1.51812	0.808099	0.318714	0.279611	1.00030
	2	1.04421	0.977936	0.036644	0.565168	
	3	1.00451	0.997680	0.003845		
	4	1.00052	0.999734	0.000359		

<table>
<tbody>
<tr><td colspan="7" align="center">Passband Ripple = 1 dB, Stopband Gain = −40 dB</td></tr>
</tbody>
</table>

N	i	A_{0i}	B_{0i}	B_{1i}	H_o/S_o	Ω_r
2	1	14.01843	0.157290	0.411245	0.100001×10^{-1}	7.04488
3	1	3.14896	0.416088	0.292413	0.445207×10^{-1} 0.349732	2.41619
4	1	8.20047	0.238719	0.591849	0.100002×10^{-1}	1.51549
	2	1.70946	0.658897	0.171046		
5	1	2.55417	0.489991	0.396962	0.425507×10^{-1}	1.21868
	2	1.28995	0.819650	0.090443	0.349732	
6	1	7.98807	0.243668	0.601988	0.100001×10^{-1}	1.09887
	2	1.55441	0.709739	0.225936		
	3	1.12899	0.909504	0.045382		
7	1	2.52698	0.494109	0.402679	0.424523×10^{-1}	1.04600
	2	1.23338	0.848902	0.117614	0.349732	
	3	1.05957	0.955761	0.022185		
8	1	7.97731	0.243925	0.602513	0.999982×10^{-2}	1.02168
	2	1.54681	0.712467	0.228870		
	3	1.10516	0.924805	0.058531		
	4	1.02799	0.978647	0.010708		
9	1	2.52561	0.494317	0.402976	0.424450×10^{-1}	1.01029
	2	1.23053	0.850439	0.119046	0.349732	
	3	1.04885	0.963393	0.028495		
	4	1.01325	0.989756	0.005137		

TABLE 9.5-3 (continued)

N	i	A_{0i}	B_{0i}	B_{1i}	H_o/S_o	Ω_r
			Passband Ripple = 1 dB, Stopband Gain = −60 dB			
2	1	44.33121	0.049740	0.232972	0.999985×10^{-3}	22.17688
3	1	6.66138	0.198506	0.216668	0.665490×10^{-2} 0.255373	5.02121
4	1	15.57508	0.123525	0.441175	0.999992×10^{-3}	2.46079
	2	2.84628	0.402671	0.163210		
5	1	4.21402	0.296236	0.356928	0.580460×10^{-2}	1.67161
	2	1.81233	0.594376	0.110961	0.255373	
6	1	14.06238	0.135110	0.468386	0.999998×10^{-3}	1.34354
	2	2.24081	0.496567	0.255540		
	3	1.40684	0.741381	0.070756		
7	1	4.01386	0.309159	0.374430	0.572054×10^{-2}	1.18547
	2	1.58609	0.668120	0.168476	0.255373	
	3	1.21724	0.841598	0.043338		
8	1	13.90612	0.136441	0.471452	0.999997×10^{-3}	1.10308
	2	2.18302	0.508082	0.266640		
	3	1.30382	0.792531	0.105321		
	4	1.12001	0.905337	0.025900		
9	1	3.99239	0.310618	0.376394	0.571125×10^{-2}	1.05822
	2	1.56295	0.676792	0.175193	0.255373	
	3	1.16520	0.874493	0.063713		
	4	1.06755	0.944252	0.015252		
			Passband Ripple = 2 dB, Stopband Gain = −20 dB			
2	1	3.60961	0.454891	0.537326	0.100103	1.94332
3	1	1.42939	0.793180	0.204089	0.254443 0.458898	1.20808
4	1	3.25882	0.486218	0.597266	0.100102	1.05569
	2	1.10765	0.935564	0.063585		
5	1	1.39116	0.807316	0.223995	0.253878	1.01567
	2	1.02976	0.981070	0.018680	0.458898	
6	1	3.25657	0.486437	0.597679	0.100102	1.00447
	2	1.09913	0.940286	0.069417		
	3	1.00845	0.994532	0.005396		
7	1	1.39096	0.807384	0.224146	0.253839	1.00128
	2	1.02749	0.982483	0.020362	0.458898	
	3	1.00242	0.998428	0.001551		
8	1	3.25736	0.486314	0.597662	0.100040	1.00037
	2	1.09913	0.940275	0.069488		
	3	1.00782	0.994938	0.005883		
	4	1.00069	0.999548	0.000446		

TABLE 9.5-3 (continued)

N	i	A_{0i}	B_{0i}	B_{1i}	H_o/S_o	Ω_r
9	1	1.39157	0.807065	0.224311	0.253438	1.00011
	2	1.02755	0.982437	0.020419	0.458898	
	3	1.00224	0.998540	0.001696		
	4	1.00020	0.999870	0.000129		

Passband Ripple = 2 dB, Stopband Gain = −40 dB						
2	1	11.43468	0.143955	0.332767	0.100001×10^{-1}	5.76107
3	1	2.76901	0.423046	0.229015	0.412982×10^{-1}	2.13923
					0.278675	
4	1	7.25202	0.212344	0.467290	0.100001×10^{-1}	1.40842
	2	1.57676	0.677934	0.127954		
5	1	2.33100	0.490174	0.302683	0.399132×10^{-1}	1.16811
	2	1.22913	0.838222	0.064276	0.278675	
6	1	7.11859	0.215746	0.473545	0.100000×10^{-1}	1.07316
	2	1.46261	0.722961	0.164483		
	3	1.09825	0.922931	0.030620		
7	1	2.31409	0.493238	0.305999	0.398550×10^{-1}	1.03262
	2	1.18820	0.862873	0.081414	0.278675	
	3	1.04352	0.964199	0.014224		
8	1	7.11297	0.215892	0.473812	0.999999×10^{-2}	1.01470
	2	1.45794	0.724945	0.166087		
	3	1.08158	0.935151	0.038502		
	4	1.01955	0.983564	0.006530		
9	1	2.31340	0.493363	0.306141	0.398506×10^{-1}	1.00665
	2	1.18649	0.863937	0.082158	0.278675	
	3	1.03626	0.969979	0.017824		
	4	1.00884	0.992494	0.002982		

Passband Ripple = 2 dB, Stopband Gain = −60 dB						
2	1	36.16031	0.045523	0.188850	0.999994×10^{-3}	18.09398
3	1	5.82458	0.202473	0.172384	0.620820×10^{-2}	4.39729
					0.201327	
4	1	13.96818	0.108944	0.348857	0.999992×10^{-3}	2.24440
	2	2.58877	0.417854	0.126933		
5	1	3.86405	0.295378	0.277502	0.551997×10^{-2}	1.56860
	2	1.69531	0.615100	0.083998	0.201327	
6	1	12.83274	0.117802	0.367723	0.999990×10^{-3}	1.28693
	2	2.09697	0.507386	0.194121		
	3	1.34376	0.761617	0.052028		

TABLE 9.5-3 (continued)

N	i	A_{0i}	B_{0i}	B_{1i}	H_o/S_o	Ω_r
7	1	3.71433	0.306340	0.289299	0.545771×10^{-2}	1.15215
	2	1.51120	0.683872	0.124582	0.201327	
	3	1.18037	0.858228	0.030942		
8	1	12.72577	0.118715	0.369634	0.999991×10^{-3}	1.08281
	2	2.05385	0.517202	0.201368		
	3	1.26045	0.807913	0.075699		
	4	1.09763	0.917695	0.017963		
9	1	3.69977	0.307452	0.290490	0.545154×10^{-2}	1.04572
	2	1.49403	0.691114	0.128833	0.201327	
	3	1.13880	0.887078	0.044501		
	4	1.05373	0.952885	0.010283		

		Passband Ripple = 3 dB, Stopband Gain = −20 dB				
2	1	3.16206	0.446651	0.451221	0.999995×10^{-1}	1.73915
3	1	1.34231	0.807114	0.157389	0.237355	1.15516
					0.394899	
4	1	2.92756	0.472487	0.493658	0.999995×10^{-1}	1.03853
	2	1.08010	0.945318	0.044619		
5	1	1.31717	0.818329	0.170012	0.237048	1.00996
	2	1.02041	0.985364	0.011943	0.394899	
6	1	2.92670	0.472578	0.493859	0.999824×10^{-1}	1.00260
	2	1.07480	0.948718	0.047988		
	3	1.00531	0.996140	0.003149		
7	1	1.31710	0.818355	0.170085	0.237014	1.00068
	2	1.01909	0.986290	0.012829	0.394899	
	3	1.00139	0.998987	0.000827		
8	1	2.92702	0.472526	0.493855	0.999560	
	2	1.07480	0.948717	0.048015	1.00018	
	3	1.00497	0.996385	0.003382		
	4	1.00036	0.999734	0.000217		
9	1	1.31754	0.818103	0.170205	0.236721	1.00005
	2	1.01913	0.986257	0.012860	0.394899	
	3	1.00131	0.999048	0.000891		
	4	1.00010	0.999930	0.000058		

		Passband Ripple = 3 dB, Stopband Gain = −40 dB				
2	1	10.01180	0.141419	0.284766	0.999991×10^{-2}	5.05584
3	1	2.54917	0.436001	0.190366	0.392774×10^{-1}	1.98022
					0.235639	
4	1	6.67811	0.203053	0.391813	0.999996×10^{-2}	1.34663
	2	1.49919	0.696465	0.102519		

TABLE 9.5-3 (continued)

N	i	A_{0i}	B_{0i}	B_{1i}	H_o/S_o	Ω_r
5	1	2.19519	0.498382	0.246744	0.382019×10^{-1}	1.13934
	2	1.19395	0.853367	0.049526	0.235639	
6	1	6.58231	0.205696	0.396137	0.999996×10^{-2}	1.05892
	2	1.40720	0.737064	0.129340		
	3	1.08088	0.932780	0.022704		
7	1	2.18313	0.500844	0.248944	0.381626×10^{-1}	1.02545
	2	1.16144	0.874707	0.061632	0.235639	
	3	1.03473	0.969917	0.010161		
8	1	6.57883	0.205793	0.396297	0.999973×10^{-2}	1.01109
	2	1.40391	0.738612	0.130362		
	3	1.06788	0.942934	0.028071		
	4	1.01510	0.986681	0.004499		
9	1	2.18274	0.500923	0.249027	0.381575×10^{-1}	1.00485
	2	1.16026	0.875502	0.062090	0.235639	
	3	1.02925	0.974534	0.012526		
	4	1.00660	0.994131	0.001982		
			Passband Ripple = 3 dB, Stopband Gain = −60 dB			
2	1	31.66062	0.044721	0.161891	0.999982×10^{-3}	15.84610
3	1	5.33757	0.209329	0.145315	0.593160×10^{-2}	4.03471
					0.168753	
4	1	12.99985	0.103391	0.292827	0.999993×10^{-3}	2.11596
	2	2.43545	0.432543	0.105070		
5	1	3.65012	0.299596	0.229883	0.533723×10^{-2}	1.50711
	2	1.62514	0.632389	0.068114	0.168753	
6	1	12.06873	0.110910	0.307139	0.999998×10^{-3}	1.25325
	2	2.00866	0.518843	0.157996		
	3	1.30601	0.777153	0.041293		
7	1	3.52768	0.309409	0.238672	0.528678×10^{-2}	1.13252
	2	1.46533	0.697339	0.099388	0.168753	
	3	1.15851	0.870254	0.024042		
8	1	11.98665	0.111628	0.308485	0.100000×10^{-2}	1.07105
	2	1.97347	0.527579	0.163278		
	3	1.23409	0.819865	0.059153		
	4	1.08453	0.926218	0.013672		
9	1	3.51659	0.310331	0.239495	0.528205×10^{-2}	1.03859
	2	1.45140	0.703663	0.102420	0.168753	
	3	1.12295	0.896257	0.034067		
	4	1.04578	0.958601	0.007671		

Source: Reprinted from Lonnie C. Ludeman, *Fundamentals of Digital Signal Processing,* Harper & Row, New York, 1986.

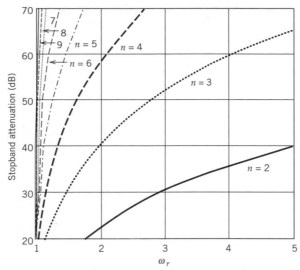

Figure 9.5-6
Elliptic filter selection graph.

EXAMPLE 9.5-5 Find the normalized transfer function for a low-pass filter with the following specifications:

(a) Cutoff frequency $f_c = 1000$ Hz

(b) Acceptable ripple of 3 dB

(c) Stopband attenuation ≥ 40 dB above 1200 Hz

SOLUTION

Solve for frequency transition sharpness,

$$\omega_r = \frac{\omega_s}{\omega_c} = \frac{2\pi(1200)}{2\pi(1000)} = 1.2.$$

Using Table 9.5-3, the factors in the equation for $\omega_r = 1.1393 < 1.2$ are

$$
\begin{aligned}
H_o &= 0.0382, & s_0 &= 0.2356, \\
B_{01} &= 0.4984, & B_{11} &= 0.2467, & A_{01} &= 2.1952, \\
B_{02} &= 0.8534, & B_{12} &= 0.0495. & A_{02} &= 1.1940,
\end{aligned}
$$

Therefore, the normalized transfer function for this filter is

$$H(\omega) = \frac{0.0382}{j\omega + 0.2346} \frac{(j\omega)^2 + 2.1952}{(j\omega)^2 + 0.2467j\omega + 0.4984} \frac{(j\omega)^2 + 1.1940}{(j\omega)^2 + 0.0495j\omega + 0.8534}.$$

The required low-pass filter response is found by using the normalized frequency,

$$\omega_n = \sqrt{\omega_c \omega_s} = 2\pi\sqrt{(1000)(1200)} = 6.88 \text{ krad/s}.$$

The required frequency response is then

$$H(\omega) = \frac{0.0382}{j\omega/\omega_n + 0.2346} \frac{(j\omega/\omega_n)^2 + 2.1952}{(j\omega/\omega_n)^2 + 0.2467(j\omega/\omega_n) + 0.4984}$$

$$\times \frac{(j\omega/\omega_o)^2 + 1.1940}{(j\omega/\omega_n)^2 + 0.0495(j\omega/\omega_n) + 0.8534}$$

$$= \frac{0.0382}{j\omega/6880 + 0.2346}$$

$$\times \frac{(j\omega/6880)^2 + 2.1952}{(j\omega/6880)^2 + 0.2467(j\omega/6880) + 0.4984}$$

$$\times \frac{(j\omega/6880)^2 + 1.1940}{(j\omega/6880)^2 + 0.0495(j\omega/6880) + 0.8534}.$$

9.5.4 OpAmp Realization of Elliptic Filters

To accommodate the relative positions of poles and zeros in second-order elliptical filter transfer functions, a "notch" filter characteristic is generalized as

$$T(j\omega) = \frac{(j\omega)^2 + \omega_z}{(j\omega)^2 + (\omega_o/Q)j\omega + \omega_o^2}. \tag{9.5-17}$$

At the notch frequency ω_z the filter transfer function response is zero, or

$$T(j\omega_z) = 0. \tag{9.5-18}$$

The relative magnitudes of ω_z and ω_o give rise to the following characteristics:

$\omega_z/\omega_o < 1$ is a high-pass characteristic.

$\omega_z/\omega_o = 1$ is a regular notch characteristic.

$\omega_z/\omega_o < 1$ is a low-pass notch characteristic.

A low-pass notch characteristic has less attenuation of signal at frequencies lower than the notch frequency (where the signal is zero) when compared to the signal at

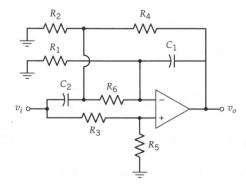

Figure 9.5-7
Boctor low-pass filter for second-order elliptic filter.

frequencies higher than the notch frequency. The high-pass notch characteristic has less attenuation of signal at frequencies higher than the notch frequency when compared to the signal at frequencies lower than the notch frequency. The regular notch characteristic has identical signal attenuation on both sides of the notch frequency.

A practical OpAmp implementation of the low-pass notch transfer function, attributed to S. A. Boctor, is the circuit shown in Figure 9.5-7.

The transfer function of the Boctor low-pass filter to implement a second-order elliptic filter is found by routine circuit analysis:

$$H(\omega) = \left(\frac{R_5}{R_3 + R_5}\right) \times$$

$$\frac{(j\omega)^2 + j\omega\left[\dfrac{1}{(R_6C_2)} + \dfrac{1}{(R_2\|R_4)} + \dfrac{1}{(R_1C_1)} - \left(\dfrac{1}{(R_6C_1)}\right)(R_3/R_5)\right] + \dfrac{(R_1 + (R_2\|R_4) + R_6)}{(R_1(R_2\|R_4)R_6C_1C_2)}}{(j\omega)^2 + \dfrac{1}{R_6C_2} + \dfrac{1}{(R_2\|R_4)C_2} + \dfrac{1}{R_4R_6C_1C_2}}.$$

$$(9.5\text{-}19)$$

The component values were determined by an algorithm found by John M. Cioffi while he was a junior at the University of Illinois in 1977. The normalized component values are

$$R_1 = \frac{2}{k_1\Omega_z^2 - 1}, \tag{9.5-20}$$

$$R_1 = \frac{2}{1 - k_1}, \tag{9.5-21}$$

$$R_3 = \frac{1}{2}\left(\frac{k_1}{Q^2} + k_1\Omega_z^2 - 1\right), \tag{9.5-22}$$

$$R_4 = \frac{1}{k_1}, \tag{9.5-23}$$

$$R_5 = R_6 = 1, \tag{9.5-24}$$

$$C_1 = \frac{k_1}{2Q}, \tag{9.5-25}$$

$$C_2 = 2Q, \tag{9.5-26}$$

where $\Omega_z = \omega_z/\omega_o$ and k_1 is chosen by the designer under the condition that $(\omega_o/\omega_z)^2 < k_1 < 1$. The normalized resistor values are multiplied by a magnitude scaling factor k_m to obtain practical element values. The capacitor values for the Boctor low-pass circuit are divided by the product of the frequency scaling factor $k_f = \omega_o$, and the magnitude scaling factor k_m to obtain practical element values. That is,

$$R_{i,\text{real}} = k_m(R_{i,\text{normalized}}), \quad i = 1, \ldots, 6, \tag{9.5-27}$$

$$C_{i,\text{real}} = \frac{C_{i,\text{normalized}}}{k_f k_m}, \quad i = 1, 2. \tag{9.5-28}$$

EXAMPLE 9.5-6 Design an OpAmp low-pass filter with the following specifications:

    a) Cutoff frequency $f_c = 1000$ Hz

    b) Acceptable ripple of 1 dB

    c) Stopband attenuation ≥ 20 dB above 2500 Hz

SOLUTION

Solve for frequency transition sharpness:

$$\omega_r = \frac{\omega_s}{\omega_c} = \frac{2\pi(2500)}{2\pi(1000)} = 2.5.$$

Using Table 9.5-3, the factors in the equation for $\omega_r = 2.3247 < 2.5$ are

$$H_o = 0.1002, \qquad A_{01} = 4.4234, \qquad B_{01} = 0.4972, \qquad B_{11} = 0.6767.$$

Therefore, the normalized transfer function for this filter is

$$H(\omega) = \frac{0.1002[(j\omega)^2 + 4.4234]}{(j\omega)^2 + 0.6767j\omega + 0.4972}.$$

The required low-pass filter response is found by using the normalized frequency:

$$\omega_n = \sqrt{\omega_c \omega_s} = 2\pi\sqrt{(1000)(2500)} = 9.93 \text{ krad/s}.$$

A second-order Boctor low-pass notch circuit is required. From the transfer function,

$$A_{VN} = 0.1002,$$

$$\omega_o = \sqrt{0.4972} = 0.7051, \qquad \omega_z = \sqrt{4.4234} = 2.1032,$$

$$Q = \frac{0.7051}{0.6767} = 1.0420.$$

The ratio $\omega_z/\omega_o = 2.983$ confirms the requirement for a low-pass notch circuit. To determine the component values of the low-pass Boctor circuit, choose the factor k_1 under the condition $(\omega_o/\omega_z)^2 < k_1 < 1$: Choose $k_1 = 0.5$.

    Scale the frequency using $k_f = 2\pi(1000) = 6283$ as required to scale to the cutoff frequency, and let the magnitude scaling factor be $k_m = 10{,}000$ using Equations 9.5-20 to 9.5-28 and substituting ω_z/ω_o for ω_z:

$$R_1 = \frac{2}{k_1(\omega_z/\omega_{o3})^2 - 1}\, k_m \approx 5.6 \text{ k}\Omega, \qquad R_1 = \frac{2}{1 - k_1}\, k_m \approx 39 \text{k}\Omega,$$

$$R_3 = \frac{1}{2}\left[\frac{k_1}{Q^2} + k_1\left(\frac{\omega_z}{\omega_o}\right)^2 - 1\right]k_m \approx 20 \text{ k}\Omega,$$

$$R_4 = \frac{1}{k_1}\, k_m \approx 20 \text{ k}\Omega, \qquad R_5 = R_6 = 1k_m = 10 \text{ k}\Omega,$$

$$C_1 = \frac{k_1}{2Q}\frac{1}{k_f k_m} \approx 3900 \text{ pF}, \qquad C_2 = 2Q\frac{1}{k_f k_m} \approx 0.033 \text{ }\mu\text{F}.$$

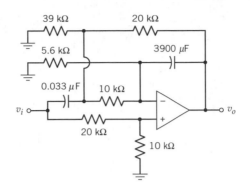

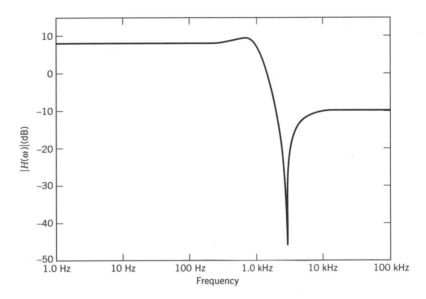

The OpAmp filter implementation that fulfills the required specification and the SPICE frequency response plot are as shown.

The high-pass Boctor notch circuit is used to implement a second-order high-pass elliptic filter (shown in Figure 9.5-8).

Analysis of the Boctor high-pass filter yields the following component relationships:

$$R_1 = \left(\frac{1 + \alpha}{\alpha\gamma\omega_z}\right)^2, \tag{9.5-29}$$

$$R_2 = 1, \tag{9.5-30}$$

$$R_3 = G\frac{(1 + \alpha)^2}{a\gamma^2}, \tag{9.5-31}$$

$$R_4 = \frac{G}{G - 1}\frac{1}{\alpha}, \tag{9.5-32}$$

$$R_5 = \frac{G}{\alpha}, \tag{9.5-33}$$

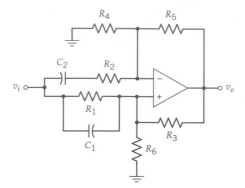

Figure 9.5-8
Boctor high-pass filter for second-order elliptic filter.

$$R_6 = \frac{G(1 + \alpha)^2}{\alpha\gamma^2[G(1 + \alpha - \alpha\omega_z^2) - 1]} \quad , \tag{9.5-34}$$

$$C_1 = C_2 = \frac{\alpha\gamma}{1 + \alpha} \quad , \tag{9.5-35}$$

where

$$\gamma = \frac{1}{Q(1 - \omega_z^2)} \quad , \tag{9.5-36}$$

$$\alpha = \frac{\gamma^2 - 1}{1 + \gamma^2\omega_z^2}. \tag{9.5-37}$$

This circuit provides a high-frequency gain of greater than 1 and applies only under the condition that $Q < [1 - (\omega_z/\omega_o)^2]^{-1}$. As in the Chebyshev filters, the transfer function for the high-pass filter is found by performing an analog-to-analog transformation.

9.6 COMPARISON OF FILTER TYPES

The Butterworth, Chebyshev, and elliptic filter types have been discussed in the previous section. The three filter types offer different performance characteristics that may be used to fulfill design specifications. To simplify the choice of filter used to fulfill a certain specification, several key characteristics are compared.

1. *Cutoff frequency.* The Butterworth filter passband is typically defined by the cutoff frequency, which is the half-power point in the frequency response. However, in Chebyshev and elliptic filters the passband, and therefore, the so-called cutoff frequency, is defined by the frequency that identifies the end of the ripple. These two frequencies are different unless $\epsilon = 1$. Therefore, it is important to keep separate the definitions of cutoff frequency and half-power points for different filter types.

2. *Ripple.* The Butterworth filter has a maximally flat response that results in a ripple-free passband and stopband. Chebyshev filters exhibit ripple in either the passband or the stopband for Chebyshev type 1 or type 2 filters; that is, passband ripple corresponds to type 1 filters and stopband ripple corresponds to type 2 filters. Elliptic filters are characterized by ripple in both the

passband and the stopband. In the Chebyshev and elliptic responses, the ripple is dependent on the factor ϵ.

3. *DC response for a low-pass filter.* In the Butterworth filter, the response is peaked at DC. However, in the Chebyshev and elliptic designs, the DC filter response may be either peaked or at the power level of the response at the cutoff frequency: That is, the DC response is peaked for odd-order filters and corresponds to the cutoff frequency power level for even-order filters.

4. *Transition band.* One of the most important filter specifications is the steepness of the transition band. Steeper transitions from the passband to stopband requires the use of higher-order filters: Higher-order filters require more electronic components to implement. Fortunately, the three filter types discussed in the previous section have varying transition band steepness given an identical filter order. The Butterworth filter is the least steep in the transition band, whereas the elliptic filter is the steepest.

For example, for an $n = 5$ filter (with $\epsilon = 1$), the Chebyshev response has 24 dB greater attenuation than the Butterworth response. For the same order, the elliptic response has approximately 27 dB greater attenuation than the Butterworth response, and approximately 3 dB greater attenuation than the Chebyshev response. In order to match the steepness of the transition band for a fifth-order elliptical filter, the order of the Butterworth filter is approximately $n = 13$. In terms of the number of OpAmps, a fifth-order elliptic filter requires three OpAmps (one first-order filter and two Boctor second-order sections) whereas a thirteenth order Butterworth filter requires seven OpAmps (one first-order filter and six Sallen and Key second-order sections). A significant savings in parts, power, and printed circuit space can be attained if the elliptic filter implementation is chosen. However, other factors such as ripple in the passband and stopband must be considered.

9.7 SWITCHED-CAPACITOR FILTERS

Using MOSFETS as switches, switched-capacitor (SC) filters can be designed in precision monolithic integrated circuits and are widely used in digital signal processing applications. The advantage of using MOSFET SC filters is derived from the simple fact that it is difficult to manufacture precision resistors in integrated circuits. OpAmps and capacitors are more easily fabricated. Although accurate values of capacitances may be difficult to achieve, precise ratios of capacitances can easily be achieved in integrated circuits.

The key features of SC filters are as follows:

- The filter can be fabricated in a precision monolithic integrated circuit.
- Since MOS technology allows high component density in integrated circuits, a single chip can be fabricated to fulfill both analog and digital signal processing.
- Metal–oxide–semiconductor devices have very low power dissipation and temperature coefficients.
- Precise capacitance ratios can be fabricated. This is particularly important since active filters depend largely on *RC* time constants: The time constant can be controlled by capacitance ratios.
- Because resistors are eliminated, power consumption is reduced.

9.7.1 MOS Switch

The FET as an analog switch was discussed in Section 4.5. An ideal enhancement-type NMOSFET switch has the following design goals:

- In the ON state ($V_{GS} > V_T$), it passes signal from the drain to source without attenuation.
- In the OFF state ($V_{GS} < V_T$), the signal is not passed from the drain to source.
- Transitions between the ON and OFF states are instantaneous.

Real NMOSFET switches have the following characteristics:

- In the ON state ($V_{GS} > V_T$), the drain–source resistance r_d is on the order of kilohms ($<$10 kΩ).[11]
- In the OFF state ($V_{GS} < V_T$), the drain–source resistance r_d is on the order of several hundred megaohms (100–1000 MΩ).

The ratio of the resistance values between the ON and OFF states of an enhancement NMOSFET is on the order or 10^5 or 100 dB. An equivalent representation of the characteristic of enhancement NMOSFET switches is shown in Figure 9.7-1. A voltage signal is applied to the gate–source junction with fast rise and fall

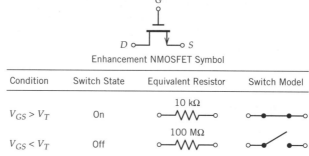

Figure 9.7-1
Circuit and switch representation of enhancement NMOSFET.

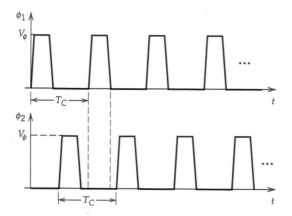

Figure 9.7-2
Biphase nonoverlapping clocks.

---

[11] The comparatively large ON resistance of these switches is due to on-chip geometries that are not present in discrete FET switches as described in Chapter 4.

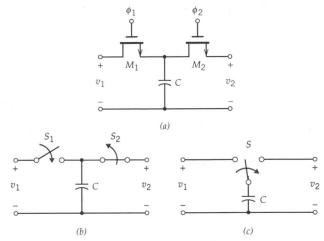

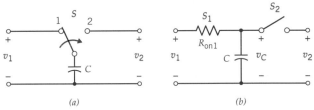

(a)

Figure 9.7-3
(*a*) NMOSFET switched capacitor; (*b*) FETs replaced by SPST switches; (*c*) simplified functional circuit representation.

(a)   *(b)*

Figure 9.7-4
(*a*) Switched capacitor with node 1 closed; (*b*) equivalent circuit with FET on resistance shown.

times and with a peak amplitude greater than V_T. Since the gate–source voltage is now a time-varying signal, it is represented as v_{gs} or more commonly as a pulse train signal ϕ. The NMOSFET switch is open or closed depending on the value of ϕ. This type of switch is known as the single-pole single-throw (SPST) switch.

The clock signal ϕ, which is a pulse train signal, is usually generated by an external digital system. The pulse train is periodic with a period of T_C with a clock frequency of $f = 1/T_C$. The clock signal is used to turn the NMOSFET ON or OFF.

A two-phase clock is shown in Figure 9.7-2. The two clock signals ϕ_1 and ϕ_2 have the same clock frequency but are out of phase; that is, when ϕ_1 is ON, ϕ_2 is OFF. The duty cycle (percentage of time ON to the period of the signal) is commonly slightly less than 50% to ensure nonoverlapping clocks.

For two NMOSFET switches in series in a T arrangement with a capacitor forms an SC circuit, shown in Figure 9.7-3*a*, with each FET with M_1 driven by ϕ_1 and M_2 by ϕ_2, one of the two FETs will always be ON. Simplified equivalent diagrams of the SC circuit are shown in Figures 9.7-3*b,c*. Figure 9.7-3*b* is derived by directly replacing the NMOSFETs with SPST switch representations, while Figure 9.7-3*c* is the functional equivalent diagram of the SC circuit. With biphase clock signal inputs ϕ_1 and ϕ_2 to FETs M_1 and M_2, respectively, it is evident that there will never be a direct connection between v_1 and v_2, as shown in Figure 9.7-3*c*. The term *switched capacitor* derives from the switching operation of the FETs on the capacitor.

For a time-varying input voltage, $v_1(t)$, with switch S_1 closed and S_2 open, the equivalent circuit is that shown in Figure 9.7-4. If $v_1(t)$ is a constant voltage, the

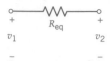

Figure 9.7-5

Approximate Equivalent circuit for switched capacitor.

voltage across the capacitor C will increase with the time constant $\tau = R_{ON1}C$, where R_{ON1} is the ON resistance of the NMOSFET M_1. For typical capacitance of $C = 1$ pF and $R_{ON1} = 10$ kΩ, the voltage across the capacitor will reach 63% of the input voltage when $\tau = R_{ON1}C = 10$ ns.

For the SC filter to operate properly, the time constant formed by $R_{ON1}C$ must be significantly small compared to the variations in the input voltage signal. If the switch position is changed from 1 to 2, the charge on the capacitor will discharge at the output, v_2. The charge transferred is

$$Q = C(v_1 - v_2) \qquad (9.7\text{-}1)$$

over a discharge time T_C. The average current through the capacitor during this discharging period in time is

$$i(t) = \frac{dQ}{dt} = \frac{\Delta Q}{\Delta t} \approx \frac{C(v_1 - v_2)}{T_c}. \qquad (9.7\text{-}2)$$

The *equivalent resistor* formed by the switched capacitor to yield the same value of current is

$$R_{eq} = \frac{v_1 - v_2}{i(t)} = \frac{T_C}{C} = \frac{1}{f_cC}. \qquad (9.7\text{-}3)$$

since $i(t) = (v_1 - v_2)/R_{eq}$, the approximate equivalent circuit for the switched capacitor of Figure 9.7-3a is Figure 9.7-5.

Consider the range of values of R_{eq}. Using $C = 1$ pF (requiring a silicon area on the chip of approximately 0.01 mm^2) and a typical switching frequency of the NMOSFETs of $f_C = 100$ kHz, R_{eq} is found to be 10 MΩ. For the SC network to be useful, the switching frequency f_C must be much larger than the signal frequencies of interest in $v_1(t)$ and $v_2(t)$.

9.7.2 Simple Integrator

The SC networks are used to implement active circuits for a variety of analog operations. A simple integrator shown in Figure 9.7-6a is described by the transfer function,

$$\frac{v_o(\omega)}{v_i(\omega)} = -\frac{1}{j\omega RC_f}. \qquad (9.7\text{-}4)$$

The resistance in the integrator of Figure 9.7-6a can be replaced with a switched capacitor to obtain Figure 9.70-6b. The integrator transfer function can be rewritten using Equation (9.7-3):

$$\frac{v_o(\omega)}{v_i(\omega)} = -\frac{f_cC_1}{j\omega C_f}. \qquad (9.7\text{-}5)$$

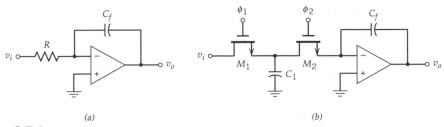

Figure 9.7-6
(a) Simple active *RC* integrator; (b) switched capacitor implementation of (a).

The transfer function of the SC implementation of the active integrator clearly shows a dependence on the clock frequency of the NMOSFET input signals and the *ratio* of the two capacitors. Since MOS integrated circuit technology can hold tight tolerances on capacitance ratios, precision integrators may be fabricated using the SC implementation.

9.7.3 Gain Stage

An SC inverting amplifier arrangement is shown in Figure 9.7-7a with two SC filter networks, each equivalent to a resistance. The gain of the circuit is

$$A_V = \frac{v_o}{v_i} = -\frac{C_2}{C_1}. \tag{9.7-6}$$

In Figure 9.7-7a, the capacitor C_1 is charged to input signal level for $\phi_1 = V_\phi$ and C_2 is discharged. When $\phi_1 = 0$, then the voltage across C_1 is applied to the OpAmp and C_2 becomes the feedback path.

A direct replacement of the resistors in the inverting OpAmp amplifier configuration with switched capacitors yields the circuit shown in Figure 9.7-7b. This circuit is unworkable since the switches used to form the feedback resistance are never closed simultaneously. Therefore, the feedback path from the output to the input of the OpAmp is only periodically closed and the OpAmp will saturate.

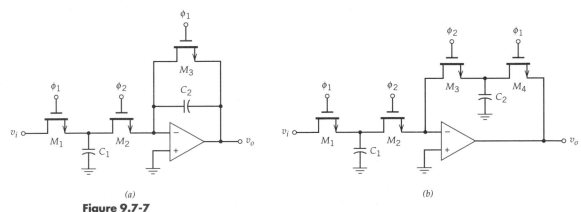

Figure 9.7-7
(a) Practical switched-capacitor inverting amplifier; (b) unworkable switched-capacitor inverting amplifier.

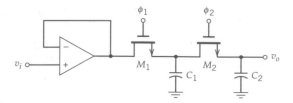

Figure 9.7-8
First-order low-pass filter.

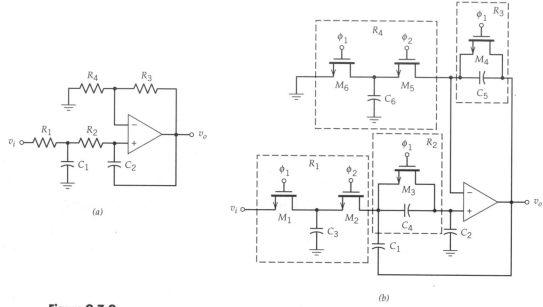

Figure 9.7-9
(*a*) Sallen and Key second-order low-pass filter section; (*b*) switched-capacitor implementation.

9.7.4 Low-Pass Filters

An SC active low-pass filter is shown in Figure 9.7-8. The NMOSFETs M_1 and M_2 and the capacitor C_1 form the equivalent resistance, R_{eq}, such that the transfer function of this circuit is

$$\frac{v_o(\omega)}{v_i(\omega)} = \frac{1}{R_{eq}C_2} \frac{1}{j\omega + 1/(R_{eq}C_2)} = \frac{f_C C_1}{C_2} \frac{1}{j\omega + f_C C_1/C_2}. \tag{9.7-7}$$

Through careful SC network arrangements, second-order SC filter circuit can be implemented. Figure 9.7-9 shows a Sallen and Key second-order all-pole low-pass filter and its SC equivalent. Note that resistors R_2 and R_3 are in the feedback path and the SC implementation replaces these resistors with the alternate SC network of a NMOSFET in parallel with the switched capacitor.

9.8 OPAMP LIMITATIONS

Thus far, designs using OpAmps have assumed that the OpAmp has very large open-loop gain ($A \geq 200,000$) over all frequencies. Filter designs have used this assumption by ignoring potential frequency effects of OpAmps in their design. In reality, Op-

Amps themselves have a frequency response: The frequency response is dependent on the closed-loop gain of the amplifier. Therefore, the OpAmp may itself limit the operational frequency range of the application.

When a large-step input voltage is applied to an OpAmp, the output waveform rises with a finite slope called the *slew rate*. The slewing behavior of the output is due to amplifier nonlinearities. Therefore, the slew rate cannot be calculated from the frequency response of the OpAmp using linear analysis. The inability of the OpAmp output to rise in voltage as quickly as linear theory predicts will also limit the frequency range of the amplifier.

9.8.1 Frequency Response of OpAmps

Returning to the definition of the OpAmp output voltage presented in Equation 1.2-1, a frequency-dependent gain is used:

$$v_o = A(j\omega)(v_2 - v_1) = A(j\omega)v_i, \tag{9.8-1}$$

where

$A(j\omega)$ = frequency-dependent large-signal gain
v_2 = noninverting input voltage
v_1 = inverting input voltage
v_o = output voltage
v_i = voltage applied between input terminals of OpAmp

For a μA741, the gain characteristics are shown in Figure 9.8-1. The μA741 is an internally compensated OpAmp, meaning that a capacitor is fabricated on the chip for stability, resulting in a low-frequency pole in the transfer function.

The curve shows an open-loop gain of between 100 and 120 dB at DC. Then at the OpAmp corner frequency, $f_{ca} = 5$ Hz, the gain falls off at -20 dB/decade, reaching 0 dB at 1 MHz. The product of the gain with the bandwidth is constant and is defined as the *gain bandwidth product (GBP)*. The GBP of the μA741 is 10^6. The open-loop gain $A(j\omega)$ of this curve is

$$A(j\omega) = \frac{A_o}{1 + j\omega/\omega_{ca}} = \frac{A_o\omega_{ca}}{j\omega + \omega_{ca}} = \frac{GBP}{j\omega + \omega_{ca}}, \tag{9.8-2}$$

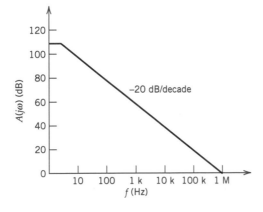

Figure 9.8-1
OpAmp open-loop gain characteristics.

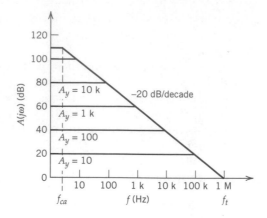

Figure 9.8-2
Closed-loop gain characteristics.

where A_o is the open-loop gain at DC, $\omega_{ca} = 2\pi f_{ca}$,

and

$$GBP = A_o\omega_{ca}. \qquad (9.8\text{-}3)$$

Referring to Figure 9.8-2, the GBP is confirmed as being constant, where the bandwidth is defined by the corner frequency (f_c) and the closed loop gain as A_v.

The output voltage v_o of an inverting OpAmp amplifier shown in Figure 9.8-3 is

$$v_o = A(j\omega)v_a, \qquad (9.8\text{-}4)$$

where v_a is the input to the OpAmp.

The input voltage to the OpAmp is

$$v_a = \frac{v_o - v_i}{R_s + R_f} R_s + v_i. \qquad (9.8\text{-}5)$$

Combining Equations 9.8-4 and 9.8-5, and knowing that $R_f/R_s = K$, the gain of the circuit is

$$A_V = \frac{v_o}{v_i} = -K\frac{A(j\omega)/(1 + K)}{1 + A(j\omega)/(1 + K)}. \qquad (9.8\text{-}6)$$

Let $A(j\omega) = GBP/j\omega$. Then the voltage gain in Equation 9.8-6 becomes

$$A_V = \frac{v_o}{v_i} = -K\frac{GBP/(1 + K)}{j\omega + GBP/(1 + K)}. \qquad (9.8\text{-}7)$$

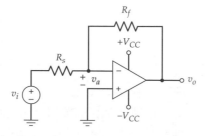

Figure 9.8-3
Inverting OpAmp amplifier.

Equation 9.8-7 is a transfer function of the inverting OpAmp amplifier taking into account the frequency response of the OpAmp. It is of the form of a single-pole transfer function with 3 dB frequency:

$$\omega_{3dB} = \frac{GBP}{1 + K}.$$

(9.8-7a)

9.8.2 OpAmp Slew Rate

Time-domain responses of OpAmp circuits require an introduction to a characteristic that is important in the specification of OpAmps. The most important of these is the slew rate. Because OpAmps have responses that are frequency dependent, the output due to a step input is not a perfect step, causing the output to be distorted.

When a step function, shown in Figure 9.8-4a, is applied to a unity-gain follower OpAmp circuit, the resulting output is that similar to Figure 9.8-4b. The output voltage is in the form of a low-pass filter:

$$A_V(\omega) = \frac{v_o(\omega)}{v_i(\omega)} = \frac{1}{1 + j\omega/\omega_t},$$

(9.8-8)

where
$$v_o(\omega) = \text{output voltage}$$
$$v_i(\omega) = \text{input voltage}$$
$$\omega_t = \text{unity-gain bandwidth}$$

The step response of the circuit is an exponentially rising waveform:

$$v_o(t) = V_1(1 - e^{-\omega_t t}) = V_1(1 - e^{-t/\tau}).$$

(9.8-9)

For a small input voltage V_1, the output will be nonsaturating, and the output waveform is shown in Figure 9.8-4b.

The initial slope of the output voltage is

$$\frac{dv_o(t)}{dt} = V_1\omega_t = V_1(GBP).$$

(9.8-10)

The preceding results are valid for linear operation, implying that the input voltage must be sufficiently small so that the OpAmp does not saturate. For a large, saturating input voltage, the output waveform is shown in Figure 9.8-4c. Note that the initial slope of the output response is lower than predicted for linear theory. The

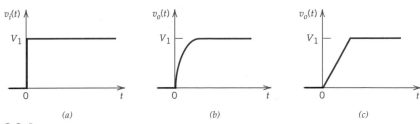

Figure 9.8-4
(a) Input voltage to OpAmp unity-gain follower; (b) output waveform with V_1 small (nonsaturating); (c) input waveform with V_1 large.

inability of the OpAmp to rise as quickly as predicted by linear theory is called *slew rate limiting*. The initial slope of the output response is called the *slew rate*, (SR) defined as

$$SR = \left.\frac{dv_o(t)}{dt}\right|_{max}.$$

(9.8-11)

The OpAmp begins slewing when the initial slope of the output waveform is less than $V_1(GBP)$. The slew rate is specified in manufacture's specifications in volts per microsecond.

The slew rate is related to power bandwidth f_p. The power bandwidth of the OpAmp is defined as the frequency at which a sinusoidal output begins to distort. The implication is that slew rate limits the bandwidth of OpAmp operation for amplifier and filter designs.

Given a output signal at the rated output voltage of the OpAmp,

$$v_o(t) = V_{out} \sin(2\pi f_p t),$$

(9.8-12)

the maximum slew rate is

$$\left.\frac{dv_o(t)}{dt}\right|_{max} = V_{out}2\pi f_p \cos(2\pi f_p t)|_{t=0} = V_{out}2\pi f_p.$$

(9.8-13)

If $V_{out}2\pi f_p > SR$, the output waveform is distorted. The power bandwidth is therefore defined by the slew rate:

$$f_p = \frac{SR}{2\pi V_{rated}},$$

(9.8-14)

where V_{rated} is the rated OpAmp output voltage.

A detailed SPICE model for the μA741, created from its equivalent models, which incorporates the bandwidth limiting nature of the OpAmp, follows:

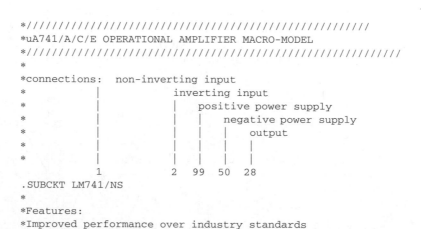

```
*/////////////////////////////////////////////////////////
*uA741/A/C/E OPERATIONAL AMPLIFIER MACRO-MODEL
*/////////////////////////////////////////////////////////
*
*connections:   non-inverting input
*              |              inverting input
*              |             |   positive power supply
*              |             |   |   negative power supply
*              |             |   |   |   output
*              |             |   |   |   |
*              |             |   |   |   |
               1             2  99  50  28
.SUBCKT LM741/NS
*
*Features:
*Improved performance over industry standards
```

```
*Plug-in replacement for LM709, LM201, MC1439, 748
*Input and output overload protection
*
***************INPUT STAGE*************
*
IOS 2  1   20N           ;Input offset current
R1  1  3   250K
R2  3  2   250K
I1  4  50  100U
R3  5  99  517
R4  6  99  517
Q1  5  2   4      QX
Q2  6  7   4      QX
*Fp2=2.55 MHz
C4  5 6 60.3614P
*
***********COMMON MODE EFFECT***********
*
I2 99 50 1.6MA               ;Supply current
EOS 7 1 POLY(1) 16 49 1E-3 1    ;Input offset voltage
R8 99 49 40K
R9 49 50 40K
*
*********OUTPUT VOLTAGE LIMITING********
V2 99 8 1.63
D1 9 8 DX
D2 10 9 DX
V3 10 50 1.63
*
**************SECOND STAGE*************
*
EH 99 98 99 49 1       ;Level shifter
G1 98 9 5 6 2.1E-3
*Fpl= 5 Hz
R5 98 9 95.493MEG
C3 98 9 333.33P
*
***************POLE STAGE**************
*
*Fp=30 MHz
G3  98 15 9 49 1E-6
R12 98 15 1MEG
C5  98 15 5.3052E-15
*
*********COMMON-MODE ZERO STAGE*********
*
*Fpcm=300 Hz
G4  98 16 3 49 3.1623E-8
L2  98 17 530.5M
R13 17 16 1K
*
**************OUTPUT STAGE*************
*
F6 50 99 POLY(1) V6 450U 1    ;Supply current correction
E1  99 23 99 15 1
R16 24 23 25                   ;Output resistance
D5  26 24 DX
V6  26 22 0.65V
R17 23 25 25
D6  25 27 DX
```

```
V7   22  27  0.65V
V5   22  21  0.63V
D4   21  15  DX
V4   20  22  0.63V
D3   15  20  DX
L3   22  28  100P                    ;Output inductor
*
**************MODELS USED*************
*
.MODEL DX D(IS=1E-15)
.MODEL QX NPN(BF=625)
*
.ENDS
*///////////////////////////////////////////////////////////////
```

9.9 CONCLUDING REMARKS

Signal filtering concepts have been described in this chapter. It was demonstrated that bandpass filters could be designed by cascading low-pass and high-pass filters. Three types of active filters employing OpAmps were shown. Design criteria such as cutoff frequency, passband ripple, and stopband attenuation are used to determine order and type of active filter that best suits the requirements.

The three filter types and their characteristics are

Butterworth filter

- Maximally flat, no ripple
- All-pole low-pass filter
- Can be implemented using Sallen and Key circuit

Chebyshev Type 1 filter

- Ripple in passband
- All-pole low-pass filter
- Can be implemented using Sallen and Key circuit
- Has a steeper transition region than Butterworth for a given order filter

Elliptic filter

- Ripple in both passband and stopband
- Poles and zeros in the low-pass filter transfer function
- Can be implemented using the Boctor circuit
- Has the steepest transition region of the three filter types discussed for a given order filter

Switched capacitor networks were discussed as a means to implement active circuits in MOS integrated circuit using capacitance ratios instead of resistors. The design advantage lies in the elimination of resistors, which require large areas when fabricated in integrated circuits. The major drawback is the frequency limitations imposed by the switching.

The frequency response limitations of OpAmps was discussed in terms of gain bandwidth and slew rate. The response of an active circuit depends not only on the components external to the OpAmp but on the OpAmp itself.

SUMMARY DESIGN EXAMPLE

Digital-to-analog (D/A) conversion of electrical signals typically requires output low-pass filtering in order to remove undesirable high-frequency signal components. As an example, digital audio systems typically require this low-pass filtering at the output so that the signals passed on to the power amplification stages are faithful reproductions of the original audio input.

Audio signals are contained to the frequency band, 40 Hz $< f <$ 20 kHz. Compact disk systems sample this audio signal at 44.1 kHz prior to analog-to-digital conversion. The digital samples are then converted to 16 binary bits of information (65,536 levels) and encoded onto the compact disk. When the disk is to be replayed, the individual words of information are sampled several times (oversampled) before D/A conversion. The output of the D/A converter contains the original signal spectrum with sidebands centered at frequencies that are multiples of the product of the sampling rate and the oversampling constant. It is these sidebands that must be eliminated.

Design a filter to eliminate the sidebands of 4× oversampled compact disk D/A conversion without distorting the frequency content of the original signal by more than 0.25 dB at any frequency.

Solution:

The human ear is particularly sensitive to variations in frequency content: The designed filter should be smooth in the passband. A Butterworth filter is a good choice (Chebyshev, type 2 might be an alternative). The sideband signals should be attenuated so that they are less than one level of digitization. Quantization into 65,536 levels implies that filter must introduce 96.33 dB attenuation at the lower frequency edge of the first sideband. The lower edge of this sideband is at

$$f_s = 4(44.1 \text{ kHz}) - 20 \text{ kHz} = 156.4 \text{ kHz}.$$

A 0.25-dB variation at the edge of the passband sets the value of ϵ for the Butterworth filter:

$$\sqrt{1 + \epsilon^2} = 10^{0.025} \Rightarrow \epsilon = 0.24342.$$

The order of the filter is then determined to be

$$n = \frac{1}{2} \frac{\log(10^{9.633} - 1) - \log(\epsilon^2)}{\log(f_s/f_c)} = 6.08.$$

At least a seventh-order Butterworth filter is necessary. The three second-order damping coefficients necessary for this filter are

$$\zeta_1 = 0.2225, \quad \zeta_2 = 0.6235, \quad \zeta_3 = 0.9010.$$

The resonant frequency for the Butterworth filter is given by

$$\omega_o = \frac{\omega_c}{\sqrt[n]{\epsilon}} = \frac{2\pi(20,000)}{\sqrt[7]{0.24343}} = 153.77 \text{ krad/s} \quad (24.473 \text{ kHz}).$$

Since gain is not a factor in the design requirements, a uniform time constant realization is chosen for simplicity and uniformity. One resistor–capacitor pair that will adequately approximate the resonant frequency [the filter resonant frequency has been chosen slightly larger (24.54 kHz) than the theoretical value] is

$$R = 1.38 \text{ k}\Omega, \qquad C = 0.0047 \text{ }\mu\text{F}.$$

One possible filter realization and a SPICE verification of the design follow.

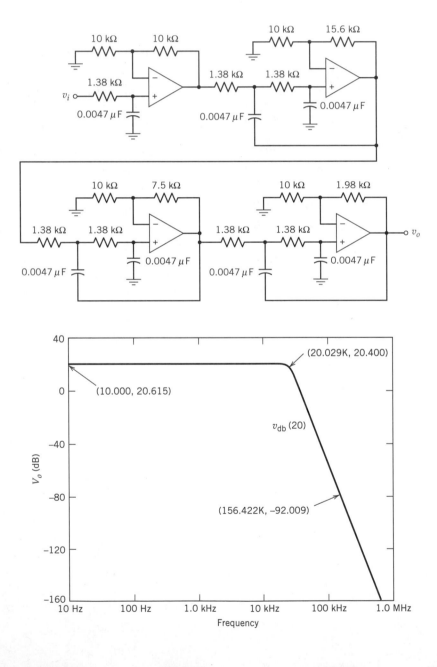

REFERENCES

Bowron, P. and Stephenson, F. W., *Active Filters for Communications and Instrumentation,* McGraw-Hill Book Company, New York, 1979.

Chen, W. K., *Passive and Active Filters: Theory and Implementations,* John Wiley and Sons, Publishers, New York, 1986.

Ghausi, M. S., *Electronic Devices and Circuits: Discrete and Integrated,* Holt, Rinehart and Winston, New York, 1985.

Ludeman, L. C., *Fundamentals of Digital Signal Processing,* Holt, Rinehart, and Winston, Philadelphia, 1986.

Millman, J., and Grabel, A., *Microelectronics,* 2nd Ed. McGraw-Hill Book Company, New York, 1987.

Millman, J. and Halkias, C. C., *Integrated Electronics: Analog and Digital Circuits and Systems,* McGraw-Hill Book Company, New York, 1972.

Nilsson, J. W., *Electrical Circuits,* 3rd Ed., Addison-Wesley, Reading, 1990.

Savant, C. J., Roden, M. S., and Carpenter, G. L., *Electronics Design: Circuits and Systems,* 2nd Ed., Benjamin/Cummings Publishing Company, Redwood City, 1991.

Sedra, A. S. and Smith, K. C., *Microelectronic Circuits,* 3rd Ed., Holt, Rinehart, and Winston, Philadelphia, 1991.

Van Valkenburg, M. E., *Analog Filter Design,* Holt, Rinehart, and Winston, Philadelphia, 1982.

PROBLEMS

9-1. A complex-conjugate pair of poles is characterized by its resonant frequency and damping coefficient:

$$\omega_o = 1 \text{ krad/s}, \qquad \zeta = 0.6.$$

  a. Determine the maximum magnitude error of the Bode straight-line magnitude approximation.

  b. Determine the maximum magnitude phase error of the ζ-dependent Bode straight-line phase approximation.

9-2. The voltage gain of an amplifier is described by the following transfer function:

$$A_V(\omega) = \frac{(j\omega)^2}{[1 + 0.9(j\omega/50) + (j\omega/50)^2]} \times \frac{1}{(1 + j\omega/10,000)(1 + j\omega/80,000)}$$

Plot the straight-line approximate Bode diagram. Compare results to a computer-generated exact plot of the magnitude and phase.

9-3. The transconductance of an amplifier is described by the following transfer function:

$$G_M(\omega) = \frac{(j\omega)^2}{(1 + j\omega/10)(1 + j\omega/40)} \times \frac{1}{1 + 0.7(j\omega/15,000) + (j\omega/15,000)^2}$$

Plot the straight-line approximate Bode diagram. Compare results to a computer-generated exact plot of the magnitude and phase.

9-4. The transresistance of an amplifier is described by flat midband region of value 20 kΩ edged by low- and high-frequency poles. The two low-frequency poles are at

$$f_{L1} = 100 \text{ Hz}, \qquad f_{L2} = 30 \text{ Hz}.$$

There are three high-frequency poles: a complex-conjugate pair described by

$$f_{Ho} = 80 \text{ kHz}, \qquad \zeta = 0.8$$

and a single pole at

$$f_{H3} = 300 \text{ kHz}.$$

  a. How many zero-frequency zeros does the expression for the transresistance contain?

b. Plot the straight-line approximate Bode diagram.

9-5. The current gain of an amplifier is described by flat midband region of value 1.8 kA/A edged by low- and high-frequency poles.

There are three low-frequency poles: a complex-conjugate pair described by

$$f_{Lo} = 70 \text{ Hz}, \qquad \zeta = 0.7$$

and a single pole at

$$f_{L3} = 20 \text{ Hz}.$$

The three high-frequency poles are at

$$f_{H1} = 14 \text{ kHz}, \qquad f_{H2} = 26 \text{ kHz},$$

$$f_{H3} = 160 \text{ kHz}.$$

a. How many zero-frequency zeros does the expression for the transresistance contain?

b. Plot the straight-line approximate Bode diagram.

9-6. Design requirements require the use of a fourteenth-order low-pass Butterworth filter. Available tables only provide the damping coefficients for the first- through eleventh-order filters. What are the damping coefficients necessary for this fourteenth-order low-pass Butterworth filter?

9-7. Design requirements require the use of a fifteenth-order high-pass Butterworth filter. Available tables only provide the damping coefficients for first- through eleventh-order filters. What are the damping coefficients necessary for this fifteenth-order high-pass Butterworth filter?

9-8. A low-pass Butterworth filter is to be designed to meet the following design criteria:

- **Passband**
  Nominal gain $A_{vo} = 1$
  $\gamma_{max} = 0.6$ dB at frequency $f_c = 150$ Hz
- **Stopband**
  $f_s = 900$ Hz
  $\gamma_{min} = 60$ dB min

Determine the order of filter necessary to achieve the design goals and the filter transfer function.

9-9. A high-pass Butterworth filter is to be designed to meet the following design criteria:

- **Passband**
  Nominal gain $A_{vo} = 1$
  $\gamma_{max} = 0.3$ dB at frequency $f_c = 150$ Hz
- **Stopband**
  $f_s = 60$ Hz
  $\gamma_{min} = 50$ dB min

Determine the order of filter necessary to achieve the design goals and the filter transfer function.

9-10. First-order filter stages can be realized with a variety of OpAmp circuit topologies. Determine the frequency response of the transfer function of the two first-order stages shown.

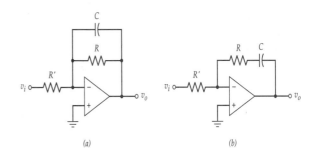

(a)   (b)

9-11. Third-order Butterworth transfer functions can be realized using a single OpAmp.

a. Determine the transfer function of the circuit shown.

b. Choose appropriate component values to achieve a third-order low-pass Butterworth filter with resonant frequency, $\omega_o = 2$ krad/s.

c. Verify the design using SPICE.

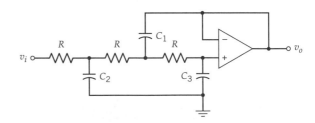

9-12. A second-order, resonant band-pass filter can be realized as shown.

a. Determine the transfer function of the circuit shown.

b. Choose appropriate component values to achieve a resonant frequency of 1 kHz with a damping coefficient of 0.5.

c. Use SPICE to simulate the circuit: Determine the bandwidth between 3 dB frequencies.

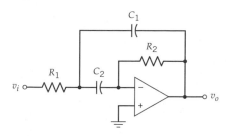

9-13. Design a unity-gain seventh-order Butterworth low-pass filter using OpAmps, resistors, and capacitors with a 3-dB frequency of 200 Hz. Verify your design with SPICE.

9-14. Design a uniform time constant (sixth-order Butterworth low-pass filter using OpAmps, resistors, and capacitors with a passband edge at 400 Hz and $\gamma_{max} = 1.5$ dB. Verify your design with SPICE.

9-15. Design a unity-gain fifth-order Butterworth high-pass filter using OpAmps, resistors, and capacitors with a 3-dB frequency of 200 Hz. Verify your design with SPICE.

9-16. Design a uniform time constant eighth-order Butterworth high-pass filter using OpAmps, resistors, and capacitors with a passband edge at 1 kHz and $\gamma_{max} = 2.2$ dB. Verify your design with SPICE.

9-17. *Scenario*: You are a junior engineer at Alcalá Engineering (a significant, but fictitious, electronics firm). The company is in the process of designing a device that has an analog multiplier as a critical component as shown. In the process of investigating many possible designs for this multiplier, the attached design was uncovered (reprinted from *EDN*, October 25, 1990). Your boss, the chief engineer, stops by your desk, comments on the inferior design of the low-pass filter, asks you to fix it, and leaves.

Your task

- Investigate the design of the current four-pole, low-pass filter.

- Prepare a design that will improve on the current design.

- Prepare a formal proposal to the chief engineer for your design. Theoretical analysis and computer simulation of designs are mandatory.

Constraints

- Alcalá Engineering is operating in a highly competitive environment. Thus the addition of a significant number of component parts is unacceptable (i.e., the filter is to remain a four-pole filter).

- The low-frequency gain of this filter is significant to the design of the analog multiplier and therefore must not be changed from that of the original design.

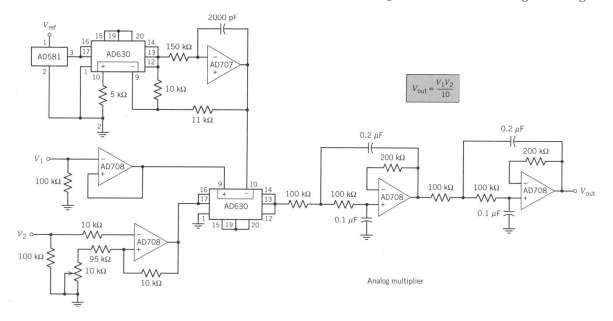

$$V_{out} = \frac{V_1 V_2}{10}$$

Analog multiplier

9-18. A high-pass filter is to be designed to meet the following specifications.

- *Passband*
  Nominal voltage gain $A_{Vo} = 25$ dB
  Passband edge $f_c = 500$ Hz
  $\gamma_{max} = 1.5$ dB maximum

- *Stopband*
  $f_s = 150$ Hz
  $\gamma_{min} = 50$ dB (gain in stopband must be less than -25 dB)

  a. What is the minimum order Butterworth filter that will meet these design goals?

  b. Design such a filter.

  c. Verify the design with computer simulation

9-19. A low-pass filter is to be designed to meet the following specifications.

- *Passband*
  Nominal voltage gain $A_{Vo} = 0$ dB
  Passband edge $f_c = 500$ Hz
  $\gamma_{max} = 1.3$ dB maximum

- *Stopband*
  $f_s = 1.6$ kHz
  $\gamma_{min} = 55$ dB

  a. What is the minimum order Butterworth filter that will meet these design goals?

  b. Design such a filter.

  c. Verify the design with computer simulation.

9-20. A Butterworth band-pass filter is to be constructed to meet the following design goals:

- *Passband*
  Nominal voltage gain $A_{Vo} = 0$ dB
  Passband edges $f_{c1} = 600$ Hz and $f_{c2} = 4$ kHz
  $\gamma_{max} = 0.4$ dB maximum

- *Stopbands*
  $\gamma_{min} = 35$ dB (both stopbands)
  Cutoff frequencies $f_{s1} = 300$ Hz and $f_{s2} = 8$ kHz

Design the filter and verify the design goals using SPICE.

9-21. A fifth-order low-pass Butterworth filter is under design. Its design goals include

$$A_{vo} = 1, \quad f_c = 1 \text{ kHz}, \quad \gamma_{max} = 1.5 \text{ dB}.$$

In order to reduce the number of OpAmps, it has been decided to use a unity-gain, third-order stage of the topology shown and a second-order stage.

The transfer function for this third-order stage has been determined to be

$$v_o = v_i[1 + j\omega R(C_2 + 3C_3)$$
$$+ (j\omega)^2 2R^2 C_3(C_1 + C_2)$$
$$+ (j\omega)^3 R^3 C_1 C_2 C_3]^{-1}.$$

Complete the filter design and verify compliance to the design goals using SPICE.

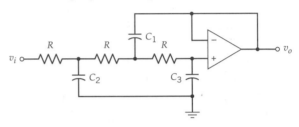

9-22. A Butterworth band-stop filter is to be constructed from a fourth-order Butterworth low-pass filter and a fourth-order Butterworth high-pass filter. All input signals in the frequency range

$$500 \text{ Hz} \leq f \leq 2 \text{ kHz}$$

are to be attenuated by at least 40 dB: Signals in the passbands are to have unity gain and the passbands are to be as wide as possible.

Design the filter and verify the design goals using SPICE.

9-23. Design a low-pass filter to meet the following specifications for Butterworth and Chebyshev filter types:

- 1 dB attenuation at 100 Hz
- >20 dB attenuation at 200 Hz

Compare the order for each of the filters.

9-24. Design a 1-kHz Chebyshev low-pass filter with the following specifications:

- Passband ripple = 2 dB
- Cutoff frequency $f_c = 1$ kHz
- Stopband attenuation >20 dB at 1.8 kHz

Confirm the design with SPICE.

9-25. Plot the transfer function of a seventh-order Chebyshev type I filter with a cutoff frequency of 4 kHz and 1-dB passband ripple. Determine the stopband attenuation at 6 kHz.

9-26. Contrast the attenuation provided by a fourth-order low-pass Chebyshev filter to a Butterworth filter of the same order at a stopband frequency twice that of the cutoff frequency. Let the ripple of the Chebyshev filter equal 1 dB.

9-27. Find the transfer function of a low-pass filter with the following specifications:

- Cutoff frequency $= f_c = 200$ Hz
- Passband ripple $= 2$ dB
- Stopband attenuation > 20 dB above 400 Hz

9-28. Find the transfer function of a high-pass filter with the following specifications:

- Cutoff frequency $f_c = 100$ Hz
- Passband ripple $= 3$ dB
- Stopband attenuation > 20 dB below 50 Hz

9-29. Design a high-pass filter with the following specifications:

- Cutoff frequency $f_c = 100$ Hz
- Passband ripple $= 3$ dB
- Stopband attenuation > 20 dB below 50 kHz

Confirm the design with SPICE.

9-30. Find the normalized transfer function for an elliptical low-pass filter with the following specifications:

- Cutoff frequency $f_c = 750$ Hz
- Acceptable ripple $= 1$ dB
- Stopband attenuation > 40 dB above 1200 Hz

9-31. Design a low-pass elliptic filter with the following specifications:

- Transition frequency $= 100$ Hz
- Acceptable ripple $= 1$ dB
- Stopband attenuation > 20 dB above 200 Hz

Compare the order of this filter with the order required for Butterworth and Chebyshev filters with the same specifications.

9-32. Design a third-order low-pass elliptic filter with the following specifications:

- Cutoff frequency $f_c = 1$ kHz
- Acceptable ripple $= 3$ dB

Confirm the design with SPICE.
    What is the stopband attenuation at 2 kHz?

9-33. A switched-capacitor integrator, shown in the figure, is designed to have as its output the following expression:

$$v_o(t) = -10 \int v_i(t) \, dt.$$

If the switching frequency is 100 kHz, what value of capacitor C is necessary?

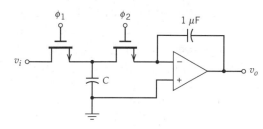

9-34. For the switched-capacitor integrator shown, what input resistance corresponds to the 2 and 12 pF for C_1 with a clock frequency of 100 kHz?

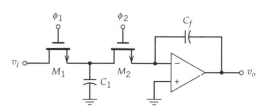

9-35. Design a switched-capacitor version of the differential integrator shown and find the transfer function.

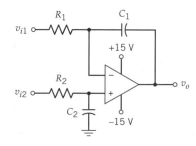

9-36. Design an inverting amplifier with a gain of -10 using the circuit shown. Assume a clock frequency of 100 kHz and $C_1 = 5$ pF. Show the clock waveforms for ϕ_1 and ϕ_2.

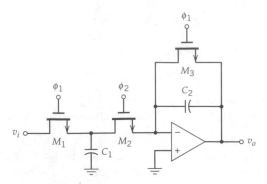

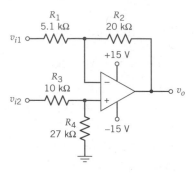

9-37. Design a first-order low-pass switched-capacitor filter with a cutoff frequency of 500 Hz with a clock frequency of 100 kHz. Use a capacitor value of 8 pF for one of the fixed capacitors. Find the transfer function of the circuit.

9-38. Design a first-order high-pass switched-capacitor filter with a cutoff frequency of 500 Hz with a clock frequency of 100 kHz. Use a capacitor value of 8 pF for one of the fixed capacitors. Find the transfer function of the circuit.

9-39. Design a switched-capacitor version of the difference amplifier shown and find the transfer function.

9-40. For an OpAmp with a slew rate of 7 V/μs used in the unity-gain configuration, what is the shortest 0- to 5-V pulse that can be used to ensure a full-amplitude output?

9-41. For an inverting amplifier using an OpAmp with a slew rate of 8 V/μs, determine the highest frequency input at which an 18-V peak-to-peak output sine wave can be generated.

CHAPTER 10

FREQUENCY RESPONSE OF TRANSISTOR AMPLIFIERS

The characterization of the performance of amplifiers as described in Section II of this text has suppressed all frequency-dependent effects. The amplifiers have been described in the so-called midband frequency range. This range of frequencies is characterized by two basic simplifying assumptions:

- The midband frequencies are high enough so that discrete circuit capacitors appear to have negligible impedance with respect to the resistances in the circuit.
- The midband frequencies are low enough so that the active elements (transistors, OpAmps, etc.) appear to have frequency-invariant properties.

The magnitude of the gain of a typical amplifier is shown in Figure 10-1. At low frequencies the circuit coupling and bypass capacitors reduce the gain.[1] At high frequencies the circuit active elements degrade in performance. This degradation in performance also causes a drop in the gain magnitude. Between these two extremes, the midband region of constant gain prevails.

In order to predict the frequency-dependent performance of amplifiers and other electronic circuitry, several options are available:

- Computer simulation
- Analytic calculations using phasor techniques and expanded transistor models on the entire circuit
- Separate high-frequency and low-frequency effects from midband performance in analytic calculations

Computer simulation alone accurately predicts circuit performance but gives the circuit designer little insight into which factors are dominant in determining the performance characteristics. Analytic calculations also give accurate performance prediction, but these calculations often involve extremely difficult algebraic maniplation that must be repeated for each individual circuit. The separation of individual effects (midband, low frequency, and high frequency) gives the designer insight into exactly which circuit elements are dominant in each performance characteristic.

---

[1] Amplifiers without bypass or coupling capacitors do not experience a degradation in performance characteristics at low frequencies. The midband region for such amplifiers extends to DC.

593

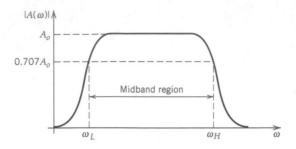

Figure 10-1
Typical amplifier gain frequency response

This chapter focuses on the separation of effects method as the primary analysis and design tool with computer simulation as an aid in final fine tuning of designs. The separation method follows the basic procedure:

1. The amplifier midband performance properties are determined as described in Section II of this text.

2. The pole locations that affect low-frequency performance are separately determined.

3. The pole locations that affect high-frequency performance are separately determined.

4. The results of steps 1–3 are combined to form the total response characteristic.

As was seen in Chapter 9, midband gain and pole location are sufficient to determine a frequency response characteristic.

After a short reminder on frequency distortion, the chapter discusses time-domain testing of amplifiers through the step response. Since amplifier performance characteristics are usually only of interest between the 3-dB frequencies, the concept of dominant poles is explored. The effects of bias and coupling capacitors on low-frequency response is discussed for both simple BJT and FET amplifier stages. After modeling the high-frequency characteristics of diodes, BJTs, and FETs, the high-frequency response of simple amplifier stages is discussed.

Multistage amplifier frequency response is determined through the cascading of the responses of the simple stages.

The frequency response of feedback amplifiers is discussed in Chapter 11.

10.1 FREQUENCY DISTORTION

In the previous chapters, transistor amplifiers were assumed to have constant gain-frequency response over the desired frequency passband. These amplifiers were described in the so-called midband frequency range where external reactive elements and the reactive components of the transistor small-signal model did not affect the frequency-domain properties of the amplifier. The gain of the amplifier outside the midband frequency range is dependent on the reactive elements, causing possible unequal gains for the frequency components of the input signal. This *frequency distortion* is reflected in the gain-frequency (Bode) and phase-frequency plots of the amplifier.

10.1.1 Gain and Phase Response

In section 9.2, a frequency response in the passband was shown to have no distortion when the circuit transfer function is constant with respect to frequency and the phase shift induced by the circuit is linear with respect to frequency. The gain of an amplifier is reduced outside of the midband. The gain-frequency curve is that of a band-pass filter.

Consider a single-pole high-pass function,

$$H_{HP}(\omega) = \frac{1}{1 - j(\omega_L/\omega)}, \tag{10.1-1}$$

where ω_L is the (lower) cutoff frequency for the high-pass transfer function.

The magnitude and phase of the single-pole transfer function of Equation 10.1-1 are

$$|H_{HP}(\omega)| = \frac{1}{\sqrt{1 + (\omega_L/\omega)^2}} \tag{10.1-2}$$

and

$$\angle H_{HP}(\omega) = \tan^{-1}\left(\frac{\omega_L}{\omega}\right). \tag{10.1-3}$$

The frequency response is shown on a Bode diagram in Figure 10.1-1.

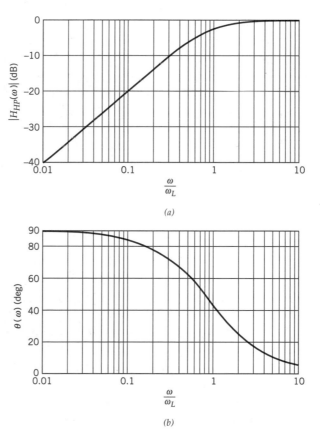

(a)

(b)

Figure 10.1-1

(*a*) Magnitude of high-pass transfer function. (*b*) Phase of high-pass transfer function.

When a single-pole system is excited by an input signal consisting of two sinusoids of frequencies $0.5\omega_L$ and $6\omega_L$ (Figure 10.1-2a), the output (Figure 10.1-2b) is distorted due to the difference in gain and phase of the two input sinusoids.

Similar analysis may be performed to show frequency distortion in systems described by single-pole low-pass transfer functions:

$$H_{LP}(\omega) = \frac{1}{1 + j(\omega/\omega_H)}, \tag{10.1-4}$$

where ω_H is the (high) cutoff frequency for the low-pass transfer function.

The magnitude and phase of the single-pole low-pass transfer function are

$$|H_{LP}(\omega)| = \frac{1}{\sqrt{1 + (\omega/\omega_H)^2}} \tag{10.1-5}$$

and

$$\angle H_{LP}(\omega) = -\tan^{-1}\left(\frac{\omega}{\omega_H}\right). \tag{10.1-6}$$

Bode diagrams of Equations 10.1-5 and 10.1-6 are mirror images of Figure 10.1-1 with a high cutoff frequency of ω_H.

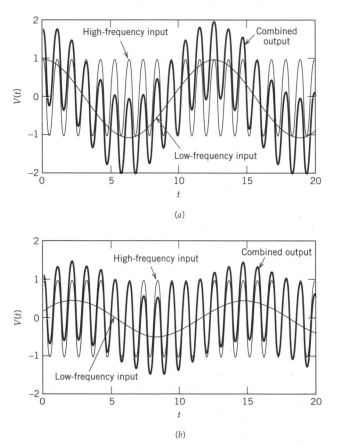

(a)

(b)

Figure 10.1-2

(a) Input signal consisting of two sinusoids. (b) Resulting distorted output due to high-pass transfer function.

10.1.2 Step and Pulse Response

Signal distortion by an amplifier can be classified into three general categories:

1. If an amplifier is linear but its amplitude response is not constant with frequency, the amplifier introduces *amplitude distortion*. The amount of amplitude distortion can be determined from the frequency response of the amplifier.

2. If an amplifier is linear but its phase shift is not a linear function of frequency, it introduces *phase,* or *delay, distortion*.

3. If an amplifier is non-linear, *non-linear distortion* results; e.g., superposition can no longer be used.

The step response provides information about all three types of distortion mentioned. Therefore, the determination of the step response (pulse response testing is a subset of step response) is a powerful test of amplifier linearity.

A step voltage, shown in Figure 10.1-3, is a voltage signal that maintains a zero value for all time $t < 0$ and a constant value V_1 for all time $t > 0$. The transition between the two voltage levels occurs at $t = 0$ and is ideally instantaneous. In real systems, the transition time from 0 to V_1 is accomplished in an arbitrarily small time interval that is significantly smaller than the response time of the circuit under test.

For a single-pole electronic circuit, the output voltage in the time domain is of the form

$$v_o = B_1 + B_2 e^{-t/\tau} \tag{10.1-7}$$

where

$$
\begin{aligned}
v_o &= \text{input voltage} \\
B_1, B_2 &= \text{constants} \\
\tau &= \text{time constant of electronic circuit}
\end{aligned}
$$

The constant B_1 is the final steady-state value of the output voltage as $t \to \infty$. If the final voltage is defined as V_f, then $B_1 = V_f$. The constant B_2 is determined by the initial voltage V_i. At $t = 0$, $v_o = V_i = B_1 + B_2$, or $B_2 = V_i - V_f$. Therefore, the general solution for a single-pole electronic circuit is given as

$$v_o = V_f + (V_i - V_f)e^{-t/\tau}. \tag{10.1-8}$$

Figure 10.1-3
Step input voltage.

Low-Pass Circuit Response

For a low-pass electronic circuit, the initial voltage $V_i = 0$ and the final voltage $V_f = V_1$.[2] Therefore, the output is given by[3]

$$v_o = V_1(1 - e^{-t/\tau}), \tag{10.1-9}$$

where

$$\tau = \frac{1}{\omega_H}. \tag{10.1-9a}$$

The low-pass circuit step response is shown in Figure 10.1-4. The rise time t_r is defined as the time it takes the voltage to rise from 0.1 to 0.9 of its final value. The time required for v_o to rise to $0.1V_1$ is approximately 0.1τ and the time to reach $0.9V_1$ is approximately 2.3τ. The difference between these two time values is the rise time of the circuit and is given by

$$t_r = (2.3 - 0.1)\tau = 2.2\tau. \tag{10.1-10}$$

Since the time constant τ of a circuit is equal to the inverse of the low-pass cutoff frequency in radians, ω_H, Equation 10.1-10 can be related to the cutoff frequency,

$$t_r = 2.2\tau = \frac{2.2}{\omega_H} = \frac{2.2}{2\pi f_H} \approx \frac{0.35}{f_H}. \tag{10.1-11}$$

Thus, the rise time is proportional to the time constant of the circuit and inversely proportional to the cutoff frequency.

A variant of the step input voltage is the pulse input shown in Figure 10.1-5. The low-pass circuit response to a pulse is identical to that of the step response given in Equation 10.1-12 for $t < t_p$. At the end of the pulse ($t = t_p$), the output voltage is at

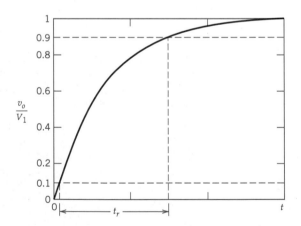

Figure 10.1-4
Low-pass circuit step response.

---

[2] In this discussion, only the time-varying portion of the signal is considered: In order to obtain the total output signal, the quiescent condition must be included.

[3] The relationship between Equations 10.1-4 and 10.1-9 is obtained through Laplace techniques.

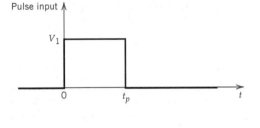

Figure 10.1-5
Pulse input voltage.

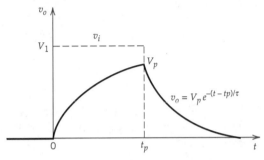

Figure 10.1-6
Low-pass circuit pulse response.

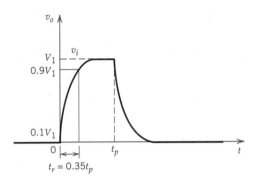

Figure 10.1-7
Low-pass pulse response with short time constant
for $f_H = 1/\tau_p$.

a peak of V_p and must decrease to zero with a time constant τ for $t > t_p$, as indicated
in Figure 10.1-6.

The output waveform is distorted when compared to the input pulse waveform.
In particular, the output waveform extends well beyond the pulse width.

To minimize pulse distortion, the rise time must be small compared to the pulse
width. If the cutoff frequency f_H is chosen to be $1/\tau_p$, then the rise time $t_r = 0.35t_p$.
The output voltage for this case is shown in Figure 10.1-7. The following rule of
thumb may be used with regards to pulse response: A pulse shape will be preserved
if the cutoff frequency is greater than the reciprocal of the pulse width.

High-Pass Circuit Response

From Equation 10.1-8, the time-domain response of the high-pass circuit has an initial
voltage $V_i = V_1$ and a final voltage $V_f = 0$. Therefore, the output voltage of the high-
pass circuit is that of a decaying exponential:

$$v_o = V_1 e^{-t/\tau},$$

(10.1-12)

where, in this case, τ is related to the lower cutoff frequency:

$$\tau = \frac{1}{\omega_L}. \tag{10.1-12a}$$

The input and output voltages are shown in Figure 10.1-8. The output voltage is 0.61 of its initial value at 0.5τ, 0.37 at 1.0τ, and 0.14 at 2.0τ. The output is nearly decayed to 5% of the peak voltage after 3.0τ, and less than 1% if $t > 5.0\tau$.

For a pulse with t_p small compared to the time constant τ, the output response is shown in Figure 10.1-9. The output at $t = t_p$ is $v_o = V_1e^{-t_p/\tau} = V_p$. Since V_p is less than V_1, the voltage becomes negative and decays exponentially to zero. For $t > t_p$, the output voltage is

$$v_o = V_1(e^{-t_p/\tau} - 1)e^{-(t-t_p)/\tau}. \tag{10.1-13}$$

Note that the distortion in the pulse caused by the high-pass circuit has resulted in a tilt to the top of the pulse and an undershoot at the end of the pulse. The percent *tilt* or *sag* in time t_p is approximately

$$v_o \approx V_1\left(1 - \frac{t}{\tau}\right). \tag{10.1-14}$$

If Figure 10.1-8 is redrawn for $t_p \ll \tau$, shown in Figure 10.1-10, the tilt in the output voltage is clear. The percent *tilt* or *sag* in time t_1 is

$$\text{sag} = \frac{V_1 - V_p}{V_1} \times 100 = \frac{t_p}{\tau} \times 100\%. \tag{10.1-15}$$

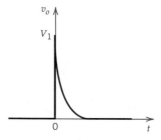

Figure 10.1-8
High-pass circuit step response.

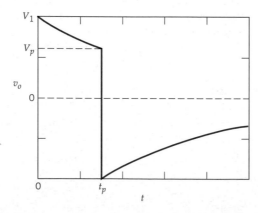

Figure 10.1-9
High-pass circuit pulse response.

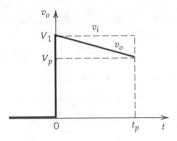

Figure 10.1-10
Tilt in step or pulse response due to high-pass circuit.

Equation (10.1-15) is valid for the tilt of each half cycle of a symmetrical square wave with peak-to-peak voltage of V_1 and a period $T = 2t_p$. If $f_{sq} = 1/T_{sq}$ is the frequency of the square wave, the sag may be expressed as

$$\text{sag} = \frac{T_{sq}}{2\tau} \times 100 = \frac{1}{2f_{sq}\tau} \times 100 = \frac{\pi f_L}{f} \times 100\%. \qquad (10.1\text{-}16)$$

The tilt is then directly proportional to the cutoff frequency f_L.

EXAMPLE 10.1-1 Find the single-pole unity-gain frequency response required for a 10% sag at 1 kHz.

SOLUTION

Using Equation 10.1-16, the cutoff frequency of the high-pass transfer function is determined:

$$f_L = \frac{\text{sag}}{100}\left(\frac{f}{\pi}\right) = \frac{10(1000)}{100\pi} = 31.85 \text{ Hz.}$$

The cutoff frequency must not exceed 31.85 Hz for the desired sag.
Therefore, the transfer function from Equation 10.1-1 is

$$H_{HP}(\omega) = \frac{1}{1 - j(\omega_L/\omega)} = \frac{1}{1 - j[2\pi(31.85)/\omega]} = \frac{1}{1 - j(200/\omega)}.$$

10.2 DOMINANT POLES

The location of the high or low cutoff frequency for electronic circuits with more than one pole (and zero) becomes increasingly difficult to find analytically as the number of poles increases. Computer software packages with an equation solver, such as MathCAD, or advanced engineering and scientific calculators with the equation solver capabilities may be required to find the cutoff frequency. However, it may be desirable to use simple formulas and approximations that yield the cutoff frequencies when initially designing a circuit.

The dominant pole approximation simply ignores second-order effects of poles (and zeros) that are far removed from the pole that dominates in the calculation for the cutoff frequency of the electronic circuit. The low cutoff frequency of a high-pass response is dominated by the largest pole in the transfer function; on the other hand, the high cutoff frequency of a low-pass response is dominated by the smallest pole. When circuits are designed with dominant poles, the analysis of the cutoff frequencies becomes considerably simpler. Nondominant poles and zeros must be carefully analyzed in some instances that may affect circuit stability, such as in feedback amplifiers and oscillators. Stability issues in feedback amplifiers are discussed in Chapter 11.

10.2.1 Low Cutoff Frequency (High Pass)

The general form of the transfer function of a low-frequency response is

$$H_{\mathrm{HP}}(\omega) = \frac{(j\omega + \omega_{z1})(j\omega + \omega_{z2}) \cdots (j\omega + \omega_{zn})}{(j\omega + \omega_{p1})(j\omega + \omega_{p2}) \cdots (j\omega + \omega_{pn})}, \tag{10.2-1}$$

where $\omega_{z1}, \ldots, \omega_{zn}$ are the frequencies for the low-frequency zeros of the response and $\omega_{p1}, \ldots, \omega_{pn}$ are the frequencies for the low-frequency poles of the response. If a midband region exists, the number of poles must equal the number of zeros. The highest pole frequency is ω_{p1} and subsequent poles are ordered such that $\omega_{p1} \geq \omega_{p2} \geq \omega_{p3} \geq \cdots \geq \omega_{pn}$. As ω approaches midband frequencies, $H_{\mathrm{HP}}(\omega)$ approaches unity.

In designing a circuit, the zeros are placed at very low frequencies so that they do not contribute appreciably to the determination of the low cutoff frequency. A dominant pole, say ω_{p1}, has a much higher frequency than the others. In this case, the transfer function in Equation 10.2-1 can be approximated as a first-order high-pass response:

$$H_{\mathrm{HP}}(\omega) \approx \frac{j\omega}{j\omega + \omega_{p1}}. \tag{10.2-2}$$

The approximate transfer function implies that the amplifier low cutoff frequency is dominated by a pole at $j\omega = -\omega_{p1}$, and that the low cutoff frequency is approximately

$$\omega_l \approx \omega_{p1}. \tag{10.2-3}$$

As a rule of thumb, the dominant pole approximation can be made if the highest frequency pole is separated from the nearest pole or zero by a factor of 4.

If a dominant pole does not exist, a complete Bode plot may have to be contructed to determine the low cutoff frequency. Alternately, approximate formulas may be used to determine the location of the low cutoff frequency. For instance, a frequency response with two poles and two zeros has a transfer function of the form

$$H_{\mathrm{HP}}(\omega) = \frac{(j\omega + \omega_{z1})(j\omega + \omega_{z2})}{(j\omega + \omega_{p1})(j\omega + \omega_{p2})}. \tag{10.2-4}$$

The squared magnitude of the transfer function is

$$|H_{\mathrm{HP}}(\omega)|^2 = \frac{(\omega^2 + \omega_{z1}^2)(\omega^2 + \omega_{z2}^2)}{(\omega^2 + \omega_{p1}^2)(\omega^2 + \omega_{p2}^2)}. \tag{10.2-5}$$

The low cutoff frequency is defined by $|H_{\text{HP}}(\omega_l)|^2 = \frac{1}{2}$. Therefore, Equation 10.2-5 becomes

$$\frac{1}{2} = \frac{(\omega_l^2 + \omega_{z1}^2)(\omega_l^2 + \omega_{z2}^2)}{(\omega_l^2 + \omega_{p1}^2)(\omega_l^2 + \omega_{p2}^2)}$$

$$= \frac{1 + (1/\omega_l^2)(\omega_{z1}^2 + \omega_{z2}^2) + (1/\omega_l^4)(\omega_{z1}^2\omega_{z2}^2)}{1 + (1/\omega_l^2)(\omega_{p1}^2 + \omega_{p2}^2) + (1/\omega_l^4)(\omega_{p1}^2\omega_{p2}^2)}. \qquad (10.2\text{-}6)$$

Equation 10.2-6 is simplified by assuming that the zeros are significantly smaller than the low cutoff frequency,

$$\frac{1}{2} \approx \frac{1}{1 + (1/\omega_l^2)(\omega_{p1}^2 + \omega_{p2}^2) + (1/\omega_l^4)(\omega_{p1}^2\omega_{p2}^2)}. \qquad (10.2\text{-}7)$$

Solving for ω_l using the quadratic equation yields

$$\omega_L = \frac{\omega_{p1}}{\sqrt{\dfrac{-(1 + k^2) + \sqrt{(1 + k^2)^2 + 4k^2}}{2}}}, \qquad (10.2\text{-}8)$$

where $\omega_{p1} = k\omega_{p2}$.

A plot of Equation 10.2-8 is shown in Figure 10.2-1, clearly showing that for $k > 4$, the dominant pole frequency is within 5.4% of the low cutoff frequency. If the zeros cannot be ignored, Equation 10.2-6 can be solved by using the approximation that ω_L is greater than all of the poles and zeros. By neglecting the terms containing $1/\omega_L^4$, the low cutoff frequency is approximately

$$\omega_L \approx \sqrt{\omega_{p1}^2 + \omega_{p2}^2 - 2\omega_{z1}^2 - 2\omega_{z2}^2}. \qquad (10.2\text{-}9)$$

Equation 10.2-9 can be extended to any number of poles and zeros; that is,

$$\omega_L \approx \sqrt{\omega_{p1}^2 + \omega_{p2}^2 + \cdots + \omega_{pn}^2 - 2\omega_{z1}^2 - 2\omega_{z2}^2 - \cdots - 2\omega_{zn}^2}. \qquad (10.2\text{-}10)$$

If a dominant pole exists, Equations 10.2-9 and 10.2-10 reduce to Equation 10.2-3.

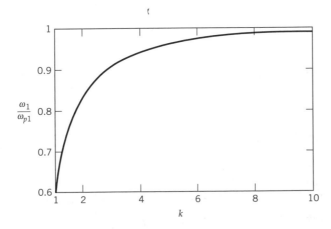

Figure 10.2-1
Low cutoff frequency as function of ratio of pole locations.

10.2.2 Low-Pass Response

The general form of the transfer function of a high-frequency response is

$$H_{LP}(\omega) = \frac{(1 + j\omega/\omega_{z1})(1 + j\omega/\omega_{z2}) \cdots (1 + j\omega/\omega_{zn})}{(1 + j\omega/\omega_{p1})(1 + j\omega/\omega_{p2}) \cdots (1 + j\omega/\omega_{pn})}, \tag{10.2-11}$$

where $\omega_{z1}, \ldots, \omega_{zn}$ are the frequencies for the high-frequency zeros of the response and $\omega_{p1}, \ldots, \omega_{pn}$ are the frequencies for the high-frequency poles of the response.

If a midband region exists, the number of poles must be greater than or equal to the number of zeros. The lowest pole frequency is ω_{p1} and subsequent poles are ordered such that $\omega_{p1} \leq \omega_{p2} \leq \omega_{p3} \leq \cdots \omega_{pn}$. As ω approaches midband frequencies, $H_{LP}(\omega)$ approaches unity.

In designing a circuit, the zeros are placed at very high frequencies so that they do not contribute appreciably to the determination of the high cutoff frequency. A dominant pole, say ω_{p1}, has a much lower frequency than the others. In this case, the transfer function in Equation 10.2-1 can be approximated as a first-order high-pass response,

$$H_{LP}(\omega) \approx \frac{1}{(1 + j\omega/\omega_{p1})}. \tag{10.2-12}$$

The approximate transfer function implies that the amplifier high cutoff frequency is dominated by a pole at $j\omega = -\omega_{p1}$, and that the high cutoff frequency is approximately

$$\omega_H \approx \omega_{p1}. \tag{10.2-13}$$

As a rule of thumb, the dominant pole approximation can be made if the lowest frequency pole is separated from the nearest pole or zero by a factor of 4.

If a dominant pole does not exist, approximate formulas may be used to determine the location of the high cutoff frequency. For instance, a frequency response with two poles and two zeros has a transfer function of the form

$$H_{LP}(\omega) = \frac{(1 + j\omega/\omega_{z1})(1 + j\omega/\omega_{z2})}{(1 + j\omega/\omega_{p1})(1 + j\omega/\omega_{p2})}. \tag{10.2-14}$$

An approximate formula for the high cutoff frequency can be derived in a manner similar to that used above to determine the low cutoff frequency. The high cutoff frequency is

$$\omega_H = \omega_{p1} \sqrt{\frac{-(1 + k^2) + \sqrt{(1 + k^2)^2 + 4k^2}}{2}}, \tag{10.2-15}$$

where $\omega_{p1} = \omega_{p2}/k$.

Figure 10.2-1 can be used to show that for $k \geq 4$, the dominant pole frequency ω_{p1} is within 5.4% of the high cutoff frequency.

If the zeros cannot be ignored, ω_u is approximately

$$\omega_H \approx \frac{1}{\sqrt{1/\omega_{p1}^2 + 1/\omega_{p2}^2 - 2/\omega_{z1}^2 - 2/\omega_{z2}^2}}. \tag{10.2-16}$$

Equation 10.2-16 can be extended to any number of poles and zeros; that is,

$$\omega_H \approx \frac{1}{\sqrt{\dfrac{1}{\omega_{p1}^2} + \dfrac{1}{\omega_{p2}^2} + \cdots + \dfrac{1}{\omega_{pn}^2} - \dfrac{2}{\omega_{z1}^2} - \dfrac{2}{\omega_{z2}^2} - \cdots - \dfrac{2}{\omega_{zn}^2}}}. \tag{10.2-17}$$

If a dominant pole exists, Equations 10.2-16 and 10.2-13 reduce to Equation 10.2-13.

10.3 EFFECT OF BIAS AND COUPLING CAPACITORS ON LOW-FREQUENCY RESPONSE

As mentioned at the beginning of this chapter, the low-frequency response is separately determined by analyzing the effects of bias and coupling capacitors on the circuit. In this section, the methods for determining BJT and FET amplifier low cutoff frequency are shown.

10.3.1 BJT Low-Frequency Response

Common-Emitter Amplifier

A single-stage common-emitter amplifier with input and output coupling capacitors is shown in Figure 10.3-1a. The small-signal equivalent model of the circuit is shown in Figure 10.3-1b.

In performing the analysis, it is convenient to assume that h_{oe}^{-1} is very large. With this assumption, the small-signal circuit shows that there is little interaction between

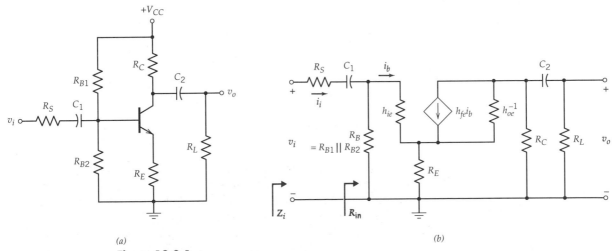

(a)   (b)

Figure 10.3-1
(a) Common-emitter amplifier with input and output coupling capacitors. (b) Small-signal equivalent circuit.

the coupling capacitors. The poles can be found by determining the output voltage and input current of the amplifier.

The input impedance is of the small-signal equivalent of Figure 10.3-1b is

$$Z_i = R_{in} + R_S + \frac{1}{j\omega C_1} \tag{10.3-1}$$

$$= [R_B||(h_{ie} + (1 + h_{fe})R_E)] + R_S + \frac{1}{j\omega C_1}.$$

The input current to the circuit is the ratio for the input voltage to the input impedance:

$$i_i = \frac{v_i}{R_{in} + R_S + 1/j\omega C_1}, \tag{10.3-2}$$

and the base current is

$$i_i = \frac{v_i R_B}{R_B + h_{ie} + (1 + h_{fe})R_E} \frac{j\omega C_1}{1 + j\omega C_1(R_{in} + R_S)}$$

$$= \frac{v_i}{h_{ie} + (1 + h_{fe})R_E} \frac{j\omega R_{in} C_1}{j\omega C_1(R_{in} + R_S) + 1} \tag{10.3-4}$$

$$= \frac{v_i R_{in} C_1}{[h_{ie} + (1 + h_{fe})R_E]\omega_{p1}} \frac{j\omega}{j\omega + \omega_{p1}},$$

where one of the pole locations is

$$\omega_{p1} = \frac{1}{C_1(R_S + \{R_B||[h_{ie} + (1 + h_{fe})R_E]\})}. \tag{10.3-5}$$

The output voltage of the amplifier is

$$v_o = -\frac{h_{fe}i_b R_C R_L}{R_C + R_L + 1/j\omega C_2} = -h_{fe}i_b R_O \frac{j\omega}{j\omega + \omega_{p2}}, \tag{10.3-6}$$

where the other pole is at

$$\omega_{p2} = \frac{1}{C_2(R_C + R_L)}. \tag{10.3-7}$$

Inspection of the pole frequencies shows that the poles are determined by the capacitor multiplied by the resistive discharge path. This is merely the time constant of the circuit. That is, the discharge path for C_1 is through the resistance $R_S + \{R_B||[h_{ie} + (1 + h_{fe})R_E]\}$ and discharge path for C_2 is through the resistance $R_C + R_L$. Therefore, the poles can be found by simply determining the time constant of that portion of the amplifier circuit being analyzed. From the bias conditions and selection of the two capacitor values, the dominant pole can be established by one of the capacitors.

Common-Emitter Amplifiers with Emitter-Bypass Capacitors

The simple analysis described above cannot be accurately used for a common-emitter amplifier with emitter-bypass capacitor shown in Figure 10.3-2a due to the interaction of the capacitors C_1 and C_E. By studying the small-signal equivalent in Figure 10.3-2b, it becomes obvious that there is some interaction between the input coupling capacitor and the emitter-bypass capacitor.

In performing the analysis, it is again convenient to assume that h_{oe}^{-1} is very large. To find the pole locations, nodal analysis is carried out using frequency-domain equivalent impedances. The current gain $A_i(\omega)$ is found by nodal analysis:

$$I_S = \frac{v_i}{R_S} = \frac{v_1}{R_S} + j\omega C_1(v_1 - v_2), \tag{10.3-8a}$$

$$j\omega C_1(v_1 - v_2) = \frac{v_2}{R_B} + \frac{v_2 - v_3}{h_{ie}}, \tag{10.3-8b}$$

$$\frac{v_2 - v_3}{h_{ie}} + h_{fe}i_b = v_3\left(j\omega C_E + \frac{1}{R_E}\right), \tag{10.3-8c}$$

$$0 = h_{fe}i_b + \frac{v_4}{R_C} + v_4\frac{j\omega C_2}{R_L C_2 + 1}. \tag{10.3-8d}$$

The small-signal base current is

$$i_b = \frac{v_2 - v_3}{h_{ie}}. \tag{10.3-9}$$

The output current is

$$i_o = \frac{j\omega C_2 v_4}{j\omega R_L C_2 + 1}. \tag{10.3-10}$$

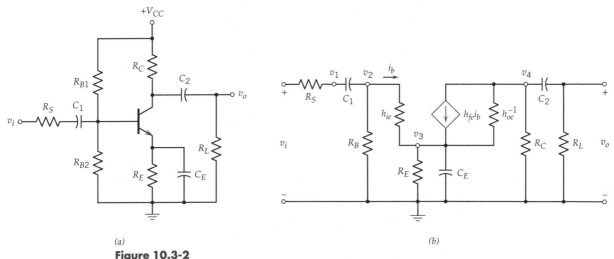

(a)   *(b)*

Figure 10.3-2
(a) Common-emitter amplifier with emitter-bypass capacitor. *(b)* Small-signal equivalent circuit.

The solution to the nodal equations in Equation 10.3-8 for the current gain is

$$A_i(\omega) = \frac{i_o}{i_S} = \frac{-A_{im}(j\omega)^2(j\omega + 1/R_E C_E)}{[j\omega + 1/[(R_L + R_C)C_2]][(j\omega)^2 + j\omega a_2 + a_1]}, \tag{10.3-11}$$

where

$$a_2 = \frac{(R_B + h_{ie})/C_1 + \{(R_B + R_S)[h_{ie} + (1 + h_{fe})R_E] + R_S R_B\}/R_E C_E}{R_B R_S + h_{ie}(R_B + R_S)}$$

$$\approx \frac{1}{C_1(R_S + h_{ie})} + \frac{h_{fe}}{C_E(R_S + h_{ie})}, \qquad R_B \gg R_S, h_{ie}, \tag{10.3-12a}$$

$$a_1 = \frac{R_B + h_{ie} + (1 + h_{fe})R_E}{R_E C_E C_1[R_B R_1 + h_{ie}(R_B + R_S)]}$$

$$\approx \frac{R_B + (1 + h_{fe})R_E}{R_B R_E (R_S + h_{ie})C_1 C_E}, \qquad R_B \gg R_S, h_{ie}, \tag{10.3-12b}$$

and A_{im} is the midband current gain with all capacitors shorted:

$$A_{im} = \frac{h_{fe}(R_S \| R_B)(R_C)}{[(R_S \| R_B) + h_{ie}](R_C + R_L)}$$

$$\approx \frac{h_{fe}R_S}{R_S + h_{ie}} \frac{R_C}{R_C + R_L}, \qquad R_B \gg R_S. \tag{10.3-12c}$$

As expected, the order of the numerator and denominator are the same: three zeros and three poles. It is evident from the above equations that, as expected, the output coupling capacitor C_2 does not interact with the emitter-bypass capacitor C_E and the input coupling capacitor C_1, since C_2 does not appear in Equation 10.3-12 and contributes a pole in Equation 10.3-11. From Equations 10.3-12a and 10.3-12b, a strong interaction between C_1 and C_E is evident.

The three capacitor values can be manipulated to design a circuit with a specific low cutoff frequency ω_l. Since C_2 contributes its own pole and does not interact with the other capacitors, the difficulty in determining the pole locations does not rest with this capacitor value. The pole location due to C_2 can be designed to be significantly lower than the chosen dominent pole. The other two poles are a complex interaction of C_1 and C_E. The zero location due to $1/R_E C_E$ must be significantly smaller than the low cutoff frequency. The two remaining pole locations are found by making the following approximations:

$$(j\omega)^2 + j\omega a_2 + a_1 = (j\omega + \omega_{p1})(j\omega + \omega_{p2}), \quad \text{where } \omega_{p1} \gg \omega_{p2},$$

$$\approx (j\omega + a_2)\left(j\omega + \frac{a_1}{a_2}\right), \quad \text{where } a_2 \gg \frac{a_1}{a_2}. \tag{10.3-13}$$

The pole locations due to the interacting C_1 and C_E are found from Equations 10.3-11 and 10.3-13:

$$\omega_{p1} \approx a_2 \approx \frac{1}{R_S + h_{ie}}\left(\frac{1}{C_1} + \frac{1 + h_{fe}}{C_E}\right), \tag{10.3-14}$$

$$\omega_{p2} \approx \frac{a_1}{a_2} \approx \frac{R_B + (1 + h_{fe})R_E}{R_B R_E[C_E + (1 + h_{fe})C_1]}. \tag{10.3-15}$$

In designing a circuit with a specific low cutoff frequency, it is suggested that the cutoff frequency equal $\omega_l \approx \omega_{p1} \approx a_2$, and select the values for C_1 and C_E. The value for C_2 can be designed to be significantly lower than the dominant pole. Since one end of the emitter-bypass capacitor C_E is ground, a large value chemical capacitor (electrolytic or tantalum) can be used. In doing so, a suggested value of C_E is given as

$$C_E \geq (h_{fe} + 1)C_1, \tag{10.3-16}$$

which allows for the solution for C_1 using Equation 10.3-14 and $\omega_l \approx \omega_{p1} \approx a_2$:

$$C_1 \approx \frac{2}{\omega_l(R_S + h_{ie})}. \tag{10.3-17}$$

With the C_1 and C_E established, the ω_{p2} can be found:

$$\omega_{p2} = \frac{R_B + (1 + h_{fe})R_E}{2h_{FE}R_BR_EC_1}. \tag{10.3-18}$$

The pole location due to C_2 is fixed at $\omega_{p3} \ll \omega_l$:

$$\omega_{p3} = \frac{1}{(R_C + R_L)C_2} \leq \frac{\omega_l}{10}. \tag{10.3-19}$$

Therefore, the output-coupling capacitor value is

$$C_2 \geq \frac{10}{\omega_l(R_C + R_L)}. \tag{10.3-20}$$

EXAMPLE 10.3-1 Complete the design for the common-emitter amplifier shown for a low cutoff frequency of 40 Hz. Find the midband gain. From the quiescent point, the following parameters were determined: $h_{fe} = 200$ and $h_{ie} = 34.7$ kΩ.

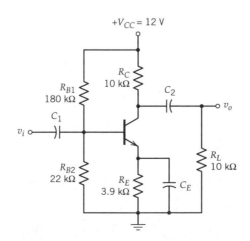

SOLUTION

The midband gain is easily found from the midband small-signal model:

$$A_{vm} = \frac{-h_{fe}(R_C||R_L)}{h_{ie}} = \frac{-200(5000)}{34,700} = -28.8.$$

Equation 10.3-17 yields the value for C_1:

$$C_1 \approx \frac{2}{\omega_l(h_{ie})} = \frac{2}{2\pi(40)(34,700)} \approx 0.22 \ \mu\text{F}.$$

Using Equation 10.3-16, the value of the emitter-bypass capacitor is

$$C_E \geq h_{fe}C_1 = 200(0.22 \times 10^{-6})$$

$$= 44 \ \mu\text{F} \Rightarrow 48 \ \mu\text{F} \quad \text{(standard value)}.$$

Lastly, Equation 10.3-20 yields

$$C_2 \geq \frac{10}{\omega_l(R_C + R_L)} = \frac{10}{2\pi(40)(20,000)}$$

$$\approx 2 \ \mu\text{F} \Rightarrow 2.2 \ \mu\text{F} \quad \text{(standard value)}.$$

The values of the capacitors are consistent with a dominant pole response.

Although the exact analysis above will yield accurate results, it may be preferable in design to use approximate relationships using a simplified approach. The simplified approach consists of assuming that the capacitances do not interact. The pole corresponding to each capacitor is determined with the other capacitors shorted: That is, the pole location corresponding to each capacitor is determined by finding the resistance in its discharge path.

For the common-emitter amplifier with coupling and bypass capacitors, the poles corresponding to C_1 and C_2 have been solved above in the analysis for the common emitter amplifier with input and output capacitors (Equations 10.3-5 and 10.3-7). The voltage gain of the circuit with C_1 and C_2 shorted yields the frequency-dependent effects of the emitter-bypass capacitor. The base current is then

$$i_b = \frac{v_i}{h_{ie} + (h_{fe} + 1)[R_E||(1/j\omega C_E)]}, \quad (10.3\text{-}21)$$

where $1/j\omega C_E$ is the impedance of the emitter-bypass capacitor. The output voltage is

$$v_o = -h_{fe}i_b(R_C||R_L). \quad (10.3\text{-}22)$$

Substituting Equation 10.3-21 into Equation 10.3-22 yields the voltage gain

$$A_v(\omega) = \frac{v_o}{v_i} = \frac{-h_{fe}(R_C||R_L)(1 + 1/j\omega R_E C_E)}{h_{ie}(1 + 1/j\omega R_E C_E) + (1 + h_{fe})/j\omega C_E}. \quad (10.3\text{-}23)$$

With the capacitor shorted, the midband gain is found:

$$A_v = \frac{v_o}{v_i} = \frac{-h_{fe}(R_C||R_L)}{h_{ie}}. \tag{10.3-24}$$

The gain is therefore

$$A_v(\omega) = \frac{A_{vm}(j\omega + 1/R_E C_E)}{j\omega + (1/R_E + (h_{fe} + 1)/h_{ie})/C_E}. \tag{10.3-25}$$

The pole due to C_E is therefore

$$\omega_{p3} = \left(\frac{1}{R_E} + \frac{h_{fe} + 1}{h_{ie}}\right)\frac{1}{C_E}. \tag{10.3-26}$$

The pole in Equation 10.3-26 is then designed to be significantly smaller than ω_{p1} and ω_{p2} in Equations 10.3-5 and 10.3-7 to complete the simplified design approach; that is, C_E is chosen so that $\omega_{p3} \le \omega_{p1}, \omega_{p2}$.

In integrated circuit design, it is undesirable to have large coupling and bypass capacitors since they require a large chip area. Since most integrated circuit amplifiers use constant current source biasing schemes and are commonly DC coupled in a differential amplifier configuration, a low cutoff frequency does not exist. This implies that DC signals are amplified by these circuits.

Common-Base Amplifier
Figure 10.3-3 shows a common-base amplifier and its small-signal equivalent. The analysis and design procedure follows that of the common emitter amplifier.

In performing the analysis, it is again convenient to assume that h_{oe}^{-1} is very large. With this assumption, the small-signal circuit shows that there is little interaction between the coupling capacitors. Therefore, the pole locations due to each capacitor can be independently calculated.

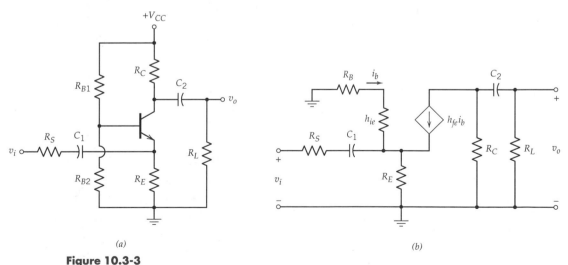

(a)   (b)

Figure 10.3-3
(a) Common-base amplifier with input and output coupling capacitors. (b) Small-signal equivalent circuit.

The discharge path for C_1 is through the resistance $R_S + [(1 + h_{fe})R_E||(h_{ie} + R_B)]$. Therefore, the pole due to C_1 is

$$\omega_{p1} = \frac{1}{C_1\{R_S + [(1 + h_{fe})R_E||(h_{ie} + R_B)]\}}. \tag{10.3-27}$$

The discharge path for C_2 is through the resistance $R_C + R_L$. Therefore, the pole due to C_2 is

$$\omega_{p2} = \frac{1}{C_2(R_C + R_L)}. \tag{10.3-28}$$

From the bias conditions and selection of the two capacitor values, the dominant pole can be established by one of the capacitors.

The simple analysis described above cannot be accurately used for a common-base amplifier with base-bypass capacitor, C_B from the base to ground, due to the interaction of the capacitors C_1 and C_B. For this circuit, the analysis similar to that performed for the common-emitter amplifier with emitter-bypass capacitor is required. However, if an exact solution to the poles is not required, a simplified approach similar to that shown for the common-emitter amplifier with emitter-bypass capacitor may be used, where the pole determined by C_B is significantly smaller than the other two poles found in Equations 10.3-27 and 10.3-28. That is, the pole due to C_B may be ignored since $\omega R_B C_B \gg 1$.

10.3.2 FET Low-Frequency Response

A single-stage common source enhancement NMOSFET amplifier with input and output coupling capacitors and source-bypass capacitor is shown in Figure 10.3-4a. The small-signal equivalent model of the circuit is shown in Figure 10.3-4b.

In performing the analysis, it is convenient to assume that r_d is large compared to the output load and that the impedance of the dependent current source is also very large. With these assumptions, the small-signal circuit shows that there is little

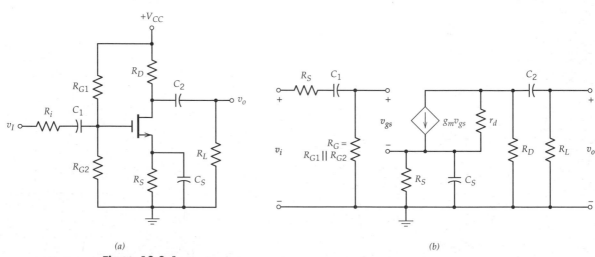

(a)   (b)

Figure 10.3-4
(a) Common-source amplifier. (b) Small-signal equivalent circuit.

interaction between the coupling and bypass capacitors due to the large input resistance inherent in FETs. Therefore, the pole locations due to each capacitor can be independently calculated.

The discharge path for C_1 is through the resistance $R_G + R_S$. Therefore, the pole due to C_1 is

$$\omega_{p1} = \frac{1}{C_1(R_G + R_S)}. \tag{10.3-29}$$

When C_2 and C_S are shorted, the transfer function of the amplifier due to C_1 is

$$A_{v1}(\omega) = \frac{A_{vm}(j\omega)}{j\omega + 1/[C_1(R_G + R_i)]}, \tag{10.3-30}$$

where A_{vm} is the midband gain of the amplifier. The midband gain of the amplifier is found by analyzing the circuit with all capacitors shorted. The midband gain is

$$A_{vm} = -\frac{g_m R_G(R_D + R_L + r_d)}{R_i + R_G}. \tag{10.3-31}$$

The discharge path for C_2 is through the resistance R_L and the parallel combination of R_D and the R_o. Therefore, the pole due to C_2 is

$$\omega_{p2} = \frac{1}{C_2[R_L + (R_D||R_o)]}$$

$$= \frac{1}{C_2(R_L + \{R_D||[r_d + (1 + g_m r_d)R_S]\})}. \tag{10.3-32}$$

When C_1 and C_S are shorted, the transfer function of the amplifier due to C_2 is

$$A_{v2}(\omega) = \frac{A_{vm}(j\omega)}{j\omega + \dfrac{1}{C_2(R_L + \{R_D||[r_d + (1 + g_m r_d)R_S]\})}}. \tag{10.3-33}$$

When C_1 and C_2 are shorted, the transfer function of the amplifier due to C_S is

$$A_{v3}(\omega) = \frac{A_{vm}(1/R_S C_S)}{j\omega + \dfrac{[r_d + (R_D||R_L) + (1 + g_m r_d)R_S]}{R_S C_S[r_d + (R_D||R_L)]}}. \tag{10.3-34}$$

Therefore, the pole location due to C_S is

$$\omega_{pS} = \frac{r_d + (R_D||R_L) + (1 + g_m r_d)R_S}{R_S C_S[r_d + (R_D||R_L)]}. \tag{10.3-35}$$

It is typical to let C_1 or C_2 establish the dominant pole. The value for C_S can be designed so that ω_{pS} is significantly lower than the dominant pole frequency. Since one end of the emmitter-bypass capacitor C_S is ground, a large value chemical capacitor (electrolytic or tantalum) can be used.

Because of the inherently large input resistance of the FET, the common-gate and common-drain amplifier analysis and design procedures follow that of the common-source amplifier: In all cases, each pole location is determined by calculating the pole location for each capacitor while the others are shorted since the capacitors do not interact to establish the poles.

10.4 HIGH-FREQUENCY MODELS OF BJT

Analysis of the response of BJT circuits at high frequencies is based on accurate modeling of the frequency-dependent performance of transistors. The dominant model used for small-signal analysis of a BJT in the forward-active region, the *h*-parameter model as presented in Chapter 3 does not contain frequency-sensitive elements and is therefore invariant with respect to changes in frequency. It is therefore necessary to introduce a new BJT model or to reinterpret an old model to include frequency-dependent terms. Once again, the Ebers–Moll model provides an excellent basis for creation of a more simple model in the forward-active region. In this modeling process, it is necessary to begin with a slight variation of the Ebers–Moll model that was presented in Chapter 3. As is shown in Figure 10.4-1, this presentation has added an additional base resistance of small value, r_b, to model the parasitic bulk resistance between the physical base terminal contact and the active base region[4] and a large output resistance r_o to model reduced output resistance due to early voltage (V_A) effects.

In the forward-active region and at low frequencies the Ebers–Moll model of Figure 10.4-1 can be replaced by the linear two-port model shown in Figure 10.4-2. This model is known as the hybrid-π model. It is similar to the *h*-parameter model

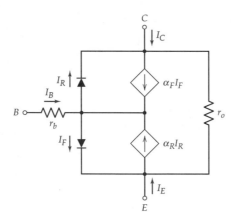

Figure 10.4-1
Ebers–Moll model of BJT.

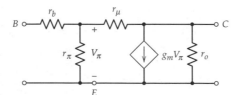

Figure 10.4-2
Low-frequency hybrid-π BJT model.

---

[4] Many models also include parasitic bulk resistances in series with the emitter and collector terminals. These two resistances typically have a much smaller effect on amplifier performance than r_b and therefore will be ignored (assumed to be zero) in the presentation provided here.

used previously in this text but has particular utility when frequency-dependent terms are included. Also of particular interest is the direct correlation between the individual hybrid-π impedances and the corresponding circuit elements in the Ebers–Moll model. The resistances r_b and r_o are directly carried from one model to the other, while r_π and r_μ are the forward resistance of the base–emitter junction and the reverse resistance of the base–collector junction, respectively.

The relationships between h-parameter and hybrid-π models can be obtained by application of the two-port parameter tests as was performed (albeit on the Ebers–Moll model) in Section 5.2. The base–collector junction reverse resistance is extremely large ($r_\mu > 20$ MΩ) and is typically ignored (left as an open circuit). The simplified hybrid-π parameters are related to the h-parameter model parameters by

$$g_m = \frac{h_{fe}}{r_\pi} = \frac{|I_c|}{\eta V_t}, \qquad r_\pi = \beta_F \frac{\eta V_t}{|I_c|} = \frac{\beta_F}{g_m},$$

$$r_o \approx \frac{1}{h_{oe}} = \frac{|V_A|}{|I_c|}, \qquad r_b = h_{ie} - r_\pi. \tag{10.4-1}$$

As will be seen in Section 10.7, the hybrid-π model is also useful in modeling FETs.

In addition to modeling these purely resistive, second-order, low-frequency effects, the diodes of the Ebers–Moll model are replaced by more complex, frequency-dependent models. The frequency-dependent component of transistor behavior is based on the capacitive component of p–n junction impedance. Once the capacitive nature of a p–n junction is known, a frequency-dependent model for a BJT can be obtained.

10.4.1 Modeling a p–n Junction Diode at High Frequencies

In the semiconductor region near a p–n junction under a voltage bias, there is a significant buildup of electrical charge on each side of the junction. Since this charge buildup is dependent on the voltage applied across the junction, there is a capacitance associated with the junction. This capacitance is strongly dependent on the doping densities of the two semiconductor regions and the geometry of the junction. It is modeled as a capacitor shunting the dynamic resistance of the junction (Figure 10.4-3).

In most electronic applications the p–n junction capacitance is dominated by the diffusion of carriers in the depletion regions. A good analytic approximation of this depletion capacitance C_j is given by

$$C_j \approx \frac{C_{j0}}{(1 - V_d/\psi_o)^m}, \tag{10.4-2}$$

where

C_{j0} = small-signal junction capacitance at zero voltage bias
ψ_o = junction built-in potential
m = junction grading coefficient ($0.2 < m \leq 0.5$)

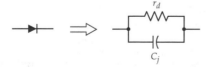

Figure 10.4-3
High-frequency model of p–n junction.

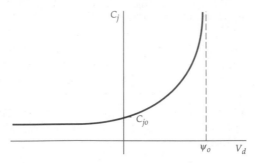

Figure 10.4-4
The p–n junction depletion capacitance.

While theoretical derivations are possible for each of the capacitance parameters, it is more common in practice to determine these parameters empirically. A plot of the junction capacitance as described by Equation 10.4-2 is shown in Figure 10.4-4. This expression is very accurate for reverse-biased junctions and for forward-biased junctions where the junction current is small. In most electronic applications, it provides an adequate level of modeling.[5]

Notice that the junction capacitance under reverse-biased conditions exhibits small variation while under forward-biased conditions it increases dramatically with bias voltage.

EXAMPLE 10.4-1 If the zero-bias capacitance of a p–n junction is 5 pF and the built-in potential ψ_o is 640 mV, determine the junction capacitance at junction voltages V_d of -20, -10, -5, -1, 0.3, and 0.6 V. Assume the junction grading coefficient has value $m = 0.40$.

SOLUTION

Equation 10.4-2 yields

$$C_j \approx \frac{5 \text{ pF}}{(1 - V_d/0.64)^{0.40}}$$

The tabulated results are

V_d (V)	C_j (pF)
-20	1.25
-10	1.62
-5	2.09
-1	3.43
0.3	6.44
0.6	15.2

The small variation in the junction capacitance for a wide range of negative bias is evident.

---

[5] High forward current modeling requires two modifications. Equation 10.4-2 indicates infinite capacitance at the built-in potential: In actuality, the depletion capacitance decreases slightly for voltages above the built-in potential. An additional capacitance term, the carrier diffusion capacitance, must also be included. This term is directly proportional to diode current.

10.4.2 Modeling the BJT at High Frequencies in the Forward-Active Region

In order to model the BJT at high frequencies, the hybrid-π model of Figure 10.4-2 is altered by shunting each p–n junction dynamic resistance with an appropriate junction capacitance. This alteration is shown in Figure 10.4-5. Here, the base–emitter junction has been modeled by a capacitor C_π in parallel with the junction forward resistance r_π. Similarly, the base–collector junction has been modeled by a capacitor C_μ ($r_\mu \approx \infty$).

Determination of the values of the two junction capacitances for a particular transistor is necessary in order to complete the modeling process. While it is possible to analytically determine these capacitances from the physical dimensions and properties of an individual transistor, it is quite common to determine the values experimentally. The base–collector junction is reverse biased when the BJT is in the forward-active region, hence the junction capacitance C_μ is *relatively* independent of quiescent conditions. Typical manufacturer data sheets provide a value of C_μ at a given reverse bias (typically, V_{CB} is 5 or 10 V): Further discussions of C_μ in this text will consider it to be constant at the manufacturer's supplied value.[6] The forward-biased base–emitter junction exhibits greater variation with bias conditions: Its junction capacitance C_π must therefore be determined with greater caution.

The value of the base–emitter junction capacitance C_π for a BJT is usually determined through a measurement of the variation with frequency of the BJT short-circuit current gain. The transistor is placed in a test fixture as shown in Figure 10.4-6. The term "short circuit" applies to the collector–emitter port of the BJT, which appears, *in an AC sense*, to be a short circuit. A plot of the current gain frequency response is then determined experimentally. In order to correlate these measurements with the

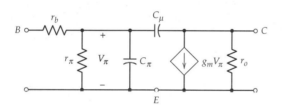

Figure 10.4-5

High-frequency hybrid-π model of BJT.

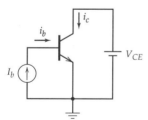

Figure 10.4-6

Measurement of short-circuit current gain.

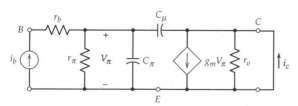

Figure 10.4-7

Short-circuit current gain AC equivalent circuit.

---

[6] Here, C_μ exhibits the same variation with junction bias as is given in Equation 10.4-2. If data are available to determine the built-in potential ψ_o and the junction grading coefficient m, the values should be used to determine a better approximation for C_μ.

hybrid-π parameters, an AC equivalent circuit of the test circuit must be created (Figure 10.4-7).

The current gain is found to be

$$A_I(\omega) = \frac{i_c(\omega)}{i_b(\omega)} = \left(\frac{i_c}{V_\pi}\right)\left(\frac{V_\pi}{i_b}\right),$$

(10.4-3)

which becomes

$$A_I(\omega) = (g_m)\left(\frac{r_\pi}{1 + j\omega r_\pi(C_\pi + C_\mu)}\right) = \frac{h_{fe}}{1 + j\omega r_\pi(C_\pi + C_\mu)}.$$

(10.4-4)

This current gain expression is in the form of a single-pole low-pass frequency response. The transistor short-circuit gain has a low-frequency value of h_{fe} and a 3-dB frequency of

$$\omega_{3dB} = \frac{1}{r_\pi(C_\pi + C_\mu)}.$$

(10.4-5)

A more common description of the results of Equation 10.4-4 depends on the frequency at which the current gain is unity, ω_T:

$$\omega_T = \frac{h_{fe} - 1}{r_\pi(C_\pi + C_\mu)} \approx \frac{g_m}{C_\pi + C_\mu}.$$

(10.4-6)

This unity-gain frequency ω_T is often referred to as the gain–bandwidth product: the product of short-circuit current gain at a particular frequency and that frequency has constant value for all frequencies greater than ω_{3dB}. Manufacturers data sheets will either provide a value for ω_T or provide the gain at some other high frequency. The gain–bandwidth product is given by

$$\omega_T = A_m \omega_m,$$

(10.4-7)

where ω_m is the frequency at which the manufacturer made the gain measurement and A_m is the gain at that frequency.

EXAMPLE 10.4-2 Given a silicon *npn* BJT with parameters

$$\beta_F = 150, \qquad V_A = 350, \qquad r_b = 30 \ \Omega,$$
$$C_\mu = 3 \ \text{pF}, \qquad f_T = 250 \ \text{MHz},$$

determine an appropriate small-signal hybrid-π model for the transistor when operating in the circuit shown.

SOLUTION

The first step in modeling any BJT is to determine the quiescent conditions. Since this is the same transistor and circuit as Example 5.2-1, the quiescent conditions have already been determined to be

$$I_C = 9.3 \ \text{mA}, \qquad V_{CE} = 7.90 \ \text{V}.$$

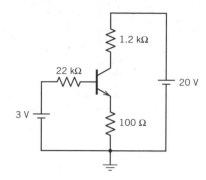

The BJT is in the forward-active region and the hybrid-π model parameters can be determined from Equations 10.4-1 and 10.4-6. Equation 10.4-1 yields the low-frequency parameters

$$g_m = \frac{|I_c|}{V_t} = \frac{9.30 \times 10^{-3}}{26 \times 10^{-3}} = 358 \text{ mS}, \qquad r_\pi = \frac{\beta_F}{g_m} = \frac{150}{0.358} = 419.4 \ \Omega,$$

$$r_o = \frac{|V_A|}{|I_c|} = \frac{350}{9.30 \times 10^{-3}} = 37.6 \text{ k}\Omega, \qquad r_b = 30 \ \Omega.$$

Equation 10.4-6 yields the capacitor values

$$C_\mu = 3 \text{ pF},$$

$$C_\pi = \frac{g_m}{\omega_T} - C_\mu = \frac{0.358}{2\pi(250 \times 10^6)} - 3 \text{ pF} = 225 \text{ pF}.$$

10.5 MILLER'S THEOREM

The process of determining the high-frequency poles for transistor amplifiers is not always a simple process. In particular, the introcuction of an impedance that connects amplifier input and output ports adds a great deal of complexity in the analysis process. One technique that often helps reduce the complexity in some circuits is the use of Miller's theorem. Miller's theorem addresses the problem introduced by the interconnection of input and output ports.

Miller's theorem applies to the process of creating equivalent circuits. This general circuit theorem is particularly useful in the high-frequency analysis of certain transistor amplifiers at high frequencies. It is based on the principle that two circuits appear equivalent if they have identical voltgage–current relationships at the ports where they interconnect with any adjoining circuitry.

Miller's theorem generally states: *Given any general linear network having a common terminal and two terminals whose voltage ratio, with respect to the common terminal, is given by*

$$V_2 = AV_1. \tag{10.5-1}$$

If the two terminals of the network are then interconnected by an impedance Z, an equivalent circuit can be formed. This equivalent circuit consists of the same general linear network and

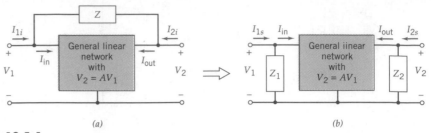

Figure 10.5-1
Miller equivalent circuits: (*a*) interconnecting impedance; (*b*) port-shunting impedances.

two impedances; each of which shunt a network terminal to common terminal. These two impedances have value (Figure 10.5-1):

$$Z_1 = \frac{Z}{1 - A}, \qquad Z_2 = \frac{AZ}{A - 1}. \tag{10.5-2}$$

Miller's theorem can be verified by showing that the voltage–current relationships at each port of the two circuits are identical. Notice that the voltage at each port of the linear network is applied in the same manner so that the port currents I_{in} and I_{out} are unchanged by the attached impedances. At the left port of Figure 10.5-1*a*, the input current I_{1i} is given by

$$I_{1i} = I_{in} + \frac{V_1 - V_2}{Z} = I_{in} + \frac{V_1 - AV_1}{Z}. \tag{10.5-3}$$

The input current of the equivalent circuit (Figure 10.5-1*b*) is given by

$$I_{1s} = I_{in} + \frac{V_1}{Z_1}. \tag{10.5-4}$$

Using the relationship given in Equation 10.5-2, the input current expression becomes

$$I_{1s} = I_{in} + \frac{V_1}{Z/(1 - A)} = I_{in} + \frac{V_1(1 - A)}{Z} = I_{1i}. \tag{10.5-5}$$

Similarly, at the right port of Figure 10.5-1*a*, the output current I_{2s} is given by

$$I_{2i} = I_{out} + \frac{V_2 - V_1}{Z} = I_{in} + \frac{V_2 - (1/A)V_2}{Z}. \tag{10.5-6}$$

The output current of the equivalent circuit (Figure 10.5-1*b*) is given by

$$I_{2s} = I_{out} + \frac{V_2}{Z_2}. \tag{10.5-7}$$

Using the relationship given in Equation 10.5-2, the output current expression becomes

$$I_{2s} = I_{out} + \frac{V_2}{AZ/(A-1)} = I_{out} + \frac{V_2 - (1/A)V_2}{Z} = I_{2i}. \qquad (10.5\text{-}8)$$

For these two Miller equivalent circuits, the individual port voltage–current relationships are identical and the two circuits appear identical to any other circuitry that may be connected at these two ports.

Replacement of the input–output interconnection is particularly useful in the analysis of common-emitter and common-source amplifiers. In these amplifiers, the voltage gain necessary to invoke Miller's theorem is easily attainable and the substitution produces significant reduction in analysis complexity.

EXAMPLE 10.5-1 The circuit shown consists of an OpAmp inverting amplifier bridged by a capacitor. Determine the Miller equivalent circuit if the inverting amplifier is considered to be the general linear network.

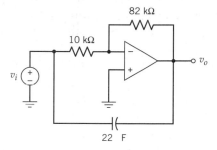

SOLUTION

The gain of the OpAmp inverter is given by

$$A = \frac{v_o}{v_i} = -\frac{82 \text{ k}\Omega}{10 \text{ k}\Omega} = -8.2.$$

The two Miller impedances are given by

$$Z_1 = \frac{Z}{1-A} = \frac{1/(j\omega(22 \times 10^{-6}))}{1 - (-8.2)} = \frac{1}{j\omega(202.4 \times 10^{-6})},$$

$$Z_2 = \frac{AZ}{A-1} = \frac{-8.2\{1/(j\omega(22 \times 10^{-6}))\}}{-8.2 - 1} = \frac{1}{j\omega(24.68 \times 10^{-6})}.$$

Here, Z_1 appears to be a 202.4-μF capacitor and Z_2 a 24.68-μF capacitor. The Miller equivalent circuit is shown.

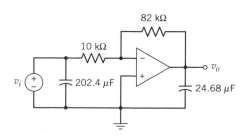

▬▬▬ 10.6 HIGH-FREQUENCY RESPONSE OF SIMPLE BJT AMPLIFIERS

At high frequencies, amplifier response is characterized by the midband gain and the high-frequency poles. Each pole affects the frequency response curves by introducing 20 dB/decade attenuation, which begins at the pole frequency. Once the midband gain of an amplifier has been determined in the usual fashion, it is only important to determine the pole locations in order to completely determine the amplifier high-frequency response.

In this section, the three single-transistor BJT amplifier types are analyzed in order to determine their high-frequency pole locations. Miller's theorem is used, where appropriate, to reduce the complexity of the analysis. As in all small-signal analysis, transistor quiescent conditions must be first calculated so that the transistor parameters can be accurately determined. In order to focus discussion on pole frequency determination, it is assumed that the quiescent analysis has been previously performed and that the transistor parameters are well known.

While the emphasis of this section is on analysis, the information gained in that analysis is significant in the practice of design. Only with a secure knowledge of the effects of circuit element values and transistor parameters on amplifier response can the designer make appropriate adjustments to meet or exceed specifications.

10.6.1 Common-Emitter Amplifier High-Frequency Characteristics

The midband AC model of typical common-emitter amplifier and its high-frequency equivalent, which uses the hybrid-π BJT model, are shown in Figure 10.6-1. The use of the analysis techniques previously developed in this text for this circuit is made difficult by the presence of the capacitor C_μ, which interconnects the input and output sections of the amplifier. Miller's theorem is particularly useful in reducing the analysis complexity for this case. In Figure 10.6-1b, the shaded two-port network has two terminals for which the voltage gain is well known:

$$A = \frac{v_o}{v_\pi} = -g_m(r_o\|R_c) = -g_m R_c',$$ (10.6-1)

where R_c' is the equivalent load resistance including the output impedance of the BJT:

$$R_c' = (r_o\|R_c).$$ (10.6-2)

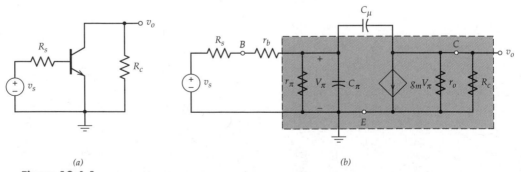

(a)   (b)

Figure 10.6-1
Common-emitter equivalent circuits: (a) midband AC equivalent; (b) high-frequency equivalent.

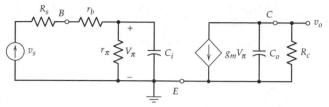

Figure 10.6-2
Miller's theorem applied to common-emitter amplifier.

The capacitor C_μ bridges the two terminals of the shaded two-port network. Miller's theorem replaces this bridging capacitor with equivalent capacitances that shunt the ports of the network (Figure 10.6-2).

The input capacitance C_i is the parallel combination of C_π and the Miller input capacitance C_1:

$$C_i = C_\pi + (1 + g_m R'_c)C_\mu \tag{10.6-3}$$

and the output capacitance C_o is the Miller output capacitance C_2:

$$C_o = C_\mu \frac{1 + g_m R'_c}{g_m R'_c}. \tag{10.6-4}$$

The voltage gain of the circuit is easily determined through typical phasor techniques. In this case, the gain is the product of a voltage division at the input and a current source–impedance product at the output:

$$A_V = \frac{v_o}{v_s} = \left(\frac{v_o}{v_\pi}\right)\left(\frac{v_\pi}{v_s}\right) = \left(\frac{-g_m R'_c}{1 + j\omega C_o R'_c}\right)\left(\frac{r_\pi/(1 + j\omega C_i r_\pi)}{r_\pi/(1 + j\omega C_i r_\pi) + R_s + r_b}\right). \tag{10.6-5}$$

Algebraic manipulation yields the desired results:

$$A_V = -g_m R'_c \frac{r_\pi}{r_\pi + R_s + r_b} \frac{1}{1 + j\omega C_o R'_c} \frac{1}{1 + j\omega C_i[r_\pi\|(R_s + r_b)]}. \tag{10.6-6}$$

This gain has two simple poles at frequencies

$$\omega_{p1} = \frac{1}{C_i[r_\pi\|(R_s + r_b)]} \tag{10.6-7}$$

$$\omega_{p2} = \frac{1}{C_o R'_c} = \frac{1}{C_\mu [(1 + g_m R'_c)/(g_m R'_c)]R'_c} = \frac{1}{C_\mu (1/g_m + R'_c)}. \tag{10.6-8}$$

Typical circuit element values imply that $\omega_{p1} < \omega_{p2}$, but very large load resistance may alter that relationship: Both pole locations should always be checked. Typically the high-frequency response of a common-emitter amplifier is limited by C_i through the multiplicative alteration of C_π using Miller's theorem. This increase in the input circuit capacitance and the resulting decrease in pole frequency magnitude is called the Miller effect.

These two pole locations, with the midband gain, determine the total high-frequency response of a common-emitter amplifier. The input impedance of the amplifier experiences a pole at the first of the two poles, ω_{p1}: The output impedance has a pole at the second, ω_{p2}.

EXAMPLE 10.6-1 Determine the high-frequency poles for the common emitter circuit shown. Assume a silicon *npn* BJT with parameters

$$\beta_F = 150, \qquad V_A = 350, \qquad r_b = 30 \ \Omega,$$
$$C_\mu = 3 \text{ pF}, \qquad f_T = 100 \text{ MHz}.$$

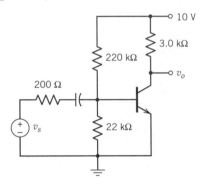

SOLUTION

The coupling capacitor is replaced by an open circuit to determine the BJT quiescent: They are found to be

$$I_B = 10.45 \ \mu\text{A}, \qquad I_C = 1.568 \text{ mA}, \qquad V_{CE} = 5.30 \text{ V}.$$

The hybrid-π parameters are then determined from these quiescent conditions:

$$g_m = 60.31 \text{ mS}, \qquad r_\pi = 2.487 \text{ k}\Omega, \qquad C_\mu = 3.0 \text{ pF},$$
$$r_o = 223.2 \text{ k}\Omega, \qquad r_b = 30 \ \Omega, \qquad C_\pi = 93.0 \text{ pF}.$$

The input Miller capacitance is given by

$$C_i = C_\pi + (1 + g_m R_c')C_\mu$$
$$= 93 \times 10^{-12} + (1 + 179.5)3 \times 10^{-12} = 634.6 \text{ pF}.$$

The source resistance R_s is the 200-Ω resistor in parallel with the two biasing resistors; thus the two pole frequencies are determined from Equations 10.6-7 and 10.6-8:

$$\omega_{p1} = \frac{1}{C_i[r_\pi||(R_s + r_b)]} = \frac{1}{634.5 \times 10^{-12}(198)}$$
$$= 7.957 \text{ Mrad/s (1.27 MHz)}$$

$$\omega_{p2} = \frac{1}{C_\mu(1/g_m + R_c')} = \frac{1}{3 \times 10^{-12}(1/0.06031 + 2960)}$$
$$= 112.0 \text{ Mrad/s (17.8 MHz)}$$

Clearly ω_{p1} is a dominant pole and the high 3-dB frequency is

$$f_H \approx 1.27 \text{ MHz}$$

More exactly, the high 3-dB frequency can be determined to be

$$\omega_H = (7.957 \text{ Mrad/s}) \left(\sqrt{\frac{-(1 + k^2) + \sqrt{(1 + k^2)^2 + 4k^2}}{2}} \right),$$

where

$$k = \frac{112.0}{7.957} = 14.08.$$

Therefore

$$\omega_H = (7.957 \text{ Mrad/s})(0.995) = 7.917 \text{ Mrad/s} \ (1.27 \text{ MHz})$$

10.6.2 Common-Collector Amplifier High-Frequency Characteristics

The midband AC model of typical common-collector amplifier and its high-frequency equivalent are shown in Figure 10.6-3. While common-collector amplifiers are often constructed without a collector resistor, it is included here for generality. Common-emitter amplifiers with an emitter resistor have the same general topology as the common-collector amplifier and therefore have virtually the same pole locations.

The analysis procedure for this circuit is once again complicated by the presence of the capacitor C_μ. Miller's theorem is one of many possible techniques that can be applied to reduce the analysis complexity for this case. In Figure 10.6-3b, the shaded

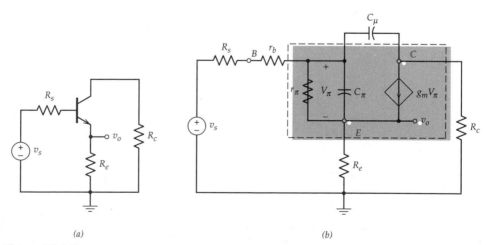

(a) (b)

Figure 10.6-3
Common-collector equivalent circuits: (a) midband AC equivalent; (b) high-frequency equivalent.

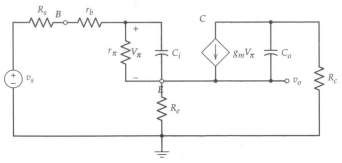

Figure 10.6-4
Miller's theorem applied to common-collector amplifier.

two-port network has two terminals for which the voltage gain, with respect to the collector, can be approximated as[7]

$$A = \frac{v_{ec}}{v_{\pi}} \approx -g_m(R_c + R_e), \tag{10.6-9}$$

where v_{ec} is the small-signal emitter–collector voltage.

The capacitance C_μ is redistributed to the input and output ports using Miller's theorem to form the circuit shown in Figure 10.6-4.

The input capacitance is the sum of C_π and the Miller input capacitance C_1:

$$C_i = C_\pi + C_1 = C_\pi + C_\mu[1 + g_m(R_c + R_e)]. \tag{10.6-10}$$

The output capacitance is the Miller output capacitance C_2:

$$C_o = C_\mu \frac{1 + g_m(R_c + R_e)}{g_m(R_c + R_e)} \approx C_\mu. \tag{10.6-11}$$

For purely resistive loads of typical value, the pole introduced by the output capacitance is typically at a very high frequency beyond the scope of realistic measurement. If the load has a capacitive component, that capacitive component will dominate the determination of the pole location. The pole introduced by the input capacitance is of more interest: The Miller effect has increased the input capacitance significantly and consequently lowered the magnitude of this pole frequency. In the equivalent circuit of Figure 10.6-4, assume that the pole introduced by C_o at a sufficiently high frequency so that C_o can be replaced by an open circuit. The source input current is given by

$$i_s = \frac{v_\pi}{Z_\pi} = \frac{v_\pi(1 + j\omega C_i r_\pi)}{r_\pi}, \tag{10.6-12}$$

where Z_π is the parallel combination of r_π and C_i.

---

[7] The complete frequency response analysis of a general common-collector amplifier is extremely complex. The techniques presented here invoke several assumptions that simplify the analysis procedure without undue loss of accuracy. Here it is assumed that the current through the collector and emitter resistors is nearly identical and that the pole in the specified Miller voltage gain is at a sufficiently high frequency (it is approximately at ω_T) so as not to alter the results significantly.

The collector current is sufficiently larger than the base current in a BJT so that the output voltage can be approximated as

$$v_o \approx (g_m v_\pi) R_e \qquad (10.6\text{-}13)$$

The source voltage is the sum of v_o, v_π, and the voltage across the input resistances:

$$v_s = v_o + v_\pi + i_s(R_s + r_b), \qquad (10.6\text{-}14)$$

which can be rewritten as

$$v_s = g_m v_\pi R_e + v_\pi + \frac{v_\pi(1 + j\omega C_i r_\pi)}{r_\pi}(R_s + r_b). \qquad (10.6\text{-}15)$$

Equation 10.6-15 can be arranged to obtain v_π in terms of v_s:

$$v_\pi = \frac{v_s r_\pi}{R_s + r_b + (1 + g_m R_e) r_\pi + j\omega C_i r_\pi (R_s + r_b)}. \qquad (10.6\text{-}16)$$

The voltage gain is the combination of Equations 10.6-13 and 10.6-16:

$$\frac{v_o}{v_s} \approx \frac{g_m r_\pi R_e}{R_s + r_b + (1 + g_m R_e) r_\pi + j\omega C_i r_\pi (R_s + r_b)}. \qquad (10.6\text{-}17)$$

This gain expression has a simple high-frequency pole located at

$$\omega_{p1} \approx \left[C_i \left(\frac{r_\pi(R_s + r_b)}{R_s + r_b + (1 + g_m R_e) r_\pi} \right) \right]^{-1}$$
$$- \left[C_i \left(r_\pi \, \middle\| \, \frac{R_s + r_b}{1 + g_m R_e} \right) \right]^{-1}. \qquad (10.6\text{-}18)$$

The complete representation of the pole location is given by replacing C_i with its value in terms of C_π and C_μ (Equation 10.6-10):

$$\omega_{p1} \approx \left[\{C_\pi + C_\mu[1 + g_m(R_c + R_e)]\} \left(r_\pi \, \middle\| \, \frac{R_s + r_b}{1 + g_m R_e} \right) \right]^{-1}. \qquad (10.6\text{-}19)$$

This single pole is usually the only pole of interest in the analysis of common collector amplifiers. More complete analysis shows that a very high frequency zero is also present along with another high-frequency pole: Both of these are usually unimportant to the amplifier designer. The approximations included in the derivation of Equation 10.6-19 make a strong case for verification of the pole frequency by computer simulation: A variation in the pole location by $\pm 10\%$ is common.

Common-emitter with emitter resistor amplifiers have the same basic topology as the common-collector amplifier anlayzed in this section. The resultant pole location remains the identical. Simulation verification is also strongly suggested. In the derivation, it is necessary to replace only the gain equation (Equation 10.6-13) by

$$v_o = -(g_m v_\pi) R_c. \qquad (10.6\text{-}20)$$

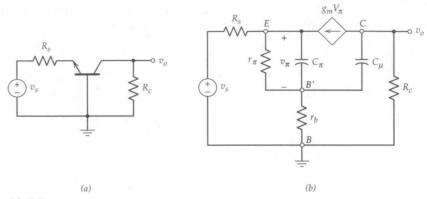

Figure 10.6-5
Common-base equivalent circuits: (*a*) midband AC equivalent; (*b*) high-frequency equivalent.

10.6.3 Common-Base Amplifier High-Frequency Characteristics

The midband AC model of typical common-base amplifier and its high-frequency equivalent are shown in Figure 10.6-5. Notice that there is no element bridging the input and output terminals. It is not necessary to invoke Miller's theorem in order to find the pole frequencies. The capacitance-increasing Miller effect is not present in common-base amplifiers: The pole magnitudes are consequently at very high frequencies.

In order to simplify the analysis of this complex circuit, notice that the voltage across the resistor r_b is very small: Both r_b and the BJT base current are relatively small quantities. It is therefore reasonable to assume the voltage at B' is essentially AC ground and the voltage at the emitter is given by

$$v_e \approx -v_\pi. \tag{10.6-21}$$

The output voltage is given by the product of the current source value and the impedance between the collector and common:

$$v_o = -g_m v_\pi \frac{1}{1 + j\omega C_\mu R_c}. \tag{10.6-22}$$

In order to find the relationship between v_π and v_s, the currents are summed at the emitter terminal of the BJT:

$$\frac{v_s + v_\pi}{R_s} + g_m v_\pi + j\omega C_\pi v_\pi + \frac{v_\pi}{r_\pi} = 0. \tag{10.6-23}$$

This expression can be solved for v_π:

$$v_\pi = \frac{-(v_s/R_s)(1/R_s + g_m + 1/r_\pi)}{1 + j\omega C_\pi(1/R_s + g_m + 1/r_\pi)}. \tag{10.6-24}$$

Equations 10.6-22 and 10.6-24 are combined to form the total gain expression:

$$v_o = -g_m \frac{-(v_s/R_s)(1/R_s + g_m + 1/r_\pi)}{1 + j\omega C_\pi(1/R_s + g_m + 1/r_\pi)} \frac{1}{1 + j\omega C_\mu R_c}. \tag{10.6-25}$$

TABLE 10.6-1 SUMMARY OF BJT AMPLIFIER HIGH-FREQUENCY POLE LOCATIONS

Amplifier Type	Pole Locations
Common-emitter	$\dfrac{1}{[C_\pi + C_\mu(1 + g_m R_c')][r_\pi \| (R_s + r_b)]} , \dfrac{1}{C_\mu(1/g_m + R_c')}$
Common-collector or common-emitter with emitter resistor	$\dfrac{1}{\{C_\pi + C_\mu[1 + g_m(R_c + R_e)]\}\{r_\pi \| [(R_s + r_b)/(1 + g_m R_e)]\}}$
Common-base	$\omega_T, \dfrac{1}{C_\mu R_c}$

NOTE: $R_c' = R_c \| r_o$.

There are two simple poles. The input pole is at frequency

$$\omega_{p1} = \frac{r_\pi + g_m r_\pi R_s + R_s}{C_\pi r_\pi R_s} = \frac{r_\pi + (1 + g_m r_\pi)R_s}{C_\pi r_\pi R_s}. \tag{10.6-26}$$

Since $g_m r_\pi = \beta_F$, the expression can be further simplified:

$$\omega_1 = \frac{r_\pi + (1 + \beta_F)R_s}{C_\pi r_\pi R_s} \approx \frac{\beta_F}{C_\pi r_\pi} = \omega_T. \tag{10.6-27}$$

the output pole is at frequency

$$\omega_{p2} = \frac{1}{C_\mu R_c}. \tag{10.6-28}$$

Both poles are at very high frequencies: Common-base amplifier stages are not usually the frequency-limiting elements in a multistage amplifier. A summary of BJT amplifier high frequency pole locations is given in Table 10.6-1.

10.7 HIGH-FREQUENCY DYNAMIC MODELS OF FET

Analysis of the response of JFET and MOSFET circuits at high frequencies is based on accurate modeling of the frequency-dependent performance of the transistors. The FET models used for signal analysis in the low-to-midband frequencies in the saturation region, as presented in Chapter 4, do not contain frequency-sensitive elements and are therefore invariant with respect to changes in frequency. It is therefore necessary to introduce the high-frequency variant of the model to include the frequency-dependent terms.

The current–voltage relationships of the JFET in Chapter 4 and the small-signal models of Chapter 5 were derived only for voltages assumed to change slowly with time. For high-frequency analysis, the relationships must be modified to include the following two effects:

1. The JFET structure acts as a parallel plate capacitor when viewed from the gate and source terminals, with the gate and channel forming the two plates. The plate capacitor separation is the width of the gate-to-channel junction depletion region. A capacitive current will flow when there is a change in the gate-to-source voltage.

2. The majority carriers require a finite transition time to cross the source-to-gate channel. If the gate voltage changes significantly during the time when the majority carriers are traversing the channel, the static expression of the drain current becomes invalid.

There is also a small capacitance in the region between the drain and source formed by the channel and the two terminal regions.[8] There is an additional large gate-to-drain resistance r_{gd} that for all practical purposes is an open circuit. The frequency-dependent components are gate-to-source capacitance C_{gs}, gate-to-drain capacitance C_{gd} (sometimes called the overlap capacitance), and drain-to-source capacitance C_{ds}. Since $C_{gs} \gg C_{ds}$, C_{ds} can usually be ignored.

The frequency-dependent effects due to charge storage in JFETs occurs mainly in the two-gate junctions. The drain–source capacitance C_{ds} is small and therefore does not appreciably affect the high-frequency response of the FET. The two remaining capacitances can be modeled as voltage-dependent capacitors with values determined by the following expressions:

$$C_{gs} = \frac{C_{gs0}}{(1 + |V_{GS}|/\psi_o)^m} \tag{10.7-1}$$

$$C_{gd} = \frac{C_{gd0}}{(1 + |V_{GD}|/\psi_o)^m}. \tag{10.7-2}$$

where

C_{gs0}, C_{gd0} = zero-bias gate–source and gate–drain junction capacitances, respectively (F)

V_{GS}, V_{DS} = quiescent gate–source and drain–source voltages, respectively

m = gate p–n grading coefficient (SPICE default = 0.5)

ψ_o = gate junction (barrier) potential, typically 0.6 V (SPICE default = 1 V)

The frequency-dependent elements for the MOSFET can be obtained in the same manner as the JFET. The gate-to-source capacitance C_{gs} is a function of the rate of change of gate charge with respect to the instantaneous gate voltage. The effect of the electrostatic coupling between the gate and the drain can be represented by the incremental gate-to-drain capacitance C_{gd}. Since both of these capacitances are effected by the gate voltage, the intrinsic capacitance formed between the gate, oxide

---

[8] A detailed discussion of the capacitances for the JFET and MOSFET high-frequency small-signal model is beyond the scope of this book. For more detailed descriptions, the reader is referred to the books (listed as references at the end of this chapter) by M. S. Tyagi, P. R. Gray, and R. G. Meyer and P. Antognetti and G. Massobrio.

layer, and the channel is of critical interest. The capacitance formed by the oxide layer at the gate is defined as

$$C_{ox} = \frac{\epsilon_{ox} WL}{t_{ox}} = C'_{ox} WL, \qquad (10.7\text{-}3)$$

where

C_{ox} = oxide capacitance formed by gate and channel
C'_{ox} = oxide capacitance per unit area
ϵ_{ox} = permittivity of oxide layer (silicon oxide, SiO_2: $3.9\epsilon_o$)
t_{ox} = thickness of oxide layer (separation between gate and channel)
W, L = width and length of channel under gate, respectively

The permittivity of vacuum is $\epsilon_o = 8.851 \times 10^{-12}$ F/m. The oxide capacitance per unit area can be calculated from physical parameters:

$$C'_{ox} = \frac{1}{\mu} \frac{2I_{DSS}}{V_{PO}^2} \quad \text{for depletion MOSFETs} \qquad (10.7\text{-}4a)$$

$$C'_{ox} = \frac{1}{\mu} (2000) \quad \text{for enhancement MOSFETs,} \qquad (10.7\text{-}4b)$$

where μ is the charge mobility (typically 600 cm^2/V-s for n-channel, 200 cm^2/V-s for p-channel.

For a MOSFET operating in saturation, the relevant capacitances for the small-signal high-frequency model are

$$C_{gs} = \tfrac{2}{3} C_{ox} + C_{gs0} W \qquad (10.7\text{-}5)$$

$$C_{gd} = C_{gd0} W, \qquad (10.7\text{-}6)$$

where C_{gs0}, C_{gd0} are the zero-bias gate-source and gate-drain capacitances, respectively (typically $C_{gs0} = C_{gd0} = 3 \times 10^{-12}$ F/m) and are related to C'_{ox}.

The capacitances in the high-frequency small-signal model of the MOSFET are relatively constant over the frequency range. Note also that the MOSFET zero-bias capacitance has dimensions of farads per meter, and in the JFET it has units of farads.

Although the values of the components are different, the JFET and MOSFET share the same small-signal model arrangement shown in Figure 10.7-1.

Since C_{ds} is small compared to C_{gs}, the drain–source capacitance may be ignored in most analysis and design situations, and the simplified model shown in Figure 10.7-2 may be used. Circuit parameters, at specific bias conditions, can be obtained from the manufacturers' data sheets. The data is usually provided in terms of y parameters. More specifically, the common source short-circuit input capacitance C_{iss}, reverse transfer capacitance C_{rss}, and output capacitance C_{oss} are provided. These

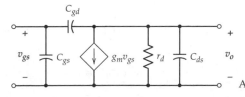

Figure 10.7-1
Accurate FET high-frequency model.

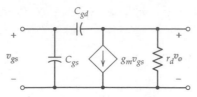

Figure 10.7-2

Simplified FET high-frequency model.

manufacturer-specified capacitances are related to the high-frequency small-signal model parameters by the following relationships:

$$C_{gd} \approx C_{rss}, \tag{10.7-7}$$

$$C_{gs} \approx C_{iss} - C_{rss}, \tag{10.7-8}$$

$$C_{ds} \approx C_{oss} - C_{rss}. \tag{10.7-9}$$

The maximum operating frequency ω_T is the frequency at which the FET no longer amplifies the input signal: That is, the dependent current source $g_m v_{gs}$ is equal to the input current.[9] Using an analysis similar to that found in 10.4 to find the BJT maximum operating frequency,

$$\omega_T = \frac{g_m}{C_{gs} + C_{ds}}. \tag{10.7-10}$$

In general, BJTs have higher maximum operating frequencies that FETs. Two factors are responsible for the lower frequency performance of FETs compared to BJTs:

- For a given area and operating current, the g_m of silicon FETs are less than half of silicon BJTs.
- In MOSFET structures, considerable capacitance is observed at the input due to the oxide layer. In JFETs, semiconductor properties and physical dimensions of the device result in long channel lengths that reduce high-frequency performance.

EXAMPLE 10.7-1   Assume as given an enhancement NMOSFET with parameters

$$K = 2.96 \text{ mA/V}^2, \qquad V_A = 150 \text{ V}, \qquad V_T = 2 \text{ V},$$

$$C_{iss} = 60 \text{ pF} \quad \text{at } V_{GS} = 0 \text{ V},$$

$$C_{oss} = 25 \text{ pF} \quad \text{at } V_{GS} = 0 \text{ V},$$

$$C_{rss} = 5 \text{ pF} \quad \text{at } V_{GS} = 0 \text{ V},$$

$$W = 30 \text{ μm}, \qquad L = 10 \text{ μm}, \qquad \mu = 600 \text{cm}^2/\text{V-s} = 0.6 \text{ m}^2/\text{V-s}$$

operating at $I_D = 5$ mA.

Determine an appropriate small-signal model for the transistor.

---

[9] The overlap resistance r_{gd} and the capacitances C_{gs} and C_{gd} allow for the existence of a significant input current at high frequencies.

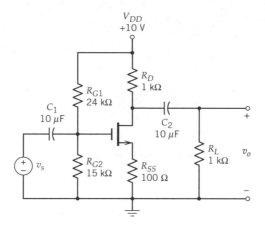

SOLUTION

The first step to modeling any FET (or any transistor circuit) is to determine the quiescent conditions. Since this is the same transistor and circuit as Example 5.8-2, the quiescent conditions have already been determined to be

$$V_{GS} = 3.3 \text{ V}, \qquad I_D = 5 \text{ mA}$$

operating in the saturation region.

The transconductance is

$$g_m = 2\sqrt{I_D K} = 2\sqrt{(5 \times 10^{-3})(2.96 \times 10^{-3})} = 7.7 \text{ mS}.$$

The drain–source resistance of the FET is

$$r_d = \frac{V_A}{I_D} = \frac{150}{5 \times 10^{-3}} = 30 \text{ k}\Omega.$$

The MOSFET capacitances provided are for zero-bias conditions. Therefore, the following zero-bias small-signal capacitors can be found:

$$C_{gd0} = \frac{C_{rss}}{W} = \frac{5}{W} \quad \text{pF/m,}$$

$$C_{gs0} = \frac{C_{iss} - C_{rss}}{W} = \frac{55}{W} \quad \text{pF/m.}$$

In saturation, the small-signal capacitances are

$$C_{gs} = \frac{2}{3}\frac{2KWL}{\mu} + C_{gs0}W = 1.97 \times 10^{-12} + 55 \times 10^{-12} \approx 57 \text{ pF,}$$

$$C_{gd} = C_{gd0}W = 5 \text{ pF.}$$

The complete small-signal model of the circuit is shown.

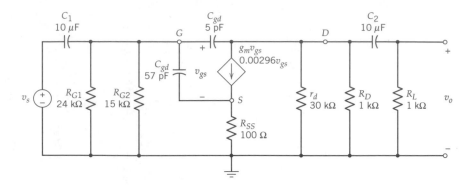

10.8 HIGH-FREQUENCY RESPONSE OF SIMPLE FET AMPLIFIERS

The high-frequency response of simple FET amplifiers is, as with BJT amplifiers, characterized by the amplifier high-frequency pole locations. In this section, common-source and common-drain amplifiers are grouped together for analysis using Miller's theorem. Common-gate amplifiers are treated separately. As in all small-signal analysis, transistor quiescent conditions must be first calculated so that the transistor parameters can be accurately determined. In order to focus discussion on pole frequency determination, it is assumed that the quiescent analysis has been previously performed and that the transistor parameters are well known.

10.8.1 Common-Source and Common-Drain Amplifier High-Frequency Characteristics

The midband AC model of typical general common-source or common-drain amplifier and its high-frequency equivalent, which uses the hybrid-π BJT model, are shown in Figure 10.8-1. The difference in the two amplifier circuit topologies depends only on where the output is taken: v_{od} is the common-source output and v_{os} is the common drain output.

The analysis technique follows the basic technique developed for the common-collector high-frequency analysis of Section 10.6.2. Miller's theorem is one of many

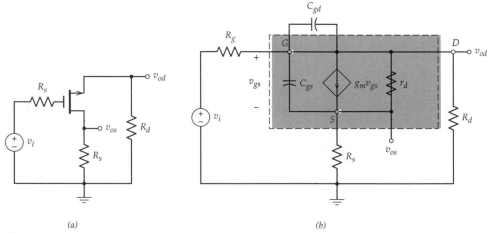

(a)   (b)

Figure 10.8-1

Common-source or common-drain equivalent circuits: (*a*) midband AC equivalent; (*b*) high-frequency equivalent.

possible techniques that can be applied to reduce the analysis complexity for this case. In Figure 10.8-1b, the shaded two-port network has two terminals for which the voltage gain, with respect to the collector, can be approximated as[10]

$$A = \frac{v_{ds}}{v_{gs}} \approx -g_m[r_d||(R_d + R_c)].$$ (10.8-1)

The capacitance C_{gd} is redistributed to the input and output ports using Miller's theorem to form the circuit shown in Figure 10.8-2.

The input capacitance is the sum of C_{gs} and the Miller input capacitance C_1:

$$C_i = C_{gs} + C_1 = C_{gs} + C_{gd}\{1 + g_m[r_d||(R_d + R_s)]\}.$$ (10.8-2)

The output capacitance is the Miller output capacitance C_2[11]:

$$C_o = C_{gd} \frac{1 + g_m[r_d||(R_d + R_s)]}{g_m[r_d||(R_d + R_s)]} \approx C_{gd}.$$ (10.8-3)

For purely resistive loads of typical value, the pole introduced by the output capacitance is typically at a very high frequency beyond the scope of realistic measurement. If the load has a capacitive component, that capacitive component will dominate the determination of the pole location. The pole introduced by the input capacitance is of more significance: The Miller effect has increased the input capacitance and consequently lowered the magnitude of the pole frequency. In the equivalent circuit of Figure 10.8-2, assume that the pole introduced by C_o at a sufficiently high frequency so that C_o can be replaced by an open circuit. The generator input current is given by

$$i_g = v_{gs}(j\omega C_i)$$ (10.8-4)

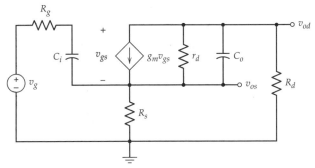

Figure 10.8-2
Miller's theorem applied to general FET amplifier.

---

[10]The complete frequency response analysis of this general FET amplifier is extremely complex. The techniques presented here invoke several assumptions that simplify the analysis procedure without undue loss of accuracy. Here it is assumed that the current through the source and drain resistors is nearly identical and that the pole in the specified Miller voltage gain is at a sufficiently high frequency ($\approx \omega_T$) so as not to alter the results significantly.

[11]Some FET models include a very small drain–source capacitance C_{ds}. In that case, the output capacitance C_o is the sum of C_{gd} and C_{ds}. The remainder of the analysis is unchanged.

The drain current is sufficiently larger than the gate current in a FET so that the voltage at the source terminal, v_{os}, can be approximated as

$$v_{os} \approx \frac{v_{gs} g_m r_d R_s}{r_d + R_S + R_d}. \tag{10.8-5}$$

The generator voltage is the sum of v_{os}, v_{gs} and the voltage across the input resistance:

$$v_g = v_{os} + v_{gs} + i_g(R_g), \tag{10.8-6}$$

which can be rewritten as

$$v_g = \frac{v_{gs} g_m r_d R_s}{r_d + R_s + R_d} + v_{gs} + v_{gs}(j\omega C_i)R_g. \tag{10.8-7}$$

Equation 10.8-7 can be rewritten to obtain v_{gs} in terms of v_g:

$$v_{gs} = \frac{v_g}{g_m r_d R_s/(r_d + R_s + R_d) + 1 + (j\omega C_i)R_g}. \tag{10.8-8}$$

The common drain voltage gain is the combination of Equations 10.8-5 and 10.8-8:

$$\begin{aligned} A_{CD} &= \frac{v_{os}}{v_g} \\ &= \frac{g_m r_d R_s}{g_m r_d R_s + (r_d + R_d + R_s) + j\omega C_i R_g(r_d + R_d + R_s)} \end{aligned} \tag{10.8-9}$$

This gain expression has a simple high-frequency pole located at

$$\omega_{p1} = \frac{1 + g_m r_d R_s/(r_d + R_d + R_s)}{(C_{gs} + C_{gd}\{1 + g_m[r_d||(R_d + R_s)]\})R_g}. \tag{10.8-10}$$

While the pole introduced by C_o has been previously ignored, it can be determined by replacing r_d with the parallel combination of r_d and C_{gd} in Equation 10.8-5:

$$v_{os} \approx \frac{v_{gs} g_m r_d R_s(1 + j\omega r_d C_{gd})}{r_d + (R_s + R_d)(1 + j\omega r_d C_{gd})}. \tag{10.8-11}$$

This expression introduces the second pole for the gain expression at

$$\omega_{p2} = \frac{1}{C_{gd}[r_d||(R_s + R_d)]}. \tag{10.8-12}$$

A very high frequency zero is also introduced. The zero is usually ignored. It is located at

$$\omega_{z1} = \frac{1}{C_{gd} r_d} \tag{10.8-13}$$

The common-source voltage gain expression is of the same form as the common-drain expression where Equation 10.8-5 is replaced with

$$v_{od} \approx \frac{-v_{gs}g_m r_d R_d}{r_d + R_s + R_d} \tag{10.8-14}$$

to obtain a voltage gain expression:

$$
\begin{aligned}
A_{CS} &= \frac{v_{od}}{v_g} \\
&= \frac{-g_m r_d R_d}{g_m r_d R_s + (r_d + R_d + R_s) + j\omega C_i R_g (r_d + R_d + R_s)} \cdot
\end{aligned}
\tag{10.8-15}
$$

This common-source gain expression has the same poles as the common-drain expression.

Two common simplifications of the pole location expressions occur when the source resistor R_s is of zero value (a common source case) or when the drain resistor R_d is zero value (a common-drain case). The pole frequencies are then as follows: *If $R_s = 0$ (common-source without source resistor)*,

$$\omega_{p1} = \frac{1}{\{C_{gs} + C_{gd}[1 + g_m(r_d || R_d)]\}R_g} \tag{10.8-16}$$

$$\omega_{p2} = \frac{1}{C_{gd}(r_d || R_d)} \cdot$$

If $R_d = 0$ (common-drain without drain resistor),

$$\omega_{p1} = \frac{1}{\{C_{gs}/[1 + g_m(r_d || R_s)] + C_{gd}\}R_g} \tag{10.8-17}$$

$$\omega_{p2} = \frac{1}{C_{gd}(r_d || R_s)} \cdot$$

EXAMPLE 10.8-1   Determine the high-frequency poles for the amplifier shown.
     The JFET parameters are

$$I_{DSS} = 6 \text{ mA}, \qquad V_{PO} = -4.7 \text{ V}, \qquad V_A = 100 \text{ V},$$

and the operational parameters are

$$V_{DS} = 4 \text{ V},$$
$$C_{iss} = 4.5 \text{ pF} \quad \text{at } V_{GS} = 0,$$
$$C_{rss} = 1.5 \text{ pF} \quad \text{at } V_{GS} = 0.$$

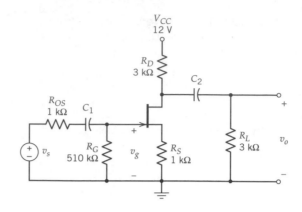

SOLUTION

Find the quiescent conditions to determine if the JFET is in the saturation region; that is, the bias condition must fulfill the following criteria:

$$V_{DS} \geq V_{GS} - V_{PO}.$$

Solve for V_{GS} by finding the voltage across R_s. The drain current I_D is given as

$$I_D = I_{DSS}\left(1 - \frac{V_{GS}}{V_{PO}}\right)^2$$

$$= I_{DSS}\left(1 - \frac{V_G - V_S}{V_{PO}}\right)^2.$$

Substituting $V_G = 0$, and $V_S = I_D R_S$ into the equation for the drain current above,

$$I_D = I_{DSS}\left(1 + \frac{I_D R_S}{V_{PO}}\right)^2.$$

Solving for I_D in the second-order equation above yields $I_D = 2$ mA. Therefore,

$$V_{GS} = V_G - V_S = 0 - I_D R_s = -0.002(1000) = -2 \text{ V}.$$

The condition for operation in the saturation is confirmed:

$$V_{DS} \geq V_{GS} - V_{PO} \quad \text{or} \quad 4 \geq -2 - (-4.7) = 2.7.$$

The small-signal parameters for the JFET are

$$g_m = \frac{-2I_D}{V_{PO} - V_{GS}} = \frac{-0.002}{-4.7 - (-2)} = 1.48 \text{ mS,}$$

$$r_d = \frac{V_A}{I_D} = \frac{100}{0.002} = 50 \text{ k}\Omega.$$

The zero-bias small-signal capacitors are

$$C_{gs0} \approx C_{iss} - C_{rss} = 4.5 \text{ pF} - 1.5 \text{ pF} = 3 \text{ pF},$$

$$C_{gd0} \approx C_{rss} = 1.5 \text{ pF}.$$

The small-signal capacitors at the quiescent point are

$$C_{gs} = \frac{C_{gs0}}{\sqrt{1 + |V_{GS}|/\psi_o}} = \frac{3 \times 10^{-12}}{\sqrt{1 + 2/0.6}} = 1.44 \text{ pF},$$

$$C_{gd} = \frac{C_{gd0}}{\sqrt{1 + |V_{GD}|/\psi_o}} = \frac{1.5 \times 10^{-12}}{\sqrt{1 + 6/0.6}} = 0.452 \text{ pF}.$$

Substituting values into the previously derived mathematical expressions for the poles of a common-source amplifier yields

$$\omega_{p1} = \frac{1 + g_m r_d R_S/[R_S + r_d + (R_D||R_L)]}{[C_{gs} + C_{gd}(1 + g_m\{r_d||[R_S + (R_D||R_L)]\})](R_G||R_{OS})}$$

$$= 692.6 \text{ Mrad/s} \Rightarrow 110.2 \text{ MHz},$$

$$\omega_{p2} = \frac{1}{C_{gd}\{r_d||[R_S + (R_D||R_L)]\}}$$

$$= 929.2 \text{ Mrad/s} \Rightarrow 147.9 \text{ MHz}$$

Since neither pole is dominant, the high cutoff frequency must be calculated using both poles:

$$\omega_H = (692.8 \text{ Mrad/s})\left\{ \sqrt{\frac{-(1 + k^2) + \sqrt{(1 + k^2)^2 + 4k^2}}{2}} \right\}$$

$$= 447.4 \text{ Mrad/s} \Rightarrow 71.2 \text{ MHz}$$

10.8.2 Common-Gate Amplifier High-Frequency Characteristics

The midband AC model of typical common-gate amplifier and its high-frequency equivalent are shown in Figure 10.8-3.

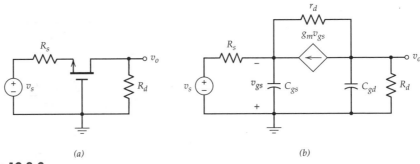

(a)   (b)

Figure 10.8-3
Common-gate equivalent circuits: (a) midband AC equivalent; (b) high-frequency equivalent.

Common-gate amplifiers are similar to common-base amplifiers in that they do not suffer from performance-reducing Miller effects. The bridging resistance r_d does present problems in the analysis of the circuit. One method of analysis that brings results fairly quickly for the common-base circuit is the node-voltage method. At the drain node of the FET, KCL is applied:

$$v_o G_d + v_o(j\omega C_{gd}) + g_m v_{gs} + (v_o + v_{gs})g_d = 0. \qquad (10.8\text{-}18)$$

The shorthand notation of replacing inverse resistance with conductance (signified by the letter g with the appropriate subscript) has been used to simplify analytic representation. At the source node of the FET a similar operation is performed:

$$(v_s + v_{gs})G_s + v_{gs}(j\omega C_{gs}) + g_m v_{gs} + (v_o + v_{gs})g_d = 0. \qquad (10.8\text{-}19)$$

These two nodal equations are combined (eliminating v_{gs}) to form the gain expression for the common-gate amplifier:

$$A_V = \frac{v_o}{v_s} = \frac{G_s}{[(G_d + g_d + j\omega C_{gd})/(g_m + g_d)](G_s + g_m + j\omega C_{gs}) - g_d}. \qquad (10.8\text{-}20)$$

In this gain expression, it should be noted that the output resistance of the FET, r_d, is typically much larger than the circuit resistors R_s and R_d (as well as the inverse of g_m). It is therefore reasonable to approximate the gain expression of Equation 10.8-20 by ignoring the small negative term in the denominator:

$$A_V \approx \frac{G_s(g_m + g_d)}{(G_d + g_d + j\omega C_{gd})(G_s + g_m + j\omega C_{gs})}.$$

This approximate gain expression for the common-gate amplifier has poles at

$$\omega_{p1} = \frac{R_s}{C_{gs}(1 + g_m R_s)}, \qquad \omega_{p2} = \frac{1}{C_{gd}(r_d||R_d)}.$$

These two poles are at very high frequencies: As with the common base amplifier, the common gate amplifier is not usually the frequency-limiting element in a multistage amplifier. A summary of FET amplifier high frequency pole location is given in Table 10.8-1.

TABLE 10.8-1 SUMMARY OF FET AMPLIFIER HIGH-FREQUENCY POLE LOCATIONS

Amplifier Type	Pole Locations	
Common-drain or common-source	$\dfrac{1 + g_m r_d R_s/(r_d + R_d + R_s)}{(C_{gs} + C_{gd}\{1 + g_m[r_d||(R_d + R_s)]\})R_g{}'}$	$\dfrac{1}{C_{gd}[r_d||(R_s + R_d)]}$
Common-gate	$\dfrac{R_s}{C_{gs}(1 + g_m R_s)}$	$\dfrac{1}{C_{gd}(r_d||R_d)}$

Multistage amplifier frequency-domain analysis is a combination of techniques shown in the previous sections of this chapter and the analysis and design techniques shown in Chapter 6. The important concepts from Chapter 6 and the previous sections of this chapter that will be used for multistage amplifier analysis are:

- The total voltage gain of a cascade-connected amplifier can be expressed as a product of gains of the individual stages and simple voltage divisions.
- Each stage presents a load to the previous stage: Its input resistance is part of the total load that is apparent to the previous stage.
- The total input resistance or total output resistance may be modified by the interaction of individual stages.

The analysis procedure used for designing multistage amplifiers is

- Perform midband gain analysis of the total circuit and of each stage taking into account the load presented by the input resistance of the next stage.
- Perform low-frequency analysis of the circuit. In most cases, careful application of the time constant approach will yield a good approximation of the low cutoff frequency.
- Perform high-frequency analysis using the high-frequency model of the circuit. When appropriate, use Miller's theorem to create a Miller's equivalent circuit of the amplifier. In most cases, the high-frequency output capacitance of the Miller's equivalent model may be ignored since the pole associated with the output capacitance is significantly higher than the pole for the Miller's equivalent input capacitance. This greatly simplifies analysis and design without sacrificing a great deal of accuracy.

The analysis of multistage amplifier high-frequency response is greatly simplified by assuming noninteracting capacitances in the high-frequency model. This assumption is particularly important when determining the gain of an amplifier stage loaded by the input impedance of a following stage—often a complex impedance. That impedance is evaluated at midband frequencies, thus eliminating the reactive component from the analysis. Using this assumption, the input impedance of the amplifier stage at the output of the stage under consideration presents a purely resistive load.

10.9.1 Multistage Amplifier with Coupling Capacitor between Stages

A common method found in multistage amplifier design makes use of coupling capacitors between stages. This method has the advantage that the bias condition of each stage is unaffected by the others. The disadvantage is the potential degradation in amplifier low-frequency response; that is, the low cutoff frequency may increase without careful selection of capacitor values. Additionally, the use of coupling capacitors between stages does not lend itself to integrated circuit implementations of the circuit since large capacitors require large areas, or "real estate,"[12] on the chip.

Figure 10.9-1 is a two-stage common-emitter cascaded amplifier.

---

[12]Areas on integrated circuits are sometimes referred to as "real estate" since a chip may be designed with a near capacity of number of components, and space may not be available.

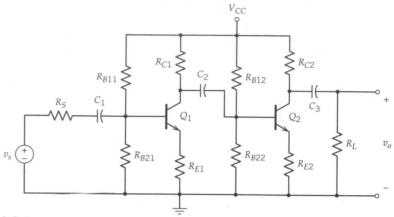

Figure 10.9-1

Two-stage common-emitter cascaded amplifier.

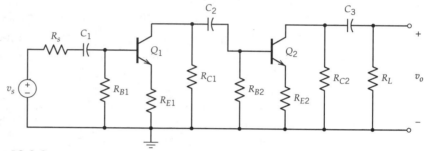

Figure 10.9-2

The AC model of Figure 10.9-1.

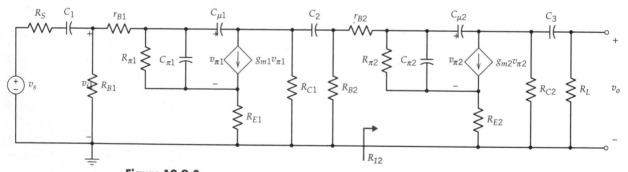

Figure 10.9-3

All-frequency hybrid-π small-signal model of amplifier in Figure 10.9-1.

The AC model of the circuit is shown in Figure 10.9-2, where $R_{B1} = R_{B11}||R_{B21}$ and $R_{B2} = R_{B12}||R_{B22}$.

The midband gain of the circuit has been found in Example 6.1-1. However, the gain expressions in Example 6.1-1 are in terms of h parameters. Since hybrid-π parameters are being used for the high-frequency analysis, the gain is found in terms of the hybrid-π parameters from the complete hybrid-π small signal model of the circuit shown in Figure 10.9-3. To find the midband gain, the bypass and coupling capacitors C_1, C_2, and C_3 are short circuited, and the high-frequency internal capacitors are open circuited.

The midband gain of the amplifier is

$$A_{Vm} = \frac{v_o}{v_s} = \left(\frac{v_{i1}}{v_s}\right)\left(\frac{v_{o1}}{v_{i1}}\right)\left(\frac{v_o}{v_{o1}}\right), \tag{10.9-1}$$

where

$$\frac{v_{i1}}{v_s} = \frac{R_{B1}}{R_{B1} + R_S}, \tag{10.9-2}$$

$$\frac{v_{o1}}{v_{i1}} = -\frac{g_{m1}r_{\pi1}[(R_{C1}||R_{B2})||R_{i2}]}{r_{b1} + r_{\pi1} + (1 + g_{m1}r_{\pi1})R_{E1}}, \tag{10.9-3}$$

$$\frac{v_o}{v_{o1}} = -\frac{g_{m2}r_{\pi2}(R_{C2}||R_L)}{R_{i2}}, \tag{10.9-4}$$

and

$$R_{i2} = r_{b2} + r_{\pi2} + (1 + g_{m2}r_{\pi2})R_{E2}.$$

The low cutoff frequency is found by setting all high-frequency transistor capacitors as open circuits and following the time constant analysis of Section 10.3. The low-frequency poles of the amplifier in Figure 10.9-1 are

$$\omega_{L1} = \frac{1}{C_1(R_S + \{R_{B1}||[r_{b1} + r_{\pi1} + (1 + g_{m1}r_{\pi1})R_{E1}]\})}, \tag{10.9-5}$$

$$\omega_{L2} = \frac{1}{C_2[R_{C1} + (R_{B2}||R_{i2})]}, \tag{10.9-6}$$

$$\omega_{L3} = \frac{1}{C_3(R_{C2} + R_L)}. \tag{10.9-7}$$

The high cutoff frequency is found by applying Miller's theorem and the equations derived in Section 10.6 (10.8 for FETs) for each stage. Equation 10.6-19 is modified to take into account the circuit in Figure 10.9-1:

$$\omega_{H1} \approx \left[(C_{\pi1} + C_{\mu1}\{1 + g_{m1}[(R_{C1}||R_{B2}||R_{i2}) + R_{E1}]\})\left(r_{\pi1}\left|\left|\frac{r_{b1} + (R_S||R_{B1})}{1 + g_{m1}R_{E1}}\right)\right.\right]^{-1}, \tag{10.9-8a}$$

$$\omega_{H2} \approx \left[(C_{\pi2} + C_{\mu2}\{1 + g_{m2}[(R_{C2}||R_L) + R_{E2}]\})\left(r_{\pi2}\left|\left|\frac{r_{b2} + (R_{C1}||R_{B2})}{1 + g_{m2}R_{E2}}\right)\right.\right]^{-1}. \tag{10.9-8b}$$

The poles associated with the Miller's equivalent output capacitances for each stage are not calculated since they are significantly higher than the poles of Equations 10.9-8a and 10.9-8b.

EXAMPLE 10.9-1 The two-stage cascade-connected amplifier of Example 6.1-1 is as shown. The two silicon BJTs have characteristic parameters:

$$h_{fe} \approx \beta_F = 150, \qquad r_b = 30\ \Omega, \qquad C_{\mu1} = C_{\mu2} = 3\ \text{pF},$$

$$\omega_{T1} = \omega_{T2} = 2\pi(400)\ \text{Mrad/s}, \qquad V_A = 350.$$

Determine the low and high cutoff frequencies of the circuit.

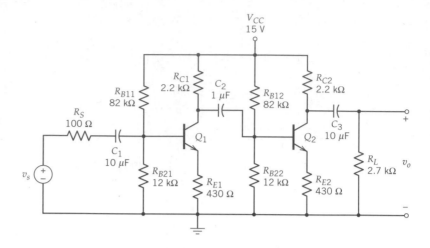

SOLUTION

The determination of the multistage amplifier performance follows the same basic steps that were derived in Chapter 6:

1. Model the transistors with the appropriate DC model.
2. Determine the circuit quiescent conditions. Verify forward-active region for BJTs or saturation region for FETs.
3. Determine transistor AC parameters from quiescent conditions.
4. Create an AC equivalent circuit.
5. Determine the midband performance of each amplifier stage by
   a. replacing the transistors by their respective AC models or
   b. using previously derived results for the circuit topology.
6. Combine stage performance quantities to obtain total midband gain.
7. Perform low-frequency analysis to determine the low cutoff frequency using the small-signal AC model with all internal (high-frequency) capacitors open circuited.
8. Perform high-frequency analysis to determine the high cutoff frequency using the small signal AC model including the internal (high-frequency) capacitors and short circuiting all external (coupling and bypass) capacitors.
9. Combine results of DC, midband, low-frequency, and high-frequency analysis to obtain total circuit performance.

The DC analysis was performed in Example 6.1-1 and the result for both collector currents and collector–emitter voltages are

$$I_C = 2.427 \text{ mA}, \qquad V_{CE} = 8.64 \text{ V}.$$

Therefore, the small signal parameters are

$$g_{m1} = g_{m2} = g_m = \frac{|I_C|}{V_t} = \frac{2.417 \times 10^{-3}}{26 \times 10^{-3}} \approx 93 \text{ mS,}$$

$$r_{\pi 1} = r_{\pi 2} = \frac{\beta_F}{g_m} = \frac{150}{0.093} = 1.61 \text{ k}\Omega,$$

$$C_{\pi 1} = C_{\pi 2} = \frac{g_m}{\omega_T} - C_\mu = \frac{0.093}{2\pi(400 \times 10^3)} - 3 \times 10^{-12} = 37 \text{ nF.}$$

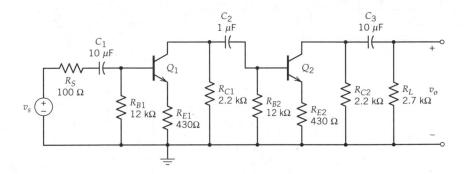

The AC model of the circuit is as shown, where $R_{B1} = R_{B11}||R_{B21}$ and $R_{B2} = R_{B12}||R_{B22}$.
Since V_A is very large, the output resistance h_{oe}^{-1} is very large and can be approximated as an infinite resistance. The all-frequency hybrid-π small-signal model of the amplifier is as shown, where $R_{i2} = r_{b2} + r_{\pi2} + (1 + g_{m2}r_{\pi2})R_{E2} = 66.6$ kΩ.

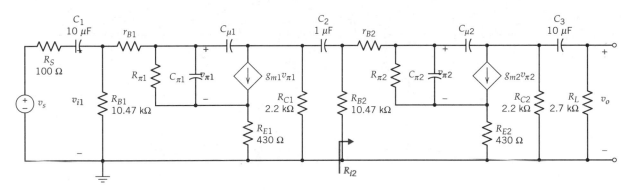

The low- and high-frequency poles are found by applying Equations 10.9-5 to 10.9-8.

$$\omega_{L1} = \frac{1}{C_1(R_S + \{R_{B1}||[r_{b1} + r_{\pi1} + (1 + g_{m1}r_{\pi1})R_{E1}]\})} \Rightarrow 9.74 \text{ rad/s} = 1.6 \text{ Hz},$$

$$\omega_{L2} = \frac{1}{C_2[R_{C1} + (R_{B2}||R_{i2})]} = 89 \text{ rad/s} \Rightarrow 14 \text{ Hz},$$

$$\omega_{L3} = \frac{1}{C_3(R_{C2} + R_L)} = 20.4 \text{ rad/s} \Rightarrow 3.2 \text{ Hz},$$

$$\omega_{H1} \approx \left[(C_{\pi1} + C_{\mu1}\{1 + g_{m1}[(R_{C1}||R_{B2}||R_{i2}) + R_{E1}]\})\left(r_{\pi1} \left\| \frac{r_{b1} + (R_S||R_{B1})}{1 + g_{m1}R_{E1}} \right)\right]^{-1}$$

$$= 8.46 \text{ Mrad/s} \Rightarrow 1.35 \text{ MHz},$$

$$\omega_{H2} \approx \left[(C_{\pi2} + C_{\mu2}\{1 + g_{m2}[(R_{C2}||R_L) + R_{E2}]\})\left(r_{\pi2} \left\| \frac{r_{b2} + R_{C1}||R_{B2}}{1 + g_{m2}R_{E2}} \right)\right]^{-1}$$

$$= 624.5 \text{ krad/s} \Rightarrow 99.4 \text{ kHz}.$$

where ω_{L2} is the dominant low-frequency pole. The lower 3-dB
frequency is 14 Hz .
ω_{H2} is the dominant high-frequency pole. The high cutoff
frequency is 99.4 kHz .

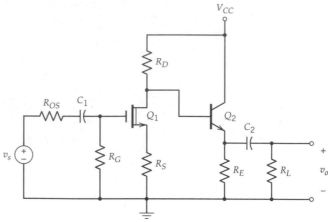

Figure 10.9-4
A DC (direct) coupled two-stage amplifier.

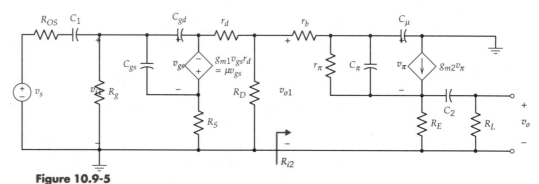

Figure 10.9-5
Complete hybrid-π small-signal model of Figure 10.9-4.

10.9.2 Multistage Amplifier with DC (Direct) Coupling between Stages

In Example 10.9-1, the coupling capacitor between the two stages established the dominant low-frequency pole and therefore the low cutoff frequency. The effect of coupling capacitors between amplifier stages on the low-frequency response can be eliminated by DC coupling the stages. The midband analysis of a DC (or direct) coupled two-stage common-source–common-collector amplifier (Figure 10.9-4) was shown in Example 6.1-2.

The complete hybrid-π small-signal model of the direct-coupled two-stage amplifier in Figure 10.9-4 is shown in Figure 10.9-5. The analysis of the midband gain, low cutoff frequency, and high cutoff frequency are found in the manner as the capacitor-coupled amplifiers.

Following the analysis of the capacitor-coupled multistage amplifier, and the results in Sections 10.6 and 10.8, the midband gain of the direct-coupled common-source–common-collector amplifier is

$$A_{Vm} = \frac{v_o}{v_s}$$

$$= -\frac{R_G}{R_G + R_{OS}} \frac{(R_D\|R_{i2})\mu}{r_d + (R_D\|R_{i2}) + (1 + \mu)R_s} \frac{R_{i2} - (r_b + r_\pi)}{R_{i2}},$$

$$(10.9\text{-}9)$$

where

$$\mu = g_{m1}r_d, \qquad R_{i2} = r_b + r_\pi + (1 + g_{m2}r_\pi)(R_E||R_L).$$

The low-frequency poles are calculated from the product of the resistive discharge paths, calculated using Thevenin equivalent resistances, and the external capacitors:

$$\omega_{l1} = \frac{1}{C_1(R_{OS} + R_G)}, \tag{10.9-10}$$

$$\omega_{l2} = \left\{ C_2 \left[R_L + \left(R_E \,\middle\|\, \frac{r_b + r_\pi + r_d + (1 + g_{m1}r_d)R_S}{1 + g_{m2}r_\pi} \right) \right] \right\}^{-1} \tag{10.9-11}$$

Using Miller's theorem, the high-frequency Miller's equivalent small-signal model is found (Figure 10.9-6).

The high-frequency capacitors are found using Miller's theorem:

$$C_{i1} = C_{gs} + C_{gd}(1 - A_{v1})$$
$$= C_{gs} + C_{gd} \left[1 + \frac{\mu(R_D||R_{i2})}{r_d + (R_D||R_{i2}) + (1 + \mu)R_S} \right], \tag{10.9-12}$$

$$C_{o1} = C_{gd}\left(1 - \frac{1}{A_{v1}}\right)$$
$$= C_{gd} \left[1 + \frac{r_d + (R_D||R_{i2}) + (1 + \mu)R_S}{\mu(R_D||R_{i2})} \right], \tag{10.9-13}$$

$$C_{i2} = C_\pi + C_\mu(1 - A_{v2})$$
$$= C_\pi + C_\mu \left[1 - \frac{R_{i2} - (r_b + r_\pi)}{R_{i2}} \right], \tag{10.9-14}$$

$$C_{o2} = C_\mu\left(1 - \frac{1}{A_{v2}}\right)$$
$$= C_\mu \left[1 - \frac{R_{i2}}{R_{i2} - (r_b + r_\pi)} \right]. \tag{10.9-15}$$

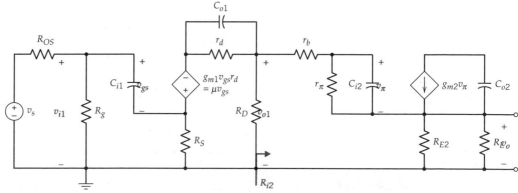

Figure 10.9-6
Miller's equivalent high-frequency model of Figure 10.9-4.

The high-frequency poles are found by applying Equation 10.8-14 to 10.8-10 and 10.8-12 for the common-source stage and Equation (10.6-18) for the common-collector stage, and making the appropriate substitutions for the circuit in Figure 10.9-6:

$$\omega_{H1} = \frac{1 + \mu(R_D||R_{i2})/[r_d + R_S + (R_D||R_{i2})]}{C_{i1}(R_{OS}||R_G)}, \tag{10.9-16}$$

$$\omega_{H2} = \frac{1}{C_{o1}\{r_d||[R_S + (R_D||R_{i2})]\}}, \tag{10.9-17}$$

$$\omega_{H3} = \left\{ C_{i2}\left[r_\pi \middle\| \left(\frac{r_b + [R_D||(r_d + R_S)]}{1 + g_{m2}R_E} \right) \right] \right\}^{-1}. \tag{10.9-18}$$

Since the high-frequency pole due to the Miller's equivalent output capacitance of the BJT is far from the other poles, the high cutoff frequency is not affected by the high-frequency pole due to Miller's equivalent output capacitance.

10.9.3 Darlington Pair

The frequency response of amplifiers using the Darlington pairs can be found using the techniques developed thus far in this chapter. Figure 10.9-7 is a common-emitter amplifier using the dual common-collector composite transistor.

The complete hybrid-π small-signal equivalent is shown in Figure 10.9-8. The analysis of this amplifier follow techniques developed in this chapter. The midband gain is found by treating all external (bypass and coupling) capacitors as short circuits and all high-frequency capacitors as open circuits. The gain expression is a little different than those that have been derived thus far. The midband gain has additive terms because the two dependent current sources, $g_{m1}v_{\pi1}$ and $g_{m2}v_{\pi2}$, add to yield the amplifier output current. The midband gain of the amplifier is

$$A_{Vm} = -\frac{v_{i1}}{v_s}(g_{m1}v_{\pi1} + g_{m2}v_{\pi2})R_O, \tag{10.9-19}$$

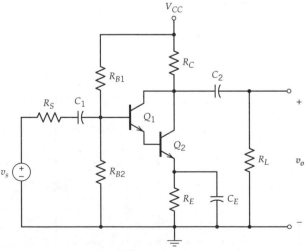

Figure 10.9-7
Common-emitter amplifier using Darlington pair.

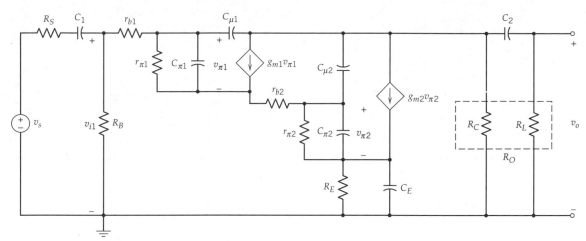

Figure 10.9-8
All-frequency hybrid-π small-signal model of Figure 10.9-7.

where

$$R_O = R_C \| R_L$$

for midband and high-frequency analysis,

$$R_B = R_{B1} \| R_{B2},$$

and

$$\frac{v_{i1}}{v_s} = \frac{R_B}{R_B + R_S}.$$

The controlling voltages for the dependent current sources are

$$v_{\pi 1} = \frac{v_{i1} r_{\pi 1}}{r_{b1} + (1 + g_{m1} r_{\pi 1})(r_{b2} + r_{\pi 2})}, \qquad (10.9\text{-}20)$$

$$v_{\pi 2} = \frac{v_{i1} r_{\pi 2}(1 + g_{m1} r_{\pi 1})}{r_{b1} + r_{\pi 1} + (1 + g_{m1} r_{\pi 1})(r_{b2} + r_{\pi 2})}. \qquad (10.9\text{-}21)$$

The complete expression for the midband gain is found by substituting Equations 10.9-20 and 10.9-21 into Equation 10.9-19,

$$A_{vm} = -\frac{R_B R_O}{R_B + R_S} \times$$

$$\left(\frac{g_{m1} r_{\pi 1}}{r_{b1} + (1 + g_{m1} r_{\pi 1})(r_{b2} + r_{\pi 2})} + \frac{g_{m2} r_{\pi 2}(1 + g_{m1} r_{\pi 1})}{r_{b1} + r_{\pi 1} + (1 + g_{m1} r_{\pi 1})(r_{b2} + r_{\pi 2})} \right).$$

$$(10.9\text{-}22)$$

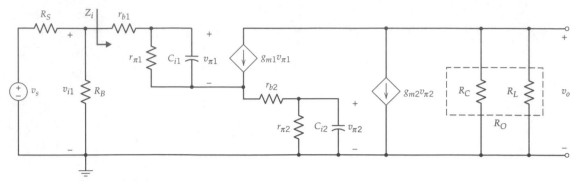

Figure 10.9-9
High-frequency Miller's equivalent circuit of Figure 10.9-7.

The high-frequency model is simplified using Miller's theorem to yield the high-frequency Miller's equivalent small-signal model shown in Figure 10.9-9. The high-frequency Miller's equivalent output capacitances are not included since the poles associated with these capacitances are not dominant.

The pole locations are found by determining the input impedance Z_i in Figure 10.9-9,

$$
Z_i = r_{b1} + \left(r_{\pi 1} \,\middle\|\, \frac{1}{j\omega C_{i1}}\right) + (1 + g_{m1})\left[r_{b2} + \left(r_{\pi 2} \,\middle\|\, \frac{1}{j\omega C_{i2}}\right)\right]
$$

$$
= \frac{\left[r_{b1}\left(\dfrac{1}{r_\pi \pi_1 C_{i1}} + j\omega\right) + \dfrac{r_{\pi 1}}{r_{\pi 1} C_{i1}}\right]\left(\dfrac{1}{r_{\pi 2} C_{i2}} + j\omega\right) + (1 + g_{m1}r_{\pi 1})\left[r_{b2}\left(\dfrac{1}{r_{\pi 2} C_{i2}} + j\omega\right) + \dfrac{r_{\pi 2}}{r_{\pi 2} C_{i2}}\right]\left(\dfrac{1}{r_{\pi 1} C_{i1}} + j\omega\right)}{\left(\dfrac{1}{r_{\pi 2} C_{i2}} + j\omega\right)\left(\dfrac{1}{r_{\pi 1} C_{i1}} + j\omega\right)}.
$$

$$(10.9\text{-}23)$$

From Equation 10.9-23, the pole locations of the Darlington circuit in Figure 10.9-7 are

$$
\frac{1}{r_{\pi 1} C_{i1}}, \qquad \frac{1}{r_{\pi 2} C_{i2}}.
$$

10.9.4 Cascode Amplifier

A cascode amplifier and its AC equivalent circuit are shown in Figure 10.9-10a and b where Q_1 forms the common-emitter stage and Q_2 forms the common-base stage. The analysis of the cascode amplifier is simplified because, as shown in Section 10.6.3, the common-base amplifier high-frequency cutoff is very high. Therefore, the common-base stage does not limit the high-frequency operation of the cascode amplifier.

The analysis and design for the low cutoff frequency follows the technique demonstrated in the previous sections of this chapter. Since the common-base stage does not affect the high-ferquency response of the cascode amplifier, the simplified high-frequency small-signal model shown in Figure 10.9-11 is used to find the dominant high-frequency pole.

The equivalent load resistance R_O of the common-emitter stage is the Thévenin input resistance of the common base stage,

$$
R_O = \frac{r_{\pi 2} + r_{b2}}{1 + g_{m2}r_{\pi 2}}. \tag{10.9-24}
$$

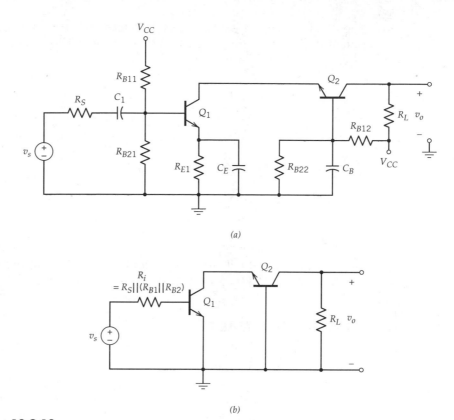

(a)

(b)

Figure 10.9-10
(a) Cascode amplifier. (b) The AC model of the cascode amplifier.

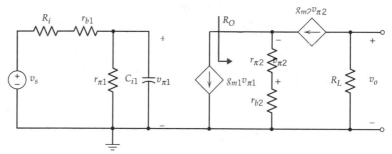

Figure 10.9-11
Simplified high-frequency model of cascode amplifier.

The midband gain of the amplifier is

$$A_{Vm} = -g_{m1}R_O = -g_{m1}\frac{r_{\pi 2} + r_{b2}}{1 + g_{m2}r_{\pi 2}}. \qquad (10.9\text{-}25)$$

The Miller's equivalent high-frequency input capacitor is

$$C_{i1} = C_{\pi 1} + C_{\mu 1}(1 + g_{m1}R_O). \qquad (10.9\text{-}26)$$

The dominant high-frequency pole is determined from the following transfer function of the input portion of the high-frequency small-signal model of Figure 10.9-11:

$$\frac{v_{\pi 1}}{v_s} = \frac{r_{\pi 1} \| (1/j\omega C_{i1})}{R_i + r_{b1} + [r_{\pi 1} \| (1/j\omega C_{i1})]}$$

$$= \frac{1}{C_1(R_i + r_{b1})} \frac{1}{j\omega + (r_{\pi 1} + R_i + r_{b1})/[C_1 r_{\pi 1}(R_i + r_{b1})]}. \tag{10.9-27}$$

Therefore, the high cutoff frequency is

$$\omega_H = \frac{r_{\pi 1} + R_i + r_{b1}}{C_1 r_{\pi 1}(R_i + r_{b1})}. \tag{10.9-28}$$

The other high-frequency poles are significantly higher in frequency than Equation 10.9-28 due to the small value of the Miller equivalent output capacitance of Q_1 and the very high cutoff frequency of the common base stage.

10.10 CONCLUDING REMARKS

Frequency response of amplifiers have been discussed in this chapter. The distortion caused by frequency limitations was explored in the time domain. A demonstration of the effect of the amplifier transfer function (gain and phase response) on the input signal confirmed that frequency distortion can dramatically alter the output signal shape when compared to its input. Low and high pass responses to a pulse input were explored and the concept of sag was introduced.

For multipole electronic systems, the difficulty in analytically finding the low and high cutoff frequencies was resolved through simple formulas and approximations of the cutoff frequencies. This technique utilized the dominant pole of electronic systems to readily determine the cutoff frequency. Dominant poles were defined as those poles that are at least a factor of 4 from the next nearest pole frequency.

The low-frequency response of an electronic amplifier was shown to be a function of the external coupling and bypass capacitors of an amplifier. Although in some instances the capacitors interacted with each other in the circuit, a method was developed for approximating the low-frequency poles by analyzing the poles associated with individual capacitors. This technique relied on the dominant pole concept.

High-frequency hybrid-π models were developed for the BJT and the FET. The high-frequency models included capacitances that determine the high-frequency operation of the devices. Miller's theorem was developed to simplify the analysis of the high-frequency models.

Single and multistage amplifiers were analyzed using the techniques developed in this chapter. The analysis and design of amplifiers followed the following procedure:

1. Model the transistors with the appropriate DC model.
2. Determine the circuit quiescent conditions. Verify forward-active region for BJTs or saturation region for FETs.
3. Determine transistor AC (hybrid-π or h) parameters from quiescent conditions.
4. Create AC equivalent circuit.

TABLE 10.10-1 CONVERSION OF BJT HIGH-FREQUENCY MODELING PARAMETERS

Source of Parameters	Conversion to Hybrid-π Model Parameters	Conversion to SPICE Parameters
Manufacturer's Data Books f_T @ I_{CT} - gain bandwidth product C_{obo} @ V_{CB} - output capacitance C_{ibo} @ V_{EB} - input capacitance	$C_\mu = C_{obo}$ $C_\pi = \dfrac{g_m}{2\pi f_T} - C_\mu$	$CJC = C_{obo}\left(1 + \dfrac{V_{CB}}{\psi}\right)^m$ $CJE = C_{ibo}\left(1 + \dfrac{V_{EB}}{\psi}\right)^m$ $TF = \dfrac{1}{2\pi f_T} - \dfrac{\eta V_t}{|I_{CT}|}\left[\dfrac{CJE}{\left(1 - \dfrac{0.7}{\psi}\right)^m} + C_{obo}\right]$
SPICE Models CJE - zero-bias base-emitter capacitance CJC - zero-bias base-collector capacitance TF - forward transit time	$C_\pi = \dfrac{CJC}{\left(1 + \dfrac{V_{CB}}{\psi}\right)^m}$ $C_\pi = g_m TF + \dfrac{CJE}{\left(1 - \dfrac{0.7}{\psi}\right)^m}$	$VJE \approx VJC \approx \psi \approx 0.75$ $MJE \approx MJC \approx m \approx 0.33$
IC Design Parameters C_{jeo} - zero-bias base-emitter capacitance $C_{\pi o}$ - zero-bias base-collector capacitance τ_F - forward transit time	$C_\mu = \dfrac{C_{jeo}}{\left(1 + \dfrac{V_{CB}}{\psi}\right)^m}$ $C_\pi = g_m \tau_F + \dfrac{C_{jeo}}{\left(1 - \dfrac{0.7}{\psi}\right)^m}$	$CJE = C_{ieo}$ $CJC = C_{\mu o}$ $TF = \tau_F$

NOTES: (1) f_T is assumed to be approximately constant in this text. It has a non-linear dependent on I_C.
(2) The values for ψ and m are device dependent. Typical values are the indicated SPICE default values:
(3) The subscripted junction voltages are for npn BJTs. The equations for pnp BJTs are identical with these subscripts reversed.

TABLE 10.10-2 CONVERSION OF JFET HIGH-FREQUENCY MODELING PARAMETERS

Source of Parameters	Conversion to Hybrid-π Model Parameters	Conversion to SPICE Parameters
Manufacturer's Data Books C_{iss} @ $V_{GS}=0$ - input short-circuit capacitance C_{rss} @ $V_{GS}=0$ - reverse transfer capacitance	$C_{gd} = \dfrac{C_{rss}}{\left(1 + \dfrac{\lvert V_{GD}\rvert}{\psi}\right)^{m}}$ $C_{gs} = \dfrac{C_{iss} - C_{rss}}{\left(1 + \dfrac{\lvert V_{GS}\rvert}{\psi}\right)^{m}}$	$CGD = C_{rss}$ $CGS = C_{iss} - C_{rss}$
SPICE Models CGD - zero-bias gate-drain capacitance CGS - zero-bias gate-source capacitance	$C_{gd} = \dfrac{CGD}{\left(1 + \dfrac{\lvert V_{GD}\rvert}{\psi}\right)^{m}}$ $C_{gs} = \dfrac{CGS}{\left(1 + \dfrac{\lvert V_{GS}\rvert}{\psi}\right)^{m}}$	$PB = \psi \approx 0.6$ $M = m \approx 0.5$
IC Design Parameters C_{gdo} - zero-bias gate-drain capacitance C_{gso} - zero-bias gate-source capacitance	$C_{gd} = \dfrac{C_{gdo}}{\left(1 + \dfrac{\lvert V_{GD}\rvert}{\psi}\right)^{m}}$ $C_{gs} = \dfrac{C_{gso}}{\left(1 + \dfrac{\lvert V_{GS}\rvert}{\psi}\right)^{m}}$	$CGD = C_{gdo}$ $CGS = C_{gso}$

NOTES: (1) The values for ψ and m are device dependent. Typical values are: $\psi = 0.6$ V and m = 0.5. The SPICE default values are: $\psi = 1.0$ V and m = 0.5.

(2) The subscripted junction voltages are for n-channel JFETs. The equations for p-channel JFETs are identical with these subscripts reversed.

TABLE 10.10-3 CONVERSION OF MOSFET HIGH-FREQUENCY MODELING PARAMETERS

Source of Parameters	Conversion to Small-Signal Parameters	Conversion to SPICE Parameters
Manufacturer's Data Books C_{iss} @ V_{GS} - input capacitance C_{rss} @ V_{GS} - reverse transfer capacitance C_{oss} @ V_{GS} - output capacitance	$C_{GS} \approx C_{iss} - C_{rss}$ $C_{GD} \approx C_{rss}$ $C_{DS} \approx C_{oss} - C_{rss}$	$TOX = \dfrac{2\epsilon_{OX}\,WL}{3(C_{iss} - C_{rss})}$ $CGDO = \dfrac{C_{rss}}{W}$
SPICE Models TOX - Gate Oxide Thickness L - Channel Length W - Channel Width CGDO - Gate-Drain Overlap Capacitance per unit channel width	$C_{GS} = \dfrac{2}{3}\left(\dfrac{\epsilon_{OX}WL}{TOX}\right)$ $C_{GD} = C_{GDO}W$ $C_{DS} = 0$	$\epsilon_{OX} = 3.9\epsilon_o = 3.9(8.51\text{pF/m})$, for SiO_2
IC Design Parameters[3] Levels 2, 3, and 4 PSpice Models	Levels 2, 3, and 4 PSpice Parameters	Levels 2, 3, and 4 PSpice Parameters

NOTES: (1) Default geometry parameters in PSpice are L = W = 100 μm. In SPICE2, the default geometry parameters are L = W = 1m. Recommend setting L = W = 100 μm or less in the MOSFET model statement. Failure to make L = W may affect the constants used to determine the DC operating point.

(2) The value for ϵ_{OX} is dependent on the gate insulation material. The value of $3.9\epsilon_o$ is for SiO_2 insulated gates only.

(3) Integrated circuit parameters require the most accurate representation of the physical realization of the device. Detailed device geometry information is required. More complex models of the MOSFET are used in specifying parameters for IC design. These models are beyond the scope of this book: the simple mocel presented in this text does not have sufficient detail to be used for IC design.

5. Determine the midband performance of each amplifier stage by:
   a. replacing the transistors by their respective AC models or
   b. using previously derived results for the circuit topology.
6. Combine stage performance quantities to obtain total midband gain.
7. Perform low-frequency analysis to determine the low cutoff frequency using the low-frequency, small-signal AC model.
8. Perform high-frequency analysis to determine the high cutoff frequency using:
   a. The small-signal hybrid-π model including the internal (high-frequency) capacitors and short circuiting all external (coupling and bypass) capacitors. Use Miller's theorem to simplify the small-signal model of the amplifier where appropriate.
   b. Previously derived expressions for the pole locations taking into account appropriate loading conditions.
9. Combine results of DC, midband, low-frequency, and high-frequency analysis to obtain total circuit performance.

As an aid to the designer faced with a variety of sources for transistor data, Tables 10.10-1 (BJT), 10.10-2 (JFET), and 10.10-3 (MOSFET) have been included. These tables provide formulas for transistor parameter conversion between the various models. The three parameter sets included in the tables are: manufacturer's data book parameters, hybrid-π model parameters, and SPICE model parameters.

SUMMARY DESIGN EXAMPLE: PREAMPLIFIER FOR A LORAN C NAVIGATIONAL SYSTEM RECEIVER

The Loran C is a standard navigational system using pulsed information in the very low frequency (VLF) range. Loran C is a form of hyperbolic navigational system that uses time differentials between signals from a pair of widely separated transmitting sites to locate the coordinates of the receiving station. The required operational frequency range is 10 Hz to 1500 kHz. The output of a "whip" antenna, with an output resistance of 2200 Ω, must be amplified by a factor of +2 to +3 by a preamplifier. This preamplified signal is applied to the 50-Ω input of the Loran C navigational receiver, which will process the incoming information. The output impedance of the preamplifier must match the input impedance of the Loran C receiver to eliminate potential signal reflections that will corrupt the signal to the Loran C receiver. It is critical that the preamplifier is a noninverting amplifier that will preserve the duty cycle of the pulsed information detected by the whip.

Design an inexpensive Loran C preamplifier to meet the stated requirements. A +9-V automotive/marine battery is used as the power source.

Solution:

The antenna signal is applied to a common-source n-JFET amplifier that will provide moderate gain and high input resistance. However, since a common-source n-JFET amplifier inverts the input signal, a unity-gain inverter is required at the output. Therefore, the output of the FET stage is applied to a unity-gain common-emitter npn BJT amplifier. To minimize signal reflections, the output resistance of the BJT

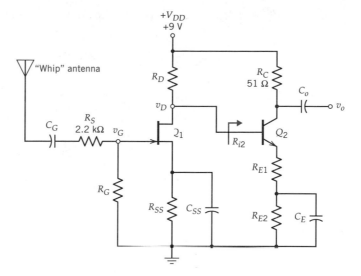

unity-gain inverting amplifier must be 50 Ω. A possible topology for the amplifier is as shown.

The BJT parameters are

$$\beta_F = 200, \qquad V_A = 160 \text{ V}, \qquad r_b = 30 \text{ }\Omega,$$

$$C_{ib0} = 8 \text{ pF} \quad \text{at } V_{EB} = 0.5 \text{ V},$$

$$C_{ob0} = 4 \text{ pF} \quad \text{at } V_{CB} = 5 \text{ V},$$

$$f_T = 300 \text{ MHz}.$$

The JFET parameters are

$$I_{DSS} = 6 \text{ mA}, \qquad V_{PO} = -4.7 \text{ V}, \qquad V_A = 100 \text{ V},$$

$$C_{iss} = 4.5 \text{ pF} \quad \text{at } V_{GS} = 0 \text{ V},$$

$$C_{rss} = 1.5 \text{ pF} \quad \text{at } V_{GS} = 0 \text{ V}.$$

Initiate the design by selecting the DC operating point and synthesizing the circuit for operation in the midband region.

Let $I_D = 1$ mA. Solve for V_{GS}:

$$I_D = I_{DSS}\left(1 - \frac{V_{GS}}{V_{PO}}\right)^2,$$

or

$$V_{GS} = V_{PO}\left(1 - \sqrt{\frac{I_D}{I_{DSS}}}\right) = -4.7\left(1 - \sqrt{\frac{1}{6}}\right) = -2.8 \text{ V}.$$

Since $V_G = 0$, $V_S = 2.8$ V. This implies that the source resistor is

$$R_{SS} = \frac{V_S}{I_D} = \frac{2.8}{0.001} = 2.8 \text{ k}\Omega.$$

The midband gain of the amplifier is

$$A_v \approx \frac{v_o}{v_g} = \left(\frac{v_d}{v_g}\right)\left(\frac{v_\pi}{v_d}\right)\left(\frac{v_o}{v_\pi}\right)$$

$$= [-g_{m1}(R_D \| R_{i2})]\left(\frac{r_\pi}{r_\pi + (1 + g_{m2}r_\pi)R_{E1}}\right)(-g_{m2}R_C),$$

where

$$g_{m1} = \frac{-2I_D}{V_{PO} - V_{GS}} = \frac{-2(0.001)}{-4.7 - (-2.8)} = 1.05 \text{ mS},$$

$$g_{m2} = \frac{|I_C|}{V_t}, \qquad r_\pi = \frac{\beta_F V_t}{|I_C|}$$

for a large r_d.

For a gain of 2–3, the parallel combination of R_D and R_{i2} is in the range

$$\frac{2}{g_{m1}g_{m2}R_C r_\pi / [r_\pi + (1 + g_{m2}r_\pi)R_{E1}]} \leq R_D \| R_{i2}$$

$$\leq \frac{3}{g_{m1}g_{m2}R_C r_\pi / [r_\pi + (1 + g_{m2}r_\pi)R_{E1}]}.$$

Since the second stage will be designed to have an inverting gain magnitude of approximately unity, the gain of the first stage can be designed to have a gain of -2.5

The second stage will be designed for a collector current of 1 mA. Therefore,

$$r_\pi = 5.2 \text{ k}\Omega, \qquad g_{m2} = 0.0385 \text{ S}.$$

for an inverting unity-gain second stage,

$$A_{v2} = \frac{v_o}{v_d} = \frac{v_\pi}{v_d}\frac{v_o}{v_\pi} \approx \frac{r_\pi}{r_\pi + (1 + g_{m2}r_\pi)R_{E1}}(-g_{m2}R_C).$$

But $g_{m2}R_C = (0.0385)(51) = 1.96$. Therefore, the expression $r_\pi / [r_\pi + (1 + g_{m2}r_\pi)R_{E1}] \approx 0.5$ for $A_{v2} = -1$. This implies that $(1 + g_{m2}r_\pi)R_{E1} \approx 5.2$ kΩ, or $R_{E1} = 26 \approx 27 \Omega$.
The input resistance of the second stage is then

$$R_{i2} = r_b + r_\pi + (1 + g_{m2}r_\pi)R_{E1} = 30 + 5200 + 5200 = 10.43 \text{ k}\Omega.$$

For an overall gain of -2.5,

$$R_D = 3.17 \text{ k}\Omega \approx 3.3 \text{ k}\Omega.$$

Checking the DC conditions of the first stages, the drain–source voltage of the JFET is

$$V_{DS} = V_{DD} - I_D(R_D + R_{SS}) = 9 - 0.001(3300 + 2800) = 2.9 \text{ V}.$$

To check that the FET is in the saturation region of operation,

$$V_{DS} \geq V_{GS} - V_{PO} = -2.8 - (-4.7) = 1.9 \text{ V},$$

and since $V_{DS} = 2.9 \text{ V} > 1.9 \text{ V}$, the FET is operating in the saturation region. The DC voltage at the drain is $V_D = 9 - 0.001(3300) = 5.7 \text{ V}$.

Now solve for R_{E2} to bias the BJT at $I_C = 1 \text{ mA}$:

$$R_{E2} = \frac{\beta_F}{I_C(\beta_F + 1)} (V_D - V_{\gamma2}) - R_{E1}$$

$$= \frac{200}{0.001(201)} (5.7 - 0.7) - 27 = 4900 \approx 4.99 \text{ k}\Omega.$$

This indicates that $V_{CE} \approx 4 \text{ V}$.

To confirm that the frequency response of the circuit is sufficient to receive the Loran C signal, the device capacitances must first be found.

For the FET,

$$C_{gs} = \frac{C_{gs0}}{\sqrt{1 + |V_{GS}|/\Psi_o}} = \frac{C_{iss} - C_{rss}}{\sqrt{1 + |V_{GS}|/\Psi_o}} = 1.3 \text{ pF},$$

$$C_{gd} = \frac{C_{gd0}}{\sqrt{1 + |V_{GS}|/\Psi_o}} = \frac{C_{rss}}{\sqrt{1 + |V_{GS}|/\Psi_o}} = 0.463 \text{ pF}.$$

For the BJT,

$$CJC = C_{obo}\left(1 + \frac{V_{CB}}{0.75}\right)^{0.33} = 7.83 \text{ pF},$$

$$CJE = C_{ibo}\left(1 + \frac{V_{EB}}{0.75}\right)^{0.33} = 9.47 \text{ pF},$$

$$TF = \frac{1}{2\pi f_T} - \frac{\eta V_t}{|I_{CT}|}\left[\frac{CJE}{(1 - 0.7/0.75)^{0.33}} + C_{obo}\right] = 460 \text{ ps},$$

$$C_{\mu2} = \frac{CJC}{(1 + V_{CB}/0.75)^{0.33}} = 4.5 \text{ pF},$$

$$C_{\pi2} = \frac{g_{m2}}{\omega_{T2}} - C_{\mu2} = 15.9 \text{ pF}.$$

The high-frequency response is found by using Equations 10.9-12 to 10.9-15 and 10.9-7.

The high-frequency capacitors are found using Miller's theorem:

$$C_{i1} = C_{gs} + C_{gd}(1 - A_{v1})$$

$$= C_{gs} + C_{gd}\left[1 + \frac{\mu(R_D||R_{i2})}{r_d + (R_D||R_{i2}) + (1 + \mu)R_{SS}}\right] = 2.1 \text{ pF},$$

$$C_{o1} = C_{gd}\left(1 - \frac{1}{A_{v1}}\right)$$

$$= C_{gd}\left[1 + \frac{r_d + (R_D||R_{i2}) + (1 + \mu)R_{SS}}{\mu(R_D||R_{i2})}\right] = 1.2 \text{ pF},$$

$$C_{i2} = C_\pi + C_\mu(1 - A_{v2})$$

$$= C_\pi + C_\mu\left[1 - \frac{R_{i2} - (r_b + r_\pi)}{R_{i2}}\right] = 18.2 \text{ pF},$$

$$C_{o2} \approx C_{\mu2} \approx 4.5 \text{ pF}.$$

The high-frequency poles are

$$\omega_{H1} = \frac{1 + \mu(R_D||R_{i2})/[r_d + R_{SS} + (R_D||R_{i2})]}{C_{i1}(R_S||R_G)}$$

$$= 528 \text{ Mrad/s} \Rightarrow 85 \text{ MHz},$$

$$\omega_{H2} = \frac{1}{C_{o1}\{r_d||[R_{SS} + (R_D||R_{i2})]\}}$$

$$= 331 \text{ Mrad/s} \Rightarrow 53 \text{ MHz}.$$

$$\omega_{H3} \approx \frac{1}{[C_{\pi2} + C_{\mu2}(1 + g_{m2}R_C)]\{r_\pi||[(r_b + R_D)/(1 + g_{m2}R_{E1})]\}}$$

$$= 27.5 \text{ Mrad/s} \Rightarrow 4.4 \text{ MHz}.$$

Therefore, the high-frequency cutoff is significantly greater than the required 1500 kHz: The high-frequency cutoff meets specification.

The specifications require a low-frequency cutoff of 10 Hz. Therefore, design all low-frequency poles at 5 Hz or less,

$$\omega_{L1} = 10\pi = \frac{1}{C_1(R_S + R_G)} \Rightarrow C_1 = 3182 \text{ pF} \approx 3300 \text{ pF},$$

and let

$$C_E = 4700 \text{ μF}.$$

The design is now complete.

■■■■■■■■ **REFERENCES**

Antognetti, P. and Massobrio, G., *Semiconductor Device Modeling with SPICE,* McGraw-Hill, New York, 1988.

Ghausi, M. S., *Electronic Devices and Circuits: Discrete and Integrated,* Holt, Rinehart and Winston, New York, 1985.

Gray, P. R. and Meyer, R. G. *Analysis and Design of Analog Integrated Circuits, 3rd Ed.,* John Wiley & Sons, Inc., New York, 1993.

Millman, J., *Microelectronics: Digital and Analog Circuits and Systems,* McGraw-Hill Book Company, New York, 1979.

Millman, J. and Halkias, C. C., *Integrated Electronics: Analog and Digital Circuits and Systems,* McGraw-Hill Book Company, New York, 1972.

Savant, C. J., Roden, M. S., and Carpenter, G. L., *Electronics Design: Circuits and Systems,* 2nd Ed., Benjamin/Cummings Publishing Company, Redwood City, 1991.

Sedra, A. S. and Smith, K. C., *Microelectronic Circuits, 3rd Ed.,* Holt, Rinehart, and Winston, Philadelphia, 1991.

Seymour, J., *Electronic Devices & Components, 2nd Ed.,* Longman Scientific & Technical, Essex, U.K., 1988.

Tyagi, M. S., *Introduction to Semiconductor Materials and Devices,* John Wiley & Sons, New York, 1991.

PROBLEMS

10-1. A rectangular pulse of 10-μs duration is the input to an amplifier described by a single high-frequency pole. Plot the amplifier output for the following locations of that single pole:

   a. 10 MHz

   b. 1 MHz

   c. 100 kHz

10-2. The output voltage response of an amplifier to a input consisting of a unit voltage step is shown. Assume the high-frequency response is characterized by a single pole. Determine:

   a. The voltage gain of the amplifier.

   b. The high 3-dB frequency of the amplifier.

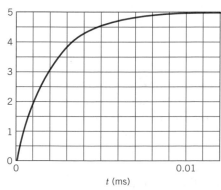

t (ms)

10-3. A rectangular pulse of duration 10 ms is the input to an amplifier described by a single low-frequency pole. Plot the amplifier output for the following locations of that single pole:

   a. 1 Hz

   b. 10 Hz

   c. 100 Hz

10-4. The response of a unity-gain buffer to a 100-Hz square wave with 5-V amplitude is shown. If the low-frequency response is characterized by a single pole, determine the low 3-dB frequency of the unity-gain buffer.

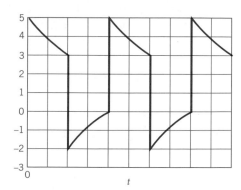

10-5. The low-frequency response of an amplifier is characterized by three poles of frequency 30, 10, and 2 Hz and three zero-frequency zeros. Calculate the lower 3-dB frequency f_L using the following techniques:

a. Dominant pole approximations

b. Root-sum of squares approximation (Equation 10.2-9)

Compare results to exact calculations.

10-6. An audio amplifier is described by a single low-frequency, $f_L = 100$ Hz, and a single high-frequency pole, $f_H = 20$ kHz.

a. Sketch the response of this amplifier to square waves of frequency

$$f_{sq} = 250 \text{ Hz}, \qquad f_{sq} = 4 \text{ kHz}.$$

b. Using simple, first-order OpAmp filters to model the amplifier, use SPICE to determine the response of the amplifier to these same square waves.

c. Compare results of parts a and b.

10-7. A particular amplifier can be described as having a midband gain of 560, three high-frequency poles at frequencies

$$f_{p1} = 25 \text{ kHz}, \qquad f_{p2} = 75 \text{ kHz},$$

$$f_{p3} = 150 \text{ kHz},$$

and a complex-conjugate pair of low-frequency poles ($\zeta = 0.6$) with resonant frequency

$$f_o = 100 \text{ Hz}.$$

a. Draw the idealized Bode magnitude and phase plots. On the graph label all slopes and use a small circle to indicate where the slope changes. Compare these straight-line approximate curves to exact curves (use a software package to generate the exact curves).

b. Use dominant-pole analysis to estimate the high and low 3-dB frequencies. Compare these estimates to exact calculations.

10-8. The BJT is the given circuit described by

$$\beta_F = 200, \qquad V_A = 150 \text{ V}.$$

a. Determine the location of the three low-frequency poles.

b. Which poles can be considered dominant?

c. Determine the low 3-dB frequency of the circuit.

d. Verify results using SPICE.

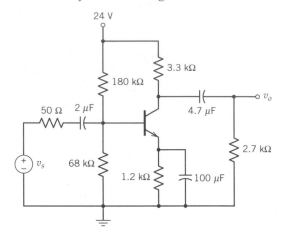

10-9. For the circuit shown, find the low 3-dB frequency and the midband voltage gain. Confirm using SPICE. The transistor is described by

$$\beta_F = 200, \qquad f_T = 600 \text{ MHz}, \qquad V_A = 200 \text{ V}.$$

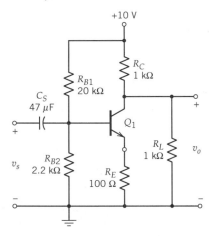

10-10. The design goals for the amplifier shown include maximum symmetrical output voltage swing and a low 3-dB frequency of 70 ± 3 Hz. The silicon BJT is described by

$$\beta_F = 150, \qquad V_A = 300 \text{ V}.$$

Complete the circuit design: Use SPICE to verify compliance with the design goals.

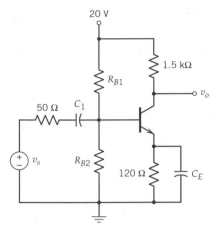

10-11. For the circuit shown, find the mid-band voltage gain and low 3-dB frequency. The transistor has $\beta_F = 100$, $f_T = 600$ MHz, and $V_A = 200$ V. Let $R_{B1} \| R_{B2} = 10$ kΩ. Confirm using SPICE.

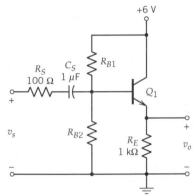

10-12. For the circuit shown, find the midband voltage gain. Complete the design to achieve a low 3-dB frequency of 40 Hz. The transistor has $\beta_F = 200$ and $V_A = 200$ V.

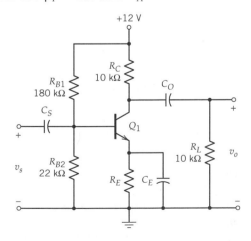

10-13. For the circuit shown, find the low 3-dB frequency and the midband voltage gain. The transistor is described by $V_{PO} = -4$ V, $I_{DSS} = 8$ mA, and $V_A = 200$ V. Confirm using SPICE.

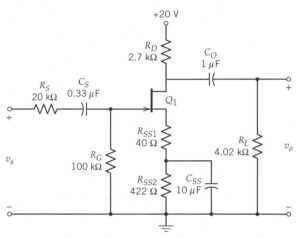

10-14. The design goals for the circuit shown include a low 3-dB frequency f_L at 80 Hz. The MOSFET has parameters

$$I_{DSS} = 5 \text{ mA}, \qquad V_{PO} = -2 \text{ V}, \qquad V_A = 150 \text{ V}.$$

    a. Complete the design by specifying capacitor values that will accomplish the design goal to within ±5%.

    b. Compare the results to SPICE simulation.

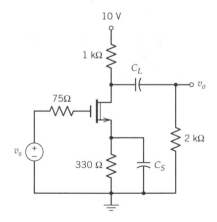

10-15. The design goals for the circuit shown include a low 3-dB frequency f_L at 50 Hz. The MOSFET has parameters

$$K = 2.5 \text{ mA/V}^2, \qquad V_T = 1.5 \text{ V},$$

$$V_A = 120 \text{ V}.$$

a. Complete the design by specifying capacitor values that will accomplish the design goal to within ±5%.

b. Compare the results to SPICE simulation.

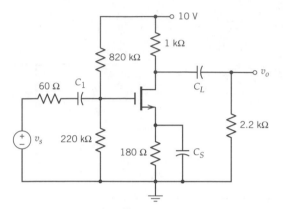

 10-16. Design an n-JFET common-source amplifier for a low 3-dB frequency of 40 Hz with a midband voltage gain of -2. The input resistance of the amplifier is 100 kΩ. The output resisance is 10 kΩ. A power supply voltage of 12 V is available. The Q point is defined by $V_{DS} = 6$ V and $V_{GS} = -1$ V. The transistor parameters are $I_{DSS} = 5$ mA, $V_{PO} = -4$ V, and $V_A = 150$ V.

(F) **10-17.** The silicon BJT in the circuit shown has parameters

$$\beta_F = 120, \qquad V_A = 200.$$

a. Determine a realistic minimum value for the capacitor C so that a 75-Hz square wave will experience no more than 5% sag.

b. Determine the midband gain for the circuit.

c. Determine the low 3-dB frequency using the capacitor chosen in part a.

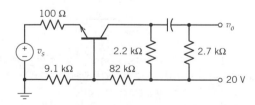

(F) **10-18.** The design goals for the circuit shown include a lower 3-dB frequency f_L at

100 Hz. The silicon BJT has parameters $\beta_F = 150$ and $V_A = 350$ V.

a. Complete the design by specifying capacitor values that will accomplish the design goal to within ±5%.

b. Sketch the Bode voltage gain plot for the amplifier.

c. Compare results to SPICE simulation.

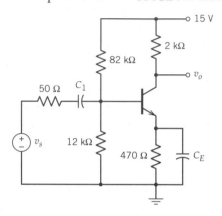

(F) **10-19.** The design goals for the circuit shown include the requirement that a 100-Hz square wave will experience no more than 4% sag. The silicon BJT has parameters $\beta_F = 160$ and $V_A = 150$ V.

a. Complete the design by specifying capacitor values that will accomplish the design goal to within ±5%.

b. Compare results to SPICE simulation.

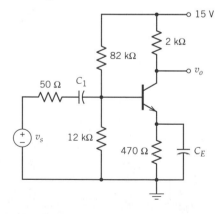

10-20. The zero-bias capacitance of a p–n junction is 2 pF and the built-in potential ψ_o is 780 mV. Assume the junction grading coefficient has value $m = 0.35$.

a. Plot the junction capacitance for the range of junction capacitances $-10 \text{ V} \leq V_D \leq 0.75 \text{ V}$.

b. If the capacitance value at $V_d = -6 \text{ V}$ is used as a reference, determine the maximum percent variation in the capacitance over the range $-10 \text{ V} \leq V_d \leq -2$ V.

10-21. A silicon BJT is described by the following parameters:

$\beta_F = 160, \quad V_A = 200 \text{ V}, \quad r_b = 10 \ \Omega,$

$C_\mu = 2.5 \text{ pF}, \quad f_T = 160 \text{ MHz}.$

Determine the high-frequency hybrid-π model for the BJT under the following bias conditions:

a. $I_c = 2 \text{ mA}$

b. $I_c = 5 \text{ mA}$

10-22. The silicon BJT in the circuit shown is described by

$\beta_F = 140, \quad V_A = 144 \text{ V}, \quad r_b = 12 \ \Omega,$

$C_\mu = 3 \text{ pF}, \quad f_T = 120 \text{ MHz}.$

a. Determine the high-frequency hybrid-π model for the BJT.

b. Determine the high 3-dB and low 3-dB frequencies for the given circuit.

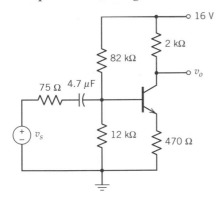

10-23. The silicon BJT in the circuit shown is described by the following parameters:

$\beta_F = 120, \quad V_A = 160 \text{ V}, \quad r_b = 25 \ \Omega,$

$C_\mu = 3.5 \text{ pF}, \quad f_T = 225 \text{ MHz}.$

Determine the high-frequency hybrid-π model for the BJT.

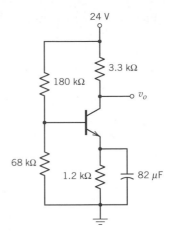

10-24. The transistors in the circuit shown are identical and have the following characteristics:

$\beta_F = 200, \quad f_T = 600 \text{ MHz}, \quad V_A = 200 \text{ V},$

$C_{obo} = 2 \text{ pF}.$

a. Find the midband voltage gain.

b. Find the high 3-dB frequency.

c. What is the largest peak-to-peak swing attainable with this circuit?

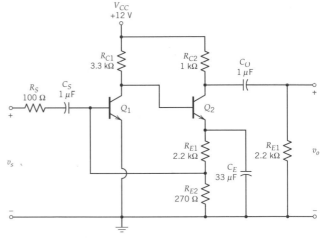

10-25. Complete the design of the circuit shown for maximum symmetrical swing in the midband frequency range. The transistor is described by

$\beta_F = 180, \quad f_T = 600 \text{ MHz}, \quad C_{obo} = 8 \text{ pF},$

$V_{CE} = 2.5 \text{ V}, \quad V_A = 200 \text{ V}.$

a. Find the high 3-dB frequency. Confirm using SPICE.

b. Compare part a to the SPICE frequency response for the small-signal model of the circuit.

c. Compare part b to the SPICE frequency response found when using the Miller's equivalent model of the transistor.

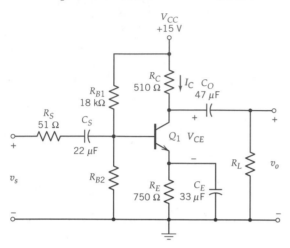

10-26. Determine the high and low 3-dB frequencies for the circuit shown. The silicon BJT is described by

$$\beta_F = 150, \quad V_A = 300 \text{ V}, \quad r_b = 20 \text{ } \Omega,$$

$$C_\mu = 3 \text{ pF}, \quad f_T = 200 \text{ MHz}.$$

Sketch a Bode voltage gain plot for the circuit.

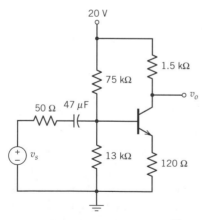

10-27. For the circuit shown, FET is described by

$$I_{DSS} = 5 \text{ mA}, \quad V_{PO} = -2 \text{ V}, \quad V_A = 150 \text{ V},$$

$$C_{rss} = 6.5 \text{ pF}, \quad C_{iss} = 35 \text{ pF}.$$

a. Determine the high 3-dB frequency for the voltage gain.

b. Remove the capacitor shunting the resistor at the FET source (the 220 μF) and determine the high 3-dB frequency.

c. Comment on results.

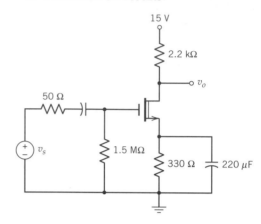

10-28. Complete the design of the amplifier shown for a Q point defined by $V_{GS} = 5$ V.

a. Find the low and high 3-dB frequencies.

b. Simulate the design using SPICE.

c. Compare simulation in part b to a simulation of the small-signal equivalent circuit of the amplifier.

d. In part c, plot the voltage across all of the capacitors (including the device capacitances) and comment on the results.

The transistor parameters are $K = 0.3$ mA/V^2, $V_T = 2$ V, $V_A = 150$ V, $C_{iss} = 12$ pF, and $C_{rss} = 2$ pF.

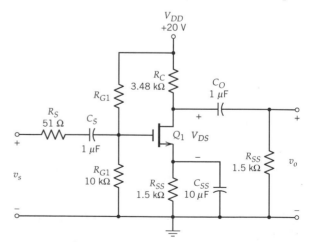

10-29. The JFET in the circuit shown is described by parameters

$V_{PO} = -1.8$ V, $I_{DSS} = 4$ mA,

$V_A = 90$ V, $C_{gs0} = 3.5$ pF,

$C_{gd0} = 1.5$ pF.

Determine the high-frequency model of the JFET for the following values of the gate voltage:

   a. $V_g = 0$ V

   b. $V_g = 1$ V

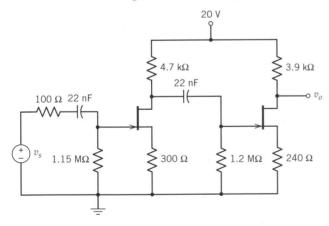

10-30. Determine the high and low 3-dB frequencies for the circuit shown. The JFET transistors have parameters

$V_{PO} = -2$ V, $I_{DSS} = 5$ mA, $V_A = 200$ V,

$C_{gs0} = 3$ pF, $C_{gd0} = 1$ pF.

Sketch a Bode voltage gain plot for the circuit, and verify results using SPICE. (Note: SPICE Parameters CGS $= C_{gs0}$ and CGD $= C_{gd0}$).

10-31. The given circuit was designed with identical transistors with characteristics

$I_{DSS} = 10$ mA, $V_{PO} = -2$ V,

$V_A = 100$ V, C_{rss} 7.3 pF,

$C_{iss} = 26.5$ pF.

Determine:

   a. Midband voltage gain

   b. Low 3-dB frequency

   c. High 3-dB frequency

Verify the results using SPICE.

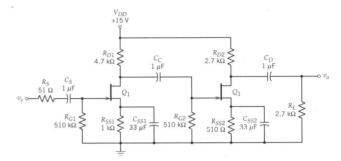

10-32. Complete the design of the circuit shown by determining a value for R_d that will achieve a quiescent output voltage of 5 V. Determine the high and low 3-dB frequencies for the completed circuit. The transistors are described by

JFET	BJT
$I_{DSS} = 2$ mA	$\beta_F = 120$
$V_{PO} = -2$ V	$V_A = 135$ V
$V_A = 120$ V	$C_{ob0} = 6$ pF
	at $V_{CB} = 5$ V
$C_{gs0} = 4$ pF	$C_{ib0} = 15$ pF
	at $V_{EB} = 0.5$ V
$C_{gd0} = 1.5$ pF	$f_T = 180$ MHz
	at $I_{CT} = 2.5$ mA

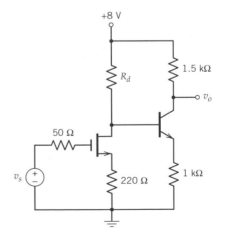

10-33. Design a three-stage *npn* BJT amplifier with an overall gain that is greater than 500 with a high 3-dB frequency of 10 MHz and a low 3-dB frequency under 100 Hz. The transistors are identical with the following characteristics:

$$\beta_F = 200, \quad f_T = 400 \text{ MHz},$$

$$V_A = 200 \text{ V}, \quad C_{obo} = 2 \text{ pF}$$

A 15-V power supply is available. Confirm the design with SPICE.

10-34. For the circuit shown the transistor parameters are

$$I_{DSS} = 10 \text{ mA}, \quad V_{PO} = -3.1 \text{ V},$$

$$V_A = 100 \text{ V}, \quad C_{rss} = 6.5 \text{ pF},$$

$$C_{iss} = 33.5 \text{ pF}$$

for the JFET and

$$\beta_F = 200, \quad f_T = 500 \text{ MHz}, \quad V_A = 200 \text{ V},$$

$$C_{obo} = 5 \text{ pF}$$

for the BJT. Determine the high and low 3-dB frequencies.

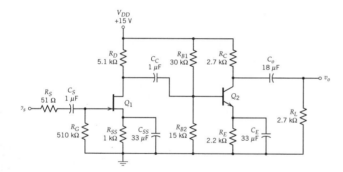

10-35. Determine the high and low 3-dB frequencies for the circuit shown. The JFET has parameters

$$V_{PO} = -2 \text{ V}, \quad I_{DSS} = 2 \text{ mA}, \quad V_A = 200 \text{ V},$$

$$C_{gso} = 3 \text{ pF}, \quad C_{gdo} = 1 \text{ pF}.$$

The BJT has parameters

$$\beta_F = 150, \quad V_A = 200 \text{ V},$$

$$C_{obo} = 7 \text{ pF} \quad \text{at } V_{BC} = 5 \text{ V},$$

$$C_{ibo} = 16 \text{ pF} \quad \text{at } V_{BE} = 0.5 \text{ V},$$

$$f_T = 160 \text{ MHz} \quad \text{at } I_{CT} = 3 \text{ mA}.$$

Compare results to SPICE simulation.

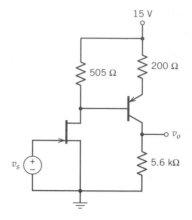

10-36. For the common-collector circuit shown, the BJT is described by

$$\beta_F = 150, \quad r_b = 30 \text{ }\Omega, \quad V_A = 200 \text{ V},$$

$$C_\mu = 0.5 \text{ pF}, \quad f_T = 200 \text{ MHz}.$$

a. Complete the design by determining the coupling capacitors so that the low 3-dB frequency is 10 ± 2 Hz.

b. What is the midband gain?

c. Determine the high 3-dB frequency.

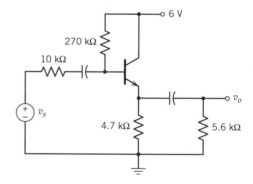

10-37. In the emitter-coupled amplifier circuit shown, it has been decided to ground one of the inputs and only use a single-ended output. The transistor parameters are

$$\beta_F = 120, \quad V_A = 250 \text{ V}, \quad r_b = 15 \text{ }\Omega,$$

$$C_\mu = 2 \text{ pF}, \quad f_T = 150 \text{ MHz}.$$

a. Determine the quiescent conditions on all transistors.

b. Determine the midband voltage gain of the circuit.

c. Determine the high and low 3-dB frequencies.

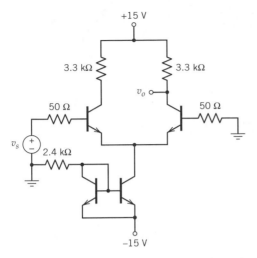

10-38. The BJT cascode amplifier shown uses identical transistors with parameters

$$\beta_F = 200, \qquad f_T = 600 \text{ MHz},$$

$$V_A = 200 \text{ V}, \qquad C_{obo} = 2 \text{ pF}.$$

a. Complete the design of the amplifier for $I_E = -3$ mA and $V_{CE1} = V_{CE2} = 3.5$ V. Find the quiescent point of all of the transistors.

b. Determine the high and low 3-dB frequencies.

c. Confirm the results using SPICE.

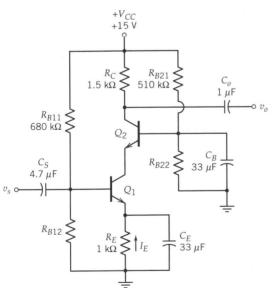

10-39. The transistors in the differential amplifier shown below are identical and have the following characteristics:

$$I_{DSS} = 10 \text{ mA}, \qquad V_{PO} = -4.5 \text{ V},$$

$$C_{rss} = 7.3 \text{ pF}, \qquad C_{iss} = 26.5 \text{ pF},$$

$$C_{oss} = 8.3 \text{ pF}, \qquad V_A = 100 \text{ V}.$$

Determine the high and low 3-dB frequencies.

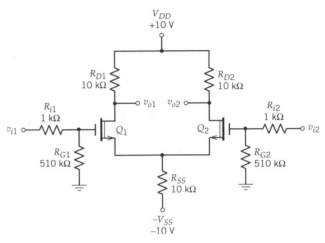

10-40. For the single-input differential amplifier shown, find the low and high 3-dB frequencies. Assume identical transistors with $\beta_F = 120$, $f_T = 600$ MHz, $V_A = 200$ V, and $C_{obo} = 2$ pF.

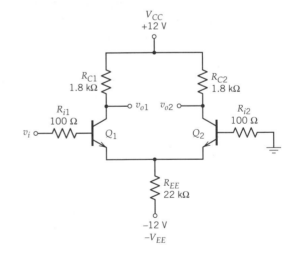

CHAPTER 11

FEEDBACK AMPLIFIER FREQUENCY RESPONSE

The basic topology of a feedback amplifier as previously described is shown in Figure 11-1.

In Chapter 8 several of the benefits of negative feedback were presented and analyzed. The mixing of a portion of the output signal with the input signal was shown to provide several benefits:

- The gain of the amplifier is stabilized against variation in the characteristic parameters of the active devices.
- The input and output impedances of the amplifier can be selectively increased or decreased.
- Non-linear signal distortion is reduced.

An additional benefit of negative feedback is the general increase in the midband frequency range. Previously described as a reduction in the variation of the gain due to *changes in frequency*, this increase in midband frequency range is exhibited by *both* an increase in the high 3-dB frequency f_H and a decrease in the low 3-dB frequency f_L.[1]

The above-listed benefits were shown to be accompanied by a drawback:

- The gain of the circuit is reduced.

An additional drawback is the possibility of oscillations. Amplifiers for which the high- and low-frequency responses are described by either single- or double-pole expressions are shown to be inherently stable and not subject to oscillations. If the amplifier high- or low-frequency response is described by three or more poles, the application of feedback brings forth the possibility of instability in the form of oscillations. This possibility of instability can be controlled either by limiting the range of applied feedback or with the addition of a compensation network to the amplifier.

In this chapter the benefit of increased bandwidth and the possibility of oscillations related to negative feedback are explored in electronic amplifiers. As in Chapter 10, the frequency response of an amplifier is broken into three regions: low-frequency, midband, and high-frequency regions. The effect of feedback on the midband region was discussed thoroughly in Chapter 8. Of interest here is the effect of

---

[1] Amplifiers for which $f_L = 0$ will, with the application of negative feedback, continue to operate with no low-frequency deterioration.

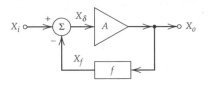

Figure 11-1
Basic negative-feedback topology.

feedback on the poles that determine the amplifier response in the low- and high-frequency regions. After deriving the pole location change in single-pole systems, the analysis is expanded to two-pole, three-pole, and many-pole systems. The concept of dominant poles is utilized to simplify system characterization when applicable.

While the early sections of this chapter focus on analysis techniques, the later sections focus on design techniques to ensure stable amplifier operation. Stability against oscillation is explored through the use of Bode diagrams and computer simulation as well as other standard stability criteria. Several common compensation networks that can alter the pole–zero characterization of the basic forward amplifier and thereby ensure amplifier stability are also explored.

11.1 EFFECT OF FEEDBACK ON AMPLIFIER BANDWIDTH (SINGLE-POLE CASE)

It has been previously shown that the gain A_f of a feedback amplifier is simply related to the forward gain A of a loaded basic forward amplifier and the feedback ratio f:

$$A_f = \frac{A}{1 + Af}. \qquad (11.1\text{-}1)$$

As in the derivations and demonstrations of Chapter 8, the amplifier gain can be expressed as a transresistance, a current gain, a voltage gain, or a transconductance depending on the topology of the mixing and sampling processes. The frequency response derivations presented here are not dependent on the mixing or sampling topology.

As was seen in Chapter 10, the small-signal sinusoidal forward gain of an amplifier is dependent on the frequency of the input. The high-frequency response of an amplifier that is characterized by a single high-frequency pole ω_H and a midband gain A_o is given by

$$A = \frac{A_o}{1 + j\omega/\omega_H}. \qquad (11.1\text{-}2)$$

When the feedback gain relationship (Equation 11.1-1) is applied to this single-pole frequency-dependent gain expression, the resultant feedback gain is given by

$$A_f = \frac{A_o/(1 + j\omega/\omega_H)}{1 + [A_o/(1 + j\omega/\omega_H)]f} = \frac{A_o}{1 + A_o f + j\omega/\omega_H} \qquad (11.1\text{-}3)$$

$$= \frac{A_o}{1 + A_o f} \frac{1}{1 + \dfrac{j\omega}{(1 + A_o f)\omega_H}}.$$

This gain is of the same form as that of a single-pole amplifier with reduced gain and increased pole frequency:

$$A_{of} = \frac{A_o}{1 + A_o f}, \qquad \omega_{Hf} = (1 + A_o f)\omega_H. \tag{11.1-4}$$

The gain has been reduced, as expected, by the return ratio ($D = 1 + A_o f$) and the *high-frequency pole value has also been increased by the return ratio.*

Similarly, if the low-frequency response of an amplifier is characterized by a single pole,

$$A = \frac{A_o}{1 + \omega_L/j\omega}, \tag{11.1-5}$$

the feedback gain can be derived to be

$$A_f = \frac{A_o/(1 + \omega_L/j\omega)}{1 + \dfrac{A_o}{(1 + \omega_L/j\omega)}f} = \frac{A_o}{1 + A_o f + \omega_L/j\omega}$$

$$= \frac{A_o}{1 + A_o f} \frac{1}{1 + \dfrac{\omega_L}{(1 + A_o f)j\omega}}. \tag{11.1-6}$$

This gain is again of the same form as that of a single low-frequency pole amplifier. The feedback amplifier low-frequency pole magnitude is given by

$$\omega_{Lf} = \frac{\omega_L}{1 + A_o f}. \tag{11.1-7}$$

The gain has been reduced, as expected, by the return ratio ($D = 1 + A_o f$) and the *low-frequency pole value has been decreased by the return ratio.*

Frequency Response

The magnitude of the gain function of a general amplifier with (A_f) and without (A) feedback is shown in Figure 11.1-1a, with frequency on a logarithmic scale. The increase in the width of the midband frequency range is evident. In Figure 11.1-1b, the Bode straight-line approximate amplitude plot is shown. It is interesting to note that the intersection of the two Bode plots occurs at $\omega = \omega_{Lf}$ and $\omega = \omega_{Hf}$. Outside the midband region of the feedback amplifier, the Bode plots are coincident. This property is a useful aid in the design of feedback amplifiers described by simple poles.

Step Response

The time-domain response of a single-pole amplifier to a unit step is a simple exponential decay to the final value:

$$X_o(t) = X_{o(\text{final})} + (X_{o(\text{initial})} - X_{o(\text{final})})e^{-t/\tau}, \tag{11.1-8}$$

where

$$\tau = \frac{2\pi}{\omega_H}. \tag{11.1-9}$$

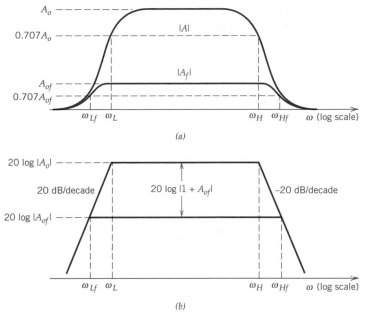

Figure 11.1-1
Changes to gain with application of feedback: (*a*) gain on a linear scale; (*b*) idealized Bode plot.

Application of feedback increases the high-frequency pole ω_H and consequently reduces the time constant of the exponential decay for the step response:

$$\tau_f = \frac{2\pi}{\omega_{Hf}} \tag{11.1-10}$$

The amplifier responds more quickly to step inputs. The tilt or sag of the amplifier is measured with a pulse train of frequency ω and is dependent on the low-frequency pole:

$$\text{sag} = \frac{\pi\omega_L}{\omega} \times 100\% = \frac{\pi f_L}{f} \times 100\%. \tag{11.1-11}$$

The application of feedback reduces ω_L and consequently reduces the tilt or sag of the pulse train response.

11.2 DOUBLE-POLE FEEDBACK FREQUENCY RESPONSE

The high-frequency response of an amplifier that is characterized by two high-frequency poles ω_1 and ω_2 and a midband gain A_o is given by

$$A = \frac{A_o}{(1 + j\omega/\omega_1)(1 + j\omega/\omega_2)}. \tag{11.2-1}$$

In order to more appropriately study the change in pole location for multiple-pole amplifiers, a change in the notation to describe frequency will be used:

$$j\omega \Rightarrow s = j\omega + \sigma. \tag{11.2-2}$$

This notation is common as one progresses from Fourier analysis to Laplace analysis and should be familiar to the reader. Under this change of variables, Equation 11.2-1 becomes

$$A = \frac{A_o}{(1 + s/\omega_1)(1 + s/\omega_2)}. \tag{11.2-3}$$

When the feedback gain relationship (Equation 11.1-2) is applied to this double-pole frequency-dependent gain expression, the resultant feedback gain is given by

$$A_f = \frac{A_o}{(1 + s/\omega_1)(1 + s/\omega_2) + A_o f}, \tag{11.2-4}$$

which is more commonly written in the form

$$A_f = \frac{A_o}{1 + A_o f}\left[1 + \frac{s}{1 + A_o f}\left(\frac{1}{\omega_1} + \frac{1}{\omega_2}\right) + \frac{s^2}{1 + A_o f}\frac{1}{\omega_1\omega_2}\right]^{-1}. \tag{11.2-5}$$

As expected, the low-frequency gain is reduced by the return ratio. However, the new locations of the two poles is not obvious. One heuristic approach to visualizing the general pole location change due to various amounts of feedback is through a plot of the denominator of Equation 11.2-4:

$$P(s) = \left(1 + \frac{s}{\omega_1}\right)\left(1 + \frac{s}{\omega_2}\right) + A_o f. \tag{11.2-6}$$

The plot of the denominator, $P(s)$, is shown in Figure 11.2-1 for four values of the loop gain $A_o f$. With unity return ratio (no feedback), the curve intersects the horizontal axis at the two poles $-\omega_1$ and $-\omega_2$ of the basic amplifier gain A. Increasing the feedback raises the curve $P(s)$ by $A_o f$ and results in a translation of the two poles, seen as the intersection of the curve with the horizontal axis, toward each other. With sufficiently large loop gain, the curve $P(s)$ no longer crosses the horizontal axis and the poles become a complex-conjugate pair.

A more rigorous approach to determining the migration of the pole locations with varying feedback casts Equation 11.2-5 into the general form

$$A_f = \frac{A_o}{(1 + A_o f)[1 + 2\zeta(s/\omega_o) + s^2/\omega_o^2]}. \tag{11.2-7}$$

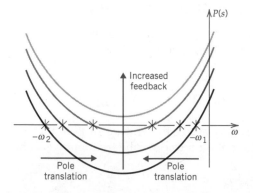

Figure 11.2-1
Heuristic interpretation of pole migration with increasing loop gain.

This format was initially presented in Chapter 9 and is standard for a two-pole system with resonant frequency

$$\omega_o = \sqrt{\omega_1\omega_2(1 + A_of)} \tag{11.2-8}$$

and damping coefficient

$$\zeta = \frac{\omega_1 + \omega_2}{2\sqrt{(1 + Af)\omega_1\omega_2}} = \frac{\omega_1 + \omega_2}{2\omega_0}. \tag{11.2-9}$$

The resonant frequency *and* the damping coefficient depend on the original pole location and return difference $D = 1 + A_of$. The resonant frequency *increases* as the square root of the return difference, while the damping coefficient *decreases* by the same factor. With the use of the quadratic formula, the two poles can be found to be located at

$$s = -\frac{\omega_1 + \omega_2}{2} \pm \frac{\omega_1 + \omega_2}{2}\sqrt{1 - \frac{1}{\zeta^2}}. \tag{11.2-10}$$

As predicted by the heuristic approach, when there is a large damping coefficient, $\zeta > 1$ (small amounts of feedback), there are two distinct, real poles: For small damping coefficient, $\zeta < 1$ (large amounts of feedback), the argument of the square root becomes negative and the poles become a complex-conjugate pair.

It is common to display the locus of pole locations with varying damping coefficient on the complex s plane. This plot, called a *root-locus plot*, is shown in Figure 11.2-2. As the damping coefficient decreases (return difference increases), the poles converge along the real axis and then split vertically. Notice that when the poles of this two-pole gain function become a complex-conjugate pair, the real pair of the pole location remains a constant. This constancy of the real part of the complex-pair pole location indicates that the poles are always located in the left-half of the complex plane: The resonant frequencies e^{st} contain an exponential decay. Consequently, resonances do not experience uncontrolled growth and the two-pole feedback amplifier is always stable.

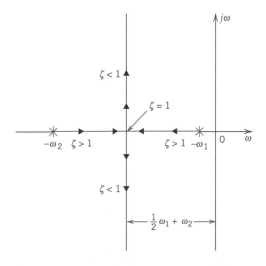

Figure 11.2-2
Root locus of two-pole gain function.

11.2.1 Frequency Response

The frequency response of a two-pole system has been discussed extensively in Section 9.1. A normalized plot of the frequency response of a high-frequency two-pole system in the general neighborhood of ω_o with the damping coefficient as a parameter ($0.2 \leq \zeta \leq 1.0$ in increments of 0.1) is shown in Figure 11.2-3.

In Chapter 9, a peak in the frequency response was found to exist for damping coefficients less than $1/\sqrt{2}$. This characteristic is particularly important in the design of amplifiers: The design requirements of good amplifiers rarely allows for significant peaks in the frequency response of the gain. The frequency at which this peak occurs is related to ω_o:

$$\omega_{\text{peak}} = \omega_o \sqrt{1 - 2\zeta^2} \tag{11.2-11}$$

and has magnitude

$$|A_f(\omega_{\text{peak}})| = \frac{|A_{of}|}{2\zeta\sqrt{1 - \zeta^2}}. \tag{11.2-12}$$

The design specifications on the midband flatness will place a lower limit on the damping coefficient (and consequently the return difference). Good amplifiers rarely have damping coefficients less than 0.5.

In amplifier design it is important to find the change in the 3-dB frequency as feedback is applied to the amplifier. In two-pole systems, the high 3-dB frequency ω_{Hf} is dependent on the resonant frequency ω_o and the damping coefficient ζ. Each of these quantities is, in turn, dependent on the original pole locations ω_1 and ω_2 and the return difference D. The calculation of the 3-dB frequency is therefore a complicated process. One approach to this process begins with the notational convention that the amplifier before the application of feedback is described by two simple poles with frequency ratio

$$\omega_2 = k\omega_1. \tag{11.2-13}$$

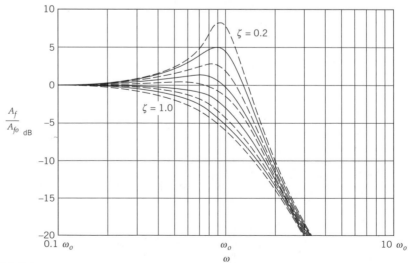

Figure 11.2-3
Normalized feedback amplifier gain.

It was previously shown in Section 10.2 that if $k > 4$, a dominant pole exists. After the application of feedback, it is also possible that a dominant pole exists. To determine the appropriate condition for a dominant pole with feedback, the ratio of the two feedback pole frequencies, k_f, given in Equation 11.2-10 is taken:

$$k_f = \frac{-\left[\dfrac{(\omega_1 + \omega_2)}{2}\right]\left(1 + \sqrt{1 - \dfrac{1}{\zeta^2}}\right)}{-\left[\dfrac{(\omega_1 + \omega_2)}{2}\right]\left(1 - \sqrt{1 - \dfrac{1}{\zeta^2}}\right)} > 4. \tag{11.2-14}$$

The condition necessary for a feedback dominant pole is the solution of Equation 11.2-14, given by

$$\left(1 + \sqrt{1 - \frac{1}{\zeta^2}}\right) > 4\left(1 - \sqrt{1 - \frac{1}{\zeta^2}}\right) \tag{11.2-15}$$

or

$$\zeta > 1.25. \tag{11.2-16}$$

If this condition on the damping coefficient is met, the 3-dB frequency of the system can be obtained with the methods described in Section 10.2. While the dominant-pole case is significant, in many cases the damping coefficient is smaller than the appropriate value ($\zeta < 1.25$) and other methods of determining the 3-dB frequency must be used. The 3-dB frequency must be obtained by analysis of the magnitude of the feedback gain equation (Equation 11.2-7). If the parameter s is replaced by its sinusoidal equivalent, $s \Rightarrow j\omega$, the gain expression becomes

$$\frac{A_f}{A_{of}} = \frac{1}{1 + 2\zeta(j\omega/\omega_o) - \omega^2/\omega_o^2}, \tag{11.2-17}$$

which, when related to relative magnitudes, reduces to

$$\left|\frac{A_f}{A_{of}}\right|^2 = \frac{1}{(1 - \omega^2/\omega_o^2)^2 + 4\zeta^2(\omega^2/\omega_o^2)} \tag{11.2-18}$$

The 3-dB frequency ω_{Hf} occurs when the gain power ratio is reduced to one-half its midband value or when the denominator of Equation 11.2-18 has value 2:

$$\left(1 - \frac{\omega_{Hf}^2}{\omega_o^2}\right)^2 + 4\zeta^2\frac{\omega_{Hf}^2}{\omega_o^2} = 2 \tag{11.2-19}$$

Application of the quadratic formula leads to a solution for ω_{Hf}.

$$\frac{\omega_{Hf}^2}{\omega_o^2} = 1 - 2\zeta^2 + \sqrt{(1 - 2\zeta^2)^2 + 1}. \tag{11.2-20}$$

Equation 11.2-8 can be recast into a form to relate ω_o to the lowest nonfeedback pole frequency:

$$\omega_o = \omega_1\sqrt{k(1 + Af)} = \omega_1\sqrt{kD}, \qquad (11.2\text{-}21)$$

which leads to a relationship between the feedback 3-dB frequency and the lowest nonfeedback pole:

$$\frac{\omega_{Hf}}{\omega_1} = \sqrt{kD[1 - 2\zeta^2 + \sqrt{(1 - 2\zeta^2)^2 + 1}]} \qquad (11.2\text{-}22)$$

Because of the dependence of ζ on the initial pole spacing k and the return difference D, the true significance of Equation 11.2-22 is disguised. One form of interpreting the significance of this expression is the ratio of 3-dB frequency to the product of the return difference and ω_1:

$$\omega_{Hf} = \left(\sqrt{\frac{k}{D}[1 - 2\zeta^2 + \sqrt{(1 - 2\zeta^2)^2 + 1}]}\right)(D\omega_1). \qquad (11.2\text{-}23)$$

Further insight can be obtained by replacing the return difference (inside the square root) with its equivalent expression in k and ζ:

$$D = \frac{(k + 1)^2}{4k\zeta^2}.$$

Equation 11.2-23 now takes the form

$$\omega_{Hf} = \left(\frac{2k\zeta}{k + 1}\sqrt{[1 - 2\zeta^2 + \sqrt{(1 - 2\zeta^2)^2 + 1}]}\right)(D\omega_1). \qquad (11.2\text{-}23a)$$

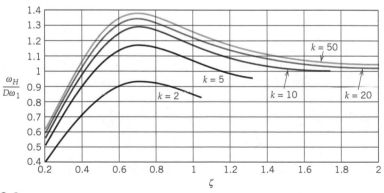

Figure 11.2-4
High 3-dB frequency as function of ζ and nonfeedback pole spacing.

The relationship is similar to the single-pole case with an additional factor:

$$\omega_{Hf} = K(\zeta, k)D\omega_1, \tag{11.2-24}$$

where the factor $K(\zeta, k)$ is given by

$$K(\zeta, k) = \frac{2k\zeta}{k+1} \sqrt{[1 - 2\zeta^2 + \sqrt{(1 - 2\zeta^2)^2 + 1}]}. \tag{11.2-25}$$

This relationship is shown in Figure 11.2-4 for a variety of initial pole ratios k.

EXAMPLE 11.2-1 A two-pole amplifier with midband gain $A_o = 1000$ and two high-frequency poles $f_1 = 100$ kHz and $f_2 = 1$ MHz has feedback applied so that the midband gain is reduced to

$$A_{of} = 80.$$

Determine the location of the new poles and the high 3-dB frequency.

SOLUTION

The ratio of the two pole frequencies is $k = 10$. The return difference is the ratio of the two gains:

$$D = \frac{A_o}{A_{of}} = \frac{1000}{80} = 12.5.$$

The resonant frequency is found to be

$$f_o = \sqrt{f_1 f_2(1 + A_o f)} = \sqrt{(0.1 \times 10^6)(1.0 \times 10^6)12.5} = 1.118 \text{ MHz}$$

and the damping coefficient is

$$\zeta = \frac{f_1 + f_2}{2f_o} = \frac{(0.1 \times 10^6 + 1.0 \times 10^6)}{2(1.118 \times 10^6)} = 0.492.$$

The high 3-dB frequency f_H can be found by obtaining $K(\zeta, 10)$ from Figure 11.2-4 (or Equation 11.2-25) and by multiplying $K(\zeta, 10), D$, and f_1:

$$f_H = K(\zeta, 10)Df_1 = (1.146)(12.5)(0.1 \times 10^6) = \textbf{1.43 MHz.}$$

The two poles are a complex-conjugate pair located at

$$s = 2\pi[-0.55 \times 10^6 \pm j(0.55 \times 10^6)(\sqrt{3.1311})]$$
$$= \textbf{-1.1}\pi \pm j\textbf{1.95}\pi \text{ Mrad/s.}$$

A plot of the amplifier gain (in decibels) with and without feedback is shown.

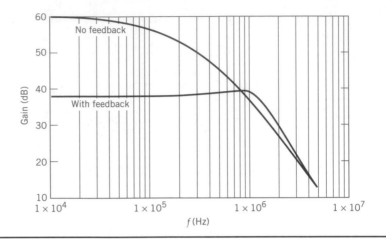

Low-Frequency Response

If the low-frequency response of an amplifier is described by two low-frequency poles ω_{1L} and ω_{2L}, the gain function can be expressed as

$$A = \frac{A_o}{(1 + \omega_{1L}/s)(1 + \omega_{2L}/s)}. \tag{11.2-26}$$

The application of feedback to the amplifier will produce a convergence of poles similar to that described for high-frequency poles. There are, however, distinct differences. The low-frequency resonant frequency ω_{oL} is given by

$$\omega_{oL} = \sqrt{\frac{\omega_{1L}\omega_{2L}}{1 + Af}}, \tag{11.2-27}$$

and the expression for the damping coefficient is

$$\zeta_L = \frac{\omega_{oL}}{2}\left(\frac{1}{\omega_{1L}} + \frac{1}{\omega_{2L}}\right). \tag{11.2-28}$$

The frequency response plots take the same form as for high-frequency poles with the single exception of being mirror images on the logarithmic frequency scale. The low 3-dB frequency ω_{Lf} can be determined in a similar manner as was the case for ω_{Hf}. For damping coefficients ζ greater than 1.25, a dominant pole exists and the methods of Chapter 10 can be used to determine ω_{Lf}. If the damping coefficient is smaller than 1.25, the methods of this chapter are more appropriate.

If the two low-frequency poles are identified by ω_{1L}, the pole closest to the midband region, and

$$\omega_{2L} = \frac{\omega_{1L}}{k}, \tag{11.2-29}$$

the low 3-dB frequency is given by

$$\omega_{Lf} = \frac{\omega_{1L}}{K(\zeta_L, k)D}, \tag{11.2-30}$$

where $K(\zeta, k)$ is defined by Equation 11.2-25.

The frequency response curve will also exhibit a peak if the low-frequency damping coefficient is less than $1/\sqrt{2}$. Peaks in the low-frequency response are usually undesirable in amplifiers, and good amplifiers rarely have damping coefficients less than 0.5. The frequency of this peak is related to ω_{oL},

$$\omega_{\text{peak}} = \frac{\omega_{oL}}{\sqrt{1 - 2\zeta^2}},$$
(11.2-31)

and has magnitude

$$|A_f(\omega_{\text{peak}})| = \frac{|A_{of}|}{2\zeta\sqrt{1 - \zeta^2}}.$$
(11.2-32)

EXAMPLE 11.2-2 A two-pole amplifier with midband gain $A_o = 1000$ and two low-frequency poles $f_1 = 100$ Hz and $f_2 = 10$ Hz has feedback applied so that the midband gain is reduced to

$$A_{of} = 80.$$

Determine the low 3-dB frequency.

SOLUTION

Here the ratio of the two pole frequencies is 10. The return difference is the ratio of the two gains:

$$D = \frac{A_o}{A_{of}} = \frac{1000}{80} = 12.5.$$

The resonant frequency is found to be

$$f_{oL} = \sqrt{\frac{f_{1L}f_{2L}}{1 + A_{of}}} = \sqrt{\frac{100(10)}{12.5}} = 8.944 \text{ Hz},$$

and the damping coefficient is

$$\zeta_L = \frac{f_{oL}}{2}\left(\frac{1}{f_{1L}} + \frac{1}{f_{2L}}\right) = \frac{8.944}{2}\left(\frac{1}{100} + \frac{1}{10}\right) = 0.492.$$

The low 3-dB frequency f_L can be found by obtaining $K(\zeta_L, 10)$ from Figure 11.2-4 (or Equation 11.2-25) and by dividing f_{1L} by the product of $K(\zeta_L, 10)$ and D:

$$f_L = \frac{f_{1L}}{K(\zeta_L, 10)D} = \frac{100}{1.146(12.5)} = \textbf{7.0 Hz}.$$

A plot of the amplifier gain (in decibels) is shown with and without feedback.

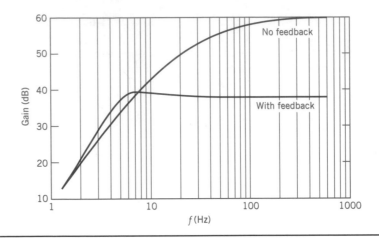

11.2.2 Step Response

The step response for an amplifier with two real high-frequency poles was discussed in Section 10.1. Feedback amplifiers with greater than unity value damping coefficients behave in the same manner: The step response is the sum of two exponential terms converging to a steady-state value. Damping coefficients less than unity value are underdamped systems and have a more complex step response.

The response of an underdamped feedback amplifier to a unit step is given by

$$X(t) = A_{of}\left[1 - \left(\frac{\zeta\omega_o}{\omega_d}\sin\omega_d t + \cos\omega_d t\right)e^{-\zeta\omega_o t}\right], \tag{11.2-33}$$

where the damped frequency ω_d is given by the expression

$$\omega_d = \omega_o\sqrt{1 - \zeta^2}. \tag{11.2-34}$$

A plot of the normalized step response of a feedback amplifier is shown in Figure 11.2-5 for several values of the damping coefficient ζ.

For small values of ζ, the step response of a feedback amplifier overshoots and oscillates about the final value before settling down to a steady-state condition. While a small amount of overshoot is often acceptable, large overshoot is often unsatisfactory for quality amplifiers. In order to quantify the response, several quantities are defined as follows:

Rise time: time to rise from 10 to 90% of the final value

Delay time: time to rise to 50% of the final value

Overshoot: peak excursion above the peak value

Damped period: time interval for one cycle of oscillation

Settling time: time for response to settle to within $\pm P$ percent of the steady-state value

These parameters are displayed for a typical underdamped step response in Figure 11.2-6.

The quantifying parameters can be obtained through careful analysis of Equation

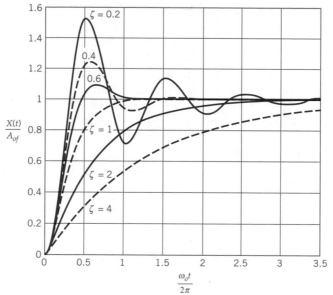

Figure 11.2-5
Step response of two-pole feedback amplifier.

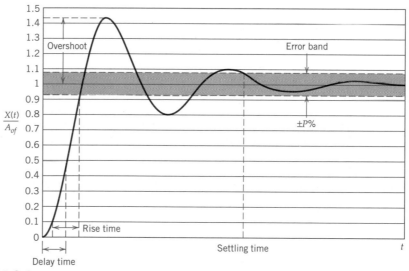

Figure 11.2-6
Step response for $\zeta = 0.25$.

11.2-32. One significant parameter, the overshoot, is obtained by setting the first derivative of Equation 11.2-32 to zero: The resultant time of the first peak is

$$t_{\text{peak}} = \frac{2\pi}{\omega_d} = \frac{2\pi}{\omega_o \sqrt{1 - \zeta^2}}. \tag{11.2-35}$$

The peak value of the step response is given by $X(t_{\text{peak}})$:

$$X(t_{\text{peak}}) = A_{of}(1 + e^{-\pi\zeta/\sqrt{1 - \zeta^2}}), \tag{11.2-36}$$

which is equivalent to

$$\text{Overshoot} = e^{-\pi\zeta/\sqrt{1-\zeta^2}} \times 100\%. \tag{11.2-37}$$

A good amplifier will have small rise, delay, and settling times and a small overshoot. Rise and delay times decrease with increased return ratio D while overshoot and settling times increase once the system becomes underdamped. The design process balances the conflicting requirements.

Tilt or Sag

It has previously been shown that the low-frequency poles tilt the constant portion of a square wave. Amplifiers described by underdamped low-frequency poles also exhibit tilted square-wave response. The response of a low-frequency underdamped amplifier to a unit step is given by

$$X_L(t) = A_{of}\left[\cos(\omega_{dL}t) - \frac{\zeta\omega_{oL}}{\omega_{dL}}\sin(\omega_{dL}t)\right]e^{-\zeta\omega_{oL}t}. \tag{11.2-38}$$

While this response is oscillatory, it can be approximated as

$$X_L(t) \approx A_{of}(1 - 2\zeta\omega_{oL}t) \tag{11.2-39}$$

for small values of $\omega_{oL}t$. Under these approximations, the percent tilt of a square wave of frequency ω is given by

$$\text{sag} \approx \frac{2\zeta\pi\omega_{oL}}{\omega} \times 100\% = \frac{2\zeta\pi f_{oL}}{f} \times 100\%. \tag{11.2-40}$$

Notice that increasing feedback reduces the resonant frequency and the damping coefficient: The percent tilt will be decreased by each of these reductions.

11.3 MULTIPOLE FEEDBACK FREQUENCY RESPONSE

It has been shown that the frequency response of an amplifier is dependent on the pole locations. The application of feedback alters the pole locations and therefore the frequency response. In first- and second-order systems, the migration of poles is relatively simple to describe: The poles move in a predictable manner and remain in the left-hand plane. As a consequence, single- or double-pole feedback amplifiers are always stable. In systems with three or more poles, hereafter identified as *multipole amplifiers*, some of the poles migrate into the right-hand plane if sufficiently large quantities of feedback (large return difference) are applied. Multiple amplifiers can become unstable.[2] If, however, the feedback is moderate, multipole amplifiers are stable and extremely useful.

The high-frequency response of an amplifier that is characterized by three high-frequency poles ω_1, ω_2, and ω_3 and a midband gain A_o is given by

$$A = \frac{A_o}{(1 + s/\omega_1)(1 + s/\omega_2)(1 + s/\omega_3)}. \tag{11.3-1}$$

[2] Stability criteria and compensation are discussed in the Sections 11.4 and 11.5.

When the feedback gain relationship (Equation 11.1-2) is applied to this triple-pole frequency-dependent gain expression, the resultant feedback gain is given by

$$A_f = \frac{A_o}{(1 + s/\omega_1)(1 + s/\omega_2)(1 + s/\omega_3) + A_o f}. \tag{11.3-2}$$

As expected, the midband gain is reduced by the return difference: The new pole locations are obscure. Figure 11.3-1 shows the heuristic approach to pole location migration. As can be seen in the figure, the two lower frequency poles ω_1 and ω_2 converge and, with sufficient feedback, become a complex-conjugate pair. With increased feedback, the highest magnitude pole frequency ω_3 increases in magnitude and *becomes more distant from the two lower frequency poles*. This separation of the poles is of particular interest as it indicates that the lower frequency poles become more dominant poles as feedback increases.

As with the two-pole case, the heuristic approach gives no insight into pole migration once poles become a complex-conjugate pair. A root-locus plot is necessary to show the pole migration in the complex plane: Figure 11.3-2 shows the migration of poles for a third-order system.

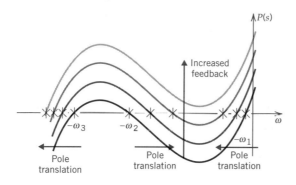

Figure 11.3-1
Heuristic interpretation of pole migration: third-order system.

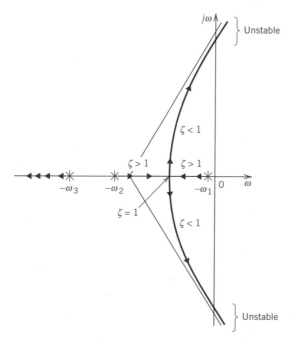

Figure 11.3-2
Root locus of three-pole gain function.

The lower frequency pole pair converges until the damping coefficient associated with the pair is of unity value. At this point, the pair diverges vertically: This migration mimics the two-pole case. However, as feedback is increased and the damping coefficient is reduced further, the pole pair migrates toward the right-hand plane. With large amounts of feedback, the pole pair crosses the axis into the right-hand plane and the system becomes unstable. Since small damping coefficients are, in general, undesirable in amplifiers, a quality amplifier with a relatively flat midband region will not sufficiently displace the pole pair to cause instability.

As the number of poles increases, it becomes more difficult to intuitively understand the migration of poles with changes in feedback. Continued application of the heuristic interpretation shows that the lowest pole pair will converge in all cases. Each successive pole pair thereafter will also converge. In odd-order systems, the final highest magnitude frequency pole will move to even higher frequencies. The migration of the poles once they have become complex-conjugate pairs is, as usual, determined through the use of a root-locus plot.

There are three simple rules to aid in the construction of an approximate root-locus plot for an amplifier described by high-frequency poles:

1. The average of the pole frequencies remains unchanged with the application of feedback. It is given as

$$\sigma_o = \frac{1}{n_p} \sum_{c=1}^{n_p} p_c. \tag{11.3-3}$$

where n_p is the number of poles and p_c are the pole frequencies.

2. When a pole pair becomes underdamped ($\zeta < 1$), the root locus for that pair departs from the real axis at an angle of $90°$. The location of the departure point is at a zero of the slope of the denominator function, $P(s)$.

3. Each pole pair root locus approaches an asymptote that intersects the real axis at σ_o and diverges from the real axis at an angle of

$$\phi = \pm \frac{180° + m360°}{n_p}, \qquad m = 1, 2, \ldots, \tfrac{1}{2}n_p. \tag{11.3-4}$$

For example, the asymptotes for a third-order system are at $\pm 60°$ and at $180°$ (seen as dotted lines on Figure 11.3-2), for a fourth-order system at $\pm 45°$ and $\pm 235°$, and for a fifth-order system at $\pm 36°$, $\pm 108°$, $180°$.

Some conclusions that can be drawn from these construction rules follow:

- The two lowest frequency poles form a pole pair that will have the greatest significance in determining amplifier high-frequency response.
- Poles that are at higher frequencies than the lowest frequency pole pair will migrate farther away from the pole pair. For example, in a third-order system, the second and third poles migrate away from each other.
- If the lowest frequency pole pair is dominant before the application of feedback, it will remain dominant.

The consequence of these conclusions is the treatment of multipole amplifiers as if they were two-pole amplifiers as long as the second and third poles are at least two octaves apart:

$$\left|\frac{\omega_3}{\omega_2}\right| \geq 4. \tag{11.3-5}$$

The low-frequency response of a feedback amplifier is also approximately determined by its dominant poles: the pole pair that is closest to the midband region.

EXAMPLE 11.3-1   A three-pole amplifier with midband gain $A_o = 1000$ and three high-frequency poles $f_1 = 100\,\text{kHz}$, $f_2 = 1\,\text{MHz}$, and $f_3 = 6\,\text{MHz}$ has feedback applied so that the midband gain is reduced to

$$A_{of} = 80.$$

Determine the high 3-dB frequency.

SOLUTION

This amplifier is the same as the amplifier of Example 11.2-1 with the addition of an additional prefeedback pole at 6 MHz. The ratio of the second and third pole frequencies is given by

$$\left|\frac{\omega_3}{\omega_2}\right| = \left|\frac{f_3}{f_2}\right| = \frac{6\,\text{MHz}}{1\,\text{MHz}} = 6 \geq 4.$$

The pole pair is dominant and the results of Example 11.2-1 are approximately valid:

$$f_H \approx \textbf{1.43 MHz}.$$

A computer simulation of the exact response (including all three poles) found the 3-dB frequency to be at $f_H = 1.489\,\text{MHz}$. The approximate result is entirely satisfactory with less than -4% variation in the 3-dB frequency from the exact result. The three feedback amplifier poles can be numerically found to be

$$p_1 = -0.4332 + j1.054\,\text{MHz},$$
$$p_2 = -0.4332 - j1.054\,\text{MHz},$$
$$p_3 = -6.233\,\text{MHz}.$$

Notice that the sum of the feedback poles is $-7.1\,\text{MHz}$, the same as the sum of the nonfeedback poles.

11.4 STABILITY IN FEEDBACK CIRCUITS

It was shown in Chapter 8 that the benefits of negative feedback were obtained at the expense of reduction in gain by the reduction factor D. An additional drawback to using feedback in circuits is the possibility of self-oscillation; that is, the amplifier may become an unstable system. In general, a stable electronic system has an output response that decays to zero with time when excited by any initial energy in the system. The initial energy can take the form of random noise in the power supply rails. A stable system in steady state does not have a time-varying output when the input is zero. An unstable or marginally stable electronic system will have some form of output when the input is zero.

Instability in electronic circuits stems from the fact that the forward gain of the amplifier and reverse gain from the feedback elements may be frequency sensitive. At low and high frequencies, the output voltage may be shifted in phase and changed in magnitude relative to the midband frequencies. At the mixing point in the feedback circuit, the input signal to the circuit and the output signal of the feedback network may, because of this additional phase shift, add rather than subtract, which can result in possible oscillation.

Unstable circuit behavior can be visualized by studying Equation 8.1-5, which is repeated here:

$$A_f(s) = \frac{A(s)}{1 + A(s)f}. \tag{11.4-1}$$

When $A(s)f = -1$, the total gain of the circuit $A_f(s) \to \infty$, a condition that is intolerable in amplifiers and represents an output that is essentially limited by the power rails.[3]

For the loop gain $A(s)f$ to be negative, the combination of phase shifts by the forward gain and the feedback ratio must cause a summation of in-phase signals at the mixing point of the feedback amplifier. Stable amplifier operation requires the magnitude of the loop gain, $|A(s)f|$, to be less than unity when its phase angle approaches 180°.[4]

11.4.1 Gain and Phase Margins

Amplifier stability can be analyzed using the magnitude and phase versus frequency plots of the loop gain. Figure 11.4-1 shows the magnitude and phase versus frequency plots of a three-pole loop gain function. The *gain margin* is the difference in gain magnitude (in decibels) between the 0-dB loop gain (unity magnitude) and the loop gain magnitude at a phase angle $\angle A(s)f$ of 180° from the midband, that is, gain margin $= -|A(s)f|_{180°}$ in decibels. A *positive value for the gain margin indicates stable amplifier operation*; if the gain margin is near zero, the amplifier is unstable and its output will oscillate without regard to its input.

The *phase margin* is the difference between the loop gain phase angle $\angle A(s)f$ at $|A(s)f| = 0$ dB (unity magnitude) and $\angle A(s)f = -180°$; that is,

$$\text{Phase margin} = \angle A(s)f|_{0dB} + 180°. \tag{11.4-2}$$

For stable operation, the phase margin must be greater than 0°.

Alternately, the stability of a feedback amplifier can be determined from the difference between the magnitude and phase versus frequency plots of the open-loop gain and the feedback ratio. If the feedback ratio is frequency independent (resistive), $20 \log|1/f|$ is a straight horizontal line on the open-loop gain magnitude versus frequency plot. The phase of f is constant for a resistive feedback network. In

---

[3] The output voltage is actually limited by the nonlinear regions of the transistor characteristics.
[4] Recall that a 180° phase shift causes a sign change in the signal.

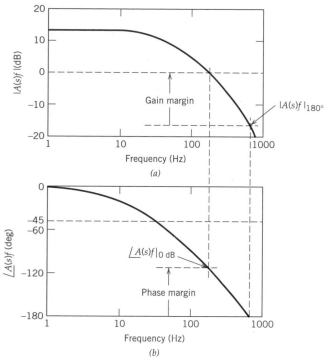

Figure 11.4-1
Definition of (*a*) gain margin and (*b*) phase margin.

the midband region, $A(s)$ and f both have the same sign: Their phases must be identical at either 0° or 180°. The magnitude of the loop gain (in decibels) can be expressed as the difference of the magnitude of the open-loop gain and the magnitude of the inverse of the feedback ratio,

$$20 \log|A(s)f| = 20 \log|A(s)| - 20 \log\left|\frac{1}{f}\right|. \tag{11.4-3}$$

Equation 11.4-3 can be used to plot the magnitude and phase versus frequency characteristics for the open-loop gain and the feedback ratio to determine amplifier gain and phase margins for varying values of f. Figure 11.4-2 shows how the plots of the open-loop gain and feedback ratio can be used to determine the gain and phase margins. The phase of the open-loop amplifier is plotted as a relative phase difference to the midband phase of $A(s)$. The figure shows a straight horizontal line (for a resistive network) representing the feedback ratio, $20 \log|1/f|$, superimposed on the magnitude versus frequency plot of the open-loop gain. The loop gain is the difference between the open-loop gain and the inverse of the feedback ratio. The gain margin is the difference (in decibels) of the inverse of the feedback ratio to the open-loop gain magnitude (in decibels) at a phase angle $\angle A(s)$ of 180°; that is,

$$\text{Gain margin (dB)} = 20 \log|1f| - 20 \log|A(s)|_{180°}. \tag{11.4-4}$$

The phase margin is the difference of the phase angle at the intersection of the magnitudes of the open-loop gain and inverse of the feedback ratio to $\angle A(s) = -180°$.

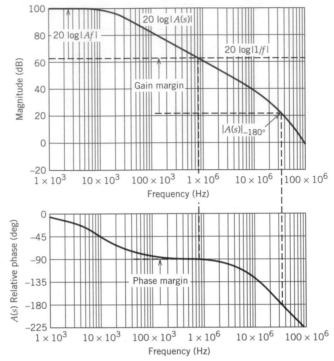

Figure 11.4-2
Alternate stability analysis technique.

It is common to specify the design of feedback amplifiers with gain and phase margins greater than 10 dB and 50°, respectively. This assures stable amplifier operation over variations in component parameter values.

The amount of phase margin has a significant effect on the shape of the closed-loop magnitude response of the feedback amplifier. Using the definitions of gain and phase margins, the frequency ω_1 where the loop gain is unity is

$$A(j\omega_1)f = 1 \cdot e^{-j\theta}, \tag{11.4-5}$$

where

$$\theta = 180° - \text{phase margin.} \tag{11.4-6}$$

The closed-loop gain at ω_1 is

$$A_f(j\omega_1) = \frac{A(j\omega_1)}{1 + A(j\omega_1)f} \tag{11.4-7}$$

$$\frac{(1/f)e^{-j\theta}}{|1 + e^{-j\theta}|}.$$

The magnitude of the closed-loop gain is therefore

$$|A_f(j\omega_1)| = \frac{1/f}{|1 + e^{-j\theta}|} = \frac{1/f}{\sqrt{(1 + \cos\theta)^2 + \sin^2\theta}}. \tag{11.4-8}$$

Using Equation 11.4-8, the magnitude of the closed-loop gain for a 50° phase margin, or $\theta = 180° - 50° = 130°$, is

$$|A_f(j\omega_1)| = \frac{1/f}{\sqrt{(1 + \cos 130°)^2 + \sin^2 130°}} = \frac{1.18}{f}. \qquad (11.4\text{-}9)$$

Equation 11.4-9 indicates that the magnitude plot of the closed-loop gain will peak by a factor of 1.18, or 1.4 dB, above the midband gain value of $1/f$. For lower phase margins, the peaking in the magnitude plot increases. For example, for a 10° phase margin, the closed-loop gain magnitude plot will peak by 5.7, or 15.1 dB, above the midband gain at frequency at ω_1. For a 90° phase margin, the magnitude plot of the closed-loop gain at ω_1 is 0.707, of -3 dB, of the midband gain.

For a two-pole system or a three-pole system that is dominated by the two lowest pole frequencies, the phase margin is related to the damping coefficient in Equation 9.1-17, which is repeated here:

$$|A_f(\omega_{\text{peak}})| = |A_f(\omega_1)| = \frac{1}{2\zeta\sqrt{1 - \zeta^2}}. \qquad (11.4\text{-}10)$$

An equation solver may be used to determine ζ in Equation 11.4-10 for a given phase margin.

EXAMPLE 11.4-1 A feedback amplifier has an open-loop transfer function

$$A(s) = \frac{100 \times 10^3}{(1 + s/2\pi(10 \times 10^3))(1 + s/2\pi(200 \times 10^3))(1 + s/2\pi(20 \times 10^6))}$$

and resistive feedback ratio $f = \frac{1}{100}$.

  a) Find the gain and phase margins.
  b) Is the amplifier stable?
  c) Plot the closed-loop gain response.

SOLUTION

(a) The gain and phase margins are found by plotting (either asymptotic or computer) the open-loop transfer function and the feedback ratio as shown.

From the frequency plots, the phase margin equals $+10°$ and the gain margin equals $+10$ dB.

(b) The feedback amplifier is just stable since both the gain margin and the phase margin are greater than zero. However, since the phase margin is less than $+50°$, variations in temperature and component parameters may cause the amplifier to become unstable.

Since the circuit frequency response has dominant-pole characteristics, the approximate analysis discussed in this section may be used to determine the amount of peaking. It was shown in this section that a 10° phase margin corresponded to an amplitude peak of 15.1 dB relative to the midband gain.

(c) The closed-loop response exhibits peaking of slightly greater than 15 dB above the midband gain.

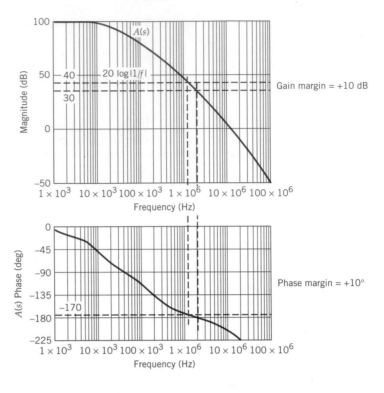

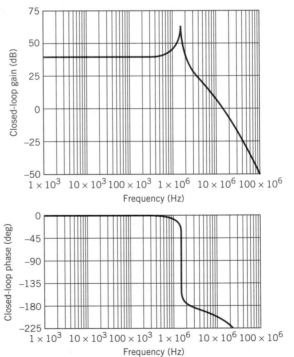

11.4.2 Nyquist Stability Criterion

An alternate method for analyzing feedback amplifiers was developed by Nyquist in 1932. The Nyquist criterion can determine whether a linear amplifier is stable.

A closed-loop feedback amplifier with gain $A_f(s)$ is stable if it has no poles with positive or zero real parts. Assuming that the open-loop gain and the feedback ratio are stable,[5] only the loop gain $A(s)f$ need be inspected for poles with positive or zero real parts.

The Nyquist diagram is simply a polar plot of $A(s)f$ for s on the contour shown in Figure 11.4-3a.

The Nyquist diagram maps the right half of the s plane, shown in Figure 11.4-3a, into the interior of the contour in the Af plane (Figure 11.4-3b). If there are zeros of $D(s)$ in the right-half plane (RHP), the Af plane contour will enclose the point $-1\angle 0°$, which is called the critical point. The number of times that the Af plane contour encircles the critical point in a counterclockwise direction is equal to the number of zeros of $D(s)$ with positive real parts.

The behavior of the closed-loop response is determined largely by the nearness of the plot of $A(j\omega)f$ to the -1 point on the Nyquist diagram. Figure 11.4-4a is a plot of a loop gain response that encircles the critical point and is unstable by the Nyquist criterion. Figure 11.4-4b is a plot of a stable amplifier whose loop gain response does not encircle the critical point.

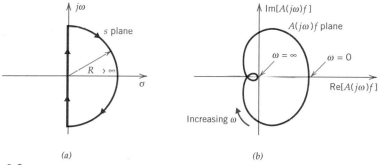

(a)   *(b)*

Figure 11.4-3
(*a*) Nyquist s-plane contour. (*b*) An $A(j\omega)f$ plane contour for loop gain with three identical poles.

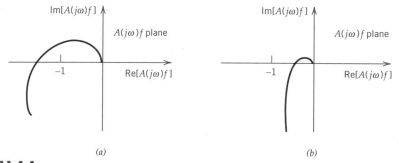

(a)   *(b)*

Figure 11.4-4
Nyquist diagram for (*a*) stable loop gain and (*b*) unstable loop gain.

---

[5] The open-loop gain in electronic amplifiers is assumed to be stable. However, if the open-loop amplifier contains feedback elements, the open-loop gain must be analyzed for stability. Passive-component feedback ratios are stable.

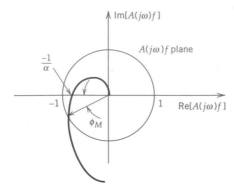

Figure 11.4-5
Gain margin and phase margin for Nyquist diagram.

The Nyquist diagram can be used to find the gain and phase margins of a feedback amplifier. Since stability is related to the amplifier gain at $-180°$ phase shift, the gain margin and phase margin are defined in a Nyquist diagram as follows (Figure 11.4-5):

Gain margin. For a stable amplifier, the ratio $1/\alpha$ is in decibels, where α is the distance from the $-180°$ crossover on the Nyquist diagram [where the plot of $A(j\omega)f$ crosses the negative real axis] to the origin, corresponds to the gain margin, that is, gain margin (dB) $= 20 \log|1/\alpha|$. If the Nyquist diagram has multiple $-180°$ crossovers, the gain margin is determined by that point that lies closest to the critical point.

Phase margin. For a stable amplifier, the phase margin is the magnitude of the minimum angle ϕ_M between a line from the origin to the point where the Nyquist diagram of $A(j\omega)f$ intersects a circle of unit radius with the center at the origin and the negative real axis.

EXAMPLE 11.4-2 A three-stage transistor amplifier with feedback was found to have a loop gain frequency response that is approximately

$$A(j\omega)f = \frac{5}{(1 + j\omega/2\pi(10^4))^3}.$$

a) Plot the loop gain on the complex plane.
b) Is the amplifier stable?
c) At what frequency does a 180° phase shift occur for the amplifier?
d) What is the maximum magnitude of the loop gain for stable operation for networks having this form of transfer function?
e) What are the gain and phase margins?

SOLUTION

(a) Using MathCAD, the given Nyquist plot was drawn.
(b) The amplifier is stable since the Nyquist plot does not encircle the $-1 + j0$ point.

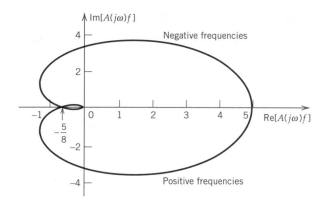

(c) From the expression of the loop gain, the intersection of the plot with the negative real axis occurs when the imaginary part of the loop gain is zero. The denominator of the loop gain is expanded to solve for the frequency where the imaginary part of the loop gain is zero,

$$\left(1 + \frac{j\omega}{\omega_o}\right)^3 = \left(1 - 3\frac{\omega^2}{\omega_o^2}\right) + j\left(3\frac{\omega^2}{\omega_o^2} - \frac{\omega^3}{\omega_o^3}\right),$$

where $\omega_o = 2\pi(10 \times 10^3)$. This expression is real when

$$3\frac{\omega^2}{\omega_o^2} - \frac{\omega^3}{\omega_o^3} = 0$$

at the freqneucy $\omega_o = 108.8$ krad/s $\Rightarrow$ 17,320 Hz.
    At 17,320 Hz, the loop gain is

$$|A(j\omega)f|_{\omega=108.8 \text{ krad/s}} = \frac{5}{1 - 3(\omega^2/\omega_o^2)}$$

$$= \frac{5}{1 - 3(108.8 \times 10^3/2\pi(10 \times 10^3))} = -\frac{5}{8}.$$

This valus is greater than -1, and as is evident from the plot of the loop gain, no encirclement of $-1 + j0$ occurs at this value of gain.

(d) The gain can be increased by $\frac{8}{5}$ before $-1 + j0$ is intersected. Therefore, the condition for absolute stability for amplifiers having this form of loop gain transfer function is

$$|A(j\omega)f| < (\tfrac{8}{5})5 = 8.$$

However, by doing so, there will no longer be any margin for stability and the amplifier transient response will become increasingly oscillatory as this limiting value of gain is approached.

(e) From part (c), $\alpha = \frac{5}{8}$ and $20\log(1/\frac{5}{8}) = 4.08$ dB is the gain margin. Data from the Nyquist plot show $A(j\omega)f = 1$ at $\omega = 87.5$ krad/s $\Rightarrow$ 13.88 kHz and $\angle A(j\omega)f = 17.3°$. The phase margin is $|\angle A(j\omega)f - 180°| = 168°$.

11.5 COMPENSATION NETWORKS

Feedback amplifiers have been shown to be unstable if they are characterized by negative phase margins. In negative-feedback amplifiers, the potential for instability is present when the amplifier open-loop (loaded basic forward) transfer function has three or more poles. Unstable amplifiers can be stabilized by the following:

- Decreasing the loop gain Af of the amplifier
- Adding a compensation network to the amplifier to shape the loop gain frequency response characteristic so that the phase and gain margins are positive and in the acceptable range (desirable phase margin $\geq 50°$ and gain margin ≥ 10 dB)

Careful design is required in each of these cases to ensure stable amplifier operation over temperature and parameter variations.

In many cases, decreasing the loop gain to achieve stability is not acceptable due to bias or amplifier gain constraints. Alternately, the designer may not have control over f, as in the case of OpAmps. Here the amplifier must operate over a wide range of feedback ratios determined by the user rather than the designer of the OpAmp. In such cases, compensation networks are added to the amplifiers to increase gain and phase margins.

The placement of compensation networks in feedback amplifiers is of some importance. The basic topology of a negative-feedback amplifier of Figure 8.1-1 is repeated in Figure 11.5-1a. The triangular symbol is a linear amplifier of gain A, and the rectangular symbol is the feedback network of feedback ratio f. *The compensation network must be placed in the signal path of the linear amplifier and the feedback network*, as shown in Figure 11.5-1b. Depending on the feedback topology, the compensation network could be placed between amplifier stages internal to a multistage linear amplifier or at the output of a linear amplifier. It is necessary that the signal pass through the amplifier, compensation network, and feedback network.

Compensation networks add poles or a combination of poles and zeros to the loop gain transfer characteristic to achieve desired gain and phase margins. The following compensation techniques and their passive-component circuit implementations are presented in this section:

- Dominant pole
- Lag–lead
- Lead
- Phantom zero

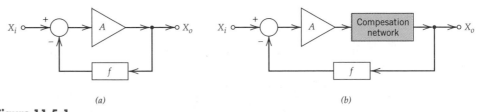

(a) *(b)*

Figure 11.5-1
(*a*) Basic negative-feedback topology. (*b*) Compensation network added to negative-feedback topology.

11.5.1 Dominant-Pole (Lag) Compensation

The simplest form of compensation adds an additional dominant (real) pole in the transfer characteristic of the open-loop amplifier gain. The transfer function of the dominant-pole compensation network is

$$H_{DP}(s) = \frac{1}{1 + s/\omega_p}, \tag{11.5-1}$$

where the pole location of the network, ω_p, is significantly smaller than the poles of the uncompensated amplifier: $\omega_p \ll \omega_{p1}, \omega_{p2}, \ldots$.

When this compensation network is added to the circuit, the open-loop gain becomes

$$A_{comp}(s) = A(s)H_{DP}(s) = \frac{A(s)}{1 + s/\omega_p}. \tag{11.5-2}$$

The dominant pole added by the compensation network is chosen so that the loop gain $|A_{comp}(s)f|$ is 0 dB at a frequency where the poles of the uncompensated open-loop gain $A(s)f$ contribute negligible phase shift. Typically, the compensation network adds a dominant pole that reduces the compensated loop gain to unity (0 dB gain) at the lowest high-frequency pole, ω_{p1}, of the open-loop amplifier. The compensation network causes a phase "lag" in the signal path. Therefore, the phase of the compensated loop gain is shifted lower in frequency, which results in increased gain and phase margins.

Figure 11.5-2 shows the method for finding the dominant pole of the compensation network. The dominant pole of the compensation network is designed so that its transfer function has a gain of 0 dB at the first pole frequency of the uncompensated loop gain, with a slope of -20 dB/decade. The dominant-pole frequency of the compensation network, ω_p, is found through the following relationship:

$$0 = 20 \log|A_{midband}| - 20 \log\left(\frac{\omega_{p1}}{\omega_p}\right), \tag{11.5-3a}$$

or simply,

$$\omega_p = \frac{\omega_{p1}}{A_{midband}}. \tag{11.5-3b}$$

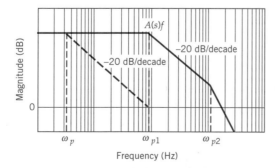

Figure 11.5-2
Construction technique for finding dominant-pole location.

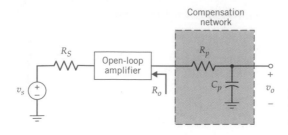

Figure 11.5-3
Dominant-pole compensation network at output terminal of open-loop amplifier.

A small-signal model of the amplifier can be used to determine the (loaded basic forward) open-loop transfer function. However, the transfer function of an open-loop amplifier becomes increasingly cumbersome for open-loop characteristics with more than three poles. Therefore, it may be preferable to use SPICE computer simulations to provide a graphical output of the compensated loop gain transfer characteristic from which the new pole locations can be found.

A simple dominant-pole compensation network, shown in the shaded area in Figure 11.5-3, is added to the open-loop amplifier. As stated previously, the compensation network could, depending on the feedback topology, reside within the open-loop amplifier so that the signal passes through the linear amplifier, compensation network, and feedback network. Here, the dominant pole of the compensated open-loop amplifier transfer function is

$$\omega_p \approx \frac{1}{(R_p + R_o)C_p},$$

(11.5-4)

where R_o is the output resistance of the open-loop amplifier.

The procedure for designing a dominant pole compensation network is as follows:

1. Find the midband loop gain.
2. Find the dominant high frequency pole, ω_{p1}, or high 3-dB bandwidth of the loop gain transfer characteristic.
3. Design the dominant pole of the compensation network such that the compensated loop gain drops to 0 dB, with a slope of -20 dB/decade, at the uncompensated high 3-dB cutoff frequency.

The dominant-pole compensation network is a simple low-pass filter with cutoff frequency at ω_{p1} that is significantly lower than the 3-dB frequency (f_H) of the open-loop amplifier. Therefore, the dominant-pole compensation network yields *greater phase margin at the expense of bandwidth.*

EXAMPLE 11.5-1 Consider the OpAmp inverting amplifier shown. The open-loop gain of the amplifier is

$$R_M = \frac{-10^4}{(1 + s/2\pi(10^6))(1 + s/2\pi(10 \times 10^6))(1 + s/2\pi(30 \times 10^6))}.$$

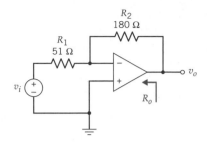

The input and output resistances of the OpAmp are $R_o = 75 \ \Omega$ and $R_i = 1 \ M\Omega$, respectively.

Find the gain and phase margins and, if necessary, compensate the circuit using a dominant-pole compensation network.

SOLUTION

Since the feedback topology of the amplifier is a shunt–shunt configuration, the feedback ratio is $f = -1/R_2 = -1/180$. The gain and phase versus frequency plots of the loop gain, $R_M f$, shown clearly indicate negative gain and phase margins. Therefore, the amplifier is unstable in its current configuration.

A dominant-pole compensation network is used to stabilize the amplifier. The pole location of the compensation network is found by following the suggested dominant-pole compensation network design procedure in this section.

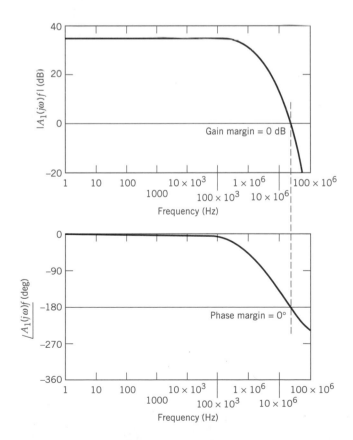

The dominant-pole frequency of the uncompensated loop gain, equal to the dominant pole of the open-loop amplifier, is 1 MHz. The midband loop gain is 34.9 dB. The compensation network is designed so that the magnitude of the loop gain at 1 MHz is 0 dB. Since the compensation network has a slope of -20 dB/decade, the location of the dominant pole, f_p, of the compensation network is easily found by using the following formula:

$$0 = 20 \log |R_{M,midband}| - 20 \log \left(\frac{10^6}{f_p} \right),$$

where $R_{M,midband} = 34.9$ dB. Solving for f_p yields $f_p = 17.9$ kHz.

The compensation circuit has a dominant pole at $f_p = 17.9$ kHz. The component values of the compensation network are found for $R_p = 0$, and $R_o < R_2$ to ensure

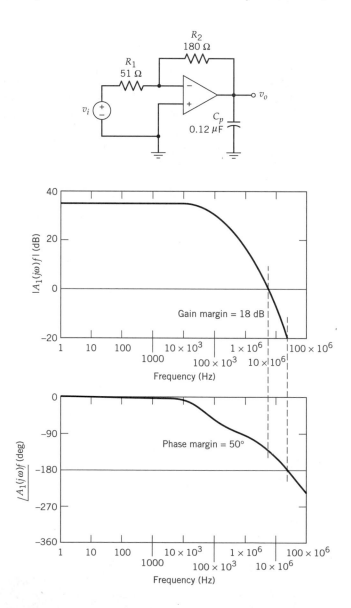

that the feedback resistor does not influence the dominant pole of the compensation network. The capacitance of the compensation network is therefore

$$C_p = \frac{1}{2\pi f_p R_o} = \frac{1}{2\pi(17.9 \times 10^3)(75)} = 0.118 \ \mu\text{F} \approx 0.12 \ \mu\text{F}.$$

The compensated amplifier is as shown.

The gain and phase versus frequency plots for the compensated amplifier are also shown.

From the plots, the gain margin is 18 dB and the phase margin is 50°. These values are acceptable design margins for stability.

EXAMPLE 11.5-2 Given the two-stage shunt–series feedback amplifier shown, find the gain and phase margins. If necessary, compensate the circuit using a dominant-pole compensation network. The capacitors C_S, C_F, C_{E1}, and C_{E2} are very large and contribute only to the lower 3-dB frequency.

The BJT SPICE model parameters are $\beta_F = 200$, CJE = 35.5 pF, CJC = 19.3 pF, and TF = 477 ps.

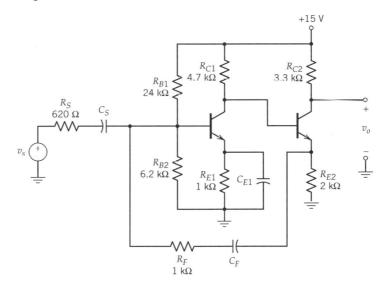

SOLUTION

Direct-current analysis shows that the bias voltages and currents of the circuit are $I_{CQ1} = 1.8$ mA, $V_{CEQ1} = 2.0$ V, $I_{CQ2} = 2.3$ mA, and $V_{CEQ2} = 5.7$ V.

The loop gain frequency characteristics are found with SPICE using the following circuit shown. The circuit is modified to take into account the loading due to the feedback elements. The input voltage source has been converted to a Norton source since the open-loop current gain is the quantity of interest in shunt–series amplifiers.

The input and output loading effects due to the feedback network are $R_{F1} = R_F + R_{E2} = 3$ kΩ and $R_{F2} = 1$ kΩ, respectively. The large capacitor C_F is used to block the DC currents and voltages in order to maintain the bias conditions.

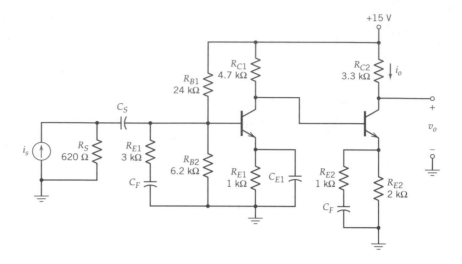

The feedback ratio is

$$f = \frac{R_{E2}}{R_{E2} + R_F} = \frac{2000}{2000 + 1000} \approx 0.67.$$

SPICE is used to determine the frequency response of the loop gain characteristics. The results are as shown below.

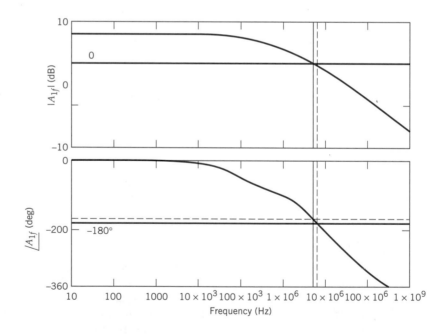

Clearly, the loop-gain characteristics show a phase margin approximately equal to zero, indicating that the amplifier is close to becoming unstable. There are two poles below 1 GHz that are approximately 66 kHz and 1.1 MHz.

A dominant-pole compensation network is added to the open-loop amplifier to achieve a phase margin of greater than 50°. The dominant-pole frequency of the uncompensated loop gain is 66 kHz. The magnitude of the midband loop gain is 42.8 dB. The compensated loop gain is designed so that the magnitude of the gain at 66 kHz is 0 dB. Since the compensation network has a slope of −20 dB/decade, the location of the dominant pole, f_p, of the compensation network is easily found by using the following formula:

$$0 = 20 \log|A_{I,\text{midband}}| - 20 \log\left(\frac{66 \times 10^3}{f_p}\right),$$

where $A_{I,\text{midband}} = 42.8$ dB. Solving for f_p yields $f_p = 478$ Hz.

The compensation circuit is designed for a dominant pole at $f_p = 478$ Hz. The output resistance of the first transistor stage is 4.7 kΩ. The input resistance R_{i2} of the second transistor stage is very large due to the presence of R_{E2} and can therefore be ignored. Series sampling presents serious obstacles to having the compensation network follow the basic amplifier. Thus a location between the two BJT stages of amplification was chosen. The component values of the compensation network are found for $R_p = 0$, and $R_o < R_{i2}$ to ensure that the frequency response due to the second stage is not significantly altered. The capacitance of the compensation network is therefore

$$C_p = \frac{1}{2\pi f_p R_o} = \frac{1}{2\pi(478)(4700)} = 71 \text{ nF} \approx 0.068 \ \mu\text{F}.$$

The compensated open-loop amplifier is as shown below.

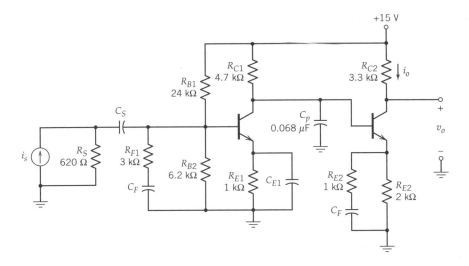

The compensated loop gain transfer characteristics are shown in the following SPICE output. The addition of the dominant-pole compensation network has increased the phase margin to approximately 85° and gain margin to approximately

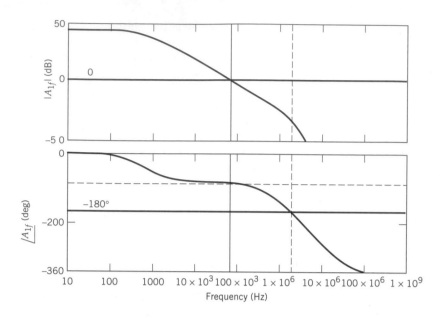

30 dB at a significant cost to bandwidth. From the SPICE plot, the bandwidth of the loop gain response is only 557 Hz.

The closed-loop amplifier circuit with dominant-pole compensation is shown.

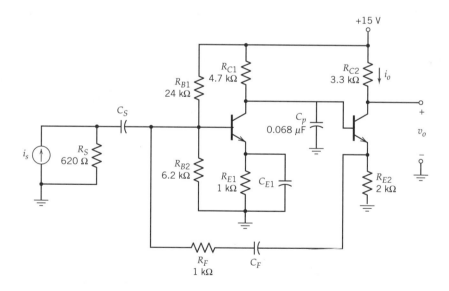

SPICE frequency response plots of the compensated closed-loop amplifiers are also shown. The 3-dB frequency of the compensated amplifier is 67 kHz, which is significantly lower than the uncompensated 3-dB frequency of 3.5 MHz. There is no gain penalty when the lag compensation network is used.

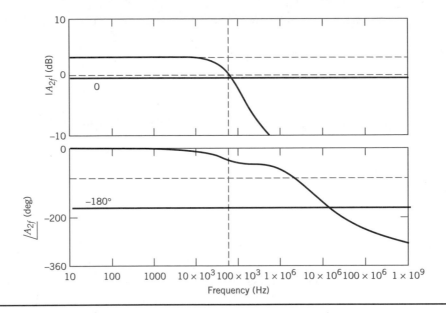

11.5.2 Lag–Lead (Pole–Zero) Compensation

Although dominant-pole (lag) compensation is successful in reducing the loop gain to 0 dB before the phase shift of the open-loop amplifier becomes excessive, the amplifier bandwidth is significantly reduced. In many cases, design specifications may call for maximum bandwidth and a specified closed-loop gain. The lag–lead compensation network, which introduces both a pole and a zero, will usually yield a wider bandwidth amplifier than a dominant-pole network. The transfer function of the lag–lead network is

$$H_{LL}(s) = \frac{1 + s/\omega_z}{1 + s/\omega_p}, \qquad \omega_p < \omega_z, \tag{11.5-5}$$

where ω_z is the zero location and ω_p is the pole location of the lag–lead network.

The lag–lead compensated open-loop gain is then

$$A_{comp}(s) = H_{LL}(s)A(s) = A(s)\frac{1 + s/\omega_z}{1 + s/\omega_p}. \tag{11.5-6}$$

The zero is chosen at the smallest high-frequency pole location of the open-loop amplifier transfer function. This has the effect of increasing the compensated bandwidth over the simple dominant-pole compensated amplifier. For example, given an open-loop gain transfer function

$$A(s) = \frac{A_{midband}}{(1 + s/\omega_{p1})(1 + s/\omega_{p2})(1 + s/\omega_{p3})}, \tag{11.5-7}$$

where ω_{p1} is the smallest pole, the lag–lead compensated transfer function becomes

$$A_{comp}(s) = H_{LL}(s)A(s)$$

$$= \frac{A_{midband}(1 + s/\omega_z)}{(1 + s/\omega_p)(1 + s/\omega_{p1})(1 + s/\omega_{p2})(1 + s/\omega_{p3})}. \quad (11.5\text{-}8)$$

The smallest pole of $A(s)$ is canceled by setting the zero of the lag–lead network at the same point; that is,

$$\omega_z = \omega_{p1}. \quad (11.5\text{-}9)$$

The open-loop gain becomes a three-pole transfer function with a dominant pole ω_p,

$$A_{comp}(s) = \frac{A_{midband}}{(1 + s/\omega_p)(1 + s/\omega_{p2})(1 + s/\omega_{p3})}. \quad (11.5\text{-}10)$$

Figure 11.5-4 shows the method for finding the dominant pole of the lag–lead compensation network. The dominant pole of the compensation-network is designed so that its transfer function has a gain of 0 dB at the second pole frequency of the uncompensated loop gain, with a slope of -20 dB/decade. The dominant-pole frequency of the compensation network, ω_p, is found through the following relationship:

$$0 = 20 \log|A_{midband}| - 20 \log\left(\frac{\omega_{p2}}{\omega_p}\right), \quad (11.5\text{-}11a)$$

or simply,

$$\omega_p = \frac{\omega_{p2}}{A_{midband}}. \quad (11.5\text{-}11b)$$

The result of the introduction of a zero, in addition to the dominant pole, in the lag–lead compensation network is an increase in open-loop (and therefore in the closed-loop) bandwidth.

A lag–lead compensation network implementation is shown in Figure 11.5-5. As stated previously, the compensation network could, depending on the feedback topology, reside within the open-loop amplifier so that the signal passes through the

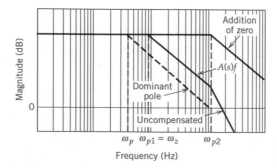

Figure 11.5-4
Construction technique for finding lag–lead compensation pole and zero locations.

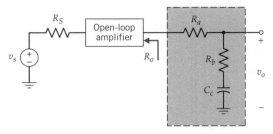

Figure 11.5-5

Lag–lead compensation network at output terminal of open-loop amplifier.

linear amplifier, compensation network, and feedback network. The dominant pole of the compensated open-loop amplifier transfer function is

$$\omega_p = \frac{1}{(R_a + R_b + R_o)C_c},$$

(11.5-12)

where R_a is the output resistance of the open-loop amplifier. The zero location of the compensation network is located at the lowest pole of the open-loop amplifier to increase the loop gain bandwidth,

$$\omega_z = \omega_{p1} = \frac{1}{R_bC_c}.$$

(11.5-13)

The Bode diagrams for the transfer function of the lag–lead compensation network are shown in Figure 11.5-6.

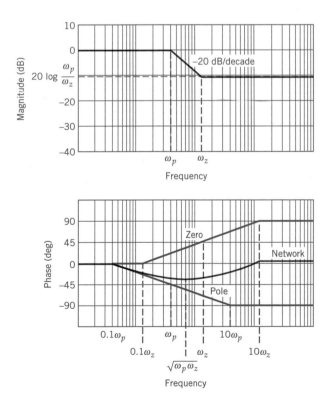

Figure 11.5-6

Bode diagrams for lag–lead compensation network.

The procedure for designing a lag–lead compensation network is as follows:

1. Find the midband loop gain.
2. Find the dominant high frequency pole, ω_{p1}, or high 3-dB bandwidth of the loop gain transfer characteristic.
3. Find the second to the lowest pole, ω_{p2}.
4. Design the compensation network so that the compensated loop gain drops to 0 dB, with a slope of -20 dB/decade, at ω_{p2} of the uncompensated loop gain characteristics.
5. Design the compensation network so that the zero of the network is at ω_{p1} of the uncompensated loop gain characteristic.
6. In some cases, the zero location may have to be changed due to interactions with the reactive components of the open-loop amplifier.

There is no loop gain magnitude penalty when using a lag–lead compensation network.

11.5.3 Lead Compensation (Equalizer)

The midband loop gain is often fixed by the specifications on the closed-loop gain and bandwidth of an amplifier. If a dominant-pole (or a lag–lead) network results in a bandwidth that is too narrow, an alternate solution must be found. Since stability depends only on the phase margin, a compensation network that would introduce phase *lead* at this point would be the desired alternate solution. A simple lead network, or equalizer, has a transfer function

$$H_{eq}(s) = \frac{s + \omega_{eq,z}}{s + \omega_{eq,p}}, \qquad \omega_{eq,p} \gg \omega_{eq,z}. \tag{11.5-14}$$

Note that the transfer function of the lead compensation network is similar to the lag–lead compensation characteristics in Equation 11.5-4, except that here the pole occurs at a higher frequency than the zero.

Using Equation 11.5-11, the lag–lead compensated open-loop gain is

$$A_{comp}(s) = H_{eq}(s)A(s) = A(s)\frac{s + \omega_{eq,z}}{s + \omega_{eq,p}}. \tag{11.5-15}$$

The zero is chosen at the *second high-frequency pole location* of the loop gain transfer function, $\omega_{eq,z} = \omega_{p2}$. The pole location is chosen to be significantly large so that $\omega_{eq,p}$ does not affect the bandwidth of the loop gain. This has the effect of increasing the compensated bandwidth over both the dominant-pole and lag–lead compensated amplifiers. However, there is a gain penalty that is incurred when using lead compensation networks. The midband gain yields

$$H_{eq}(0) = \frac{\omega_{eq,z}}{\omega_{eq,p}}. \tag{11.5-16}$$

This attenuation must be taken into account since it directly impacts midband gain. Figure 11.5-7 shows the method for finding the zero of the lead compensation

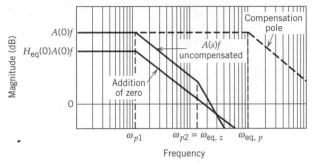

Figure 11.5-7
Construction technique for finding lead compensation pole and zero locations.

network. The zero of the compensation network is designed so that its transfer function has zero at the second-pole frequency of the uncompensated loop gain, with a slope of -20 dB/decade, $\omega_{eq,z} = \omega_{p2}$. The zero frequency of the compensation network is found through the following relationship:

$$0 = 20 \log|A_{\text{midband}}| - 20 \log\left(\frac{\omega_{p1}}{\omega_z}\right), \tag{11.5-17a}$$

or simply,

$$\omega_z = \frac{\omega_{p1}}{A_{\text{midband}}}. \tag{11.5-17b}$$

For example, if an open-loop transfer function has the form

$$A(s) = \frac{A_{\text{midband}}}{(1 + s/\omega_{p1})(1 + s/\omega_{p2})(1 + s/\omega_{p3})}, \tag{11.5-18}$$

where ω_{p1} is the smallest pole, the lead compensated loop gain transfer function becomes

$$A_{\text{comp}}(s)f = \frac{A_{\text{midband}}H_{\text{eq}}(0)f(1 + s/\omega_{eq,z})}{(1 + s/\omega_{eq,p})(1 + s/\omega_{p1})(1 + s/\omega_{p2})(1 + s/\omega_{p3})}. \tag{11.5-19}$$

The second lowest pole of $A(s)$ is canceled by setting the zero of the lead network at the same point; that is,

$$\omega_{eq,z} = \omega_{p2}. \tag{11.5-20}$$

The lead network cancels the second lowest amplifier pole. The pole in the lead compensation network is chosen to be large enough so that it has little effect on the phase margin. The open-loop gain becomes the three-pole transfer function,

$$A_{\text{comp}}(s)f = \frac{A_{\text{midband}}H_{\text{eq}}(0)f}{(1 + s/\omega_{eq,p})(1 + s/\omega_{p1})(1 + s/\omega_{p3})}, \tag{11.5-21}$$

where $\omega_{eq,p} \gg \omega_{p3} > \omega_{p1}$ or, if ω_{p1} is dominant, $\omega_{p3} \gg \omega_{eq,p} \gg \omega_{p1}$.

The introduction of a zero in the lead compensation network results in an increase in open-loop (and therefore in closed-loop) bandwidth.

A lead compensation network implementation is shown in Figure 11.5-8. As stated previously, the compensation network could, depending on the feedback topology, reside within the open-loop amplifier so that the signal passes through the linear amplifier, compensation network, and feedback network. The zero of the compensation network is located at the second lowest pole frequency of the open-loop amplifier characteristics to increase bandwidth,

$$\omega_{eq,z} = \frac{1}{R_a C_c} .$$ (11.5-22)

The pole location of the compensation network,

$$\omega_{eq,p} = \frac{1}{(R_a \parallel R_b)C_c},$$ (11.5-23)

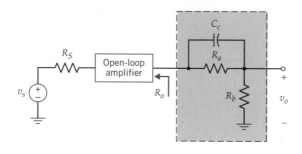

Figure 11.5-8
Lead compensation network at output terminal of open-loop amplifier.

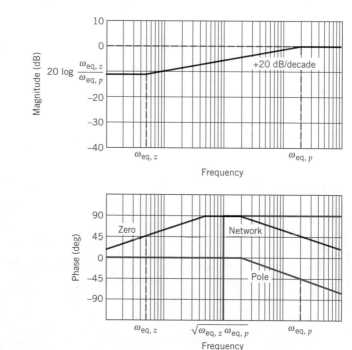

Figure 11.5-9
Bode diagrams for lead compensation network.

is chosen to be significantly large so as to have minimal effect on the loop gain crossover point (0-dB point). Using Equations 11.5-16, 11.5-22, and 11.5-23 yields the expression for the attenuation due to the compensation network,

$$H(0) = \frac{R_b}{R_a + R_b}.$$

(11.5-24)

Taking into account the output resistance of the open-loop amplifier, the actual output signal attenuation is

$$H'(0) = \frac{R_b}{R_o + R_a + R_b}.$$

(11.5-25)

The resistor values must be carefully selected so that the closed-loop amplifier gain is acceptable.

The Bode diagrams for the transfer function of the lead compensation network are shown in Figure 11.5-9.

EXAMPLE 11.5-3 Apply a lead compensation network to the two-stage shunt–series feedback circuit of Example 11.5-1.

SOLUTION

A lead compensation network is added to the open-loop amplifier to achieve a phase margin of greater than 50°. The resistor R_a is chosen as 47 kΩ, which is much larger than the $R_o = 4.7$ kΩ of the first transistor stage. This choice of R_a is used to find the capacitance value C_c of the compensation network,

$$\omega_{\text{eq,z}} \approx \frac{1}{R_a C_c} \Rightarrow C_c \approx \frac{1}{2\pi(1.1 \times 10^6)(47 \times 10^3)} = 3 \text{ pF}.$$

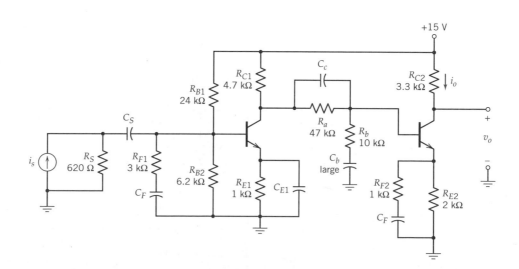

The pole frequency for the compensation network is chosen as $\omega_{eq,p} = 6.4$ MHz. Then R_b is given by

$$R_b = \frac{1}{\omega_{eq,p}C_c - 1/R_a}$$

$$= \frac{1}{2\pi(6.4 \times 10^6)(3 \times 10^{-12}) - 1/(47 \times 10^3)} \approx 10 \text{ k}\Omega.$$

The compensated loop gain transfer characteristics are shown below in the SPICE output. The addition of the lead compensation network has increased the phase margin to well above 50°. As expected, the midband gain is attenuated from that of the uncompensated midband loop gain.

The closed-loop amplifier circuit with lead compensation is shown below. The 3-dB frequency of the lead compensated amplifier is 955 kHz, which is lower than

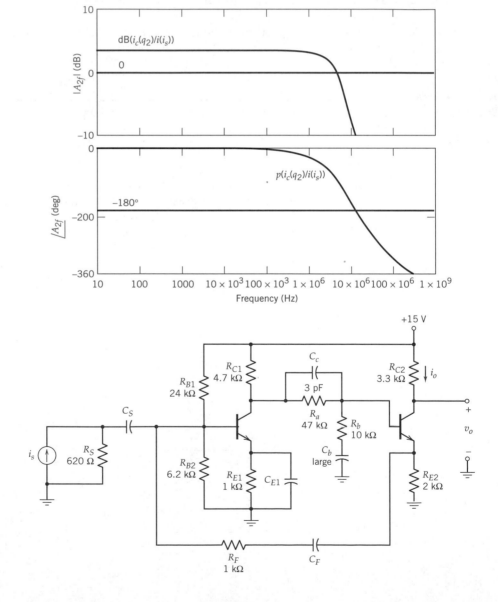

the 3.5 MHz of the uncompensated amplifier but is greater than the dominant-pole compensated amplifier frequency of 67 kHz.

SPICE frequency response plots of the compensated amplifier are shown.

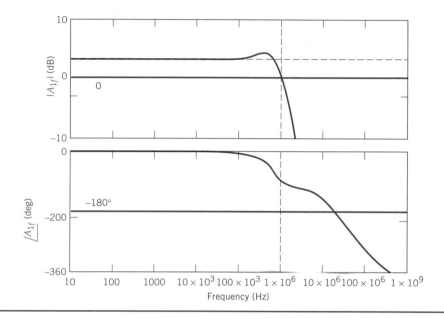

11.5.4 Phantom-Zero Compensation

Instead of shaping the open-loop transfer characteristics of an amplifier, the loop gain may be altered by the addition of reactive elements in the feedback network, f, for compensating a feedback amplifier. The use of reactive elements in the feedback circuit is essentially equivalent to adding lead compensation: The locations of the zero and the pole for a phantom compensation network is the same as that of the lead compensation network. Figure 11.5-10 shows phantom-zero compensation for a shunt–shunt feedback amplifier. The shaded area in the figure constitutes the phantom compensation network. Note that removal of the capacitor C_f results in a simple resistive shunt–shunt feedback amplifier.

Another slightly different version of phantom-zero compensation is the Miller compensation method shown in Figure 11.5-11. In the miller compensation method, a capacitor C_f is placed in shunt between the collector and base (drain and gate for

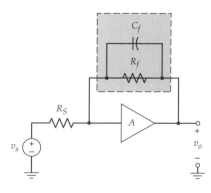

Figure 11.5-10
Phantom-zero compensation for shunt–shunt feedback amplifier.

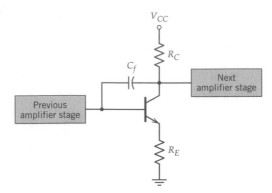

Figure 11.5-11
Miller's compensation method using shunting capacitor.

FETs) of an internal transistor amplifier stage that establishes the amplifier's lowest high-frequency pole ω_{p1}. The effect will be an increase in the input capacitance due to Miller's effect, which in turn reduces ω_{p1} and results in a dominant pole.

The advantage of Miller's compensation is the reduction of the capacitance value compared to the techniques shown in Sections 11.5.1–11.5.3. The small value of C_f is multiplied by the Miller effect factor (amplifier gain) resulting in a much larger capacitance value. Since small values of C_f can be used, Miller's compensation may be used in integrated circuit design.

11.6 CONCLUDING REMARKS

Frequency response of feedback amplifiers is discussed in this chapter. The effects of feedback on the poles in the low- and high-frequency regions were investigated. It was shown that pole locations changed with feedback. In particular, the low-frequency pole decreased with increased return ratio and the high-frequency pole increased with increased return ratio. This "pole-splitting" phenomenon was discussed for single-, double-, and multipole feedback frequency responses.

Time-domain effects of feedback were also shown to be dependent on the return ratio. The rise and delay times (with step inputs) decrease with increased return ratio, while overshoot and settling times increase once the amplifier becomes underdamped. Increased feedback proved to reduce the damping coefficient in double-pole feedback amplifiers, resulting in a decrease in the percent tilt.

The root-locus method was introduced as a tool to analyze the pole behavior with feedback. A simple set of rules was established to aid in the construction of approximate root-locus plots of amplifiers described by high-frequency poles. The concept of dominant poles was used to simplify amplifier frequency characteristics when applicable.

The stability of amplifiers was investigated. Stable systems were defined and the conditions for instability discussed. Design goals for stability were related to frequency response analysis (or Bode plots). Gain and phase margins were defined and used to quantify stability. An alternate method developed by Nyquist for determining stability was also discussed.

Methods to compensate for unstable amplifiers were established. These methods include the addition of electrical networks to shift the pole locations of the amplifiers. The trade-offs for each compensation technique were presented.

SUMMARY DESIGN EXAMPLE: FIBER-OPTIC TRANSIMPEDANCE PREAMPLIFIER

Fiber-optic receivers consist of a photodiode, which can be modeled as a current source and a parallel capacitor, and amplifying electronics. The first stage of amplification, which is often called the preamplifier, is critical in determining the performance of a fiber-optic receiver. A moderately low input resistance amplifier is often used as the preamplifier. The low-input resistance allows for the RC time constant of the amplifier and the photodiode shunt capacitance to be small, allowing for large-bandwidth operation. Since the input signal can be modeled as a current source, output voltages are desired, and large transresistances are required to amplify the low input currents, a transimpedance (transresistance) amplifier is often used.

Low-cost transistor components can be used for low-bandwidth applications. Low-data-rate (corresponding to moderate-bandwidth) fiber-optic links are often used in simple computer networks. For a 32-kbit/s link, a high-frequency cutoff of approximately 22.4 kHz is required. With proper digital encoding, the low cutoff frequency is approximately 1 kHz. The minimum midband transresistance requirement is -2000. A photodiode with a 10-pF shunt capacitance and responsivity of 0.9 A/W is used. A +6-V power supply is available. The *npn* BJT parameters are

$$\beta_F = 150, \qquad V_A = 160 \text{ V}, \qquad r_b = 30 \ \Omega,$$
$$C_{ibo} = 8 \text{ pF} \quad \text{at } V_{EB} = 0.5 \text{ V},$$
$$C_{obo} = 4 \text{ pF} \quad \text{at } V_{CB} = 5 \text{ V},$$
$$f_T = 300 \text{ MHz}.$$

Design the fiber-optic preamplifier to meet the required specifications.

Solution

The transimpedance amplifier topology is as shown.

Since the transresistance must be at least -2000, let $R_f = 3.3 \text{ k}\Omega$, which will allow the amplifier to exceed the required specifications.

Also let $I_C = 3.8$ mA and $V_{CE} = 0.785$ V. This allows for a small value for $r_\pi = 1.02 \text{ k}\Omega$ and $g_m = 0.146$.

Let the value of the emitter degenerative resistor R_E be 27 Ω; then the collector resistor $R_C = 3.9 \text{ k}\Omega$.

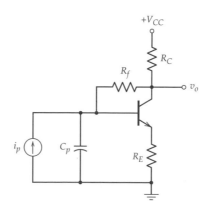

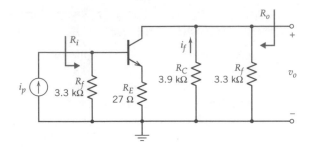

The midband loaded forward amplifier is as shown.
The feedback quantity is

$$f = \frac{i_f}{v_o} = \frac{-1}{3300}.$$

The input resistance is

$$R_i = 3300 \parallel [r_\pi + r_b + (1 + g_m r_\pi)27] = 2 \text{ k}\Omega.$$

The output resistance is

$$R_o = 3900 \parallel 3300 = 1.79 \text{ k}\Omega.$$

The amplifier forward transresistance is

$$R_M = \frac{v_o}{i_p} = A_v R_i = -105.2 \text{ k}\Omega.$$

The feedback quantities of interest are

$$D = I + Af = 32.9,$$

$$R_{if} = \frac{R_i}{D} = 60.8 \ \Omega, \qquad R_{of} = \frac{R_o}{D} = 54.3 \ \Omega,$$

$$R_{Mf} = \frac{R_M}{D} = -3200$$

which meets the midband transresistance specification.
Find the frequency response of the transimpedance amplifier by determining the
capacitances for the BJT and the frequency response of the loaded forward amplifier:

$$\text{CJC} = C_{obo}\left(1 + \frac{V_{CB}}{0.75}\right)^{0.33} = 7.83 \text{ pF},$$

$$\text{CJE} = C_{ibo}\left(1 + \frac{V_{EB}}{0.75}\right)^{0.33} = 9.47 \text{ pF},$$

$$\text{TF} = \frac{1}{2\pi f_T} - \frac{\eta V_t}{|I_{CT}|}\left[\frac{\text{CJE}}{(1 - 0.7/0.75)^{0.33}} + C_{obo}\right] = 460 \text{ ps},$$

and

$$C_\mu = \frac{\text{CJC}}{(1 + V_{CB}/0.75)^{0.33}} = 7.6 \text{ pF}, \qquad C_\pi = \frac{g_m}{\omega_T} - C_\mu = 70 \text{ pF}.$$

The high-frequency capacitors are found using Miller's theorem:

$$C_{i2} = C_p + C_\pi + C_\mu(1 - A_v) = C_p + C_\pi + C_\mu\left(1 - \frac{R_M}{3300}\right) = 331 \text{ pF},$$

$$C_o \approx C_\mu \approx 7.6 \text{ pF}.$$

The loaded forward amplifier high-frequency pole is

$$\omega_H \approx \left\{ C_{i2}\left[r_\pi \;\Big\|\; \left(\frac{r_b + R_f}{1 + g_m R_E}\right) \right] \right\}^{-1} \Rightarrow 6.4 \text{ MHz}$$

The loaded forward amplifier high-frequency pole corresponding to the C_o is significantly higher.

Therefore, the high-frequency cutoff of the feedback amplifier is

$$\omega_{Hf} = \frac{\omega_H}{D} = 200 \text{ krad/s} \Rightarrow 32 \text{ kHz},$$

which fulfills the specification for the high-frequency cutoff.

Since the preamplifier does not use external capacitors, the low-frequency response extends to DC.

REFERENCES

D'Azzo, J. J., and Houpis, C. H., *Linear Control System Analysis and Design: Conventional and Modern,* 3rd ed., McGraw-Hill, New York, 1988.

Fitchen, F. C., *Transistor Circuit Analysis and Design,* Van Nostrand, Princeton, NJ, 1966.

Ghausi, M. S., *Electronic Devices and Circuits: Discrete and Integrated,* Holt, Rinehart and Winston, New York, 1985.

Gray, P. R., and Meyer, R. G., *Analysis and Design of Analog Integrated Circuits,* 3rd ed., Wiley, New York, 1993.

Millman, J., *Microelectronics: Digital and Analog Circuits and Systems,* McGraw-Hill, New York, 1979.

Millman, J., Halkias, C. C., *Integrated Electronics: Analog and Digital Circuits and Systems,* McGraw-Hill, New York, 1972.

Nise, N. S., *Control Systems Engineering,* Benjamin/Cummings, Redwood City, California, 1992.

Rosenstark, S., *Feedback Amplifier Principles,* Macmillan, New York, 1986.

Sedra, A. S., and Smith, K. C., *Microelectronic Circuits,* 3rd ed., Holt, Rinehart, and Winston, Philadelphia, 1991.

Schilling, D. L., and Belove, C., *Electronic Circuits,* 3rd ed., McGraw-Hill, New York, 1989.

■■■■ PROBLEMS

11-1. The voltage gain of an amplifier is described by the following quantities: midband gain 1000, low 3-dB frequency 100 Hz, and high 3-dB frequency 50 kHz.

 a. Assume that the high- and low-frequency responses are characterized by single poles. Feedback is to be applied to increase the bandwidth of the amplifier by a factor of 10. Assuming the feedback network does not load the initial amplifier, what return difference is necessary to accomplish the design goal? Determine the new midband gain and the high and low 3-dB frequencies after the application of feedback.

 b. Upon a more careful analysis of the original amplifier (prior to application of feedback), it was discovered that a second high-frequency pole exists at 1 MHz. What effect will this second pole have on the feedback amplifier specified by the results of part a?

11-2. An audio amplifier is described by two high-frequency poles at 25 and 60 kHz. It is desired to increase the bandwidth with the application of feedback without significantly destroying the step response of the amplifier. A design goal of no more than 2% overshoot is established. Assuming that the feedback network will not change the pole location of the original amplifier, determine the following:

 a. Maximum high 3-dB frequency that can be obtained

 b. Rise time before and after application of feedback

11-3. The response to an OpAmp noninverting amplifier to a unit step voltage input is shown. Assume the amplifier frequency response is characterized by a single high-frequency pole.

 a. Determine the high 3-dB frequency.

 b. Determine the gain bandwidth product (GBP) of the amplifier.

 c. If the OpAmp has an open loop gain of 110 dB, what is the 3-dB frequency of the OpAmp open-loop gain?

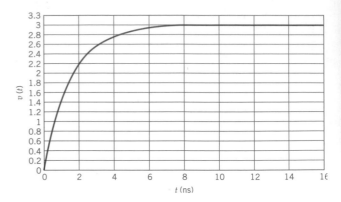

11-4. An amplifier is described by two low-frequency poles at

$$f_{L1} = 50 \text{ Hz}, \qquad f_{L2} = 4 \text{ Hz}.$$

 a. Determine the low 3-dB frequency of this amplifier.

 b. Estimate the sag that a 1-kHz square wave will experience passing through this amplifier.

 c. It is desired to reduce the sag to 1% with the application of feedback. What return ratio D will result in that amount of sag?

 d. What is the new 3-dB frequency of the resultant feedback amplifier?

11-5. The current gain of an amplifier is described by the following quantities: midband gain 10 A/mA, low 3-dB frequency 120 Hz, and high 3-dB frequency 2.4 MHz.

 a. Assume the high- and low-frequency responses are characterized by single poles. Feedback is to be applied to increase the high 3-dB frequency to 20 MHz. Assuming the feedback network does not load the initial amplifier, what return difference is necessary to accomplish the design goal? Determine the new midband gain and the low 3-dB frequency after the application of feedback.

 b. Upon a more careful analysis of the original amplifier (prior to the application of feedback), it was discovered that a second low-frequency pole exists at $f_{L2} = 10$ Hz and a second high-frequency pole exists at $f_{H2} = 48$ MHz. What effect will these initially nondominant poles have on the feedback amplifier specified by the results of part a?

11-6. An audio amplifier is described by a midband gain of 420, a single low-frequency pole at

$$f_{L1} = 100 \text{ Hz},$$

and two high-frequency poles at

$$f_{H1} = 14 \text{ kHz}, \qquad f_{H2} = 26 \text{ kHz}.$$

The amplifier is to be redesigned as a feedback amplifier so that the midband region will be extended to cover frequencies in the range 40 Hz $\leq f \leq$ 20 kHz. Assuming that the application of feedback does not alter the original amplifier, determine the following properties of the redesigned amplifier:

a. Minimum feedback ratio f that will accomplish the new design goals
b. Feedback return difference D
c. High and low 3-dB frequencies
d. Midband gain
e. Are there any peaks in the frequency response? If so, determine the peak characteristics.

11-7. An audio amplifier is described by a midband gain of 251, two low-frequency poles at

$$f_{l1} = 60 \text{ Hz}, \qquad f_{l2} = 20 \text{ Hz},$$

and two high-frequency poles at

$$f_{h1} = 40 \text{ kHz}, \qquad f_{h2} = 100 \text{ kHz}.$$

The amplifier is to be redesigned as a feedback amplifier so that the midband region will be as large as possible without introducing any peaks in the frequency response. Assuming that the application of feedback does not alter the original amplifier, determine the following properties of the redesigned amplifier:

a. Minimum feedback ratio f that will accomplish the new design goals
b. Feedback return difference D
c. High and low 3-dB frequencies
d. Midband gain

11-8. A designer is attempting to create a feedback amplifier with a midband voltage gain of 40. The basic forward gain elements of this amplifier are two amplifier stages (each described by single high- and low-frequency poles) with the following properties:

	1	2
Midband voltage gain	15	20
High 3-dB frequency	30 kHz	20 kHz
Low 3-dB frequency	40 Hz	30 Hz

Assume that series interconnection of the amplifier stages does not change the above stated properties.

a. Apply global feedback to the amplifier so that a gain of 40 is obtained and determine the resultant frequency response.
b. Unfortunately the technique of part a results in significant peaks in the frequency response. An alternate approach is to propose an attenuator placed in the forward gain path, as shown. Determine the values of the attenuation factor g and feedback ratio f that will result in maximum bandwidth, no peaks in the frequency response, and a voltage gain of 40. Note, for this configuration, the feedback gain is given by

$$A_f = \frac{A_1 g A_2}{1 + f A_1 g A_2}.$$

c. What are the high and low 3-dB frequencies for the amplifier designed in part b?

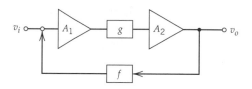

11-9. Another circuit topology that might be used to accomplish the design goals of the previous problem incorporates both local and global feedback. For the topology shown, determine the local and global feedback ratios f_{local}, and f_{global} that will maximize the bandwidth and produce a voltage gain of 40 without introducing any peaks in the frequency response. The individual, noninteracting amplifier stages are described by the following

	1	2
Midband voltage gain	15	20
High 3-dB frequency	30 kHz	20 kHz
Low 3-dB frequency	40 Hz	30 Hz

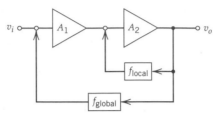

11-10. A macromodel for a μA741 bipolar OpAmp is provided in the appendices (it is also in the PSpice library **eval.lib**).

a. Use SPICE to determine the frequencies of the first two high-frequency poles of the μA741 OpAmp. Assume a resistive load of 820 Ω.

b. Design a simple inverting amplifier with a voltage gain of −10. Use SPICE to determine the high 3-dB frequency of this amplifier.

c. Compare the results of part b to those predicted by feedback theory.

11-11. A macromodel for an LM324 bipolar OpAmp is provided in the appendices (it is also in the PSpice library **eval.lib**).

a. Use SPICE to determine the frequencies of the first two high-frequency poles of the LM324 OpAmp. Assume a resistive load of 560 Ω.

b. Design a simple noninverting amplifier with a voltage gain of +8. Use SPICE to determine the high 3-dB frequency of this amplifier.

c. Compare the results of part b to those predicted by feedback theory.

11-12. A macromodel for an LF411 JFET input OpAmp is provided in the appendices (it is also in the PSpice library **eval.lib**).

a. Use SPICE to determine the frequencies of the first two high-frequency poles of the LF411 OpAmp. Assume a resistive load of 1 kΩ.

b. Design a simple noninverting amplifier with a voltage gain of +15. Use SPICE to

determine the high 3-dB frequency of this amplifier.

c. Compare the results of part b to those predicted by feedback theory.

11-13. The transistors in the given feedback amplifier circuit are 2N2222 (description is given in the PSpice libraries). The quiescent conditions have been previously determined to be

$$I_{c1} \approx I_{c2} \approx 2.0 \text{ mA}.$$

a. Determine the midband voltage gain and approximate 3-dB frequencies of the amplifier.

b. Compare hand calculations to PSpice simulation.

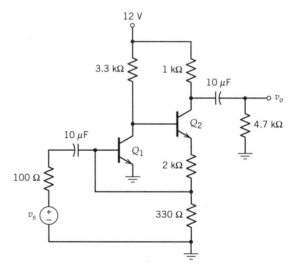

11-14. For the single-stage feedback amplifier shown, determine the high and low 3-dB frequencies. The FET parameters are

$$K = 1 \text{ mA/V}^2, \quad C_{rss} = 1.3 \text{ pF}, \quad V_T = 1.5 \text{ V},$$
$$C_{iss} = 5.0 \text{ pF}, \quad V_A = 160 \text{ V}.$$

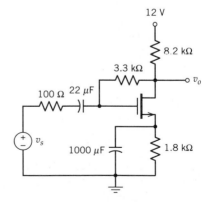

11-15. The transistors in the given feedback amplifier circuit are 2N3904 (description is given in the PSpice libraries). The quiescent conditions have been previously determined to be

$$I_{c1} \approx I_{c2} \approx 1.32 \text{ mA}.$$

a. Determine the midband voltage gain and approximate 3-dB frequencies of the amplifier.

b. Compare hand calculations to PSpice simulation.

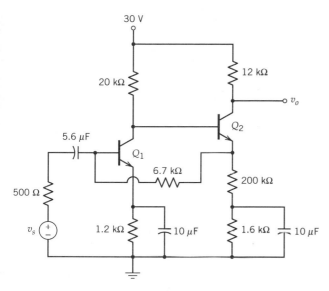

11-16. For the amplifier shown, determine the high and low 3-dB frequencies. The JFETs are described by

$$V_{PO} = -2 \text{ V}, \qquad I_{DSS} = 4 \text{ mA}, \qquad V_A = 200 \text{ V},$$

$$C_{rss} = 1.0 \text{ pF}, \qquad C_{iss} = 5.0 \text{ pF}.$$

Verify the analysis using SPICE.

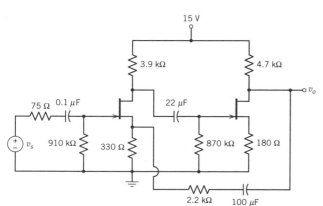

11-17. The transistor in the given feedback amplifier circuit is a 2N3904 (description is given in the PSpice libraries).

a. Complete the amplifier design by determining a realistic value for the capacitor C that will result in a low 3-dB frequency of approximately 100 Hz.

b. Determine the high 3-dB frequency of the amplifier.

c. Use PSpice to determine the high and low 3-dB frequencies of the final design. Compare the simulation results to hand calculations. Comment.

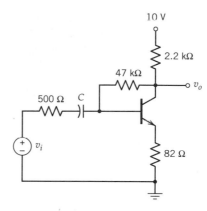

11-18. Determine the high and low 3-dB frequencies for the feedback amplifier shown. Assume the following circuit parameters:

BJTs	2N2222
Coupling capacitors	10 μF
Bypass capacitor	1000 μF

Compare results to SPICE simulation.

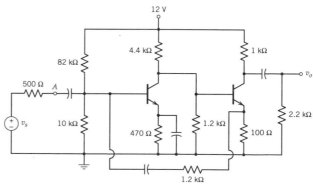

11-19. Compute the high and low 3-dB frequencies for the amplifier of the previous problem if the mixing point is moved to location A. Compare results to SPICE simulation.

11-20. For the feedback amplifier shown, assume the following circuit parameters:

BJTs	2N2222
Coupling capacitors	20 µF
Bypass capacitor	2200 µF

a. Determine the high and low 3-dB frequencies.

b. Compare results to SPICE simulation.

c. The frequency response for this circuit is not flat in the midband. Modify the circuit so that the midband region is flat. Use SPICE to verify that the design modifications improve the midband region.

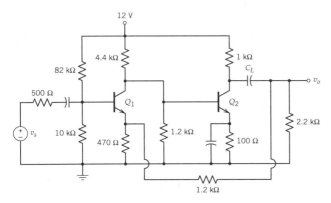

11-21. In the previous problem the load capacitor C_L is included within the feedback amplifier due to the sampling point of the feedback network. Assume the sampling point is moved to the collector of Q_2 and a 100-µF capacitor is inserted in series with the 1.2-kΩ feedback resistor. This change essentially removes C_L from the feedback amplifier.

a. Determine the high and low 3-dB frequencies of the modified circuit.

b. Compare results to SPICE simulation.

11-22. A particular series–shunt feedback amplifier can be described as having a midband voltage gain of 10,000 with three high-frequency poles at the following frequencies:

$$f_{p1} = 1 \text{ MHz}, \quad f_{p2} = 4 \text{ MHz},$$
$$f_{p3} = 25 \text{ MHz}.$$

a. Write an expression for A_v as a function of frequency.

b. Draw the idealized Bode magnitude and phase plots. On the graph label all slopes and use a small circle to indicate where the slope changes.

c. At what frequency will this feedback amplifier oscillate (if it oscillates at all)?

d. In order to have at least 45° phase margin, what is the amount of feedback that can be added? Is this a maximum or a minimum for stable amplifier operation?

e. What is the gain margin for the amount of feedback determined in part c?

11-23. An amplifier can be described by a midband gain of 251, two low-frequency poles at

$$f_{l1} = 60 \text{ Hz}, \quad f_{l2} = 20 \text{ Hz},$$

and four high-frequency poles at

$$f_{h1} = 40 \text{ kHz}, \quad f_{h2} = 100 \text{ kHz},$$
$$f_{h3} = 200 \text{ kHz}, \quad f_{h4} = 1 \text{ MHz}.$$

a. Draw the idealized Bode magnitude and phase plots. On the graph label all slopes and use a small circle to indicate where the slope changes. Compare these straight-line approximate curves to exact curves (use a software package to generate the exact curves).

b. At what frequency will this feedback amplifier oscillate (if it oscillates at all)?

c. In order to achieve a gain margin of at least 5 dB, what is the amount of feedback that can be added? Is this a maximum or a minimum for stable amplifier operation?

d. What is the phase margin for the amount of feedback determined in part c?

11-24. The loop gain of an inverting amplifier is

$$A(j\omega)f = \frac{-250}{(1 + j\omega)(1 + j0.01\omega)^2}.$$

Determine whether the amplifier is stable using:

a. The Nyquist plot

b. The gain and phase plot of the loop gain

11-25. The loop gain of an inverting amplifier is

$$A(j\omega)f = \frac{-K}{(1 + j0.01\omega)^3}.$$

a. If $K = 5$, is the amplifier stable?

b. What value of K defines a gain margin of greater than 10 dB and phase margin of greater than 50°?

11-26. The loop gain of an amplifier is

$$A(j\omega) = \frac{24,000}{[1 + j\omega/(2 \times 10^5)]^2(1 + j\omega/10^5)}$$

with a feedback factor $f = -5$.

a. Plot the magnitude and phase of the loop gain.

b. Determine whether the amplifier is stable. If the amplifier is stable, determine the gain and phase margins.

11-27. An OpAmp with an open loop gain of

$$A(j\omega) = \frac{10^5}{(1 + j\omega/10^5)^2(1 + j\omega/10^2)}$$

is used to design an inverting amplifier with a gain of -125. Determine the gain and phase margin of the inverting amplifier using:

a. The Nyquist plot

b. The gain and phase plot of the loop gain

11-28. Sketch the Nyquist plot of the loop gain for a three-pole amplifier with a DC open-loop gain $A_o = -1100$ and open-loop poles at 500 kHz, 1.1 MHz, and 1.8 MHz. Determine whether the amplifier is stable with the following feedback factors:

a. $f = -0.005$

b. $f = -0.02$

Determine the maximum value of f for which the amplifier is stable.

11-29. A three-pole amplifier has an open-loop DC gain of -1000 and poles located at 1.1 MHz, 12 MHz, and 28 MHz. Dominant-pole compensation is applied to the amplifier.

a. Find the location of the dominant pole so that the open-loop gain is first a constant and then falls to 0 dB at a rate of -20 dB/ decade for frequencies less than 1.1 MHz.

b. Determine the maximum value of the feedback factor f for which this compensated amplifier is marginally stable.

c. Determine the value of the feedback factor f for which this compensated amplifier has a gain margin greater than 10 dB and phase margin greater than 50°?

11-30. Lag–lead compensation is used with an amplifier with a DC gain of -1200 and poles at 1 MHz, 10 MHz, and 220 MHz. The zero of the compensation network is selected to cancel the 1-MHz pole of the uncompensated amplifier.

a. Find the pole of the lag–lead compensation network so that the amplifier is stable with a phase margin of 50° when the feedback factor $f = -0.01$.

b. Determine the bandwidth of the compensated feedback amplifier.

11-31. An inverting OpAmp inverting amplifier is to be stabilized using lead compensation, as shown. The OpAmp input and output resistances are $R_i = 1$ MΩ and $R_o = 75$ Ω. The open-loop gain of the OpAmp is

$$A(j\omega) = \frac{24,000}{[1 + j\omega/(2 \times 10^5)]^2(1 + j\omega/10^5)}.$$

a. Complete the design for a phase margin of 50°.

b. Confirm the result with SPICE.

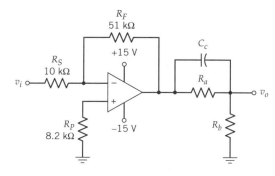

11-32. An inverting OpAmp inverting amplifier is to be stabilized using dominant-pole compensation, as shown. The OpAmp input and output resistances are

$$R_i = 1 \text{ M}\Omega, \qquad R_o = 75 \text{ }\Omega.$$

The open-loop gain of the OpAmp is

$$A(j\omega) = \frac{24,000}{[1 + j\omega/(2 \times 10^5)]^2(1 + j\omega/10^5)}.$$

a. Complete the design for a phase margin of 50°.

b. Confirm the result with SPICE.

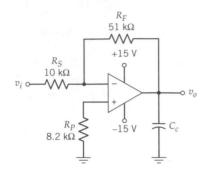

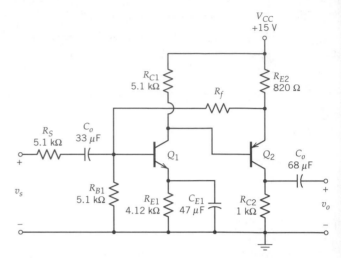

(F) 11-33. The lag–lead compensated amplifier shown uses an OpAmp with input and output resistances of $R_i = 1$ MΩ and $R_o = 75$ Ω, respectively. The open-loop gain of the OpAmp is

$A(j\omega)$

$$= \frac{1000}{[1 + j\omega/(2\pi \times 10^6)][1 + j\omega/(2\pi \times 10^6)]}.$$

a. Complete the design for a phase margin of 50°.

b. Confirm the result with SPICE.

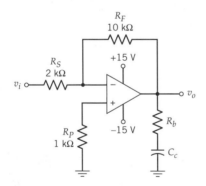

11-34. For the shunt–series feedback amplifier shown, the transistor characteristics are: $\beta_F = 200$, $V_A = 120$ V, CJC = 7.31 pF, CJE = 22.01 pF, and TF = 411 ps for *npn* transistors and $\beta_F = 200$, $V_A = 100$ V, CJC = 14.76 pF, CJE = 19.82 pF, and TF = 603.7 ps for *pnp* transistors.

a. Determine the value of the feedback resistor R_f for unstable amplifier operation.

b. Using the value of R_f found in part a, design a stable dominant-pole compensated amplifier.

c. Confirm the results using SPICE.

(F) 11-35. Design a lead compensated amplifier using the result found in part a of problem 11-34. Confirm the results using SPICE.

(F) 11-36. Design a lag–lead compensated amplifier using the result found in part a of problem 11-34. Confirm the results using SPICE.

11-37. For the differential amplifier shown, assume identical transistors with

$$\beta_F = 120, \qquad V_A = 175 \text{ V}, \qquad \text{CJE} = 19 \text{ pF},$$

$$\text{CJE} = 36 \text{ pF}, \qquad \text{TF} = 3.3 \text{ ns}.$$

a. Determine the value of the feedback resistor R_{f1} for unstable amplifier operation.

b. Using the value of R_{f1} found in part a, design a stable dominant-pole compensated amplifier.

c. Confirm the results using SPICE.

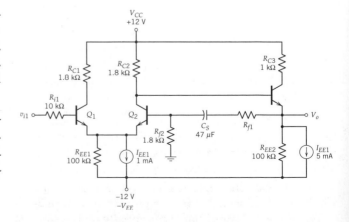

11-38. Design a lead compensated amplifier using the result found in part a of problem 11-37. Confirm the results using SPICE.

11-39. The amplifier shown uses identical NMOSFETs with the following characteristics:

$$K = 1.25 \text{ mA/V}^2, \quad V_T = 2 \text{ V}, \quad V_A = 150 \text{ V},$$

$$C_{iss} = 12 \text{ pF}, \qquad C_{rss} = 2 \text{ pF}.$$

a. Determine the value of the feedback resistor R_f for unstable amplifier operation.

b. Using the value of R_f found in part a, design a stable dominant-pole compensated amplifier.

c. Confirm the results using SPICE.

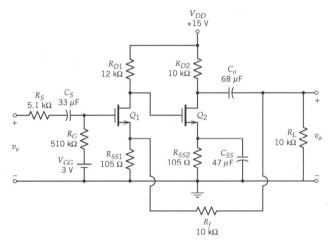

11-40. Design a lead compensated amplifier using the result found in part a of problem 11-39. Confirm the results using SPICE.

11-41. Design an analog amplifier to meet the following specifications:

Input	A differential amplifier
Input voltage	At least ±1 mV
Midband voltage gain	1000 ± 10
Midband input resistance	At least 1 MΩ
Midband output resistance	Less than 100 Ω
Low 3-dB frequency	100 Hz, maximum
High 3-dB frequency	1 MHz, minimum

Design limitations are:

- Coupling capacitors are to be limited to 1 μF maximum.
- Bypass capacitors are to be limited to 1000 μF maximum.
- At least one BJT and one JFET must be used.

Active devices are limited to the following types (descriptions are available in the PSpice library **eval.lib**):

2N2222A	*npn* bipolar transistor
2N2907A	*pnp* bipolar transistor
2N3904	*npn* bipolar transistor
2N3906	*pnp* bipolar transistor
2N3819	*n*-channel Junction field effect transistor
2N4393	*n*-channel Junction field effect transistor

A report on the design is required. Minimum requirements for the report include:

- A complete theoretical analysis of the design
- SPICE verification of design specifications
- Comparisons, conclusions, and comments

11-42. Design an analog feedback amplifier to meet the following specifications:

Midband voltage gain	100 ± 5 (driving a 1-kΩ load)
Midband input resistance	At least 40 kΩ
Midband output resistance	Less than 50 Ω
Low 3-dB frequency	30 Hz, maximum
High 3-dB frequency	10 MHz, minimum
Output voltage swing	±4 V (from quiescent value), minimum

Design limitations are:

- Coupling capacitors are to be limited to 1 μF maximum.
- Bypass capacitors are to be limited to 1000 μF maximum.

Active devices are limited to the following types (descriptions are available in the PSpice library **eval.lib**):

2N2222A	*npn* bipolar transistor
2N2907A	*pnp* bipolar transistor
2N3904	*npn* bipolar transistor
2N3906	*pnp* bipolar transistor
2N3819	*n*-channel JFET
2N4393	*n*-channel JFET

A report on the design is required. Minimum requirements for the report include:

- A complete theoretical analysis of the design
- SPICE verification of design specifications
- Comparisons, conclusions, and comments

12

OSCILLATOR CIRCUITS

An electronic oscillator is a circuit that produces a periodic output without an input signal. A harmonic (often called linear) oscillator, which is the topic of this chapter, is a subset of electronic oscillators that produce an output signal that is approximately sinusoidal. The oscillation is based on a resonant circuit often designed using inductors and capacitors. Crystals may be used to closely control the oscillation frequency.

Modern applications of oscillators include audio and electronic communication systems. These systems often contain several oscillators including crystal-controlled reference oscillators, voltage-controlled oscillators (VCOs), and voltage-controlled crystal oscillators. Although many integrated circuits exist for generating periodic signals, discrete oscillator designs have significant advantages over many integrated solutions. In many instances, integrated circuit oscillators cannot meet the high-frequency and low-noise requirements of communication systems. Discrete oscillators are also used in high-quality audio systems, which require high stability and low noise.

The basic feedback amplifier topologies, presented in previous chapters, are used to analyze and design oscillators. These circuits can be used to generate essentially sinusoidal waveforms by carefully designing the amplifier to operate at the critical point where the loop gain Af is -1. When the loop gain is at the critical point, the circuit is oscillatory and delivers a sinusoidal waveform without an externally applied input signal.

Harmonic oscillator designs that incorporate feedback amplifier topology with the loop gain at the critical point are used in a variety of electronic systems. These circuits are commonly called linear oscillators since they generate the waveforms through resonance phenomenon; that is, a frequency-selective feedback circuit is used to amplify the frequency of interest. An alternate approach using non-linear signal wave-shaping electronic circuits is often employed in oscillator circuit design. These non-linear circuits, called multivibrators, are discussed in Chapter 13.

12.1 LINEAR ANALYSIS

The basic topology of a harmonic oscillator, shown in Figure 12.1-1, is identical to the feedback amplifier topology introduced in Chapter 8. The open-loop amplifier gain $A(s)$ and feedback return ratio $f(s)$ are frequency selective. Figure 12.1-1 shows an input signal $X_i(s)$ for the sake of analysis; oscillators do not require an input signal to generate a sinusoidal output waveform. In oscillator circuits, $X_o(s)$ is nonzero for $X_i(s) = 0$.

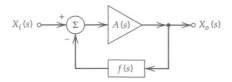

Figure 12.1-1

Basic topology of harmonic oscillator with injected input signal.

The closed-loop gain of the circuit shown in Figure 12.1-1 is

$$A_f(s) = \frac{A(s)}{1 + A(s)f(s)},$$ (12.1-1)

where $A_f(s) = X_o(s)/X_i(s)$.

The output signal is therefore

$$X_o(s) = \frac{A(s)}{1 + A(s)f(s)} X_i(s).$$ (12.1-2)

In electronic oscillators, the circuit is not excited by an external time-varying source, that is, the input signal $X_i(j\omega) = 0$. Therefore, for a nonzero output $X_o(s)$, Equation 12.1-2 must yield a loop gain $A(j\omega_o)f(j\omega_o) = -1$ at the frequency of oscillation ω_o. From the discussion in Chapter 11, this loop gain condition causes instability, resulting in amplifier oscillation.

The condition of unity loop gain with a 180° phase shift is called the *Barkhausen criterion*. In order to achieve the requisite phase shift, reactive elements must be used in the feedback loop. Since the feedback element is reactive, the signal phase shift is invariably a function of frequency. This implies that there is only one frequency where the Barkhausen criterion is satisfied. The circuit oscillates at the frequency where the Barkhausen criterion is satisfied.

When the oscillator is designed for $A(j\omega_o)f(j\omega_o) = -1$, there is a chance that oscillation will cease due to variations in the characteristics of the active elements of the amplifier caused by slight temperature variations, power supply noise, and so on, that may force $|A(j\omega)f(j\omega)| < 1$. If the magnitude of the loop gain becomes smaller than unity, the oscillation will decay and then cease. Therefore, an oscillator with $A(j\omega_o)f(j\omega_o)$ exactly equal to -1 is not realizable in practice. If $|A(j\omega)f(j\omega)| > 1$, the signal that is fed back into the mixing point (the summing node) from the feedback network will be smaller than at the input to $f(s)$. Therefore, the output will appear larger than that of the previous trip around the loop. This larger voltage at the output will then reappear as a still larger voltage on the next trip around the loop. It therefore appears that, for $|A(j\omega)f(j\omega)| > 1$, the amplitude of the oscillation increases without limit. In fact, the output signal amplitude is limited by the power supply rails and the onset of nonlinear operation of the active devices of the amplifier. As the oscillations increase in amplitude, the non-linearity of the circuit becomes more apparent. The onset of nonlinearity limits the amplitude of the oscillation.

In practical oscillators, it is necessary to design $|A(j\omega)f(j\omega)|$ to be slightly greater than unity. To ensure that the magnitude of the loop gain will not fall below unity with electronic noise or variations in transistor and circuit parameters, $|A(j\omega)f(j\omega)| > 1.05$. As stated before, the amplitude of oscillation in practical harmonic oscillators is limited by the onset of non-linearity. In many cases, the excursion into non-linear regions of operation by oscillators is small, thereby allowing linear theory to be used to design the circuits. Additionally, the reactive feedback network is frequency selective and serves a dual purpose in feedback oscillator design. First, it ensures that the Barkhausen criterion is satisfied and that the conditions of oscillation are met only at the desired frequency, which is significantly lower than the high cutoff frequency of the amplifier. Second, the reactive feedback network removes the harmonics of the distorted signal caused by the onset of non-linearity so that a relatively pure sinusoid at the fundamental oscillation frequency appears at both the output (sampling point) and the mixing point.

12.1.1 Design of Practical Oscillators

The design of oscillators is more an art than an exact science since the steady-state operating conditions cannot be accurately predicted by simple mathematical techniques. At the onset of non-linearity, the linear models used for the analysis of amplifiers are not representative of all aspects of circuit behavior. However, linear analysis is useful in predicting many aspects of oscillator operation.

In oscillators, the output of the circuit must be fed back into the input with a gain slightly greater than unity and with a phase shift of 0° (or some multiple of 360°). This condition of oscillation only occurs at the frequency of the sinusoidal oscillation. If the phase shift through the reactive feedback network is assumed to be independent of the amplifier operating conditions, the steady-state frequency of oscillation will be identical to the initial oscillation frequency in the transient state at the beginning of oscillation. For independent amplifier operation, accurate predictions of the initial oscillation frequency can be made using linear small-signal analysis. Linear small-signal analysis yields the minimum required amplifier gain and operating frequency of an oscillator. In most oscillator designs, the reactive network has an effect on amplifier operation. To counteract the loading effect, the magnitude of the loop gain at the oscillation frequency is $1.3 > |A(j\omega)f(j\omega)| > 1.05$ to ensure generation of a sinusoid with minimum harmonic distortion. Higher values of $|A(j\omega)f(j\omega)|$ results in an oscillatory waveform with higher harmonic content. The transient time prior to steady-state oscillation is shorter for higher loop gains at the oscillation frequency.

Another factor that complicates oscillator design is the frequency dependence of the reactive components used in the feedback network. In particular, for radio frequency (RF) applications, a capacitor larger than a few hundred picofarads tends to appear inductive above 10 MHz. An inductor may appear capacitive at higher frequencies due to stray capacitance between the windings of the inductor. These so-called "parasitic" effects are difficult to model using conventional circuit theory. In fact, the parasitic effects of the reactive components may cause the circuit to oscillate at a frequency other than that predicted by small-signal analysis; that is, an oscillator circuit that appears to meet required specifications may oscillate not only at the designed frequency of oscillation, but at a lower frequency (a phenomenon called "motor-boating") and at one or more frequencies higher than the designed oscillating frequency. In most cases, these effects can be mitigated by employing high-quality inductors and by connecting small (100–300-pF) capacitors in parallel with all bypass and coupling capacitors. These smaller capacitors provide an effective short circuit at frequencies where the larger capacitors become inductive.

The design process of an oscillator begins with the determination of the circuit topology and the value of its elements. The circuit element values determine the desired oscillating frequency. However, such an analysis does not predict circuit power output, efficiency, waveform purity, frequency stability, or temperature and power supply variation sensitivities. These effects are most often resolved in the design by computer simulation and hardware prototyping to adjust circuit component values until the desired overall performance is achieved. In practice, many designers use a few "pet" oscillator circuits that are adapted to fulfill the required performance specifications.

12.1.2 Frequency Stability

Frequency stability is expressed in terms of the amount of frequency change with respect to a change in a particular circuit parameter, such as the small-signal transistor current gain h_{fe}. Assume that κ is one of these circuit parameters, so that a change in $\Delta\kappa$ about its equilibrium point causes a change in the loop gain of a circuit. That is,

$$\Delta(Af) = \frac{\partial(Af)}{\partial\kappa} \Delta\kappa. \tag{12.1-3}$$

Stable frequency operation requires that the loop gain remain invariant. From Equation 12.1-1, the condition for stable oscillator operation implies that $\partial(Af)/\partial\kappa \to 0$.

For variations in κ, the loop gain of an oscillator is expressed as

$$Af = 1 + \Delta(Af) \tag{12.1-4}$$

This corresponds to a new oscillating frequency and amplitude of oscillation:

$$\omega = \omega_o + \Delta\omega \tag{12.1-5a}$$

$$v = v_o + \Delta v, \tag{12.1-5b}$$

where ω_o and v_o are the initial equilibrium oscillating frequency and amplitude, respectively.

The difference between the new and old equilibrium points are

$$\Delta(Af) = \frac{\partial(Af)}{\partial\kappa} \Delta\kappa + \frac{\partial(Af)}{\partial v} \Delta v + \frac{\partial(Af)}{\partial\omega} \Delta\omega = 0. \tag{12.1-6}$$

Solving Equation 12.1-6 for $\Delta\omega$,

$$\Delta\omega = \frac{[\partial(Af)/\partial\kappa] \, \Delta\kappa + [\partial(Af)/\partial v] \, \Delta v}{\partial(Af)/\partial\omega}. \tag{12.1-7}$$

Three methods for stabilizing oscillators, using Equation 12.1-7, are as follows:

1. The numerator of Equation 12.1-7 is set equal to zero. That is,

$$\Delta v = \frac{\partial(Af)/\partial\kappa}{\partial(Af)/\partial v} \Delta\kappa. \tag{12.1-8}$$

Equation 12.1-8 states that a rate of change in the circuit parameter κ must be counterbalanced by a change in another circuit parameter.

2. The denominator of Equation 12.1-7 must approach infinity. In order for $\Delta\omega$ to approach zero, $\partial(Af)/\partial\omega$ must approach infinity. An alternate expression that will allow for the solution to this condition is

$$\frac{\partial(Af)}{\partial\omega} = f\frac{\partial A}{\partial\omega} + A\frac{\partial f}{\partial\omega}. \qquad (12.1\text{-}9)$$

To simplify the analysis, assume that A is independent of ω. Then Equation 12.1-9 simply becomes

$$\frac{\partial(Af)}{\partial\omega} = A\frac{\partial f}{\partial\omega}. \qquad (12.1\text{-}10)$$

Since the feedback factor f describes a reactive feedback network, it is complex. Therefore, the amplitude and phase relationship must obey the relationship

$$A\frac{\partial f}{\partial\omega} \rightarrow \infty. \qquad (12.1\text{-}11)$$

Therefore, the magnitude and phase of f are $A(\partial|f|/\partial\omega) \rightarrow \infty$ and $A(\partial\theta_f/\partial\omega) \rightarrow \infty$. Assuming that A is independent of ω,

$$\frac{\partial(Af)}{\partial\omega} = A\frac{\partial f}{\partial\omega}.$$

Since the slope $\partial|f|/\partial\omega$ is directly proportional to the quality factor Q of the oscillator circuit, the fulfillment of the requirement of Equation 12.1-11 requires that a high-quality-factor (Q) circuit is required for stability.

3. The open-loop gain of the oscillator circuit can be used to stabilize the circuit. In this case, A must be made very large. Therefore, f must be small to meet the oscillation criterion of $Af = -1$. However, the derivative of f with respect to ω must be large to fulfill the requirement of Equation 12.1-11.

12.2 *RC* OSCILLATORS

Simple *RC* oscillators are commonly used in audio frequency applications, which span the frequency range from several hertz to several tens of kilohertz. The two most commonly used oscillator circuits are the *RC* phase-shift and Wien bridge oscillators.

12.2.1 Phase-Shift Oscillator

The phase-shift oscillator is one of the simplest oscillators to design and construct in the audio frequency range. The oscillator exemplifies the simple principles and conditions of oscillation discussed in Section 12.1. A simple OpAmp-based phase-shift oscillator is shown in Figure 12.2-1.

In this circuit, an inverting OpAmp amplifier is followed by an *RC* "ladder" network consisting of three cascaded arrangements of a resistor *R* and a capacitor *C*. The three resistors and three capacitors in the feedback network have identical val-

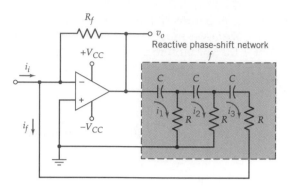

Figure 12.2-1
OpAmp-based phase-shift oscillator.

ues. The output of the reactive phase-shift ladder network is returned to the input of the inverting OpAmp amplifier. This feedback configuration is the shunt–shunt topology. Therefore, the analysis used for shunt–shunt feedback topologies is used. The amplifier gain is

$$A = R_M = -R_f. \tag{12.2-1}$$

The feedback ratio is

$$f = \frac{i_f}{v_o} = -\frac{i_3}{v_o}. \tag{12.2-2}$$

Using the Barkhausen criterion,

$$Af = -1 = -R_f\left(-\frac{i_3}{v_o}\right) = \frac{R_f i_3}{v_o}. \tag{12.2-3}$$

If the loading of the reactive phase-shift network on the inverting amplifier can be neglected, the output of the inverting amplifier shifts the input signal by 180°. At a particular frequency, the RC network shifts the phase by an additional 180°, resulting in a total phase shift of 0° from the output of the OpAmp amplifier to the input. The circuit oscillates at this frequency, provided that the amplifier gain is sufficiently large.

In order to solve for the loop gain, mesh equations for mesh 1, mesh 2, and mesh 3 are formulated:

$$\text{Mesh 1:} \quad v_o = i_1(R - jX) - i_2 R + i_3 0,$$
$$\text{Mesh 2:} \quad 0 = -i_1 R + i_2(2R - jX) - i_3 R, \tag{12.2-4}$$
$$\text{Mesh 3:} \quad 0 = i_1 0 - i_2 R + i_3(2R - jX),$$

where

$$X = \frac{1}{\omega C}, \quad \alpha = \frac{X}{R} = \frac{1}{\omega RC} = \frac{\omega_o}{\omega}, \quad \omega_o = \frac{1}{RC}.$$

Solve for the current i_3 using Cramer's rule,

$$i_3 = \frac{-R^2 v_o}{R^3[1 - 5\alpha^2 + j(\alpha^3 - 6\alpha)]}. \tag{12.2-5}$$

Therefore, the loop gain at the frequency of oscillation is

$$Af = \frac{R_f i_3}{v_o} = \frac{-R_f}{R[1 - 5\alpha^2 - j(6\alpha - \alpha^3)]} = -1. \tag{12.2-6}$$

By setting the quantity

$$\alpha^3 - 6\alpha = 0, \tag{12.2-7}$$

the frequency of oscillation is found by solving for α,

$$\alpha = \sqrt{6} = \frac{\omega}{\omega_o} \Rightarrow \omega = \frac{\omega_o}{\sqrt{6}}. \tag{12.2-8}$$

The frequency of oscillation is

$$\omega = \frac{1}{RC\sqrt{6}} \tag{12.2-9}$$

At the frequency of oscillation, the feedback factor $f = \frac{1}{29}$. For the magnitude of the loop gain to be slightly greater than unity, the magnitude of the open-loop gain of the oscillator must be slightly greater than 29. This condition states that an inverting amplifier must have a gain $|A| > 29$. Since an inverting is used, the resistors R_f and R are related by the required gain of the amplifier,

$$\left| -\frac{R_f}{R} \right| = |A| > 29. \tag{12.2-10}$$

EXAMPLE 12.2-1. Design of OpAmp Phase-Shift Oscillator

Design an OpAmp phase-shift oscillator to produce a 1 kHz ± 10% sinusoidal wave. Use a μA741 OpAmp and power supply voltages of ±15 V. Verify the design using SPICE.

SOLUTION

The circuit topology of the oscillator is identical to Figure 12.2-1.

Resistance values on the order of 10 kΩ are most practical in OpAmp circuits. The components for the feedback network can be determined using Equation 12.2-9:

$$\omega = 2\pi(1000) = \frac{1}{RC\sqrt{6}}.$$

Since there are more standard resistor value choices available to the designer than capacitor values, choose a reasonable and common capacitor value to initiate the design. For this design, a capacitor value of 0.01 μF is selected:

$$C = 0.01 \text{ μF.}$$

Using the chosen capacitor value, solve for the resistor in the RC ladder network:

$$R = \frac{1}{2\pi(1000)(0.01 \times 10^{-6})\sqrt{6}} = 6497 \approx 6.49 \text{ kΩ.}$$

The magnitude of the gain of the inverting OpAmp amplifier at midband must be greater than 29 by at least 5%. Choose a gain of $1.2 \times 29 = 34.8$. Then the feedback resistor R_f of the inverting amplifier is given as

$$R_f = -(A_{vo}R) = -[-34.8(6490)] = 225.9 \text{ k}\Omega \approx \textbf{226 k}\Omega.$$

The complete circuit design of the phase-shift oscillator is as shown with the SPICE analysis nodes.

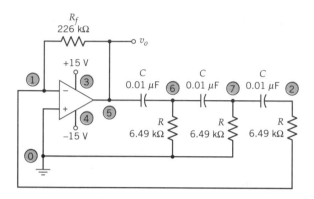

SPICE Simulation

The netlist for the phase-shift oscillator circuit is listed:

```
OpAmp Phase-Shift Oscillator
Rf     1    5      226k
VCC    3    0      +15
VEE    4    0      -15
X1  0 1 3 4 5 uA741
*-----------------------------------------------------------------------
*connections: non-inverting input
*             | inverting input
*             | | positive power supply
*             | | | negative power supply
*             | | | | output
*             | | | | |
.subckt uA741  1  2  3  4  5
*
cl    11 12 8.661E-12
c2     6  7 30.00E-12
dc     5 53 dx
de    54  5 dx
dlp   90 91 dx
dln   92 90 dx
dp     4  3 dx
egnd  99  0 poly(2) (3,0) (4,0) 0 .5 .5
fb     7 99 poly(5) vb vc ve vlp vln 0 10.61E6-10E6 10E6 10E6-10E6
ga     6  0 11  12  188.5E-6
gcm    0  6 10  99  5.961E-9
iee   10  4 dc  15.16E-6
hlim  90  0 vlim 1K
ql    11  2 13 qx
q2    12  1 14 qx
r2     6  9  100.0E3
rcl    3 11  5.305E3
rc2    3 12  5.305E3
```

```
re1   13   10   1.836E3
re2   14   10   1.836E3
ree   10   99   13.19E6
ro1   8    5    50
ro2   7    99   100
rp    3    4    18.16E3
vb    9    0    dc 0
vc    3    53   dc 1
ve    54   4    dc 1
vlim  7    8    dc 0
vlp   91   0    dc 40
vln   0    92   dc 40
.model dx D(Is=800.0E-18 Rs=1)
.model qx NPN(Is=800.0E-18 Bf=93.75)
.ends
*$
C1    5    6    0.01u IC=0
R1    6    0    6.49k
C2    6    7    0.01u IC=0
R2    7    0    6.49k
C3    7    8    0.01u IC=0
R3    2    1    6.49k
*In order to run the oscillator, the UIC (Use Initial Condition)
*command must be used. Note also, the IC (Initial Condition)
*specification on the capacitor values.
.TRAN 10u 50m 0 UIC
.OP
.PROBE
.END
```

The output of the oscillator is as shown. Note the transient time prior to steady-state oscillation. The transient time is approximately 45 ms. A larger loop gain will shorten the transient time but will also increase harmonic distortion of the steady-state sinusoid.

In the accompanying figure, the oscillation frequency is shown to be 1.08 kHz, which meets the design requirement.

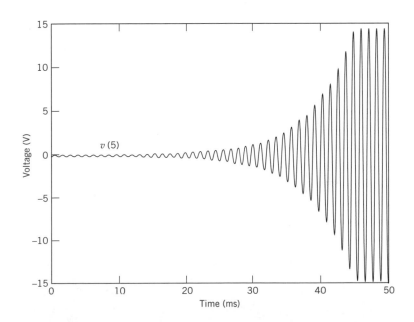

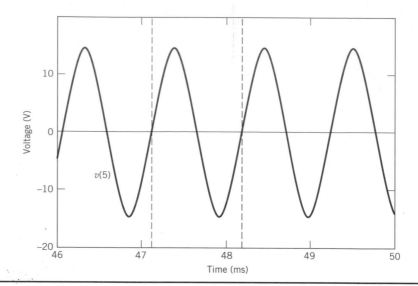

BJT Realization of Phase-Shift Oscillator

Discrete phase-shift oscillators can be designed using a single BJT or FET amplifier. Figure 12.2-2 shows a BJT-based phase-shift oscillator. The design procedure for the BJT-based phase-shift oscillator is identical to that of the OpAmp oscillator where the frequency of oscillation is determined by the R and C values of the reactive ladder network, according to Equation 12.2-9. One difference lies in the value of the resistor R' in the RC ladder network that connects the reactive feedback network to the base of the BJT.

The value of R' is adjusted so that

$$R = R' + (R_B \| h_{ie}), \qquad (12.2\text{-}11)$$

where $R_B = R_{B1} \| R_{B2}$.

It is evident that the RC phase-shift network is not independent of the BJT amplifier. Therefore, a complete analysis of the circuit is required. Assuming operation in the midband frequency range, the small-signal behavior of the BJT-based phase-

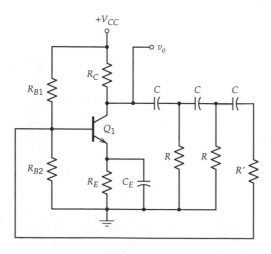

Figure 12.2-2
BJT-based phase-shift oscillator.

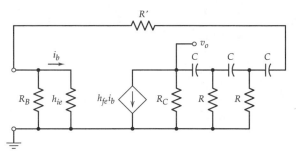

Figure 12.2-3
Small-signal equivalent circuit of BJT-based phase-shift oscillator.

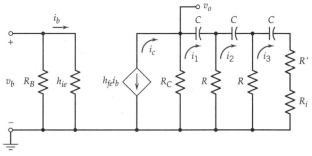

Figure 12.2-4
Equivalent circuit used to calculate loop gain.

shift oscillator in Figure 12.2-2 can be analyzed using the small-signal equivalent model in Figure 12.2-3. Assume that h_{oe}^{-1} is very large and can be ignored.

The small-signal equivalent circuit of the BJT-based phase-shift oscillator shown in Figure 12.2-3 is modified in Figure 12.2-4 to allow calculation of the loop gain of the circuit. The equivalent resistance R_i is given as

$$R_i = R_B \| h_{ie}. \tag{12.2-12}$$

The loop gain for the circuit in Figure 12.2-4 is most conveniently represented in terms of currents,

$$Af = \frac{i_3}{i_b}. \tag{12.2-13}$$

The loop gain is found by using the method of analysis shown in the phase-shift oscillator circuit. When the oscillator is designed so that $R_B \gg h_{ie}$ (i.e., $R_i = R_B \| h_{ie} \approx h_{ie}$), Equation 12.2-13 yields

$$A(j\omega)f(j\omega) = \frac{h_{fe}}{3 + \dfrac{R}{R_c} - j\dfrac{4}{RX} - j\dfrac{6}{R_CX} - \dfrac{1}{R_C^2X^2} - \dfrac{5}{RR_CX^2} + j\dfrac{1}{R^2R_CX^3}}, \tag{12.2-14}$$

where $X = 1/(\omega C)$.

The Barkhausen criterion for oscillation requires that $A(j\omega)f(j\omega) = 1\angle180° = -1$. Therefore, oscillation occurs when the imaginary part of the denominator in Equation 12.2-14 is zero,

$$0 = j\left(\frac{4}{RX} - \frac{6}{R_C X} + \frac{1}{R^2 R_C X^3}\right)$$
$$= j\left[\frac{1}{\omega}\left(\frac{4}{RC} - \frac{6}{R_C C}\right) + \frac{1}{\omega^3 R^2 R_C C^3}\right]. \tag{12.2-15}$$

Equation 12.2-15 simplifies to

$$\frac{1}{\omega^2 R R_C C^2} = 4 + \frac{6R}{R_C}. \tag{12.2-16}$$

Solving for the oscillation frequency ω in Equation 12.2-16 results in the expression

$$\omega = \frac{1}{RC\sqrt{\dfrac{4R_C}{R+6}}}. \tag{12.2-17}$$

The real part of the loop gain $A(j\omega)f(j\omega) = -1$. Under this condition, the imaginary component of the denominator of Equation 12.2-14 is zero. Therefore, at the oscillation condition,

$$-1 = \frac{h_{fe}}{3 + R/R_C - (1/\omega^2)\left(\dfrac{1}{R_C^2 C^2} + \dfrac{5}{R R_C C^2}\right)}. \tag{12.2-18}$$

The required current gain h_{fe} for the fulfillment of the Barkhausen criterion is found by substituting Equation 12.2-16 into Equation 12.2-18 and solving for h_{fe},

$$h_{fe} = 23 + \frac{29}{R_C/R} + 4\frac{R_C}{R}. \tag{12.2-19}$$

To find the resistor ratio R_C/R for the minimum h_{fe} required to fulfill the Barkhausen criterion, the derivative of h_{fe} with respect to R_C/R is found and set to zero,

$$0 = \frac{dh_{fe}}{d(R_C/R)} = \frac{d}{d(R_C/R)}\left(23 + \frac{29}{R_C/R} + 4\frac{R_C}{R}\right). \tag{12.2-20}$$

Equation 12.2-20 yields the resistor ratio R_C/R,

$$\frac{R_C}{R} = \sqrt{\frac{29}{4}} = 2.7. \tag{12.2-21}$$

Therefore, to fulfill the oscillation condition, the BJT current gain must be

$$h_{fe} \geq 23 + \frac{29}{2.7} + 4(2.7) = 44.5. \tag{12.2-22}$$

The result of Equation 12.2-22 is that the circuit meets the Barkhausen criterion and will oscillate if $h_{fe} \geq 44.5$. Most small-signal BJTs used in oscillator design have current gains well in excess of 44.5. To ensure oscillation, the loop gain is adjusted to be slightly greater than unity by adjusting the collector resistor R_C.

MOSFET Realization of Phase-Shift Oscillator

A depletion NMOSFET-based phase-shift oscillator, shown in Figure 12.2-5, is simpler to analyze than BJT-based oscillators since the gate (input) resistance is very high. Therefore, the analysis of the FET phase-shift oscillator is similar to the OpAmp oscillator.

When the FET-based phase-shift oscillator is design so that $R \gg R_D \| r_d$, the loading of the reactive feedback network on the amplifier can be neglected. The equation for the loop gain is found using the methods described for the BJT-based and OpAmp-based phase-shift oscillators. The loop gain of the NMOSFET-based phase-shift oscillator in Figure 12.2-5 is

$$A(j\omega)f(j\omega) = \frac{jg_m(R_D\|r_d)\omega^3 R^3 C^3}{(1 - 6\omega^2 R^2 C^2) + j\omega RC(5 - \omega^2 R^2 C^2)}, \tag{12.2-23}$$

where g_m is the mutual conductance of the FET.

At the oscillation condition,

$$\omega_o = \frac{1}{RC\sqrt{6}} \Rightarrow f_o = \frac{1}{2\pi RC\sqrt{6}}. \tag{12.2-24}$$

Substituting Equation 12.2-24 into Equation 12.2-23 yields the expression for the loop gain at the oscillation frequency,

$$A(j\omega)f(j\omega) - \frac{g_m(R_D\|r_d)}{29}. \tag{12.2-25}$$

From Equation 12.2-25, the gain of the FET amplifier must be $g_m R_L = -29$. This condition implies that the magnitude of the gain of the FET amplifier must be at least 29. This conclusion is identical to that of the OpAmp phase-shift oscillator.

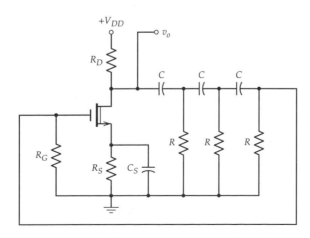

Figure 12.2-5
Depletion NMOSFET-based phase-shift oscillator; bias resistor $R_G \gg R$.

12.2.2 Wien Bridge Oscillator

The Wien bridge oscillator circuit uses a balanced bridge as its reactive feedback network. The OpAmp implementation of the Wien bridge oscillator is shown in Figure 12.2-6 where the bridge network is clearly depicted. When using an OpAmp as its gain element, the oscillator is assumed to operate at a frequency significantly lower than the unity-gain frequency of the OpAmp.

The feedback configuration of the Wien bridge oscillator is the series–shunt topology. To simplify the analysis of the circuit, Figures 12.2-7a and b are used. To find the loop gain of the oscillator circuit, the feedback path at the output of the OpAmp is broken and an external voltage v_L is applied, as shown in Figure 12.2-7b. Auxiliary voltages v_i, v_1, and v_2 are indicated to simplify the analysis.

The open-loop gain of the OpAmp is A_V (very large positive gain), and the output voltage is $v_o = A_V v_i$. The loop gain is therefore

$$Af = -\frac{v_o}{v_L} = -\frac{v_i}{v_L} A_V. \tag{12.2-26}$$

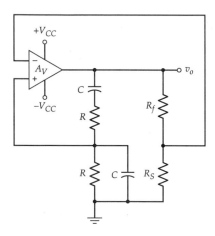

Figure 12.2-6
OpAmp Wien bridge oscillator circuit.

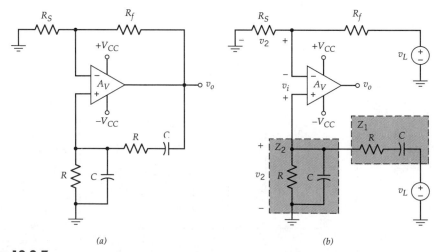

(a)   (b)

Figure 12.2-7
(a) OpAmp Wien bridge oscillator (redrawn). (b) Circuit used to calculate loop gain.

Noting that $v_i = v_2 - v_1$ and $A = A_V$, the feedback factor is

$$f = -\frac{v_i}{v_L} = -\frac{v_2 - v_1}{v_L} = -\left(\frac{Z_2}{Z_1 + Z_2} - \frac{R_S}{R_S + R_f}\right). \qquad (12.2\text{-}27)$$

Therefore, the loop gain for the circuit is

$$Af = -\frac{v_o}{v_L} = -A_V\left(\frac{Z_2}{Z_1 + Z_2} - \frac{R_S}{R_S + R_f}\right)$$

$$= -A_V\left(\frac{R/(1 + j\omega RC)}{R/(1 + j\omega RC) + R - j\left(\dfrac{1}{\omega C}\right)} - \frac{R_S}{R_S + R_f}\right) \qquad (12.2\text{-}28)$$

$$= -A_V\left(\frac{j\omega/\omega_o}{j\omega/\omega_o + (1 + j\omega/\omega_o)^2} - \frac{R_S}{R_S + R_f}\right),$$

where $\omega_o = 1/RC$.

Using the condition for oscillation $Af = -1$, Equation 12.2-28 yields

$$0 = 1 - \left(\frac{\omega}{\omega_o}\right)^2 + j\,\frac{\omega}{\omega_o}\left[3 - \frac{A_V(R_S + R_f)}{R_S(1 + A_V) + R_f}\right]. \qquad (12.2\text{-}29)$$

Equating the real and imaginary parts to zero, Equation 12.2-29 yields

$$\omega = \omega_o = \frac{1}{RC} \qquad (12.2\text{-}30)$$

for the oscillation frequency and

$$\frac{R_f}{R_S} = \frac{2A_V + 3}{A_V - 3} \approx 2 \qquad (12.2\text{-}31)$$

for the ratios of the gain resistors R_f and R_S.

12.3 *LC* OSCILLATORS

The *LC* oscillators are commonly used in RF applications, which span the frequency range from several hundred kilohertz to several hundred megahertz. The two most commonly used oscillator circuits are the Colpitts and Hartley oscillators. These oscillators are configured in the general form shown in Figure 12.3-1. The amplifier may be an OpAmp, BJT, or FET amplifier. The amplifier gain is A_v, the input resistance to the amplifier is R_i, output resistance is R_o, and complex impedances are Z_1, Z_2, and Z_3. The feedback topology of the circuit in Figure 12.3-1 is a series–shunt feedback.

The load impedance of the circuit is

$$Z_L = [(Z_1||R_i) + Z_3]||Z_2. \qquad (12.3\text{-}1)$$

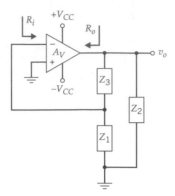

Figure 12.3-1
General form of many oscillator circuits.

The open-loop gain is

$$A = \frac{-A_v Z_L}{Z_L + R_o}.$$

(12.3-2)

The feedback factor is

$$f = \frac{Z_1 \| R_i}{(Z_1 \| R_i) + Z_3}.$$

(12.3-3)

The loop gain is found by multiplying Equations 12.3-2 and 12.3-3,

$$Af = \frac{A_v(Z_1 \| R_i)Z_2}{R_o[(Z_1 \| R_i) + Z_2 + Z_3] + Z_2[(Z_1 \| R_i) + Z_3]}.$$

(12.3-4)

If the impedances are purely reactive (inductive or capacitive), then $Z_1 = jX_1$, $Z_2 = jX_2$, and $Z_3 = jX_3$, where $X = \omega L$ for inductors and $X = -1/(\omega C)$ for capacitors. The loop gain is

$$Af = \frac{-A_v[(X_1^2 R_i + jX_1 R_i^2)/(R_i^2 + X_1^2)]X_2}{jR_o[X_1 R_i^2/(R_i^2 + X_1^2) + X_2 + X_3] - X_2[X_1^2 R_i/(R_i^2 + X_1^2) + X_3]}.$$

(12.3-5)

For the loop gain to be real with no phase shift and $R_i \gg X_1$, the imaginary component of Equation 12.3-5 is set equal to zero,

$$0 = X_1 + X_2 + X_3,$$

(12.3-6)

and therefore,

$$Af = \frac{A_v X_1 X_2}{X_2(X_1 + X_3)} = \frac{A_v X_1}{(X_1 + X_3)}.$$

(12.3-7)

From Equation 12.3-6, the circuit oscillates at a resonant frequency corresponding to the series combination of X_1, X_2, and X_3.

Equation 12.3-7 is simplified by using the relationship in Equation 12.3-6,

$$Af = \frac{A_v X_1}{X_2}.$$

(12.3-8)

That the loop gain $|Af|$ must have a magnitude of at least unity and X_1 must have the same sign as X_2, that is, they must be the same type of reactance, either both capacitive or inductive, implies that X_3 must be inductive if X_1 and X_2 are capacitive, or vise versa.

A Colpitts oscillator is a circuit where Z_1 and Z_2 are capacitors and Z_3 is an inductor. A Hartley oscillator is a circuit where Z_1 and Z_2 are inductors and Z_3 is a capacitor. In the latter case, mutual inductance between Z_1 and Z_2 will alter the relationships derived above.

The *LC* oscillators are commonly used in RF applications in the frequency range between 100 kHz and several hundred megahertz. In this range of frequencies, the simple audio frequency small-signal models of active devices (BJTs and FETs) cannot be used. High-frequency transistor models must be used in the design and analysis. The *LC* oscillators are found extensively in communication electronics. For example, in AM and FM receivers, station tuning is accomplished by varying a capacitance or an inductance.

12.3.1 Colpitts Oscillator

A BJT-based Colpitts oscillator circuit is shown in Figure 12.3-2. The capacitors C_B and C_E are very large coupling and bypass capacitors, respectively. The small-signal model at the output of the BJT-based Colpitts oscillator is shown in Figure 12.3-3.

The equivalent device output capacitance C_o is not included in Figure 12.3-3 since the high-frequency response of BJT amplifiers is limited by $C_i = C_\pi + [1 + g_m(R||R_C)]C_\mu$. The resistor $R = (r_b + r_\pi)||R_B$, and $C_2 = C_2' + C_i$. The output resistance of the BJT, r_o, is assumed to be very high.

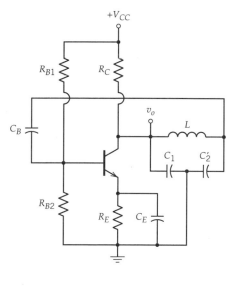

Figure 12.3-2
BJT-based Colpitts oscillator.

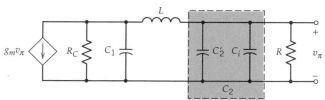

Figure 12.3-3
Output equivalent circuit for Colpitts oscillator in Figure 12.3-2.

The loop gain is found by breaking open the loop at the base of the BJT and calculating the signal that is returned via the feedback network. The loop gain is

$$Af = g_m\{R_C||-jX_{C1}||[jX_L + (-jX_{C_2}||R)]\}\frac{-jX_{C_2}||R}{(-jX_{C_2}||R) + jX_L} \quad (12.3\text{-}9)$$

$$= \frac{g_m(R||R_C)}{j\omega(R||R_C)C_1 + [1 + j\omega(R||R_C)C_2](1 - \omega^2LC_1)}.$$

For the loop gain to be real (180° phase shift),

$$\omega_o^2 = \frac{C_1 + C_2}{LC_1C_2}, \quad (12.3\text{-}10)$$

$$1 = \frac{-g_m(R_C||R)}{1 - \omega_o^2LC_1}. \quad (12.3\text{-}11)$$

The condition for oscillation is found by substituting Equation 12.3-10 into Equation 12.3-11,

$$1 + g_m(R_C||R) > 1 + \frac{C_1}{C_2}. \quad (12.3\text{-}12)$$

If the oscillator is designed so that $R_B \gg r_b + r_\pi$, then $R \approx r_b + r_\pi$. Since r_b is small compared to r_π, $R \approx r_\pi = \beta_F/g_m$. Substituting this simplified expression for R into Equation 12.3-12 yields the required capacitance ratio

$$\frac{\beta_F R_C}{R_C + r_\pi} > \frac{C_1}{C_2}. \quad (12.3\text{-}13)$$

The frequency of oscillation is found from Equation 12.3-10 as

$$f_o = \frac{1}{2\pi}\sqrt{\frac{C_1 + C_2}{LC_1C_2}}. \quad (12.3\text{-}14)$$

EXAMPLE 12.3-1.
BJT-Based Colpitts Oscillator

Complete the design of the given circuit for an oscillation frequency of 10.7 MHz. Assume that the transistor parameters are

$$\beta_F = 200, \qquad V_A = 150\ \text{V}, \qquad r_b = 30\ \Omega,$$

$$C_\mu = 3\ \text{pF}, \qquad f_T = 250\ \text{MHz}.$$

SOLUTION

First determine the bias condition of the oscillator. Applying KVL to the base–emitter loop,

$$0 = \frac{V_{CC}R_{B2}}{R_{B1} + R_{B2}} - \frac{I_C}{\beta_F}(R_{B1}||R_{B2}) - V_\gamma - I_C\frac{\beta_F + 1}{\beta_F}R_E.$$

Substituting in the appropriate values yields the collector current

$$I_C = 1\ \text{mA}.$$

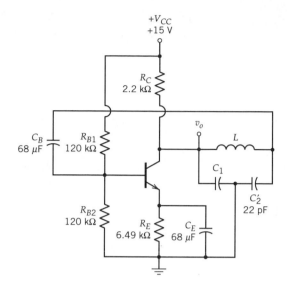

Checking the collector–emitter loop results in $V_{CE} = 6.3$ V.

The hybrid-π parameters of interest are

$$g_m = \frac{|I_C|}{V_t} = \frac{0.001}{0.026} = 38.5 \text{ mS}, \qquad r_\pi = \frac{\beta_F}{g_m} = \frac{200}{0.0385} = 5.2 \text{ k}\Omega.$$

The output resistance of the BJT, r_o, is significantly high enough to ignore in the design. Also, $R_B \gg r_\pi$.

The capacitance C_π is given by

$$C_\pi = \frac{g_m}{\omega_T} - C_\mu - \frac{0.0385}{2\pi(250 \times 10^6)} - 3 \times 10^{-12} = 21.5 \text{ pF}.$$

Miller's equivalent input capacitance of the BJT is

$$C_i = C_\pi + C_\mu(1 + g_m R) = C_\pi + C_\mu\{1 + g_m[R_{B1}\|R_{B2}\|R_C\|(r_b + r_\pi)]\}$$

$$= (21.5 + 3\{1 + 0.0385[120 \times 10^3\|120 \times 10^3\|2.2 \times 10^3\|(30 + 5.2 \times 10^3)]\}) \times 10^{-12}$$

$$= 199 \text{ pF}.$$

Then

$$C_2 = C_2' + C_i = 22 + 199 = 221 \text{ pF}.$$

If the magnitude of the closed-loop gain is designed for 1.25, then

$$\frac{\beta_F R_C}{1.25(R_C + r_\pi)} = \frac{C_1}{C_2},$$

$$C_1 = \frac{\beta_F R_C C_2}{1.25(R_C + r_\pi)} = \frac{200(2200)(221 \times 10^{-12})}{1.25[2200 + (200/0.0385)]} = 10.4 \text{ nF}.$$

Select $C_1 = \mathbf{0.010\ \mu F}$.

The frequency of oscillation is 10.7 MHz, expressed as

$$f_o = \frac{1}{2\pi} \sqrt{\frac{C_1 + C_2}{LC_1C_2}}.$$

Solving for L,

$$L = \frac{C_1 + C_2}{(2\pi f_o)^2 C_1 C_2} = \frac{(0.01 \times 10^{-6} + 221 \times 10^{-12})}{[2\pi(10.7 \times 10^6)]^2 (0.01 \times 10^{-6})(221 \times 10^{-12})}$$

$$= 1.02\ \mu H \approx \mathbf{1.0\ \mu H}.$$

SPICE Results

```
10.7 MHz BJT Colpitts Oscillator
VCC   1   0   15V
RC    1   4   2.2k
RB1   1   3   120k
CB    2   3   68u IC=0
RB2       3   0   120k
Q1    4   3   5 npnbjt
RE    5   0   6.49k
CE    5   0   68u IC=0
C1    4   0   0.01u IC=0
*C2 includes the input capacitance of the BJT
C2    2   0   221p IC=0
L1    4   2   1.0u
.MODEL npnbjt NPN(IS=1p BF=200 VA=150)
.OP
.TRAN 1n 2u 0 1n UIC
.PROBE
.END
```

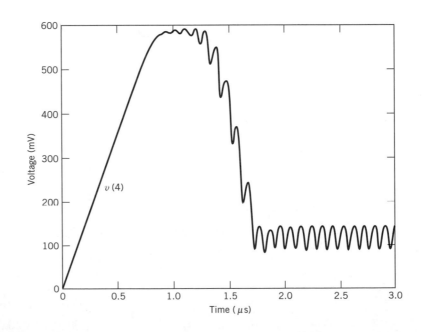

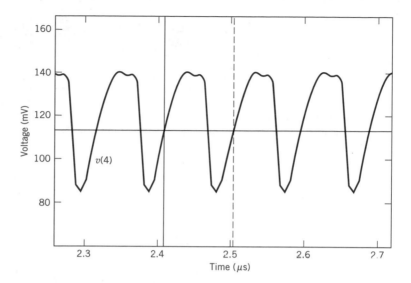

The resulting output at the collector of the BJT is as shown. Note the finite transient time to steady-state oscillation.

The steady-state output of the oscillator is also shown. The period of oscillation is 92.9 ns, which corresponds to 10.76 MHz. The oscillation frequency of the simulated circuit is within 1% of the desired oscillating frequency of 10.7 MHz.

A FET-based Colpitts oscillator, shown in Figure 12.3-4, is somewhat simpler to design and analyze since the input (gate) resistance is very large.

The capacitors C_G and C_S are very large coupling and bypass capacitors, respectively. If the circuit is designed with a large value of R_G, the small-signal model at the output of the FET-based Colpitts oscillator is shown in Figure 12.3-3.

The equivalent device output capacitance C_o is not included in Figure 12.3-5, since the high-frequency response of FET amplifiers is limited by $C_i = C_{gs} + [1 + g_m R]C_{gd}$. The resistor R is given as $R = r_d \| R_D$, and $C_2 = C_2' + C_i$.

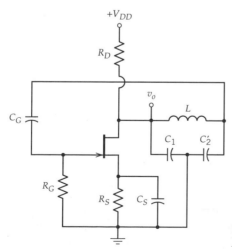

Figure 12.3-4
JFET-based Colpitts oscillator.

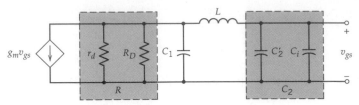

Figure 12.3-5
Output equivalent circuit for Colpitts oscillator in Figure 12.3-4.

The loop gain is found by breaking the loop open at the gate of the FET and calculating the signal that is returned via the feedback network. The loop gain is

$$Af = g_m\{R\|[-jX_{C_1}\|(jX_L - jX_{C2})]\}\frac{-jX_{C_2}}{jX_L - jX_{C2}}. \tag{12.3-15}$$

For simplicity, let $C_1 = C_2$. Then the loop gain is

$$Af = \frac{-g_m R X_C^2}{R(jX_L - 2jX_C) - jX_C(jX_L - jX_C)}. \tag{12.3-16}$$

For the loop gain to be real (180° phase shift),

$$\omega_o = \frac{\sqrt{2}}{\sqrt{LC}}, \tag{12.3-17}$$

$$1 = g_m R. \tag{12.3-18}$$

The condition for oscillation indicates that the circuit will oscillate if

$$g_m R > 1. \tag{12.3-19}$$

An OpAmp-based Colpitts oscillator is shown in Figure 12.3-6. The small output resistance of the OpAmp is included in the analysis.

The oscillation condition occurs at

$$\omega_o = \frac{\sqrt{2}}{\sqrt{LC}}. \tag{12.3-20}$$

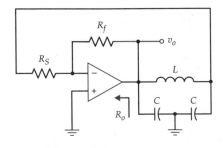

Figure 12.3-6
OpAmp-based Colpitts oscillator.

The oscillator circuit should be designed so that

$$\frac{R_f}{R_S} > 1. \tag{12.3-21}$$

The advantage of using Colpitts oscillators comes from the fact that tuning a single inductor allows for variation in the oscillation frequency. This is especially advantageous since inductors can easily be tuned by introducing ferrite material in through the cross section of the inductor coil.

12.3.2 Hartley Oscillator

The Hartley oscillator circuit is similar to the Colpitts oscillator with one major difference: The capacitors and inductors have be "swapped." A Colpitts oscillator can be mapped into a Hartley oscillator by exchanging jX_L and $-jX_C$. A FET-based Hartley oscillator circuit is shown in Figure 12.3-7, where C_D is a large coupling capacitor required to isolate the DC and AC models. The inductors in the circuit are identical, and for simplicity, there is no mutual inductance.

The loop gain is found by breaking the loop open at the gate of the FET and calculating the signal that is returned via the feedback network. The loop gain is

$$\Lambda f = \frac{-g_m R X_L^2}{R(-jX_C + 2jX_L) + jX_L(jX_L - jX_C)}. \tag{12.3-22}$$

For the loop gain to be real (180° phase shift),

$$\omega_o^2 = \frac{1}{\sqrt{2LC}}, \tag{12.3-23}$$

$$1 = g_m R. \tag{12.3-24a}$$

The condition for oscillation indicates that the circuit will oscillate if

$$g_m R > 1. \tag{12.3-24b}$$

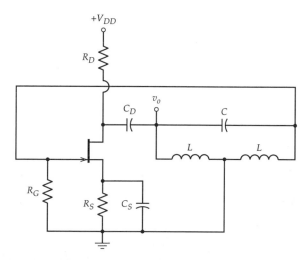

Figure 12.3-7
FET-based Hartley oscillator.

The advantage of using Hartley oscillators comes from the fact that tuning a single capacitor allows for variation in the oscillation frequency. This is especially advantageous since a single-voltage variable capacitor called a varactor diode can be used to easily tune the circuit.

12.4 CRYSTAL OSCILLATORS

Crystals are three-dimensional, mechanical oscillating components that oscillate in many different modes. The crystal oscillations are governed by the crystal piezoelectric properties and the arrangement and shape of the electrodes attached to the crystal. Crystals are fabricated so that several oscillating and harmonic modes can be used in the design of a circuit. Crystals are available in a wide range of discrete frequencies.

For stable frequency operation, the oscillator should be designed so that a crystal is the controlling element for the oscillation. The crystal oscillator is often a critical component in communications systems and in digital signal processing applications. The circuit symbol for the piezoelectric crystal is shown in Figure 12.4-1a. The electrical equivalent circuit representing the resonant nature of the piezoelectric crystal is shown in Figure 12.4-1b. The equivalent circuit is similar to the familiar RLC passive resonant circuit.

The capacitance C_p is on the order of 10 pF and includes the capacitance associated with the mechanical package, and C_s is on the order of 0.05 pF. The crystal equivalent inductance L is very large for quartz crystals, on the order of several tens of henrys. The internal losses of the crystal is represented by R_s, which is typically small. The resistive loss R_s is related to the quality factor Q of the crystal, since the energy lost during any periodic signal is associated with the dissipative resistance. Here, Q is defined as the ratio of maximum energy stored to the amount lost per periodic cycle. It also determines the bandwidth of resonant circuits. The bandwidth is calculated from Q and the resonant frequency by

$$\text{BW} = \frac{f_o}{Q}. \tag{12.4-1}$$

The quality factor is related to the dissipative resistance and inductance through the relationship

$$Q = \frac{\omega_o L}{R_s}. \tag{12.4-2}$$

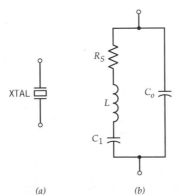

Figure 12.4-1
(a) Circuit symbol for a crystal. (b) Equivalent circuit for a crystal.

Crystals are inherently high-Q devices. When used in oscillator circuits, a crystal increases the oscillation stability. This was shown in Section 12.1, where

$$\frac{\partial(Af)}{\partial\omega} = A\frac{\partial f}{\partial\omega}.$$

Since the slope $\partial|f|/\partial\omega$ is directly proportional to the quality factor Q of the oscillator circuit, the high-Q characteristic of crystals allows the fulfillment of the requirement of Equation 12.1-10 that $A(\partial f/\partial\omega) \rightarrow \infty$. Therefore, crystal oscillators are inherently very stable.

Figure 12.4-1b provides a simplified equivalent circuit of a crystal at one oscillating frequency. In reality, the crystal has many oscillatory modes (and frequencies). Therefore, a more accurate model of a crystal depicting its many oscillatory frequencies and harmonics is shown in Figure 12.4-2.

The circuit in Figure 12.4-2 contains several series-resonant circuits whose frequencies are nearly the odd harmonics of the fundamental oscillatory frequency. The higher resonant frequencies are called overtones of the fundamental. Typical circuit parameters for fundamental, third, and fifth overtone crystals are given in Table 12.4-1.

If $R_s = 0$, crystal input resistance in a narrow frequency range around f_o is given as

$$Z(j\omega) = \frac{(j\omega C_o)^{-1}[j\omega L + (j\omega C_1)^{-1}]}{j\omega L + (j\omega C_1)^{-1} + (j\omega C_o)^{-1}}. \qquad (12.4\text{-}3)$$

The impedance $Z(j\omega)$ will be zero when L and C_1 are in resonance, that is,

$$f_s = \frac{1}{2\pi\sqrt{LC_1}}, \qquad (12.4\text{-}4)$$

where f_s is the series-resonant frequency of the crystal. Conversely, the crystal will have infinite input impedance at the frequency

$$f_a = \frac{1}{2\pi\sqrt{L[C_oC_1/(C_o + C_1)]}}, \qquad (12.4\text{-}5)$$

where f_a is the antiresonant frequency of the crystal. An ideal crystal (one with $R_s = 0$) behaves as both a series-resonant and a parallel-resonant circuit with infinite Q.

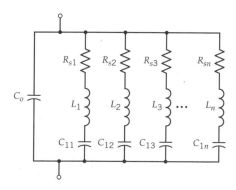

Figure 12.4-2
Electric circuit equivalent of crystal showing many oscillatory states.

TABLE 12.4-1 TYPICAL CRYSTAL DATA

f, MHz	Oscillation Mode	R_s, Ω	C_o, pF	C_1, fF	Q
1.0	Fundamental	250	4.0	9.0	65,000
2.0	Fundamental	70	3.5	10.0	110,000
5.0	Fundamental	15	6.0	24.0	86,000
10.0	Fundamental	12	6.0	24.0	50,000
20.0	Fundamental	12	6.0	24.0	25,000
45.0	Third overtone	25	5.0	1.5	90,000
100.0	Fifth overtone	40	5.0	0.3	130,000

Realistically, all crystals have some series resistance R_s. For nonzero R_s, the input impedance of a crystal is

$$Z(j\omega) = \frac{(j\omega C_o)^{-1}[j\omega L + R_s + (j\omega C_1)^{-1}]}{j\omega L + R_s + (j\omega C_1)^{-1} + (j\omega C_o)^{-1}}. \tag{12.4-6}$$

In practice, the input impedance of a crystal at resonance with nonzero R_s is the same as one with $R_s = 0$. The effect of R_s is primarily in the reduction of Q.

Crystal oscillators can be implemented in a variety of topologies. For example, a Pierce oscillator, shown in Figure 12.4-3, is simply a Colpitts oscillator where the crystal has replaced the inductor in the feedback path. The Pierce oscillator depends on the inductive component of the crystal to provide a feedback of the proper phase. The crystal will oscillate at a frequency between f_a and f_S.

The circuit in Figure 12.4-3 is designed and analyzed in the same manner as the LC Colpitts oscillator. Feedback is provided through the crystal, which is operating near the series-resonant mode frequency. For oscillators operating above 20 MHz,

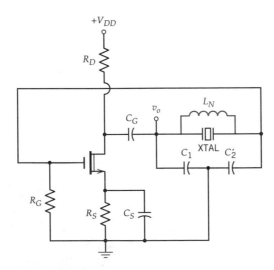

Figure 12-4.3
Pierce crystal oscillator.

an inductor L_N is placed in parallel with the crystal. The purpose of L_N is to neutralize the package capacitance C_o, where

$$L_N \approx \frac{1}{(2\pi f_o)^2 C_o}. \tag{12.4-7}$$

The Pierce oscillator is most commonly used with fundamental-mode crystals.

A Pierce oscillator designed from a CMOS inverter as an amplifier is shown in Figure 12.4-4. The reactive components of the feedback network consist of C_1, C_2, and XTAL (the crystal).

The CMOS amplifier has a high-gain inverting amplifier. The reactive feedback network reverses the phase again so that there is positive feedback. This CMOS oscillator is often used as a simple timing circuit.

The effect of adding series or parallel elements to the crystal is quite complex. The resonant frequency is altered, and the dissipative resistance is scaled. Losses due to the added components will also alter the composite Q of the oscillator. To complicate things further, all of the effects mentioned vary as the circuit is tuned or adjusted.

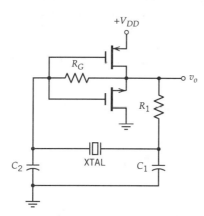

Figure 12.4-4
CMOS Pierce oscillator.

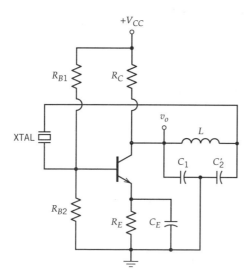

Figure 12.4-5
Improved crystal Colpitts oscillator.

Figure 12.4-5 shows an improved crystal Colpitts oscillator circuit. The schematic shows a series crystal in the feedback path of the oscillator.

Here the crystal appears as a series resistance R_s (crystal resistance) at the series-resonant frequency. Since R_s is small, it can usually be ignored in designing the circuit. The design procedure is identical to an LC Colpitts oscillator with addition of a crystal in the feedback path.

12.5 CONCLUDING REMARKS

Electronic harmonic oscillator circuits were discussed in this chapter. Although several different configurations were presented, all of the harmonic oscillators shared the Barkhausen criterion for oscillation, which simply states that the loop gain of the circuit has unity magnitude gain and a phase of 180°.

Oscillators were designed from BJTs, FETs, and OpAmps. The advantage of using FETs and OpAmps lies in the large input resistance of the devices. Because of their large input resistances, the device does not affect the impedance of the reactive feedback network. However, at higher frequencies, OpAmps cannot be used since the high-frequency response is limited by the relatively low gain–bandwidth product. Therefore, for high-frequency RF applications, BJT and FET oscillators are used.

At high frequencies, LC reactive networks are commonly used. This is due to the inductive nature of resistors at high frequencies, which prohibits the use of RC phase-shift oscillators at these frequencies.

For good frequency stability, crystals are used in the oscillator circuit. Because crystals are inherently reactive, they can be used in place of inductors in oscillator circuits. At higher frequencies, however, a small inductor should be placed in parallel with the crystal to counteract the package capacitance.

SUMMARY DESIGN EXAMPLE: COLOR REFERENCE OSCILLATOR FOR COLOR TELEVISION

The color seen on a typical television is emitted from special phosphors. The color cathode-ray tube (CRT) has three separate phosphors placed on the screen. When struck by electrons, one phosphor emits red, one green, and one blue. The three phosphors are arranged very close to each other in alternating vertical stripes or as tiny points in a matrix pattern. The color that is projected to the user is a combination of the three phosphor emissions.

The electronic signal required to produce color on an otherwise black and white picture signal is called chrominance or chroma. The frequency of the chroma signal is 3.5795 MHz. The chroma signal is a vector with the phase angle indicating the hue (tint, or type of color) and magnitude indicating the saturation level or color intensity. A red fire engine is a saturated red object.

There are several oscillator designs that can be considered for the 3.5795-MHz oscillator:

- Phase-shift oscillator
- Wien bridge oscillator
- Colpitts oscillator
- Hartley oscillator
- Pierce oscillator

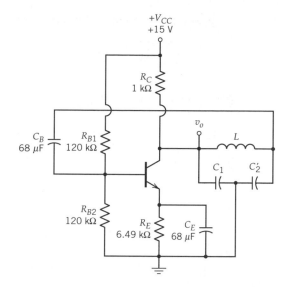

Each of the oscillator configurations listed can be designed with OpAmps, BJTs, and FETs. OpAmp implementations are not practical in this case since the frequency is higher than most low-cost OpAmp gain–bandwidth products. For low cost, it is also common to use inexpensive BJTs for discrete electronic circuits. A common device is the 2N3904 npn BJT. The BJT specifications are

$$\beta_F = 200, \qquad V_A = 150 \text{ V}, \qquad r_b = 30 \ \Omega,$$

$$C_{ib0} = 8 \text{ pF} \quad \text{at } V_{EB} = 0.5 \text{ V},$$

$$C_{ob0} = 4 \text{ pF} \quad \text{at } V_{CB} = 5 \text{ V},$$

$$f_T = 300 \text{ MHz}.$$

The oscillator chosen is the BJT Colpitts oscillator. Bias the transistor at a modest collector current. Choose $I_C = 1$ mA with a large Thévenin base resistance to simplify analysis. The circuit configuration chosen is as shown.

In order to solve for the reactive feedback elements, the effective input and output capacitances of the BJT must be found. The parameters of interest are CJE and CJC for SPICE simulations and C_π and C_μ for analytical design.

From Chapter 10, the relationship between SPICE and analytical parameter values are

$$\text{CJC} = C_{ob0}\left(1 + \frac{V_{CB}}{0.75}\right)^{0.33}, \tag{12.5-1}$$

$$\text{CJE} = C_{ib0}\left(1 + \frac{V_{EB}}{0.75}\right)^{0.33}, \tag{12.5-2}$$

$$\text{TF} = \frac{1}{2\pi f_T} - \frac{\eta V_t}{|I_{CT}|}\left[\frac{\text{CJE}}{(1 - 0.7/0.75)^{0.33}} + C_{ob0}\right]. \tag{12.5-3}$$

These equations yield CJE = 9.47 pF, CJC = 7.83 pF, and TF = 460 ps.

The hybrid-π parameters of interest are

$$g_m = \frac{|I_C|}{V_t} = \frac{0.001}{0.026} = 38.5 \text{ mS}, \qquad r_\pi = \frac{\beta_F}{g_m} = \frac{200}{0.0385} = 5.2 \text{ k}\Omega.$$

The capacitance C_μ is given as

$$C_{\mu 1} = \frac{\text{CJC}}{\left(1 + \dfrac{V_{CB}}{0.75}\right)^{0.33}}$$

$$= \frac{\text{CJC}}{\left(1 + \dfrac{15 - I_C R_C - [V_{CC}R_B/(R_{B2} + R_{B1}) - I_C(R_{B1}\|R_{B2})/\beta_F]}{0.75}\right)^{0.33}}.$$

The resulting value of C_μ is 3.7 pF.

The output resistance of the BJT, r_o, is significantly high enough to ignore in the design. Also, $R_B \gg r_\pi$.

The capacitance C_π is given as

$$C_\pi = \frac{g_m}{\omega_T} - C_\mu = \frac{0.0385}{2\pi(300 \times 10^6)} - 3.7 \times 10^{-12} = 16.8 \text{ pF}.$$

Miller's equivalent input capacitance of the BJT is

$$C_i = C_\pi + C_\mu(1 + g_m R) = C_\pi + C_\mu\{1 + g_m[R_{B1}\|R_{B2}\|R_C\|(r_b + r_\pi)]\}$$
$$= (16.5 + 3.7\{1 + 0.0385[120 \times 10^3\|120 \times 10^3\|10^3\|(30 + 5.2 \times 10^3)]\}) \times 10^{-12}$$
$$= 139 \text{ pF}.$$

Let $C_2' = 22$ pF. Then,

$$C_2 = C_2' + C_i = 22 + 139 = 161 \text{ pF}.$$

If the magnitude of the closed-loop gain is designed for 1.25, then

$$\frac{C_1}{C_2} = \frac{\beta_F R_C}{1.25(R_C + r_\pi)},$$

$$C_1 = \frac{\beta_F R_C C_2}{1.25(R_C + r_\pi)} = \frac{200(1000)(161 \times 10^{-12})}{1.25[1000 + (200/0.0385)]} = 4200 \text{ pF}.$$

Select $C_1 = \mathbf{3900}$ **pF**. The frequency of oscillation is 3.5795 MHz, expressed as

$$f_o = \frac{1}{2\pi}\sqrt{\frac{C_1 + C_2}{LC_1C_2}}.$$

Solving for L,

$$L = \frac{C_1 + C_2}{(2\pi f_o)^2 C_1 C_2} = \frac{3900 \times 10^{-12} + 161 \times 10^{-12}}{[2\pi(3.5795 \times 10^6)]^2(3900 \times 10^{-12})(161 \times 10^{-12})}$$
$$= 12 \text{ }\mu\text{H} \approx \mathbf{10 }\text{ }\boldsymbol{\mu}\textbf{H}.$$

Because standard values are used, the actual oscillation frequency is $f_o = 3.72$ MHz. A simple method for ensuring stable and accurate oscillation frequency is to use a 3.5795-MHz crystal. A crystal Colpitts oscillator can be designed by simply replacing the inductor in the reactive feedback network using the calculated capacitor values.

REFERENCES

Ghausi, M. S., *Electronic Devices and Circuits: Discrete and Integrated,* Holt, Rinehart and Winston, New York, 1985.

Millman, J., and Halkias, C. C., *Integrated Electronics: Analog and Digital Circuits and Systems,* McGraw-Hill, New York, 1972.

Sedra, A. S., and Smith, K. C., *Microelectronic Circuits,* 3rd ed., Holt, Rinehart and Winston, Philadelphia, 1991.

Schilling, D. L., and Belove, C., *Electronic Circuits,* 3rd ed., McGraw-Ilill, New York, 1989.

Young, P. H., *Electronic Communication Techniques,* 3rd ed., Merrill, New York, 1994.

PROBLEMS

12-1. For the network shown, show:

a. When used with an OpAmp to form an oscillator, show that the resonant frequency is $f = 2\pi RC$ and the gain must exceed 3.

b. Use SPICE to verify the result of part a.

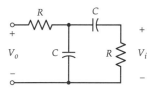

12-2. Design an OpAmp-based phase-shift oscillator circuit at 8 kHz. Assume an ideal OpAmp.

12-3. Show that a four-section RC network reduces the required network voltage gain for a phase-shift oscillator.

a. Determine the minimum required voltage gain of the amplifier.

b. Determine the resonant frequency of a four-section RC phase-shift oscillator.

12-4. Determine the resonant frequency of a phase-shift oscillator with three RC sections with capacitor to ground as shown.

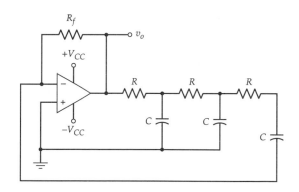

12-5. For a four-section RC phase-shift network similar to problem 12-27 with capacitor to ground, for a phase-shift oscillator:

a. Determine the minimum required voltage gain of the amplifier.

b. Determine the resonant frequency of the phase-shift oscillator.

12-6. Assume the following FET parameters: $V_{PO} = -3.5$ V, $I_{DSS} = 10$ mA, and $V_A = 200$ V.

a. Complete the given design for an oscillation frequency of 5 kHz. Make any necessary assumptions.

b. Simulate the circuit using SPICE and confirm the oscillation frequency. How is the oscillation frequency affected when the temperature of the FET is changed to 50°C?

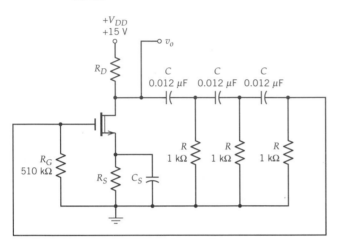

12-7. Assume the following BJT parameters of interest: $\beta_F = 200$ and $V_A = 200$ V.

a. Complete the design of the BJT-based phase-shift oscillator shown. What is the frequency of oscillation? Make any necessary assumptions.

b. Simulate the circuit using SPICE and confirm the oscillation frequency. How is the oscillation frequency affected when the temperature of the BJT is changed to 50°C?

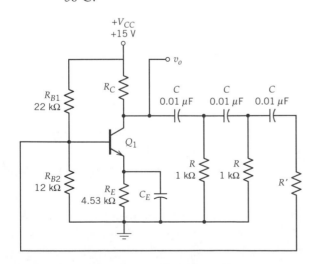

12-8. Complete the given design of an OpAmp-based Wien bridge oscillator for 19.2 kHz. Use $C = 1\ \mu F$ and $R_f = 5.1\ k\Omega$.

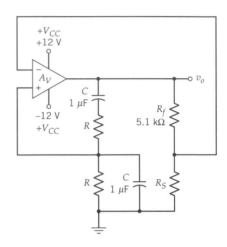

12-9. Using an ideal OpAmp, design a Colpitts oscillator at 1 kHz using an inductor value of $L = 10\ \mu H$.

12-10. Design a BJT-based common-emitter configured Colpitts oscillator at 1 MHz. The BJT parameters of interest are

$$\beta_F = 200, \qquad V_A = 150\ V,\ r_b = 30\ \Omega,$$

$$C_{ib0} = 8\ pF \quad \text{at } V_{EB} = 0.5\ V,$$

$$C_{ob0} = 4\ pF \quad \text{at } V_{CB} = 5\ V,$$

$$f_T = 300\ MHz.$$

A +15-V power supply is available. Simulate the circuit using SPICE and confirm the oscillation frequency. How is the oscillation frequency affected when the temperature of the BJT is changed to 50°C?

12-11. Assume the following BJT parameters:

$$\beta_F = 200, \qquad V_A = 150\ V, \qquad r_b = 30\ \Omega,$$

$$C_{ib0} = 8\ pF \quad \text{at } V_{EB} = 0.5\ V,$$

$$C_{ob0} = 4\ pF \quad \text{at } V_{CB} = 5\ V,$$

$$f_T = 300\ MHz.$$

a. Design a BJT-based common-emitter configured Colpitts oscillator at 10 MHz.

Bias the transistor at $I_C = 1$ mA and $V_{CE} = 8$ V. A power supply voltage of +24 is available.

b. Redesign the oscillator using the parameters found in Table 12.4-1 for a 10-MHz fundamental crystal.

c. Simulate both circuits with SPICE and comment on the results.

12-12. Design an NMOSFET-based common-source configured Colpitts oscillator at 2 MHz. The FET parameters of interest are

$$I_{DSS} = 6 \text{ mA}, \qquad V_{PO} = -4.7 \text{ V}, \qquad V_A = 100 \text{ V},$$

$$C_{iss} = 4.5 \text{ pF} \quad \text{at } V_{GS} = 0 \text{ V},$$

$$C_{rss} = 1.5 \text{ pF} \quad \text{at } V_{GS} = 0 \text{ V}.$$

A +15-V power supply is available. Bias the transistor at $I_D = 1$ mA and $V_{DS} = 5$ V. Simulate the circuit using SPICE and confirm the oscillation frequency. How is the oscillation frequency affected when the temperature of the FET is changed to 50°C?

12-13. Determine the relationship for the frequency of oscillation for the common-base configured Colpitts oscillator in terms of C_1, C_2, and L using small-signal hybrid-π analysis and appropriate assumptions. Assume that $r_o^{-1} \ll C_1 C_2'(r_b + r_\pi)$ and C_C and C_B are large-valued capacitors.

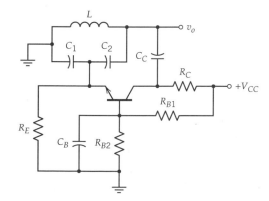

12-14. Determine the relationship for the frequency of oscillation for the common-gate configured Colpitts oscillator in terms of C_1, C_2,

and L using small-signal analysis and appropriate assumptions. The terms C_D and C_G are large-valued capacitors.

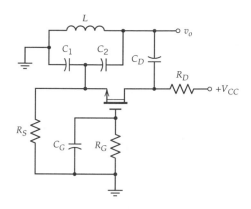

12-15. Design a BJT-based common-base configured Colpitts oscillator at 1 MHz. The BJT parameters of interest are

$$\beta_F = 200, \qquad V_A = 150 \text{ V}, \qquad r_b = 30 \ \Omega,$$

$$C_{ibo} = 8 \text{ pF} \quad \text{at } V_{EB} = 0.5 \text{ V},$$

$$C_{obo} = 4 \text{ pF} \quad \text{at } V_{CB} = 5 \text{ V},$$

$$f_T = 300 \text{ MHz}.$$

A +15-V power supply is available. Bias the transistor at $I_C = 1$ mA and $V_{CE} = 5$ V. Simulate the circuit using SPICE and confirm the oscillation frequency. How is the oscillation frequency affected when the temperature of the BJT is changed to 50°C?

12-16. Design an NMOSFET-based common-gate configured Colpitts oscillator at 2 MHz. The FET parameters of interest are

$$I_{DSS} = 6 \text{ mA}, \qquad V_{PO} = -4.7 \text{ V}, \qquad V_A = 150 \text{ V},$$

$$C_{iss} = 4.5 \text{ pF} \quad \text{at } V_{GS} = 0 \text{ V},$$

$$C_{rss} = 1.5 \text{ pF} \quad \text{at } V_{GS} = 0 \text{ V}.$$

A +15-V power supply is available. Bias the transistor at $I_D = 1$ mA and $V_{DS} = 5$ V. Simulate the circuit using SPICE and confirm the oscillation frequency. How is the oscillation frequency affected when the temperature of the FET is changed to 50°C?

12-17. Design a NJFET-based common-source configured Hartley oscillator at 1 MHz. The FET parameters of interest are

$I_{DSS} = 10$ mA, $V_{PO} = -3.5$ V, $V_A = 150$ V,

$C_{iss} = 6$ pF   at $V_{GS} = 0$ V,

$C_{rss} = 2$ pF   at $V_{GS} = 0$ V.

A +15-V power supply is available. Bias the transistor at $I_D = 1$ mA and $V_{DS} = 5$ V.

12-18. Determine the relationship for the frequency of oscillation for the transformer coupled common-base configured Hartley oscillator shown in terms of C, L, and the transformer turns ratio using small-signal hybrid-π analysis and appropriate assumptions. Assume that C_C and C_B are large-valued capacitors.

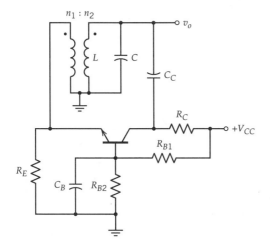

12-19. In many instances it is preferable to use autotransformers, which are easily fabricated, instead of regular transformers for coupling signals in feedback configurations. The autotransformer and its equivalent circuit are as shown.

Determine the relationship for the frequency of oscillation for the autotransformer coupled common-base configured Hartley oscillator shown in terms of C, L, and the autotransformer turns ratio using small-signal hybrid-π analysis and appropriate assumptions. Assume that C_C and C_B are large-valued capacitors.

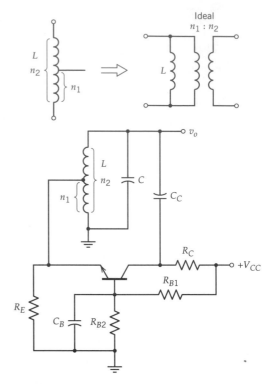

12-20. What is the relationship for the frequency of oscillation for the autotransformer coupled common-base configured Hartley oscillator shown in problem 12-16 when R_E is eliminated $(R_E \rightarrow \infty)$?

12-21. Design a BJT-based autotransformer coupled common-base configured Hartley oscillator at 1 MHz. The BJT parameters of interest are

$\beta_F = 200$, $V_A = 150$ V, $r_b = 30\ \Omega$,

$C_{ib0} = 8$ pF   at $V_{EB} = 0.5$ V,

$C_{ob0} = 4$ pF   at $V_{CB} = 5$ V,

$f_T = 300$ MHz.

A +15-V power supply is available. Bias the transistor at $I_C = 1$ mA and $V_{CE} = 5$ V. Simulate the circuit using SPICE and confirm the oscillation frequency. How is the oscillation frequency affected when the temperature of the BJT is changed to 50°C?

12-22. Determine the relationship for the frequency of oscillation for the transformer coupled common-source configured Hartley oscillator shown in terms of C, L, and the autotransformer turns ratio using small-signal analysis and appropriate assumptions. Assume that C_C and C_G are large-valued capacitors.

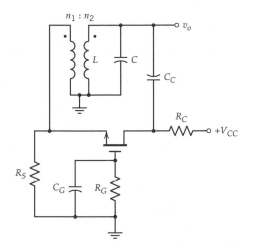

12-23. Determine the relationship for the frequency of oscillation for the autotransformer coupled common-source configured Hartley oscillator shown in terms of C, L, and the autotransformer turns ratio using small-signal analysis and appropriate assumptions. Assume that C_C and C_G are large-valued capacitors.

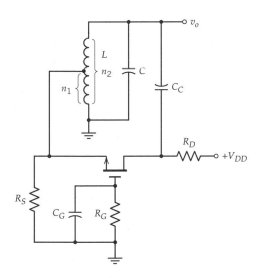

12-24. What is the relationship for the frequency of oscillation for the autotransformer coupled common-gate configured Hartley oscillator shown in problem 12-19 when R_S is eliminated ($R_S \rightarrow \infty$)?

12-25. Design a NMOSFET-based autotransformer coupled common-gate configured Hartley oscillator at 2 MHz. The FET parameters of interest are

$$I_{DSS} = 6 \text{ mA}, \qquad V_T = 2.0 \text{ V}, \qquad V_A = 150 \text{ V},$$

$$C_{iss} = 4.5 \text{ pF} \quad \text{at } V_{GS} = 0 \text{ V},$$

$$C_{rss} = 1.5 \text{ pF} \quad \text{at } V_{GS} = 0 \text{ V}.$$

A +15-V power supply is available. Bias the transistor at $I_D = 1$ mA and $V_{DS} = 5$ V. Simulate the circuit using SPICE and confirm the oscillation frequency. How is the oscillation frequency affected when the temperature of the FET is changed to 50°C?

12-26. Determine the relationship for the frequency of oscillation for the autotransformer coupled common-collector configured Hartley oscillator shown in terms of C, L, and the autotransformer turns ratio using small-signal hybrid-π analysis and appropriate assumptions. Assume that C_C and C_B are large-valued capacitors.

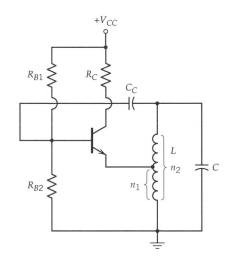

12-27. If the temperature stability of the crystal is 5.0 ppm/°C, determine the percent change in oscillator frequency after a rise in temperature of 50°C for a Pierce oscillator. Assume that the temperature variation only affects the series equivalent capacitance of the crystal.

12-28. A small inductance is added in parallel with a crystal operating in the series mode in a crystal oscillator. Will the frequency of the oscillation increase or decrease? Explain. Use simulations where appropriate.

12-29. A small "trimmer" capacitance is added in parallel with a crystal operating in the antiresonant mode in a crystal oscillator. Will the frequency of the oscillation increase or decrease? Explain. Use simulations where appropriate.

12-30. Design a test circuit to confirm the resonance frequency of a 2-MHz crystal with the parameters given in Table 12.4-1. Confirm its operation by simulating the circuit using SPICE.

12-31. Design a 1-MHz CMOS Pierce oscillator using the crystal data in Table 12.4-1. The MOS parameters are $K_n = 100 \ \mu A/V^2$, $K_n = 100 \ \mu A/V^2$, $V_{Tn} = |V_{Tn}| = 1$ V, $C_{iss} = 50$ pF at $V_{GS} = 0$ V, and $C_{rss} = 5$ pF at $V_{GS} = 0$ V.

12-32. Find the condition under which the circuit shown will oscillate. Assume $V_A = 180$ V.

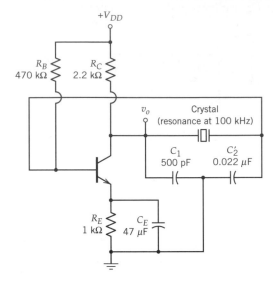

13

WAVEFORM GENERATION AND WAVESHAPING

In addition to sinusoidal waveforms, electronic systems often have need for signals with other wave shapes. Common waveforms include single pulses of fixed duration as well as periodic square waves and triangular waves. Fixed-duration pulses are used primarily for timing of events in communication and control systems. Periodic square waves are formed by a regular series of pulses: One typical use is as a clock for digital systems. If pulse timing, duration, or amplitude can be externally controlled, these altered square waves become the foundation of many digital communication systems. Triangular waves are particularly significant in scanning an electron beam across a CRT screen (as in television or oscilloscope applications), in precise time measurements, and in time modulation.

Electronic circuits that generate nonsinusoidal waveforms, such as pulse, square, and triangular waveforms, are typically based upon electronic multivibrators. These multivibrators are characterized by a very rapid transition between two distinct output states and can be grouped into three basic categories based on the time stability of these output states. A bistable multivibrator will rest indefinitely in either output state until triggered to change state. A monostable multivibrator has one stable state: the other state is of fixed, finite duration (a quasi-stable state) that can only be activated with a triggering signal. A constant, periodic switching between quasi-stable states characterizes an astable multivibrator. Astable, periodic switching is also known as non-linear oscillation.

In this chapter, multivibrator circuits intended for precision analog applications are presented. These circuit designs are based on a slight variation of the operational amplifier, typically known as the *comparator*. Circuits that generate single pulses of varying time duration or square and triangular waveforms of variable time symmetry using multivibrators are described. Similar circuits utilizing an integrated circuit multivibrator, in the form of an integrated circuit timer, are also described. A square-wave VCO is presented as a special case of non-linear oscillation.

Arbitrary, periodic waveforms are often derived from triangular waveforms. Two techniques to generate sinusoids from triangular waveforms are described as examples of waveform alteration. The primary technique is based on piecewise linear amplification using the forms of the basic diode clipping circuits described in Chapter 2.

Multivibrator applications based on OpAmps and comparators are best suited for electronic applications from a fraction of a hertz up to a few megahertz. High-frequency, analog multivibrator applications are beyond the scope of this discussion. High-frequency multivibrators intended for digital applications are presented in Chapter 16.

▬▬▬▬ 13.1 MULTIVIBRATORS

Multivibrator circuits are fundamental to many waveshaping and wave generation circuits. In the most common form they have two output states, each of which may be either stable or quasi-stable. As such, multivibrator circuits can be grouped into three classifications based on the stability of the output states:

- *Bistable multivibrator—two stable output states.* Bistable circuits require a triggering signal to transition between output states. Once in a particular output state, the circuit remains in that stable state indefinitely until triggered for a transition to the other stable output state.

- *Astable multivibrators—two quasi-stable output states.* Without any external triggering, the astable multivibrator transitions periodically between two quasi-stable output states.

- *Monostable multivibrator—one stable and one quasi-stable output state.* In monostable circuits, a triggering signal is required to induce an output transition from the stable state to the quasi-stable state. After the circuit remains in the quasi-stable state for a fixed time, typically long in comparison with the time of transition between states, it returns to the stable state without external triggering.

13.1.1 Bistable Multivibrators

One of the most simple forms of a bistable circuit is the comparator. A comparator shares many characteristics with high voltage gain OpAmp: It has two inputs and a single output determined by the difference of two input signals. As seen in Figure 13.1-1, the circuit symbol for a comparator is identical with that of an OpAmp and the terminals are identified in the same manner. If the noninverting input is at a higher potential than the inverting input, the output of the comparator is the HIGH voltage output state: Reversing the sense of the inputs yields the LOW voltage output state. The HIGH and LOW output voltage states are often determined by the power supply rails but may be internally controlled to be other voltage levels (5 and 0 V are a common pair when interfacing to digital circuitry). While OpAmps can be used as comparators, a comparator is not typically used with a wide range of negative feedback. Therefore the comparator can be optimized for rapid transition between states at the expense of linearity and feedback stability. As such, a comparator will typically transition between states much more quickly than an OpAmp of similar design. Comparators suffer from the same nonideal characteristics as OpAmps. Slew rate and frequency response are, arguably, the most problematic of these nonideal properties.[1]

While comparators perform well in a noiseless environment, the noise content

Figure 13.1-1
Comparator circuit symbol.

---

[1] Nonideal characteristics of OpAmps are discussed in Section 1.5. The discussion is also valid for comparators.

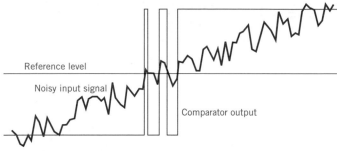

Figure 13.1-2
Output of comparator due to noisy input.

of a typical signal can create false triggering or multiple triggering of the output of a comparator. In a typical application, a comparator is used to determine when the noninverting input voltage exceeds the voltage level at the inverting input. As shown in Figure 13.1-2, the addition of noise to an increasing input signal at the noninverting input can cause several transitions of the comparator output when the input voltage nears the voltage reference level present at the inverting input. Usually a single transition of the output is desired in such a comparison. Multiple triggering of the output of a comparator can be eliminated with the application of *positive* feedback. Positive feedback retains the rapid output transition of a comparator but alters the trigger level so that two separate trigger levels exist: one for positive-slope signals and another for negative-slope signals. The resultant transfer relationship exhibits hysteresis as shown in Figure 13.1-3. Any input signal below the negative-slope transition voltage V_T^- results in a LOW output V_L. If the output state is LOW, it will not transition to the HIGH state unless the input is greater than the positive-slope transition voltage V_T^+. Similarly, any input above V_T^+ results in a HIGH output V_H that will not transition to the LOW state unless the input falls below V_T^-. Thus signals, after crossing a threshold, do not respond to input signal changes unless the variation is large enough to cross the deadband.

An example of a comparator with positive feedback is shown in Figure 13.1-4. This form of circuit is called a *Schmitt trigger*. Aside from being especially useful in converting a slowly varying or noisy signal into a clean, pulsed form with sharp transitions, the Schmitt trigger is particularly useful in converting sine-wave input

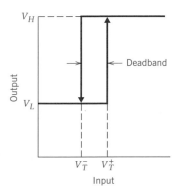

Figure 13.1-3
Schmitt trigger transfer characteristic hysteresis.

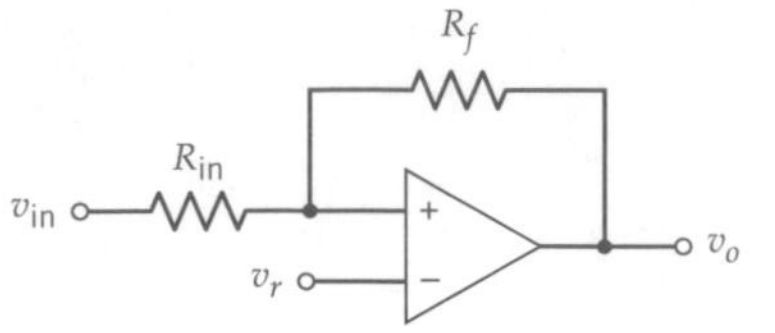

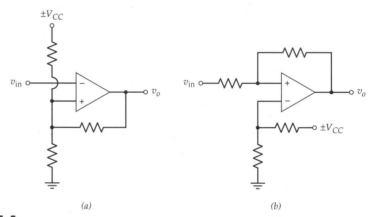

Figure 13.1-4
Simple Schmitt trigger circuit.

into a pulse-train output. Variation of the pulse-train duty cycle in this application is accomplished by varying the triggering voltage levels V_T^+ and V_T^-.

The Schmitt trigger can be realized in many configurations. When comparators are used as the basic active element,[2] inverting and noninverting forms are typically realized, as is shown in Figure 13.1-5. The additional resistors shown provide stable reference voltage v_r. If a more precise reference voltage is necessary, precision voltage reference circuitry may be utilized instead of simple resistor networks.[3]

Figure 13.1-5
Typical Schmitt trigger circuits: (a) inverting Schmitt trigger; (b) noninverting Schmitt trigger.

EXAMPLE 13.1-1 The simple Schmitt trigger circuit of Figure 13.1-4 is configured so that the two stable output voltages are +10 V. The reference voltage v_r is at ground potential. Complete the design by choosing R_{in} and R_f so that the output transition to +10 V occurs at the threshold voltage, $v_{in} = V_T^+ = 0.5$ V. At what input voltage level does the transition to −10 V occur (V_T^-)?

SOLUTION

The output will transition to +10 V when the noninverting input reaches the reference voltage $v_r = 0$ V. The noninverting input voltage can be obtained from the input

---

[2] Bipolar and CMOS realization of Schmitt trigger circuits for digital applications are presented in Section 16.5.
[3] Precision voltage references are discussed in Section 14.2.1.

voltage $v_{in} = 0.5$ V and the LOW output voltage $v_o = -10$ V using the voltage division technique:

$$v_r = -10 + \frac{R_f}{R_f + R_{in}}[0.5 - (-10)] = 0.0$$

or

$$\frac{R_f}{R_f + R_{in}} = \frac{10}{10.5} \Rightarrow R_f = 20R_{in}.$$

Many resistor pairs will fulfill this simple requirement. As with OpAmps, it is important to choose resistors that are small compared to the input resistance of the comparator and large compared to the output resistance. The following choice is one pair of values that will meet the design goals using standard resistance values:

$$R_f = 20 \text{ k}\Omega, \qquad R_{in} = 1 \text{ k}\Omega.$$

The transition to -10 V is an exact mirror image of the transition to $+10$ V: It occurs at

$$v_{in} = -0.5 \text{ V}.$$

Thus the HIGH output state ($+10$ V) is triggered when the input exceeds 0.5 V and remains until the input falls below -0.5 V.

13.1.2 Astable Multivibrators

Astable multivibrators continuously transition between quasi-stable output states without the aid of an external triggering input. The output of an astable circuit therefore becomes a square wave with the waveform duty cycle as a possible variable. One such astable circuit is shown in Figure 13.1-6. Here the two inputs to a comparator are coupled to the comparator output through two different type networks. The noninverting input is connected through a resistive voltage divider: As the output toggles between quasi-stable states, this input similarly toggles between states. The inverting input is connected to the comparator output by an RC charging network: This input exponentially transitions toward the output voltage. When the inverting input voltage matches the voltage at the noninverting input, the output voltage toggles to the other quasi-stable state. This output toggling causes successive exponential change and toggling.

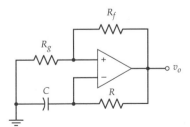

Figure 13.1-6
Astable multivibrator.

The two quasi-stable output voltage states of the astable multivibrator are at fixed voltages V_H and V_L. When the output is in one of these states, the voltage state at the noninverting ("+") comparator node is given by

$$v_H^+ = \frac{R_g}{R_g + R_f} V_H \quad \text{or} \quad v_L^+ = \frac{R_g}{R_g + R_f} V_L. \tag{13.1-1}$$

Just prior to a HIGH to LOW transition, for example, the output voltage is V_H, the noninverting input is at v_H^+, and the inverting input is in exponential transition from its initial value v_L^+ toward V_H. This exponential transition will continue until the voltage at the noninverting input matches that at the inverting input, v_H^+, at which time the output will toggle to V_L. That is,

$$v^-(t) = V_H - (V_H - v_L^-)e^{-t/RC}. \tag{13.1-2}$$

The total exponential transition time is the solution to the expression

$$v^-(t) = v_H^+ \tag{13.1-3}$$

or

$$V_H - \left(V_H - \frac{R_g}{R_g + R_f} V_L\right)e^{-t/RC} = \frac{R_g}{R_g + R_f} V_H. \tag{13.1-4}$$

The transition time is given by

$$t_{HL} = RC \ln\left(\frac{(R_g + R_f)V_H - R_g V_L}{R_f V_H}\right)$$

$$= RC \ln\left(1 + \frac{R_g(V_H - V_L)}{R_f V_H}\right). \tag{13.1-5}$$

A LOW to HIGH transition time is determined in the same manner: The expression is the same as Equation 13.1-5 with V_H and V_L interchanged. As such, the duty cycle of the output square wave can be altered somewhat. The period of this repeated output toggling becomes the sum of the two transition times:

$$\tau = RC\left[\ln\left(1 + \frac{R_g(V_H - V_L)}{R_f V_H}\right) + \ln\left(1 + \frac{R_g(V_L - V_H)}{R_f V_L}\right)\right]. \tag{13.1-6}$$

The complexity of these expressions for transition time can be simplified greatly by assuming a particular comparator operational configuration. Typically an astable multivibrator is operated with symmetric voltage limits, $V_L = -V_H$. This operation practice results in identical transition times and a 50% duty cycle:

$$t_{HL} = t_{LH} = RC \ln\left(1 + 2\frac{R_g}{R_f}\right). \tag{13.1-7}$$

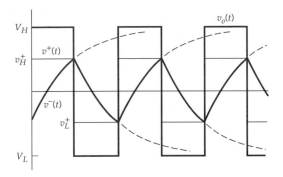

Figure 13.1-7
Astable multivibrator waveforms.

In this case the period of the repeated output toggling is twice the transition time:

$$\tau = 2t_{HL} = 2RC \ln\left(1 + 2\frac{R_g}{R_f}\right). \tag{13.1-8}$$

In practice, the slew rate limitations of the circuit comparator will lengthen each of the transition times. This has the effect of lowering the frequency of operation somewhat. The waveforms associated with a symmetric astable multivibrator are shown in Figure 13.1-7.

13.1.3 Monostable Multivibrators

A monostable multivibrator produces a single output pulse, typically of precise amplitude and duration, each time a trigger signal is applied to the input. As such, monostable multivibrators are useful in transforming a train of pulses with variable amplitude and/or duration into a train of pulses with standard amplitude and duration. The output during the single pulse is quasi-stable; between pulses the output is stable.

An astable multivibrator can be transformed into a monostable multivibrator by stabilizing one of the quasi-stable output states. The circuit of Figure 13.1-8 has its negative output V_L stabilized by the introduction of the stabilizing diode D_S. This diode prevents the inverting input $v^-(t)$ from becoming sufficiently negative to fall below the lower noninverting state:

$$v^-(t) > v_L^+ = \frac{R_g}{R_g + R_f} V_L. \tag{13.1-9}$$

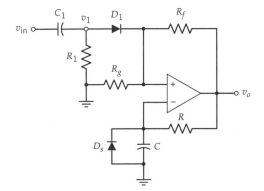

Figure 13.1-8
Monostable multivibrator.

Therefore the comparator will not spontaneously toggle to the positive output state V_H. In order for an external signal to trigger a pulse, the capacitive input circuit formed by C_1, R_1, and D_1 is added. This input circuit allows a fast rise time pulse to momentarily lift the noninverting input above the inverting input, which triggers a HIGH output.

As with the astable circuit, the monostable multivibrator output states are given by V_H (quasi-stable) and V_L (stable). When the circuit is in the stable state, the two inputs to the comparator are

$$v_L^+ = \frac{R_g}{R_g + R_f} V_L, \qquad v_L^- = -V_\gamma \approx -0.7 \text{ V}. \tag{13.1-10}$$

A positive pulse applied to the input, v_{in}, will raise the noninverting input above the inverting input and force the comparator output to the HIGH state, V_H. The noninverting input toggles and the inverting input begins an exponential transition toward V_H.

$$v_H^+ = \frac{R_g}{R_g + R_f} V_H, \qquad v^-(t) = V_H - (V_H - v_L^-)e^{-t/RC}. \tag{13.1-11}$$

The exponential transition will continue until the voltage at the noninverting input matches that at the inverting input, v_H^+, at which time the output will toggle to V_L. The duration of the positive pulse is given by the time of this exponential transition:

$$t_{\text{pulse}} = RC \ln\left(\frac{(R_g + R_f)(V_H - v_L^-)}{R_f V_H}\right). \tag{13.1-12}$$

After the output toggles to LOW, the noninverting input returns to its stable state, v_L^+, and the inverting input begins an exponential transition toward V_L:

$$v^-(t) = V_L - (V_L - v_H^+)e^{-t/RC}. \tag{13.1-13}$$

This exponential transition is halted by the diode D_S before the inverting input is sufficiently negative to toggle the comparator to HIGH. Consequently, the LOW output state is stable. The time to return to the stable state is given by

$$t_{\text{recovery}} = RC \ln\left(\frac{V_L - v_H^+}{V_L - v_L^-}\right). \tag{13.1-14}$$

Once the monostable multivibrator returns to its stable state, it can be retriggered to output another single pulse. The characteristic of a single output pulse for each triggering leads to the common alternate identification of a monostable multivibrator as a *one-shot*. Typical monostable multivibrator waveforms are shown in Figure 13.1-9. In order to ensure clean, single pulses, design guidelines suggest that the input circuit time constant be small compared to the pulse duration.

One-shot circuits find greatest use in analog or asynchronous digital circuitry. Use in synchronous digital circuitry is discouraged due to a variety of problems.

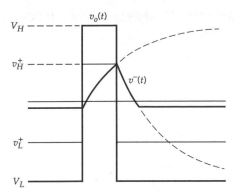

Figure 13.1-9
Monostable multivibrator waveforms.

EXAMPLE 13.1-2 The monostable multivibrator circuit of Figure 13.1-8 is configured so that the two stable output voltages are ± 10 V. Complete the design so that the circuit responds to a pulse input with an output pulse of duration 10 ms. What is the recovery time for the design?

SOLUTION

The pulse duration is given by Equation 13.1-12:

$$t_{\text{pulse}} = RC \ln\left(\frac{(R_g + R_f)(V_H - v_L^-)}{R_f V_H}\right) = RC \ln\left(\frac{(R_g + R_f)(10.7)}{R_f(10)}\right).$$

In order to reduce the effects of stray noise, it is good practice to keep within the first two time constants of the exponential decay. While many sets of component values will satisfy the constraints, the following set is chosen.

Choose $RC = t_{\text{pulse}}$ (one time constant) = 10 ms.

One pair of standard components meeting this choice is

$$R = \textbf{10 k}\Omega, \qquad C = \textbf{1 }\boldsymbol{\mu}\textbf{F};$$

a further consequence of this choice is

$$1 = \ln\left(\frac{(R_g + R_f)(10.7)}{R_f(10)}\right).$$

One pair of standard value resistors that will meet this requirement is

$$R_g = \textbf{15.4 k}\Omega, \qquad R_f = \textbf{10 k}\Omega.$$

The recovery time is determined by Equation 13.1-14. With the above choices for component values, it is

$$t_{\text{recovery}} = 10{,}000(1 \times 10^{-6})\left(\frac{-10 - 5}{-10 - (-0.7)}\right) = 4.78 \text{ ms}.$$

▬▬▬ 13.2 GENERATION OF SQUARE AND TRIANGULAR WAVEFORMS USING ASTABLE MULTIVIBRATORS

In electronic applications, the three most useful waveforms are as follows:

- Sinusoidal waveforms
- Square waveforms
- Triangular waveforms

Precision generation of these waveforms is vital to the proper operation of a wide variety of electronic devices and electronic test equipment. Square waves can be generated either by passing a sinusoid through a bistable circuit, such as a Schmitt trigger, or directly with an astable multivibrator. Triangular waveforms can be obtained by integrating a square waveform. Simultaneous generation of all three waveform types presents a challenge. Fortunately one form of astable multivibrator is capable of simultaneous generation of square and triangular waveforms; sinusoids can be derived from the triangular waveform using non-linear waveshaping techniques.[4]

13.2.1 Precision Square-Wave Generation

The astable multivibrator discussed in Section 13.1.2 is a near-ideal generator of square waves. With symmetric output voltage levels, a uniform square wave is generated with period

$$\tau = 2RC \ln\left(1 + 2\frac{R_g}{R_f}\right). \tag{13.2-1}$$

A precision square-wave generator based on this astable multivibrator is shown in Figure 13.2-1. In this circuit the symmetric output voltage levels are ensured through the use of a back-to-back, matched pair of Zener diodes.

Alteration of the time symmetry of the waveform so that the HIGH output time is not equal to the LOW output time is often desired in such a circuit. Two possible

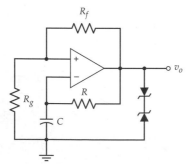

Figure 13.2-1
Astable multivibrator square-wave generator.

---

[4] These non-linear techniques are discussed in Section 13.3.

techniques used to alter the symmetry are significant in any discussion of multivibrators:

- A monostable multivibrator (one-shot) is connected in series with the output of the circuit of Figure 13.2-1. The monostable multivibrator is adjusted to output a pulse of varying duration without varying the period of the waveform.
- The astable multivibrator circuit is modified so that the HIGH and LOW output times are unequal.

Due to the significant increase in the complexity of the circuit, including the addition of another comparator, the addition of a one-shot is usually not the best choice. Minor modification of the astable multivibrator is simple and effective.

While it has been shown that nonsymmetric HIGH and LOW voltage states will lead to nonsymmetric waveforms, it is more common to vary the square-wave time symmetry through control of the exponential RC time constant. The circuit shown in Figure 13.2-2 is one possible multivibrator realization that provides different charging and discharging time constants. During the HIGH output state, diode D conducts and diode D' is off. The circuit reduces to that of Figure 13.2-1; the time constant is given by the product of RC. The addition of a diode in the negative-feedback path reduces the voltage apparent to the charging network by V_γ. Since the charging time is not dependent on this voltage, the HIGH state duration time is as previously derived:

$$t_H = RC \ln\left(1 + 2\frac{R_g}{R_f}\right). \tag{13.2-2}$$

When the output is in the LOW state, the operational modes of the two diodes D and D' are interchanged and the alternate negative-feedback resistor R' acts in the exponential decay. The resultant LOW output-state duration time is similarly given by

$$t_L = R'C \ln\left(1 + 2\frac{R_g}{R_f}\right). \tag{13.2-3}$$

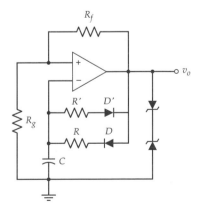

Figure 13.2-2
Astable, nonsymmetric square-wave generator.

If the two negative-feedback resistors are not equal in value, $R \neq R'$, the HIGH and LOW duration times are unequal and the square wave is asymmetric. The period of the asymmetric square wave is given by the sum of the two duration times:

$$\tau = t_H + t_L = (R + R')C \ln\left(1 + 2\,\frac{R_g}{R_f}\right). \tag{13.2-4}$$

The duty cycle is typically defined as the ratio of the high-state duration t_H to the period τ:

$$\text{Duty cycle} = \frac{t_H}{\tau} = \frac{R}{R + R'}. \tag{13.2-5}$$

EXAMPLE 13.2-1   Design a circuit using an astable multivibrator to produce a 10-kHz square wave with a 35% HIGH state duty cycle.

SOLUTION

When basing time on RC decay, it is best to stay approximately within the first two time constants. Choosing $R_f = R_g$ will satisfy this guideline. It is also important to choose resistors that will meet standard comparator resistance guidelines. Therefore, within these guidelines, arbitrarily choose

$$R_g = R_f = \textbf{10 k}\boldsymbol{\Omega}.$$

Then

$$\tau = \frac{1}{f} = 100 \;\mu\text{s} = (R + R')C(1.0986)$$

or

$$(R + R')C = 91.024 \;\mu\text{s}.$$

Arbitrarily choose a convenient value of C that will keep the resistor values R and R' within comparator resistance constraints. Here choose

$$C = \textbf{0.01 }\boldsymbol{\mu}\textbf{F} \Rightarrow R + R' = 9.1024 \text{ k}\Omega.$$

A 35% duty cycle implies

$$\frac{R}{R + R'} = 0.35 \Rightarrow R = 0.53846R'.$$

These two constraints result in the final two standard-value resistors:

$$R' = 5.916 \text{ k}\Omega \approx \textbf{5.90 k}\boldsymbol{\Omega}, \qquad R = 3.1858 \text{ k}\Omega \approx \textbf{3.20 k}\boldsymbol{\Omega}.$$

13.2.2 Simultaneous Square and Triangular Wave Generation

Astable multivibrators, as described previously (and shown in Figure 13.2-1), provide a square-wave output and a pseudotriangular waveform as an internal signal. Unfortunately, this internal waveform has voltage transitions that are exponential rather than linear. Transition linearity can be achieved by replacing RC capacitive charging. Two distinct possibilities for linearization of the signal are dominant in typical realizations:

- current source fed capacitive charging or
- integration of the square wave.

In the first possibility, the negative-feedback resistor R is replaced by a constant-current device. This current source typically has as its primary element, a FET operating in the saturation region. The second possibility uses an OpAmp integrator to create a linearly varying signal. Since OpAmp integrators invert the signal, the output of the integrator is fed back to the noninverting terminal of the comparator, as shown in Figure 13.2-3. In this form, the comparator takes the form of a simple Schmitt trigger.

This astable circuit also has two quasi-static states. They exist for the square-wave output v_{sq} at fixed voltages V_H and $-V_H$ (once again, the symmetry is ensured by back-to-back, matched Zener diodes). The Schmitt trigger toggles between output states when the voltage at its noninverting input is at ground potential. This toggle requirement reflects back on the integrator output v_{tr} as

$$v_{\text{tr(toggle)}} = -\frac{R_{\text{in}}}{R_f} v_{sq} = \mp \frac{R_{\text{in}}}{R_f} V_H. \tag{13.2-6}$$

As previously stated, the time-dependent output $v_{tr}(t)$ is an inverted integral of the difference between $v_{sq}(t)$ and the symmetry control voltage V_s (a constant):

$$v_{tr}(t) = \frac{-1}{RC} \int [v_{sq}(t) - V_s] \, dt. \tag{13.2-7}$$

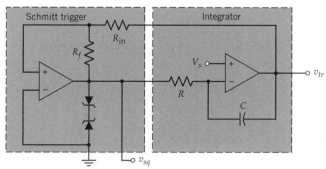

Figure 13.2-3
Triangle/square wave generator.

When v_{sq} is in the HIGH state, v_{tr} is linearly decreasing between its toggle values:

$$v_{tr}(t) = \frac{-1}{RC} \int (V_H - V_s) \, dt = \frac{V_s - V_H}{RC} t + \text{a HIGH constant}. \quad (13.2\text{-}8)$$

The "HIGH constant" of integration need not be evaluated for this discussion. The time for the decreasing voltage transition is given by the difference in toggle values divided by the slope of the linear transition:

$$2\left(\frac{R_{in}}{R_f} V_H\right) = \frac{V_H - V_s}{RC} t_- \Rightarrow t_- = \frac{2R_{in}RCV_H}{R_f(V_H - V_s)}. \quad (13.2\text{-}9)$$

When v_{sq} is in the LOW state, v_{tr} is linearly increasing between its toggle values with a different slope:

$$v_{tr}(t) = \frac{-1}{RC} \int -(V_H + V_s) \, dt = \frac{V_H + V_s}{RC} t + \text{a LOW constant}. \quad (13.2\text{-}10)$$

While the difference in toggle values remains the same, the change in slope results in a different transition time for the positive transition:

$$2\left(\frac{R_{in}}{R_f} V_H\right) = \frac{V_H + V_s}{RC} t_+ \Rightarrow t_+ = \frac{2R_{in}RCV_H}{R_f(V_H + V_s)} \quad (13.2\text{-}11)$$

The signal will continuously repeat the transitions. The period for the total waveform is the sum of the transition times:

$$\tau = t_- + t_+ = \frac{4R_{in}RC}{R_f} \frac{V_H^2}{V_H^2 - V_s^2}. \quad (13.2\text{-}12)$$

The positive-slope duty cycle is given by the ratio of the positive transition time to the period:

$$\text{Duty cycle} = \frac{t_+}{\tau} = \frac{1}{2}\left(1 - \frac{V_s}{V_H}\right). \quad (13.2\text{-}13)$$

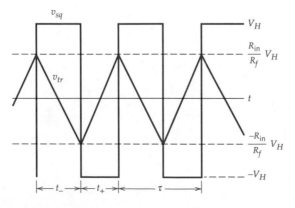

Figure 13.2-4
Triangle/square-wave generator signals.

Symmetric waveforms (those with a 50% duty cycle) are obtained when V_s is at ground potential. Positive V_s results in the positive-slope waveform segment having shorter duration than the negative-slope segment (positive-slope duty cycle <50%); negative V_s reverses the relationship. In this circuit, the frequency of oscillation is also dependent on V_s: It is maximized when $V_s = 0$ and decreases in a nonlinear fashion as the magnitude of V_s increases. The generator waveforms v_{tr} and v_{sq} are shown in Figure 13.2-4 for positive V_s.

13.2.3 Voltage-Controlled Frequency of Oscillation

Control of the frequency of oscillation of a waveform generator is often important. The frequency of oscillation for the waveform generator of Section 13.2-2 can be controlled by

- variation of component values (typically R or C) or
- variation of the symmetry signal V_s.

A manually controlled variable resistor or capacitor can be an adequate frequency control device in many applications. Other design constraints may require that the frequency be controlled by electrical, rather than mechanical, inputs. In particular, the need for a VCO is significant in many demodulation circuits for communication purposes.

While the circuit of Section 13.2-2 meets the voltage control requirement, there are two significant drawbacks to using this circuit as a VCO:

- A change in the frequency produces a change in the duty cycle.
- The frequency depends on the control voltage in a non-linear fashion: It is dependent on $1 - (V_s^2 - V_H^2)$.

Square and triangle waves, with voltage-controlled frequency, can be generated simultaneously with a circuit of the basic topology shown in Figure 13.2-5. This circuit is an astable multivibrator where the rate of linearly charging a capacitor is decoupled from the amplitude of the output of the square wave. This decoupling is ac-

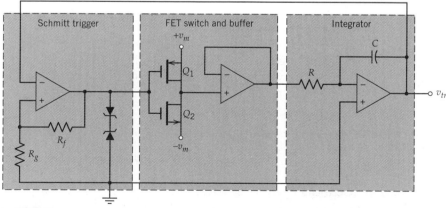

Figure 13.2-5
Voltage-controlled oscillator.

complished by the insertion of a FET single-pole, double-throw (SPDT) switch between the Schmitt trigger and the integrator. The SPDT switch is controlled by the output of the Schmitt trigger: It alternately couples the input of the integrator to either the frequency control voltage v_m or its negative, $-v_m$ (usually obtained from an OpAmp unity-gain amplifier). It is required that v_m be positive and of sufficient magnitude so that the switch operates properly.

The output of the integrator, v_{tr}, depends on v_m and v_{sq}:

$$v_{tr}(t) = \frac{-1}{RC} \int \pm v_m \, dt = \frac{\mp v_m}{RC} t + \text{a constant of integration.} \qquad (13.2\text{-}14)$$

When v_{sq} is HIGH, the output of the FET switch is connected to $-v_m$; when v_{sq} is LOW the output of the switch is connected to $+v_m$. As in the previously discussed waveform generator, the triangular wave transitions are linear with time. The duration of each transition is determined from the toggle voltages and the slope of the transition. The toggle values for the Schmitt trigger are obtained from a voltage division of its two output states: V_H and $-V_H$:

$$v_{tr(toggle)} = \frac{R_g}{R_f + R_g} v_{sq} = \pm \frac{R_g}{R_f + R_g} V_H. \qquad (13.2\text{-}15)$$

The time for the voltage transition is given by the difference in toggle values divided by the slope of the linear transition:

$$\frac{2R_g}{R_f + R_g} V_H = \frac{v_m}{RC} t \Rightarrow t = \frac{2R_g RC}{R_f + R_g} \frac{V_H}{v_m}. \qquad (13.2\text{-}16)$$

Circuit symmetry implies that positive and negative transitions are of equal duration; the period of oscillation is double that of a single transition:

$$\tau = 2t = \frac{4R_g RC}{R_f + R_g} \frac{V_H}{v_m}. \qquad (13.2\text{-}17)$$

The frequency of oscillation is the inverse of the period:

$$f = \frac{1}{\tau} = \frac{R_f + R_g}{4R_g RC} \frac{v_m}{V_H}. \qquad (13.2\text{-}18)$$

The results of Equation 13.2-18 indicate the desired linear dependence of frequency with input voltage v_m. Typically, the linearity extends over three or more decades. If greater frequency variation is necessary, it must be accomplished through resistor switching (typically resistor R is switched).

EXAMPLE 13.2-2 Design a VCO that will output a 15-V peak-to-peak square wave of variable frequency from 100 Hz to 2 kHz for an input voltage range of 0.1–2 V.

SOLUTION

The output of the circuit shown in Figure 13.2-5 will satisfy all design requirements with proper choice of components and component values. A square wave can be obtained at the v_{sq} terminal of the circuit. The 15-V peak-to-peak output voltage requirement is obtained by careful choice of the Zener diode matched pair. The pair is chosen so that $2(V_z + V_\gamma) = 15$ V. This requirement will result in diodes with a Zener voltage

$$V_z \approx 6.8 \text{ V}.$$

The frequency variation is obtained from Equation 13.2-18. Using the lower range of frequency and voltage, the resultant component constraint is

$$100 = \frac{R_f + R_g}{4R_g RC} \frac{0.1}{7.5} \Rightarrow \frac{R_f + R_g}{R_g RC} = 30{,}000.$$

Many sets of component values will satisfy this constraint and keep within the standard comparator resistance guidelines. One set is

$$R_f = R_g = 10 \text{ k}\Omega \Rightarrow RC = 66.67 \times 10\text{-6},$$
$$C = 4700 \text{ pF} \Rightarrow R = 14.2 \text{ k}\Omega.$$

All these values are standard component values.

13.3 NON-LINEAR WAVEFORM SHAPING

Arbitrarily shaped, periodic signals can be derived from triangular waveforms through a two-port network with an appropriate, non-linear transfer function. The non-linear property of the network is usually derived using a piecewise linear approximation through the introduction of a series of breakpoints. Typically the breakpoints are realized with diode[5] or transistor switching networks. While selectively filtering the harmonic content of the output of a multivibrator may seem to be an effective, alternate means of waveshaping, this technique is usually not a practical solution. The steep transition regions required to filter the various harmonics imply a very complex, high-order filter design. In most situations the design goals can be more easily met by piecewise linear approximation. Digitally generated waveforms are another alternative that is particularly popular in arbitrary waveform generators. (The design of digital waveform generators is beyond the scope of this discussion.)

A general waveshaping network using diodes is shown in Figure 13.3-1. A triangle wave, v_{tr}, enters the network, is shaped by the piecewise linear two-port network, and then is buffered by the unity-gain amplifier. The shaped output is v_o. The network is composed of an array of elements consisting of a series-connected diode, resistor, and voltage sources. If the voltages $\{V_i^+\}$ consist of a progressively increasing set of positive voltages, this portion of the array will progressively flatten the waveform as it increases. Similarly, an increasing magnitude set of negative voltages $\{V_i^-\}$ will progressively flatten the negative portion of the waveform. Reversing a diode in either the upper or lower portion of the array will have the opposite effect on the waveform at the breakpoint.

The primary difficulty encountered in this form of array is provision for many independent voltage sources. Surprisingly, design solutions are simplified by the switching nature of the array. One possibility is the use of forward-biased diodes to specify the breakpoint voltages. In order to demonstrate the utility of such an array, a simple triangle-to-sinusoid converter will be demonstrated.

---

[5] Diode switching networks to shape waveforms are initially introduced in Section 2.6.

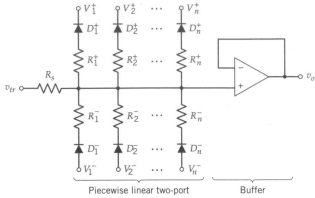

Figure 13.3-1
Piecewise linear wave-shaping circuit.

EXAMPLE 13.3-1 In order to produce a sinusoid of magnitude V_s from a triangle wave of magnitude $V_{tr(max)}$, the transfer function for the triangle-to-sinusoid converter must be given by the expression

$$v_o = V_s \sin\left(\frac{\pi}{2} \frac{v_{tr}}{V_{tr(max)}}\right).$$

It has been decided to approximate this transfer function with the seven-segment diode array shown. Here the voltage sources are provided by forward-biased diodes. This array clips all inputs so that the maximum possible output is $\pm 3V_\gamma \approx \pm 1.8$ V. There are four other breakpoints (progressively introduced by diode conduction) at ± 0.6 and ± 1.2 V. The slopes of the transfer function in the various regions are given by[6]

$$m_1 = 1, \qquad\qquad\qquad -0.6 \text{ V} < v_o < 0.6 \text{ V},$$

$$m_2 = \frac{R_1}{R_1 + R_s}, \qquad\qquad 0.6 \text{ V} < |v_o| < 1.2 \text{ V},$$

$$m_3 = \frac{R_1 R_2}{R_1 R_2 + R_1 R_s + R_2 R_s}, \quad 1.2 \text{ V} < |v_o| < 1.8 \text{ V}.$$

---

[6] These equations assume that the forward dynamic resistance of the diodes is zero valued. If the dynamic resistance can be reasonably approximated, the resistor values R_1 and R_2 should be reduced by that approximation.

The analytic expression for the slope of the transfer function is given by

$$\frac{dv_o}{dv_{\text{tr}}} = \frac{\pi}{2} \frac{V_s}{V_{\text{tr(max)}}} \cos\left(\frac{\pi}{2} \frac{v_{\text{tr}}}{V_{\text{tr(max)}}}\right).$$

Matching the slope at the various breakpoints yields

$$m_1\big|_{v_{\text{tr}}=0} = 1 \Rightarrow \frac{V_s}{V_{\text{tr(max)}}} = \frac{2}{\pi}.$$

For a 1.8-V magnitude sinusoid this expression implies that $V_{\text{tr(max)}} = 2.825$ V. At the first breakpoint $v_o = v_{\text{tr}} = \pm 0.6$ V,

$$m_2 = 0.945 \Rightarrow R_1 - 17.17 R_s.$$

At the second breakpoint $v_o = 1.2$ V, $v_{\text{tr}} = 1.27$ V,

$$m_3 = 0.7613 \Rightarrow R_2 = 2.690 \ R_s.$$

Reasonable choices for the resistors using standard values are

$$R_s = 1 \text{ k}\Omega, \qquad R_1 = 17.2 \text{ k}\Omega, \qquad R_2 = 2.7 \text{ k}\Omega.$$

A PSpice analysis of the circuit was performed. The input and output waveforms are shown. Total harmonic distortion of the near-sinusoidal output of this circuit was calculated to be about 1.0%—a reasonable value for such a simple circuit. Typical converters usually have at least six breakpoints on each side of ground.

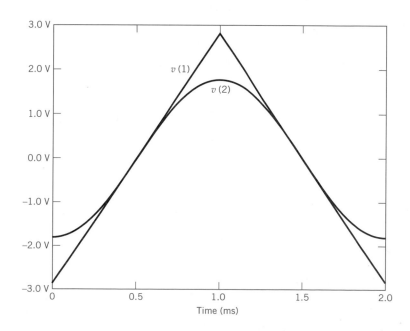

```
Triangle-to-sinusoid converter
VS    1    0PWL(0 −2.825 1M 2.825 2M −2.825)
RS    1    2    1K
D1P   2    10   D1N4148
D1N   10   2    D1N4148
R1    10   0    17.2K
D2P1  2    21   D1N4148
D2N1  22   2    D1N4148
D2P2  21   20   D1N4148
D2N2  20   22   D1N4148
R2    20   0    2.7K
D3P1  2    31   D1N4148
D3N1  33   2    D1N4148
D3P2  31   32   D1N4148
D3N2  34   33   D1N4148
D3P3  32   0    D1N4148
D3P3  0    34   D1N4148
.TRAN .01M 2M 0 0.01M
.LIB NOM.LIB
.PROBE
.FOUR 500 V(2)
.END
```

Another similar form of piecewise linear two-port network uses BJTs as the switching device rather than diodes. A seven-segment array demonstrating this technique applied to triangle-to-sinusoid conversion is shown in Figure 13.3-2.[7] This design was based on the breakpoints and slopes derived in Example 13.3-1. The BJT-based circuit has several advantages over the diode-based circuit. Primary among the advantages is the flexibility concerning the amplitude of v_{tr}. As long as

$$V_{CC} \approx V_{tr(max)}, \qquad 1.8 < V_{tr(max)} < 15 \text{ V},$$

this circuit performs well with less than 1.5% total harmonic distortion. One particular drawback is the rather strong dependence on the forward-biased dynamic resistance of the base–emitter junctions of the various transistors. Performance is greatly dependent on these values (in the design shown, the 2.7-kΩ resistor was reduced to 2 kΩ and the 17.2-kΩ resistor was reduced to 14.3 kΩ due to dynamic resistance considerations).

Another technique useful in the conversion of triangular waves to sinusoids is non-linear amplification. A particularly useful circuit that employs this technique is shown in Figure 13.3-3. The design utilizes logarithmic amplification obtained with an overdriven differential gain stage. Input triangle waves alternately force one of the two BJTs to the verge of saturation.

In this amplifier the triangle wave is amplified linearly near the zero crossing; in the regions near the peaks of the triangle the amplification is logarithmic. For a well-controlled triangle voltage input, the resultant transfer function is near ideal and produces a low-distortion sinusoid output v_o. Optimal performance occurs with

$$52 \text{ mV} < I_1R < 86 \text{ mV}, \qquad V_{tr(max)} \approx 95 \text{ mV}.$$

---

[7] A more complete discussion can be found in Grebene, 1984, pp. 592–595.

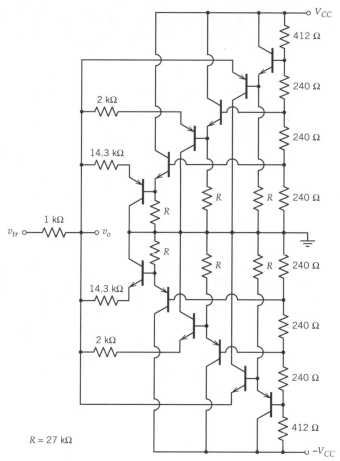

Figure 13.3-2
Seven-segment triangle to sinusoidal converter.

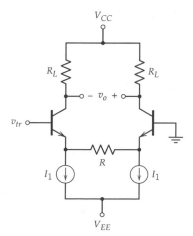

Figure 13.3-3
Differential pair triangle-sinusoidal nonlinear amplifier.

A logarithmic triangle-to-sinusoid converter is particularly useful at high frequencies where the higher voltage input requirement of piecewise linear converters may be difficult to produce with sufficient accuracy.

13.4 INTEGRATED CIRCUIT MULTIVIBRATORS

An alternative to the realization of multivibrators with discrete elements exists in the form of an integrated circuit package. These integrated circuit packages, commonly identified as *integrated circuit timers*, contain the primary components needed to implement monostable and astable multivibrators: A minimum number of external components is necessary to complete the multivibrator design. Among the integrated circuit timers currently available, the 555 timer has gained widest acceptance due to its versatility and low cost.

The basic functional block diagram of a 555 timer is shown in Figure 13.4-1. It consists of two comparators, a resistive network that sets the trigger levels of the comparators, a transistor that acts as a switch, and a set–reset (SR) flip-flop. The three internal resistors, labeled R, divide the input voltage V_{CC} so that the voltage trigger levels of the two comparators CP_1 and CP_2 are at $2V_{CC}/3$ and $V_{CC}/3$, respectively. These trigger levels can be altered by applying an external voltage at the *control* input terminal. The outputs of the comparators control the state of the SR flip-flop. An SR flip-flop is a form of bistable circuit:[8] It is a level retention circuit with complementary outputs Q and $\overline{Q}$. The output Q transitions to match the S input when only one input is HIGH. When both inputs are LOW, the flip-flop retains its last value of Q and holds it until at least one input transitions to HIGH. The terms *set* and *reset* refer to the action of the output Q. Here, Q *sets* (transitions to HIGH) when the set input S

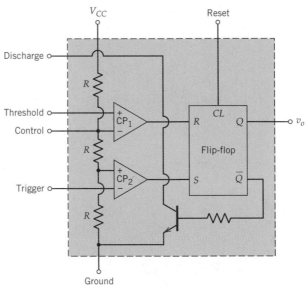

Figure 13.4-1
Simplified functional diagram of 555 integrated circuit timer.

---

[8] More information on the characteristics of flip-flops can be found in Section 16.5.

is HIGH; Q *resets* (transitions to LOW) when the reset input R is HIGH. The transistor switch shorts the *discharge* input to ground when $\overline{Q}$ is HIGH.

The 555 timer is available in both bipolar and CMOS technologies. In various forms it is capable of producing timing signals with a duration that ranges from microseconds to hours. Astable oscillation up to a few megahertz is possible. While many functional operations can be performed with this circuit, discussion will be limited to a monostable multivibrator (one-shot) and an astable multivibrator (non-linear oscillator).

13.4.1 555 Timer Monostable Multivibrator

Timing operations using a 555 timer are typically based on the charging and discharging of external networks. Most fundamental of these timing operations is that of a monostable multivibrator. The 555 timer implementation of a monostable multivibrator is shown in Figure 13.4-2. Here the timed interval is controlled by a single external RC network. The basic connections are as follows:

- A capacitor C is connected from the *threshold* input to *ground*.
- A resistor R_A is connected from the *threshold* input to the *positive power* V_{CC}.
- The *discharge* and *threshold* inputs are shorted together.
- The *reset* input is held HIGH.
- The *control* input is left open (or connected through a small capacitor to ground).

The stable state of this circuit exists when the output Q of the flip-flop is LOW. In order to achieve that state, the input v_{in} must be greater than the trigger level of CP_2. In this state, the complementary output of the flip-flop, $\overline{Q}$ is HIGH. This causes the transistor switch to activate, forcing a rapid discharge of the capacitor voltage to

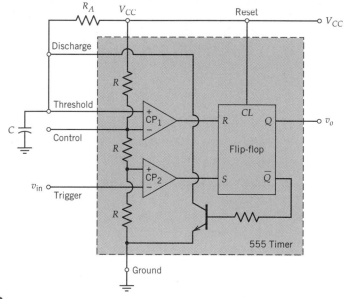

Figure 13.4-2
A 555 timer monostable multivibrator.

essentially zero. The output of CP$_1$ is LOW. The output of each comparator is therefore LOW: the LOW inputs to the flip-flop retain the LOW output until an input change occurs.

The quasi-stable state occurs when the input voltage v_{in} momentarily drops below the trigger level of CP$_2$:

$$v_{in} < V_{CC}/3. \tag{13.4-1}$$

This change in input level forces the output of CP$_1$ to a HIGH state, setting the flip-flop (Q, HIGH, and $\overline{Q}$, LOW). The $\overline{Q}$ output of the flip-flop deactivates the transistor switch and allows the capacitor to begin charging toward the positive power voltage V_{CC}:

$$v_c(t) = V_{CC}\left(1 - e^{-t/R_AC}\right). \tag{13.4-2}$$

The charging will continue until $v_c(t)$ exceeds the trigger voltage of CP$_1$. The duration of the charging is given by the solution to

$$\tfrac{2}{3}V_{CC} = V_{CC}\left(1 - e^{-t/R_AC}\right) \Rightarrow t = R_AC \ln(3). \tag{13.4-3}$$

When $v_c(t)$ reaches the trigger level, the output of CP$_1$ goes HIGH. If the input signal $v_i(t)$ has returned HIGH, the flip-flop changes state and returns to Q = LOW: A single HIGH pulse of duration $R_AC \ln(3)$ is formed. If $v_i(t)$ is still LOW, the HIGH pulse continues until $v_i(t)$ goes HIGH, at which time the pulse terminates.

13.4.2 555 Timer Astable Multivibrator

An astable multivibrator where the frequency of oscillation and the duty cycle can be independently controlled with two external resistors and a single external capacitor is shown in Figure 13.4-3. In this configuration, the HIGH duty cycle is limited to the range of 50–100%. The basic connections are as follows:

- A capacitor C is connected from the *trigger* input to *ground*.
- A resistor R_A is connected from the *discharge* input to the *positive power* V_{CC}.
- A resistor R_B is connected from the *threshold* input to the *discharge* input.
- The *trigger* and *threshold* inputs are shorted together.
- The *reset* input is held HIGH.
- The *control* input is left open (or connected through a small capacitor to ground).

In this astable configuration, the capacitor voltage $v_c(t)$ transitions exponentially between the trigger levels of the two comparators (established by the resistors labeled R):

$$\tfrac{1}{3}V_{CC} \le v_c(t) \le \tfrac{2}{3}V_{CC}. \tag{13.4-4}$$

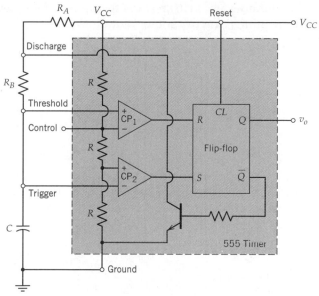

Figure 13.4-3
A 555 timer astable multivibrator.

Both the direction and rate of capacitor voltage transition are controlled by the switching transistor. When $v_c(t)$ is *between* the trigger levels, both comparators will have a LOW output. This output state signals the flip-flop to retain its last output state. When $v_c(t)$ reaches the upper trigger level, the output of CP_1 momentarily changes to HIGH. This action activates the reset state of the flip-flop (Q, LOW, and $\overline{Q}$, HIGH). The output of the flip-flop activates the transistor switch, which forces the capacitor to begin an exponential discharge through the resistor R_B to ground:

$$v_c(t) = \tfrac{2}{3}V_{CC}\, e^{-t/R_B C}.\tag{13.4-5}$$

As soon as the voltage drops below the upper trigger level, the output of CP_1 returns to LOW; however, the flip-flop retains its output state and the discharge continues. It continues until $v_c(t)$ reaches the lower trigger level. The duration of this discharging transition, t_d, occurs at the solution to

$$\tfrac{1}{3}V_{CC} = \tfrac{2}{3}V_{CC}\, e^{-t_d/R_B C} \Rightarrow t_d = R_B C\, \ln(2).\tag{13.4-6}$$

When $v_c(t)$ reaches the lower trigger level, the output of CP_2 momentarily changes to a HIGH. This action activates the set state of the flip-flop (Q, HIGH, and $\overline{Q}$, LOW). The output of the flip-flop deactivates the transistor switch, which forces the capacitor to begin exponential charging through the resistors R_A and R_B toward V_{CC}:

$$v_c(t) = V_{CC} - \tfrac{2}{3}V_{CC}\, e^{-t/(R_A + R_B)C}.\tag{13.4-7}$$

As soon as the voltage rises above the lower trigger level, the output of CP_2 returns to LOW; however, the flip-flop retains its output state and the charging continues.

It continues until $v_c(t)$ reaches the upper trigger level. The duration of this charging transition, t_c, occurs at the solution to

$$\tfrac{2}{3}V_{CC} = V_{CC} - \tfrac{2}{3}V_{CC}\,e^{-t_c/(R_A+R_B)C} \tag{13.4-8}$$

or

$$t_c = (R_A + R_B)C \ln(2). \tag{13.4-9}$$

The transitions continue indefinitely, producing a non-linear oscillation. A HIGH output occurs during the charging transition and a LOW output occurs during the discharging transition. The period of oscillation is given by the sum of the transition times:

$$
\begin{aligned}
\tau &= t_c + t_d \\
&= (R_A + R_B)C \ln(2) + R_B C \ln(2) \\
&= (R_A + 2R_B)C \ln(2)
\end{aligned}
\tag{13.4-10}
$$

or

$$\tau \approx 0.693(R_A + 2R_B)C. \tag{13.4-11}$$

The HIGH duty cycle is given by the ration of t_c to τ:

$$\text{Duty cycle} = \frac{t_c}{\tau} = \frac{R_A + R_B}{R_A + 2R_B}. \tag{13.4-12}$$

Since it is not possible to set $R_A = 0$ and still have a functioning astable multivibrator, the HIGH duty cycle will always be greater than 0.5 (50%). The output voltage and capacitor voltage waveforms for a 60% duty cycle 555 timer astable multivibrator are shown in Figure 13.4-4.

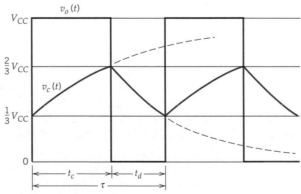

Figure 13.4-4
Waveforms for 555 timer astable multivibrator (60% duty cycle).

13.5 CONCLUDING REMARKS

A variety of circuits that generate nonsinusoidal waveforms has been presented in this chapter. These circuits are based on electronic multivibrators. Multivibrators are classified into three basic types based on the stability of the output states:

- Bistable multivibrator: two stable output states
- Astable multivibrator: two quasi-stable output states
- Monostable multivibrator: one stable and one quasi-stable output state

Bistable multivibrators provide two distinct output states when appropriately triggered. One of the most useful of these circuits is the Schmitt trigger. A Schmitt trigger performs the particularly useful task of noise elimination through hysteresis in its transfer function. Astable multivibrators are non-linear oscillators that provide periodic square and/or triangular waveforms. Monostable multivibrators provide a single-output pulse of fixed duration when triggered. Monostable and astable multivibrators can be realized through the use of an integrated circuit timer. Most dominant among integrated circuit timers is the 555 timer family.

Arbitrary, periodic waveforms can be derived from triangular waveforms. A technique to generate arbitrary waveforms using diode clipping circuitry is commonly used. Low harmonic content sinusoids can be generated using this technique or by non-linear amplification.

SUMMARY DESIGN EXAMPLE

The transmission of digitial computer signals over standard telephone lines requires the use of a modulator–demodulator unit known as a modem. Typically modems transmit 2400 digital pulses per second. In the construction of such a device, a local square-wave oscillator is often required at that frequency. In order to interface with other digital circuits, the output of the oscillator must be TTL compatible. These design goals reduce to the following specifications:

$$f = 2400 \text{ Hz}, \qquad V_H \approx 5 \text{ V}, \qquad V_L \approx 0 \text{ V}.$$

Design such a device so that the square-wave duty cycle lies between 40 and 60%.

Solution:

The obvious design alternatives are as follows:

- Linear oscillator
- Discrete-element astable multivibrator
- Integrated circuit timer astable multivibrator

The two astable multivibrator types have a distinct advantage in complexity, size, and cost over a linear oscillator at this low frequency. In addition the TTL-compatible output voltage levels lead to a 555 timer realization as an extremely advantageous choice. Therefore, the 555 timer circuit topology shown in Figure 13.4-3 is chosen as the basis for this design.

The specifications lead to specific parameter values needed in this design. If V_{CC} = 5 V, the output voltage levels will meet specification. The frequency of oscillation requirement leads to

$$(R_A + 2R_B)C \ln(2) = \frac{1}{2400} = 416.7 \ \mu s.$$

Arbitrarily choose the capacitor to be a convenient standard value:

$$C = 0.027 \ \mu F \Rightarrow R_A + 2R_B = 22.26 \ k\Omega.$$

A 555 timer can only have duty cycle greater than 50%. Arbitrarily choose the duty cycle to lie easily within the design goals at 55%. This choice leads to a ratio of resistor values:

$$R_A + R_B = 0.55(R_A + 2R_B) \Rightarrow R_B = 4.5R_A.$$

Combining the two constraining equations for resistance values leads to standard-value resistors:

$$R_A = 2.23 \ k\Omega, \qquad R_B = 10.0 \ k\Omega.$$

The design is complete. If exact-value components are used, the frequency is 2403.7 Hz (0.15% error) with a duty cycle of 55.02% (within specifications).

DESIGN VERIFICATION USING PSPICE:

The evaluation version of PSpice is supplied with a macromodel of a CMOS version of the 555 timer. The simple astable multivibrator oscillator in this design example can be modeled as follows:

```
Chapter 13 Summary Design Example
*2400 HZ Oscillator with 55% duty cycle using 555 timer
X1 0 2 3 8 5 2 7 8 555D
*the terminals of the 555D are Gnd, Trg, Vo, Rst, Ctl, Thd, Dis, Vcc
C1 2 0 0.027U
RB 2 7 10K
RA 7 8 2.23K
*The following is a load resistor at the output
R1 3 0 10K
VCC 8 0 5
.lib nom.lib
.tran 10u 1m
.probe
.end
```

A plot of the output and capacitor voltages is shown below. The simulation yield a square wave of frequency 2378 Hz (−0.92%) with 54.5% duty cycle (within specifications).

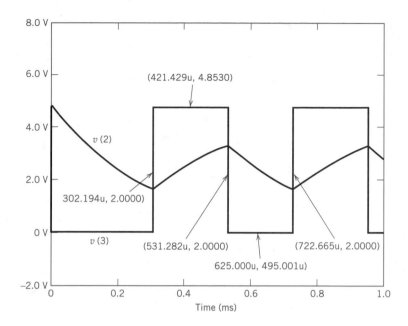

(421.429u, 4.8530)

v (2)

302.194u, 2.0000)

v (3)

(531.282u, 2.0000)

(722.665u, 2.0000)

625.000u, 495.001u)

Time (ms)

REFERENCES

Linear Circuits Data Book, Vol. 3: Voltage Regulators/Supervisors, Comparators, Special Functions, and Building Blocks, Texas Instruments, Dallas, TX, 1992.

Grebene, A.B., *Bipolar and MOS Analog Integrated Circuit Design,* Wiley, New York, 1984.

Hambley, A.R., *Electronics, A Top-Down Approach to Computer-Aided Circuit Design,* Macmillan, New York, 1994.

Millman, J., *Microelectronics, Digital and Analog Circuits and Systems,* McGraw-Hill, New York, 1979.

Millman, J., and Taub, H., *Pulse, Digital, and Switching Waveforms,* McGraw-Hill, New York, 1965.

Savant, C. J., Roden, M.S., and Carpenter, G.L., *Electronic Circuit Design, An Engineering Approach,* Benjamin/Cummings, Menlo Park, 1987.

Sedra, A.S., and Smith, K.C., *Microelectronic Circuits,* 3rd. ed. Saunders College Publishing, Philadelphia, 1991.

Wojslaw, C.F., and Moustakas, E.A., *Operational Amplifiers,* Wiley, New York, 1986.

PROBLEMS

 13-1. A simple Schmitt trigger has the following design requirements:

- Stable output voltages at ±5 V
- Reference voltage $v_r = 0$ V
- Threshold voltages $V_T^+ = ±1$ V

Design a circuit to meet these requirements to within ±0.05 V.

13-2. A sinusoidal input voltage $v(t) = A \sin(1000\pi t)$ is the input to the simple Schmitt trig-

ger circuit created in Example 13-1-1. Quantitatively described the output voltage if:

a. The input sinusoidal amplitude is 1.0 V (i.e., $A = 1.0$)

b. The input sinusoidal amplitude is 0.4 V (i.e., $A = 0.4$)

Assume the output voltage is zero valued at $t = 0$.

 13-3. A noninverting Schmitt trigger has the following design requirements:

- Stable output voltage at ± 10 V
- Positive-slope threshold voltage $V_T^+ = 2$ V
- Negative-slope threshold voltage $V_T^- = 0$ V

Design a circuit to meet these requirements to within ± 0.05 V.

13-4. An inverting Schmitt trigger has the following design requirements:

- Stable output voltage at ± 10 V
- Positive-slope threshold voltage $V_T^+ = 1.5$ V
- Negative-slope threshold voltage $V_T^- = -3$ V

Design a circuit to meet these requirements to within ± 0.02 V.

13-5. Another form of OpAmp Schmitt trigger is shown.

a. Determine the transfer characteristic as a function of the circuit parameters. Assume $V_{\text{ref}} < V_{CC} - 2$.

b. If the circuit is constructed with the following circuit element values, what are the threshold and output voltages?

$$R_g = 8.2\text{k}\Omega, \quad R = 1\text{ k}\Omega,$$
$$V_{ref} = 4 \text{ V}, \quad V_{CC} = 15 \text{ V}.$$

c. Use SPICE to verify the results of part b.

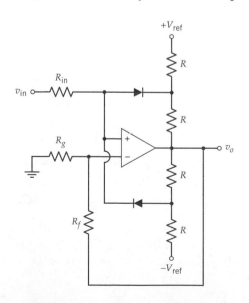

13-6. The astable multivibrator circuit shown is constructed with the following component properties:

$$R_f = 2 \text{ k}\Omega, \quad D_1: \; V_\gamma = 0.6 \text{ V}, \quad V_z = 6.4 \text{ V},$$
$$R_g = 1 \text{ k}\Omega \quad D_2: \; V_\gamma = 0.6 \text{ V}; \quad V_z = 3.4 \text{ V}.$$

a. What are V_H and V_L?

b. Determine the HIGH duty cycle (t_{HL}/τ).

c. Complete the design by choosing component values R and C so that the circuit will oscillate at 1 kHz.

d. Verify the results using SPICE.

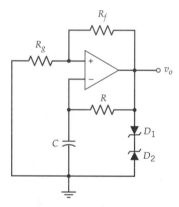

13-7. Design an astable multivibrator to produce a symmetric square wave of frequency 2 kHz. Verify the design using SPICE.

13-8. Design an astable multivibrator to produce a symmetric square wave with an amplitude of 6 $V_{\text{p-p}}$ and frequency of 1.2 kHz. Verify the design using SPICE.

13-9. Design an astable multivibrator to produce a square wave of frequency 500 Hz with a HIGH state duty cycle of 40%. It is required that $V_H = 5$ V; V_L may be varied. Verify the design using SPICE.

13-10. Design an astable multivibrator to produce a square wave of frequency 500 Hz with a HIGH state duty cycle of 40%. It is required that $V_H = -V_L = 5$ V. *Hint*: In order to have different time constants for the two transitions, diodes may be placed in the discharge path. Verify the design using SPICE.

13-11. Design a one-shot circuit that responds to a pulse input with an output pulse of duration 3 ms. Verify the design using SPICE (any OpAmp macromodel will suffice in modeling a comparator in this application).

13-12. Design a one-shot circuit that responds to a pulse input with an output pulse of duration 1 ms. Verify the design using SPICE (any OpAmp macromodel will suffice in modeling a comparator in this application).

13-13. A stable 120-Hz pulse train of amplitude 5 V is required. Since commercial power is extremely stable, it has been decided to use the 60-Hz, 110-V power line voltage as a triggering source for this pulse train. Design a system that will produce the required pulse train using a simple one-shot based on the simple design topology shown. Verify the design using SPICE.

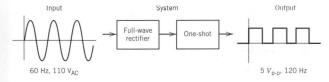

Input   System   Output

Full-wave rectifier → One-shot

60 Hz, 110 V_{AC} 5 V_{p-p}, 120 Hz

13-14. A stable 60-Hz pulse train of amplitude 10 V is required. Using commercial power line input and a design topology similar to that described in the previous problem, design a system to produce the required pulse train. Verify the design using SPICE.

13-15. Design an astable multivibrator to produce a symmetric triangle wave with a peak-to-peak amplitude of 8 V and frequency of 3.4 kHz. Verify the design using SPICE.

13-16. Design an astable multivibrator to produce a triangle wave with the following characteristics:

- Frequency 500 Hz
- Amplitude 10 V peak-to-peak
- Positive-slope duty cycle 40%

Verify the design using SPICE.

13-17. It is suggested that a device to sound the orchestral tuning note ($A = 440$ Hz) could be inexpensively mass produced using an astable multivibrator as the tone oscillator. The presence of harmonics of the 440-Hz tone in the output is desirable.

    a. Design an astable multivibrator to produce that frequency.

    b. Trained musicians are easily capable of determining pitch (frequency) error of ±3 cents.

A cent is defined as 1% of a semitone, where a pitch transition of a semitone produces a change in frequency by a factor of the twelfth root of 2. Thus pitches (when compared to the ideal pitch frequency f_o) that are considered "in tune" must lie within the frequency range

$$(^{12}\sqrt{2})^{-3/100} f_o < f < (^{12}\sqrt{2})^{3/100} f_o$$
$$\Rightarrow 2^{-3/1200} f_o < f < 2^{3/1200} f_o.$$

Comment on the practicality of using the design of part a for this mass-produced device.

13-18. Design a VCO that will output a 10-V peak-to-peak square wave of variable frequency from 50 Hz to 1 kHz for an input voltage range of 0.2–4 V.

13-19. Design a VCO that will output a 10-V peak-to-peak triangle wave of variable frequency from 50 Hz to 4kHz for an input voltage range of 0.1–8 V.

13-20. The diode network shown is a proposed triangle-to-sinusoid converter. It is said to be capable of taking a 20-V peak-to-peak triangular wave input and produce a sinusoid with less than 3% total harmonic distortion. Unfortunately the design does not include the proper value for the reference voltage $\pm V_{ref}$.

    a. Determine the breakpoint voltages as a function of V_{ref}. (Assume $V_{ref} > 2V_\gamma$.)

    b. Choose an appropriate value for V_{ref} and determine the total harmonic distortion of the output using SPICE.

    c. Try another value for V_{ref}. Compare the resultant output total harmonic distortion with that of part b.

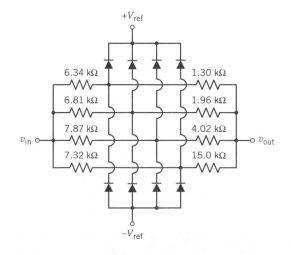

13-21. One common use of non-linear waveshaping occurs in digital telephony. Analog-to digital (A/D) conversion introduces the same quantization noise for small-amplitude signals than for large-amplitude signals. Thus small signals appear to have more *relative* noise due to the A/D conversion than large signals. In order to equalize the signal-to-noise ratio, the input message is amplified in a non-linear fashion prior to A/D conversion. This process is called *compressing*. Upon receipt, the message is digital-to-analog (D/A) converted and non-linearly amplified to restore linearity. This process is called *expanding*. Together the two processes are called *companding*. Using the basic topology shown in Figure 13.3-1, design a seven-segment compressing wave-shaping network with the following characteristics:

$$v_o = v_{in}, \qquad\qquad |v_{in}| < 1$$
$$v_o = 0.5(v_{in} + 1), \qquad 1 < v_{in} < 3$$
$$v_o = 0.25(v_{in} + 5), \qquad 3 < v_{in} < 7$$
$$v_o = 0.125(v_{in} + 17), \qquad 7 < v_{in} < 15$$
$$v_o = 0.5(v_{in} - 1), \qquad -1 > v_{in} > -3$$
$$v_o = 0.25(v_{in} - 5), \qquad -3 > v_{in} > -7$$
$$v_o = 0.125(v_{in} - 17), \qquad -7 > v_{in} > -15$$

13-22. Using the basic topology shown in Figure 13.3-1 (some diodes must be reversed), design a seven-segment expanding wave-shaping network for use in the receiver of the digital telephony system described above. The desired design characteristics are:

$$v_o = v_{in}, \qquad\qquad |v_{in}| < 1$$
$$v_o = 2v_{in} - 1, \qquad 1 < v_{in} < 2$$
$$v_o = 4v_{in} - 5, \qquad 2 < v_{in} < 3$$
$$v_o = 8v_{in} - 17, \qquad 3 < v_{in} < 4$$
$$v_o = 2v_{in} + 1, \qquad -1 > v_{in} - 2$$
$$v_o = 4v_{in} + 5, \qquad -2 > v_{in} > -3$$
$$v_o = 8v_{in} + 17, \qquad -3 > v_{in} > -4$$

Hint: An amplifier is needed in addition to the passive diode network.

13-23. Design a monostable multivibrator using a 555 timer with a pulse output of duration 10 ms. The input is a 5-V signal that drops to 0 V for a duration of 1 ms to trigger the multivibrator.

13-24. Series connection of integrated circuit timers can produce an output consisting of a delayed pulse of fixed duration. The first timer fixes the delay, with its negative transition triggering the second timer. The second timer sets the duration of the output pulse. The two timers are capacitively coupled with a pull-up resistor on the input of the second timer. The *RC* timer constant of this coupling circuit must typically be less than 50 μs (for a 555 timer). Design such a two-integrated-circuit timer that will produce an output pulse that is delayed from an input triggering pulse. The design specifications are:

Input trigger	5 V to 0 V for 10 μs
Output delay	2 ms ± 10 μs
Output duration	0.5 ms ± 10 μs

Use SPICE to verify the design.

13-25. Verify the waveforms shown in Figure 13.4-4 by designing an 555 timer astable multivibrator to oscillate at 2 kHz with 60% HIGH duty cycle. Use PSpice to display the waveforms.

13-26. The expression for the output waveform duty cycle of an integrated circuit timer oscillator is given by Equation 13.4-12. A 50% duty cycle implies that $R_A = 0$ (the discharge and V_{CC} terminals are shorted together). Explain why the circuit will not oscillate with a 50% duty cycle. Use PSpice simulation to verify that the circuit will not oscillate.

13-27. The circuit shown purports to use an integrated circuit (IC) timer to produce square-wave oscillation with a 50% duty cycle. The design parameters are

$$f = \frac{1}{1.386 R_1 C_1}, \qquad R_2 > 10 R_1.$$

a. Use this circuit topology to design a 540-Hz oscillator.

b. Use PSpice and the macromodel of a 555 timer to check the design for frequency and duty cycle compliance with specifications.

c. If design parameters are not met, explain any mechanisms that cause the variation.

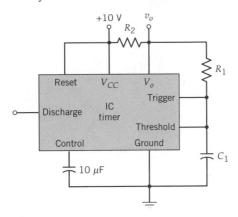

13-28. The circuit shown is a voltage-to-frequency converter based upon linear charging of a capacitor by a constant current source. Use PSpice and the macromodels for a 741 OpAmp and 555 timer to determine the linearity of the voltage-to-frequency conversion over the range $0.5\ V < v_{in} < 5\ V$.

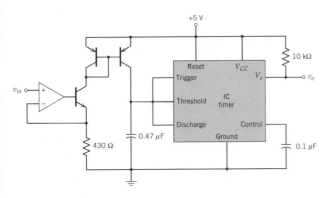

13-29. One-shot circuits that use an integrated circuit timer have a very rapid recovery time. This rapid recovery creates opportunities for unusual applications of the timer. One application is the use of an integrated circuit timer monostable multivibrators (one-shot) to divide the frequency of a pulse train. In order to divide the frequency by a factor N, the pulse duration of the one-shot is chosen to lie in the range

$$(N - DC)\tau < t < N\tau,$$

where τ is the period of the original pulse train and DC is the HIGH-state duty cycle given as a fraction rather than as a percentage. The trigger terminal serves as the input to the frequency divider. Using a 555 timer, design a circuit that will divide the frequency of a 10-kHz square wave (DC = 0.5) by a factor of 3 ($N = 3$). Verify the design using PSPICE.

13-30. The circuit shown may be used to detect irregularities in a train of pulses. As long as the pulse spacing is shorter than the timing interval of the timing circuit, the monostable circuit is continuously triggered. Pulse spacing greater than the timing interval or the termination of the pulse train allows completion of the timing interval and the generation of an output pulse.

a. Assume a pulse train input at 1 kHz with a LOW duty cycle of 20%. Complete the design so that a missing LOW pulse will be detected (the pulse spacing is effectively increased by the absence of a pulse).

b. Test the design of part a using SPICE.

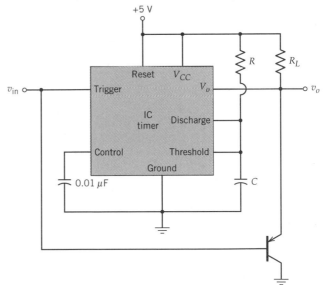

3,746,968
AMPLITUDE-TO-FREQUENCY CONVERTER
Robert A. Pease, Wilmington, Mass., assignor to Teledyne, Inc., Los Angeles, Calif.
Filed Sept. 8, 1972, Ser. No. 287,575
Int. Cl. H02m *5/00*; 321 *60;18*

U.S. Cl. 321—60

9 Claims

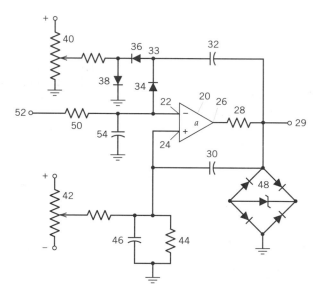

An amplitude-to-frequency converter is described which uses a single operational amplifier having both regenerative and degenerative feedback loops. The degenerative feedback loop includes a diode and capacitor in series, the inverting input of the amplifier is shunted to ground through an input capacitor. The output of the amplifier is connected to ground through a bilateral voltage limiting circuit. The amplifier will provide a rectangular wave output at a repetition rate linearly related to the amplitude of the input signal, the only precision values required being supplied by the voltage limiting circuit and the degenerative feedback capacitors.

Application-Specific Electronics

This final section of this work focuses on three significant branches of electronic circuitry that are not necessarily in the direct path taken by the three previous sections. Each is particularly significant in its own right. Discussion here provides an introduction to these specialized topics:

- Power circuit discussions focus on DC power.
- Communication circuit discussions focus on high-frequency operation and the frequency translation of signals.
- Digital circuit discussions focus on the non-linear operation of digital gates and the speed at which they operate.

Electronic power circuitry is vital in the operation of all electrical apparatus. The focus of discussion on power electronics presented here is directed at providing clean, reliable power to electronic circuitry. Central to the discussion is the design, operation, and use of voltage regulators. Another topic of particular interest to the electronics designer, transient suppression along with overvoltage protection, is presented. The thyristor family of electronic devices is introduced.

Communication circuits typically translate baseband signals to a higher frequency range for transmission. In order to accomplish this task and reverse it at the receiver, a variety of fundamental building blocks are necessary. An introduction to these building blocks is presented here. Voltage-controlled osillactors and mixers form the primary building blocks discussed for transmission modulators, while mixers and phase-lock loops are presented as common in receiver demodulators. A discussion of the trade-offs between active and passive filters complements previous discussion in Chapter 9. Analog-to-digital conversion is introduced.

While an introduction to the basic operation of a few logic gates was presented in Chapters 3 and 4, discussion there focused on transistor operation rather than gate operation. Here a closer look at bipolar and FET gate operation, including gate speed, is taken. Fundamental problems encountered when transistors transition between regions of operation is analyzed. Several design alternatives to reduce these problems are discussed. Comparisons between gate families are made, but not emphasized. A short introduction to gallium arsenide logic is presented.

CHAPTER 14

POWER CIRCUITS

All the electronic circuits described in the first three sections of this book have as their main purpose the modification of an input signal so as to perform useful work on the load. This signal modification may take the form of signal amplification, frequency filtering, digital logical operations, or a variety of other possibilities. Inherent in all designs is an electrical power source. Most often this source is a DC voltage source; in some instances, the DC is derived from an AC source. The proper operation of all electronic circuitry depends on the application of uniform electrical power in the form of these sources. Power electronic circuits are responsible for converting the available electric energy into a form with appropriate uniformity.

Regardless of the form of the input electrical energy (AC, DC, or a combination) the basic form of an electrical power system remains consistent. A typical electrical power system is shown in Figure 14-1. In such a system the input energy is filtered, converted to a new form, or shifted to a new level by an electrical power circuit, and again filtered. Electronic (or possibly mechanical) observation of the load conditions is an important factor in control of all operations.

Previous chapters have described the principles of AC-to-DC conversion through the use of diode half-wave or full-wave rectifiers followed with a simple passive low-pass filter (active filtering is extremely inefficient in power applications). The benefits of feedback on stability have also been discussed. This chapter focuses on combining previously explored devices and principles into power circuits useful to the electronic designer. While many of these devices are commercially available in integrated circuit packages, knowledge of the principles of operation is valuable to the prudent device user or circuit designer.

Within many power circuits there exists the need for a high-power switch. BJTs and FETs are extremely useful as switching elements and have previously been discussed. Both of these transistor switches find common use in power circuits. Thyristors form another family of semiconductor devices that are extremely useful in switching applications. Most common among this family are the silicon-controlled rectifier, typically used in DC applications, and the triac for AC applications.

Voltage regulators are a common device available for providing a stable DC output voltage for a wide range of input voltage and output current. As realized in a three-pin integrated circuit package, these regulators are extremely effective, easy to implement, and inexpensive. The internal design of both linear and switching regulators is discussed. While the design of DC power supplies is not specifically addressed in this chapter, many of the design principles of simple voltage regulators can be applied to their more powerful relative.

Protection against high-voltage transient or the accidental application of an over-voltage is another major concern of the electronic designer. Several types of tran-

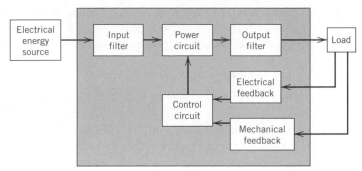

Figure 14-1
Typical electronic power system.

sient suppressors are discussed. Similarly, several overvoltage protection circuits are described.

14.1 THYRISTORS

Thyristors are a form of solid-state switch that is activated by a triggering signal. Prior to activation, a thyristor acts as an extremely high impedance path. Once activated, a thyristor acts as a low-impedance path and *remains activated* until the switched current falls below a minimum value, the holding level, at which time it deactivates. Once a thyristor is activated, the triggering signal is no longer necessary to continue activation unless low switched current deactivates the thyristor. The most common members of the thyristor family include the following:

- Silicon-controlled rectifiers (SCRs)
- Triacs
- Programmable unijunction transistors (PUTs)
- Silicon bilateral switches (SBSs)
- Sidacs

Silicon-controlled rectifiers are most commonly used as power control elements, triacs are bidirectional switches and are most useful in AC power applications under 40 A, PUTs are most often used in timer circuits, SBSs are most commonly used as gate trigger devices for the power control elements, and Sidacs are a high-voltage bilateral trigger device. The thyristor family also includes the gate turn-off (GTO) thyristor, a device capable of being turned off, as well as on, with the application of different polarity gate current. The GTO thyristor typically is used in power inverters (DC-to-AC power converters).

Thyristors have several distinct advantages over mechanical switches. Primary among these advantages over mechanical switches are the following:

- High switching speed
- Low-energy switch triggering
- Automatic debounce characteristics
- Zero-current deactivation that avoids contact arcing in inductive circuits

While each type of thyristor has its particular uses, discussion in this chapter will be limited to the two most commonly used thyristor types, the SCR and the triac. Several SCR applications will be presented in later sections of this chapter.

14.1.1 Silicon-Controlled Rectifier

The Silicon controlled rectifier (SCR) is the foundation element of the thyristor family. Its circuit symbol is shown in Figure 14.1-1. The three SCR terminals take their names from diode terminology and are identified as anode A, cathode K, and gate G. The SCR can conduct currents in excess of 50 A from the anode to the cathode and is capable of blocking voltages up to 800 V. The gate terminal is the input for the triggering signal.

SCRs have a four-layer structure of alternating p-type and n-type semiconductor material. Conceptually, this structure takes the form of Figure 14.1-2a. The structure functionally acts as two complementary BJTs connected as shown in Figure 14.1-2b.

Analysis of the SCR equivalent circuit leads to its modes of operation. The SCR has two fundamental operational states:

- An OFF state where only extremely small leakage currents pass from the anode to the cathode. In this state transistor Q_1 is cut off and Q_2 is in the forward-active region, albeit with extremely low current flow.

- An ON state where current flows freely between anode and cathode. Here both transistors are in saturation.

In order to understand the transition between modes, it is necessary to examine the basic transistor current relationships:

$$I_{C1} = -\alpha_1 I_{E1} + I_{CO1} = \alpha_1 I_K + I_{CO1}, \tag{14.1-1}$$

$$I_{C2} = -\alpha_2 I_{E2} + I_{CO2} = -\alpha_2 I_A + I_{CO2}. \tag{14.1-2}$$

Here the two quantities I_{CO1} and I_{CO2} are the collector leakage currents of each transistor[1] (with an open emitter) and α_1 and α_2 are the collector–emitter current ratios. Note that in this formulation, α_i is transistor region dependent and is only equal to α_{Fi} in the forward-active region. Kirchhoff's current law applied to the SCR equivalent circuit of Figure 14.1-2b yields expressions for the currents into the anode and out of the cathode of the SCR:

$$I_A = I_{C1} - I_{C2} \tag{14.1-3}$$

$$I_K = I_A + I_G = I_{C1} - I_{C2} + I_G. \tag{14.1-4}$$

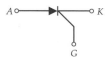

Figure 14.1-1
The SCR circuit symbol.

---

[1] Since Q_1 is an *npn* BJT, the collector leakage current I_{CO1} is a positive quantity. Here, Q_2 is an *pnp* BJT and has negative I_{CO2}.

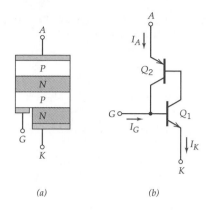

(a)   (b)

Figure 14.1-2

The SCR structure: (a) typical medium-power SCR semiconductor structure; (b) equivalent circuit.

The above four equations can be combined to determine the SCR anode current as a function of the transistor parameters and the gate current:

$$I_A = [\alpha_1(I_A + I_G) + I_{CO1}] - (-\alpha_2 I_A + I_{CO2}), \qquad (14.1\text{-}5)$$

which leads to

$$I_A = \frac{\alpha_1 I_G + I_{CO1} - I_{CO2}}{1 - (\alpha_1 + \alpha_2)}. \qquad (14.1\text{-}6)$$

When the control transistor Q_2 is OFF, $\alpha_1 \approx 0$ and $\alpha_2 < 1$: The SCR anode current is a small multiple of the sum of the magnitude of the leakage currents. As current is applied to the gate of the SCR and Q_1 begins to turn on, α_1 increases until the denominator of Equation 14.1-6 becomes zero. This singularity in the expression for the anode current I_A is physically realized by a rapid increase in the anode current until it is limited by the external circuitry to which the SCR is connected. At that current-limiting point, I_G may be removed and the two BJT leakage currents will suffice to latch the SCR in the ON state. SCR deactivation will only occur when the anode current drops to a zero value due to external circuit circumstances. Reactivation of the SCR will occur when V_{AK} is again positive and a current pulse enters the gate. For the case of sinusoidal V_{AK} and a pulse train at the SCR gate SCR conduction current is shown in Figure 14.1-3.

While a conducting SCR presents very low dynamic resistance to the conduction current, there is a voltage drop across the anode–cathode terminals. The voltage

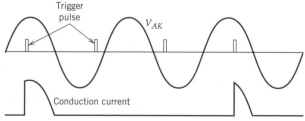

Figure 14.1-3

The SCR conduction triggered by a pulse train.

across the SCR in the ON state is approximately given by the sum of a base–emitter drop, a collector–emitter drop, and an ohmic loss:

$$V_{AK} \approx V_{BE\text{(sat)}} + V_{CE\text{(sat)}} + R_B I_A.$$

Here R_B is the bulk resistance associated with the SCR. The ON voltage drop has a minimum value in the range of 1 V and may be as high as 2–3 V for high-current SCRs. The actual voltage drop for any particular SCR is best determined from manufacturer's specifications or experimental evaluation.

In addition to the desired SCR activation with positive gate current, there are four false SCR activation mechanisms that must be avoided:

- High rate of change of anode–cathode voltage dV_{AK}/dt
- High anode–cathode voltage V_{AK}
- High device junction temperature
- Energy injection into device semiconductor junctions, principally in an optical fashion

Reasonable design caution can avoid false SCR activation by the last three mechanisms: Opaque packaging avoids optical activation, a proper heat sink avoids high temperatures, and selection of an SCR with a correct voltage-blocking rating avoids an avalanche breakdown. Fast variation of the anode–cathode voltage V_{AK} is the most likely significant false activation mechanism in most applications.

A voltage applied across the anode–cathode terminals of an SCR induces a current through the SCR proportional to the derivative of the applied voltage. This current is due to the junction capacitances between the layers of semiconductor:[2]

$$i = C\,\frac{dV_{AK}}{dt}.$$

If this capacitance-charging current exceeds the gate-triggering current, the SCR is subject to false triggering. Sensitivity to rate-of-change false activation can be reduced by a resistance shunting the gate–cathode junction or, more commonly, controlled with a snubber circuit. Snubber circuits limit the rate of change of the voltage across a SCR (or more generally any thyristor). In their most simple form, a snubber circuit consists of a series-connected resistor and capacitor that shunt the SCR, as shown in Figure 14.1-4. The RC time constant of the snubber circuit limits the rise time of the anode–cathode voltage and thereby reduces the possibility of false activation. On occasion the resistor in the snubber circuit may be shunted or connected in series

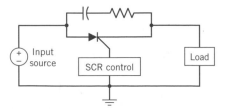

Figure 14.1-4
The SCR application with snubber circuitry.

---

[2] Junction capacitance is discussed in Sectin 10.4.1.

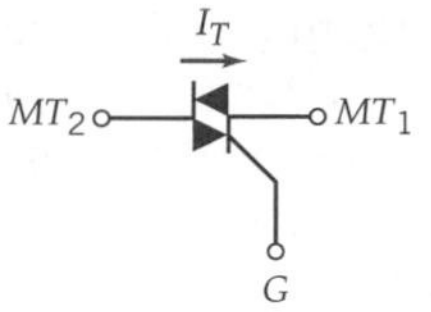

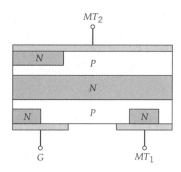

Figure 14.1-5
Triac circuit symbol.

with a diode. This diode is added to aid in suppression of transient voltages that may occur.

14.1.2 Triac

The triac is also a three-terminal semiconductor switch: Its circuit symbol is shown in Figure 14.1-5. The triac switch differs from the SCR in that it is capable of conduction currents in either direction. The current-carrying terminals are simply identified at MT_1 and MT_2 (main terminals 1 and 2). The triggering terminal, G, remains identified as the gate. Positive conduction current is identified as flowing from MT_2 through the triac to MT_1. Positive gate current flows into the triac.

The triac is a five-layer semiconductor device as shown in Figure 14.1-6. In many ways it can be thought of as two complementary SCRs (a p–n–p–n device and an n–p–n–p device) connected in parallel. There are, however, properties of the triac that cannot be described by such a simple model. In particular is the ability of the triac to be triggered into conduction in *either direction* by either a positive or negative current into the gate. This bipolar triggering ability can be accounted for by noting that the region between the gate and MT_1 is essentially two complementary diodes. Either a positive or negative gate current will bring one of these diodes into conduction, triggering the same transistor action found in a SCR.

The four possible regions of triac operation are identified in Figure 14.1-7: The regions are identified by the polarity of the main terminal voltage difference and the polarity of the gate current. While it is possible to activate the triac in any region, the activation sensitivity to the gate current varies with region.

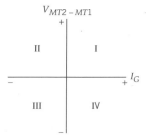

Figure 14.1-6
Typical triac semiconductor structure.

Figure 14.1-7
Triac regions of operation.

Region I (positive gate current and positive main terminal voltage difference) is the most sensitive of the four regions. Regions II and III are only slightly less sensitive to gate current turn-on. Region IV typically requires as much as four times the gate activation current as region I; this region of operation is avoided whenever possible. In many AC applications, the polarity of the gate current is automatically reversed as the polarity of the input voltage reverses so that operation is always in regions I and III.

The triac is subject to the same false activation mechanisms as the SCR. Many of the same precautions must be taken to ensure that the only activation mechanism is through the gate current. One additional precaution that must be taken involves the necessity of bidirectional snubber circuitry. While simple *RC* snubber circuitry is effective for both conduction current polarities (assuming proper choice of nonpolarized capacitor), more complex SCR snubber circuitry may not be bidirectional and therefore inappropriate for triac use.

14.2 VOLTAGE REGULATOR DESIGN

The performance of electronic circuitry often depends on the application of stable DC power to the circuitry. Zener diode regulators, as discussed in Chapter 2, are one possible method for providing relatively stable DC power. However, Zener diode regulators have significant drawbacks, the most significant of which is the dependence of the output voltage on load current and temperature. A better alternative to the use of a Zener diode in efficiently ensuring stable power is the use of an integrated circuit voltage regulator. Such devices are highly effective, widely available, and relatively inexpensive. An integrated circuit voltage regulator provides specific, stable DC power over a wide range of load current and input voltage conditions and has relatively small variation in output with temperature.

An integrated circuit voltage regulator consists of three basic elements, as shown in Figure 14.2-1:

- A voltage reference elements that provides a known stable voltage level V_{ref} that is essentially independent of temperature and the input voltage
- An error amplifier that compares the output voltage or some fraction of the output voltage to the reference voltage V_{ref}
- A power control element that converts, as indicated by the error amplifier, the input voltage to the desired output voltage over varying load conditions

Each of these elements can be realized with several circuit topologies. The following discussion provides a sampling of the element topologies common in typical voltage

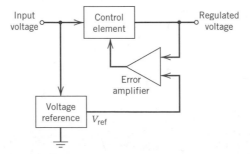

Figure 14.2-1
Typical voltage regulator block diagram.

regulators. While all examples shown here assume that the voltage to be regulated is positive, regulation of both voltage polarities is common and accomplished with the same general circuit topology. In addition, tracking regulators provide a symmetric pair of regulated voltages (i.e., ±10 V) for applications that need matched, regulated power. One such application is the dual power bus needed for OpAmps.

14.2.1 Voltage Reference

The design goal of the voltage reference element is to provide a known, stable reference voltage V_{ref}. There are three basic circuit topologies that provide a reference voltage within common voltage regulators. Two of these circuits are based on the Zener diode reverse breakdown voltage V_z and the third is based on the base–emitter junction voltage of a BJT. Simplest among the three circuits is a basic Zener diode reference, as shown in Figure 14.2-2.

In this circuit the output voltage V_{ref} is dependent on the diode Zener breakdown voltage V_z, the diode zener resistance r_z, the input voltage V_{in}, and the resistor value R:

$$V_{\text{ref}} = V_z \frac{R}{r_z + R} + V_{\text{in}} \frac{r_z}{r_z + R}. \tag{14.2-1}$$

The basic Zener diode reference can be satisfactory in applications where the input voltage V_{in} is relatively stable. However, the susceptibility of this circuit to variation in input voltage and load-current-induced temperature variation may make it a poor choice in many applications.[3]

An improvement in the basic Zener diode voltage reference can be obtained by making the diode current independent of input voltage V_{in}. One typical circuit topology that reduces diode current variation by driving the Zener diode with a constant-current source is shown in Figure 14.2-3.

Here the current in the Zener diode is the sum of the base current in Q_1 and the current through resistor R_{SC}:

$$I_z = I_{B1} + I_{R_{\text{SC}}}. \tag{14.2-2}$$

In most applications the base current can be ignored:

$$I_z \approx I_{R_{\text{SC}}} = \frac{V_{BE1}}{R_{\text{SC}}} \tag{14.2-3}$$

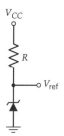

Figure 14.2-2
Basic Zener diode voltage reference.

---

[3] The Zener voltage for integrated circuit Zener diodes varies with temperature by approximately +2.2 mV/°C.

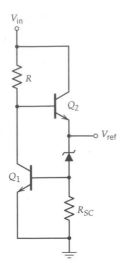

Figure 14.2-3
Constant-current Zener diode voltage reference.

This gives a stable reference voltage V_{ref}:

$$V_{\text{ref}} = V_z + I_z r_z + V_{BE1} \approx V_z + V_{BE1}\left(1 + \frac{r_z}{R_{\text{SC}}}\right). \tag{14.2-4}$$

Temperature variation of the reference voltage can be minimized by balancing the *positive* temperature variation of the Zener voltage V_z with the *negative* temperature variation of the base–emitter voltage V_{BE1}. The major disadvantage of the constant-current Zener diode reference voltage is the need for the input voltage V_{in} to be relatively large: It must remain, depending on the exact circuit design, at least 1.5 V more than the Zener voltage.[4] This restriction on the input voltage limits the minimum voltage application of such a regulator.

A third common reference voltage circuit, the bandgap voltage reference, is shown in Figure 14.2-4. This design allows for a minimum input–output voltage difference as small as 0.6 V: The input and output vary only by the voltage across the input resistor R.

The output of the bandgap voltage reference is based on the highly predictable base–emitter voltage of a BJT in the forward-active region. Here the output reference voltage V_{ref} is given by

$$V_{\text{ref}} = V_{BE3} + I_3 R_3. \tag{14.2-5}$$

The current I_3 is essentially the output of a Widlar current source[5] formed by transistors Q_1 and Q_2 and resistors R_1 and R_2. The current source has an output current I_{C2} given by the solution to the transcendental equation

$$I_{C2} R_2 = \frac{\beta_F}{\beta_F + 1} \eta V_T \ln\left(\frac{I_{C1}}{I_{C2}}\right). \tag{14.2-6}$$

---

[4] Manufacturers specify the minimum voltage difference between input and output voltage: For this design topology, it typically lies in the 2–3-V range.
[5] Widlar current sources are discussed in Section 6.4.5.

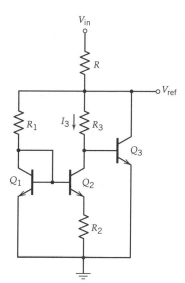

Figure 14.2-4
Bandgap voltage reference.

If the base current of Q_3 is assumed to be small compared to the collector current of Q_2, the output voltage can be derived from Equations 14.2-5 and 14.2-6:

$$V_{\text{ref}} = \eta V_T \frac{R_3}{R_2} \ln\left(\frac{I_{C1}}{I_{C2}}\right) + V_{BE3}. \tag{14.2-7}$$

It can be seen from Equation 14.2-7 that the bandgap reference circuit has an output, V_{ref}, that is independent of the input voltage V_{in}. An additional benefit of this circuit topology is its relative insensitivity to temperature variation. If the circuit is built in integrated circuit form, then all circuit elements are essentially at the same temperature. The ratio of the collector currents I_{C1} and I_{C2} remains basically constant over temperature changes, and the variation in the reference voltage with temperature is given by

$$\frac{\Delta V_{\text{ref}}}{\Delta T} \approx \eta \frac{R_3}{R_2} \ln\left(\frac{I_{C1}}{I_{C2}}\right) \frac{\Delta V_T}{\Delta T} + \frac{\Delta V_{BE3}}{\Delta T}. \tag{14.2-8}$$

In Chapter 2, the voltage equivalent temperature, V_T, was defined as

$$V_T = \frac{kT}{q} \approx \frac{T}{11{,}600}. \tag{14.2-9}$$

Consequently V_T has a positive variation with temperature. In Chapter 3, it was shown that the variation in V_{BE} with temperature is negative. Thus it is possible to have essentially no temperature variation in V_{ref} if the two temperature-dependent terms in Equation 14.2-8 cancel. The necessary conditions for cancellation are

$$\frac{R_3}{R_2} \ln\left(\frac{I_{C1}}{I_{C2}}\right) = -\frac{11{,}600}{\eta} \frac{\Delta V_{BE3}}{\Delta T}. \tag{14.2-10}$$

A judicious choice of the resistors R_1, R_2, and R_3 allows the designer to produce a wide variety of reference voltages that are relatively insensitive to temperature and input voltage variation. A typical reference voltage value used in many regulators is 2.5 V.

All of the described voltage reference circuits find uses other than in voltage regulators: They can be used whenever an independent voltage reference is necessary. A precision reference voltage is usually achieved with either the bandgap or constant-current Zener reference circuits. These precision voltage reference circuits are used in a wide variety of electronic applications. Most common among these is as reference for precision analog-to-digital conversion. A few of the common applications for voltage reference circuits are as follows:

- Analog-to-digital (A/D) conversion
- Digital-to-analog (D/A) conversion
- Digital multimeter applications
- Voltage-to-frequency conversion, VCO
- Frequency-to-voltage conversion, FM detection

14.2.2 Error Amplifier

The error amplifier in integrated circuit voltage regulators is basically a differential amplifier that compares the output voltage (or a fraction of the output voltage) to the reference voltage. They may take as simple a form as an emitter-coupled or source-coupled pair but are more often similar to an OpAmp circuit.[6] These differential amplifiers must have a high common-mode rejection ratio (CMRR), low offset currents and voltages, and a high power supply rejection ratio (PSRR) to be effective in voltage regulation applications.

14.2.3 Linear Voltage Regulators

The inputs to the power control element of a voltage regulator are the supply voltage and the control signal from the error amplifier. The output of this element must be a constant, specific voltage over a wide range of load currents and impedances. While the voltage reference and error amplifier elements can be virtually identical for a wide range of voltage regulators, the power control element varies widely depending on the type of desired regulation. The three basic types of power control elements are as follows:

- Series
- Shunt
- Switching

Voltage regulators are often classified by the type of control unit employed. A voltage regulator with a series or shunt power control element is classified as a *linear voltage regulator*; a regulator with a switching power control element is a *switching regulator*.

---

[6] Differential amplifiers are discussed in several sections of this book. See Chapters 1 and 6 for basic discussions.

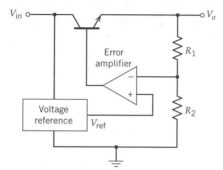

Figure 14.2-5
Basic series regulator topology.

Linear regulators have a distinct noise advantage over switching regulators. Due to energy conversion efficiency considerations, linear regulators are typically found in low-power electronic applications; switching regulators find greater use in high-power applications where they act as regulated power supplies.

Series Regulators

The series regulator is best suited for medium-load-current applications where the input–output voltage difference is not large. The commonly used, three-terminal integrated circuit voltage regulator found in many electronic designs is usually of this design. Safety features such as input overvoltage and output short circuit protection can also be provided in series regulators.

For a series regulator, the error amplifier regulates the output voltage V_o through an active series element, usually a transistor, as shown in Figure 14.2-5. If the output voltage falls below

$$V_o \leq \frac{R_2}{R_2 + R_1} V_{\text{ref}},$$

(14.2-11)

the error amplifier will supply a positive voltage signal to the BJT. The transistor base–emitter junction becomes more forward biased and increases the current to the load. The output voltage is thereby increased until an appropriate balance is achieved. Large input–output voltage differences imply that the active control element must dissipate significant power. The internal losses lead to a typical working efficiency for a series regulator of 40–50%. A high input–output voltage differential can reduce this figure further. The potentially large internal power loss is the greatest disadvantage of a series regulator.

Shunt Regulators

While the shunt regulator is usually the most inefficient of all regulator topologies, it can be a good choice in some applications. It is relatively insensitive to input voltage variation, protects the source from load current transients, and is inherently rugged against an accidental load short circuit. It is the simplest of all regulators: For example, the Zener diode regulator, as described in Chapter 2, is a passive form of shunt regulator.

The basic circuit topology of an active shunt regulator is shown in Figure 14.2-6. Here, the active series pass element of the series regulator has been replaced by a resistor R_S, and the changes in load current are neutralized by shunting excess cur-

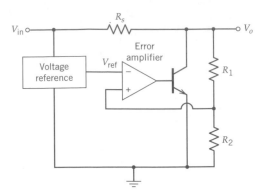

Figure 14.2-6
Basic shunt regulator topology.

rent through an active element (usually a transistor) to ground. When the output voltage falls below

$$V_o \leq \frac{R_2}{R_2 + R_1} V_{ref},$$ (14.2-12)

the error amplifier reduces voltage to the base of the BJT, thereby reducing the shunt current. Since the current through the series pass element R_S must remain constant, more current is available for the load and V_o is increased until an appropriate balance is achieved.

14.2.4 Switching Voltage Regulators

The switching regulator is most often used in relatively high power applications where power conversion efficiency is of high concern. Switching regulators have a typical operating efficiency of 60–90%. In addition, they provide good regulation over a wide range of input voltage and maintain high efficiency over a wide range of load current. It is also possible to provide regulated voltages *larger* than the input voltage. A switching regulator has a significant size and weight advantage over other regulators in high-power applications. Typically a switching regulator is smaller by a factor of between 4 and 8.

The basic topology of a switching regulator, as shown in Figure 14.2-7, is similar to a linear regulator except the active element is modulated in an ON/OFF mode of operation by a switch control unit. Typically the control unit uses the output of the error amplifier to pulse-width modulate the output of a constant-frequency oscillator. The pulse-width-modulated square wave varies the ON/OFF duty cycle of the active element, similarly varying the average value of the passed voltage. The addition of a rectifying low-pass filter after the series active element smoothes the output voltage to the desired value. Another realization of the switch control unit consists of a VCO with nonvarying ON time. A major drawback of this realization is the added complexity of the rectifying filter due to variable-frequency inputs. Consequently the VCO realization is not typically found in modern switching voltage regulators.

The added complexity of a switching regulator in the form of complex switch control circuitry and the need for a high-frequency active switching element increase the cost of the regulator. As a result, switching regulators can only compete economically with other regulators in high-power applications (greater than ~20 W). Switching regulators typically have somewhat higher output ripple and can be more sus-

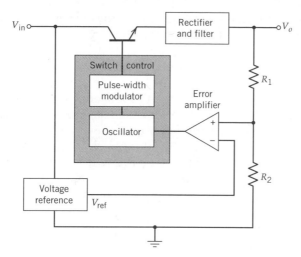

Figure 14.2-7
Typical switching regulator topology.

ceptible to load current transients than linear regulators. In addition, high-rate (20–500 kHz), square-wave switching generates significant electromagnetic interference (EMI) and radio-frequency interference (RFI). It is possible to successfully diminish both EMI and RFI with proper filtering.

The rectifying filtering section of a switching voltage regulator can take several different topologies. Most common among these topologies are the following:

- Buck
- Boost
- Buck–boost

The basic topology of each rectifying filter section is shown in Figure 14.2-8. The majority of other rectifying filters are direct derivatives of these three types.

The output voltage V_o of the *buck* filter configuration (Figure 14.2-8a) is always less than the input voltage V_{in}. Here, a switch is placed in series between the input voltage and the input to an LC low-pass filter. When the switch is conducting, current flows through the inductor to the load. When the switch opens, the magnetic field within the inductor maintains the current flow to the load, pulling current through

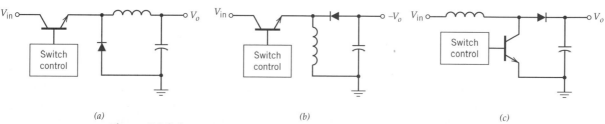

(a)   *(b)*   *(c)*

Figure 14.2-8
Switching regulator filtering configurations: (*a*) buck (step down); (*b*) buck–boost filtering (step up or down); (*c*) boost filtering (step up). The active element switch is shown here as a BJT. In many switching regulator designs, it is a much more complex element. Many designs use a class A amplifier with a transformer output as the high-frequency switch.

the diode from ground. In a buck circuit, the output voltage is proportional to the product of the input voltage and the switch duty cycle:

$$V_o \approx V_{\text{in}} \times \text{duty cycle.} \tag{14.2-13}$$

Peak switch current in the buck circuit is proportional to the load current.

The *boost* filter configuration (Figure 14.2-8c) has the unique property of providing an output voltage that is always *larger* than the input voltage. In this circuit, the switch is placed within the *LC* filter so that it can shunt current to ground. With the switch in its conducting state, the inductor current increases. When the switch opens, the output voltage V_o is the sum of the input voltage and the voltage across the inductor (a positive voltage due to decreasing inductor current). Boost regulators deliver a fixed amount of power to the load:

$$P_L = \tfrac{1}{2}LI^2 f_o, \tag{14.2-14}$$

where I is the peak inductor current and f_o is the switch operating frequency. In order to determine the output voltage, the load resistance R_L must be known:

$$V_o = \sqrt{P_o R_L} = I\sqrt{\tfrac{1}{2}LR_L f_o}. \tag{14.2-15}$$

Of course, the peak inductor current I is proportional to the duty cycle of the switch operation. Boost circuits are particularly useful in charging capacitive circuits (as in a capacitive-discharge automotive ignition system) and make good battery chargers.

The *buck–boost* rectifying filter (Figure 14.2-8b) provides the possibility of output voltages that are either higher or lower than the input voltage. The circuit operates in much the same fashion as the boost circuit with the exception that the output voltage is simply the voltage across the inductor. The buck–boost circuit also delivers constant power to the load, independent of the load resistance. Hence, Equation 14.2-14 and 14.2-15 are valid. The buck–boost circuit has the distinct feature of providing a negative-voltage output. This change of polarity is often an advantage, sometimes · a drawback. Isolation of the input from the output through transformer coupling avoids any problems.

For all these configurations, transient changes in the load conditions may cause problems. If the load suddenly becomes a very high impedance or suddenly become disconnected, the energy stored in the inductor has no path for dissipation. In a worse-case scenario, arcing across the load may occur. Switching power regulators are currently in a constant state of change due to major innovations in component design. It appears that major improvements in design are near.

14.3 VOLTAGE REGULATOR APPLICATIONS

The three-terminal, linear, integrated circuit voltage regulator is the most common voltage regulator in electronic applications with power requirements less than ~20 W and voltages less than 50 V. Its small size, high quality, and cost-effective properties make it an extremely useful device. In its most simple form, the three-terminal regulator requires a minimum number of external components for proper operation. A typical connection of a positive, fixed-output regulator is shown in Figure 14.3-1. The only external components necessary for this particular regulator are two small-value, high-frequency capacitors to improve stability and transient response.

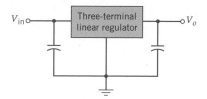

Figure 14.3-1
Typical fixed integrated circuit voltage regulator connection.

Whenever using an integrated circuit voltage regulator, the circuit designer is faced with several design choices based on the properties and limitations of these regulators. Linear regulators are categorized by their regulated output voltage. The most basic categories are based on the following properties:

- Output voltage polarity
- Fixed- or variable-output voltage
- Dual-tracking output voltages

The maximum output current is one additional constraint that must be considered. Integrated circuit voltage regulators typically come with maximum current ranging from 100 mA to 3 A. It is possible to extend this maximum output current with the addition of external circuit pass elements (Section 14.3.1).

The *polarity* of the input and output voltage usually determines the use of a positive or negative regulator: Positive regulators typically are used to regulate positive voltages; negative regulators typically regulate negative voltages. This is particularly true in systems where the input and output share a common ground. However, in systems where the ground reference can be floating at either the input or the output, the positive and negative regulators may be interchanged (Figure 14.3-2). In this special case a positive regulator can be used to regulate negative voltages and a negative regulator can be used to regulate positive voltages.

Fixed-output voltage regulators are available in a variety of output voltage values and current ratings. They provide an inexpensive, simple means of regulating output voltage and have several advantages:

- Ease of use
- Few external components
- Reliable performance
- Internal thermal and short circuit protection

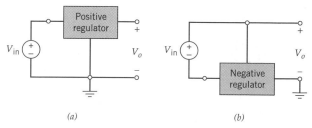

(a)   (b)

Figure 14.3-2
Voltage regulation alternatives: (*a*) positive output using positive regulator; (*b*) positive output using negative regulator.

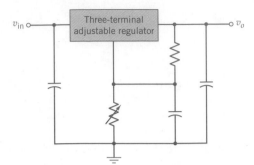

Figure 14.3-3
Typical adjustable voltage regulator connection.

The main disadvantage of a fixed-output voltage regulator lies in the inability to precisely adjust its output. The variation in output voltage may be as large as ±5% for any specified value. A similar problem exists due to the limited selection of output voltage values that are available.

Adjustable-output voltage regulators are best suited for applications requiring high-precision voltage regulation and/or regulation at a nonstandard voltage level. In addition, the regulated voltage may be sensed at a location remote from the output of the regulator. This feature allows for compensation due to losses in a distributed load or external pass components. Additional features often found on adjustable regulators include adjustable short circuit current limiting, access to the reference voltage V_{ref}, and overload protection. Adjustable regulators typically require a few more external components than fixed regulators (Figure 14.3-3). The capacitors improve stability and transient response

It is possible to accomplish the performance of an adjustable-voltage regulator using a fixed-output regulator and an OpAmp. One circuit topology that has an adjustable output is shown in Figure 14.3-4. In this circuit the variable output is limited to values larger than the specified, regulated output of the fixed regulator. While designs of this type are effective, the additional components required often make them economically impractical.

Many systems require balanced, dual-polarity power. An OpAmp that requires $\pm V_{\text{CC}}$ about a common ground is such a system. An obvious solution to dual-polarity applications is two independent regulators, one positive and one negative, paired together. Two problems arise with such a solution: power-on latch-up and under-voltage output imbalance. Latch-up is due to the intolerance of each individual regulator to reverse voltages applied at its output. In dual-polarity systems with a single

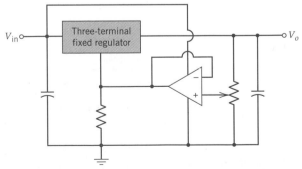

Figure 14.3-4
Variable regulated output using a fixed-output voltage regulator.

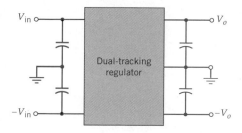

Figure 14.3-5
Typical dual-tracking regulator connection.

load across both outputs, reverse voltages may appear during the power-on operation and cause latch-up of the output of one of the regulators. This condition can be avoided by placing diodes, from input to output and output to ground of each regulator, to avoid significant reverse-voltage application. In many systems that require balanced power of opposite polarity, the application of unbalanced power will offset the output. If, for a variety of reasons, the imbalance is not constant, there will be a time-varying offset in the signal output: Time-varying outputs are usually interpreted as information signals.

Dual-tracking regulators provide a solution to both problems. Latch-up is internally controlled and no additional external components are necessary. In order to avoid an imbalance in output voltages, the control system within a dual-tracking regulator monitors both the positive and negative power outputs. If either output falls out of regulation, the tracking regulator will respond by varying the other output to match: A decrease in the magnitude of the positive output will result in an equal decrease in the magnitude of the negative output. A typical dual-tracking connection is shown in Figure 14.3-5. As is the case with most linear regulators, output capacitors improve ripple and transient performance. Input capacitors may be necessary if the source is particularly noisy or if the regulator is placed too far from the unregulated power supply.

Limitations as to the input voltage range and output voltage and current ranges over which regulation will occur apply to all voltage regulators. The *safe operating area* (SOA) defines the limits of these ranges. Exceeding the limits can result in catastrophic failure, temporary device shutdown, or failure to properly regulate the output. The SOA is defined by manufacturer's specifications relating to the input voltage, the output current, maximum power dissipation, and in the case of variable-voltage regulators, the output voltage. These specifications are described as follows:

$V_{\text{in(max)}}$	Absolute maximum input voltage with respect to regulator ground terminal
$(V_{\text{in}} - V_o)_{\text{min}}$	Minimum input–output voltage difference at which regulation can be maintained; also called dropout voltage
$(V_{\text{in}} - V_o)_{\text{max}}$	Maximum input–output voltage difference
$I_{L\text{(max)}}$	Maximum current deliverable to load from regulator
$P_{D\text{(max)}}$	Maximum power that can be dissipated by regulator
$V_{o\text{(min)}}$	For adjustable regulators, minimum output voltage that can be regulated
$V_{o\text{(max)}}$	For adjustable regulators, maximum output voltage that can be regulated

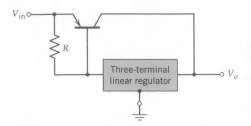

Figure 14.3-6
Series pass element to boost output current.

Of these specifications, $V_{in(max)}, I_{L(max)}$, and $P_{D(max)}$ can result in catastrophic failure if proper protection is not provided. Often this protection is within the regulator itself; in some cases it must be provided with external circuitry. The other specifications are functional limits that, if exceeded, imply a failure in the regulation ability of the device.

14.3.1 Extending Capabilities of Simple Voltage Regulators

It is often desirable to extend the SOA of a regulator with external components so as to exceed one of the specified limits. Extending the maximum values of output current, input voltage, and output voltage is necessary in many voltage regulator applications. While many varied uses of simple integrated circuit voltage regulators exist, a constant-current source provides a good example of one possibility.

Increasing Maximum Output Current

The effective output current of a simple voltage regulator can be increased through the use of an external bypass element, as shown in Figure 14.3-6.

In this realization the bypass element is a power *pnp* BJT: It must be chosen with the capability to provide ample output current and to dissipate enough power across its collector–emitter terminals. The resistor R must be chosen so that the internal bias current of the voltage regulator, I_{bias}, does not turn on the external bypass BJT. This condition is met if

$$R \leq \frac{V_{BE(on)}}{I_{bias}}.$$

Special care must be taken in this realization to protect against a possible output short circuit. Under this condition, the bypass element often must dissipate large quantities of power and may fail.

Increasing Input and/or Output Voltage Levels

It is often necessary to provide a regulated voltage larger than an available linear regulator. One possible circuit topology that will increase the regulated voltage for a particular regulator is shown in Figure 14.3-7. This circuit uses a Zener diode in between the ground terminal of the regulator and the actual circuit ground. The new regulated output voltage is the sum of the regulator output voltage and the series Zener voltage:

$$V_{o(new)} = V_{o(reg)} + V_{z1}.$$

Also shown in Figure 14.3-7 is an overvoltage protection circuit (shaded area and the diode D_P). Under normal operation the input BJT, Q_P, is in saturation and the

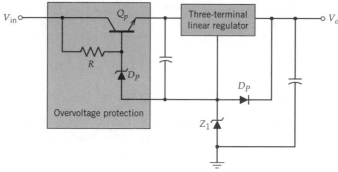

Figure 14.3-7
High-output voltage output circuit topology.

circuit operates as if the protection circuit was not present. However, if the input voltage exceeds the sum of the two Zener voltages,

$$V_{in} > V_{z1} + V_{zP},$$

the input BJT, Q_P, will enter the forward-active region and begin to dissipate power. The voltage drop across the BJT collector–emitter terminals will protect the input of the regulator from overvoltage, and excess current will be shunted through R_p, Z_p, and Z_1. The protection diode D_p protects the regulator against an output short circuit. While this protection circuit is shown in conjunction with the Zener diode realization of extending the output voltage, it can be used alone. With the absence of Zener diode Z_1, the diode D_P is also excluded. All elements in this overvoltage protection circuit will dissipate large quantities of power when activated and must be rated for that occurrence.

Using Voltage Regulator to Provide Constant Current

Another common regulated source necessary in many electronic application is a constant-current source. Three terminal positive-voltage regulators can be effectively used to provide this source, as shown in Figure 14.3-8.

In this configuration the output current I_o can be adjusted to any value from the minimum regulator bias current (≈ 8 mA) to the maximum current deliverable be the regulator, $I_{L(max)}$. The output current is the sum of current through the variable resistor and the bias current through the ground terminal of the regulator:

$$I_o = \frac{V_{reg}}{R} + I_{IB}.$$

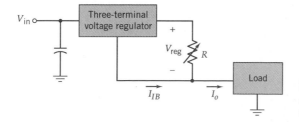

Figure 14.3-8
Adjustable constant-current source regulator configuration.

The input voltage for this configuration must always be greater than the sum of the minimum input–output voltage for the regulator, the regulator voltage, and the voltage at the load:

$$V_{in} \geq (V_{in} - V_o)_{min} + V_{reg} + V_o.$$

The choice of regulator is a balancing of conflicting requirements. Small V_{reg} is desirable to reduce power dissipation and allow for large variation in V_o. Large V_{reg} gives a more precise setting of the load current through the variable resistor. Five-volt fixed regulators often provide a good compromise.

14.4 TRANSIENT SUPPRESSION AND OVERVOLTAGE PROTECTION

A common problem in most electronic applications comes in the form of an inappropriately large input voltage. Protection must be provided so that erroneous voltage application connection at the input or transient overvoltages do not damage the circuitry. Transient high voltages are usually the result of the release of stored electric energy in the form of a current or voltage pulse. In order to ensure proper circuit protection against improper operation or actual damage, the energy within the transient must be dissipated within an added suppressor. Transients may result from sources within a circuit or have external sources. One typical internal source is inductive switching (as in motor control circuitry). External sources of transients include line voltage variations and EMI. One particularly unfriendly environment (in terms of transients) for electronic applications occurs in automotive applications.

Common transient suppressors and/or overvoltage protection circuits limit the peak AC or DC voltage and include the following:

- Carbon block spark gap suppressors
- Zener diodes
- Varistors
- Selenium rectifiers
- Overvoltage "crowbar" circuits

The first four items on this list are well suited as protection against short-duration high-voltage transients. In addition, Zener diodes and selenium rectifiers have good capability to protect against longer duration overvoltages so long as their average power ratings are not exceeded. Overvoltage crowbar circuits are best used as protection against long-duration overvoltages or misapplication of electrical power. Each of these protection circuits is connected as a shunt to the circuit to be protected.

A *carbon block spark gap suppressor* is an effective means of keeping transient voltages below a level of 300–1000 V. It is commonly found in telecommunication and power distribution systems. The suppressor is simply two carbon electrodes, one connected through a resistor to ground, separated by an air gap of approximately 0.1 mm. When a high-voltage transient is encountered, the energy in the transient is passed to ground through the arc in the air gap between the electrodes. The main drawbacks to carbon block spark gap suppressor use is its relatively short lifetime and inherent variation in arcing voltage.

Zener diodes, discussed in Chapter 2, exhibit excellent voltage-limiting properties. They are perhaps the most often used transient suppressor. Often a "back-to-back"

Figure 14.4-1
Zener diodes connected for bipolar overvoltage protection.

connection, as shown in Figure 14.4-1, is used to suppress transient overvoltages that are bipolar in nature. Many Zener transient voltage suppressors are available commercially packaged as a back-to-back pair. While typical Zener diodes are designed to operate at less than their maximum power rating, Zener transient suppressors are designed to effectively limit large, short-duration power pulses. The design change is accomplished by increasing junction area so as to withstand the high-energy surge of a transient. When Zener diodes fail, they typically fail to a short circuit. This feature, coupled with the shunt connection of Zener protection circuits, ensures that the circuit is protected even though the protection circuit has failed.

A *varistor* is a non-linear resistor that has electrical properties similar to a back-to-back Zener diode connection (Figure 14.4-2). A varistor can therefore provide an alternative to Zener dioes in many applications. The device is constructed of metal (often zinc) oxides sintered into a polycrystalline structure or silicon carbide sintered into a suitable ceramic binder forming a hard ceramiclike material. As a result of the sintering process, highly resistive intergranular boundaries are distributed throughout the volume of the device. The boundaries are the source of the voltage-dependent non-linear behavior of the varistor. The distributed nature of the boundaries make the varistor well suited for high-power transient suppression. Compared to Zener diodes, varistors are inexpensive. Varistors also share the attractive fail-to-short characteristic with Zener diodes.

The non-linear behavior of a varistor can be described by its volt–ampere transfer relationship:

$$V = \frac{C}{I}\,|I|^{(2-a)}, \tag{14.4-1}$$

where the passive sign convention relates the voltage and current polarities[7] and the two varistor property-dependent constants are given by C, the voltage across the varistor at $I = 1$ A, and a a factor describing the nonlinearity of the V–I curve. Typically $0.5 \leq a \leq 0.83$; as a increases, the sharpness of the curve increases. In Figure 14.4-3, the volt–ampere transfer relationship of a typical varistor ($a = 0.7$) is compared to that of a back-to-back Zener diode connection. The difference in the sharpness of the cutoff is evident.

The major drawbacks to varistors can be significant. Varistors do not clamp voltages as effectively as Zener diodes and may therefore by unsuitable in many applications where an absolute maximum voltage limit must be maintained. Also, they

Figure 14.4-2
Varistor circuit symbol.

---

[7] The volt–ampere transfer relationship for a varistor is symmetric. The literature typically provides a relationship that assumes current flow is always positive:

$$V = CI^{(1-a)}.$$

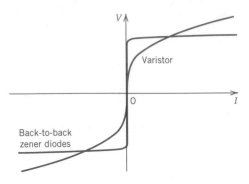

Figure 14.4-3
Volt–ampere relationship for varistor compared to back-to-back Zener diode pair.

can only dissipate relatively small amounts of average power and degrade significantly when stressed near their maximum ratings. They can, however, offer a significant cost advantage over Zener diodes. Varistors are best used in applications for protection against rarely occurring transients.

Selenium rectifiers, in the reverse breakdown mode of operation, can clamp voltages in much the same manner as Zener diodes. An attractive features of many selenium rectifiers is the ability to survive a limited number of surges greater than their maximum typical rating. Unfortunately they do not have as sharp a "knee" in the V–I curve as the Zener diode (though sharper than the varistor) and the ON resistance is somewhat greater. Therefore the use of selenium rectifiers as overvoltage protection devices is diminishing as the demands for protection become more stringent.

Overvoltage "crowbar" circuits protect loads with a switchable shunt element. This element is often realized with a SCR that is activated if overvoltage conditions exist. A simple form of overvoltage crowbar circuit, utilizing a Zener diode as the overvoltage sensing element, is shown in Figure 14.4-4. In this circuit, an input voltage V_{in} that exceeds the SCR gate activation voltage plus the Zener breakdown voltage will force current into the SCR gate activating the SCR. The SCR will then shunt current away from the load protecting it from damage. One drawback of SCR crowbar circuits is that they will not deactivate unless the source current goes to zero. Deactivation of the SCR is usually accomplished by a series circuit breaker or fuse incorporated in series with the source, V_{in}.

While a Zener diode–SCR crowbar circuit is relatively inexpensive and easy to use, there are several drawbacks to its use. These drawbacks are mainly due to the properties of Zener diodes. In particular, the Zener voltage values commercially available are limited, often have insufficient tolerances, and may not activate sharply enough (the knee of the diode curve may be too rounded). These drawbacks are particularly significant when the voltage protection limit must be fairly small; many digital circuits require overvoltage protection at power voltages less than 10 V. A

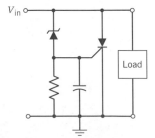

Figure 14.4-4
Simple SCR crowbar circuit.

good solution to these problems involves the use of an integrated circuit sensing circuit.

An overvoltage crowbar circuit using sensing integrated circuit is shown in Figure 14.4-5. The shaded portion of the figure highlights the sensing circuit. This integrated circuit package contain two OpAmps, a voltage reference circuit, a Zener diode, a current source, and three BJTs. The operation of the circuit is reliable and accurate.

In the normal, deactivated state, the input voltage to the protection circuit, V_{in}, is appropriately small:

$$V_{in} < \frac{R_2 + R_1}{R_2} V_{ref}. \tag{14.4-2}$$

This deactivated state assures that the first OpAmp, OA_1, provides a high voltage to the base of Q_1, putting it into the saturation region. Then OA_2 provides a low voltage to Q_2 and Q_3, which are in cutoff. No current is supplied to the gate of the SCR, which is therefore not activated. Should V_{in} increase so that the input voltage fails to meet the constraint of Equation 14.4-2, OA_1 will turn off Q_1. The current source will then pass its current through the Zener diode, raising the positive terminal of OA_2 higher than V_{ref}. The BJT combination Q_2 and Q_3 will turn on and activate the SCR, shunting all the current away from the load.

Using an overvoltage sensing integrated circuit to activate the SCR allows the circuit designer to provide a temperature-independent voltage reference and adjust the crowbar voltage using the two resistors R_1 and R_2. The sensing circuits are readily available from a wide range of manufacturers and are usually found within the "power supply supervisory" listings. It is also common to package a circuit equivalent to that shown in Figure 14.4-5 in a single integrated circuit package. These crowbars are available in a range of voltages and short circuit current ratings.

While it is impossible to demonstrate the operation of every protection circuit, the protection methods discussed here provide a good sampling of typical techniques. Zener diodes provide appropriate transient protection in most small-scale electronic applications. Higher power applications that need protection from longer duration overvoltages usually use crowbar devices. Crowbar circuits are also often

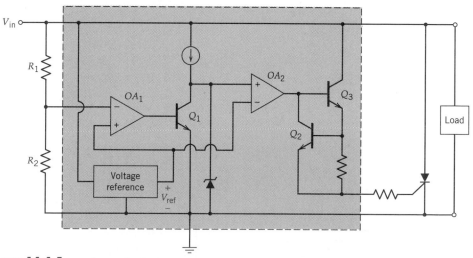

Figure 14.4-5
Crowbar protection using overvoltage sensing circuit.

activated by temperature sensors (often a thermistor) to protect against damage to a circuit due to overheating. Integrated circuit overvoltage sensors occasionally are used to deactivate a semiconductor switch in series with the load rather than the shunt SCR as described here.

14.5 CONCLUDING REMARKS

The study of power electronics should, more properly, be covered in an entire book rather than in one short chapter. Here, the discussion has been restricted to a few common devices that an electronic circuit designer will find necessary in order have clean, constant DC power as an input to circuits with other major electronic functions. Thyristors were shown to be a useful family of triggered switches dominated by the SCR and Triac. Both linear and switched voltage regulators are commonly used: Linear regulators dominate low-power applications, while switched regulators find greater use in high-power applications. Several forms of active and passive protection circuits were also discussed.

SUMMARY DESIGN EXAMPLE

In the design of A/D and D/A converters there is a need for a precision voltage reference that is relatively invariant with input voltage and temperature variation.

Design a 2.5-V precision voltage reference that will operate for input voltages ranging from 4 to 10 V.

Solution:

The three voltage reference topologies available are as follows:

- Zener diode
- Constant-current Zener diode
- Bandgap

The basic Zener diode reference is somewhat dependent on input voltage and must therefore be discarded. The constant-current Zener diode reference is a possibility,

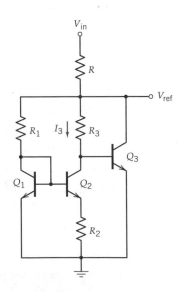

but the relatively small possible difference between input and output voltage may pose a problem. Therefore, the bandgap reference seems the best choice and will form the basis of the chosen design.

As described in Section 14.2, the bandgap reference output voltage is given by

$$V_{ref} \approx \eta V_T \frac{R_3}{R_2} \ln\left(\frac{I_{C1}}{I_{C2}}\right) + V_{BE3}.$$

Assuming that $V_{BE2} = V_\gamma = 0.7$ V, this expression reduces to

$$2.5 = 0.026 \frac{R_3}{R_2} \ln\left(\frac{I_{C1}}{I_{C2}}\right) + 0.7$$

or

$$\frac{R_3}{R_2} \ln\left(\frac{I_{C1}}{I_{C2}}\right) = 69.231.$$

A few arbitrary design choices must be made to continue. Choose

$$I_{C2} = 0.5 \text{ mA} \Rightarrow R_3 = \frac{2.5 - 0.7}{I_{C2}} = \textbf{3.6 k}\boldsymbol{\Omega}.$$

Also choose

$$I_{C1} = 1.0 \text{ mA} \Rightarrow R_2 = \textbf{36.0 }\boldsymbol{\Omega},$$

$$\Rightarrow R_1 = \frac{2.5 - 0.7}{I_{C1}} = \textbf{1.8 k}\boldsymbol{\Omega}.$$

The resistor R must be small enough so that at least 1.5 mA $(I_{C1} + I_{C2})$ will flow at the smallest input voltage. The upper limit on the input voltage merely puts a restriction on the power dissipation. Therefore

$$R < \frac{4 \text{ V} - 2.5 \text{ V}}{1.5 \text{ mA}} = 1 \text{ k}\Omega \Rightarrow \text{Choose } R = \textbf{910 }\boldsymbol{\Omega}.$$

The maximum power dissipated in the resistor R is

$$P_{R(\text{max})} = \frac{(10 - 2.5)^2}{910} = 61.8 \text{ mW}.$$

A $\frac{1}{4}$-W resistor will suffice.

Any high-β_F BJTs can be used for this design. The ideal temperature variation for the BJTs is given by the relationship of Equation 14.2-10. This expression reduces to

$$\frac{\Delta V_{BE3}}{\Delta T} = \frac{11{,}600}{\eta}\left(\frac{R_3}{R_2} \ln\left(\frac{I_{C1}}{I_{C2}}\right)\right) = -5.98 \text{ mV/K}.$$

This is a rather large value. More realistic BJTs will provide a voltage with small but nonzero temperature variation.

■ REFERENCES

Linear Circuits Data Book, Vol. 3: *Voltage Regulators and Supervisors,* Texas Instruments, Dallas, TX, 1989.

Linear/Switchmode Voltage Regulator Handbook, 4th ed., Motorola, Phoenix, AZ, 1989.

Thyristor Device Data Manual, Motorla, Phoenix, AZ, 1992.

Baliga, J.B., Chen, D.Y. (Ed.), *Power Transistors: Device Design and Applications,* Institute of Electrical and Electronics Engineers, New York, 1984.

Cherniak, S., *A Review of Transients and Their Means of Suppression,* Motorola, Phoenix, AZ, 1991.

Fisher, M.J., *Power Electronics,* PWS-Kent Publishing, Boston, 1991.

Horowitz, P. and Hill, W., *The Art of Electronics,* 2nd ed., Cambridge University Press, Cambridge, MA, 1989.

Kassakian, J.G., Schlecht, M.F., and Verghese, G.C., *Principles of Power Electronics,* Addison-Wesley, Reading, MA, 1991.

Millman, J., *Microelectronics, Digital and Analog Circuits and Systems,* McGraw-Hill, New York, 1979.

Mitchell, D.M., *DC–DC Switching Regulator Analysis,* McGraw-Hill, New York, 1988.

■ PROBLEMS

14-1. Model the action of an SCR using the equivalent circuit of Figure 14.1-2*b* and SPICE. Use BJTs with $\beta_F = 100$. Apply a voltage V_s of 18 $V_{\text{p-p}}$ at 60 Hz in series with 100 Ω across the anode–cathode terminals. Into the gate of the SCR model inject 10-mA current pulses of 50 μs duration from a source with an output resistance of 100 kΩ. Verify the operation of the SCR model as shown in Figure 14.1-3.

14-2. A macromodel for the 2N1595 SCR is available in the PSpice model libraries. Its terminals are ordered in the model call as {**anode gate cathode**}. Repeat the functional test of problem 14-1 (the minimum gate trigger current is 2 mA; increase the magnitude of the current pulses so that the SCR will properly trigger).

14-3. A macromodel for the 2N5444 Triac is available in the PSpice model libraries. Its terminals are ordered in the model call as {**MT2 gate MT1**}. Repeat the functional test of problem 14-1 on a 100-$V_{\text{p-p}}$ sinusoid to show that *positive* gate current pulses trigger conduction in both directions (the minimum magnitude gate trigger current is 70 mA; increase the magnitude of the current pulses so that the SCR will properly trigger).

14-4. A macromodel for the 2N5444 Triac is available in the PSpice model libraries. Its terminals are ordered in the model call as {**MT2 gate MT1**}. Repeat the functional test of problem 14-1 on a 100-$V_{\text{p-p}}$ sinusoid to show that *negative* gate current pulses trigger conduction in both directions (the minimum gate magnitude trigger current is 70 mA; increase the magnitude of the current pulses so that the SCR will properly trigger).

14-5. A power conversion system is under design. The input to this system is a standard 110 V AC at 60 Hz. The output is variable-voltage DC supplied to a 100-Ω resistive load. It has been decided to achieve the design goals using a variable-delay one-shot and an SCR as shown.

a. Complete the design using a 2N1595 SCR and a capacitor that will provide no more than 8% ripple.

b. Use PSpice to verify proper operation when the SCR is triggered over 20% of the input sinusoidal waveform.

c. Use PSpice to verify proper operation when the SCR is triggered over 80% of the input sinusoidal waveform.

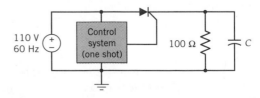

14-6. The power conversion system of problem 14-5 is to be redesigned using a Triac rather than an SCR. This redesign requires the addition of a full-wave rectifier bridge but allows greater efficiency of power conversion.

a. Complete the design using a 2N5444 Triac and a capacitor that will provide no more than 8% ripple.

b. Use PSpice to verify proper operation when the Triac is triggered over 160% of the input sinusoidal waveform.

14-7. The basic Zener diode voltage reference circuit of Figure 14.2-2 is proposed as a voltage reference circuit with an output of 4 V. Assume the Zener diodes are characterized by a Zener voltage $V_z = 4$ V and a Zener resistance $r_z = 40\ \Omega$. The resistor R in the voltage reference circuit has value of 10 kΩ.

a. Determine the nominal output voltage V_{ref} of the circuit to an input $V_{in} = 16$ V.

b. Assume that V_{in} experiences an AC ripple of amplitude 2 V_{p-p}. What is the AC ripple in the output V_{ref}?

14-8. The circuit shown is proposed as a voltage reference circuit with an output of 4 V. Assume the Zener diodes are characterized by a Zener voltage $V_z = 4$ V and a Zener resistance $r_z = 40\ \Omega$.

a. Determine the nominal output voltage V_{ref} of the circuit to an input $V_{in} = 16$ V.

b. Assume that V_{in} experiences an AC ripple of amplitude 2 V_{p-p}. What is the AC ripple in the output V_{ref}?

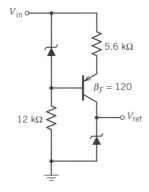

14-9. Design a constant-current voltage reference with an output of 2.0 V. Assume an input voltage ranging from 4 to 6 V, silicion BJTs, and Zener diodes with the following properties:

Minimum Zener current	1 mA
Zener resistance	5 Ω

14-10. A constant-current voltage reference is designed with the basic topology identified in Figure 14.2-3 and the following circuit elements (parameters are found in the PSpice libraries):

BJTs, 2N2222   Zener diode, 1N750
$R_{sc} = 680\ \Omega$ $R = 1.6\ k\Omega$

a. Estimate the nominal output voltage.

b. Assume the input voltage V_{in} varies between 8 and 12 V; use SPICE to determine the variation in the output voltage V_{ref}.

c. Compare SPICE results to theory.

14-11. Design a bandgap voltage reference with an output of 1.2 V. Assume an input voltage of about 5 V. What is the ideal temperature variation of the transistor base–emitter junction voltage?

14-12. Design a bandgap voltage reference with an output of 1.4 V. Assume an input voltage of approximately 12 V. What is the ideal temperature variation of the transitor base–emitter junction voltage?

14-13. Design a 5-V series voltage regulator using the basic topology shown in Figure 14.2-5. The input voltage falls in the range 6 V $< V_{in} < 10$ V. Assume the following components are available:

Precision voltage reference	1.2 V
Comparator	LM 111
BJT	$\beta_F = 100$
Resistors	Any standard value

Verify correct operation using SPICE for a load of 100 Ω. *Note*: The macromodel for the LM111 comparator has an open-collector output; a pull-up resistor is required for HIGH output.

14-14. Design a 3.3-V series voltage regulator using the basic topology shown in Figure 14.2-5. The input voltage falls in the range

$6\ V < V_{in} < 12\ V$. Assume the following components are available:

Precision voltage reference	1.1V
Comparator	LM111
BJT	$\beta_F = 120$
Resistors	Any standard value

Verify correct operation using SPICE for a loads of 100 Ω and 1 kΩ. *Note*: The macromodel for the LM111 comparator has an open-collector output; a pull-up resistor is required for HIGH output.

14-15. Design a 5-V shunt voltage regulator using the basic topology shown in Figure 14.2-6. The input voltage falls in the range $6\ V < V_{in} < 10\ V$. Assume the following components are available:

Precision voltage reference	1.2V
Comparator	LM111
BJT	$\beta_F = 100$
Resistors	Any standard value

Verify correct operation using SPICE for a load of 100 Ω. *Note*: The macromodel for the LM111 comparator has an open-collector output; a pull-up resistor is required for HIGH output.

14-16. Design a 3.3-V shunt voltage regulator using the basic topology shown in Figure 14.2-6. The input voltage falls in the range $6\ V < V_{in} < 10\ V$. Assume the following components are available:

Precision voltage reference	1.2 V
Comparator	LM111
BJT	$\beta_F = 100$
Resistors	Any standard value

Verify correct operation using SPICE for a load of 100 Ω. *Note*: The macromodel for the LM111 comparator has an open-collector output; a pull-up resistor is required for HIGH output.

14-17. Use PSpice and the macromodel for the LM7805C to verify that a +5-V regulator can correctly regulate to achieve +5 V output.

a. Over what range of output current will the LM7805C macromodel provide correct regulaton?

b. What are the minimum input voltages that provide correct regulation for output currents of 10 and 100 mA?

14-18. Use PSpice and the macromodel for the LM7815C to verify that a +15-V regulator can correctly regulate to achieve +15 V output.

a. Over what range of output current will the LM7815C macromodel provide correct regulation?

b. What are the minimum input voltages that provide correct regulation for output currents of 10 and 100 mA?

14-19. Use PSpice and the macromodel for the LM7805C to verify that a +5-V regulator can correctly regulate to achieve −5 V output. Assume the input voltage varies in the range $-7 < V_i < -15$ and a load current of 150 mA.

14-20. Design a 7.5-V voltage regulator using a 5-V, three-terminal regulator and the basic topology shown in Figure 14.3-4. The input voltage falls in the range $9\ V < V_{in} < 15\ V$. Assume the following components are available:

5-V fixed-voltage regulator	LM7805 C (a macromodel is provided in Appendix)
OpAmp	μA741
Resistors	Any standard value

Verify correct operation using PSpice for a load of 100 Ω.

14-21. Design a 16-V voltage regulator using a 12-V, three-terminal regulator and the basic topology shown in Figure 14.3-4. The input voltage falls in the range $20\ V < V_{in} < 30\ V$. Assume the following components are available:

12-V fixed-voltage regulator	LM7812C (a macromodel is provided in Appendix)
OpAmp	μA741
Resistors	Any standard value

Verify correct operation using PSpice for a load of 200 mA.

14-22. An automotive application requires a current of 10 ± 0.1 mA to be delivered to a variable load ($470\ \Omega > R_{load} > 100\ \Omega$). The input voltage to the system varies between 11 and 16 V. Design a constant-current source using a LM7805 5V regulator. Verify proper operation of the design using SPICE.

14-23. The 78XX series of voltage regulators is limited to approximately 1.0 A output current (with appropriate heat sinks). The current boost configuration of Figure 14.3-6 has been selected for a design requiring a regulated output voltage at an output current $I_o \approx 5.0$ A The input voltage is $V_{in} \approx 10$ V. Complete the design by specifying proper resistance values and component power ratings. Assume $\beta_F = 100$.

14-24. The 78XX series of voltage regulators is limited to approximately 1.0 A output current (with appropriate heat sinks). The current boost configuration shown has been selected for a design requiring a regulated output voltage at an output current $I_o \approx 5.0$ A The input voltage is $V_{in} \approx 10$ V. Complete the design by specifying proper resistance values and component power ratings. Assume $\beta_F = 100$.

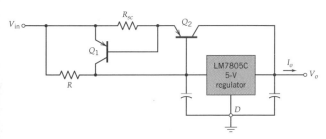

14-25. Complete the design of the overvoltage protection circuit shown by specifying the power ratings necessary for Q_p and Z_p if the input voltage is limited to 50 V. Use SPICE to verify proper operation. Assume $\beta_F = 60$ and a maximum load current of 250 mA.

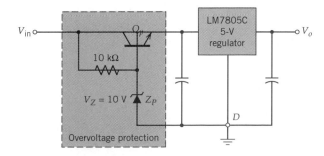

14-26. It is necessary to provide a regulated voltage at 10 V with a maximum current of 100 mA. The input voltage lies in the range 15 V $< V_i <$ 55 V. An LM7805 5-V regulator is the only readily available component. While this regulator can supply adequate current, the improper output voltage and an input voltage limitation of 35 V demand that additional circuitry be added to the design.

    a. Using the basic circuit topology of Figure 14.3-7, design a voltage regulator that will meet specifications. Be sure all component specifications include power ratings. Assume $\beta_F = 60$.

    b. Verify proper operation of the design using PSpice.

14-27. A varistor that has a non-linearity factor, $a = 0.80$ and allows 1 A of current at 50 V is being compared to back-to-back Zener diodes with SPICE parameters IS = 10 nA, IBV = 50 mA, and BV = 15 V.

    a. At what voltage does each voltage protection system allow the same current?

    b. What is the value of that current?

    c. If the current is increased 20%, what voltage appears across each system?

14-28. Back-to-back 1N750 Zener diodes are proposed as a transient protection device for a 5-V circuit. The device being protected is essentially resistive and draws a nominal current of 50 mA. Use SPICE to determine the following:

    a. The maximum voltage that will be applied to the load if an input current twice the nominal value is applied to the protected device

    b. The maximum voltage that will be applied to the load if a current spike of magnitude 150 mA and duration 0.1 μs (rise and fall times $\approx$ 0.1 ns) is applied to the load

14-29. An approximate model of a varistor can be generated in SPICE using a non-linear voltage-controlled current source. The controlling voltage for the source is simply the voltage across the source. The non-linear properties of the varistor can be approximated as an odd-order polynomial in the voltage across the varistor.

a. For a varistor that has a nonlinearity factor $a = 0.80$ that allows 1 A of current at 50 V, determine an approximate expression for the current through the varistor expressed as a fifth-order polynomial of the voltage across the varistor.

b. Use SPICE to plot the $V–I$ transfer relationship of this model.

c. Compare this $V–I$ transfer relationship to that of back-to-back Zener diodes with SPICE parameters IS = 10 nA, BV = 15 V, and IBV = 50 mA.

d. Compare the voltage across each system when subjected to a current consisting of 2 mA constant current that momentarily (duration 1 μs) jumps to 4 mA.

14-30. Design a simple SCR crowbar overvoltage protection circuit that activates at an input voltage of +10 V. Use the following parts

Zener diode	Any specified Zener voltage
Resistors	Any standard value
SCR	2N1595 (gate turn-on voltage ≈ 0.7 V, turn-on current ≈ 2 mA)

Verify correct crowbar operation using PSpice for a load of 100 Ω. Comment on the accuracy of the crowbar voltage and what variation in design might be necessary.

14-31. Design a simple SCR crowbar overvoltage protection circuit that activates at an input voltage of +40 V. Use the following parts:

Zener diode	Any specified Zener voltage
Resistors	Any standard value
SCR	2N1595 (gate turn-on voltage ≈ 0.7 V, turn-on current ≈ 2 mA)

Verify correct crowbar operation using PSpice for a load of 1 kΩ. Comment on the accuracy of the crowbar voltage and what variation in design might be necessary.

CHAPTER 15

COMMUNICATION CIRCUITS

The rapid development of computer technology has been followed by an explosion in the demand for high-speed, low-cost telecommunication products to complement the growing demand for information transfer. New communication systems have been exploited to satisfy the telecommunication needs of the public. For example, the rapid growth in private wireless communication, most notably the cellular telephone, has increased the demand for high-quality, low-cost electronic circuits.

As this chapter will show, many subsystems that make up modern communication systems are designed using the basic principles established in the previous chapters. By integrating functional electronic blocks consisting of simple amplifiers or filters, application-specific communication electronic subsystems can be designed. Although a complete discussion of communication electronics requires volumes of books, this chapter presents an overview of some of the concepts used in designing those circuits.

The discussion in this chapter is limited to circuits operating at modest frequencies (below 20 MHz). In the RF range of frequencies, a different set of two port parameters, called *S parameters*, are used to quantify active and passive devices and circuits. *S* parameters are based on the devices or circuit behavior when terminating the network ports with a known impedance rather than open or short circuits. Another reason for using *S* parameters lies in the dependence of RF and microwave circuit analysis on the reflection (or scattering, hence *S* parameter) of electromagnetic waves.

The topics in this chapter representative the type of circuits commonly used in communications systems. Overviews are presented on analog-to-digital conversion, voltage-controlled oscillators, mixers, phase-lock loops, filter concepts, modulator/demodulator design, and issues important to good receiver design.

15.1 ANALOG-TO-DIGITAL CONVERSION

Most of the chapters in this text have been concerned exclusively with digital or exclusively analog topics. However, many systems use mixed-mode signaling: Mixed-mode systems process both digital and analog signal characteristics. Consequently, the conversion of analog-to-digital signals is a common practice in modern electronic systems.

For example, audio signals in the form of music are converted to digital signals using sophisticated analog-to-digital converters (ADCs) and stored on compact disks (CDs) for sale to the public. Analog telephone voice signals are converted to digital signals for transmission on fiber-optic systems. The future high-definition television (HDTV) signals will probably use some or all digital signaling to transmit high-resolution video images.

As technology advances, more signals from various transducers (microphones, cameras, medical sensors, and other sensors) are converted from analog to digital form for efficient and reliable transmission.

There are many different ADC circuits. Many incorporate digital integrated circuits, which have not been discussed in this book. Therefore, the discussion in this chapter is limited to the Flash ADC circuit. However, prior to describing the Flash ADC, an introduction to analog-to-digital conversion principles is in order.

15.1.1 Quantizing and Sampling

An analog signal voltage range is defined as the maximum peak-to-peak voltage excursion of the signal, as shown in Figure 15.1-1*a*. When the peak-to-peak amplitude of the analog signal is divided into 2^N equal quantization levels and sampled at times $T_0, T_1, T_2, \ldots, T_n$, discrete values of the signal voltage at the sampled times can be found. The result is a digitized form of the analog signal. The analog signal with its associated quantization levels and sampling times is shown in Figure 15.1-1*b*.

The resulting digital reconstruction of the analog signal in Figure 15.1-1 using $N = 3$ quantization levels at sampling times $T_0, T_1, T_2, \ldots, T_n$ is shown in Figure 15.1-2.

When the analog signal is sampled, the voltage level corresponding to the sampling time is held at that quantization level until the next sample causes the level to shift. Also notice that each sampled quantization level has been forced to take a discrete value corresponding to the discrete quantizing levels (in Figure 15.1-2 there are eight levels). Upon completion of the sampling process, the ADC generates a binary code representing the sampled quantization levels. In Figure 15.1-2, the number of discrete quantization levels is 2^N, where $N = 3$, so the ADC generates a binary code of 3 bits/sample. Each discrete signal quantization level in the case of $N = 3$ is $\frac{1}{8}(2V_{peak})$.

15.1.2 Quantizing and Sampling Frequency

The waveform in Figure 15.1-2 is clearly not an exact duplicate of the analog waveform shown in Figure 15.1-1. By low-pass filtering the digitized waveform, the high-frequency components of the signal can be eliminated and the square corners rounded to improve its likeness to the original signal. However, because the sampled signal is forced to take on a discrete, quantized value, the original voltage level,

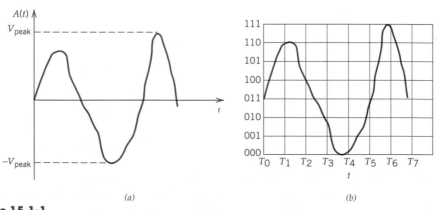

(a)   (b)

Figure 15.1-1
(*a*) Analog signal in time domain. (*b*) Quantizing times and sampling levels.

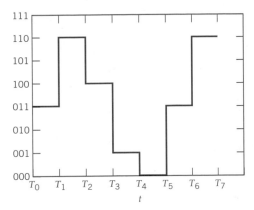

Figure 15.1-2
Digital signal reconstruction of analog signal.

which may lie between the discrete levels, may not be reproduced. In communications systems, this type of distortion is called quantizing error.

The quantizing error can be reduced by generating more discrete voltage levels and reducing the sampling interval. For example, a 4-bit ADC converts an analog signal to 16 discrete quantization levels; 14-bit ADCs converts an analog signal to 16,384 discrete quantization levels. Although advantageous in many respects, high resolution (large number of quantization intervals, 2^N) creates a significant design problem when samping small input signals due to random thermal noise.

Another common form of distortion occurs when the incoming analog waveform is sampled at too low a frequency, corresponding to an unacceptably large sampling interval. This problem is corrected by sampling at frequencies equal to or greater than required by the Nyquist sampling criterion. The Nyquist criterion states that if the highest frequency component of a waveform is f_n, then the waveform must be sampled at least at the Nyquist frequency $2f_n$.

15.1.3 Flash Analog-to-Digital Converters

Common ADC topologies used in modern communications systems include the single-ramp ADC, dual-ramp ADC, charge redistribution ADC, and Flash ADC. All but the Flash ADC require digital logic implementations. The basic strategy of the single-ramp ADC is to convert the analog input voltage to a measurable time interval and transform that time interval to a digital word. The dual-ramp ADC is a higher resolution form of the single-ramp ADC. The operation of the charge redistribution ADC is based on the cycling of charge about a binary weighted array of capacitors.

One of the simplest ADC circuits involves a parallel set of voltage comparators used to compare the incoming signal waveform voltage to a reference voltage. This form of ADC circuit is commonly called a *Flash ADC* and is shown in Figure 15.1-3. The input waveform has been processed by a sample-and-hold circuit like the one shown in Figure 15.1-4.

The function of the sample-and-hold circuit is to track the input and hold the input voltage level to provide a stable output signal value to the ADC during the sampling interval. The basic storage device that "holds" the voltage in Figure 15.1-4 is the capacitor C. The capacitor charges rapidly when the switch is closed. When the switch, usually a FET that is controlled by digital circuitry, is open, the large time constant of the capacitor and the input resistance of the output buffer OpAmp prevents the capacitor from discharging. This provides a stable voltage during the actual conversion process. For the circuit to operate properly, the charging time must be

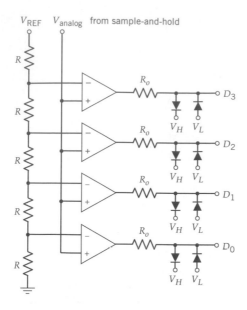

Figure 15.1-3

Four-bit flash ADC with diode clamps to limit output voltage.

very short with respect to the input waveform variations. The charging and discharging times associated with the RC circuit is designed to achieve the desired design accuracy and sampling time interval.

The operation of the parallel bank of comparators in a Flash ADC is based on the processing of the waveform in relation to the known reference voltage from a precision voltage reference.[1] This reference voltage is applied to a series of resistors of equal value that divide the reference voltage into the desired number of quantization levels.

A number of comparators, corresponding to the number of quantization levels, is used to compare the incoming analog waveform to the voltages generated by the precision reference. There is an OpAmp comparator for each voltage reference level. The analog input waveform is applied to the parallel bank of comparators. When the reference voltage for a particular comparator is greater than the input waveform, the output will be negative.

The high parallelism of the Flash ADC has the potential for very high speed analog-to-digital conversion. However, the difficulties presented by this type of ADC include the large number of resistors and comparators required for a large number of quantization levels. Additionally, the output must be encoded into a binary code, which requires additional circuitry. These encoders can be complex and may offset the high-speed advantages of the Flash ADC.

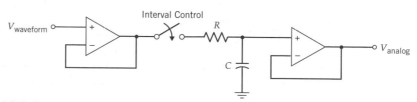

Figure 15.1-4

Sample-and-hold circuit.

---

[1] Precision voltage references are presented in Chapter 14.

Despite the potential drawbacks, the Flash ADC is in wide use and is considered a good design option for many systems where high-speed conversions are required. With the availability of inexpensive OpAmps and high-speed programmable digital logic, Flash ADCs will most likely remain as a good design option for many applications.

15.2 VOLTAGE-CONTROLLED OSCILLATORS

In communications circuits, it is often desirable to have the ability to change the oscillation frequency of an oscillator by applying a voltage to the circuit. Oscillators that allow for control of the frequency of oscillation by an applied voltage are called voltage-controlled oscillators (VCOs).[2]

In practice, the Colpitts oscillator topology is most commonly used to design VCOs in communication electronics. Several phenomena can be exploited to allow voltage control of the oscillation frequency of an electronic oscillator. They include the following:

- Bias control to alter amplifier gain and device capacitance
- Voltage-variable capacitors to alter the circuit resonant frequency
- Voltage-variable resistors to alter gain

The first two of these techniques are discussed in this section. Both arrangements alter the reactive LC feedback network of the oscillator as a function of an input voltage to alter the oscillation frequency. The use of voltage-variable resistors (e.g., FETs operating in the ohmic region) is not commonly found in VCOs for communication electronics applications. They are, however, used in phase-shift oscillator-type VCOs in low-frequency applications.

Bias-Controlled VCO

In Chapter 12, it was shown that the capacitance values of the LC reactive network of a BJT Colpitts oscillator, shown in Figure 15.2-1, is dependent on the gain of the circuit.

In order to achieve a particular frequency of oscillation, the capacitance values C_1 and C_2' are highly dependent on the gain of the BJT circuit. The value of $C_2 = C_2' + C_i$, where C_i is the BJT equivalent input capacitance, is

$$
\begin{aligned}
C_2 &= C_2' + C_i \\
&= C_2' + C_\pi + C_\mu(1 + g_m R) \\
&= C_2' + C_\pi + C_\mu\{1 + g_{m1}[R_{B1} \parallel R_{B2} \parallel R_C \parallel (r_b + r_\pi)]\}.
\end{aligned}
\tag{15.2-1}
$$

The small-signal capacitance values associated with the BJT are

$$
C_\mu = \frac{CJC}{(1 + V_{CB}/0.75)^{0.33}},
\tag{15.2-2}
$$

$$
C_\pi = \left(\frac{g_m}{2\pi f_T}\right) - C_\mu.
\tag{15.2-3}
$$

---

[2] A very simple low-frequency VCO using multivibrators is presented in Section 13.2.

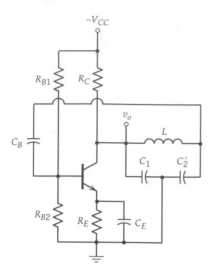

Figure 15.2-1
Colpitts oscillator circuit.

It is evident from Equations 15.2-1 to 15.2-3 that the capacitance C_2 is highly dependent on the small-signal mutual conductance g_m of the transistor. Since the mutual conductance of a BJT is a function of the bias condition, C_2 is a also a function of the bias current. By changing C_2, the oscillation frequency $f_o = (1/2\pi)\sqrt{(C_1 + C_2)/(LC_1C_2)}$ can be altered while maintaining a constant value for C_1 and L.

Therefore, one method of varying the oscillation frequency of a Colpitts oscillator is to change the bias condition of the transistor. Figure 15.2-2 shows one implementation of a Colpitts bias-controlled VCO. The transistor Q_2 is a constant-current source. The bias current is controlled by V_m. The collector current through Q_1 is increased by increasing the control voltage V_m in Q_2. An increase in the bias current

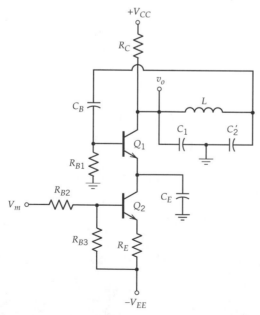

Figure 15.2-2
Bias-controlled BJT Colpitts VCO.

of Q_1 will result in a corresponding increase in the capacitor C_2. The capacitor C_E acts as an emitter bypass capacitor for Q_1. The resistor R_{B1} is a bias resistor for Q_1.

EXAMPLE 15.2-1 Bias-Controlled Colpitts Oscillator Tuning Range

Find the frequency tuning range as a function of the bias control voltage V_m for the BJT Colpitts VCO shown. The bias control voltage range is $0 < V_m < 3$ V. The BJTs have identical parameters.

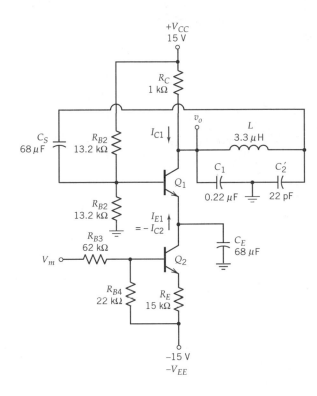

The transistor parameters are

$$\beta_F = 200, \qquad V_A = 200 \text{ V}, \qquad r_b = 30 \ \Omega,$$

$$\text{CJC} = 14 \text{ pF}, \qquad f_T = 250 \text{ MHz}.$$

SOLUTION

Determine the DC condition of the circuit by solving for I_{C1}, V_{CE1}, and V_{CE2} over the tuning voltage range V_m. The quiescent collector current for Q_1 is

$$I_{C1} = \frac{\beta_{F1}}{\beta_{F1} + 1} \frac{\beta_{F2}}{\beta_{F2} + 1} \frac{\dfrac{V_m R_{B4}}{(R_{B3} + R_{B4}) + V_{EE} - V_{\gamma 1}}}{\dfrac{R_E + (R_{B4} \parallel R_{B3})}{(\beta_{F2} + 1)}}.$$

The ranges for the collector current of Q_1, V_{CE1}, and V_{CE2} for bias voltage input of $0 < V_m < 3$ V are

$$0.99 \text{ mA} < I_{C1} < 0.94 \text{ mA},$$

$$7.24 \text{ V} < V_{CE1} < 7.29 \text{ V},$$

$$6.19 \text{ V} < V_{CE2} < 7.69 \text{ V}.$$

These voltages and collector current ranges confirm that the transistors are in the forward-active region of operation.

The mutual conductance and input resistance of Q_1 are

$$g_{m1} = \frac{|I_{C1}|}{V_t} = \frac{|I_{C1}|}{0.026}, \qquad R_i = r_{b1} + r_{\pi 1} \approx r_{\pi 1}$$

$$= \frac{\beta_{F1}}{g_{m1}} = \frac{\beta_F V_t}{|I_{C1}|} = \frac{0.026 \beta_{F1}}{|I_{C1}|}.$$

The capacitance $C_{\mu 1}$ of Q_1 is given as

$$C_{\mu 1} = \frac{\text{CJC}}{(1 + V_{CB1}/0.75)^{0.33}}$$

$$= \frac{\text{CJC}}{\left(1 + \dfrac{15 - I_{C1} R_C - [V_{CC} R_{B2}/(R_{B2} + R_{B1}) - I_{C1}(R_{B1} \| R_{B2})/\beta_{F1}]}{0.75}\right)^{0.33}}.$$

The capacitance $C_{\pi 1}$ of Q_1 is given by

$$C_{\pi 1} = \frac{g_{m1}}{2\pi f_T} - C_{\mu 1}.$$

One of the capacitors, C_2, is dependent on the bias condition of Q_1,

$$C_2 = C_2'' + \frac{g_{m1}}{2\pi f_T} - C_{\mu 1} + C_{\mu 1}[1 + g_{m1}(R_i \| R_{B1} \| R_{B2} \| R_C)],$$

and the frequency of oscillation is

$$f_o = \frac{1}{2\pi} \sqrt{\frac{C_1 + C_2}{LC_1 C_2}}.$$

All of the values above are dependent on the bias condition of Q_1. Using the equations above, the MathCAD routine shown below is used to find the frequency of oscillation as a function of the bias control voltage:

```
C1: = 0.22 · 10⁻⁶   L: = 3.3 · 10⁻⁶
```

$$\text{frequency}(C1, C2, L): \; = \frac{1}{2 \cdot \pi} \cdot \sqrt{\frac{C1 + C2}{L \cdot C1 \cdot C2}}$$

$$i: = 0 \ . \ . \ . \ 30$$

$$Vm_i: = if \ \left(i<1, 0, \frac{90}{i}\right) \quad Vm_i: = 0.1 \cdot i$$

$$IC1_i: = \frac{200^2}{201^2} \cdot \frac{15 + \dfrac{Vm_i 22}{62 + 22} - 0.7}{15000 + \dfrac{16200}{201}} \quad gm_i: = \frac{IC1_i}{0.026}$$

$$C\mu_i: = \frac{14 \cdot 10^{-12}}{\left[\left[1 + \left[\dfrac{15 - IC1_i: \left(1000 - \dfrac{16200}{200}\right)}{0.75}\right]\right]^{0.33}\right]}$$

$$C2_i: = 22 \cdot 10^{-12} + \frac{gm_i}{(2 \cdot \pi \cdot 250 \cdot 10^6)} - C\mu_i + C\mu_i$$

$$\cdot \left[1 + gm_i \cdot \left(\frac{1}{5600} + \frac{1}{1000} + \frac{IC1_i}{2000.026}\right)^{-1}\right]$$

$$fo_i: = frequency \ (C1, C2_i, L)$$

The resulting graph is as shown.

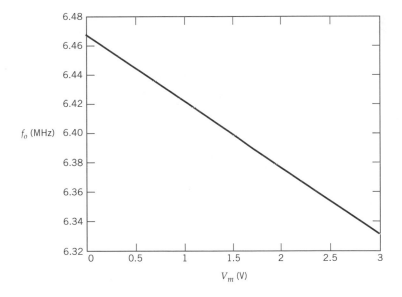

The graph indicates that the VCO tuning range is approximately 6.33 MHz $< f_o <$ 6.47 MHz for $0 < V_m <$ 3V and is quite linear.

Since the mutual conductance g_m varies only by 5%, the gain of the circuit is, for all practical purposes, unchanged. Therefore, the condition for oscillation is satisfied over the bias control voltage range.

Varicap VCO

By far the most common method for controlling the oscillation frequency of an oscillator is with a voltage using a varactor diode. The capacitance of a varactor diode

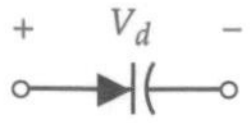

$$+ \quad V_d \quad -$$

Figure 15.2-3
Circuit symbol for varactor diode.

(sometimes called a tuning diode, voltage-variable capacitance diode, or varicap) is dependent on the reverse-bias voltage across the diode. The depletion capacitance of a diode was presented in Chapter 10 and is repeated here for convenience,

$$C_j = \frac{C_{jo}}{(1 - V_d/\psi_o)^m},$$ (15.2-4)

where

$C_{jo} \equiv$ small-signal junction capacitance at zero voltage bias (SPICE parameter CJO)

$\psi_o \equiv$ junction built-in potential (SPICE parameter VJ)

$m \equiv$ junction grading coefficient, $0.2 < m \leq 0.5$ (SPICE parameter M)

The symbol for a varactor diode is shown in Figure 15.2-3.

A common varactor diode is the MV2201. The parameters of interest for this device is $\psi_o = 0.75V$, $C_{jo} = 14.93$ pF, and $m = 0.4261$. A graph of the capacitance as a function of the reverse-bias voltage is shown in Figure 15.2-4.

A BJT-based Colpitts VCO can be designed by connecting a varactor diode in parallel (for AC conditions) to C_2', as shown in Figure 15.2-5. This has the effect of increasing the value of C_2 by adding the varicap capacitance with C_2'.

Increasing C_2 increases the oscillation frequency approximately by the square root of the increase in C_2. Care must be taken so that the voltage at the anode of the diode, V_m, is smaller than the collector voltage V_C of the BJT so as not to affect the bias conditions of the transistor.

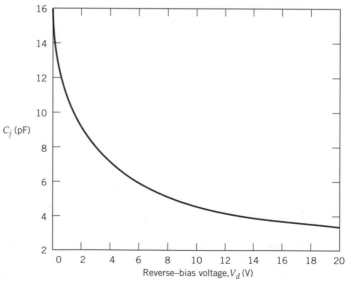

Figure 15.2-4
Capacitance as function of reverse-bias voltage for MV2201 varactor diode.

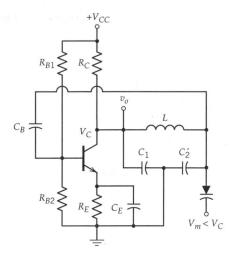

Figure 15.2-5
Varactor diode-based VCO.

EXAMPLE 15.2-2 Tuning Voltage of VCO with Varactor Diode

For the given circuit, what is the range of the tuning voltage V_m so that the VCO will tune over the frequency range of 10.5 MHz $\leq f_o \leq$ 10.7 MHz?

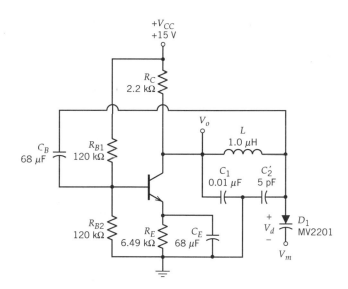

Assume that the transistor parameters are

$$\beta_F = 200, \qquad V_A = 150 \text{ V}, \qquad r_b = 30 \ \Omega,$$
$$C_\mu = 3 \text{ pF}, \qquad f_T = 250 \text{ MHz}.$$

SOLUTION

The DC conditions of this circuit were extablished in Example 12.3-1. The capacitance C_2 was found to be 221 pF.

The oscillation frequency of the VCO without the varactor diode is

$$f_o = \frac{1}{2\pi} \sqrt{\frac{C_1 + C_2}{LC_1C_2}}$$

$$= \frac{1}{2\pi} \sqrt{\frac{0.01 \times 10^{-6} + 221 \times 10^{-12}}{10^{-6}(0.01 \times 10^{-6})(221 \times 10^{-12})}} = 10.82 \text{ MHz}.$$

This implies that for small varicap capacitances, the oscillation frequency approaches 10.82 MHz.

In order to meet the specification for the maximum VCO frequency of 10.7 MHz, $C_2 = 226$ pF. The capacitance value of the varicap must be 226 pF − 221 pF = 5 pF. From the MV2201 characteristic graph of capacitance as a function of reverse-bias voltage in Figure 15.2-4, 5 pF corresponds to $V_d = 9$ V.

For the minimum VCO frequency specification of 10.5 MHz, $C_2 = 235$ pF. The capacitance value of the varicap must be 235 pF − 221 pF = 14 pF. From the MV2201 characteristic graph of capacitance as a function of reverse-bias voltage in Figure 15.2-4, 14 pF corresponds to $V_d = 0$ V.

Since $I_C = 1$ mA (found in Example 12.3-1), the DC voltage at the collector of the BJT is

$$V_C = V_{o,\text{DC}} = V_{\text{CC}} - I_C R_C = 15 - (0.001)(2200) = 12.8 \text{ V}.$$

For a slowly varying voltage V_m, the reverse-bias voltage across the varactor diode is

$$V_d = V_{o,\text{DC}} - V_m.$$

The tuning voltage is then

$$V_m = V_{o,\text{DC}} - V_d.$$

The tuning voltage for a 10.5-MHz oscillation frequency is then

$$V_m = V_{o,\text{DC}} - V_d = 12.8 - 0 = 12.8 \text{ V}.$$

The tuning voltage for a 10.7-MHz oscillation frequency is then

$$V_m = V_{o,\text{DC}} - V_d = 12.8 - 9 = 3.8 \text{ V}.$$

Therefore, the calculated tuning voltage range is $0 \text{ V} < V_m < 3.8 \text{ V}$.

15.3 MIXERS

A mixer uses the nonlinearity of a device to produce intermodulation[3] products. In most applications, a mixer is used to generate the difference frequency between the

---

[3] Intermodulation distortion was discussed in Chapter 7. Intermodulation products are those frequency components that are produced at the output of a device with nonlinear characteristics with two inputs of different frequency.

input signal (commonly called the RF in radio frequency applications) and the local oscillator (called the LO; this is just a stable oscillator circuit). Consider the output from a device with nonlinear characteristics,

$$v_o = V_{DC} + a_1 v_i + a_1 v_i^2 + a_1 v_i^3 + \ldots \qquad (15.3\text{-}1)$$

When two sinusoids are mixed, the input voltage can be represented as

$$v_i = X_1 \cos \omega_{RF} t + X_2 \cos \omega_{LO} t. \qquad (15.3\text{-}2)$$

Substituting Equation 7.5-15 into Equation 7.5.4 yields

$$\begin{aligned} v_o = V_{DC} &+ a_1(X_1 \cos \omega_{RF} t + X_2 \cos \omega_{LO} t) \\ &+ a_2(X_1 \cos \omega_{RF} t + X_2 \cos \omega_{LO} t)^2 + a_3(X_1 \cos \omega_{RF} t \\ &+ X_2 \cos \omega_{LO} t)^3 + \ldots. \end{aligned} \qquad (15.3\text{-}3)$$

By using trigonometric identities, Equation 15.3-3 yields the magnitude and frequency components of the output signal that includes the difference frequency $\omega_{RF} - \omega_{LO}$,

$$v_o = \ldots + a_2 X_1 X_2 \cos(\omega_{RF} - \omega_{LO}) t + \ldots. \qquad (15.3\text{-}4)$$

This desired difference frequency can be recovered using a band-pass filter.

In typical applications, two signals, RF and LO, are applied to the mixer, from which the difference (or other) intermodulation distortion product is selected. In a receiver, the known signal is provided by the LO; the unknown signal is the RF. These two signals are mixed and filtered. This process of intentionally creating intermodulation products is called *heterodyning*. The output of the mixing and filtering process in a receiver is the intermediate frequency (IF).

Heterodyning has many advantages. One of the advantages of heterodyning in a receiver lies in the reduction of operating frequency from the RF. That is, the RF signal is typically a very high frequency signal. By mixing the RF with an LO of a specified frequency, the IF is significantly lower in frequency than the RF, making the circuit design less complex.

The symbol for a mixer is shown in Figure 15.3-1.

15.3.1 BJT Mixers

Typically, the fundamental building block of mixer designs includes small-signal RF amplifiers. Therefore, an understanding of small-signal RF amplifier design is critical in the design of mixers. A small-signal BJT RF amplifier will be used to demonstrate this design technique.

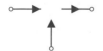

Figure 15.3-1
Symbol for mixer.

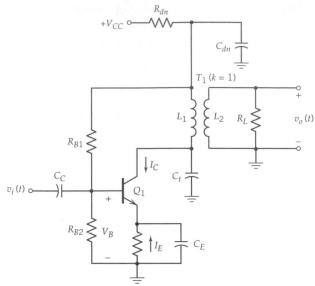

Figure 15.3-2
Small-signal RF amplifier, self-bias configuration.

Small-Signal RF Amplifier Design
As in all electronic circuit design, the design of small-signal RF amplifier involves a great number of choices and exercise in good judgment. Logically derived rules-of-thumb may be applied to narrow the options available to the designer.

For example, a single-transistor RF amplifier may utilize a self-biasing scheme for maximum temperature and device variance stability, as shown in Figure 15.3-2.

The design process for this amplifier is as follows:

- Establish stable quiescent conditions.
- Optimize collector bias currents for maximum output power.
- Decouple the power supplies.
- Establish the value of the DC blocking capacitor.
- Determine the tuning capacitor value to select the desired signal product.
- Determine the effective load at the BJT collector using the transformer Q.

For quiescent point stability defined by a 1% (or less) change in collector current for a 10% change in β_F, the biasing rule-of-thumb is

$$\frac{R_B}{R_E} \leq \frac{\beta_F}{9} - 1, \tag{15.3-5}$$

where

$$R_B = R_{B1} \parallel R_{B2}. \tag{15.3-6}$$

Another rule of thumb sets the voltage across R_E to 10% of V_{CC},

$$R_E = -\frac{V_{CC}}{10 I_E}. \tag{15.3-7}$$

A convenient value for R_{B2} is given as

$$R_{B2} = \frac{V_B}{10I_B} = \frac{V_B\beta_F}{10I_C}, \tag{15.3-8}$$

where $V_B = -I_E R_E + V_{BE}$ and $V_{BE} \approx 0.7$ V.

Optimum collector bias current is based on the maximum output (RF) power P_o and power supply voltage V_{CC}. A transformer-coupled class A amplifier can achieve a maximum efficiency of 50% at full output voltage swing. If V_E is 10% of V_{CC}, then the power supply is providing

$$P_{DC} = 0.9V_{CC}I_C \tag{15.3-9}$$

to the collector of the transistor. For a 50% efficiency,

$$P_{DC} = \frac{P_O}{\eta} = 2P_O, \tag{15.3-10}$$

such that

$$I_C = \frac{2P_O}{0.9V_{CC}}. \tag{15.3-11}$$

For example, if $P_O = 10$ mW and $V_{CC} = 12$ V, $I_C = 20$ mW/$(0.9 \times 12$ V$) = 1.85$ mA. The emitter resistance is then $R_E = 0.1(12$ V$)/1.85$ mA $= 649$ Ω.

Here, R_{dn} and C_{dn} form a low-pass filter called a decoupling network that provides an AC low-impedance point between the collector and base. The decoupling network also isolates the amplifier from the power supply rail from possible feedback to other amplifiers. The RF chokes are not recommended as a replacement for R_{dn} because of potential resonance effects with C_{dn}. A value of $R_{dn} = 100$ Ω is typical for low-power amplifiers and will result in only a few tenths of volts coupled away from the amplifier. The bypass capacitor C_{dn} is chosen for a reactance of one order of magnitude less than the resistance. For example, if the amplifier is used at 455 kHz and $R_{dn} = 100$ Ω, then $X_C \leq 10$ Ω so that $C_{dn} \geq 0.035$ μF. The parameter C_{dn} is in series with C_E (AC path). Because C_E typically has a reactance of a few ohms, it is appropriate to make $C_{dn} = C_E$.

The emitter bypass capacitor C_E provides an AC low impedance for the transistor emitter. It is typically sufficient in RF circuit design (low-impedance circuits) to make X_{CE} an order of magnitude less than the inverse of the transconductance of the transistor,

$$X_{CE} = \frac{1}{10g_m} = \frac{V_t}{10I_C}. \tag{15.3-12}$$

The value of the DC blocking capacitor C_C is determined in the same way as a bypass capacitor except that its reactance should be an order of magnitude less than the amplifier input impedance; that is, $X_{CC} = \frac{1}{10}Z_{in}$, where Z_{in} is the amplifier input impedance. The Z_{in} of an RF amplifier must include all input capacitances.

The capacitor C_t is a tuning capacitance to tune to the proper Q and bandwidth at the IF frequency,

$$C_t = \frac{1}{(2\pi f_{IF})^2 L_1},$$ (15.3-13)

where f_{IF} is the IF frequency.

At the output, the effective quality factor Q_{eff} of the transformer must be determined to determine the load reflected onto the collector of the transistor. First, the reactance of the primary of the transformer, X_{L1}, is determined. Knowing the desired bandwidth of f_{IF}, BW_{IF}, $Q_{eff} = f_{IF}|BW_{eff}$. Then $R'_{L1} = X_{L1} \times Q_{eff}$ and $n_p/n_s = \sqrt{R'_{L1}/R_L}$.

Active Mixer Design

The design of an active mixer is the same as that of the small-signal RF amplifier. The only difference is an additional input for the LO, as shown in Figure 15.3-3.

The coupling capacitor C_{LO} is given as

$$C_{LO} = \frac{1}{2\pi f_{LO} R_{LO}},$$ (15.3-14)

where R_{LO} is the LO source resistance.

15.3.2 Dual-Gate FET Mixer

A dual-gate FET is an n-channel depletion-type FET (commonly a GaAs MESFET) with two independently insulated gate terminals. The MESFETs operate essentially

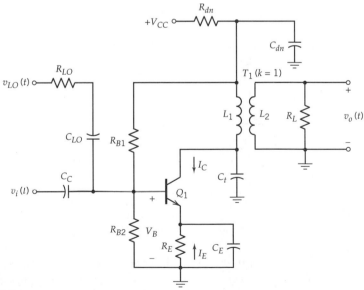

Figure 15.3-3
Active mixer.

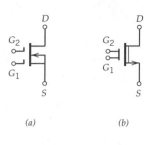

Figure 15.3-4
(*a*) Dual-gate *n*-channel depletion MESFET symbol. (*b*) Simplified dual-gate MESFET circuit symbol.

(*a*)   (*b*)

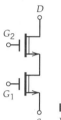

Figure 15.3-5
Dual-gate *n*-channel depletion MESFET equivalent representation.

like a MOSFET but can be used at very high frequencies, on the order of several gigahertz. These FETs have a series arrangement of two separate channels, with each channel having independent gate control. The circuit symbol for the *n*-channel depletion-type dual-gate MESFET is shown in Figure 15.3-4*a*, with the simplified symbol in Figure 15.3-4*b*. Dual-gate MESFETS are used at very high frequencies and are commonly used in mixer, modulator, and automatic gain control (AGC) circuits. When gate 2 is at AC ground, the dual-gate FET may be represented as a common source–common gate (CS–CG) pair, referred to as the cascode circuit, shown in Figure 15.3-5.

In mixer-modulator applications, the LO is applied to gate 1, and the output is taken from the drain (common source configuration). The baseband signal is applied to gate 2 to modulate the LO signal.

A dual-gate *n*-channel depletion MESFET mixer makes use of the isolated gates for good isolation between the LO, and the baseband is shown in Figure 15.3-6. The IF output is transformer coupled for impedance matching and isolation.

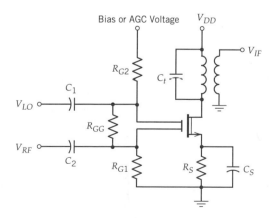

Figure 15.3-6
Dual-gate *n*-channel depletion-type MESFET mixer.

15.4 PHASE-LOCK LOOPS

The phase-lock loop (PLL) is used to track the phase and frequency of a signal. It is often used in the receiver in both AM and FM systems. In addition to AM and FM receiver applications, PLLs are used in control applications, such as in a CD player, to track the rotational speed of the CD.

The PLL is a feedback system as shown in Figure 15.4-1. In the PLL, the feedback signal is intended to follow the phase and frequency of the input signal. However, in the case where the input and feedback signals are not equal, the difference between the two, called the error signal, will change the feedback signal. The feedback signal will continue to change until it again matches the input signal. The feedback quantity that is compared with the input signal is a generalized phase angle $\theta(t)$. A generalized phase angle is composed of a frequency and phase component:

$$\theta(t) = \omega_c t + \phi_c(t), \tag{15.4-1}$$

where ω_c is a set frequency, often called the carrier frequency in communication systems, and $\phi_c(t)$ is a time-varying phase.

The frequency of the output signal of the VCO is adjusted until its generalized phase angle is close to that of the input signal of the PLL. When the generalized phase angles of the input signal and the output of the VCO are nearly identical, the two signals are synchronized. There is, however, a constant phase difference between the two signals due to signal delay through the PLL.

Although several different analog and digital implementations of the PLL are widely used, one of the most common configurations uses a mixer as a phase comparator, a loop filter with a response $K_a H(s)$, and a VCO. The error or difference voltage V_d is the output of the loop filter, which is the controlling voltage to the VCO. For a quiescent VCO frequency of ω_c, the error voltage $V_d = 0$, and the loop is said to be in lock.

Phase Detector

When the PLL is in lock, the output voltage of the phase detector is the difference frequency signal with phase difference, or static phase error, $\phi_e = \phi_c - \phi_o$. If the input signal to the mixer phase detector is $v_c(t) = V_c \sin(\omega_c t + \phi_c)$, the reference signal from the VCO is $v_o(t) = V_o \sin(\omega_o t + \phi)$, and $\omega_c = \omega_o$, then the output signal of the mixer phase detector is

$$V_p(t) = v_c(t)v_o(t) \tag{15.4-2}$$
$$= \tfrac{1}{2}(V_c V_o)K \cos \phi_e \ \tfrac{1}{2}(V_c V_o)K \cos(2\omega_o t + \phi_e),$$

where K is the mixer gain.

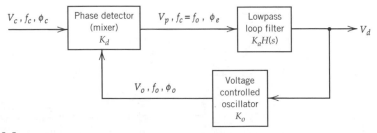

Figure 15.4-1
Block diagram of phase-lock loop.

Since the low-pass loop filter eliminates the second harmonic term from the output of the mixer phase detector, only the first term of Equation 15.4-2 is considered:

$$V_p = \tfrac{1}{2}(V_cV_o)K \cos \phi_d. \tag{15.4-3}$$

For the case where the input frequency f_s is equal to the free-running quiescent frequency of the VCO, f_o, $\phi_e = \tfrac{1}{2}\pi$, and the difference voltage $V_d = 0$. Therefore, the output voltage of the phase detector is also zero. The error signal is proportional to phase differences about 90°. For small changes in phase $\Delta\phi_e$,

$$\phi_e \approx \tfrac{1}{2}\pi + \Delta\phi_e, \tag{15.4-4}$$

and the mixer phase detector output is

$$\begin{aligned} V_p &= \tfrac{1}{2}(V_cV_o)K \cos (\tfrac{1}{2}\pi + \Delta\phi_e) = \tfrac{1}{2}(V_cV_o)K \sin (\Delta\phi_e) \\ &\approx \tfrac{1}{2}(V_cV_o)K \, \Delta\phi_e. \end{aligned} \tag{15.4-5}$$

For small phase perturbation $\Delta\phi_e$,

$$V_p \approx \tfrac{1}{2}(V_cV_oK) \, \Delta\phi_e, \tag{15.4-6}$$

under the assumption that

$$V_p = K_d(\phi_c - \phi_o), \tag{15.4-7}$$

where K_d is the phase detector scale factor, defined as

$$K_d = \tfrac{1}{2}V_cV_oK.$$

The gain of the each of the components of the PLL must be defined in order to find the closed-loop transfer function. When the loop is in lock, the gain factor of the phase detector is

$$K_d = \frac{V_p}{\sin \Delta\phi_e} \text{ V/rad.} \tag{15.4-8}$$

In PLLs the phase $\Delta\phi_e$ is usually designed to be small so that a pulse of noise will not throw the loop out of lock.

Loop Filter

The low-pass loop filter can be passive or active. When passive filters are used, an amplifier with gain K_a is usually required to increase the amplitude of the filtered difference signal. Two low-pass passive filters are shown in Figure 15.4-2. A first- or second-order Butterworth low-pass filter may be used as the active loop filter in the PLL.

For the simple low-pass filter without gain in Figure 15.4-2a, the transfer function is

$$H(s) = \frac{1}{1 + sR_1C}. \tag{15.4-9}$$

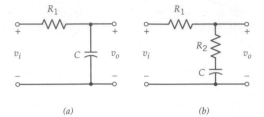

Figure 15.4-2

Two passive low-pass loop filter configurations.

(a)   (b)

For the lag–lead loop filter without gain shown in Figure 15.4-2b, the transfer function is

$$H(s) = \frac{1 + sR_2C}{1 + s(R_1 + R_2)C}. \tag{15.4-10}$$

Voltage-Controlled Oscillator

The output frequency of the VCO is expressed as

$$f_o = f_f + \frac{K_o}{2\pi} v_d, \tag{15.4-11}$$

where

f_o = oscillator output frequency in hertz

f_f = quiescent frequency of VCO in hertz

K_o = VCO voltage to frequency relationship in radians per volt

Equation 15.4-11 can be written in terms of radian frequency.

$$\omega_o = \omega_f + K_o v_d. \tag{15.4-12}$$

The total generalized phase angle of the output of the VCO is

$$\theta(t) = \int_0^t (\omega_f + \Delta\omega)\, dt = \omega_f t + \phi_o(t), \tag{15.4-13}$$

and $\Delta\omega$ is the frequency deviation from ω_f.

Therefore, the phase term is defined as

$$\phi_o(t) = \int_0^t \Delta\omega\, dt, \tag{15.4-14}$$

or

$$\frac{d\phi_o(t)}{dt} = \Delta\omega = K_o v_d. \tag{15.4-15}$$

The different voltage v_d is a DC voltage when the loop is in lock. When the PLL is not in lock, v_d is a voltage corresponding to the difference frequency $(f_c - f_o)$ that draws the VCO into synchronization with the input signal. When the PLL is in lock, the VCO output frequency equals that of the input signal. However, there is a phase difference detected by the phase detector between the VCO and input signals. This difference is called the static phase error ϕ_e. The static phase error is used to maintain the necessary control voltage on the VCO to maintain the required frequency to keep the PLL in lock.

In the s domain, Equation 15.4-15 is

$$\Phi_o(s) = K_o \frac{V_d(s)}{s}, \tag{15.4-16}$$

which clearly shows that the VCO performs as an integrator for phase errors. As an integrator, the VCO helps maintain phase lock through momentary disturbances.

Closed-Loop Transfer Function

The closed-loop transfer function of a PLL is found by determining the ratio $\Phi_o(s)/\Phi_c(s)$, where $\Phi_o(s)$ and $\Phi_c(s)$ are frequency-domain representations of $\phi_o(t)$ and $\phi_c(t)$, respectively. This quantity provides a measure of the loop response to changes in the input phase or frequency. The VCO input signal is

$$V_d(s) = V_e(s)H(s)K_a, \tag{15.4-17}$$
$$V_e(s) = K_d\Phi_e(s). \tag{15.4-18}$$

The DC loop gain is

$$K_v = K_dK_aK_o = \frac{V_e(s)}{\Phi_e(s)} \frac{V_d(s)}{V_e(s)} \frac{\Delta\omega}{V_d(s)} = \frac{\Delta\omega}{\Phi_e(s)}. \tag{15.4-19}$$

Substituting Equations (15.4-17) through 15.4-19 into Equation 15.4-16 yields

$$\Phi_o(s) = K_v \, \Omega_e(s) \, \frac{H(s)}{s}. \tag{15.4-20}$$

The open-loop transfer function is defined as

$$P(s) = \frac{\Phi_o(s)}{\Phi_e(s)} = K_v \frac{H(s)}{s}, \tag{15.4-21}$$

and the closed-loop transfer function is

$$F(s) = \frac{\Phi_o(s)}{\Phi_c(s)} = \frac{P(s)}{1 + P(s)} = \frac{K_vH(s)}{s + K_vH(s)}. \tag{15.4-22}$$

From Equation 15.4-22, it is apparent that the transfer function of the loop filter is a major factor in determining the loop performance. When the filter bandwidth is reduced, the response time to changes in phase or frequency is increased and helps maintain loop lock against momentary disturbances of the input signal.

Other types of phase detectors are also used in commercial PLLs. The type of phase detector circuit selected depends on many factors, including, cost, size, speed, noise performance, and manufacturability. Mixer phase detectors are most commonly used in applications where little VCO frequency deviation from the free-running state is required. For high-speed performance, digital phase detectors using emitter-coupled logic (ECL) circuitry are usually preferred and are commercially available in single-chip PLL packages.

EXAMPLE 15.4-1 Loop Characteristics

For the PLL shown, determine the following:

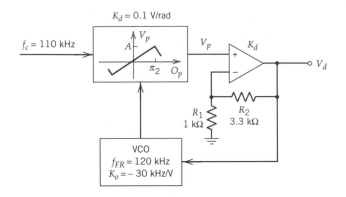

(a) Gain of OpAmp amplifier

(b) Loop gain in units of reciprocal seconds and in decibels at $\omega = 1$ rad/s

(c) VCO output frequency when PLL is phase locked

(d) Static phase error ϕ_e and V_d when PLL is phase locked

SOLUTION

(a) The Op/Amp amplifier is in the noninverting configuration. Therefore the gain of the amplifier is

$$K_a = 1 + \frac{R_2}{R_1} = 1 + \frac{3300}{1000} = \textbf{4.3.}$$

(b) For the loop gain at $\omega = 1$ rad/s

$$K_v = K_d K_a K_o = (0.1 \text{ rad/V})(4.3)(-30 \text{ kHz/rad}) = -12.9 \text{ kHz/V.}$$

Converting -12.9 kHz/V to units reciprocal seconds yields

$$K_v = (-12.9 \text{ kcycles/s-rad})(2\pi \text{ rad/cycle}) = \textbf{81,053 s}^{-1}.$$

Expressed in decibels,

$$K_{v,\text{dB}} = 20 \log(|K_v|) = 20 \log(81{,}053) = \textbf{98.2 dB at 1 rad/s.}$$

(c) When the PLL is locked, the VCO frequency is, by definition $f_o = f_i = 110$ kHz. Only a phase difference can exist between the input signal and the VCO output at the phase decector. This phase difference is called the static phase error, which yields a different voltage output from the loop filter and provides sufficient frequency control voltage to the VCO to maintain frequency lock.

(d) When the PLL is phase locked, the VCO control voltage V_d is given as

$$V_d = \frac{f_c - f_o}{K_o} = \frac{110 \text{ kHz} - 120 \text{ kHz}}{-30 \text{ kHz/V}} = \textbf{0.33 V}.$$

Since the gain of the OpAmp amplifier is 4.3, the output of the phase decector/filter V_p is given by

$$V_p = \frac{0.33 \text{V}}{4.3} = 0.077 \text{ V}.$$

The static phase error is

$$\phi_e = \frac{V_p}{k_d} = \frac{0.077 \text{ V}}{0.1 \text{ V/rad}} - \textbf{0.77 rad}.$$

Hold-In Range

The range of frequencies over which the loop maintains lock is called the *hold-in* range. For a PLL where the amplifier does not saturate and the VCO has a wide frequency range, the phase detector charactcristic limits the hold-in range. As static phase error increases due to increasing input frequency f_c, a limit for the output of the phase detector is reached beyond which the phase detector cannot supply additional corrective control voltage to the VCO. If the phase detector cannot produce more than $V_{p,\max}$, the total range of the phase detector output is $\pm V_{p,\max}$; for a total range the static phase error $\phi_e = \phi_c - \phi_o$ is π radians. The hold-in frequency is the minimum-to-maximum input frequency range $f_{c,\max} - f_{c,\min} = \Delta f_H$,

$$\Delta f_H = \tfrac{1}{2} K_v, \tag{15-4-23}$$

where K_v is in radians per second.

15.5 ACTIVE AND PASSIVE FILTER DESIGN

Signal filtering is often central to the design of many communication subsystems. The isolation or elimination of information contained in frequency ranges is of critical importance. In simple AM radio receivers, for example, the user selects one radio station using band-pass filter techniques. Other radio stations occupying frequencies close to the selected radio station are eliminated. The standard AM broadcast band in North America is from 535 to 1605 kHz.

In Chapter 9, active filter concepts using OpAmps were introduced. One of the advantages of using active filters included the addition of some gain. However, due to their limited gain–bandwidth product, active filters using OpAmps see little use in communication system design.

The two types of frequency-selective circuit configurations most commonly used in communication systems are the passive *LC* filter (low-, high-, and band-pass responses) and the tuned amplifier (band-pass response). In this section, an overview of these two types of frequency-selective circuit configurations for use in band-pass applications is provided.

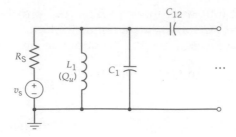

C_{12}

Figure 15.5-1
First resonator of finite-Q band-pass filter.

15.5.1 Passive *LC* Filters

The capacitor-coupled constant-*k* low- and band-pass filter configuration is shown in Figure 15.5-1. This type of filter consists simply of *LC* resonators connected in series: L_1 and C_1 form the first resonator with capacitive coupling provided by C_{12} to the second resonator. The determination of component values begins with the specifications and procedures outlined in Chapter 9, number of resonators or poles required, and required normalized end-section resonator Q of the inductor and source (or load).

Butterworth and Chebyshev band-pass filters for equal input and output terminations are designed using Table 15.5-1. The value q_1 is the required normalized end section Q of the combination of the inductor and source. The value k_{xy}, where $x = 1, 2, 3 \ldots, y = 2, 3, 4, \ldots,$ and $y = x + 1$, is the normalized coupling factor between the resonators that are in series. For example, k_{12} is the coupling factor between the first and second resonators.

The inductor has a finite quality factor Q_u, the unloaded Q of the coil. The resistance of the coil is given by

$$R_{\text{coil}} = X_L Q_u, \tag{15.5-1}$$

where

$$\text{Load reactance} \equiv X_L = \frac{R}{q_1 Q_L}, \tag{15.5-2}$$

$$Q_L \equiv \text{overall circuit loaded } Q = \frac{\text{center frequency}}{\text{bandwidth}} \tag{15.5-3}$$

$$= \frac{f_o}{\text{BW}},$$

$$R = R_S \parallel R_{\text{coil}}. \tag{15.5-4}$$

In practice, since inductors smaller than 50 nH are difficult to handle, the resistance R must be limited by

$$R \geq 2\pi f_o Q_L q_1 (50 \times 10^{-9}). \tag{15.5-5}$$

Since inductors are not ideal and have a finite Q, the circuit of Figure 15.5-1 can be modified to include the effect of the finite Q of the inductor as a coil resistance R_{coil}, as shown in Figure 15.5-2.

TABLE 15.5-1 CIRCUIT DESIGN CONSTANTS FOR CAPACITIVELY COUPLED BUTTER-
WORTH AND CHEBYSHEV FILTERS WITH EQUAL INPUT AND OUTPUT TER-
MINATIONS

n	$q_1 q_n$	k_{12}	k_{23}	k_{34}	k_{45}	k_{56}	k_{67}	k_{78}
			Butterworth (Zero Ripple)					
2	1.414	0.707						
3	1.000	0.707	0.707					
4	0.765	0.841	0.541	0.841				
5	0.618	1.000	0.556	0.556	1.000			
6	0.518	1.169	0.605	0.518	0.605	1.169		
7	0.445	1.342	0.667	0.527	0.527	0.667	1.342	
8	0.390	1.519	0.736	0.554	0.510	0.554	0.736	1.519
			0.01-dB Chebyshev					
2	1.483	0.708						
3	1.181	0.682	0.682					
4	1.046	0.737	0.541	0.737				
5	0.977	0.780	0.540	0.540	0.780			
6	0.937	0.809	0.550	0.518	0.550	0.809		
7	0.913	0.829	0.560	0.517	0.517	0.560	0.829	
8	0.897	0.843	0.567	0.520	0.510	0.520	0.567	0.843
			0.1-dB Chebyshev					
2	1.638	0.711						
3	1.433	0.662	0.662					
4	1.345	0.685	0.542	0.685				
5	1.301	0.703	0.536	0.536	0.703			
6	1.277	0.715	0.539	0.518	0.539	0.715		
7	1.262	0.722	0.542	0.516	0.516	0.542	0.722	
8	1.251	0.728	0.545	0.516	0.510	0.516	0.545	0.728
			0.5-dB Chebyshev					
2	1.950	0.723						
3	1.864	0.647	0.647					
4	1.826	0.648	0.545	0.648				
5	1.807	0.652	0.534	0.534	0.652			
6	1.796	0.655	0.533	0.519	0.533	0.655		
7	1.790	0.657	0.533	0.516	0.516	0.533	0.657	
8	1.785	0.658	0.533	0.515	0.511	0.515	0.533	0.658

TABLE 15.5-1 (continued)

n	$q_1\, q_n$	k_{12}	k_{23}	k_{34}	k_{45}	k_{56}	k_{67}	k_{78}
				1-dB Chebyshev				
2	2.210	0.739						
3	2.210	0.645	0.645					
4	2.210	0.638	0.546	0.638				
5	2.210	0.633	0.535	0.538	0.633			
6	2.250	0.631	0.531	0.510	0.531	0.631		
7	2.250	0.631	0.530	0.517	0.517	0.530	0.631	

Source: Arthur B. Williams, *Electronic Filter Design Handbook* (New York: McGraw-Hill, 1981).

In typical communication electronic applications, R_S and the load resistance R_L are specified and inductors of a known Q_u are used. In this case, the load reactance is

$$X_L = R_S\left(\frac{1}{q_1 Q_L} - \frac{1}{Q_u}\right). \tag{15.5-6}$$

Low-Pass Filter Design

The constant-k configuration is a series of T sections where L_1 is the first component or π sections where C_1 is the first component connected to the source. As expected, the relationships that define the component values for the Butterworth filter are different from those of the Chebyshev filter.

For Butterworth low-pass filters, the component values of the kth section are

$$C_k = \frac{2\,\sin(2k-1)}{\dfrac{2n}{2\pi f_c R}}, \tag{15.5-7}$$

$$L_k = \frac{2R\,\sin(2k-1)}{\dfrac{2n}{2\pi f_c}}, \tag{15.5-8}$$

where $k = 1, \ldots, n$ and f_c is the 3-dB cutoff frequency.

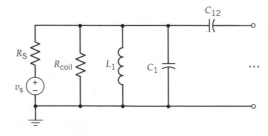

Figure 15.5-2
First resonator of band-pass filter with coil resistance.

The calculations required for the filter component values for Chebyshev filters are more complex than for Butterworth filters. The kth section component values for Chebyshev low-pass filters are found using the following coefficients:

$$g_1 = \frac{2a_1}{\sinh\left(\dfrac{E}{2n}\right)}, \tag{15.5-9}$$

$$g_k = \frac{4(a_{k-1}a_k)}{b_{k-1}g_{k-1}}, \qquad k = 2, 3, \ldots, \tag{15.5-10}$$

$$a_k = \sin\left[\frac{\pi(2k-1)}{2n}\right], \qquad k = 1, 2, \ldots, n, \tag{15.5-11}$$

$$r_{dB} \equiv \text{ripple amplitude} - 20 \log\left(\frac{1}{1+\epsilon^2}\right), \tag{15.5-12}$$

$$E = \ln\left[\coth\left(\frac{r_{dB}}{17.37}\right)\right], \tag{15.5-13}$$

$$b_k = \sinh^2\left(\frac{E}{2n}\right) + \sin^2\left(\frac{k\pi}{n}\right), \qquad k = 1, 2, \ldots, \tag{15.5-14}$$

$$\omega_c = 2\pi f_c \cosh\left[\frac{1}{n}\cosh^{-1}\left(\frac{1}{\epsilon}\right)\right], \tag{15.5-15}$$

$$C_k = \frac{g_k}{\omega_c R}, \tag{15.5-16a}$$

$$L_k = \frac{g_k R}{2\pi\omega_c}. \tag{15.5-16b}$$

Band-Pass Filter Design

The Butterworth and Chebyshev band-pass configurations share the same set of equations used to determine the component values of the filter using coefficients found in Table 15.5-1.

To find the number of resonators required, calculate the shape factor (SF) of the filter response. The shape factor is defined as the ratio of the bandwidth for a given signal attenuation, BW_{xdB}, over the 3-dB bandwidth, BW_{3dB},

$$SF = \frac{BW_{xdB}}{BW_{3dB}} = \frac{f_{xH} - f_{xL}}{f_{3H} - f_{3L}} = \frac{f_o}{BW_{3dB}}\left(\frac{f_{xH}}{f_o} - \frac{f_o}{f_{xH}}\right), \tag{15.5-17}$$

where the subscript H indicates the high frequency corresponding to the particular bandwidth figure and L indicates the low-frequency point of the particular bandwidth.

The inductor values are

$$L = \frac{R}{2\pi f_c q_1 Q_L}. \tag{15.5-18}$$

Since the coupling capacitor must resonate with the inductor at the center frequency of the band-pass filter,

$$C_{xy} = \frac{k_{xy}}{Q_L} C_{node},$$ (15.5-19a)

where

$$C_{node} = \frac{1}{(2\pi f_o)^2 L},$$ (15.5-19b)

and k_{xy} is found in Table 15.5-1. The value of the capacitor C_i is resonant with the inductor with which it is in parallel at the center frequency of the band-pass filter (with all other inductors shorted to ground). Under this condition, the coupling capacitors on either side of the inductor are in parallel with C_i. The value of C_i is therefore the difference between the total capacitance C_{node} at the node and the coupling capacitors. For example, the capacitance values for a three-resonator band-pass filter are

$$C_1 = C_{node} - C_{12},$$ (15.5-20)

$$C_2 = C_{node} - C_{12} - C_{23},$$ (15.5-21)

$$C_3 = C_{node} - C_{23} = C_1.$$ (15.5-22)

EXAMPLE 15.5-1   Design a capacitor-coupled constant-k band-pass filter centered at 455 ± 10 kHz, 1dB maximum ripple, 25 ± 5 kHz 3-dB bandwidth, and 30 dB attenuation at 520 kHz. The source and load resistors are 50 Ω and the available inductors have a quality factor of 60.

SOLUTION

The first task is to find the required order of the filter that directly corresponds to the number of resonators in the filter. The required order is

$$n = \text{SF} = \frac{f_o}{\text{BW}_{3dB}} \left(\frac{f_{xH}}{f_o} - \frac{f_o}{f_{xH}} \right)$$

$$= \frac{455{,}000}{25{,}000} \left(\frac{510{,}000}{455{,}000} - \frac{455{,}000}{510{,}000} \right) = \lceil 4.4 \rceil = 5,$$

where the partial bracket $\lceil \ \rceil$ is a symbol for the operation to round up to the next integer. Using Table 15.5-1 for $n = 5$ and 1 dB ripple, $q_1 = 2.210$, $k_{12} = 0.633$, $k_{23} = 0.535$, $k_{34} = 0.538$, and $k_{45} = 0.633$.

Now find a practical value of R. Use $q_1 = 2.210$ and the loaded circuit Q,

$$Q_L = \frac{f_o}{\text{BW}} = \frac{455{,}000}{25{,}000} = 18.2.$$

Then from Equation 15.5-5,

$$R \geq 2\pi f_o Q_L q_1 (50 \times 10^{-9}) = 5.8 \ \Omega.$$

If the inductors had infinite Q, then $R = 50 \ \Omega$. However, since the inductors used in this design has finite unloaded Q_u, the reactance of the load is calculated using Equation 15.5-6,

$$X_L = R_S\left(\frac{1}{q_1 Q_L} - \frac{1}{Q_u}\right) = 50\left(\frac{1}{2.210(18.2)} - \frac{1}{60}\right) = 0.41.$$

The value of R is then

$$R = R_S \parallel R_{\text{coil}} = \left(\frac{1}{R_S} + \frac{1}{X_L Q_u}\right)^{-1} = \left(\frac{1}{50} + \frac{1}{0.41(60)}\right)^{-1} = 16.5 \ \Omega.$$

Use Equation 15.5-17 to find all inductor values,

$$L = \frac{R}{2\pi f_c q_1 Q_L} = \frac{16.5}{2\pi(455 \times 10^3)(2.210)(18.2)}$$
$$= 0.143 \ \mu\text{H} \approx 0.15 \ \mu\text{H}.$$

To find the capacitors, C_{node} must be calculated using Equation 15-5-19b,

$$C_{\text{node}} = \frac{1}{(2\pi f_o)^2 L} = \frac{1}{[2\pi(455 \times 10^3)]^2(0.15 \times 10^{-6})}$$
$$= 0.816 \ \mu\text{F} \approx 0.82 \ \mu\text{F}.$$

The values of the coupling capacitors are

$$C_{12} = \frac{k_{12}}{Q_L} C_{\text{node}} = \frac{0.633}{18.2}(0.82 \times 10^{-6}) = 0.029 \ \mu\text{F} \approx 0.027 \ \mu\text{F},$$

$$C_{23} = \frac{k_{23}}{Q_L} C_{\text{node}} = \frac{0.535}{18.2}(0.82 \times 10^{-6}) = 0.024 \ \mu\text{F} \approx 0.027 \ \mu\text{F},$$

$$C_{34} = \frac{k_{34}}{Q_L} C_{\text{node}} = \frac{0.538}{18.2}(0.82 \times 10^{-6}) = 0.024 \ \mu\text{F} \approx 0.027 \ \mu\text{F},$$

$$C_{45} = C_{12} = 0.029 \ \mu\text{F} \approx 0.027 \ \mu\text{F}.$$

The values of the tuning capacitors are

$$C_1 = C_{\text{node}} - C_{12} = 0.82 \times 10^{-6} - 0.027 \times 10^{-6}$$
$$= 0.793 \times 10^{-6} \approx 0.82 \ \mu\text{F},$$
$$C_2 = C_{\text{node}} - C_{12} - C_{23}$$
$$= 0.82 \times 10^{-6} - 0.027 \times 10^{-6} - 0.027 \times 10^{-6}$$
$$= 0.766 \times 10^{-6} \approx 0.82 \ \mu\text{F},$$
$$C_3 = C_{\text{node}} - C_{23} - C_{34}$$
$$= 0.82 \times 10^{-6} - 0.027 \times 10^{-6} - 0.027 \times 10^{-6}$$
$$= 0.766 \times 10^{-6} \approx 0.82 \ \mu\text{F},$$
$$C_4 = C_{\text{node}} - C_{34} - C_{45}$$
$$= 0.82 \times 10^{-6} - 0.027 \times 10^{-6} - 0.027 \times 10^{-6}$$
$$= 0.766 \times 10^{-6} \approx 0.82 \ \mu\text{F},$$
$$C_5 = C_{\text{node}} - C_{45} = 0.82 \times 10^{-6} - 0.027 \times 10^{-6}$$
$$= 0.793 \times 10^{-6} \approx 0.82 \ \mu\text{F}.$$

The Chebyshev five-pole band-pass filter design results in the circuit shown.

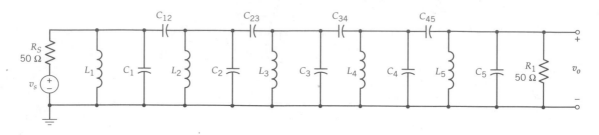

15.5.2 Tuned Amplifiers

Tuned amplifiers are designed to amplify a narrow band of frequencies centered about the center frequency f_o. They can be used as amplifying band-pass filters, in contrast to the passive LC filters previously discussed. Unlike wide-band amplifiers that were presented in previous chapters, the design of tuned amplifiers requires careful design to avoid oscillations due to the reactive load and the internal reactive feedback elements of the transistor. Since tuned amplifiers only operate over a narrow band of frequencies ($\pm 5\%$ of the center frequency), the two-port parameters of the active device at the center frequency are used in the design.

The simplest, and most common, form of tuned amplifier is the single-tuned amplifier. The single-tuned amplifier is designed using a FET, BJT, or OpAmp. Naturally, the OpAmp implementation suffers from bandwidth limitations due to the OpAmp gain–bandwidth product.

The one common denominator found in different types of single-tuned amplifiers is the use of a passive parallel resonant circuit, shown in Figure 15.5-3a using a nonideal finite-Q (unloaded) inductor and capacitor. The nonideal inductor can be modeled as an ideal inductor with a series resistor r, as shown in Figure 15.5.3b.

The relationship between the unloaded Q_u of the inductor, resonant frequency, and the inherent series resistance r is

$$Q_u = \frac{\omega_o L}{r}.$$ (15.5-23)

The admittance of the circuit is

$$Y = j\omega C + \frac{1}{r + j\omega L}$$
$$= \frac{r}{r^2 + \omega^2 L^2} + j\left(\omega C - \frac{\omega L}{r^2 + \omega^2 L^2}\right).$$ (15.5-24)

At the resonant frequency, $\omega = \omega_o$ and the imaginary component goes to zero. Therefore,

$$\omega_o C = \frac{\omega_o L}{r^2 + \omega_o^2 L^2}.$$ (15.5-25)

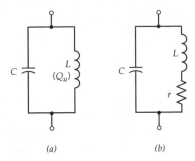

Figure 15.5-3

(a) (b) Passive parallel resonant circuit.

Solving for ω_o yields the expression for the resonant frequency in terms of the passive components,

$$\omega_o = \sqrt{\frac{1}{LC} - \left(\frac{r}{L}\right)^2}. \tag{15.5-26}$$

Using Equation 15.5-23, Equation 15.5-26 can be rewritten in terms of the unloaded Q of the inductor,

$$\omega_o = \sqrt{\frac{1}{LC}\left(\frac{Q_u^2}{Q_u^2 + 1}\right)}. \tag{15.5-27}$$

A finite-Q inductor has the effect of reducing the resonant natural frequency of an ideal LC circuit by a factor of $Q_u^2/(Q_u^2 + 1)$. Therefore, the series rL branch in Figure 15.5-3b can be replaced with a resistor R_p and L_p in parallel, creating a parallel RLC circuit, as shown in Figure 15.5-4.

The parallel equivalent circuit parameters have the values

$$R_p = r(Q_u^2 + 1), \tag{15.5-28}$$

$$L_p = L_s \frac{Q_u^2 + 1}{Q_u^2}. \tag{15.5-29}$$

If $r \ll \omega L$, then

$$R_p = \frac{\omega^2 L}{r} = \omega L Q_u = r Q_u^2. \tag{15.5-30}$$

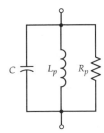

Figure 15.5-4

Parallel equivalent circuit of LC circuit using nonideal inductors.

The impedance of the parallel equivalent circuit is

$$Z(\omega) = \frac{R_p}{1 + jQ_u(\omega/\omega_o - \omega_o/\omega)}. \tag{15.5-31}$$

A single-tuned amplifier uses the parallel LC resonant circuit (commonly called the tank circuit) in place of a load resistor. The small resistance inherent in the nonideal inductor must be taken into account when biasing the circuit. Since the impedance is highest at the resonant frequency f_o, the gain of the circuit peaks at that frequency. For frequencies far from f_o, the load impedance is small, which has the effect of reducing gain.

FET Single-Tuned Amplifier

A simple single-tuned NMOSFET amplifier is shown in Figure 15.5-5. The resonant components of this tuned amplifier are $R_D \parallel r_d$, $C_T \parallel C_o$, and L_T, where C_o is the effective output capacitance of the NMOSFET. The inductor is nonideal with an unloaded Q factor specified by Q_u.

Using the parallel equivalent model of an LC tank circuit, shown in Figure 15.5-4, the small-signal model of the single-tuned NMOSFET amplifier of Figure 15.5-5 is shown in Figure 15.5-6. The resistance R_p is the resistance associated with the inductor.

The voltage transfer characteristic of the circuit is

$$\begin{aligned} A_v = \frac{v_o}{v_i} &= -\frac{g_m}{G + sC + 1/(sL_p)} \\ &= -\frac{g_m}{C}\frac{s}{s^2 + s(G/C) + 1/(L_pC)}, \end{aligned} \tag{15.5-32}$$

where

$$G = \frac{1}{R_p \parallel r_d}, \qquad C = C_o + C_T.$$

The magnitude of the gain is

$$|A_v(j\omega)| = \frac{g_m}{C}\frac{\omega}{\sqrt{\dfrac{(G/C)\omega^2 + 1}{L_pC - \omega^2)^2}}}. \tag{15.5-33}$$

Equation 15.5-33 is maximum when the circuit is operating at the resonant frequency ω_o, with bandwidth ω_{3dB} defined as

$$\omega_o = \frac{1}{\sqrt{L_pC}}, \qquad \omega_{3dB} = \frac{1}{RC}. \tag{15.5-34}$$

The gain at resonance is

$$|A_v(j\omega_o)| = \frac{g_m}{G} = -g_m(r_d \parallel R_p). \tag{15.5-35}$$

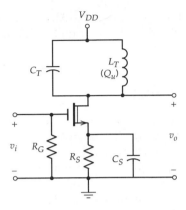

Figure 15.5-5
Single-tuned NMOSFET amplifier.

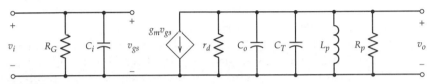

Figure 15.5-6
Small-signal equivalent circuit of NMOSFET single-tuned amplifier.

BJT Single-Tuned Amplifier

A BJT tuned amplifier is shown in Figure 15.5-7. Note that the resonant circuit is at the input of the transistor amplifier.

The current gain of the circuit shown in Figure 15.5-7 is

$$
\begin{aligned}
A_v = \frac{i_o}{i_i} &= -\frac{g_m}{G + sC + 1/(sL_p)}, \\
&= -\frac{g_m}{\dfrac{1 + sRC + R}{sI_p}},
\end{aligned}
\tag{15.5-36}
$$

where

$$
C = C_T + C_i, \qquad R = R_{B1} \parallel R_{B2} \parallel R_P, \qquad G = \frac{1}{R}.
$$

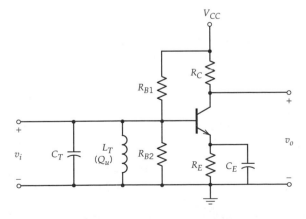

Figure 15.5-7
BJT-tuned amplifier.

Equation 15.5-36 is of the same form as Equation 15.4-32. Therefore, the expression for the resonant frequency and bandwidth of the circuit is identical to that of the FET tuned amplifier.

15.6 MODULATOR/DEMODULATOR DESIGN

Modulation is the process of combining information with a separate waveform to allow efficient transmission. Demodulation is the process of recovering information from a waveform that has been altered to carry that information. There are a variety of modulator and demodulator design options. Modulators can be designed from combinations of mixers, nonlinear amplifiers, VCOs, diode circuits, and filters. Similarly, demodulators can be designed from combinations of electronic circuits. Many of the design options can be explored through discussion of AM and FM.

Amplitude Modulation

In AM systems, it is typical to use mixers in the modulator to shift the baseband (or "raw" information such as audio) signal to a higher frequency to allow for transmission on an assigned carrier frequency. In AM radio receivers, the Federal Communications Commission approved radio frequency range is 540–1600 kHz. Each radio station is allocated a frequency in that frequency range at 10-kHz increments.

In AM modulation, the baseband information alters the amplitude of a sinusoid at a significantly higher frequency than the highest baseband frequency. The high-frequency sinusoid is called the carrier frequency. The heterodyning process is used to demodulate AM signals by shifting the AM radio frequency signal down to a lower IF frequency. Amplitude-modulated radio receivers typically use 455-kHz IF electronics to process the modulated signal.

A simple double-sideband-suppressed carrier (DSBSC) form of amplitude modulation using a mixer is shown in Figure 15.6-1.

In many instances, it is desirable not only to transmit the baseband information that has been up converted to a higher frequency, but also to send the carrier signal. Common examples of this type of AM are AM radio transmission where the station carrier frequency is transmitted with the up-converted information and television signals. One possible method of transmitting a DSB plus a carrier (DSB + C) signal is shown in Figure 15.6-2.

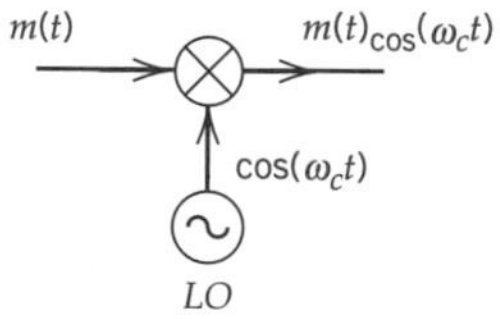

$m(t)$ $m(t)_{\cos(\omega_c t)}$

$\cos(\omega_c t)$

LO

Figure 15.6-1
A DSB-SC mixer modulator.

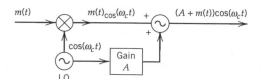

Figure 15.6-2
A DSB + C modulator.

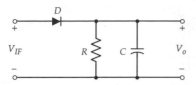

Figure 15.6-3
Envelope detection for AM.

Amplitude Demodulation

An envelope detector is the most common circuit used for converting AM signals to baseband signals. An envelope detector is simply a low-pass filter allowing only the baseband signal to pass while eliminating the IF component generated after a mixer down converts the modulated signal from the high carrier frequency. A simple envelope detector is shown in Figure 15.6-3. The diode allows only those signal greater than 0 V to pass through the detector. The input signal to the envelope detector is AM with carrier signal,

$$V_{IF}(t) = [A + m(t)]\cos \omega_{IF}t, \qquad (15.6\text{-}1)$$

where
$$A = \text{constant DC voltage}$$
$$m(t) = \text{baseband signal}$$
$$\omega_{IF} = \text{IF frequency (e.g., 455 kHz)}$$

An alternate technique to demodulate AM signals uses synchronous detection. The AM signal of Equation 15.6-1 is demodulated by multiplying the signal (using a mixer) by an LO signal of the same carrier frequency as shown in Figure 15.6-4, where $\omega_{IF} = \omega_c$.

The output of the mixer for $\omega_{IF} = \omega_c$ is

$$
\begin{aligned}
v_d(t) &= 2B[A + m(t)]\cos(\omega_{IF}t)\cos(\omega_{IF}t + \phi_o) \\
&= B[A + m(t)][\cos(\omega_{IF}t - \omega_{IF}t - \phi_o) \\
&\quad + \cos(\omega_{IF}t + \omega_{IF}t + \phi_o)] \\
&= B[A + m(t)][\cos \phi_o + \cos(2\omega_{IF}t + \phi_o)].
\end{aligned}
\qquad (15.6\text{-}2)
$$

The mixer output consists of a baseband signal, $B[A + m(t)]\cos \phi_o$, and a modulated signal at the second harmonic frequency of the carrier, $B[A + m(t)]\cos(2\omega_{IF}t + \phi_o)$. The low-pass filter removes the high-frequency component so that the resulting output is the demodulated baseband signal with some gain and DC bias,

$$V_d(t) = B[A + m(t)]\cos \phi_o. \qquad (15.6\text{-}3)$$

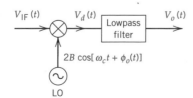

Figure 15.6-4
Synchronous direct conversion amplitude demodulator.

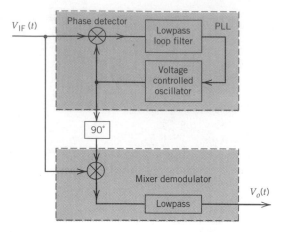

Figure 15.6-5
Phase-locked coherent amplitude demodulator.

Unfortunately, since the input phase angle of the signal is not known, the output voltage can by very small. A phase difference $\phi_o = 0$ yields the maximum output voltage. This implies that the LO must be phase locked to the carrier signal. In this case, the demodulation is truly one of coherent detection, which has superior signal-to-noise performance over noncoherent detection methods such as with the use of envelope detectors. To ensure that the LO is phase locked to the input carrier signal, a PLL can be used. The complete diagram of the PLL amplitude demodulator is shown in Figure 15.6-5. Since the phase detector causes the loop to lock with the VCO 90° out of phase with the input, a 90° phase shifter is placed between the output of the VCO and the mixer demodulator.

The RF frequency phase-shift networks that shift the phase by 90° at a single frequency are usually designed as *LC* circuits.

Frequency Modulation

In FM transmission, the baseband information alters the frequency of a sinusoid at a significantly higher frequency than the highest baseband frequency, whereas in phase-modulation (PM) transmission, the baseband information alters the phase of a sinusoid. One of the most common methods used to generate FM and PM signals is to use a VCO. By applying a time-varying signal to reverse bias a varactor diode in a VCO, an angle-(generalized angle, either phase or frequency) modulated signal can be generated. One cannot distinguish the difference between an FM or PM signal by merely observing its time-domain characteristics.

Phase modulation is characterized by

$$\varphi_{PM}(t) = A(\cos \omega_c t - k_p \sin \omega_c t), \tag{15.6-4}$$

and FM is characterized by

$$\varphi_{FM}(t) = A\left(\cos \omega_c t - k_f \sin \omega_c t \int m(\alpha)\, d\alpha\right), \tag{15.6-5}$$

where φ = voltage or current signal

k_f, k_p = FM and PM constants, respectively

A = a constant

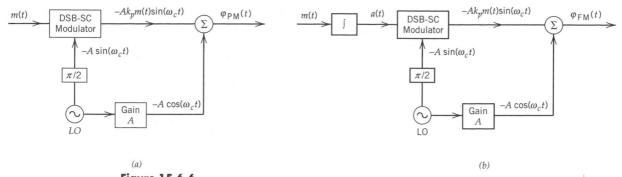

(a)   (b)

Figure 15.6-6

(a) Phase modulator using DSB-SC modulator. (b) Frequency modulator using DSBSC modulator.

Other methods for generating PM and FM signals are shown in Figure 15.6-6, in which DSB-SC modulators are used for generation of the angle-modulated signals.

Frequency Demodulation

There are a number of different methods for demodulating FM signals after down conversion to the IF. One common method uses a differentiator in series with an envelope detector, as shown in Figure 15.6-7. The output signal from an envelope detector is low-pass filtered to provide an output signal proportional (by a constant k_f) to the original baseband signal.

Another common type of frequency demodulator is the discriminator circuit. This frequency demodulator is often called a slope detector since the principle of detection is based on the detection of the signal amplitude variation as it traverses the slope of the resonant response of the circuit, as shown in Figure 15.6-8.

Although conceptually simple, discriminator circuits are sensitive to amplitude variations in the input signal. Variations in the input modulated signal will produce a falsely demodulated baseband signal. One solution is to apply the input modulated signal to a limiter prior to demodulation with a discriminator. However, this requirement has eliminated the use of discriminators in most production entertainment receiver circuits in favor of ratio detectors. In fact, although many integrated circuit receivers incorporate what are called discriminators, they are actually ratio detectors.

The ratio detector, shown in Figure 15.6-9, is similar in operation to discriminators except for its balanced peak-detector configuration, which makes it less susceptible to amplitude variations in the input modulated signal.

The ratio detector operates similarly to a half-wave rectifier with two diodes in series and an RC filter. The radio frequency choke (RFC) is placed in the circuit for DC isolation. The filtered ouput is v_3, which is proportional to $\eta|v_2|$, where v_2 is the IF phasor voltage on the secondary of the transformer and η is the diode empirical scaling constant. The voltages $|v_2|$ and v_3 are essentially independent of frequency over the operating bandwidth of the detector. Since v_3 is divided across the two R_1 resistors and is symmetrical with respect to ground, v_c and v_d are of equal amplitude and

$$v_c = v_d = \tfrac{1}{2}v_3, \tag{15.6-6}$$
$$v_3 = v_c + v_d. \tag{15.6-7}$$

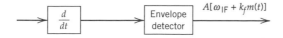

Figure 15.6-7

Frequency demodulation by direct differentiation.

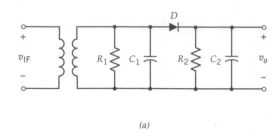

(a)

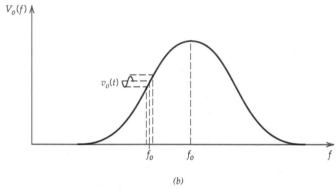

(b)

Figure 15.6-8
(a) Discriminator (slope detection) circuit. (b) Method of demodulating FM signals.

The phasor IF voltage v_a' between the anode of the top diode and ground is

$$v_a' = v_{IF} + v_2, \tag{15.6-8}$$

and the phasor IF voltage v_b' between the cathode of the bottom diode and ground is

$$v_b' = v_{IF} - v_2. \tag{15.6-9}$$

The rectified voltages across the two resistors of value R are

$$v_a = \eta|v_{IF} + v_2|, \tag{15.6-10}$$

$$v_b = \eta|v_{IF} - v_2|. \tag{15.6-11}$$

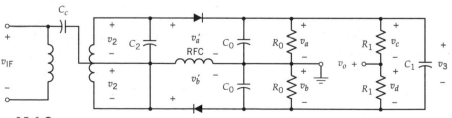

Figure 15.6-9
Ratio detector.

The output voltages across the resistor bridge and across the capacitor C_1 are

$$v_o = v_d - v_b,$$ (15.6-12)
$$v_3 = v_a + v_b.$$ (15.6-13)

Substituting Equations 15.6-6 and 15.6-13 into Equation 15.6-12 yields the output voltage with respect to the voltages across the two resistors of value R

$$v_o = \tfrac{1}{2}v_3 - v_b = \tfrac{1}{2}v_a + \tfrac{1}{2}v_b - v_b = \tfrac{1}{2}(v_a - v_b).$$ (15.6-14)

In steady state, the diodes conduct during a small portion of each IF cycle. During conduction, the capacitors are recharged. The diodes will conduct if the instantaneous IF voltages v_a' and v_b' are greater than the instantaneous voltage v_a and v_b. The peak values of v_a' and v_b' drop and the diode conduction period is cut short when the input signal level drops. For a given FM deviation, $\Delta\omega$, the Q of the circuit on the secondary of the transformer increases, making $\angle v_1/v_2$ smaller. This increases the ratio detector output as $\Delta\omega$ increases.

A PLL can be used to demodulate FM signals. If the PLL is locked onto an input frequency, the control voltage for the VCO from the loop filter/amplitude is proportional to the VCO's shift in frequency from its free-running frequency. The control voltage shifts with a shifting input signal to the PLL. If the input to the PLL is an FM signal, the VCO control voltage from the loop filter/amplifier yields the demodulated output. The PLL is able to demodulate FM signals with a higher degree of linearity than other frequency demodulation techniques.

15.7 RECEIVER DESIGN ISSUES

Many factors are considered to rate receiver performance. The following specifications are commonly applied to a wide variety of communications receivers.

Sensitivity

Sensitivity is a measure of the weakest received signal that allows acceptable reproduction of the original signal. Ultimately, the sensitivity is limited by the noise generated by the receiver electronic circuits. Therefore, the receiver output noise is an important factor in quantifying sensitivity. Sensitivity is defined as the minimum carrier signal input voltage that will produce a specified signal-to-noise power ratio (SNR) at the output of the IF section.

Noise Figure

The noise figure (NF) of a two-port network is a measure of the degradation of the SNR between the input and output terminals. A two-port network with noise is shown in Figure 15.7-1 with input signal power P_{si} and input noise power P_{ni} and corresponding output signal power P_{so} and output noise power P_{no}.

Figure 15.7-1
Two-port network with noisy input and output.

The noise figure as defined over a specified bandwidth is

$$\text{NF} = \frac{\text{input SNR}}{\text{output SNR}} = \frac{P_{si}/P_{ni}}{P_{so}/P_{no}} = \frac{P_{no}}{G_a P_{ni}} = 1 + \frac{P_{ne}}{G_a P_{ni}}, \qquad (15.7\text{-}1)$$

where P_{ne} is the noise power generated by the two-port network.

The value of NF is most commonly expressed in term of decibels:

$$\text{NF}_{\text{dB}} = 10 \log(\text{NF}). \qquad (15.7\text{-}2)$$

For a noise-free network, the input and output SNRs are equal and NF = 1 = 0 dB.

Selectivity

Selectivity is a measure of the receiver's ability to capture a desired station and eliminate unwanted signals. This quality is determined in large part by the frequency response of the frequency-selective circuits of the receiver.

Image Rejection

Image rejection is a measure of the attenuation of unwanted sum and difference frequencies produced by the mixer. Large values are desirable. Typical values are about 50 dB.

Intermediate-Frequency Rejection

The ratio of inputs at the IF and the desired carrier frequency that produce an equal output from the mixer is defined as the IF rejection ratio.

Intermodulation Distortion

Intermodulation distortion (IMD) refers to distortion products created by the non-linear response of the electronic circuit when excited by two (or more) sinusoidal inputs. It produces unwanted signals that may interfere with and corrupt the desired signal.

The common theme that prevails in the receiver qualities mentioned are as follows:

- Frequency selectivity
- Noise
- Distortion

By increasing frequency sensitivity and decreasing noise and distortion in each subsystem, a receiver with good performance can be designed.

15.8 CONCLUDING REMARKS

A sampling of communication electronic circuits were presented in this chapter. Many variants of the circuit discussed here are being used in a variety of communication system applications.

The circuits and concepts discussed in this chapter clearly show that most of the circuits used can be designed and analyzed by the methods discussed thus far in this book. An analog-to-digital converter used OpAmp comparators to yield a digital representation of the time-varying input signal. A transistor mixer circuit is merely a small-signal amplifier with two inputs that take advantage of the nonlinear characteristics of the active device. A VCO was shown to be an oscillator circuit with a voltage-variable reactive network. A PLL is a subsystem composed of a mixer, loop filter/amplifier, and a VCO. The selection of filter configurations was shown to depend on the frequency range of operation, with passive filters used when the frequencies are higher than the audio range. Communications modulators and demodulators are designed using mixers, diode circuits, and PLLs.

SUMMARY DESIGN EXAMPLE: CABLE TELEVISION CONVERTER

It is common to transmit the modulated information on RF and microwave frequencies. Therefore, the received signal must be down converted prior to demodulation. A down converter is required by consumers with non-cable-ready television sets in order to receive cable television signals. Cable television signals typically range in frequency from 55 MHz (channel 2) to 300 MHz to 1 GHz, depending on the number of standard television signals transmitted by the local cable television operator.[4] Since the cable television frequency assignments for channels greater than 13 are different from the broadcast channel frequency assignments, non-cable-ready television sets will not be able to receive channels from the cable that are greater than 13.

To solve this problem, a converter unit is provided to the customer. The purpose of the converter units is to down convert all cable-transmitted channels to channel 3. The output of the converter unit is applied to the television set with its channel selector tuned to channel 3. Cable channel selection is performed by the converter unit. Other functions are built into the cable converter unit including descrambling capability to block reception of premium entertainment channels. The down conversion to channel 3 (61.25-MHz carrier) is accomplished by a mixer. The allocated bandwidth per channel is 6 MHz. The frequency range allocated to channel 3 is 60–66 MHz.

Several mixer topologies can be considered. The two mixer topologies considered for this design are as follows:

- BJT active mixer
- Dual-gate FET mixer

For low cost, an inexpensive BJT active mixer may be most appropriate. Typical RF BJT specifications are

$$\beta_F = 50, \qquad V_A = 150 \text{ V}, \quad r_b = 30\Omega,$$

$$C_{ib0} = 8 \text{ pF} \quad \text{at } V_{EB} = 0.5 \text{ V},$$

$$C_{ob0} = 1 \text{ pF} \quad \text{at } V_{CB} = 5 \text{ V},$$

$$f_T = 5 \text{ GHz} \quad \text{at } 30 \text{ mA}.$$

To down convert channel 7 at a carrier frequency (RF frequency) of 174.25 MHz, the VCO LO is set at $f_{LO} = 61.25 \text{ MHz} + 174.25 \text{ MHz} = 235.5 \text{ MHz}$, resistance $R_{LO} = 200 \ \Omega$. The IF is 61.26 MHz.

---

[4] The frequency range depends on the number of standard National Television System Committee format channels. Operational frequency range may be significantly different when HDTV signals are transmitted.

The mixer is to deliver 10 mW to a 50-Ω load with an efficiency of 50%. The output RF transformer has a primary inductance of 1.0 μH. A 12-V DC power supply is available.

The BJT mixer configuration is as shown.

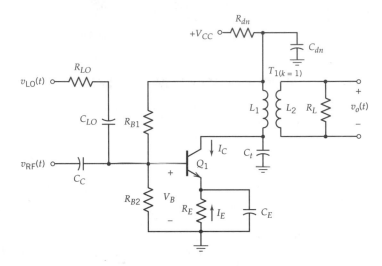

DC DESIGN:

For $V_E = 0.1V_{CC}$,

$$I_C = \frac{2P_O}{0.9V_{CC}} = \frac{2(0.01)}{0.9(12)} = 1.85 \text{ mA}.$$

For stable BJT operation,

$$R_E = \frac{V_{CC}}{10I_E} = \frac{V_{CC}(\beta_F + 1)}{10\beta_F I_C} = \frac{12(51)}{10(50)(0.00185)} = 661 \approx 680\Omega,$$

$$R_{B2} = \frac{V_B}{10I_B} = \frac{V_B\beta_F}{10I_C} = \frac{\beta_F(-I_E R_E + V_{BE})}{10I_C} = \frac{50[0.00189(680) + 0.7]}{0.0185}$$

$$= 5370 \approx 5.6 \text{ k}\Omega.$$

For quiescent point stability defined by a 1% (or less) change in collector current for a 10% change in β_F, the biasing rule-of-thumb is

$$R_B \leq R_E\left(\frac{\beta_F}{9} - 1\right) = 680\left(\frac{50}{9} - 1\right) = 3100 \approx 3.0 \text{ k}\Omega,$$

where $R_B = R_{B1} \parallel R_{B2}$.

Solving for R_{B1} yields $R_{B1} \approx 6.2$ kΩ.

AC DESIGN:

Let the decoupling network resistance $R_{\text{dn}} = 100$ Ω. The reactance of the capacitor C_{dn} must therefore be $X_C < 10$ Ω for 61.25 MHz. Therefore, the decoupling capacitor is $C_{\text{dn}} \geq 260$ pF or $C_{\text{dn}} = 330$ pF.

Since C_E is in series with C_{dn} in the AC path, let $C_E = C_{dn}$.

The emitter bypass capacitor C_E provides an AC low impedance for the transistor emitter,

$$X_{CE} = \frac{1}{10g_m} = \frac{V_t}{10I_C} = \frac{0.026}{10(0.00185)} = 1.4 \ \Omega.$$

Therefore

$$C_E = \frac{1}{\omega_{IF}X_{CE}} = \frac{1}{2\pi(61.25 \times 10^6)(1.4)} = 1850 \text{ pF} \approx 2200 \text{ pF}.$$

The value of the DC blocking capacitor C_C is determined in the same way as a bypass except that its reactance should be an order of magnitude less than the amplifier input impedance; that is, $X_{CC} = \frac{1}{10}Z_{in}$, where Z_{in} is the amplifier input impedance:

$$X_{CC} = \frac{Z_{in}}{10} \approx \frac{R_B \parallel (r_b + r_\pi)}{10} = \frac{3000 \parallel (30 + 703)}{10} = 59 \ \Omega.$$

Therefore,

$$C_C = \frac{1}{\omega_{RF}X_{CC}} = \frac{1}{2\pi(174.25 \times 10^6)(59)} = 15.4 \text{ pF} \approx 15 \text{ pF}.$$

The coupling capacitor C_{LO} is given by

$$C_{LO} = \frac{1}{2\pi f_{LO}R_{LO}} = \frac{1}{2\pi(235.5 \times 10^6)(200)} = 3.3 \text{ pF}.$$

The capacitor C_t is a tuning capacitance to tune to the proper Q and bandwidth at the IF frequency,

$$C_t = \frac{1}{(2\pi f_{IF})^2 L_1} = \frac{1}{[2\pi(61.25 \times 10^6)]^2(1 \times 10^{-6})} = 6.8 \text{ pF}.$$

Since video signals are contained in a nonsymmetric frequency range about the carrier frequency, with the carrier signal positioned at 1.25 HMz above the lower frequency corresponding to the lowest frequency of the channel range. The highest frequency corresponding to the channel range is 4.75 MHz above the carrier. Therefore, the design of the IF mixer bandwidth should contain $f_{IF} \pm 4.75$ MHz. In order to contain those frequencies, a bandwidth of ± 14.25 MHz, which is wider than ± 4.75 MHz, is used. The output from the mixer will be applied to a filter to appropriately shape the video signal.

Knowing the desired bandwidth of f_{IF} and BW_{IF}, the effective quality factor Q_{eff} of the transformer is given as

$$Q_{eff} = \frac{f_{IF}}{BW_{eff}} = \frac{61.25 \times 10^6}{14.25 \times 10^6} = 4.3.$$

Then the reactance of the primary of the transformer X_{L1} is

$$R_{L1} \, X_{L1} Q_{\text{eff}} = 2\pi(61.25 \times 10^6)(10^{-6})(4.3) = 1.65 \text{ k}\Omega.$$

The transformer turns ratio is

$$\frac{n_p}{n_s} = \sqrt{\frac{R'_{L1}}{R_L}} = \sqrt{\frac{1650}{50}} = 5.7.$$

The input impedance of the filter that follows the mixer is 50 Ω.

REFERENCES

Ghausi, M. S., *Electronic Devices and Circuits: Discrete and Integrated,* Holt, Rinehart and Winston, New York, 1985.

Hayward, W. H., *Introduction to Radio Frequency Design,* Prentice-Hall, Englewood Cliffs, NJ, 1982.

Krauss, H. L., Bostian, C. W., and Raab, F. H., *Solid State Radio Engineering,* Wiley, New York, 1980.

Lathi, B. P., *Modern Digital and Analog Communication Systems,* 2nd ed., Holt Rinehart, and Winston, Philadelphia, 1989.

Millman, J., and Halkias, C. C., *Integrated Electronics: Analog and Digital Circuits and Systems,* McGraw-Hill, New York, 1972.

Roden, M. S., *Analog and Digital Communication Systems,* 3rd ed., Prentice-Hall, Englewood Cliffs, NJ, 1991.

Sedra, A. S., and Smith, K. C., *Microelectronic Circuits,* 3rd. ed., Holt, Rinehart, and Winston, Philadelphia, 1991.

Schilling, D. L., and Belove, C., *Electronic Circuits,* 3rd ed., McGraw-Hill, New York, 1989.

Smith, J., *Modern Communication Circuits,* McGraw-Hill, New York, 1986.

Young, P. H., *Electronic Communication Techniques,* 3rd ed., Merrill Publishing, New York, 1994.

PROBLEMS

15-1. Design an 4-bit Flash ADC to digitize an analog signal with a peak-to-peak voltage of 5 V. Assume ideal OpAmps powered by a ±12-V power supply. Include the sample-and-hold circuit.

15-2. A sample-and-hold circuit has a holding capacitor of 50 pF, and the leakage current in the HOLD mode is 1 nA. If the HOLD interval is 50 μs, find the percentage of output decay rate (called *droop*).

15-3. A sample-and hold circuit has a holding capacitor of 100 pF, and the equivalent leakage resistance in the HOLD mode is 15 GΩ. Es-

timate the percentage of output decay rate (droop) if the hold interval is 100 μs.

15-4. In the circuit of Figure 15.1-4, let $R = 15$ kΩ and $C = 500$ pF. The input bias current of the output OpAmp is 300 nA. Estimate the percentage of output decay rate (droop) if the HOLD interval is 1 V.

15-5. Consider an 8-bit Flash ADC. If the voltage supply V_{CC} consists of a DC voltage V_+ and a ripple voltage with peak value $\pm\Delta$, find Δ to ensure that the error produced by Δ affects no bits other than the lowest significant bit (LSB).

15-6. Design a 1-MHz BJT-bias controlled Colpitts 1-MHz VCO with a ± 10-kHz tuning range using a 3.3-μH inductor in the reactive feedback path. Find the frequency tuning range as a function of the bias control voltage V_m for the oscillator. The BJTs have identical parameters:

$$\beta_F = 200, \qquad V_A = 200 \text{ V}, \qquad r_b = 30\Omega,$$

$$CJC = 14 \text{ pF}, \qquad f_T = 250 \text{ MHz}.$$

15-7. Generate the graph of the MV2201 characteristics for reverse-bias voltages of 0–20 V, as shown in Figure 15.2-4. The varactor parameters are $\psi_o = 0.75$ V, $C_{jo} = 14.93$ pF, and $m = 0.4261$.

15-8. Complete the design of the given circuit for a VCO tuning range of $4.95 \text{ MHz} \leq f_o \leq 5.05 \text{ MHz}$. Determine the required range of the tuning voltage V_m.

Assume that the transistor parameters are

$$\beta_F = 200, \qquad V_A = 150 \text{ V}, \qquad r_b = 30 \text{ }\Omega,$$

$$C_{ibo} = 6.5 \text{ pF} \quad \text{at } V_{EB} = 0.5 \text{ V},$$

$$C_{obo} = 3.3 \text{ pF} \quad \text{at } V_{CB} = 5 \text{ V},$$

$$f_T = 250 \text{ MHz}.$$

Simulate the circuit using SPICE and confirm the oscillation frequency.

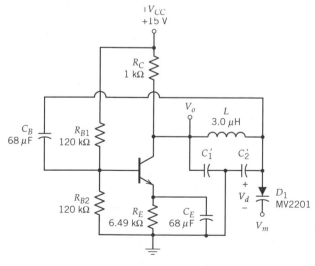

15-9. Design an NMOSFET-based Colpitts VCO with a tuning rage of $1.97 \text{ MHz} \leq f_o \leq 2.03 \text{ MHz}$ using an MV2201

varactor diode. Determine the required range of the tuning voltage V_m. The FET parameters of interest are

$$I_{DSS} = 6 \text{ mA}, \qquad V_{PO} = -4.7 \text{ V}, \qquad V_A = 100 \text{ V},$$

$$C_{iss} = 4.5 \text{ pF} \quad \text{at } V_{GS} = 0 \text{ V},$$

$$C_{rss} = 1.5 \text{ pF} \quad \text{at } V_{GS} = 0 \text{ V}.$$

A +15-V power supply is available.

Simulate the circuit using SPICE and confirm the oscillation frequency. How is the oscillation frequency affected when the temperature of the FET is changed to 50°C?

15-10. Design a 1-MHz Hartley 1-MHz VCO with a ± 10-kHz tuning range using an autotransformer and an MV2201 varactor diode. Find the frequency tuning range as a function of the diode tuning voltage V_m. The BJT parameters are

$$\beta_F = 200, \qquad V_A = 200 \text{ V}, \qquad r_b = 30\Omega,$$

$$CJC = 14 \text{ pF}, \qquad f_T = 250 \text{ MHz}.$$

15-11. Design an active mixer using a BJT with the parameters

$$\beta_F = 200, \qquad V_A = 150 \text{ V}, \qquad r_b = 30\Omega,$$

$$C_{ibo} = 3 \text{ pF} \quad \text{at } V_{EB} = 0.5 \text{ V},$$

$$C_{obo} = 1 \text{ pF} \quad \text{at} V_{CB} = 5 \text{ V},$$

$$f_T > 750 \text{ MHz}.$$

The mixer is to deliver 6 mW to a 50-Ω load with an efficiency of 40%. The RF input frequency is 100 MHz and the IF is 10.7 MHz. The primary inductance of the output transformer is 0.3 mH (infinite Q). The IF bandwidth is 500 kHz. Use SPICE to confirm the operation of the mixer (inspect the frequency spectra).

15-12. Design an active mixer using a BJT with the parameters

$$\beta_F = 200, \qquad V_A = 150 \text{ V}, \qquad r_b = 30\Omega,$$

$$C_{ibo} = 7 \text{ pF} \quad \text{at } V_{EB} = 0.5 \text{ V},$$

$$C_{obo} = 4 \text{ pF} \quad \text{at } V_{CB} = 5 \text{ V},$$

$$f_T = 350 \text{ MHz}.$$

The mixer is to deliver 15 mW to a 75-Ω load with an efficiency of 40%. The RF input frequency is 2 MHz and the IF is 455 kHz. The primary inductance of the output transformer is 0.3 mH (infinite Q). The IF bandwidth is 25 kHz. Use SPICE to confirm the operation of the mixer (inspect the frequency spectra).

15-13. For the dual-gate FET mixer shown, determine:

a. Bias voltage at both gates

b. Drain and source bias voltages for $I_D = 2$ mA

c. Value of L in order to use 10-pF (maximum) variable capacitor at midcapacity if stray and FET drain capacitances total 3 pF for IF of 45 MHz

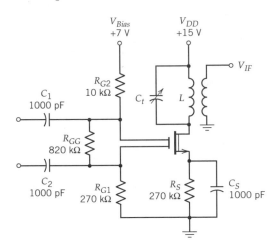

15-14. Define *phase locked*.

15-15. A phase comparator with triangular transfer characteristics has a maximum output voltage of 4 V. Determine the gain in volts per radian and volts per degree of phase.

15-16. An analog phase detector with two input generators (no VCO) has a beat frequency output of 4 V peak-to-peak at 100 Hz. Determine the phase detector gain (sensitivity) in volts per radian and volts per degree of phase.

15-17. A VCO is linear between 250 and 330 kHz. The corresponding input voltages are 220 and −220 mV, respectively.

a. Determine the VCO gain (sensitivity).

b. Determine the free-running frequency.

15-18. For a PLL with $K_d = 0.5$ V/rad, $K_a = -4$, and $K_o = -30$ kHz/V, a VCO free-running

frequency of 200 kHz and a triangular characteristic:

a. Determine the PLL loop gain.

b. Determine the VCO input voltage for the PLL locked to a 180-kHz input signal.

c. Determine the maximum voltage output from the phase detector.

d. Determine the hold-in frequency range.

15-19. Complete the system design of the PLL shown for a loop gain of 100 dB at 1 rad/s. Determine the static phase error ϕ_e and V_d when the PLL is phase locked.

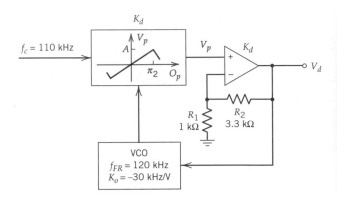

15-20. Design a capacitor-coupled constant-k band-pass filter centered at 10.7 MHz ± 200 kHz, 1 dB maximum ripple, 800 kHz ± 100 kHz 3 dB bandwidth, and 25 dB attenuation at 11.5 MHz. The source and load resistors are 75 Ω, and the available inductors have a quality factor of 70. Confirm the design using SPICE.

15-21. A high-selectivity AM receiver has a 455-kHz IF with a 10-kHz, 3-dB bandwidth requirement. Interference at 427.5 coming through the mixer and RF amplifier must be reduced by 48 dB. Assume a 50-Ω system. The available inductors have a $Q_u = 100$. Design the filter and confirm its operation using SPICE.

15-22. Perform a SPICE simulation of Example 15.5-1.

15-23. Design a capacitor-coupled band-pass filter with a minimum number of resonators, ripple ≤ 0.3 dB, center frequency 2 MHz, 3-dB bandwidth 100 kHz, and 36 dB attenuation at 2169 kHz in a 1-kΩ system. The available inductors have $Q_u = 100$. Confirm the design using SPICE.

15-24. A 2.11-GHz, 28.6-MHz bandwidth satellite receiver must reject an adjacent channel transmitter by 60 dB. Carriers are 36 MHz apart. The filter passband ripple must not exceed 0.5 dB, and the insertion loss must not exceed 4.4 dB. A cavity filter with $Q_u = 1000$ is used.

    a. How many resonators are required?

    b. What is the actual ripple?

    c. What will be the insertion loss?

15-25. Design a single-tuned FET amplifier with a center frequency at 500 kHz, $Q = 50$, and gain $|A_v| > 5$ using an inductor $L = 10$ μII with $Q_u = 200$. The source and load resistors are 1 kΩ. The FET parameters are $I_{DSS} = 5$ mA, $V_{PO} = -2$ V, $V_A = 150$ V, $C_{rss} = 6.5$ pF, $C_{iss} = 35$ pF. Confirm the design using SPICE.

15-26. Design a single-tuned BJT amplifier with a center frequency at 220 kHz, $Q = 50$, and gain $|A_v| > 5$ using an inductor $L = 10$ μH with $Q_u = 200$. The source and load resistors are 1 kΩ. The BJT parameters are $\beta_F = 200$, $f_T = 400$ MHz, $V_A = 200$ V, and $C_{obo} = 2$ pF. Confirm the design using SPICE.

15-27. Complete the design of the envelope detector shown for an IF frequency of 455 kHz, $V_\gamma = 0.7$ V, and $P = 1$ kΩ.

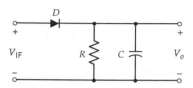

15-28. Design an LC $\pi/2$ phase shifter at an IF frequency of 10.7 MHz. What is the signal attenuation at the IF?

15-29. Design a direct differentiation frequency demodulator for an IF frequency of 1 MHz.

15-30. Derive the total system noise figure NF_{sys} for the two-amplifier system shown. The terms P_{na1} and P_{na2} are the noise powers generated by the two amplifiers. The amplifier gains are G_{a1} and G_{a2}, with corresponding noise figures N_{F1} and N_{F2}.

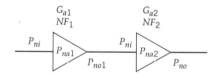

CHAPTER 16

DIGITAL CIRCUITS

A digital electronic circuit is a device that operates on single or multiple input signals to produce an output that is limited to one of a few possibilities. The most common circuits are binary digital circuits: those circuits that have a single output limited to only two output states. The two-state behavior of digital circuits leads to referring to two-state a circuit as a "gate": It is either open or closed, ON or OFF. Gates are often connected in series with other gates. Gates that drive others are "master" gates; gates that are driven are "slave" gates. A single gate can perform both master and slave operations to individual surrounding gates.

Short introductions to the operation of selected binary digital circuits are presented in the introductory chapters on transistor functionality. Chapter 3 discusses the essential operating principles of two bipolar logic families using simple linear models of the BJT. Chapter 4 discusses MOS logic inverters using the principles of load lines. While it is assumed that those chapters are prerequisites to this chapter, several digital circuit operating principles are presented that warrant repetition. Among the most important are the following:

- Logic voltage levels
- Noise margin
- Fan-out

The output of a digital circuit is characterized by two voltage levels: a logic HIGH voltage and a logic LOW voltage. These voltages are symbolized as V_{oH} and V_{oL}, respectively. The input is also characterized by two voltage levels: the level above which all inputs are a logic HIGH, $V_{iH(\text{min})}$, and the level below which all inputs are a logic LOW, $V_{iL(\text{max})}$. For purposes of noise immunity it is important that the output and input voltage levels not be the same. Specifically,

$$V_{oH} > V_{iH(\text{min})}, \qquad V_{oL} < V_{iL(\text{max})}.$$

The measures of this noise immunity are the *noise margins* NM(HIGH) and NM(LOW):

$$\text{NM(HIGH)} = V_{oH} - V_{iH(\text{min})},$$
$$\text{NM(LOW)} = V_{iL(\text{max})} - V_{oL}.$$

A descriptive diagram of the noise margins as well as slave input and master output voltage levels is shown in Figure 16-1.

Fan-out is a measure of the number of similar slave gates that a master gate can

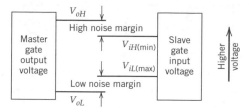

Figure 16-1
Physical interpretation of noise margin.

drive without producing logical errors. Typically current loading determines the fan-out of a gate, but as in the case of MOS gates, gate speed can be the determining factor. Fan-in is another term often found in the literature: It identifies the number of gate inputs.

While it is impossible to completely describe all types of digital circuits and all aspects of their operation in a single chapter, this chapter strives to present the *essential aspects* of the major digital circuits commonly in use. The fundamentals of the speed of digital logic transitions begins the discussion of both bipolar and MOS gates. Three families of bipolar gates and two families of MOS gates are discussed extensively. Regenerative logic circuits (latches, flip-flops, and Schmitt triggers) are presented as are the fundamentals of memory circuits. The chapter ends with a descriptive section on gallium arsenide logic circuits.

16.1 SWITCHING SPEED OF BJTS

Bipolar digital logic gates depend on the transition of the output of a BJT from one logic level to another. In the case of classic TTL logic gates this transition takes place between the cutoff and saturation regions of the output transistor. The speed at which any gate operates is limited by the transition speeds of its constituent transistors. Similarly, the maximum clock rate at which a synchronous digital system will operate is limited by transistor transition speed. The transition speed of a transistor is a function of the physical parameters of the transistor and of the components in the circuit surrounding the transistor. While it is impossible to analyze the effect of all possible circuit topologies on switching speed, an analysis of the switching speed of the simple BJT inverter provides much insight into all BJT switching.

As shown in Figure 16.1-1, a general BJT inverter consists of a common-emitter BJT with Thévenin sources connected to the base and collector terminals. These Thévenin sources represent the surrounding logic gate circuitry.

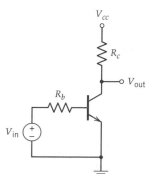

Figure 16.1-1
Simple BJT inverter.

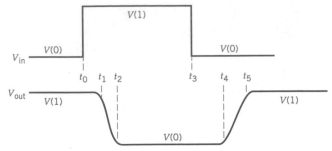

Figure 16.1-2
Response of inverter to digital pulse.

As the input voltage changes between logic levels, the output will change to the opposite logic levels. This transition cannot take place instantaneously; various delays must occur. Conceptually, the response of a simple BJT inverter to a rectangular pulse is shown in Figure 16.1-2.

The pulse response consists of regions of constancy and regions of transition. In the regions of transition, there are four significant time periods:

- Delay time $t_d = t_1 - t_0$
- Rise time $t_r = t_2 - t_1$
- Storage time $t_s = t_4 - t_3$
- Fall time $t_f = t_5 - t_4$

The delay time is the time between the pulse transition and when the response transitions 10% of the distance between HIGH and LOW states. The rise time is the time for a 10–90% transition from HIGH to LOW. The term *rise time* refers to the BJT collector current change: As the voltage transitions from HIGH to LOW, the collector current rises from a minimum to a maximum value. The storage time and fall time measure the equivalent time periods in the LOW-to-HIGH transition.

Rise Time
Perhaps the easiest transition region to analyze is the rise time. Here the BJT is in the forward-active region of operation. In this region of operation, the transistor speed is most often described by the forward time constant τ_F or by its frequency-domain equivalent, the unity-gain frequency ω_T:[1]

$$\tau_F \approx \frac{1}{\omega_T}. \qquad (16.1\text{-}1)$$

While the forward time constant is a useful parameter (e.g., it is necessary in SPICE analysis), a more relevant parameter for gate speed calculations is the forward-active region time constant τ_f:

$$\tau_f = \beta_F \tau_F \approx \frac{\beta_F}{\omega_T} = \frac{1}{\omega_{3dB}}. \qquad (16.1\text{-}2)$$

---

[1] The unity-gain frequency is defined as the radian frequency at which the common emitter gain is unity. It is discussed fully in Section 10.4.

Transitions between two steady-state levels X_i and X_f in the forward-active region are described by a simple exponential relationship with this time constant:

$$X(t) = X_f - (X_f - X_i)e^{-(t-t_0)/\tau_f}.$$ (16.1-3)

It can easily be shown that the rise time of a simple exponential transition is

$$\text{Rise time} \approx 2.2\tau_f.$$ (16.1-4)

Unfortunately, a transition entirely in the forward-active region is not the type of transition typical of BJT logic gates. In a BJT gate, the LOW logic level is characterized by the saturation region of the BJT. Thus the apparent final steady-state level is not the same as the actual final level. The BJT collector rise is consistent with a final collector current $I_{cf} = \beta_F I_b$ and the time constant τ_f. The transition is completed when the collector current reaches the saturation collector current $I_{c(\text{sat})} < \beta_F I_b$. A significantly reduced rise time results. Figure 16.1-3 is a graphical interpretation of the reduction in rise time.

The rise time can be calculated by determining the 90 and 10% times t_2 and t_1, respectively. Assume the external circuit parameters define the two currents $I_{c(\text{sat})}$ and I_{b1}:

$$I_{c(\text{sat})} = \frac{V_{cc} - V_{ce(\text{sat})}}{R_c},$$ (16.1-5)

$$I_{b1} = \frac{V_i - V_{be}}{R_b}.$$ (16.1-6)

For a saturated LOW output, define the current ratio N_1 as

$$N_1 = \frac{\beta_F I_{b1}}{I_{c(\text{sat})}}.$$ (16.1-7)

Here, N_1 is an indicator of the degree in which the BJT has been driven into the saturation region. It is called the *saturation overdrive factor*. Values of N_1 near unity

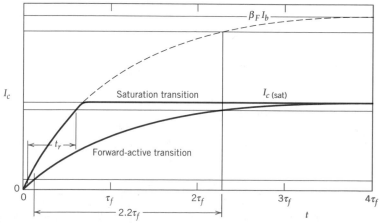

Figure 16.1-3
Saturated transition rise time.

indicate that the BJT is barely into the saturation region; larger values indicate a large excess of base current over what is necessary to saturate the BJT. The time for the collector current to achieve 90% of the saturated collector current is the solution to the expression

$$0.9I_{c(\text{sat})} = N_1 I_{c(\text{sat})}(1 - e^{-t_2/\tau_f}). \tag{16.1-8}$$

Similarly, the time to achieve 10% of the saturated collector current is the solution to

$$0.1I_{c(\text{sat})} = N_1 I_{c(\text{sat})}(1 - e^{-t_1/\tau_f}). \tag{16.1-9}$$

The rise time t_r is the difference between t_2 and t_1:

$$t_r = t_2 - t_1 = \tau_f \ln\left(\frac{N_1 - 0.1}{N_1 - 0.9}\right). \tag{16.1-10}$$

As N_1 increases, the rise time is made smaller. This seems to indicate that a strongly saturated BJT is desirable in terms of gate speed. Other factors will show that strong saturation is not desirable.

Delay Time

The delay time indicates the time between the input signal transition and the start of the rise time. Its primary components[2] are the following:

- t_{d1}, the time for minority carriers to transit the base and reach the collector
- t_{d2}, the time for the collector current to rise to 10% of its final value

Statistical analysis shows that the first of these factors is given by

$$t_{d1} \approx \frac{\tau_F}{3} = \frac{\tau_f}{3\beta_F}. \tag{16.1-11}$$

The second factor is a portion of the exponential collector current change and can be derived in the same fashion. The 0–10% portion of the collector current rise takes place in time:

$$t_{d2} = \tau_f \ln\left(\frac{N_1}{N_1 - 0.1}\right). \tag{16.1-12}$$

The delay time is the sum of these factors:

$$t_d = t_{d1} + t_{d2} \approx \tau_f\left[\ln\left(\frac{N_1}{N_1 - 0.1}\right) + \frac{1}{3\beta_F}\right]. \tag{16.1-13}$$

---

[2] Another factor is the time required to charge the base–emitter junction to the cut-in voltage. Here it is assumed that this factor is small compared to the other delay time factors.

Fall Time

The fall time t_f is analogous to the rise time. It indicates the time for the BJT collector current to fall from 90 to 10% of its saturated value. This fall in collector current occurs as the transistor transitions in the forward-active region from saturation to cutoff. The expression for fall time is therefore

$$t_f = t_4 - t_3 = \tau_f \ln\left(\frac{N_2 - 0.9}{N_2 - 0.1}\right). \tag{16.1-14}$$

Here a reverse overdrive factor, N_2, is defined as the ratio of β times the instantaneous base turn-off current to the saturation collector current:

$$N_2 = \frac{\beta_F I_{b2}}{I_{c(\text{sat})}}, \tag{16.1-15}$$

where

$$I_{b2} = \frac{-V_{be(\text{active})}}{R_b}. \tag{16.1-16}$$

Notice that N_2 is a *negative* quantity. Large-magnitude N_2 indicates a short fall time.

Storage Time

The storage time indicates the time between the input signal transition and the start of the fall time. Its primary components[3] are as follows:

- t_{s1}, the time for minority carriers to transit the base and reach the collector
- t_{s2}, the time for the collector current to rise to 10% of its final value
- t_{s3}, the time to dissipate the excess charge stored in the base of the saturated BJT

The first two factors are direct analogs of similar components in the delay time:

$$t_{s1} \approx \frac{\tau_F}{3} = \frac{\tau_f}{3\beta_F}, \tag{16.1-17}$$

$$t_{s2} = \tau_f \ln\left(\frac{N_2 - 1}{N_2 - 0.9}\right). \tag{16.1-18}$$

The third factor, t_{s3}, is related to an exponential decay in the base charge when the BJT is in the saturation region. This decay has a time constant τ_s that is a function of the forward-active region time constant τ_f and its inverse-active region counterpart τ_r:

$$\tau_s \approx \tau_f + \beta_R(\tau_f + \tau_r). \tag{16.1-19}$$

---

[3] Another factor is the time required to charge the base–emitter junction to the cut-in voltage. Here it is assumed that this factor is small compared to the other delay time factors.

This dual dependence is due to the forward-bias condition of both the base–emitter and the base–collector junctions of a saturated BJT. As a consequence, the saturation time constant is significantly longer than either of the other time constants. The time for base charge dissipation is given by

$$t_{s3} = \tau_s \ln\left(\frac{N_1 - N_2}{1 - N_2}\right). \tag{16.1-20}$$

The total storage time is given by the sum of the individual components:

$$t_s \approx \frac{\tau_f}{3\beta_F} + \tau_f \ln\left(\frac{N_2 - 1}{N_2 - 0.9}\right) + \tau_s \ln\left(\frac{N_1 - N_2}{1 - N_2}\right). \tag{16.1-21}$$

Large-magnitude N_2 (a negative quantity) will decrease the storage time. However, large N_1 will significantly *increase* the storage time. This is especially significant since the storage time constant is larger than the forward-active region time constant $\tau_s > \tau_f$. If the transistor does not enter the saturation region ($N_1 = 1$), $t_{s3} = 0$ and the storage time is composed of only the first two components of Equation 16.1-21.

Summary

While the switching speed of a BJT is largely dependent on the physical parameters of the transistor itself, the surrounding circuit parameters also have significant effect. Rise time and fall time are strongly dependent on the overdrive factors N_1 and N_2, respectively. In each case, an increase in the magnitude of the overdrive factor reduces the respective time. Delay time is also decreased by large N_1, as is storage time by large-magnitude N_2. Unfortunately storage time can be greatly *increased* by large N_1 (saturated BJTs). The propagation delay of a TTL inverter can be described in terms of the transition times derived. Its component terms are

$$t_{PHL} \approx t_d + \tfrac{1}{2}t_r, \qquad t_{PLH} \approx t_s + \tfrac{1}{2}t_f. \tag{16.1-22}$$

The average propagation delay is given by

$$t_{PD} = \tfrac{1}{2}(t_{PHL} + t_{PLH}) \approx \tfrac{1}{2}(t_d + t_s) + \tfrac{1}{4}(t_r + t_f). \tag{16.1-23}$$

16.2 BIPOLAR DIGITAL GATE CIRCUITS

Bipolar logic gates are divided into four fundamental logic gate families: diode–transistor logic (DTL), transistor–transistor logic (TTL), emitter-coupled logic (ECL), and integrated injection logic (I^2L). While the original form of DTL is now obsolete, these four families of gates form the foundation for all modern bipolar logic gates. Advancements in the design of these gates that have taken place are generally for the purpose of improving gate performance in one or more of the following areas:

- Speed of operation
- Power consumption
- Noise rejection
- Fan-out

Unfortunately, an improvement in one aspect of gate performance may degrade performance in another. Good design is a balance of often-conflicting design constraints: Modern logic gate design is an example of the balancing of these constraints. It is the purpose of this section to explore some of the designs common in digital gates.

The operation principles of several basic digital logic gates that use BJTs are presented, using quasistatic analysis techniques, in Section 3.5. Power consumption, fan-out, and many aspects of noise rejection can be analyzed using these simple, linear techniques. The speed at which a bipolar logic gate operates can be determined using the basic techniques presented in Section 16.1. Since TTL found its origins in DTL, modern design in the two logic families will be discussed together. Emitter-couple logic and I^2L will be discussed separately.

16.2.1 TTL and DTL Logic Gates

The output voltage levels of a common DTL and TTL logic circuit are developed as the output BJT switches between two regions of operation: cut-off and saturation. It is these two stable states that make these families of gates reliable and predictable.

The NAND gate forms the fundamental logical unit in both DTL and TTL logic. All other gate logic operations can be derived from this fundamental unit. The basic circuit topology of a simple TTL NAND logic gate is shown in Figure 16.2-1. This gate is typically operated with a supply voltage of 5 V and consists of several resistors and three BJTs. The input BJT, Q_1, is constructed with multiple emitters (three emitters are shown; other numbers are common) that serve as the individual logic inputs to the gate. When any of the logic inputs is a logic LOW, Q_1 enters the saturation region. There is an insufficiently high voltage at the base of Q_2 to forward bias the base–emitter junctions of Q_2 and Q_3; thus Q_2 and Q_3 are OFF and a logic HIGH output is produced. A LOW logic output occurs when all inputs are HIGH. When this input condition occurs, Q_1 enters the inverse-active region.[4] Then Q_2 turns ON and forces Q_3 into the saturation region, producing a logic LOW. Depending on exact resistor values and BJT parameters, Q_2 typically enters the saturation region, al-

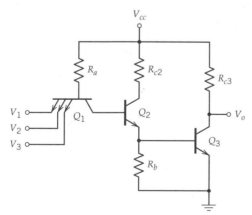

Figure 16.2-1
Basic TTL NAND gate.

---

[4] In many situations (large fan-out), the master gate cannot supply enough current to sustain the inverse-active region of Q_1 in the slave gate. In that case Q_1 is in an *inverse* saturation state: Both junctions are forward biased, but the base–collector junction is *more* forward biased than the base–emitter junction.

though not as strongly as Q_3. The nominal output logic levels for this simple, un-loaded gate are[5]

$$V_{oH} = 5 \text{ V}, \qquad V_{oL} = 0.2 \text{ V}.$$

The extremes of the input levels are

$$V_{iH(\min)} = 1.8 \text{ V}, \qquad V_{IL(\max)} = 1.1 \text{ V}.$$

These levels result in noise margins of

$$\text{NM(HIGH)} = V_{oH} - V_{iH(\min)} = 5 - 1.8 \approx 3.2 \text{ V},$$

$$\text{NM(LOW)} = V_{iL(\max)} - V_{oL} = 1.1 - 0.2 \approx 0.9 \text{ V}.$$

While this simple gate topology operates well, it has several properties that are undesirable for current integrated circuit realizations. Most problematic are its relatively high power consumption and low speed of operation.

The low operational speed of simple TTL and DTL logic gates stems from junction charge buildup in BJTs that enter the saturation region. As was seen with the simple bipolar inverter, strongly saturated BJTs exhibit a relatively long storage time. This long storage time slows the digital transitions necessary in a logic gate. In the TTL NAND gate under analysis, both the logic LOW and logic HIGH states have BJTs in saturation. A logic LOW implies that Q_1 is in saturation, and a logic HIGH implies that Q_2 and Q_3 are in saturation.

Each junction must dissipate its stored charge when the gate transitions its output between logic levels. The speed of the charge dissipation strongly depends upon the Thévenin resistance apparent to the junction. It is in this Thévenin resistance that a conflict between two TTL gate design goals is most apparent. Any attempt to increase the speed of the gate by reducing the resistance values results in an increase in the gate average power consumption. Similarly, reducing the average power consumption by increasing the resistor values results in slower gate performance. With this basic gate topology, power consumption and gate speed can only be simultaneously improved by lowering the supply voltage. Unfortunately this action reduces the HIGH noise margin, NM(HIGH).

Gate performance can be improved in several areas simultaneously only with changes in the basic topology of the gate. Historically these changes have taken place gradually and have resulted in a series of TTL gate families. The changes have been centered on two basic design techniques:

- Active charge dissipation
- Transistor saturation control

While it is not feasible to extensively discuss each variation of gate topology in this text, a brief look at several TTL gate alterations is instructive in the study of logic gate speed.

Active Charge Dissipation
In typical TTL gates the output BJT, Q_3, is the most strongly saturated transistor. In addition, the active elements surrounding Q_1 and Q_2 provided relatively low impedance paths for rapid dissipation of built-up charge. Therefore the greatest benefit

---

[5] Logic levels, noise margins, and fan-out for this gate are calculated in Section 3.5.

is obtained by focusing efforts at increasing the gate speed about the output transistor. For a LOW output, Q_3 is saturated: Both base–emitter and base–collector junctions are forward biased. While the charge buildup is largest in the base–emitter junction, it is the base–collector junction that must undergo the greatest change in charge distribution as Q_3 switches state from saturation to cutoff. In order to quickly remove the charge from the base–collector junction, the collector resistor is replaced by a low-impedance active load, as shown in Figure 16.2-2. This particular circuit topology is referred to as an active pull-up or sometimes a "totem-pole" output.

With active pull-up output, the saturated transistor Q_3 sees a Thévenin load at the collector consisting of the dynamic resistance of a diode in series with the output resistance of a common-collector amplifier:

$$R_{\text{th}} \approx r_d + \frac{R_{c2}}{\beta_{F4} + 1}.$$

This apparent resistance is much smaller than that of the collector resistor in the simple TTL topology while the Thévenin voltage has not significantly changed. These changes have the effect of reducing the overdrive factor N_1 during discharge:

$$N_{1(\text{discharge})} = \frac{\beta_F I_b}{I_{c(\text{eff})}},$$

where

$$I_{c(\text{eff})} = \frac{(V_{cc} - V_{be(\text{active})} - V_\gamma) - V_{ce(\text{sat})}}{R_{\text{th}}} > I_{c(\text{sat})}.$$

The storage time for Q_3 is significantly reduced, thereby reducing the collector current fall time and increasing the gate speed. Since the actual value of the collector current at saturation, $I_{c(\text{sat})}$, remains essentially unchanged, active pull-up has little effect on the collector current rise time.

Gate operation with active pull-up is much the same as with a resistor as the output collector load. The input circuitry of the TTL gate remains unchanged; therefore the input logic voltage levels remain unchanged. The output LOW voltage remains the same at $V_{CE(\text{sat})} \approx 0.2$ V; only the output HIGH voltage changes. For a

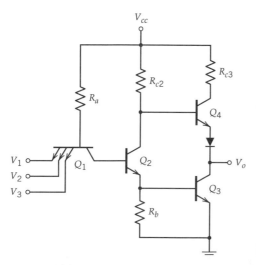

Figure 16.2-2
TTL gate with active pull-up.

logic HIGH output, Q_2 and Q_3 are cut off. It is assumed that a load attached to the output of the gate draws some current; thus Q_4 is in the forward-active region. The nominal output voltage level is the source voltage reduced by the voltage drops across the resistor R_{c2}, the base–emitter junction of Q_4, and the diode:

$$V_{oH} = V_{cc} - (V_{Rc2} + V_{BE4(on)} + V_\gamma) \approx V_{cc} - 1.5 \approx 3.5 \text{ V}.$$

The reduction in the nominal output HIGH level decreases the HIGH noise margin, NM(HIGH):

$$\text{NM(HIGH)} = V_{oH} - V_{iH(min)} = 3.5 - 1.8 \approx 1.7 \text{ V}.$$

This form of active pull-up is the output stage found in 74XX/54XX series TTL gates. Added benefits of this active pull-up circuit topology are a decrease in average power consumption of 10–20% over the simple resistive pull-up configuration and a very slight improvement in fan-out. Both improvements are due to Q_4 being in the cutoff region of operation for a LOW output.

EXAMPLE 16.2-1   Determine the fan-out of the active pull-up NAND gate shown in Figure 16.2-2. The pertinent circuit parameters are

$$V_{cc} = 5 \text{ V}, \qquad R_a = 3.9 \text{ k}\Omega, \qquad R_b = 1.0 \text{ k}\Omega,$$
$$R_{c2} = 1.5 \text{ k}\Omega, \qquad R_{c3} = 130 \text{ }\Omega.$$

The physical realization is in silicon with BJT parameters:

$$\beta_F = 50, \qquad \beta_R = 2.$$

SOLUTION

The determination of fan-out is much the same as that of the resistive pull-up circuit (Example 3.5-2). Fan-out in TTL gates is determined by a master gate with a LOW output driving a slave gate. The input current for a slave gate with a low input (0.2 V) is found to be

$$I_{in} = \frac{5.0 - 0.2 - 0.8}{3900} = 1.026 \text{ mA}.$$

If the master gate is driven by other gates of the same type, it is unreasonable to assume that a large amount of current is entering the emitter of Q_1 (it would draw the input voltage below zero). For fan-out calculations it is safer to assume the worst-

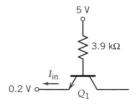

case scenario where the input current to the master gate is approximately zero. Under that scenario

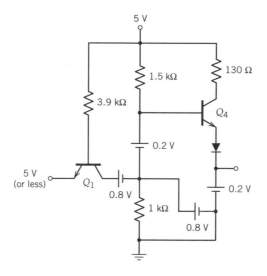

$$I_{B2} \approx I_{B1},$$

$$I_{B2} \approx \frac{5 - (0.8 + 0.8 + 0.7)}{3900} = 692 \ \mu A.$$

In this state, there is insufficient voltage between the base of Q_4 and the output terminal to forward bias both the base–emitter junction of Q_4 and the diode. Therefore, Q_4 in the master gate is OFF. In order to determine the ratio of collector to base currents in the output transistor Q_3, the collector current of Q_2 must be calculated:

$$I_{C2} = \frac{5 - 0.8 - 0.2}{1500} = 2.667 \ \text{mA}.$$

The base current of Q_3 is therefore

$$I_{B3} = I_{B2} + I_{C2} - \frac{0.8}{1000} = 2.559 \ \text{mA}.$$

The no-load collector current of Q_3, $I_{C3(nl)}$, in the master gate is zero due to Q_4 being in the cutoff region. Thus the master gate fan-out is determined from the saturation conditions on Q_3:

$$I_{C3} < \beta_F I_{B3}$$

or

$$I_{C3(nl)} + N(I_{in}) < \beta_F I_{B3},$$

or

$$0 \ \text{mA} + N(1.026 \ \text{mA}) < 50(2.559 \ \text{mA}) \Rightarrow N < 124.7.$$

The fan-out of this gate is 124 gates of similar construction. This is an increase of only one gate over the fan-out for the resistive pull-up gate discussed in Example 2.5-3.

Additional changes in the circuit topology can bring further improvements. The circuit of Figure 16.2-3 shows two such topological changes. The active pull-up circuit in this TTL NAND gate realization consists of two BJTs connected as a Darlington pair[6] rather than the BJT–diode connection previously discussed. Also shown is an active pull-down circuit connected to the base of the output transistor Q_3.

The Darlington active pair pull-up formed by Q_4 and Q_5 presents a very low Thévenin resistance to the collector of Q_3 and further reduces the saturation over-drive factor N_1:

$$R_{\text{th}} \approx \frac{1}{\beta_{F4} + 1}\left(R_{b4} \left\| \frac{R_{c2}}{\beta_{F5} + 1}\right.\right) \approx \frac{R_{c2}}{(\beta_{F4} + 1)(\beta_{F5} + 1)}.$$

This Darlington pull-up topology was first commercially seen in the 74HXX/54HXX high-speed series of TTL gates. The active pull-down circuit also shown in Figure 16.2-3 is formed by two resistors, R_{b3} and R_{b6}, and a transistor, Q_6. The effect of this configuration is to present a low Thévenin resistance to the base of Q_3:

$$R_{b3\text{th}} \approx \frac{R_{b6}}{\beta_{F6} + 1}.$$

Lowering the apparent base resistance increases the pull-down base current, thereby increasing the reverse overdrive factor N_2. An increased reverse overdrive factor lowers both the storage and fall times for Q_3.

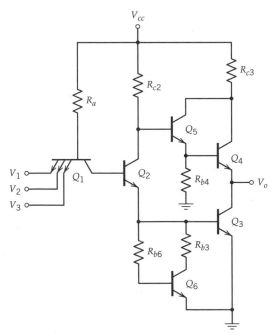

Figure 16.2-3
TTL gate with active pull-down and improved pull-up.

---

[6] Darlington pairs are discussed in Section 6.2.

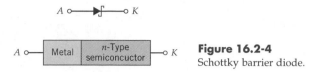

Figure 16.2-4
Schottky barrier diode.

Saturation Control

Any attempt to quickly remove charge stored in the junctions of a saturated transistor by reducing various Thévenin resistances has physical limits. The next logical step in improving gate performance is centered on limiting the charge buildup by controlling the region of operation of the transistor. Specifically, if the transistors never enter the saturation region, the component of the storage time related to the transistor storage time constant τ_s will become insignificant. Since this component can easily be the greatest contributor to slow transition times, gate speed will be dramatically improved. An exclusion from the saturation region can be accomplished if the base–collector junction is shunted by a low-V_γ diode that is less subject to charge storage effects than the transistor base–collector junction. The diode shunting the base–collector junction will not allow that junction to become sufficiently forward biased, thereby keeping the BJT in the forward-active region. While a germanium diode ($V_\gamma \approx 0.3$ V) seems ideal for the purpose of BJT saturation control,[7] fabrication of silicon BJTs and germanium diodes on the same integrated circuit chip is not practical. Schottky-barrier diodes are the ideal alternative.

Schottky-barrier diodes are formed with a metal–semiconductor junction rather than the usual p–n semiconductor junction. A representation of the metal–semiconductor junction and the circuit symbol for a Schottky-barrier diode are shown in Figure 16.2-4. At the junction between a metal and a semiconductor, the metal acts as a p-type impurity. If the semiconductor is n-type, the junction acts as a diode. The V–I characteristic of a Schottky-barrier diode is indistinguishable from that of a p–n junction except that the cut-in voltage V_γ is somewhat smaller. Depending on the doping of the semiconductor and the metal used, V_γ ranges between 0.2 and 0.5 V with silicon as the semiconductor. A typical integrated circuit realization of a Schottky-barrier diode using aluminum results in $V_\gamma \approx 0.4$ V. One particular advantage of a Schottky-barrier diode is the extremely small charge storage time (equivalently, a small junction capacitance) associated with the junction. This storage time is typically at least an order of magnitude smaller than an equivalently sized p–n junction.

Transistors that incorporate a Schottky-barrier diode shunting the base–collector junction are referred to as Schottky transistors. The circuit symbol for a Schottky npn transistor and a conceptual integrated circuit realization of this transistor are shown in Figure 16.2-5. The metallic base electrode bridges the p-type material that is the base and the n-type material that forms the collector, allowing it to serve dual purposes: to be the base contact and also to form the base–collector shunting Schottky diode.[8]

If all saturating transistors in a TTL or DTL gate are replaced by Schottky transistors, the speed of the gate will be significantly improved. The 74SXX/54SXX series

---

[7] The shunting of the base–collector junction of a silicon BJT with a germanium diode is referred to as *Baker clamping* a silicon transistor.
[8] The collector region of this BJT is shown with both an n and n^+ region. This process prevents the formation of a Schottky barrier diode at the metal–semiconductor junction.

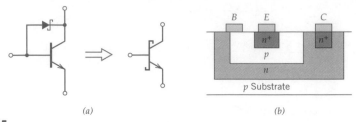

Figure 16.2-5
Schottky transistor: (*a*) circuit symbol; (*b*) IC realization.

of TTL gates have the same topology as the gate shown in Figure 16.2-3 with all BJTs *except* Q_4 replaced by Schottky transistors (Q_4 is not a saturating transistor). Other than reduced BJT storage time, the change to Schottky transistors has little effect on the performance of the NAND gate. The greatest change is in the nominal logic levels. The HIGH output level remains unchanged at $V_{oH} \approx 3.5$ V, but the LOW level is increased slightly. Since the output transistor Q_3 no longer saturates with a logic LOW output, the new low output is given by

$$V_{oL} = V_{BE3(active)} - V_\gamma = 0.7 - 0.4 \approx 0.3 \text{ V.}$$

Similarly the extremes of the input logic levels are given by

$$V_{iH(min)} = V_{BE3(active)} + V_{BE2(active)} + (V_{BE1(active)} - V_\gamma) \approx 1.7 \text{ V,}$$

$$V_{iL(max)} = V_{BE3(cut-in)} + V_{BE2(active)} - (V_{BE1(active)} - V_\gamma) \approx 0.9 \text{ V.}$$

The change in input and output logic levels also alters the noise margins. The noise margins for a Schottky TTL gate (74SXX) are given by

$$\text{NM(HIGH)} = V_{oH} - V_{iH(min)} = 3.5 - 1.7 \approx 1.8 \text{ V,}$$

$$\text{NM(LOW)} = V_{iL(max)} - V_{oL} = 0.9 - 0.3 \approx 0.5 \text{ V.}$$

Further improvements in integrated circuit fabrication techniques have allowed designers to return to a DTL input to reduce power consumption. The 74LSXX/54LSXX series of low-power Schottky gates (Figure 16.2-6) is an example of the return to DTL gate topology with active pull-up and pull-down improvements.

The return to DTL technology was facilitated by the Schottky transistor Q_2. Since this BJT no longer enters the saturation region, it is not necessary to have active elements to speed the charge dissipation during logic transitions. Transistor Q_1 was one such active element: It can return to a DTL topology without significant loss of gate speed. Another common gate design feature is shown at the input terminals of the TTL gate in Figure 16.2-6. The inputs of this series of gates are protected against transient current surges with a reverse-biased Schottky diode shunting each input to ground.

Open-Collector Outputs

Transistor–transistor logic gates are also available with no internal provision for either active or passive pull-up of the output transistor. Such a gate is identified as

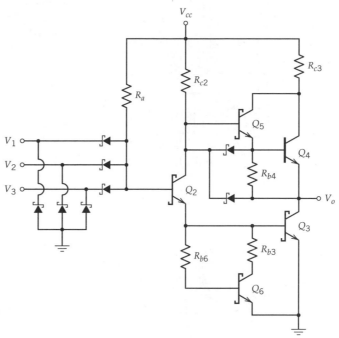

Figure 16.2-6
Low-power Schottky DTL (LS) gate with active pull-up and pull-down.

having an *open-collector output*. The two primary advantages of an open-collector output are as follows:

- Wired-AND operations can be simply created.
- Simple external loads are easily driven.

Open-collector outputs are also useful in driving a simple load such as a LED or relay. An example of wired-AND is shown in Figure 16.2-7. Here, the outputs of two open-collector NAND gates are connected through an external resistor to positive power. The system output can only go HIGH if both NAND gate output BJTs are OFF. Since the output BJT OFF state is associated with a HIGH output, the logical operation is an AND operation on the outputs of the two gates.

In order to ensure proper operation of a wired AND, the external resistor R_L has a minimum value. It must be chosen so that

$$R_L > \frac{V_{CC} - V_{OL(\text{max})}}{I_{OL(\text{max})} - NI_{IL}}.$$

The denominator terms are defined as follows: $I_{OL(\text{max})}$ is the maximum output current that the output BJT of a gate can sink ($\beta_F I_B$), and NI_{IL} is the current drawn by N gates with a LOW input. Gates with internal passive pull-up resistors typically violate the constraints on R_L and therefore must not be connected in this manner. If gates with active pull-up are connected in this manner, excessive power will be dissipated in the output stage of an individual gate leading to gate failure.

Open-collector gates, when properly terminated, have passive pull-up. As such, they exhibit relatively long propagation delays compared to gates with active pull-up.

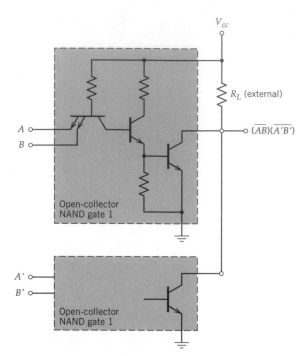

Figure 16.2-7
Two open collector NAND gates with outputs forming wired-AND.

16.2.2 ECL Logic Gates

The basic functional topology of all ECL gates is shown in Figure 16.2-8. All ECL gates use this basic topology only with variations in the individual components. The current switch is always in the form of several BJTs sharing a common connection at the emitter terminal. This configuration, essentially a very high gain differential amplifier, ensures that the emitter-coupled BJTs never enter the saturation region. Switching is between the forward-active region and cutoff. As such, ECL gates have a short storage time and are inherently the quickest form of bipolar logic. The bias network provides stable voltage references and the output driver buffers the output of the current switch to increase fan-out and match impedance for optimal transmission.

The significant advantages of the ECL gate family over other bipolar logic families include the following:

- *Complementary outputs.* Most logic elements offer both the logic function and its complement (i.e., OR/NOR). Additional logic inverters are eliminated from designs, reducing timing delays and power consumption.

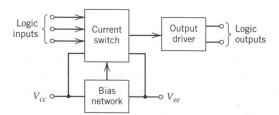

Figure 16.2-8
Basic ECL gate topology.

- *Constant supply current.* The current drain remains essentially constant regardless of gate logic state. The power supply design requirements are therefore simplified.
- *Low switching noise.* The differential amplifier and common-collector amplifier switching sections of ECL logic have very low current transients. In addition, the low voltage variation in logic states ensures that charging of any stray capacitance results in small current variation.
- *Low crosstalk.* Crosstalk in digital systems is proportional to the derivative of voltage signals. Low logic-level voltage differences, as well as greatly reduced BJT storage times, allow for effective control of signal rise and fall times without significantly slowing gate speed. Voltage rise and fall rates can be reduced in ECL by a factor of 5 (or more) over typical TTL rates.
- *High fan-out.* The pairing of high input impedance with low output impedance allows for large fan-out.

The disadvantages are primarily higher power consumption, lower noise margins, and additional design constraints in the integrated circuit environment. Emitter-coupled logic is operated with a negative power supply to reduce the effects of noise and power supply variation on gate output.

The basic operation of an early ECL OR gate[9] was described in Section 3.5. In that basic circuit the bias network is an externally supplied reference voltage. Also, the output common-collector buffer has a pull-down resistor.[10] Later ECL gate families incorporate several improvements. One series of gates that incorporates the significant changes in ECL topology is the ECL 10 K series. The topology of an ECL 10K OR/NOR gate is shown in Figure 16.2-9.

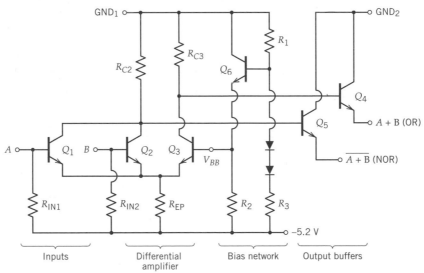

Figure 16.2-9
Basic ECL 10K OR/NOR gate.

---

[9] The gate described in Section 3.5 has the topology of an MECL I circuit. MECL II has identical topology except the bias network is included in the integrated circuit realization.

[10] A pull-down resistor at the emitter of the output BJT ensures that the output will be pulled appropriately LOW.

The circuit topology of the 10K series differs from the basic gate in several ways:

- *Internal bias network.* A bias network is included in the integrated circuit realization of the gate to provide a reference voltage to the differential amplifier.
- *Ground separation.* Two separate ground terminals are provided so that power supply transients due to switching in the differential amplifier are not passed through the ground to the output buffers.
- *Relocation of pull-down.* The relatively low-value, output-emitter, pull-down resistors have been eliminated and replaced by high-value resistors on the input BJTs. Benefits include reduced power consumption, more reliable gate operation with floating inputs, and increased gate life.

Logic levels remain unchanged at

$$V_{oH} \approx -0.72 \text{ V}, \qquad V_{oL} \approx -1.60 \text{ V}.$$

One problem that is common in ECL gates is the variation of the logic levels due to variation in the power supply voltage or due to temperature changes. Various advanced ECL circuits have addressed these issues at the expense of circuit complexity. One such advanced circuit topology is that of the ECL 100K series shown in Figure 16.2-10. Notice that the increase in power consumption due to added circuitry has been compensated for by a reduction in the power supply voltage magnitude from 5.2 to 4.5 V.

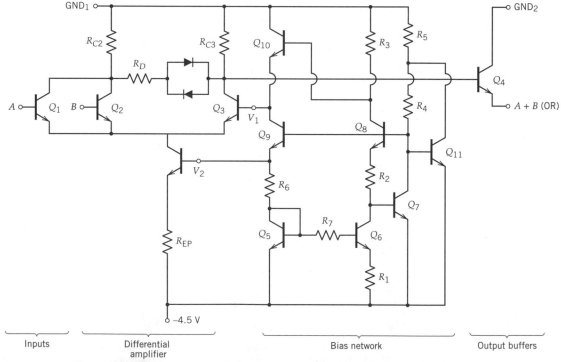

Figure 16.2-10
ECL 100 K OR gate.

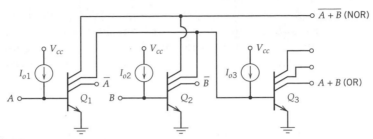

Figure 16.2-11
Simple I^2L OR/NOR logic gate.

16.2.3 I^2L Logic Gates

While TTL and ECL circuits are a good choice for small-scale (SSI) and medium-scale (MSI) integrated circuits, they have limited use in large-scale (LSI) and are not practical for very large scale (VLSI) applications. The primary reasons for these limitations on the use of TTL and ECL circuits are the following:

- Relatively high power consumption
- Relatively large surface area of integrated circuit realization

Integrated injection logic[11] (I^2L) blends high speed with high surface area density and low power consumption. In addition, the simple gate structure of I^2L provides multifunction outputs. For example, the simple two-input OR/NOR gate shown in Figure 16.2-11 provides four logical operations as outputs: the complement of each input, $\overline{A}$ and $\overline{B}$, and the OR/NOR functions $A + B$ and $\overline{A + B}$.

The basic element of an I^2L gate is a multiple-collector *npn* BJT driven at the base with a simple *pnp* BJT current source (Figure 16.2-12). The unique feature of this element is its single input and multiple outputs.

The operation of the basic I^2L element is simple if it is remembered that at least one of the multiple collector outputs must be connected to the input of another element in a master–slave pairing. If the master gate input A is HIGH, Q_1 enters the saturation region and the master gate output at each of the multiple collectors is

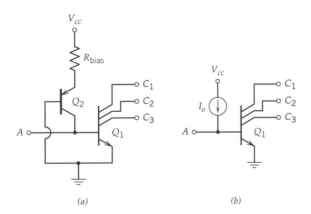

(a)   *(b)*

Figure 16.2-12
Basic I^2L digital gate element: (*a*) circuit diagram; (*b*) equivalent circuit.

---

[11]I^2L is also known as *merged transistor logic* (MTL).

$V_{CE(\text{sat})}$. If the master gate input is LOW, Q_1 enters the cutoff region. The slave gate sees a high-impedance input from the master gate. The current source forces the slave gate Q_1 into saturation, which in turn forces the slave gate input *and* the master gate output to $V_{BE(\text{sat})}$. All other collectors of the master gate will also be raised to that voltage level. The output logical voltage levels are therefore

$$V_{oH} = V_{BE(\text{sat})} \approx 0.8 \text{ V}, \qquad V_{oL} = V_{CE(\text{sat})} \approx 0.2 \text{ V}.$$

The input logical levels are the levels at which a BJT can be considered to be ON or OFF:

$$V_{iH(\text{min})} = V_\gamma \approx 0.7 \text{ V}, \qquad V_{iL(\text{max})} = V_{BE(\text{on})} \approx 0.6 \text{ V}.$$

These voltage levels are all within ~0.6 V with the transition region only ~0.1 V wide. Obviously I^2L gates are not appropriate for use in a noisy environment. The noise margins are

$$\text{NM(HIGH)} = V_{oH} - V_{iH(\text{min})} = 0.8 - 0.7 \approx 0.1 \text{ V},$$

$$\text{NM(LOW)} = V_{iL(\text{max})} - V_{oL} = 0.6 - 0.2 \approx 0.4 \text{ V}.$$

Fan-out considerations are particularly simple to visualize in I^2L gates. A HIGH output implies no interaction between master and slave gate: There is no limit on the number of slave gates imposed by a logic HIGH output. For a LOW output, the master gate output BJT must be in saturation and the slave gate output BJT in cutoff. The slave gate therefore supplies I_o to the collector of the master gate output BJT. The condition for saturation is

$$\beta_F I_B \geq I_C \Rightarrow \beta_F I_o \geq N I_o \Rightarrow \text{fan-out} = \beta_F.$$

It must be noted that the physical structure of the multiple-collector BJT limits the forward current gain β_F. In most cases β_F for a multiple-collector BJT is much smaller than that of modern, simple BJTs. Consequently, the fan-out of an I^2L gate is typically *less than ten* gates of similar construction. If the output to the final gate of a logical operation is to be taken off the chip, it is necessary to passively pull up the collector of the output BJT. This can be accomplished with a resistor connected between any of the output BJT collector terminals and a positive-voltage supply.

The interconnection of the two BJTs that make up an I^2L gate simplifies integrated circuit realization of the gate structure. The p-type base of the current source is directly connected to the p-type emitter of the inverter. Similarly, the n-type collector of the current source connects to the n-type base of the inverter. These interconnections lead to shared regions in the integrated circuit realization, as shown in Figure 16.2-13. A single resistor is usually used for all I^2L gates on a chip. A multiple-

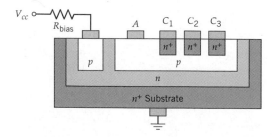

Figure 16.2-13
Integrated circuit realization of I^2L gate with three collectors.

collector structure can be constructed in a similar fashion to the multiple-emitter structure of the input transistor of TTL gates.

16.3 DYNAMIC PROPERTIES OF MOS TRANSISTORS

Metal–oxide–semiconductor digital logic gates depend on the transition of the output of a MOSFET from one logic level to another. As in bipolar transitions, the speed at which any gate operates is limited by the physical parameters of its constituent transistors and of the components in the circuit surrounding the transistor. While it is impossible to analyze the effect of all possible circuit topologies on switching speed, an analysis of the switching speed of the simple CMOS inverter provides much insight into all MOS switching.[12] As shown in Figure 16.3-1, a general CMOS inverter consists of two *complementary* MOS transistors: The input to the inverter is at the connection of the two FET gate terminals and the output is at the connection of the drain terminals. Here it is assumed that the master inverter shown drives a similar-topology slave inverter. This slave inverter is represented by its input impedance: a capacitor in parallel with a very large resistance. As the input resistance of a FET is essentially infinite, it is assumed to have no significant effect on any further calculations.

The response of the CMOS inverter to a logic LOW-to-HIGH input transition is shown in Figure 16.3-2. For a CMOS inverter these logic levels are

$$V_{iL} \approx 0 \text{ V}, \qquad V_{iH} \approx V_{DD}.$$

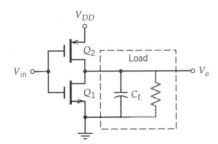

Figure 16.3-1
Typical CMOS inverter, capacitively loaded.

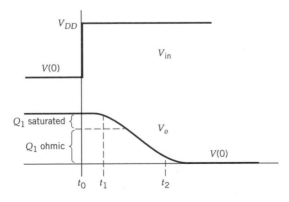

Figure 16.3-2
Transient response of CMOS inverter.

---

[12]This digital CMOS inverter is initially discussed in Section 3.4.

As in the case of a bipolar inverter, the digital transition cannot be instantaneous and must experience some delay. Due to the symmetry of the CMOS circuit, the *mathematical expressions* for the rise and fall time will be functionally identical. The rise and fall times themselves will vary with the transistor characteristic. Specifically, the rise and fall times are dependent on the *n*-channel and *p*-channel FET transconductance parameters K_n and K_p, respectively, as well as the respective threshold voltages.

The respone to a digital step consists of regions of constancy and a region of transition. In the region of transition there are two significant time periods:

- Delay time $t_d = t_1 - t_0$
- Rise time $\quad t_r = t_2 - t_1$

The delay time is the time between the input pulse transition and when the response transitions 10% of the distance between HIGH and LOW states. The rise time is the time for a 10–90% transition from HIGH to LOW. The term *rise time* is used to match the definitions in bipolar circuitry. In NMOS gate circuits, it refers to the FET drain current change: As the voltage transitions from a HIGH to LOW, the drain current rises from a minimum to a maximum value. It is a misnomer for CMOS circuits: The drain current is zero for both logic states.

During the delay time ($t_0 \leq t < t_1$), the *n*-channel MOSFET is in the saturation region of operation and is described by the voltage–current relationship:[13]

$$I_D = K(V_{GS} - V_T)^2. \tag{16.3-1}$$

Since the input voltage to the inverter is a constant value during this time period ($V_{\text{in}} = V_{DD}$), it can be seen through Equation 16.3-1 that the drain current is constant during the delay time. The load capacitance can therefore be assumed to discharge linearly:

$$I_D \, \Delta t = -C_L \, \Delta V_o \Rightarrow t_d = \frac{-C_L \, \Delta V_o}{I_D}. \tag{16.3-2}$$

The change in the output voltage, ΔV_o, is 10% of the supply voltage: $\Delta V_o = -0.1 V_{DD}$. The total delay time can therefore be easily computed:

$$t_d = \frac{0.1 C_L V_{DD}}{K(V_{DD} - V_T)}. \tag{16.3-3}$$

The rise time has two components:

$$t_r = t_{r1} + t_{r2}, \tag{16.3-4}$$

where t_{r1} is the portion of the rise time where Q_1 is in the saturation region and t_{r2} is the portion where Q_1 is in the ohmic region. The transition between the saturation and ohmic regions of an enhancement region MOSFET occurs when

$$V_{DS} = V_{GS} - V_T \Rightarrow V_o = V_{DD} - V_T. \tag{16.3-5}$$

---

[13]The voltage–current relationships for FETs of all types are presented in Table 4.4-1.

As during the delay time, the drain current is constant for a saturation region FET and the load capacitor discharges linearly during t_{r1}:

$$t_{r1} = \frac{-C_L \, \Delta V_o}{I_D} \tag{16.3-6}$$

or

$$t_{r1} = \frac{-C_L[(V_{DD} - V_T) - 0.9V_{DD}]}{K(V_{DD} - V_T)} = \frac{C_L(V_T - 0.1V_{DD})}{K(V_{DD} - V_T)}. \tag{16.3-7}$$

During the second portion of the rise time, Q_1 is in the ohmic region and can be described by the expression

$$I_D = K[2(V_{GS} - V_T)V_{DS} - V_{DS}^2]. \tag{16.3-8}$$

The load capacitor discharge is described by the differential form of its voltage–current relationship:

$$I_D = -C_L \frac{dV_o}{dt}. \tag{16.3-9}$$

Simultaneously solving Equation 16.3-8 and 16.3-9 leads to an integral expression for t_{r2}:

$$t_{r2} = \frac{C_L}{2K(V_{DD} - V_T)} \int_{V_{DD}-V_T}^{0.1V_{DD}} \frac{dV_o}{V_o^2/2(V_{DD} - V_T) - V_o}. \tag{16.3-10}$$

Evaluation of the integral gives the expression for the second portion of the rise time:

$$t_{r2} = \frac{C_L}{2K(V_{DD} - V_T)} \ln\left(20 \frac{V_{DD} - V_T}{V_{DD}} - 1 \right). \tag{16.3-11}$$

The total rise time for the CMOS inverter is the sum of the rise time components:

$$t_r = \frac{C_L}{K(V_{DD} - V_T)} \left[(V_T - 0.1V_{DD}) + \frac{1}{2} \ln\left(20 \frac{V_{DD} - V_T}{V_{DD}} - 1 \right) \right]. \tag{16.3-12}$$

A comparison of the expressions for delay time (Equation 16.3-3) and rise time (Equation 16.1-13) shows that the rise time dominates the delays inherent in digital CMOS switching. In addition, both the rise time and the delay time are directly proportional to the capacitance of the load. This dependence on load capacitance is the determining factor in MOS gate fan-out. When many slave gates are connected to the output of a single master gate, the input capacitance of the slave gates add. The rise and delay times increase directly in proportion to the number of slave gates attached to the output of the master gate. Transition speed requirements put an upper limit on this number.

As with the bipolar inverter, propagation delay is a useful descriptor of the gate speed. The propagation delay of a CMOS inverter can be described in terms of the transition times derived. Its component terms are

$$t_{PHL} \approx t_d + \tfrac{1}{2}t_r \quad \text{(n-channel FET parameters)}. \tag{16.3-13}$$

The LOW-to-HIGH propagation time has the same mathematical form but the transconductance parameter of the p-channel FET must be used:

$$t_{PLH} \approx t_d + \tfrac{1}{2}t_r \quad \text{(p-channel FET parameters).} \tag{16.3-14}$$

The average propagation time is given by

$$t_P = \tfrac{1}{2}(t_{PHL} + t_{PLH}). \tag{16.3-15}$$

EXAMPLE 16.3-1 A CMOS inverter is fabricated using a 5-V supply and MOSFETs with the following properties:

$$K_N = 0.1 \text{ mA/V}^2 \quad (n\text{-channel}), \qquad |V_T| = 1,$$
$$K_P = 0.25 \text{ mA/V}^2 \quad (p\text{-channel}).$$

Determine the averge propagation delay time if it is driving a capacitive load of 5 pF.

SOLUTION

Equation 16.3-3 yields the delay times

$$t_d = \frac{0.1 C_L V_{DD}}{K(V_{DD} - V_T)} = \frac{0.625 \times 10^{-12}}{K}.$$

The HIGH-to-LOW transition uses K_N while the Low-to-High transition uses K_P:

$$t_{dHL} = 6.25 \text{ ns}, \qquad t_{dHL} = 2.5 \text{ ns}.$$

Similarly the rise times can be calculated from Equation 6.3-12:

$$t_r = \frac{2.318 \times 10^{-12}}{K},$$

$$t_{rHL} = 23.18 \text{ ns}, \qquad t_{rLH} = 9.27 \text{ ns}.$$

The individual propagation delays are given by

$$t_{PHL} \approx t_{dHL} + \tfrac{1}{2}t_{rHL} = 17.84 \text{ ns},$$
$$t_{PLH} \approx t_{dLH} + \tfrac{1}{2}t_{rLH} = 7.14 \text{ ns}.$$

The average propagation delay is the average of the individual propagation delays:

$$t_P = \tfrac{1}{2}(t_{PHL} + t_{PLH}) \approx \textbf{12.5 ns}.$$

An NMOS inverter is similar to a CMOS inverter but fabricated with an n-channel active load rather than the p-channel switch. Transition time calculations are similar to those described for the CMOS inverter, with the exception that the switching transistor Q_1 is always in the ohmic region. The expansion of NMOS inverter calculations to cover NMOS gates is complicated further due to the dependence of the

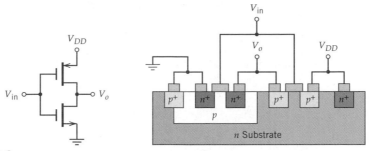

Figure 16.3-3
CMOS inverter and conceptual integrated circuit realization.

LOW output voltage V_{oL} on the exact state of the multiple switching transistors inherent to MOS gates. The CMOS and NMOS gates operate at *essentially* the same speed if comparable FETs are used.

FET Latch-Up

The integrated circuit realization of MOS gate structures produces, in addition to the MOS structures, parasitic bipolar structures (Figure 16.3-3). While the *npn* and *pnp* structures that produce parasitic BJTs are usually benign, *pnpn* structures can produce a parasitic silicon-controlled rectifier (SCR).[14] An SCR is a form of latching switch that is activated by the proper injection of a small current or by a high-derivative voltage pulse. Once the SCR structure is activated, the high currents that result, combined with the latching property of the SCR, lead to catastrophic failure of the MOS device. Modern integrated circuit design of MOS structures includes transient protection structures so that the possibility of parasitic SCR latch-up is minimized in all but the noisiest of environments. Still, MOS integrated circuits are particularly sensitive to damage by static discharge.

16.4 MOS DIGITAL GATE CIRCUITS

The two basic logic operations in MOS gates are NAND and NOR. In order to create these logic operations, MOSFETs are used essentially as switches. The two logic operations can be conceptualized (Figure 16.4-1) as a resistive load in series with either a series connection of several switches to ground (NAND) or a parallel connection of several switches to ground (NOR). In the NAND configuration, the output will be LOW only if both of the series switches are closed. In the NOR configuration, the output is LOW if either of the parallel switches is closed. In MOS realizations of such logic gates, the FET can be used for all components of the gate: the switches as well as the active resistive loads. The extremely low current draw at the controlling terminal of a FET switch, the gate terminal, facilitates cascading these simple logic gates without many of the problems caused by loading.

MOS logic gate circuits can be constructed using MOS transistors of the same type or transistors of mixed types. NMOS circuits use *n*-channel FETs exclusively, albeit often with a mix of depletion- and enhancement-mode FETs. PMOS is the complementary form of NMOS using *p*-channel FETs only. CMOS (complementary

---

[14]SCRs are discussed extensively in Section 14.1.

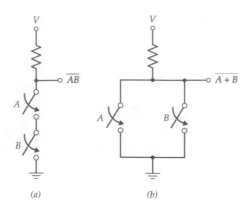

Figure 16.4-1
Two-input logic gates using simple controlled
switches: (*a*) NAND; (*b*) NOR.

MOS) uses a mixture of complementary, enhancement-mode, *n*-channel and *p*-channel FETs. Of these three types, NMOS and CMOS are the most common. Discussion here is restricted to the fundamental gates common to NMOS and CMOS logic.

NMOS Gates

NMOS gates are used extensively in LSI and VLSI microprocessors, memories, and other circuitry but are not commonly available as individually packaged circuits. All elements of basic NMOS logic gates are fabricated from *n*-channel FETs: The switches are enhancement-mode NMOS FETs and the active resistive load can be either an enhancement-mode or a depletion-mode NMOS FET. The logic gate action is essentially the same for each type of active load. For simplicity, discussion here will be limited to enhancement-type active loads. The circuit diagrams for basic, two-input NMOS NAND and NOR gates, using enhancement-mode active loads, are shown in Figure 16.4-2. It should be noted that the geometry of all the switching FETs is identical, but the active load FET usually has different characterizing parameters.

While the fundamentals of NMOS gate operation are fairly easily understood, the specifics can be more complicated. Analysis of a NOR gate is mathematically the least complicated. Discussion will begin with the NOR gate and progress to the NAND gate. The short-circuit connection between the drain and gate terminals of the active load ensures that the load FET is always in the saturation region of operation: $V_{GS} = V_{DS}$. The equation that relates the load FET drain current to its gate–source voltage is found in Table 4.4-1:

$$I_{DR} = K_R(V_{GSR} - V_T)^2. \tag{16.4-1}$$

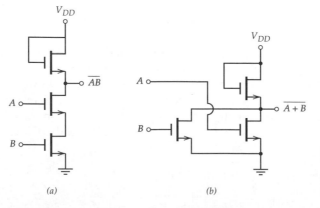

Figure 16.4-2
Two-input NMOS logic gates
(enhancement mode active load):
(*a*) NAND; (*b*) NOR.

An additional subscript is added to the FET parameters to specify the FET under consideration: R indicates the active resistive load and S indicates the switching FETs.[15]

In a two-input NOR gate, two LOW inputs produce simple results. For the switching FETs, $V_{GS} < V_T$ implies that the drain current I_{DS} is zero valued. Consequently, the drain current of the load FET must be zero valued. Substitution into Equation 16.4-1 yields

$$0 = K_R(V_{GSR} - V_T)^2 \Rightarrow V_{GSR} = V_T. \tag{16.4-2}$$

The output of the gate goes to a HIGH state, V_{oH}:

$$V_{oH} = V_{DD} - V_{GSR} = V_{DD} - V_T. \tag{16.4-3}$$

One or more HIGH inputs to the NOR gate implies that some current flows through the FETs. Any switching FET with a HIGH input is forced into the ohmic region. Solving for currents and voltages requires the use of additional FET characteristic equations. The current characteristic equation for enhancement-mode FETs operating in the ohmic region is also found in Table 4.4-1:

$$I_{DS} = K_S[2(V_{GSS} - V_T)V_{DSS} - V_{DSS}^2]. \tag{16.4-4}$$

Since there exists the possibility for N switching FETs to be simultaneously ON, the relationship between switch and load drain currents is given by

$$I_{DR} = NI_{DS}, \tag{16.4-5}$$

or

$$K_R(V_{GSR} - V_T)^2 = NK_S[2(V_{GSS} - V_T)V_{DSS} - V_{DSS}^2]. \tag{16.4-6}$$

Kirchhoff's laws relate V_{GSR} to V_{DSS}:

$$V_{GSR} = V_{DD} - V_{DSS}. \tag{16.4-7}$$

Equation 16.4-6 becomes a quadratic equation in the LOW output voltage level, $V_{oL} = V_{DSS}$, as a function of the number of HIGH inputs, N, the voltage supply V_{DD}, the FET threshold voltage V_T, the FET transconductance factors K_R and K_S, and the input voltage V_{GSS}:

$$K_R(V_{DD} - V_{DSS} - V_T)^2 = NK_S[2(V_{GSS} - V_T)V_{DSS} - V_{DSS}^2]. \tag{16.4-8}$$

The quadratic functional form of Equation 16.4-8 obscures intuition. Of most significance is a decrease in the output voltage V_{DSS} as NK_S/K_R increases. For most designs, it is important to set a maximum value on this LOW output. The maximum value will occur for only one switch ON ($N = 1$). A particular design goal will restrict the minimum ratio of transconductance factors, K_S/K_R. The usual implication of this restriction is switching FETs that are significantly wider than the load FET.

---

[15]It is assumed in this discussion that the threshold voltage V_T is the same for all FETs in the circuit. In integrated circuit fabrication this is a realistic design assumption.

EXAMPLE 16.4-1 Given a 5-V power supply. Design a three-input NOR gate to have the following output logic levels when driven by a gate of the same design:

$$V_{oH} = 3.8 \text{ V}, \qquad V_{oL(\text{max})} = 0.9 \text{ V}.$$

SOLUTION

An NMOS circuit topology similar to Figure 16.4-2*b* but with three inputs and three switching FETs will satisfy the design requirements as shown. All that remains to be determined is an acceptable set of FET parameters, V_T, K_R, and K_S.

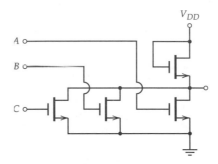

The HIGH output level is given by Equation 16.4-3:

$$V_{oH} = V_{DD} - V_{T},$$

or

$$3.8 = 5 - V_T \Rightarrow V_T = \mathbf{1.2 \text{ V}}.$$

The maximum LOW output level is determined with only one switching FET ON. For this condition the input voltage $V_{GSS} = V_{oH}$ and the output voltage $V_{DSS} = V_{oL(\text{max})}$:

$$K_R(5 - 0.9 - 1.2)^2 = 1K_S[2(3.8 - 1.2)(0.9) - (0.9)^2]$$

or

$$K_R = 0.460K_S \quad \text{or more exactly} \quad K_R \leq 0.460K_S.$$

In order to keep drain currents reasonable, the value of the transconductance factors must be kept reasonably small. Since there are no specific design constraints on FET transconductance factors, a drain current of approximately 1.93 mA will result in

$$K_R = \mathbf{230 \text{ } \mu A/V^2}, \qquad K_S = \mathbf{500 \text{ } \mu A/V^2}.$$

Other similar values will produce acceptable results. The LOW output voltage logic levels can be found to be 0.9 V for a single HIGH input, 0.527 V for two HIGH inputs, and 0.373 V for three HIGH inputs.

The analysis of an NMOS NAND gate is similar to that of the NOR gate but complicated by the series connection of the switching FETs. The HIGH output voltage is determined in the same manner as for a NOR gate and produces the same result:

$$V_{oH} = V_{DD} - V_T. \tag{16.4-9}$$

The series connection of switching FETs adds complexity to the low-output-voltage calculation. While the gate–source voltage V_{GS1S} of the grounded-source FET is simply the input voltage, subsequent switching FETs have V_{GS} reduced by the drain–source voltages of any intervening FETs. In the two-input NAND gate circuit of Figure 16.4-2a, the upper switching FET has a gate–source voltage (assuming the gate is driven by a similar gate)

$$V_{GS2S} = V_{in} - V_{DS1S} = V_{DD} - V_T - V_{DS1S}, \tag{16.4-10}$$

while the grounded-source switching FET has gate–source voltage

$$V_{GS1S} = V_{in} = V_{DD} - V_T. \tag{16.4-11}$$

Of course the NAND gate output voltage for a logic LOW is given by the sum of the switching FET drain–source voltages:

$$V_{oL} = \Sigma_{i=1}^{N} V_{DSiS} = V_{DS1S} + V_{DS2S}, \tag{16.4-12}$$

where N is the number of series switching FETs. The series connection of N identical switching FETs creates N equations relating to the drain currents:

$$K_R(V_{DD} - V_{oL} - V_T)^2 = K_S[2(V_{GSiS} - V_T)V_{DiSS} - V_{DiSS}^2], \, i = 1, 2, \ldots, \tag{16.4-13}$$

The simultaneous solution of Equations 16.4-10 and 16.4-11 determines the LOW output voltage level.

EXAMPLE 16.4-2 Given a 5-V power supply, design a three-input NAND gate to have the following output logic levels when driven by a gate of the same design:

$$V_{oH} = 3.8 \text{ V}, \qquad V_{oL(\text{max})} = 0.5 \text{ V}.$$

 SOLUTION

An NMOS circuit topology similar to Figure 16.4-2a but with three inputs and three switching FETs will satisfy the design requirements as shown. All that remains to be determined is an acceptable set of FET parameters, V_T, K_R, and K_S.

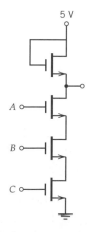

The HIGH output level is given by Equation 16.4-3:

$$V_{oH} = V_{DD} - V_T$$

or

$$3.8 = 5 - V_T \Rightarrow V_T = \textbf{1.2 V}.$$

The maximum LOW output level is determined by simultaneously solving Equations 16.4-12 and 16.4-13 (four total equations):

$$0.5 = V_{DS1S} + V_{DS2S} + V_{DS3S},$$

$$K_R(V_{oL} - 1.2)^2 = 1K_S[2(3.8 - 1.2)(V_{DS1S}) - (V_{DS1S})^2],$$

$$K_R(V_{oL} - 1.2)^2 = 1K_S[2(3.8 - 1.2 - V_{DS1S})(V_{DS2S}) - (V_{DS2S})^2],$$

$$K_R(V_{oL} - 1.2)^2 = 1K_S[2(3.8 - 1.2 - V_{DS1S} - V_{DS2S})(V_{DS3S}) - (V_{DS3S})^2].$$

There are no easy, closed-form techniques for solving four nonlinear, simultaneous equations. A MathCAD solution to find the ratio of the transconductance factors for a three-input NAND gate follows:

```
Vₜ:=1.2   V_oL:=0.5
Guess values for the ratio and the three drain-source voltages
ratio:=1   V1:=.2 V2:=.2 V3:=.3
Given
```

V1+V2+V3=V_oL $(V_{oL}-V_T)^2$=ratio·[2·(3.8−1.2−V1)·V2−V2²]

$(V_{oL}-V_T)^2$=ratio·[2·(3.8−1.2)·V1−V1²] $(V_{oL}-V_T)^2$=ratio·[2·(3.8−1.2−V1−V2)·V3−V3²]

$$\text{Find(V1,V2,V3,ratio)} = \begin{vmatrix} 0.155 \\ 0.166 \\ 0.179 \\ 0.626 \end{vmatrix}$$

The minimum ratio of transconductance factors for a 0.5-V LOW output voltage is therefore

$$\frac{K_S}{K_R} \geq 0.626.$$

Choosing the same load FET K_R as Example 16.4-1 implies that the NAND gate will draw the same power for a LOW output as the NOR gate in that example. This design choice leads to

$$K_R = \textbf{230 } \boldsymbol{\mu}\textbf{A/V}^2, \qquad K_S \geq 144 \text{ } \mu\text{A/V}^2.$$

Choosing the same switch FET K_S as Example 16.4-1 reduces the output LOW voltage to $V_{oL} \approx 0.25$ V and greatly simplifies integrated circuit layout. For these reasons, a reasonable choice might be

$$K_S = \textbf{500 } \boldsymbol{\mu}\textbf{A/V}^2.$$

Other choices will fulfill the design goals adequately. For example, identical FETs result in $V_{oL} \approx 0.4$ V.

As was previously stated, the active load for an NMOS gate can be either an enhancement-mode or depletion-mode FET. When NMOS gates use a depletion-mode load, the gate and source terminals of the active load are shorted together. This connection ensures that the depletion-mode load is always in the saturation region of operation. In the depletion-mode case, the integrated circuit fabrication procedures are more complex but typically lead to faster switching speeds. Analysis techniques, while not discussed here, are similar to those of the enhancement-mode active load case.

Rather than switches mixed with active, resistive loads, it is possible to produce NAND and NOR logic operation using only controlled switches. In this type of realization, the resistive load is replaced by a group of *oppositely* controlled switches. As shown in Figure 16.4-3, each input simultaneously produces an action on one of the positively controlled switches and the opposite action on its counterpart in the negatively controlled switches. CMOS logic gates are founded on this realization of logical switching.

CMOS Gates

Complementary MOS logic gates are available in SSI packages and are found in many LSI and VLSI applications such as calculators and watches. The very popular 74HCXX series of logic gates is an example of SSI CMOS logic. As the physical scale of integrated circuit realization becomes smaller, CMOS is becoming the most significant form of MOS gate in VLSI applications. Part of this rise in CMOS circuitry is due to its very low power consumption.

In standard CMOS each input is connected to an individual NMOS FET *and* a PMOS FET. The complementary channel FETs act as opposite-acting switches with this connection. A logic HIGH signal closes the NMOS switch and opens the PMOS switch; a LOW signal produces the reverse actions. An example of this dual connection, a two-input NAND gate, is shown in Figure 16.4-4a. This NAND gate consists of two NMOS transistors in series connected to two PMOS transistors in parallel.

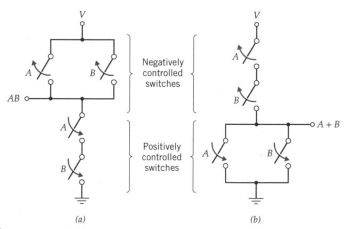

(a)   (b)

Figure 16.4-3
Conceptual switch-only two-input logic gates: (*a*) NAND; (*b*) NOR.

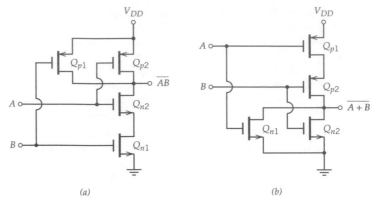

Figure 16.4-4
Two-input CMOS logic gates: (*a*) NAND; (*b*) NOR.

Whenever an input turns one of the NMOS FETs ON, the corresponding PMOS FET is turned OFF (i.e., Q_{n1} on implies Q_{p1} OFF). Thus, the output is switched to its LOW state only if *both* NMOS transistors are on (both inputs are HIGH) but is connected to V_{DD} when *either* PMOS is ON (either input is LOW). This is the ideal form for a NAND gate. A two-input NOR gate is the exact dual structure of the NAND gate: There are two parallel NMOS FET transistors connected to two series PMOS FETs, as shown in Figure 16.4-4*b*. Each additional input adds two FETs: one of each type connected in series or parallel, as appropriate, to its matching-type FETs.

Complementary MOS gates are characterized by very stable logic voltage levels. Each FET switches between the cutoff and ohmic regions of operation. In the cutoff region of operation, no drain current flows through the FET. In the CMOS gate, ohmic region FETs are, in all stable states, in series with a cutoff FET (or combinations of cutoff FETs) and must also have no drain current. Consequently, CMOS gates consume power only during the switching transient. During this transient, a short-duration current pulse flows through the FETs, leading to low total power consumption. Unfortunately, it also generates significant electrical noise.

The transfer characteristic of a CMOS gate can be approached by analyzing the states of the individual FETs. An ohmic region FETs is described by

$$I_D = K_S[2(V_{GS} - V_T)V_{DS} - V_{DS}^2]. \tag{16.4-14}$$

If the drain current must be zero valued, the implication of this simple application of KCL is

$$V_{DS} = 0.$$

The logic voltage levels of a CMOS gate are therefore limited by the supply voltage and ground:

$$V_{oH} = V_{DD}, \qquad V_{oL} = 0.$$

The voltage at which the logic transitions occur is not easily characterized for CMOS gate circuits. It is a function of the *p*-channel and *n*-channel FET transconductance

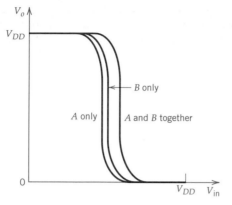

Figure 16.4-5
Two-input CMOS NAND gate transfer relationship.

factors K_N and K_P as well as which inputs are in transition. A typical transfer relationship for a two-input CMOS NAND gate is shown in Figure 16.4-5. When only input A is in transition, the transition occurs at the lowest voltage; only B in transition occurs at a slightly higher input voltage. When both inputs are simultaneously in transition, the transition occurs at its highest voltage level.

The spread in the transition region can be investigated by observing the midpoint of the output voltage transition ($V_o = \frac{1}{2}V_{DD}$) for the two extreme cases. For simplicity of discussion, assume the output is in transition from logic LOW to HIGH. For a logic LOW, both inputs to a NAND gate are high and the FETs are in the following regions of operation:

Q_{n1} and Q_{n2} in ohmic region

Q_{p1} and Q_{p2} in cutoff

The lowest voltage transition occurs when the input voltage V_{in} is connected to terminal A. At the midpoint of the transition for this case, Q_{n2} and Q_{p2} enter the saturation region while the two other FETs remain in their previous states. The currents through the FETs are

$$I_{Dp2} = -K_p(V_{in} - V_{DD} + V_T)^2, \tag{16.4-15}$$

$$I_{Dn2} = K_n(V_{in} - V_{DS1} - V_T)^2, \tag{16.4-16}$$

$$I_{Dn1} = K_n[2(V_{DD} - V_T)V_{DS1} - V_{DS1}^2]. \tag{16.4-17}$$

Simultaneous solution of these three equations, knowing that the currents must have the same magnitude, leads to the value of the input voltage V_{in} at which the transition occurs. If both inputs transition at the same time, Q_{p1} is also in the saturation region at the midpoint of the transition. This difference in FET state leads to an increased n-channel drain current since both p-channel FETs are contributing to the total current.

As with NMOS gates, CMOS gates are optimized for size and speed as well as voltage levels. Often these design goals lead to differently characterized (and sized) n-channel and p-channel FETs.

EXAMPLE 16.4-3 Determine the range of input voltages at which the midpoint of a logic transition occurs for a two-input CMOS NAND gate that uses FETs described by

$$|V_T| = 1 \text{ V}, \qquad \frac{K_n}{K_p} = 2,$$

and a 5-V power supply.

SOLUTION

The midpoint of the logic transition is found by the simultaneous solution of Equations 16.4-15 through 16.4-17. There are actually three input voltages at which the midpoint of the transition occurs: Each depends on which input is in transition. Simultaneous solution of several nonlinear equations is best accomplished numerically. A MathCAD solution for two cases (input A only and both A and B) follows. The simultaneous change is described by noting that both p-channel FETs have the same terminal voltages applied:

```
Two-input CMOS NAND gate transition solution
Vdd:=5          Vt:=1
Kn:=2           KP:=1
Guess Values
Vin:=2          Vds1:=1
Given    (Solve Block)
(Vin − Vds1 − Vt)²=2·(Vdd − Vt)·Vds1 − Vds1²
Kn·(Vin − Vds1 − Vt)²=N·Kp·(Vin − Vdd + Vt)²
''N'' indicates the number of p-channel FETs in saturation
Ans(N):=find(Vin,Vds1)
N:=1 . . . 2
Ans(N)₀       Ans(N)₁
```

Ans(N)$_0$	Ans(N)$_1$
2.345	0.175
2.622	0.245

The lowest single-input transition midpoint occurs at an input voltage of 2.345 V while the simultaneous double transition occurs at an input voltage of 2.622 V. Slight alteration of the program to model a single transition of input B (Equations 16.4-16 and 16.4-17 need slight modification) results in an input transition voltage of 2.388 V.

Transmission Gates

A transmission gate has an output signal that duplicates its input signal when a third signal, the Enable signal, is present. When the Enable signal is in its other state (often called the Inhibit state) the transmission gate is opened. A very simple CMOS realization of a transmission gate is shown in Figure 16.4-6. The transmission path is A to C, and the Enable/Inhibit signal is applied with opposite polarity to the two MOSFET gate terminals. Digital transmission gates are very similar to the parallel CMOS analog switch discussed in Section 4.5, but optimized for single-direction transmission.

Transmission gates are often used with a clock signal entering the Enable terminal. As such, the transmission gate serves to gate signal on or off. These gates are

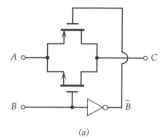

B	$\bar{B}$	A	C
0	1	0	open
0	1	1	open
1	0	0	0
1	0	1	1

(a)   (b)

Figure 16.4-6
CMOS digital transmission gate: (*a*) simplified circuit diagram; (*b*) truth table.

commonly found in multiplexers and other digital devices requiring signal switching.

16.5 BISTABLE LOGIC CIRCUITS

Electronic circuits constructed in such a manner so that two distinct, stable output states exist are commonly identified with the term *bistable*. Bistable logic circuits perform many important functions in digital circuitry. Level latches, counters, shift registers, and memories all depend on bistable circuitry. The most significant property of a bistable circuit is its ability to maintain a given stable output until an external input is applied. Application of an appropriate external input will cause a bistable circuit to change state in a predictable manner. The bistable circuit then holds the new output state until another appropriate input is applied. The most significant examples of bistable circuits are as follows:

- Latches
- Flip-flops
- Schmitt triggers

A *latch* is the simplest form of bistable circuit. This circuit "latches" its output to be the same logic level as its last valid input. The latch then holds the output at the logic level until another valid input forces a change in the latch state. As such, the latch is a very simple form of memory circuit. Latches are especially significant on shared data buses where values must be held while the bus transmits other data. *Flip-flops* are typically derived from latches. Most significant among the changes is the requirement that a clock pulse be present in order for a flip-flop to switch states. While many latches have an indeterminate output state, the output of a flip-flop is always determinate. The output of a flip-flop depends not only on the inputs but also on the current state of its output. In that sense, it also is a form of memory circuit. The *Schmitt trigger* finds greatest use in speeding the rise and fall times of digital signals that, for various reasons, have level transitions that are too slow for accurate logical manipulations. It is characterized by an input/output characteristic displaying hysteresis. Detailed presentation of bistable circuitry is beyond the scope of this discussion. Only a few common circuits are presented for demonstration of the principles.

Set–Reset Latch
The set–reset (SR) latch is a very common form of single-bit retention circuit. The SR latch is formed by cross-coupling the outputs of a pair of NOR gates into the inputs

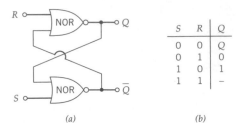

S	R	Q
0	0	Q
0	1	0
1	0	1
1	1	–

(a)   (b)

Figure 16.5-1
Basic SR latch: (*a*) logic diagram; (*b*) characteristic table.

of the opposite member of the pair, as shown in Figure 16.5-1. The output Q of the latch transitions to match the S input when only one input is HIGH. When both inputs are LOW, the latch retains its last value of Q and holds it until at least one input transitions to HIGH. The terms *set* and *reset* refer to the action of the output Q. Here Q sets (transitions to HIGH) when the set input S is HIGH; Q resets (transitions to LOW) when the reset input R is HIGH. Unfortunately the simple SR latch has a state that must be avoided: When *both inputs* are HIGH, the output is indeterminate.

A common realization of an SR latch using two NOR gates is shown in CMOS form is Figure 16.5-2.

Other integrated circuit gate families may use a different realization of the latch. In particular, it is more efficient in I^2L gate realizations to use a NAND form of the SR latch (Figure 16.5-3). This NAND realization produces the same logic characteristic as the NOR realization.

JK Flip-Flop

Flip-flops are an augmentation of a basic latch[16] that removes the indeterminate state present when both inputs are HIGH. One common flip-flop is the JK flip-flop, logically realized in Figure 16.5-4*a*. The addition of two three-input AND gates and an addition feedback path removes the ambiguity in the logic table so that if both inputs

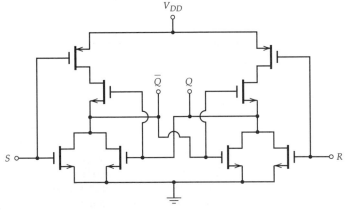

Figure 16.5-2
CMOS SR latch (NOR realization).

---

[16]Terminology has not been effectively standardized for bistable circuitry. Many sources prefer to consider the SR latch a primitive form of flip-flop; however, integrated circuit terminology usually reserves the term for the more complex circuitry.

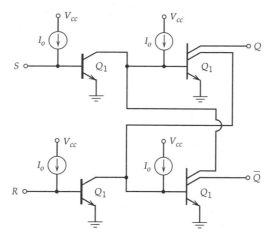

Figure 16.5-3
I^2L SR latch (NAND realization).

are HIGH, the output Q inverts. The addition of a clock signal avoids many of the problems associated with noisy input signals.

A JK flip-flop will only transition between states during the presence of a clock pulse. During that clock pulse the inputs and outputs are combined to form the logic table of Figure 16.5-4*b*. This table is the same as for an SR latch with the single exception that two HIGH inputs toggle the output to its complement in a JK flip-flop where that state resulted in an indeterminate state in the SR latch.

A characteristic of all JK flip-flops is that the output will toggle (change to the opposite state) when clocked in the presence of a HIGH signal at both inputs. Operated in that mode the circuit becomes a *T flip-flop* and is particularly useful in digital counters.

Since latches and flip-flops have standard logic gates as basic functional components, they are subject to many of the same speed restrictions: Gates with short delay times lead to fast latches or flip-flops. Typically the speed of a latch or flip-flop is specified through the *maximum clock frequency*. The maximum clock frequency is simply the highest rate at which the clock input of a bistable circuit can be driven while maintaining proper operation. Other significant operational parameters for these circuits are as follows:

- *Setup and hold times*: It is necessary that the input data arrive a short time before the triggering edge of the clock pulse and remain a short time after. These times are the setup time and hold time, respectively.

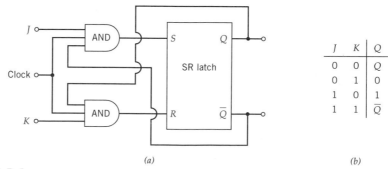

J	K	Q
0	0	Q
0	1	0
1	0	1
1	1	$\overline{Q}$

(a)   *(b)*

Figure 16.5-4
Basic JK flip-flop: (*a*) logic diagram; (*b*) characteristic table.

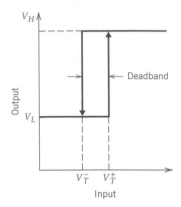

Figure 16.5-5
Schmitt trigger transfer characteristic hysteresis.

- *Clock HIGH and LOW pulse widths*: the minimum time that the clock must remain in its HIGH and LOW states for proper gate operation.

Schmitt Trigger

The output of a Schmitt trigger is bistable and has very steep transition regions. The characterizing feature of the Schmitt trigger transfer function is the presence of two separate transition regions, one for positive-slope and one for negative-slope signals, separated by a deadband region. The resultant transfer relationship exhibits hysteresis, as shown in Figure 16.5-5. Any input signal below the negative-slope transition voltage V_T^- results in a low output V_L. If the output state is LOW, it will not transition to the HIGH state unless the input is greater than the positive-slope transition voltage V_T^+. Similarly, any input above V_T^+ results in a HIGH output, V_H, that will not transition to the LOW state unless the input falls below V_T^-. Thus signals, after crossing a threshold, do not respond to input signal changes unless the variation is large enough to cross the deadband.

The Schmitt trigger is especially useful in converting slowly varying or a noisy signal into a clean digital form with sharp transitions. Another common usage is converting sine-wave input into a pulse-train output.[17] The dependence of the output on both level and level derivative is unique.

As with all bistable circuits, realizations of Schmitt triggers are possible using both FETs and BJTs. Two common realizations are shown in Figure 16.5-6. Similar circuits are available in most transistor gate families.

EXAMPLE 16.5-1 The bipolar Schmitt trigger circuit of Figure 16.5-6a is constructed with component values

$$V_{cc} = 5 \text{ V}, \qquad R_1 = 3.5 \text{ k}\Omega, \qquad R_2 = 2.6 \text{ k}\Omega, \qquad R_e = 1 \text{ k}\Omega$$

and BJTs described by

$$\beta_F = 50.$$

Determine the trigger voltages for positive and negative slope signals V_T^+ and V_T^-.

---

[17] Analog Schmitt trigger applications are presented in Section 13.1.

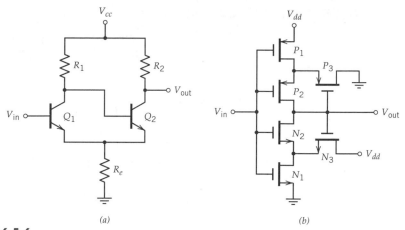

Figure 16.5-6
Typical Schmitt trigger circuits: (*a*) bipolar; (*b*) CMOS.

SOLUTION

When V_{in} is LOW, Q_1 is cut off and Q_2 is in saturation. Two loop equations can be written:

$$5 - 3500I_{b2} - 0.8 - 1000(I_{b2} + I_{c2}) = 0,$$
$$5 - 2600I_{c2} - 0.2 - 1000(I_{b2} + I_{c2}) = 0.$$

The solution to this pair of equations is

$$I_{b2} = 679 \ \mu A, \qquad I_{c2} = 1.145 \ mA \Rightarrow V_{out}$$
$$= 2.024 \ V, \qquad V_{e2} = 1.824 \ V.$$

The change of state will begin when Q_1 begins to turn on:

$$V_T^+ \approx V_{e2} + V_{\gamma 1} = 1.824 + 0.5 = \textbf{2.33 V}.$$

When V_{in} is HIGH, Q_1 is in saturation and Q_2 is cut off: $V_{out} = 5$ V. The low threshold voltage can be found by determining when Q_2 begins to turn on. For this to happen, Q_1 must enter the forward-active region and have a collector–emitter voltage equal to the cut-in voltage of Q_2, $V_{\gamma 2}$. The collector–emitter voltage of Q_1 (with Q_2 OFF) is given by

$$5 - 3500 \frac{\beta_F}{\beta_F + 1} I_{e1} - 1000I_{e1} = V_\gamma = 0.5.$$

The solution to this equation is

$$I_{e1} = 1.015 \ mA, \qquad V_T^- = V_{in} = 1000I_{e1} + 0.7 = \textbf{1.72 V}.$$

16.6 SEMICONDUCTOR MEMORIES

Memories are devices that are capable of the storage and retrieval of large amounts of digital data often required in modern digital systems. A wide variety of memory systems are available:

- Semiconductor memories
- Magnetic core memories
- Magnetic bubble memories
- Moving-surface memories (magnetic tape, disc, etc.)

Discussion here will be focused on the first of these possibilities: semiconductor memories. In addition, the discussion will be restricted to *random-access memory* (RAM) and exclude *read-only memory* (ROM).

Semiconductor RAM is typically composed of large arrays of identical, single-bit memory cells that are accessed through row and column decoding electronics. The general topology for such an array is shown in Figure 16.6-1. Each cell must be capable of reading data from an external source, storing the data until needed, and then writing the data to an external element. Consequently, the memory array includes both address/command lines as well as data lines. Since memory addressing is based on binary codes, the number of rows and columns are integral powers of 2.

The topology of the individual memory cells distinguishes the various types of semiconductor RAM. A RAM cell typically stores information in either a digital latching circuit or in the charge on a capacitor. Latches can hold their state indefinitely (assuming no loss of electrical power), while the charge in a capacitor-based storage cell gradually dissipates and must be periodically refreshed by external circuitry. The term *static random access memory* (SRAM) applies to latch-based cells, while the term *dynamic random access memory* (DRAM) applies to cells that must be refreshed.

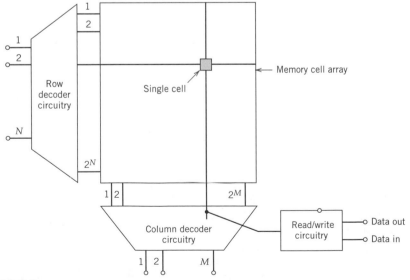

Figure 16.6-1
Semiconductor RAM array topology.

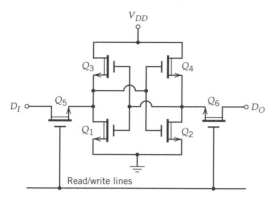

Figure 16.6-2
6-T SRAM CMOS memory cell.

Static Random Access Memory

Typical static memories are based upon the simple digital latches discussed in Section 16.5. While any of the bipolar or MOS latch realizations discussed can be used to form static memories, the MOS realizations are currently the dominant technology. Metal–oxide–semiconductor SRAM memory cells are characterized by high speed, low power consumption, and high reliability. A six-FET (6-T) CMOS memory cell based on a CMOS SR latch is shown in Figure 16.6-2. In this cell, input data D_I enters the latch through the FET switch formed by Q_5 and exits the latch through Q_6 as output data D_O. A logic HIGH signal on the read/write line activates the two FET switches, enabling the entrance or exit of data. A logic LOW signal on the read/write line disconnects the latch from external circuitry; without any external stimulation, the latch holds its current state. As with all CMOS circuitry, the current flow in this cell is extremely low except when the cell is in transition between memory states. The resultant low power consumption is extremely attractive for memory systems dependent on limited capacity power sources such as batteries.

An NMOS version of the 6-T cell is also available. In the NMOS cell, Q_3 and Q_4 are replaced by active loads, typically enhancement-mode FETs. The NMOS realization of the 6-T memory cell draws a more consistent, albeit higher, current from the power supply. The price of less electrical noise than the CMOS realization is increased power consumption.

Another variation of this cell is the four-FET (4-T) SRAM cell. This variation is essentially the same circuit as the NMOS realization. Here the active load FETs are replaced by polysilicon resistors. The variation allows for smaller cell size but increases power consumption and decreases reliability somewhat over the other two realizations.

Dynamic Random Access Memory

In many situations, the packing density of memory cells can be an important factor in memory design. The dependence of static RAM on latches prohibits a size reduction beyond the 4-T size. In addition, the control lines to a single SRAM cell can be as many as six. Dynamic RAM can break these barriers by basing data memory on capacitor charge storage. The disadvantage of DRAM lies in the need to refresh the capacitor charge before leakage can degrade the data storage process. The need to keep the memory storage capacitors small in size results in a refresh process cycling every 2–4 ms. Even though this seems frequent, for very large memory arrays, DRAM can have a distinct advantage over SRAM.

Two basic memory cells dominate DRAM design: a three-FET (3-T) and a single-

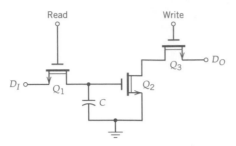

Figure 16.6-3
Typical 3-T DRAM memory cell.

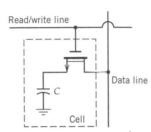

Figure 16.6-4
Simplified 1-T DRAM memory cell.

FET (1-T) design. The 3-T design is shown in Figure 16.6-3. It is basically a capacitor, two switches, and an output buffer. Incoming data D_I enters through switch Q_1 and is stored in the capacitor. Switch Q_2 acts as an inverting buffer so that the capacitor is not significantly discharged when the output switch Q_3 is activated to write the output D_O. The inversion of the output signal due to the buffer is usually compensated in either the read or write path with a single inverter. The 3-T cell requires separate read and write lines, but the data lines may be combined at the expense of more complex data encoding/decoding electronics.

The simplest of all DRAM cells is the 1-T cell shown in Figure 16.6-4. Here the data storage capacitor is connected to a single data line through a single switch. Incoming data charges the capacitor through the switch. When data is to be read, the same switch connects the capacitor to the data line. This connection completely discharges the capacitor. It is then necessary to immediately refresh that data by imposing an amplified duplicate back onto the data line. Of course this immediate-refresh-after-read is in addition to the normal refresh process necessary in all DRAM. The cell size reduction possible in the 1-T warrants, in many situations, the added complexity of the drive electronics.

16.7 GALLIUM ARSENIDE LOGIC

Gallium arsenide (GaAs) integrated circuits have a distinct speed advantage over similarly sized silicon integrated circuits. Gate operating speeds in excess of 1 gigabit per second are currently available and the speed limit appears not to have been reached. In addition, GaAs circuits typically dissipate less power than silicon circuits. Unfortunately, only two logic functions, an inverter or a NOR gate, can currently be constructed as a single logic stage. All more complex functions must be created from arrays of gates. In addition, the processing technology of GaAs circuits is not at the same stage of maturity as silicon integrated circuit processing. Currently, GaAs wafers are significantly smaller than silicon wafers. It is also currently difficult to produce transistors within a chip with uniform electrical properties compared to silicon MOS device variation. This parameter variation, coupled with a relatively high thermal coefficient for GaAs and relatively high power dissipation, places severe restrictions on noise margins and the reliability of logical operations. Still, it appears that GaAs circuits have a significant future.

Since a native oxide of GaAs does not exist, MOS-like structures are not possible in GaAs integrated circuits. Therefore other transistor structures have been developed. Three basic transistor structures have currently been shown to be useful in GaAs circuits:

- Metal–semiconductor field-effect transistors (MESFETs)
- Heterojunction field-effect transistors (HFETs)
- Heterojunction bipolar transistors (HBTs)

MESFETs are the current dominant GaAs transistor structure. Operation of a MESFET is similar to a JFET where a metal–semiconductor junction takes the place of the p–n junction of the JFET. As seen in Section 16.2, appropriate metal–semiconductor junctions form Schottky barrier diodes. Interestingly, MESFETs can be fabricated as either depletion-mode or enhancement-mode FETs. Voltage–current relationships for GaAs MESFETs are essentially the same as for silicon JFETs with parameters in the ranges

$$-2.5 \text{ V} < V_P < -0.2 \text{ V} \quad \text{or} \quad 0.1 \text{ V} < V_T < 0.3 \text{ V}.$$

Typical GaAs MESFET Schottky-barrier voltage is in the range of 0.8 V. HFETs are characterized by voltage–current relationships similar to MESFETs and are therefore not specifically discussed. HBTs are functionally similar to BJTs. The discussion here is limited to MESFET gate structures.

There are two basic families of GaAs MESFET logic circuits:

- Enhancement–depletion logic
- Source-coupled logic

The enhancement–depletion family bears considerable similarity to NMOS logic, and source-coupled logic is similar to the bipolar ECL. Because of these similarities, discussion here will be descriptive, rather than quantitative, in scope. Dominance among the GaAs logic families is not yet firmly established.

Enhancement–Depletion Logic

Gallium arsenide enhancement–depletion (ED) logic circuits share the same topology with NMOS logic circuits. As an example, the topology of the GaAs NOR gate shown in Figure 16.7-1 is essentially the same as the NMOS NOR gate shown in Figure 16.4-2b. In this realization, Q_1 and Q_2 are enhancement-mode FETs and Q_3 is a depletion-mode FET used as an active load.[18] However, the functional differences between a GaAs MESFET and a silicon MOSFET restricts direct comparisons. In particular, the Schottky-barrier diode inherent to the gate of a GaAs device allows gate current to flow when gate–source voltages exceed the Schottky barrier voltage. Consequently voltage swing must be kept small in order to avoid this detrimental gate current condition.

---

[18]The NMOS gate in Figure 16.4-2b uses an enhancement-mode FET as an active load rather than a depletion-mode FET. While GaAs requires a depletion-mode active load, NMOS can be fabricated in either form; operation of the gate is the same.

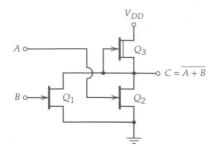

Figure 16.7-1
Simple GaAs enhancement–depletion NOR gate.

The principal differences between GaAs ED logic and NMOS logic are as follows:

- For a HIGH output, any slave gate attached to the output will clamp the output voltage to the Schottky barrier voltage ($\approx$0.8 V). The output voltage in NMOS is limited to the much larger value, $V_{DD} - V_T$.

- For a HIGH output, the current in the active load is not zero: Power is dissipated for both HIGH and LOW output levels. The NMOS dissipates essentially zero power for a HIGH output.

- When the input is HIGH, the output is LOW unless the input voltage exceeds the Schottky barrier voltage. Further increases in the input voltage increase the output voltage. If sufficiently high input voltage is applied, logical errors will occur. NMOS does not have this potential problem.

- Voltage swings are limited to the Schottky barrier voltage ($\approx$0.8 V). The NMOS swings can be much larger.

- The power supply voltage for GaAs ED logic needs to be only slightly larger than the Schottky barrier voltage. This supply voltage can therefore be significantly smaller than that of NMOS logic.

- The threshold voltage for GaAs enhancement FETs must be less than the turn-on voltage of the Schottky diode. The NMOS FETs do not have this restriction.

While it is possible to construct a GaAs NAND gate using the same topology as an NMOS NAND gate, the very small differences in voltage levels between a logic HIGH and LOW reduce the noise margin to unacceptable levels. The "stacking" of enhancement FETs to create the logic NAND function increases the logic LOW and is responsible for this reduction in noise margins. Gallium arsenide NAND gates are not commercially available.

Source-Coupled Logic

Gallium arsenide source-coupled (SC) logic is based upon a FET differential amplifier[19] in the same manner as ECL is based on a BJT differential amplifier. A simple two-input OR gate is shown in Figure 16.7-2. If either input A or B is a logic HIGH, Q_3 will enter cutoff and the output C will become a logic high. Conversely, only if both inputs are a logic LOW will Q_3 enter saturation and the output go to a logic LOW. The logic voltage levels for an unloaded gate are

$$V_{oH} \approx V_{DD} - V_T, \qquad V_{oL} \approx V_T.$$

This logic swing is significantly larger than other forms of GaAs logic. Noise margins are also significantly improved.

---

[19]Source-coupled differential amplifiers are discussed in Section 6.3.

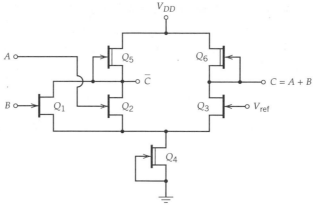

Figure 16.7-2
Simple GaAs source-coupled OR/NOR gate.

One particularly attractive feature of SC logic is its insensitivity to transistor parameter variation. Unfortunately, SC logic consumes significantly more power than other forms of GaAs logic. A simple SC OR gate is roughly twice the size of the same gate in GaAs ED logic. Gate speed is essentially equivalent in the two logic families.

16.8 CONCLUDING REMARKS

The *essential aspects* of the electronic operation of the major digital circuits commonly in use have been presented in this chapter. Discussion has been limited to the basic building blocks of each of the major logic families commonly in use. A short comparison of a four commercial logic gate families is shown in Table 16.8-1.

TABLE 16.8-1 COMPARISON OF COMMERCIAL LOGIC GATE PROPERTIES

Property	TTL		ECL, Series 100K	CMOS, Series HC
	Series H	Series LS		
Basic gate	NAND	NAND	OR/NOR	NOR or NAND
Fan-out	10	20	25	>50
Power dissipation	22.5 mW	2 mW	40 mW	1.75 mW at 1 MHz
Noise margin Low	0.4 V	0.3 V	0.14 V	1.25 V
High	0.4 V	0.7 V	0.14 V	1.25 V
Propagation delay	6 ns (C_L = 25 pF)	9.5 ns (C_L = 15 pF)	0.75 ns (C_L = 50 pF)	8 ns (C_L = 50 pF)
Maximum clock frequency (D flip-flop)	50 MHz	33 MHz	350 MHz	50 MHz

NOTE: All characteristics are typical limits for safe performance.

SUMMARY DESIGN EXAMPLE

An existing TTL (LS) digital circuit is to be interfaced to a new system using CMOS (HC) digital gates. Both systems are to be operated from a 5-V power source. The nominal input and output specifications of the two logic families for 5-V operation are known to be as follows:

	LS TTL	HC CMOS
$V_{OH(min)}$	2.7 V	4.4 V
$V_{IH(min)}$	2.0 V	3.15 V
$V_{OL(max)}$	0.5 V	0.1 V
$V_{IL(max)}$	0.8 V	0.9 V
$I_{OL(max)}$	8 mA	4 mA
$I_{IL(max)}$	−0.4 mA	−1 μA
$I_{OH(max)}$	−0.4 mA	−4 mA
$I_{IH(max)}$	20 μA	1 μA
Propagation delay	9 ns	8 ns

The small difference in propagation delay presents no significant problems in design of an interface between the two circuits. There is, however, a mismatch in acceptable logic levels. The CMOS output logic levels are compatible with the TTL input levels, but the TTL HIGH output ($V_{OH(min)} = 2.7$ V) is not compatible with the CMOS HIGH input requirements ($V_{IH(min)} = 3.15$ V). Interface circuitry to alleviate this incompatibility is to be designed.

Solution:

The output level of LS TTL gates is determined by the active pull-up circuitry. The voltage drop across this circuitry, even with minimal load current, does not allow the output voltage to *consistently* rise above the levels necessary for HCMOS circuitry. A variety of possible interfaces can be considered:

- Buffer the TTL output with a TTL gate that has passive pull-up.
- Buffer the TTL output with an HCT gate (these are HCMOS gates specifically modified to accept TTL levels at the input.
- Operate the CMOS portion of the circuit at a lower voltage level, that is, 3 V.
- Shunt the TTL active pull-up with an external, passive pull-up.

Buffering can provide a good solution. The only significant drawbacks are an addition of components and the additional propagation delay induced by the buffer. The TTL gates with passive pull-up may introduce a large additional propagation delay and are therefore discarded as an alternative. If the new CMOS system can be implemented with HCT gates as the input gates, this is probably the best solution. Operation of the CMOS portion of the circuit at 3 V will lower the required HIGH-input-voltage requirement below the TTL HIGH output. Unfortunately this also reduces noise margin and violates the constraints in the design requirements. If HCT input gates are not possible, shunting the TTL active pull-up with an external, passive pull-up as shown seems a viable design alternative.

An external resistor that shunts the output of the LS TTL gate will force the HIGH level at the interface near 5 V. The minimum value of the pull-up resistor R_p

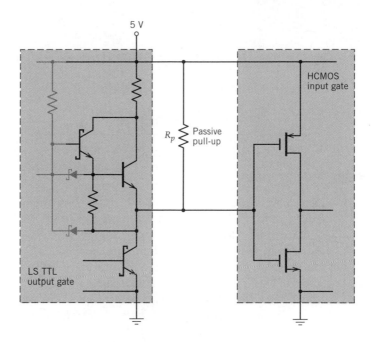

is determined by the current sinking capability of the TTL gate: The LOW voltage must remain within specifications. For a logic LOW, the maximum current through the resistor is given by the sum of the maximum TTL output current and N CMOS input currents. This LOW voltage requirement restricts R_p to

$$R_p > \frac{V_{CC} - V_{OL(max)}^{TTL}}{I_{OL(max)}^{TTL} - NI_{IL}^{HC}} \approx 563 \; \Omega.$$

The maximum value of R_p is determined by the output voltage rise time. This voltage rise is a complex process. It rises to $V_{OH(min)}$ very quickly (≈ 9 ns) due to the active pull-up. Thereafter, the rise will be exponential due to the RC time constant formed by the input capacitance of the CMOS gates and R_p. The time period for the voltage to exceed the CMOS $V_{IH(min)}$ is given by the solution to

$$v(t) = V_{CC} - (V_{CC} - V_{OH(min)}^{TTL})e^{-t/R_pC} = V_{IH(min)}^{HC},$$
$$v(t) = 5 - 2.3e^{-t/R_pC} = 3.15 \Rightarrow t = 0.22 \, R_pC.$$

If a design using a shunt, passive pull-up resistor is to be better than a design using a buffer gate, the additional rise time must be shorter than the propagation delay due to a buffer. That is,

$$t < 9 \text{ ns} \Rightarrow R_p < \frac{9 \text{ ns}}{0.22C} = \frac{40.9 \text{ ns}}{C}.$$

The total capacitance C is given by the input capacitance of the total number of CMOS gates being driven by the TTL gate. A typical value is $C \approx 10$ pF. This capacitive assumption leads to

$$R_p < 4.09 \text{ k}\Omega.$$

Thus

$$563 \ \Omega < R_p < 4.09 \ \text{k}\Omega.$$

A reasonable, standard-value choice is

$$R_p = 1.5 \ \text{k}\Omega.$$

REFERENCES

High-Speed CMOS Logic Data Book, Texas Instruments, Dallas, TX, 1988.
Buchanan, J., *CMOS/TTL Digital Systems Design*, McGraw-Hill, New York, 1990.
Glasford, G., *Digital Electronic Circuits*, Prentice-Hall, Englewood Cliffs, NJ, 1988.
Haznedar, H., *Digital Microelectronics*, Benjamin/Cummings, Redwood City, 1991.
Hodges, D., and Jackson, H., *Analysis and Design of Digital Integrated Circuits*, 2nd ed., McGraw-Hill, New York, 1988.
Shoji, M., *CMOS Digital Circuit Technology*, Prentice-Hall, Englewood Cliffs, NJ, 1988.
Taub, H., and Schilling, D., *Digital Integrated Electronics*, McGraw-Hill, New York, 1977.
Wing, O., *Gallium Arsenide Digital Circuits*, Kluwer Academic, Boston, 1990.

PROBLEMS

16-1. The BJT in the simple inverter shown is described by

$$\beta_F = 75, \qquad f_T = 200 \ \text{MHz}.$$

If the input voltage V_{in} has logic levels 0 and 5 V, determine:

    a. Rise time

    b. Delay time

    c. Fall time

    d. Storage time

    e. Average propagation delay

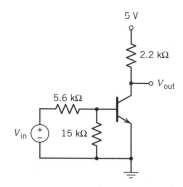

16-2. Assume the power supply for the simple inverter of problem 16-1 is reduced to 3.3 V and the input logic levels become 0 and 3.3 V. The transistor parameters are unchanged. Determine:

    a. Rise time

    b. Delay time

    c. Fall time

    d. Storage time

    e. Average propagation delay

16-3. In an attempt to decrease the propagation delay of the simple inverter described in problem 16-1, the collector resistor is reduced to 1.5 kΩ. Comment *quantitatively* on the advisability of this design change.

16-4. The BJTs in the inverter with active pull-up shown are described by

$$\beta_F = 75, \qquad f_T = 200 \ \text{MHz}.$$

If the input voltage V_{in} has logic levels 0 and 5 V, determine:

    a. Rise time

    b. Delay time

c. Fall time

d. Storage time

e. Average propagation delay

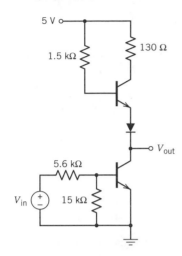

16-5. The BJTs in the inverter with active pull-up shown are described by

$$\beta_F = 75, \qquad f_T = 200 \text{ MHz.}$$

If the input voltage V_{in} has logic levels 0 and 5 V, determine:

a. Rise time

b. Delay time

c. Fall time

d. Storage time

e. Average propagation delay

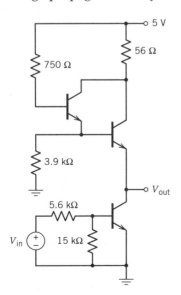

16-6. The BJTs in the inverter with active pull-up and pull-down as shown are described by

$$\beta_F = 75, \qquad f_T = 200 \text{ MHz.}$$

If the input voltage V_{in} has logic levels 0 and 5 V, determine:

a. Rise time

b. Delay time

c. Fall time

d. Storage time

e. Average propagation delay

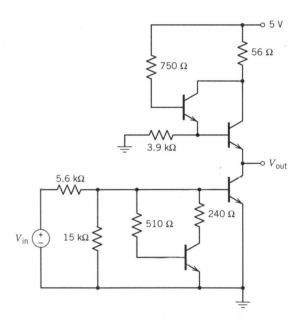

16-7. For the simple bipolar inverter shown:

a. Use SPICE to determine the gate speed. Assume the BJT has the same properties as a 2N2222 *npn* bipolar transistor.

b. Model a Schottky inverter by shunting the base–collector junction of the given BJT with a diode. Repeat the SPICE gate speed determination of part 1. Assume the diode has properties

$$I_s = 2 \text{ nA}, \qquad \eta = 1.1.$$

c. Comment on any gate speed differences.

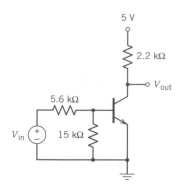

16-8. The circuit shown is a form of simplified TTL gate.

a. Determine the logical operation the gate performs on the four inputs $A,B,C,$ and D.

b. What are the logic voltage levels at the output V_{out}?

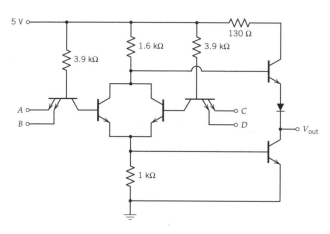

16-9. Compare the average power consumption of the simple ECL OR gate of Figure 3.5-5 to that of a 10K OR gate (Figure 16.2-9) with the following parameters:

$$R_{C2} = 218 \ \Omega, \qquad R_{C3} = 246 \ \Omega,$$

$$R_{EP} = 777 \ \Omega, \qquad R_1 = 909 \ \Omega,$$

$$R_2 = 6.12 \ k\Omega, \qquad R_3 = 4.99 \ k\Omega,$$

$$R_{IN1} = R_{IN2} = 51 \ k\Omega.$$

Assume a pull-down resistor at the output of 1.5 kΩ and BJTs with $\beta_F = 100$.

16-10. A CMOS inverter is fabricated using a 3.3-V supply and MOSFETs with the following properties:

$$K_N = 0.1 \ mA/V^2,$$

$$K_P = 0.25 \ mA/V^2, \quad |V_T| = 1.$$

a. Determine the average propagation delay time if it is driving a capacitive load of 5 pF.

b. Compare the results of part a to those found in Example 16.3-1.

16-11. A CMOS inverter is fabricated using a 5-V supply and MOSFETs with the following properties:

$$K_N = 0.4 \ mA/V^2,$$

$$K_P = 0.8 \ mA/V^2, \quad |V_T| = 1.2.$$

Determine the average propagation delay time if it is driving a capacitive load of 5 pF.

16-12. A CMOS inverter is fabricated using a 3-V supply and MOSFETs with the following properties:

$$K_N = 0.06 \ mA/V^2,$$

$$K_P = 0.15 \ mA/V^2, \quad |V_T| = 0.6 \ V.$$

Determine the average propagation delay time if the gate is driving a capacitive load of 15 pF.

16-13. A CMOS inverter is fabricated using transistors with the following properties:

$$K_N = K_P = 0.1 \ mA/V^2, \qquad |V_T| = 1.5 \ V.$$

Use SPICE to determine the voltage transfer characteristic for the following power supply conditions:

a. $V_{DD} < V_T$
b. $V_{DD} = V_T$
c. $V_{DD} = 2V_T$
d. $V_{DD} = 3V_T$

Comment on the results.

16-14. Given a 5-V power supply, design a three-input NMOS NOR gate to have the following output logic levels when driven by a gate of the same design:

$$V_{oH} = 4.0 \ V, \qquad V_{oL(max)} = 0.8 \ V.$$

What are the values of V_{oL} for one, two, and three logic high inputs?

16-15. Given a 3.3-V power supply, design a three-input NMOS NOR gate to have the following output logic levels when driven by a gate of the same design:

$$V_{oH} = 2.7 \text{ V}, \qquad V_{oL(max)} = 0.7 \text{ V}.$$

What are the values of V_{oL} for one, two, and three logic high inputs?

16-16. Given a 5-V power supply, design a three-input NMOS NAND gate to have the following output logic levels when driven by a gate of the same design:

$$V_{oH} = 4.0 \text{ V}, \qquad V_{oL(max)} = 0.8 \text{ V}.$$

16-17. Given a 3.3-V power supply, design a three-input NMOS NAND gate to have the following output logic levels when driven by a gate of the same design:

$$V_{oH} = 2.7 \text{ V}, \qquad V_{oL(max)} = 0.7 \text{ V}.$$

16-18. Determine a logical expression for the output Y of the CMOS circuit shown as a function of the three inputs A, B, and C. Use SPICE to verify the logical expression. Assume the MOSFETs are described by

$$|V_T| = 1 \text{ V}, \qquad K = 0.4 \text{ mA/V}^2.$$

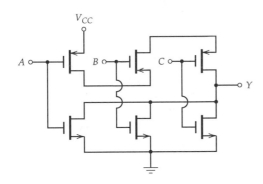

16-19. Given a 3-V power supply, design a three-input NMOS NAND gate to have the following output logic levels when driven by a gate of the same design:

$$V_{oH} = 2.0 \text{ V}, \qquad V_{oL(max)} = 0.5 \text{ V}.$$

16-20. Determine the range of input voltages at which the midpoint of a logic transition occurs for a two-input CMOS NAND gate that uses FETs described by

$$|V_T| = 0.5 \text{ V}, \qquad \frac{K_n}{K_p} = 2,$$

and a 3-V power supply.

16-21. Determine the range of input voltages at which the midpoint of a logic transition occurs for a two-input CMOS NAND gate that uses FETs described by

$$|V_T| = 1.0 \text{ V}, \qquad \frac{K_n}{K_p} = 4,$$

and a 5-V power supply. Compare this voltage range to that found in Example 16.4-3.

16-22. It is possible to form bipolar gates that operate similarly to MOS gates. One such circuit is shown. Determine the logic function implemented by this circuit.

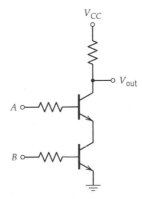

16-23. Use SPICE to implement the CMOS SR latch of Figure 16.5-2. Assume a 5-V power supply and MOSFETs described by

$$K_N = 0.1 \text{ mA/V}^2,$$

$$K_P = 0.25 \text{ mA/V}^2, \quad |V_T| = 0.6 \text{ V}.$$

Verify correct operation of the latch.

16-24. Design a bipolar Schmitt trigger using BJTs characterized by $\beta_F = 75$ and a 5-V power supply to meet the following design criteria:

- Threshold voltages: $V_T^+ = 2.2$ V and $V_T^- = 1.5$ V.

- Power consumption: Power supplied by the 5-V source must not exceed 5 mW in any stable state. Any current entering the input terminal of the Schmitt trigger is excluded from this calculation.

16-25. Design a bipolar Schmitt trigger using BJTs characterized by $\beta_F = 100$ and a 3.3-V power supply to meet the following design criteria:

- Threshold voltages: $V_T^+ = 2.0$ V and $V_T^- = 1.2$ V.

- Power consumption: Power supplied by the 3.3-V source must not exceed 3 mW in any stable state. Any current entering the input terminal of the Schmitt trigger is excluded from this calculation.

16-26. Use SPICE to determine the transfer characteristic for the given silicon bipolar Schmitt trigger. The BJTs are characterized by $\beta_F = 50$. Compare SPICE results to those obtained by the simple hand analysis of Example 16.5-1. Over what range of input voltages are the two transition regions? (*Hint*: In transient analysis SPICE may have difficulty converging for switching circuits. This can usually be eliminated by setting ITL4 = 40 in an **.OPTIONS** statement.)

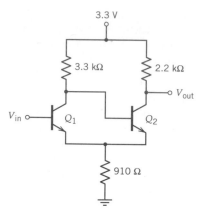

16-28. Use SPICE to determine the voltage transfer characteristic for the given MOSFET Schmitt trigger. The MOSFETs are characterized by

$$V_T = 1 \text{ V}, \qquad K = 0.2 \text{ mA/V}^2.$$

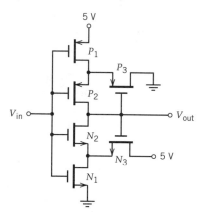

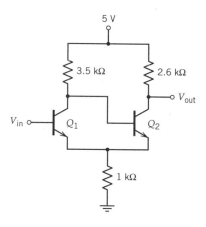

16-27. In the Schmitt trigger circuit shown, the BJTs are characterized by $\beta_F = 50$.

a. Using simple hand analysis, determine the positive- and negative-slope trigger voltages V_T^+ and V_T^-.

b. Use SPICE to validate results obtained by the simple hand analysis.

(*Hint*: In transient analysis SPICE may have difficulty converging for switching circuits. This can usually be eliminated by setting ITL4 = 40 in an **.OPTIONS** statement.)

16-29. An existing TTL (LS) digital system is to be interfaced to a new ECL digital system. The interface is unidirectional with the TTL system driving the ECL system. It has been suggested that that interface can be realized using a common-base amplifier. The nominal input and output specifications of the two logic families are known to be the following:

	LS TTL	10K ECL
$V_{oH(min)}$	2.7 V	−0.98
$V_{iH(min)}$	2.0 V	−1.105
$V_{oL(max)}$	0.5 V	−1.85
$V_{iL(max)}$	0.8 V	−1.475
$I_{oL(max)}$	8 mA	22.3 mA
$I_{iL(max)}$	−0.4 mA	65 μA
$I_{oH(max)}$	−0.4 mA	5.4 mA
$I_{iH(max)}$	20 μA	130.5 μA
Propagation delay	9 ns	2.5 ns

Complete the interface design and verify proper operation using SPICE.

16-30. It has been suggested that the TTL–CMOS interface described in the Summary Design Example could be improved upon using an active logic-level interface circuit. One such interface circuit is shown. Choose appropriate resistor values and compare the operation of this interface circuit to that of the Summary Design Example. Assume BJTs are described by $\beta_F = 50$.

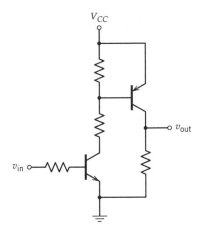

STANDARD RESISTANCE VALUES

0.5% 1%	5% 10%	0.5% 1%	5% 10%	0.5% 1%	5% 10%	0.5% 1%	5% 10%	0.5% 1%	5% 10%
10.0	10	16.4		26.4		40.7		63.4	
10.1		16.5		26.7		41.2		64.2	
10.2		16.7			27	41.7		64.9	
10.4		16.9		27.1		42.2		65.7	
10.5		17.2		27.4		42.7		66.5	
10.6		17.4		27.7			43	67.3	
10.7		17.6		28.0		43.2			68
10.9		17.8		28.4		43.7		68.1	
11.0	11	18.0	18	28.7		44.2		69.0	
11.1		18.2		29.1		44.8		69.8	
11.3		18.4		29.4		45.3		70.6	
11.4		18.7		29.8		45.9		71.5	
11.5		18.9			30	46.4		72.3	
11.7		19.1		30.1		47.0	47	73.2	
11.8		19.3		30.5		47.5		74.1	
12.0	12	19.6		30.9		48.1		75.0	75
12.1		19.8		31.2		48.7		75.9	
12.3		20.0	20	31.6		49.3		76.8	
12.4		20.3		32.0		49.9		77.7	
12.6		20.5		32.4		50.5		78.7	
12.7		20.8		32.8			51	79.6	
12.9		21.0			33	51.1		80.6	
13.0	13	21.3		33.2		51.7		81.6	
13.2		21.5		33.6		52.3			82
13.3		21.8		34.0		53.0		82.5	
13.5			22	34.4		53.6		83.5	
13.7		22.1		34.8		54.2		84.5	
13.8		22.3		35.2		54.9		85.6	
14.0		22.6		35.7		55.6		86.6	
14.2		22.9			36		56	87.6	
14.3		23.2		36.1		56.2		88.7	
14.5		23.4		36.5		56.9		89.8	
14.7		23.7		37.0		57.6		90.9	
14.9		24.0	24	37.4		58.3			91
15.0	15	24.3		37.9		59.0		92.0	
15.2		24.6		38.3		59.7		93.1	
15.4		24.9		38.8		60.4		94.2	
15.6		25.2			39	61.2		95.3	
15.8		25.5		39.2		61.9		96.5	
16.0	16	25.8		39.7			62	97.6	
16.2		26.1		40.2		62.6		98.8	

Note: Standard resistance values are obtained from the decade table by multiplying by powers of 10.

DEVICE SPICE PARAMETERS

A number of the SPICE examples and problems in this text use device parameters that are included in the evaluation version of the MicroSim PSpice parts libraries. They are included here so that users of other versions of SPICE can utilize the same set of parameters:

Part Name	Part Type
D1N750	Zener diode
MV2201	Voltage-variable capacitance diode
D1N4148	Switching diode
MBD101	Switching diode
Q2N2222	*npn* Bipolar transistor
Q2N2907A	*pnp* Bipolar transistor
Q2N3904	*npn* Bipolar transistor
Q2N3906	*pnp* Bipolar transistor
J2N3819	*n*-Channel junction field-effect transistor
J2N4393	*n*-Channel junction field-effect transistor
IRF150	*n*-Type power MOS field-effect transistor
IRF9140	*p*-Type power MOS field-effect transistor

Not included in the PSpice evaluation version libraries are macromodels of voltage regulators. These macromodels are provided here along with component values for six common voltage regulators so that all users can model these devices:

LM7805C	5-V positive-voltage regulator
LM7812C	12-V positive-voltage regulator
LM7815C	15-V positive-voltage regulator
LM7905C	5-V negative-voltage regulator
LM7912C	12-V negative-voltage regulator
LM7915C	15-V negative-voltage regulator

This text also takes advantage of other MicroSim/PSpice models that are not easily transported to other SPICE versions. For that reason the following device models *are not provided here* (PSpice simulation is recommended):

2N1595	Silicon-controlled rectifier
2N5444	Triac
555D	555 Timer subcircuit

DIODE MODEL PARAMETERS

```
.model D1N750     D(Is=880.5E-18 Rs=.25 IKf=0 N=1 Xti=3 Eg=1.11 Cjo=175p
+        M=.5516 VJ=.75 Fc=.5 Isr=1.859n Nr=2 Bv=4.7 Ibv=20.245m
+        Nbv=1.6989 Ibvl=1.9556m Nbvl=14.976 Tbvl=-21.277u)
*        Motorola    pid=1N750   case=DO-35
*        89-9-18 gjg
*        Vz=4.7 @ 20mA, Zz=300 @ 1mA, Zz=12.5 @ 5mA, Zz=2.6 @ 20mA
$
.model MV2201     D(Is=1.365p Rs=1 Ikf=0 N=1 Xti=3 Eg=1.11 Cjo=14.93p M=.4261
+        Vj=.75 Fc=.5 Isr=16.02p Nr=2 Bv=25 Ibv=10u)
*        Motorola    pid=MV2201 case=182-03
*        88-09-22 bam creation
$
.model D1N4148     D(Is=0.1p Rs=16 CJO=2p Tt=12n Bv=100 Ibv=0.1p)
*        85-??-??    Original library
$
.model MBD101     D(Is=192.1p Rs=.1 Ikf=0 N=1 Xti=3 Eg=1.11 Cjo=893.8f
+        M=98.29m Vj=.75 Fc=.5 Isr=16.91n Nr=2 Bv=5 Ibv=10u)
*        Motorola pid=MBD101 case=182-03
*        88-09-22 bam creation
$
```

BIPOLAR JUNCTION TRANSISTOR MODEL PARAMETERS

```
.model Q2N2222     NPN(Is=14.34f Xti=3 Eg=1.11 Vaf=74.03 Bf=255.9 Ne=1.307
+        Ise=14.34f Ikf=.2847 Xtb=1.5 Br=6.092 Nc=2 Isc=0 Ikr=0 Rc-1
+        Cjc=7.306p Mjc=.3416 Vjc=.75 Fc=.5 Cje=22.01p Mje=.377 Vje=.75
+        Tr=46.91n Tf=411.1p Itf=.6 Vtf=1.7 Xtf=3 Rb=10)
*        National    pid=19     case=T018
*        88-09-07 bam creation
$
.model Q2N2907A   PNP(Is=650.6E-18 Xti=3 Eg=1.11 Vaf=115.7 Bf=231.7 Ne=1.829
+        Ise=54.81f Ikf=1.079 Xtb=1.5 Br=3.563 Nc=2 Isc=0 Ikr=0 Rc=.715
+        Cjc=14.76p Mjc=.5383 Vjc=.75 Fc=.5 Cje=19.82p Mje=.3357 Vje=.75
+        Tr=111.3n Tf=603.7p Itf=.65 Vtf=5 Xtf=1.7 Rb=10)
*        National    pid=63     case=TO18
*        88-09-09 bam creation
$
.model Q2N3904     NPN(Is=6.734f Xti=3 Eg=1.11 Vaf=74.03 Bf=416.4 Ne=1.259
+        Ise=6.734f Ikf=66.78m Xtb=1.5 Br=.7371 Nc=2 Isc=0 Ikr=0 Rc=1
+        Cjc=3.638p Mjc=.3085 Vjc=.75 Fc=.5 Cje=4.493p Mje=.2593 Vje=.75
+        Tr=239.5n Tf=301.2p Itf=.4 Vtf=4 Xtf=2 Rb=10)
*        National    pid=23     case=TO92
*        88-09-08 bam creation
$
.model Q2N3906     PNP(Is=1.41f Xti=3 Eg=1.11 Vaf=18.7 Bf=180.7 Ne=1.5 Ise=0
+        Ikf=80m Xtb=1.5 Br=4.977 Nc=2 Isc=0 Ikr=0 Rc=2.5 Cjc=9.728p
+        Mjc=.5776 Vjc=.75 Fc=.5 Cje=8.063p Mje=.3677 Vje=.75 Tr=33.42n
+        Tf=179.3p Itf=.4 Vtf=4 Xtf=6 Rb=10)
*        National    pid=66     case=TO92
*        88-09-09 bam creation
$
```

JUNCTION FIELD-EFFECT TRANSISTOR MODEL PARAMETERS

```
.model J2N3819    NJF(Beta=1.304m Betatce=-.5 Rd=1 Rs=1 Lambda=2.25m Vto=-3
+           Vtotc=-2.5m Is=33.57f Isr=322.4f N=1 Nr=2 Xti=3 Alpha=311.7
+           Vk=243.6 Cgd=1.6p M=.3622 Pb=1 Fc=.5 Cgs=2.414p Kf=9.882E-18
+           Af=1)
*           National    pid=50      case=TO92
*         88-08-01 rmn BVmin=25
$
.model J2N4393    NJF(Beta=9.109m Betatce=-.5 Rd=1 Rs=1 Lambda=6m Vto=-1.422
+           Vtotc=-2.5m Is=205.2f Isr=1.988p N=1 Nr=2 Xti=3 Alpha=20.98u
+           Vk=123.7 Cgd=4.57p M=.4069 Pb=1 Fc=.5 Cgs=4.06p Kf=123E-18
+           Af=1)
*           National    pid=51      case=TO18
*         88-07-13 bam BVmin=40
$
```

METAL OXIDE FIELD-EFFECT TRANSISTOR MODEL PARAMETERS

```
.model IRF150     NMOS(Level=3 Gamma=0 Delta=0 Eta=0 Theta=0 Kappa=0 Vmax=0
+           Xj=0 Tox=100n Uo=600 Phi=.6 Rs=1.624m Kp=20.53u W=.3 L=2u
+           Vto=2.831 Rd=1.031m Rds=444.4K Cbd=3.229n Pb=.8 Mj=.5 Fc=.5
+           Cgso=9.027n Cgdo=1.679n Rg=13.89 Is=194E-18 N=1 Tt=288n)
*           Int'l Rectifier pid=IRFC150 case=TO3
*           88-08-25 bam creation
$
.model IRF9140     PMOS(Level=3 Gamma=0 Delta=0 Eta=0 Theta=0 Kappa=0 Vmax=0
+       Xj=0 Tox=100n Uo=300 Phi=.6 Rs=70.6m Kp=10.15u W=1.9 L=2u Vto=-3.67
+           Rd=60.66m Rds=444.4K Cbd=2.141n Pb=.8 Mj=.5 Fc=.5 Cgso=877.2p
+           Cgdo=369.3p Rg=.811 Is=52.23E-18 N=2 Tt=140n)
*           Int'l Rectifier pid=IRFC9140 case=TO3
*           88-08-25 bam creation
```

VOLTAGE REGULATOR (POSITIVE) MACROMODEL

```
.subckt X_VREG1 Input Output Ground params:
+     Av_feedback=1665, R1_value=1020
*
*Band-gap voltage source:
*
*    The source is off when Vin<3V and fully on when Vin>3.7V
*    Line regulation is set with Rreg=0.5 * dVin/dVbg. The
*    temperature dependence of this circuit is a quadratic fit to
*    the following points:
*                  T     Vbg(T)/Vbs(nom)
*                 ----   ------------
*                  0         .999
*                 3.75     1
*                 125       .990
*
*    The temperature coefficient of Rbg is set to 2 * the band gap
*    temperature coefficient. Tnom is assumed to be 27 deg. C and
*    Vnom is 3.7V
*
Vbg 100 0 DC 7.4V
Sbg (100,101)(Input,Ground)Sbg1
Rbg 101 0 1 TC=1.612E-5,2.255E-6
Ebg(102,0)(Input,Ground)1
```

```
Rreg 102 101 180
.model Sbg1 VSWITCH(RON=1 ROFF=1MEG VON=3.7 VOFF=3)
*
*Feedbackstage
*
*    Diodes D1,D2 limit the excursion of the amplifier
*    outputs to being near the rails. Rfb, Cfb Set the
*    corner frequency for roll-off of ripple rejection.


*
*    The opamp gain is given by: AV=(Fores/Freg)*(Vout/Vbg)
*    where Fores=output impedance corner frequency
*            with Cl=0 (typical value about 1 MHz)
*       Freg=corner frequency in ripple rejection
*            (typical value about 600 Hz)
*       Vout=regulator output voltage (5,12,15V)
*       Vbg=bandgap voltage (3.7V)
*
*    Note: AV is constant for all output voltages, but the
*    feedback factor changes. If Av= 2250, then the 3rd coeff.
*    below is:
*                    Vout   AV*Feedback factor
*                    -----  ------------------
*                     5     1665
*                    12     694
*                    15     555
*
Rfb 9 8 1MEG
Cfb 8 Ground 265PF
*Eopamp 105 0 value={2250*v(101,0)+Av_feedback*v(Ground,8)}
Vgainf200 0 {Av_feedback}
Rgainf200 0 1
Eopamp 105 0 Poly(3)(101,0)(Ground,8)(200,0)0 2250 0 0 0 0 0 0 1
Ro 105 106 1K
D1 106 108 Dlim
D2 107 106 Dlim
.model Dlim D(VJ=0.7)
V11 102 108 DC 1
V12 107 0 DC 1
*
*Quiescent current modelling
*
*    Quiescent current is set by Gq, which draws a current
*    proportional to the voltage drop across the regulator and
*    R1 (temperature coefficient .1%/deg C). R1 must change
*    with output voltage as follows: R1=R1(5v)*Vout/5v.
*
Gq Input Ground Input 9 1.67e-5
R1 9 Ground {R1_Value} TC=0.001
*
*Output Stage
*
*    The current limit is: Vbe=Io*Rsc+(Vi-Vo-6.3V)*230/10K.
*    This is modelled with a voltage controlled resistor which
*    turns on when: Io*Rsc+0.023(Vi-Vo)=0.985V.
*
Glim 0 112 value={v(7,9)+0.023*v(Input,9)}
Glim 0 112 Poly (2)(7,9)(Input,9)0 1 0.023
Rxx 112 0 1 TC=2.795e-3,-7.24e-7
*
Q1 Input 5 6 Npn1
```

```
Q2 Input 6 7 Npn1 10
.model Npn1 NPN (BF=50 IS=1E-14)
*Efb Input 4 value={v(Input,Ground)+v(0,106)}
Efb Input 4 Poly(2)(Input,Ground)(0,106)0 1 1
RB 4 10 1K TC=0.003
Slim(5,10)(112,0)Slim 1
.model Slim 1 VSWITCH (VOFF=0.6 VON=1.02 ROFF=1 RON=200K)
Re 6 7 2K
Rsc 7 9 0.26
Rbond 9 Output 0.008
.ends
*
*-------------------------------------------------------------LM7805C
.subckt LM7805 Input Output Ground
  x1 Input Output Ground x_VREG1 params:
+    AV_feedback=1665,R1_Value=1020
.ends
*
*-------------------------------------------------------------LM7812C
.subckt LM7812C Input Output Ground
  x1 Input Output Ground x_VREG1 params:
+    Av_feedback=694,R1_Value=2448
.ends
*
*-------------------------------------------------------------LM7815C
.subckt LM7815 Input Output Ground
  x1 Input Output Ground x_VREG1 params:
+    Av_feedback=555,R1_Value=3060
.ends
*
```

VOLTAGE REGULATOR (NEGATIVE) MACROMODEL

```
*
*Band-gap voltage source:
*
Vbg 100 0 DC −7.4V
Sbg (100,101)(Ground,Input) Sbg1
Rbg 101 0 1 TC=1.612E-5,−2.255E-6
Ebg(102,0)(Input,Ground)1
Rreg 102 101 180
.model Sbg1 VSWITCH (RON=1 ROFF=1MEG VON=3.7 VOFF=3)
*
*Feedback stage
*
Rfb 9 8 1MEG
Cfb 8 Ground 265PF
*Eopamp 105 0 value={2250*v(101,0)+Av_feedback*v(Ground,8)}
Vgainf200 0 {Av_feedback}
Rgainf200 0 1
Eopamp 105 0 Poly(3),(101,0)(Ground,8)(200,0)0 2250 0 0 0 0 0 1
Ro 105 106 1K
D1 108 106 Dlim
D2 106 107 Dlim
.model Dlim D(VJ=0.7)
Vl1 108 102 DC 1
Vl2 0 107 DC 1
*
*Quiescent current modelling
```

```
*
Gq Ground Input 9 Input 0.7e-5
R1 9 Ground {R1_Value} TC=0.001
*
*Output Stage
*
*Glim 112 0 value={v(Input,7)+0.023*v(Input,9)}
Glim 112 0 Poly(2)(Input,7)(Input,9)0 1 0.023
Rxx 112 0 1 TC=2.795e-3,−7.24e-7
*
Q1 9 5 6 Npn1
Q2 9 6 7 Npn1 10
.model Npn1 NPN (BF=50 IS=1E-14)
*Efb 4 9 value={v(Input,Ground)+v(0,106)}
Efb 4 9 Poly(2)(Input,Ground)(0,106)0 1 1
Rb 4 10 1K TC=0.003
Slim(5,10)(112,0)Slim 1
.model Slim 1 VSWITCH ( VOFF=0.6 VON=1.02 ROFF=1 RON=200K)
Re 6 7 2K
Rsc 7 Input 0.28
Rbond 9 Output 0.008
.ends
*
*------------------------------------------------------------LM7905C
.subckt LM7905C Input Output Ground
  x1 Input Output Ground x_VREG2 params:
+    Av_feedback=1665, R1_Value=4080
.ends
*
*------------------------------------------------------------LM7912C
.subckt LM7912C Input Output Ground
  x1 Input Output Ground x_VREG2 params:
+    Av_feedback=694,R1_Value=9792
.ends
*
*----------------------------------------------------------- -LM7915C
.subckt LM7915C Input Output Ground
  x1 Input Output Ground x_VREG2 params:
+    Av_feedback=555,R1_Value=12240
.ends
*
```

APPENDIX III

OP AMP MACROMODELS

BIPOLAR OPAMP MODELING

Complex analog integrated circuits are modeled in SPICE through the use of a macromodel: an accurate representation of the terminal characteristics of the device. The macromodels for two bipolar OpAmps (μa741 and LM324) are included in the evaluation version of PSpice. The macromodel is schematically shown in Figure A3-1. The component values are extracted from the MicroSim linear subcircuit library.

SPICE NODE LIST FOR BIPOLAR OPAMP MACROMODELS

```
*connections:   non-inverting input
*               | inverting input
*               | | positive power supply
*               | | | negative power supply
*               | | | | output
*               | | | | |
.subckt LM324  1 2 3 4 5
```

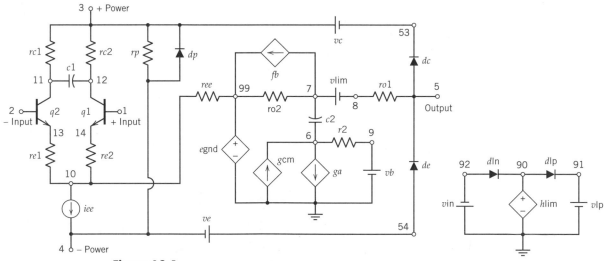

Figure A3-1
Bipolar OpAmp macromodel schematic.

```
c1      11  12  2.887E-12
c2       6   7  30.00E-12
dc       5  53  dx
de      54   5  dx
dlp     90  91  dx
dln     92  90  dx
dp       4   3  dx
egnd    99   0  poly(2)(3,0)(4,0)0 .5 .5
fb       7  99  poly(5) vb vc ve vlp vln 0 21.22E6 −20E6 20E6 20E6 −20E6
ga       6   0  11  12   188.5E-6
gcm      0   6  10  99   59.61E-9
iee      3  10  dc  15.09E-6
hlim    90   0  vlim  1K
q1      11   2  13  qx
q2      12   1  14  qx
r2       6   9  100.0E3
rc1      4  11  5.305E3
rc2      4  12  5.305E3
re1     13  10  1.845E3
re2     14  10  1.845E3
ree     10  99  13.25E6
ro1      8   5  50
ro2      7  99  25
rp       3   4  9.082E3
vb       9   0  dc  0
vc       3  53  dc  1.500
ve      54   4  dc  0.65
vlim     7   8  dc  0
vlp     91   0  dc  40
vln      0  92  dc  40
.model dx D(Is=800.0E-18 Rs=1)
.model qx PNP(Is=800.0E-18 Bf=166.7)
.ends
*$
*--------------------------------------------------------------------------
-
*connections:  non-inverting input
*              | inverting input
*              | | positive power supply
*              | | | negative power supply
*              | | | | output
*              | | | | |
.subckt uA741  1 2 3 4 5
c1      11  12  8.661E-12
c2       6   7  30.00E-12
dc       5  53  dx
de      54   5  dx
dlp     90  91  dx
dln     92  90  dx
dp       4   3  dx
egnd    99   0  poly(2)(3,0)(4,0)0 .5 .5
fb       7  99  poly(5) vb vc ve vlp vln 0 10.61E6 −10E6 10E6 10E6 −10E6
ga       6   0  11  12   188.5E-6
gcm      0   6  10  99   5.961E-9
iee     10   4  dc  15.16E-6
hlim    90   0  vlim  1K
q1      11   2  13  qx
q2      12   1  14  qx
r2       6   9  100.0E3
rc1      3  11  5.305E3
rc2      3  12  5.305E3
```

```
rel     13   10   1.836E3
re2     14   10   1.836E3
ree     10   99   13.19E6
ro1      8    5   50
ro2      7   99   100
rp       3    4   18.16E3
vb       9    0   dc 0
vc       3   53   dc 1
ve      54    4   dc 1
vlim     7    8   dc 0
vlp     91    0   dc 40
vln      0   92   dc 40
.model dx D(Is=800.0E-18 Rs=1)
.model qx NPN(Is=800.0E-18 Bf=93.75)
.ends
```

JFET INPUT OPAMP MODELING

Modeling of Op Amps with a JFET input stage is very similar to that of bipolar OpAmps. Only one such JFET OpAmp (LF411) is included in the evaluation version of PSpice. A schematic of this macromodel is shown in Figure A3-2. The component values for the LF411 are also from the MicroSim linear subcircuit library.

SPICE NODE LIST FOR JFET INPUT OPAMP MACROMODELS

```
*connections:  non-inverting input
*              | inverting input
*              | | positive power supply
*              | | | negative power supply
*              | | | | output
*              | | | | |
.subckt LF411  1 2 3 4 5
 c1      11   12   4.196E-12
```

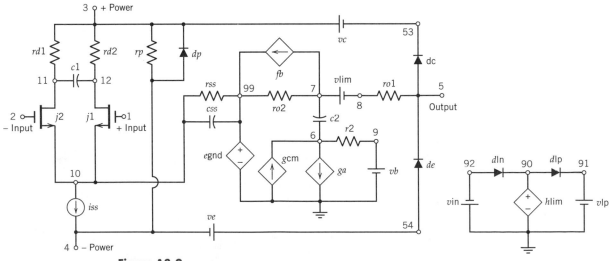

Figure A3-2
JFET input OpAmp macromodel schematic.

```
c2      6    7   10.00E-12
css    10   99   1.333E-12
dc      5   53   dx
de     54    5   dx
dlp    90   91   dx
dln    92   90   dx
dp      4    3   dx
egend  99    0   poly(2)(3,0)(4,0)0 .5 .5
fb      7   99   poly(5) vb vc ve vlp vln 0 31.83E6 —30E6 30E6 30E6 —30E6
ga      6    0   11   12   251.4E-6
gcm     0    6   10   99   2.514E-9
iss    10    4   dc  170.0E-6
hlim   90    0   vlim  1K
j1     11    2   10  jx
j2     12    1   10  jx
r2      6    9   100.0E3
rd1     3   11   3.978E3
rd2     3   12   3.978E3
ro1     8    5   50
ro2     7   99   25
rp      3    4   15.00E3
rss    10   99   1.176E6
vb      9    0   dc 0
vc      3   53   dc  1.500
ve     54    4   dc  1.500
vlim    7    8   dc  0
vlp    91    0   dc  25
vin     0   92   dc  25
.model dx D(Is=800.0E-18 Rs=1m)
.model jx NJF(Is=12.50E-12 Beta=743.3E-6 Vto=—1)
.ends
*$
*-------------------------------------------------------------------------- ---------
 —
```

VOLTAGE COMPARATOR MACROMODELS

Due to the basic functional operation of a comparator, the macromodel can be made more simple. A schematic of this macromodel is shown in Figure A3-3. The com-

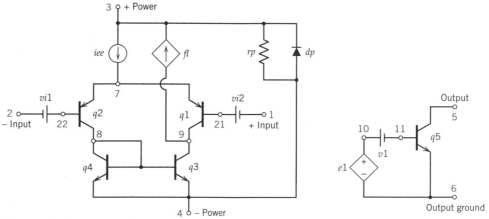

Figure A3-3
Comparator macromodel schematic.

ponent values for a LM311 comparator are also from the MicroSim linear subcircuit library.

```
*connections:   non-inverting input
*               | inverting input
*               | | positive power supply
*               | | | negative power supply
*               | | | | open collector output
*               | | | | | output ground
*               | | | | | |
.subckt LM111  1 2 3 4 5 6
 f1    9    3    v1   1
 iee   3    7    dc   100.0E-6
 vi1   21   1    dc   .45
 vi2   22   2    dc   45
 q1    9    21   7    qin
 q2    8    22   7    qin
 q3    9    8    4    qmo
 q4    8    8    4    qmi
.model qin PNP(Is=800.0E-18 Bf=833.3)
.model qmi NPN(Is=800.0E-18 Bf=1002)
.model qmo NPN(Is=800.0E-18 Bf=1000 Cjc=1E-15 Tr=118.8E-9)
 e1    10   6    9    4    1
 v1    10   11   dc   0
 q5    5    11   6    qoc
.model qoc NPN(Is=800.0E-18 Bf=34.49E3 Cjc=1E-15 Tf=364.6E-12 Tr=79.34E-9)
 dp    4    3    dx
 rp    3    4    6.122E3
 .model dx D(Is=800.0E-18 Rs=1)
 .ends
```

APPENDIX IV

DATA SHEETS

MOTOROLA
SEMICONDUCTOR ▬▬▬▬▬
TECHNICAL DATA

Quad Low Power Operational Amplifiers

The LM124 series are low-cost, quad operational amplifiers with true differential inputs. These have several distinct advantages over standard operational amplifier types in single supply applications. The quad amplifier can operate at supply voltages as low as 3.0 V or as high as 32 V with quiescent currents about one fifth of those associated with the MC1741 (on a per amplifier basis). The common mode input range includes the negative supply, thereby eliminating the necessity for external biasing components in many applications. The output voltage range also includes the negative power supply voltage.

* Short Circuited Protected Outputs
* True Differential Input Stage
* Single Supply Operation: 3.0 V to 32 V
* Low Input Bias Currents: 100 nA Max (LM324A)
* Four Amplifiers Per Package
* Internally Compensated
* Common Mode Range Extends to Negative Supply
* Industry Standard Pinouts
* ESD Clamps on the Inputs Increase Ruggedness without Affecting Device Operation

---

LM124, LM224, LM324, LM324A, LM2902

QUAD DIFFERENTIAL INPUT OPERATIONAL AMPLIFIERS

SILICON MONOLITHIC INTEGRATED CIRCUIT

J SUFFIX
CERAMIC PACKAGE
CASE 632

N SUFFIX
PLASTIC PACKAGE
CASE 646
(LM224, LM324, LM2902 Only)

D SUFFIX
PLASTIC PACKAGE
CASE 751A
(SO-14)

PIN CONNECTIONS

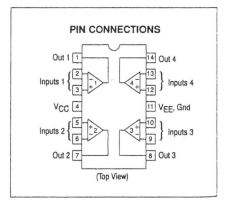

(Top View)

MAXIMUM RATINGS (T_A = +25°C, unless otherwise noted.)

Rating	Symbol	LM124 LM224 LM324,A	LM2902	Unit
Power Supply Voltages Single Supply Split Supplies	V_{CC} V_{CC}, V_{EE}	32 ±16	26 ±13	Vdc
Input Differential Voltage Range (1)	V_{IDR}	±32	±26	Vdc
Input Common Mode Voltage Range	V_{ICR}	−0.3 to 32	−0.3 to 26	Vdc
Output Short Circuit Duration	I_{SC}	Continuous		
Junction Temperature Ceramic Package Plastic Packages	T_J	175 150		°C
Storage Temperature Range Ceramic Package Plastic Packages	T_{stg}	−65 to +150 −55 to +125		°C
Operating Ambient Temperature Range	T_A	−55 to +125 −25 to +85 0 to +70 —	— — — −40 to +105	°C

NOTE: 1. Split Power Supplies.

ORDERING INFORMATION

Device	Temperature Range	Package
LM124J	−55° to +125°C	Ceramic DIP
LM2902D	−40° to +105°C	SO-14
LM2902N		Plastic DIP
LM2902J	−40° to +85°C	Ceramic DIP
LM224D	−25° to +85°C	SO-14
LM224J		Ceramic DIP
LM224N		Plastic DIP
LM324AD	0° to +70°C	SO-14
LM324AN		Plastic DIP
LM324D		SO-14
LM324J		Ceramic DIP
LM324N		Plastic DIP

LM124, LM224, LM324,A, LM2902

ELECTRICAL CHARACTERISTICS (V_{CC} = 5.0 V, V_{EE} = GND, T_A = 25°C, unless otherwise noted.)

Characteristics	Symbol	LM124/LM224 Min	Typ	Max	LM324A Min	Typ	Max	LM324 Min	Typ	Max	LM2902 Min	Typ	Max	Unit
Input Offset Voltage V_{CC} = 5.0 V to 30 V (26 V for LM2902), V_{ICR} = 0 V to V_{CC} −1.7 V, V_O = 1.4 V, R_S = 0 Ω T_A = 25°C T_A = T_{high} to T_{low} (Note 1)	V_{IO}	— —	2.0 —	5.0 7.0	— —	2.0 —	3.0 5.0	— —	2.0 —	7.0 9.0	— —	2.0 —	7.0 10	mV
Average Temperature Coefficient of Input Offset Voltage T_A = T_{high} to T_{low} (Note 1)	$\Delta V_{IO}/\Delta T$	—	7.0	—	—	7.0	30	—	7.0	—	—	7.0	—	µV/°C
Input Offset Current T_A = T_{high} to T_{low} (Note 1)	I_{IO}	— —	3.0 —	30 100	— —	5.0 —	30 75	— —	5.0 —	50 150	— —	5.0 —	50 200	nA
Average Temperature Coefficient of Input Offset Current T_A = T_{high} to T_{low} (Note 1)	$\Delta I_{IO}/\Delta T$	—	10	—	—	10	300	—	10	—	—	10	—	pA/°C
Input Bias Current T_A = T_{high} to T_{low} (Note 1)	I_{IB}	— —	−90 —	−150 −300	— —	−45 —	−100 −200	— —	−90 —	−250 −500	— —	−90 —	−250 −500	nA
Input Common Mode Voltage Range (Note 2) V_{CC} = 30 V (26 V for LM2902) V_{CC} = 30 V (26 V for LM2902), T_A = T_{high} to T_{low}	V_{ICR}	0 0	— —	28.3 28	0 0	— —	28.3 28	0 0	— —	28.3 28	0 0	— —	24.3 24	V
Differential Input Voltage Range	V_{IDR}	—	—	V_{CC}	—	—	V_{CC}	—	—	V_{CC}	—	—	V_{CC}	V
Large Signal Open-Loop Voltage Gain R_L = 2.0 kΩ, V_{CC} = 15 V, for Large V_O Swing, T_A = T_{high} to T_{low} (Note 1)	A_{VOL}	50 25	100 —	— —	25 15	100 —	— —	25 15	100 —	— —	25 15	100 —	— —	V/mV
Channel Separation 10 kHz ≤ f ≤ 20 kHz, Input Referenced	CS	—	−120	—	—	−120	—	—	−120	—	—	−120	—	dB
Common Mode Rejection R_S ≤ 10 kΩ	CMR	70	85	—	65	70	—	65	70	—	50	70	—	dB
Power Supply Rejection	PSR	65	100	—	65	100	—	65	100	—	50	100	—	dB
Output Voltage—High Limit (T_A = T_{high} to T_{low}) (Note 1) V_{CC} = 5.0 V, R_L = 2.0 kΩ, T_A = 25°C V_{CC} = 30 V (26 V for LM2902), R_L = 2.0 kΩ V_{CC} = 30 V (26 V for LM2902), R_L = 10 kΩ	V_{OH}	3.3 26 27	3.5 — 28	— — —	3.3 26 27	3.5 — 28	— — —	3.3 22 23	3.5 — 28	— — —	3.3 22 23	3.5 — 24	— — —	V
Output Voltage — Low Limit V_{CC} = 5.0 V, R_L = 10 kΩ, T_A = T_{high} to T_{low} (Note1)	V_{OL}	—	5.0	20	—	5.0	20	—	5.0	20	—	5.0	100	mV
Output Source Current (V_{ID} = +1.0 V, V_{CC} = 15 V) T_A = 25°C T_A = T_{high} to T_{low} (Note 1)	I_O +	20 10	40 20	— —	20 10	40 20	— —	20 10	40 20	— —	20 10	40 20	— —	mA
Output Sink Current (V_{ID} = −1.0 V, V_{CC} = 15 V) T_A = 25°C T_A = T_{high} to T_{low} (Note 1) (V_{ID} = −1.0 V, V_O = 200 mV, T_A = 25°C)	I_O −	10 5.0 12	20 8.0 50	— — —	10 5.0 12	20 8.0 50	— — —	10 5.0 12	20 8.0 50	— — —	10 5.0	20 8.0	— —	mA µA
Output Short Circuit to Ground (Note 3)	I_{SC}	—	40	60	—	40	60	—	40	60	—	40	60	mA
Power Supply Current (T_A = T_{high} to T_{low}) (Note 1) V_{CC} = 30 V (26 V for LM2902), V_O = 0 V, R_L = ∞ V_{CC} = 5.0 V, V_O = 0 V, R_L = ∞	I_{CC}	— —	— —	3.0 1.2	— —	1.4 0.7	3.0 1.2	— —	— —	3.0 1.2	— —	— —	3.0 1.2	mA

NOTES: 1. T_{low} = −55°C for LM124 T_{high} = +125°C for LM124
 = −25°C for LM224 = +85°C for LM224
 = 0°C for LM324, A = +70°C for LM324,A
 = −40°C for LM2902 = +105°C for LM2902

2. The input common mode voltage or either input signal voltage should not be allowed to go negative by more than 0.3 V. The upper end of the common mode voltage range is V_{CC} −1.7 V.

3. Short circuits from the output to V_{CC} can cause excessive heating and eventual destruction. Destructive dissipation can result from simultaneous shorts on all amplifiers.

uA741C, uA741I, uA741M
GENERAL-PURPOSE OPERATIONAL AMPLIFIERS

D920, NOVEMBER 1970 — REVISED JANUARY 1992

- Short-Circuit Protection
- Offset-Voltage Null Capability
- Large Common-Mode and Differential Voltage Ranges
- No Frequency Compensation Required
- Low Power Consumption
- No Latch-Up
- Designed to Be Interchangeable With Fairchild μA741

description

The uA741 is a general-purpose operational amplifier featuring offset-voltage null capability.

The high common-mode input voltage range and the absence of latch-up make the amplifier ideal for voltage-follower applications. The device is short-circuit protected and the internal frequency compensation ensures stability without external components. A low potentiometer may be connected between the offset null inputs to null out the offset voltage as shown in Figure 2.

The uA741C is characterized for operation from 0°C to 70°C. The uA741I is characterized for operation from −40°C to 85°C. The uA741M is characterized for operation over the full military temperature range of −55°C to 125°C.

symbol

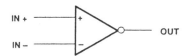

uA741M . . . J PACKAGE
(TOP VIEW)

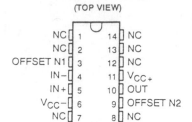

uA741M . . . JG PACKAGE
uA741C, uA741I . . . D OR P PACKAGE
(TOP VIEW)

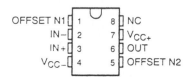

uA741M . . . U FLAT PACKAGE
(TOP VIEW)

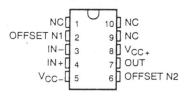

uA741M . . . FK PACKAGE
(TOP VIEW)

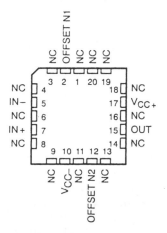

NC–No internal connection

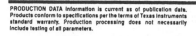

POST OFFICE BOX 655303 • DALLAS, TEXAS 75265

uA741C, uA741I, uA741M
GENERAL-PURPOSE OPERATIONAL AMPLIFIERS

AVAILABLE OPTIONS

T_A	PACKAGE					
	SMALL OUTLINE (D)	CHIP CARRIER (FK)	CERAMIC DIP (J)	CERAMIC DIP (JG)	PLASTIC DIP (P)	FLAT PACK (U)
0°C to 70°C	uA741CD				uA741CP	
−40°C to 85°C	uA741ID				uA741IP	
−55°C to 125°C		uA741MFK	uA741MJ	uA741MJG		uA741MU

The D package is available taped and reeled. Add the suffix R (e.g., uA741CDR).

schematic

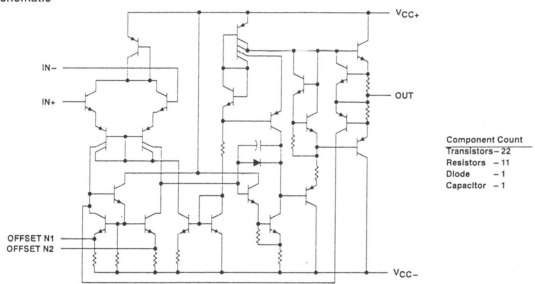

Component Count
Transistors – 22
Resistors – 11
Diode – 1
Capacitor – 1

absolute maximum ratings over operating free-air temperature range (unless otherwise noted)

		uA741C	uA741I	uA741M	UNIT
Supply voltage V_{CC+} (see Note 1)		18	22	22	V
Supply voltage $V_{CC−}$ (see Note 1)		−18	−22	−22	V
Differential input voltage (see Note 2)		±15	±30	±30	V
Input voltage any input (see Notes 1 and 3)		±15	±15	±15	V
Voltage between either offset null terminal (N1/N2) and $V_{CC−}$		±15	±0.5	±0.5	V
Duration of output short circuit (see Note 4)		unlimited	unlimited	unlimited	
Continuous total power dissipation		See Dissipation Rating Table			
Operating free-air temperature range		0 to 70	−40 to 85	-55 to 125	°C
Storage temperature range		−65 to 150	−65 to 150	−65 to 150	°C
Case temperature for 60 seconds	FK package			260	°C
Lead temperature 1,6 mm (1/16 inch) from case for 60 seconds	J, JG, or U package			300	°C
Lead temperature 1,6 mm (1/16 inch) from case for 10 seconds	D or P package	260	260		°C

NOTES: 1. All voltage values, unless otherwise noted, are with respect to the midpoint between V_{CC+} and $V_{CC−}$.
    2. Differential voltages are at the noninverting input terminal with respect to the inverting input terminal.
    3. The magnitude of the input voltage must never exceed the magnitude of the supply voltage or 15 V, whichever is less.
    4. The output may be shorted to ground or either power supply. For the uA741M only, the unlimited duration of the short circuit applies at (or below) 125°C case temperature or 75°C free-air temperature.

TEXAS
INSTRUMENTS

uA741C, uA741I, uA741M
GENERAL-PURPOSE OPERATIONAL AMPLIFIERS

DISSIPATION RATING TABLE

PACKAGE	$T_A \leq 25°C$ POWER RATING	DERATING FACTOR	DERATE ABOVE T_A	$T_A = 70°C$ POWER RATING	$T_A = 85°C$ POWER RATING	$T_A = 125°C$ POWER RATING
D	500 mW	5.8 mW/°C	64°C	464 mW	377 mW	N/A
FK	500 mW	11.0 mW/°C	105°C	500 mW	500 mW	275 mW
J	500 mW	11.0 mW/°C	105°C	500 mW	500 mW	275 mW
JG	500 mW	8.4 mW/°C	90°C	500 mW	500 mW	210 mW
P	500 mW	N/A	N/A	500 mW	500 mW	N/A
U	500 mW	5.4 mW/°C	57°C	432 mW	351 mW	135 mW

electrical characteristics at specified free-air temperature, $V_{CC}\pm = \pm15$ V

PARAMETER		TEST CONDITIONS	T_A†	UA741C MIN	UA741C TYP	UA741C MAX	UA741I, UA741M MIN	UA741I, UA741M TYP	UA741I, UA741M MAX	UNIT
V_{IO}	Input offset voltage	$V_O = 0$	25°C		1	6		1	5	mV
			Full range			7.5			6	
$\Delta V_{IO(adj)}$	Offset voltage adjust range	$V_O = 0$	25°C		±15			±15		mV
I_{IO}	Input offset curent	$V_O = 0$	25°C		20	200		20	200	nA
			Full range			300			500	
I_{IB}	Input bias current	$V_O = 0$	25°C		80	500		80	500	nA
			Full range			800			1500	
V_{ICR}	Common-mode input voltage range		25°C	±12	±13		±12	±13		V
			Full range	±12			±12			
V_{OM}	Maximum peak output voltage swing	$R_L = 10$ kΩ	25°C	±12	±14		±12	±14		V
		$R_L \geq 10$ kΩ	Full range	±12			±12			
		$R_L = 2$ kΩ	25°C	±10	±13		±10	±13		
		$R_L \geq 2$ kΩ	Full range	±10			±10			
A_{VD}	Large-signal differential voltage amplification	$R_L \geq 2$ kΩ	25°C	20	200		50	200		V/mV
		$V_O = \pm10$ V	Full range	15			25			
r_i	Input resistance		25°C	0.3	2		0.3	2		MΩ
r_o	Output resistance	$V_O = 0$, See Note 5	25°C		75			75		Ω
C_i	Input capacitance		25°C		1.4			1.4		pF
CMRR	Common-mode rejection ratio	$V_{IC} = V_{ICR}$ min	25°C	70	90		70	90		dB
			Full range	70			70			
k_{SVS}	Supply voltage sensitivity ($\Delta V_{IO}/\Delta V_{CC}$)	$V_{CC} = \pm9$ V to ±15 V	25°C		30	150		30	150	µV/V
			Full range			150			150	
I_{OS}	Short-circuit output current		25°C		±25	±40		±25	±40	mA
I_{CC}	Supply current	No load, $V_O = 0$	25°C		1.7	2.8		1.7	2.8	mA
			Full range			3.3			3.3	
P_D	Total power dissipation	No load, $V_O = 0$	25°C		50	85		50	85	mW
			Full range			100			100	

† All characteristics are measured under open-loop conditions with zero common-mode input voltage unless otherwise specified. Full range for the uA741C is 0°C to 70°C, the uA741I is –40°C to 85°C, and the uA741M is –55°C to 125°C.
NOTE 5: This typical value applies only at frequencies above a few hundred hertz because of the effects of drift and thermal feedback.

TEXAS INSTRUMENTS
POST OFFICE BOX 655303 • DALLAS, TEXAS 75265

uA741C, uA741I, uA741M
GENERAL-PURPOSE OPERATIONAL AMPLIFIERS

operating characteristics, $V_{CC}\pm = \pm15$ V, $T_A = 25°C$

PARAMETER		TEST CONDITIONS		uA741C			uA741I, uA741M			UNIT
				MIN	TYP	MAX	MIN	TYP	MAX	
t_r	Rise time	$V_I = 20$ mV,	$R_L = 2$ kΩ,		0.3			0.3		μs
	Overshoot factor	$C_L = 100$ pF,	See Figure 1		5%			5%		
SR	Slew rate at unity gain	$V_I = 10$ V, $C_L = 100$ pF,	$R_L = 2$ kΩ, See Figure 1		0.5			0.5		V/μs

---

PARAMETER MEASUREMENT INFORMATION

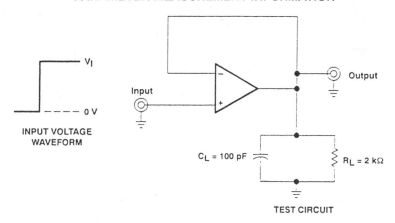

Figure 1. Rise Time, Overshoot, and Slew Rate

APPLICATION INFORMATION

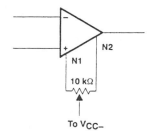

Figure 2. Input Offset Voltage Null Circuit

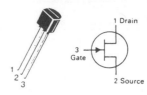

2N5484
2N5486★

CASE 29-04, STYLE 5
TO-92 (TO-226AA)

1 Drain

3 Gate

2 Source

JFET
VHF/UHF AMPLIFIERS

N-CHANNEL — DEPLETION

★These are Motorola
designated preferred devices.

MAXIMUM RATINGS

Rating	Symbol	Value	Unit
Drain-Gate Voltage	V_{DG}	25	Vdc
Reverse Gate-Source Voltage	V_{GSR}	25	Vdc
Drain Current	I_D	30	mAdc
Forward Gate Current	$I_{G(f)}$	10	mAdc
Total Device Dissipation @ $T_C = 25°C$ Derate above 25°C	P_D	350 2.8	mW mW/°C
Operating and Storage Junction Temperature Range	T_J, T_{stg}	−65 to +150	°C

ELECTRICAL CHARACTERISTICS ($T_A = 25°C$ unless otherwise noted.)

Characteristic		Symbol	Min	Typ	Max	Unit
OFF CHARACTERISTICS						
Gate-Source Breakdown Voltage ($I_G = -1.0\ \mu Adc$, $V_{DS} = 0$)		$V_{(BR)GSS}$	−25	−	−	Vdc
Gate Reverse Current ($V_{GS} = -20$ Vdc, $V_{DS} = 0$) ($V_{GS} = -20$ Vdc, $V_{DS} = 0$, $T_A = 100°C$)		I_{GSS}	− −	− −	−1.0 −0.2	nAdc μAdc
Gate Source Cutoff Voltage ($V_{DS} = 15$ Vdc, $I_D = 10$ nAdc)	2N5484 2N5486	$V_{GS(off)}$	−0.3 −2.0	− −	−3.0 −6.0	Vdc
ON CHARACTERISTICS						
Zero-Gate-Voltage Drain Current ($V_{DS} = 15$ Vdc, $V_{GS} = 0$)	2N5484 2N5486	I_{DSS}	1.0 8.0	− −	5.0 20	mAdc
SMALL-SIGNAL CHARACTERISTICS						
Forward Transfer Admittance ($V_{DS} = 15$ Vdc, $V_{GS} = 0$, $f = 1.0$ kHz)	2N5484 2N5486	$|y_{fs}|$	3000 4000	− −	6000 8000	$\mu mhos$
Input Admittance ($V_{DS} = 15$ Vdc, $V_{GS} = 0$, $f = 100$ MHz) ($V_{DS} = 15$ Vdc, $V_{GS} = 0$, $f = 400$ MHz)	2N5484 2N5486	$Re(y_{is})$	− −	− −	100 1000	$\mu mhos$
Output Admittance ($V_{DS} = 15$ Vdc, $V_{GS} = 0$, $f = 1.0$ kHz)	2N5484 2N5486	$|y_{os}|$	− −	− −	50 75	$\mu mhos$
Output Conductance ($V_{DS} = 15$ Vdc, $V_{GS} = 0$, $f = 100$ MHz) ($V_{DS} = 15$ Vdc, $V_{GS} = 0$, $f = 400$ MHz)	2N5484 2N5486	$Re(y_{os})$	− −	− −	75 100	$\mu mhos$
Forward Transconductance ($V_{DS} = 15$ Vdc, $V_{GS} = 0$, $f = 100$ MHz) ($V_{DS} = 15$ Vdc, $V_{GS} = 0$, $f = 400$ MHz)	2N5484 2N5486	$Re(y_{fs})$	2500 3500	− −	− −	$\mu mhos$

2N5484 2N5486

ELECTRICAL CHARACTERISTICS (continued) (T_A = 25°C unless otherwise noted.)

Characteristic	Symbol	Min	Typ	Max	Unit
Input Capacitance (V_{DS} = 15 Vdc, V_{GS} = 0, f = 1.0 MHz)	C_{iss}	–	–	5.0	pF
Reverse Transfer Capacitance (V_{DS} = 15 Vdc, V_{GS} = 0, f = 1.0 MHz)	C_{rss}	–	–	1.0	pF
Output Capacitance (V_{DS} = 15 Vdc, V_{GS} = 0, f = 1.0 MHz)	C_{oss}	–	–	2.0	pF
FUNCTIONAL CHARACTERISTICS					
Noise Figure	NF				dB
(V_{DS} = 15 Vdc, V_{GS} = 0, R_G = 1.0 Megohm, f = 1.0 kHz)		–	–	2.5	
(V_{DS} = 15 Vdc, I_D = 1.0 mAdc, 2N5484 R_G ≈ 1.0 k ohm, f = 100 MHz)		–	–	3.0	
(V_{DS} = 15 Vdc, I_D = 1.0 mAdc, 2N5484 R_G ≈ 1.0 k ohm, f = 200 MHz)		–	4.0	–	
(V_{DS} = 15 Vdc, I_D = 4.0 mAdc, 2N5486 R_G ≈ 1.0 k ohm, f = 100 MHz)		–	–	2.0	
(V_{DS} = 15 Vdc, I_D = 4.0 mAdc, 2N5486 R_G ≈ 1.0 k ohm, f = 400 MHz)		–	–	4.0	
Common Source Power Gain	G_{ps}				dB
(V_{DS} = 15 Vdc, I_D = 1.0 mAdc, f = 100 MHz) 2N5484		16	–	25	
(V_{DS} = 15 Vdc, I_D = 1.0 mAdc, f = 200 MHz) 2N5484		–	14	–	
(V_{DS} = 15 Vdc, I_D = 4.0 mAdc, f = 100 MHz) 2N5486		18	–	30	
(V_{DS} = 15 Vdc, I_D = 4.0 mAdc, f = 400 MHz) 2N5486		10	–	20	

POWER GAIN

FIGURE 1 – EFFECTS OF DRAIN CURRENT

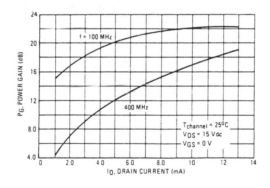

MAXIMUM RATINGS

Rating	Symbol	2N2219 2N2222	2N2218A 2N2219A 2N2222A	Unit
Collector-Emitter Voltage	V_{CEO}	30	40	Vdc
Collector-Base Voltage	V_{CBO}	60	75	Vdc
Emitter-Base Voltage	V_{EBO}	5.0	6.0	Vdc
Collector Current — Continuous	I_C	800	800	mAdc

		2N2218A 2N2219,A	2N2222,A	
Total Device Dissipation @ T_A = 25°C Derate above 25°C	P_D	0.8 4.57	0.4 2.28	Watt mW/°C
Total Device Dissipation @ T_C = 25°C Derate above 25°C	P_D	3.0 17.1	1.2 6.85	Watts mW/°C
Operating and Storage Junction Temperature Range	T_J, T_{stg}	−65 to +200		°C

THERMAL CHARACTERISTICS

Characteristic	Symbol	2N2218A 2N2219,A	2N2222,A	Unit
Thermal Resistance, Junction to Ambient	$R_{\theta JA}$	219	437.5	°C/W
Thermal Resistance, Junction to Case	$R_{\theta JC}$	58	145.8	°C/W

2N2218A,2N2219,A★ 2N2222,A★

2N2218, A/2N2219,A
CASE 79-04
TO-39 (TO-205AD)
STYLE 1

A/2N2222,A
CASE 22-03
TO-18 (TO-206AA)
STYLE 1

GENERAL PURPOSE TRANSISTORS

NPN SILICON

★2N2219A and 2N2222A
are Motorola designated
preferred devices.

ELECTRICAL CHARACTERISTICS (T_A = 25°C unless otherwise noted.)

Characteristic		Symbol	Min	Max	Unit
OFF CHARACTERISTICS					
Collector-Emitter Breakdown Voltage (I_C = 10 mAdc, I_B = 0) Non-A Suffix A-Suffix		$V_{(BR)CEO}$	30 40	— —	Vdc
Collector-Base Breakdown Voltage (I_C = 10 μAdc, I_E = 0) Non-A Suffix A-Suffix		$V_{(BR)CBO}$	60 75	— —	Vdc
Emitter-Base Breakdown Voltage (I_E = 10 μAdc, I_C = 0) Non-A Suffix A-Suffix		$V_{(BR)EBO}$	5.0 6.0	— —	Vdc
Collector Cutoff Current (V_{CE} = 60 Vdc, $V_{EB(off)}$ = 3.0 Vdc) A-Suffix		I_{CEX}	—	10	nAdc
Collector Cutoff Current (V_{CB} = 50 Vdc, I_E = 0) Non-A Suffix (V_{CB} = 60 Vdc, I_E = 0) A-Suffix (V_{CB} = 50 Vdc, I_E = 0, T_A = 150°C) Non-A Suffix (V_{CB} = 60 Vdc, I_E = 0, T_A = 150°C) A-Suffix		I_{CBO}	— — — —	0.01 0.01 10 10	μAdc
Emitter Cutoff Current (V_{EB} = 3.0 Vdc, I_C = 0) A-Suffix		I_{EBO}	—	10	nAdc
Base Cutoff Current (V_{CE} = 60 Vdc, $V_{EB(off)}$ = 3.0 Vdc) A-Suffix		I_{BL}	—	20	nAdc
ON CHARACTERISTICS					
DC Current Gain (I_C = 0.1 mAdc, V_{CE} = 10 Vdc)	2N2218A 2N2219,A, 2N2222,A	h_{FE}	20 35	— —	—
(I_C = 1.0 mAdc, V_{CE} = 10 Vdc)	2N2218A 2N2219,A, 2N2222,A		25 50	— —	
(I_C = 10 mAdc, V_{CE} = 10 Vdc)(1)	2N2218A 2N2219,A, 2N2222,A		35 75	— —	
(I_C = 10 mAdc, V_{CE} = 10 Vdc, T_A = −55°C)(1)	2N2218A 2N2219,A, 2N2222,A		15 35	— —	
(I_C = 150 mAdc, V_{CE} = 10 Vdc)(1)	2N2218A 2N2219,A, 2N2222,A		40 100	120 300	

2N2218A 2N2219,A 2N2222,A

ELECTRICAL CHARACTERISTICS (continued) (T_A = 25°C unless otherwise noted.)

Characteristic		Symbol	Min	Max	Unit
(I_C = 150 mAdc, V_{CE} = 1.0 Vdc)(1)	2N2218A		20	—	
	2N2219,A, 2N2222,A		50	—	
(I_C = 500 mAdc, V_{CE} = 10 Vdc)(1)	2N2219, 2N2222		30	—	
	2N2218A		25	—	
	2N2219A, 2N2222A		40	—	
Collector-Emitter Saturation Voltage(1)		$V_{CE(sat)}$			Vdc
(I_C = 150 mAdc, I_B = 15 mAdc)	Non-A Suffix		—	0.4	
	A-Suffix		—	0.3	
(I_C = 500 mAdc, I_B = 50 mAdc)	Non-A Suffix		—	1.6	
	A-Suffix		—	1.0	
Base-Emitter Saturation Voltage(1)		$V_{BE(sat)}$			Vdc
(I_C = 150 mAdc, I_B = 15 mAdc)	Non-A Suffix		0.6	1.3	
	A-Suffix		0.6	1.2	
(I_C = 500 mAdc, I_B = 50 mAdc)	Non-A Suffix		—	2.6	
	A-Suffix		—	2.0	

SMALL-SIGNAL CHARACTERISTICS

Characteristic		Symbol	Min	Max	Unit
Current Gain — Bandwidth Product(2)		f_T			MHz
(I_C = 20 mAdc, V_{CE} = 20 Vdc, f = 100 MHz)	All Types, Except		250	—	
	2N2219A, 2N2222A		300	—	
Output Capacitance(3)		C_{obo}	—	8.0	pF
(V_{CB} = 10 Vdc, I_E = 0, f = 1.0 MHz)					
Input Capacitance(3)		C_{ibo}			pF
(V_{EB} = 0.5 Vdc, I_C = 0, f = 1.0 MHz)	Non-A Suffix		—	30	
	A-Suffix		—	25	
Input Impedance		h_{ie}			kohms
(I_C = 1.0 mAdc, V_{CE} = 10 Vdc, f = 1.0 kHz)	2N2218A		1.0	3.5	
	2N2219A, 2N2222A		2.0	8.0	
(I_C = 10 mAdc, V_{CE} = 10 Vdc, f = 1.0 kHz)	2N2218A		0.2	1.0	
	2N2219A, 2N2222A		0.25	1.25	
Voltage Feedback Ratio		h_{re}			X 10^{-4}
(I_C = 1.0 mAdc, V_{CE} = 10 Vdc, f = 1.0 kHz)	2N2218A		—	5.0	
	2N2219A, 2N2222A		—	8.0	
(I_C = 10 mAdc, V_{CE} = 10 Vdc, f = 1.0 kHz)	2N2218A		—	2.5	
	2N2219A, 2N2222A		—	4.0	
Small-Signal Current Gain		h_{fe}			—
(I_C = 1.0 mAdc, V_{CE} = 10 Vdc, f = 1.0 kHz)	2N2218A		30	150	
	2N2219A, 2N2222A		50	300	
(I_C = 10 mAdc, V_{CE} = 10 Vdc, f = 1.0 kHz)	2N2218A		50	300	
	2N2219A, 2N2222A		75	375	
Output Admittance		h_{oe}			μmhos
(I_C = 1.0 mAdc, V_{CE} = 10 Vdc, f = 1.0 kHz)	2N2218A		3.0	15	
	2N2219A, 2N2222A		5.0	35	
(I_C = 10 mAdc, V_{CE} = 10 Vdc, f = 1.0 kHz)	2N2218A		10	100	
	2N2219A, 2N2222A		15	200	
Collector Base Time Constant		$rb'C_c$	—	150	ps
(I_E = 20 mAdc, V_{CB} = 20Vdc, f = 31.8 MHz)	A-Suffix				
Noise Figure		NF	—	4.0	dB
(I_C = 100 μAdc, V_{CE} = 10 Vdc, R_S = 1.0 kohm, f = 1.0 kHz)	2N2222A				
Real Part of Common-Emitter High Frequency Input Impedance		$Re(h_{ie})$	—	60	Ohms
(I_C = 20 mAdc, V_{CE} = 20 Vdc, f = 300 MHz)	2N2218A, 2N2219A 2N2222A				

(1) Pulse Test: Pulse Width ≤ 300 μs, Duty Cycle ≈ 2.0%.
(2) f_T is defined as the frequency at which $|h_{fe}|$ extrapolates to unity.
(3) 2N5581 and 2N5582 are Listed C_{cb} and C_{eb} for these conditions and values.

MAXIMUM RATINGS

Rating	Symbol	Value	Unit
Collector-Emitter Voltage	V_{CEO}	40	Vdc
Collector-Base Voltge	V_{CBO}	60	Vdc
Emitter-Base Voltage	V_{EBO}	6.0	Vdc
Collector Current — Continuous	I_C	200	mAdc
Total Device Dissipation @ T_A = 25°C Derate above 25°C	P_D	625 5.0	mW mW/°C
*Total Device Dissipation @ T_C = 25°C Derate above 25°C	P_D	1.5 12	Watts mW/°C
Operating and Storage Junction Temperature Range	T_J, T_{stg}	−55 to +150	°C

*THERMAL CHARACTERISTICS

Characteristic	Symbol	Max	Unit
Thermal Resistance, Junction to Ambient	$R_{\theta JA}$	200	°C/W
Thermal Resistance, Junction to Case	$R_{\theta JC}$	83.3	°C/W

*Indicates Data in addition to JEDEC Requirements.

2N3903
2N3904★

**CASE 29-04, STYLE 1
TO-92 (TO-226AA)**

3 Collector

2 Base

1 Emitter

1
2
3

GENERAL PURPOSE TRANSISTORS

NPN SILICON

★This is a Motorola
designated preferred device.

ELECTRICAL CHARACTERISTICS (T_A = 25°C unless otherwise noted.)

Characteristic		Symbol	Min	Max	Unit
OFF CHARACTERISTICS					
Collector-Emitter Breakdown Voltage(1) (I_C = 1.0 mAdc, I_B = 0)		$V_{(BR)CEO}$	40	—	Vdc
Collector-Base Breakdown Voltage (I_C = 10 µAdc, I_E = 0)		$V_{(BR)CBO}$	60	—	Vdc
Emitter-Base Breakdown Voltage (I_E = 10 µAdc, I_C = 0)		$V_{(BR)EBO}$	6.0	—	Vdc
Base Cutoff Current (V_{CE} = 30 Vdc, V_{EB} = 3.0 Vdc)		I_{BL}	—	50	nAdc
Collector Cutoff Current (V_{CE} = 30 Vdc, V_{EB} = 3.0 Vdc)		I_{CEX}	—	50	nAdc
ON CHARACTERISTICS					
DC Current Gain(1)		h_{FE}			—
(I_C = 0.1 mAdc, V_{CE} = 1.0 Vdc)	2N3903 2N3904		20 40	— —	
(I_C = 1.0 mAdc, V_{CE} = 1.0 Vdc)	2N3903 2N3904		35 70	— —	
(I_C = 10 mAdc, V_{CE} = 1.0 Vdc)	2N3903 2N3904		50 100	150 300	
(I_C = 50 mAdc, V_{CE} = 1.0 Vdc)	2N3903 2N3904		30 60	— —	
(I_C = 100 mAdc, V_{CE} = 1.0 Vdc)	2N3903 2N3904		15 30	— —	
Collector-Emitter Saturation Voltage(1) (I_C = 10 mAdc, I_B = 1.0 mAdc) (I_C = 50 mAdc, I_B = 5.0 mAdc)		$V_{CE(sat)}$	— —	0.2 0.3	Vdc
Base-Emitter Saturation Voltage(1) (I_C = 10 mAdc, I_B = 1.0 mAdc) (I_C = 50 mAdc, I_B = 5.0 mAdc)		$V_{BE(sat)}$	0.65 —	0.85 0.95	Vdc
SMALL-SIGNAL CHARACTERISTICS					
Current-Gain — Bandwidth Product (I_C = 10 mAdc, V_{CE} = 20 Vdc, f = 100 MHz)	2N3903 2N3904	f_T	250 300	— —	MHz

Rev 2

2N3903 2N3904

ELECTRICAL CHARACTERISTICS (continued) (T_A = 25°C unless otherwise noted.)

Characteristic		Symbol	Min	Max	Unit
Output Capacitance (V_{CB} = 5.0 Vdc, I_E = 0, f = 1.0 MHz)		C_{obo}	—	4.0	pF
Input Capacitance (V_{EB} = 0.5 Vdc, I_C = 0, f = 1.0 MHz)		C_{ibo}	—	8.0	pF
Input Impedance (I_C = 1.0 mAdc, V_{CE} = 10 Vdc, f = 1.0 kHz)	2N3903 2N3904	h_{ie}	1.0 1.0	8.0 10	k ohms
Voltage Feedback Ratio (I_C = 1.0 mAdc, V_{CE} = 10 Vdc, f = 1.0 kHz)	2N3903 2N3904	h_{re}	0.1 0.5	5.0 8.0	X 10^{-4}
Small-Signal Current Gain (I_C = 1.0 mAdc, V_{CE} = 10 Vdc, f = 1.0 kHz)	2N3903 2N3904	h_{fe}	50 100	200 400	—
Output Admittance (I_C = 1.0 mAdc, V_{CE} = 10 Vdc, f = 1.0 kHz)		h_{oe}	1.0	40	μmhos
Noise Figure (I_C = 100 μAdc, V_{CE} = 5.0 Vdc, R_S = 1.0 k ohms, f = 1.0 kHz)	2N3903 2N3904	NF	— —	6.0 5.0	dB

SWITCHING CHARACTERISTICS

			Symbol	Min	Max	Unit
Delay Time	(V_{CC} = 3.0 Vdc, V_{BE} = 0.5 Vdc, I_C = 10 mAdc, I_{B1} = 1.0 mAdc)		t_d	—	35	ns
Rise Time			t_r	—	35	ns
Storage Time	(V_{CC} = 3.0 Vdc, I_C = 10 mAdc, I_{B1} = I_{B2} = 1.0 mAdc)	2N3903 2N3904	t_s	— —	175 200	ns
Fall Time			t_f	—	50	ns

(1) Pulse Test: Pulse Width $\leq$ 300 μs, Duty Cycle $\leq$ 2.0%.

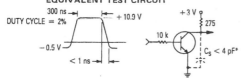

FIGURE 1 – DELAY AND RISE TIME EQUIVALENT TEST CIRCUIT

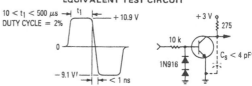

FIGURE 2 – STORAGE AND FALL TIME EQUIVALENT TEST CIRCUIT

*Total shunt capacitance of test jig and connectors

TYPICAL TRANSIENT CHARACTERISTICS

— T_J = 25°C --- T_J = 125°C

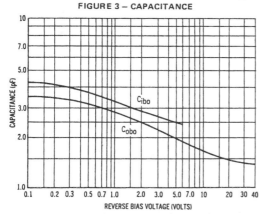

FIGURE 3 – CAPACITANCE

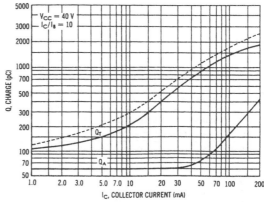

FIGURE 4 – CHARGE DATA

MAXIMUM RATINGS

Rating	Symbol	Value	Unit
Collector-Emitter Voltage	V_{CEO}	40	Vdc
Collector-Base Voltage	V_{CBO}	40	Vdc
Emitter-Base Voltage	V_{EBO}	5.0	Vdc
Collector Current — Continuous	I_C	200	mAdc
Total Device Dissipation @ T_A = 25°C Derate above 25°C	P_D	625 5.0	mW mW/°C
Total Power Dissipation @ T_A = 60°C	P_D	250	mW
Total Device Dissipation @ T_C = 25°C Derate above 25°C	P_D	1.5 12	Watts mW/°C
Operating and Storage Junction Temperature Range	T_J, T_{stg}	−55 to +150	°C

*THERMAL CHARACTERISTICS

Characteristic	Symbol	Max	Unit
Thermal Resistance, Junction to Ambient	$R_{\theta JA}$	200	°C/W
Thermal Resistance, Junction to Case	$R_{\theta JC}$	83.3	°C/W

2N3905
2N3906★

**CASE 29-04, STYLE 1
TO-92 (TO-226AA)**

3 Collector

2 Base

1 Emitter

GENERAL PURPOSE
TRANSISTORS

PNP SILICON

★This is a Motorola
designated preferred device.

ELECTRICAL CHARACTERISTICS (T_A = 25°C unless otherwise noted.)

Characteristic		Symbol	Min	Max	Unit
OFF CHARACTERISTICS					
Collector-Emitter Breakdown Voltage (1) (I_C = 1.0 mAdc, I_B = 0)		$V_{(BR)CEO}$	40	—	Vdc
Collector-Base Breakdown Voltage (I_C = 10 μAdc, I_E = 0)		$V_{(BR)CBO}$	40	—	Vdc
Emitter-Base Breakdown Voltage (I_E = 10 μAdc, I_C = 0)		$V_{(BR)EBO}$	5.0	—	Vdc
Base Cutoff Current (V_{CE} = 30 Vdc, V_{EB} = 3.0 Vdc)		I_{BL}	—	50	nAdc
Collector Cutoff Current (V_{CE} = 30 Vdc, V_{EB} = 3.0 Vdc)		I_{CEX}	—	50	nAdc
ON CHARACTERISTICS(1)					
DC Current Gain		h_{FE}			—
(I_C = 0.1 mAdc, V_{CE} = 1.0 Vdc)	2N3905 2N3906		30 60	— —	
(I_C = 1.0 mAdc, V_{CE} = 1.0 Vdc)	2N3905 2N3906		40 80	— —	
(I_C = 10 mAdc, V_{CE} = 1.0 Vdc)	2N3905 2N3906		50 100	150 300	
(I_C = 50 mAdc, V_{CE} = 1.0 Vdc)	2N3905 2N3506		30 60	— —	
(I_C = 100 mAdc, V_{CE} = 1.0 Vdc)	2N3905 2N3906		15 30	— —	
Collector-Emitter Saturation Voltage (I_C = 10 mAdc, I_B = 1.0 mAdc) (I_C = 50 mAdc, I_B = 5.0 mAdc)		$V_{CE(sat)}$	— —	0.25 0.4	Vdc
Base-Emitter Saturation Voltage (I_C = 10 mAdc, I_B = 1.0 mAdc) (I_C = 50 mAdc, I_B = 5.0 mAdc)		$V_{BE(sat)}$	0.65 —	0.85 0.95	Vdc
SMALL-SIGNAL CHARACTERISTICS					
Current-Gain — Bandwidth Product (I_C = 10 mAdc, V_{CE} = 20 Vdc, f = 100 MHz)	2N3905 2N3906	f_T	200 250	— —	MHz
Output Capacitance (V_{CB} = 5.0 Vdc, I_E = 0, f = 1.0 MHz)		C_{obo}	—	4.5	pF

Rev 2

ELECTRICAL CHARACTERISTICS (continued) (T_A = 25°C unless otherwise noted.)

Characteristic		Symbol	Min	Max	Unit
Input Capacitance (V_{EB} = 0.5 Vdc, I_C = 0, f = 1.0 MHz)		C_{ibo}	–	10.0	pF
Input Impedance (I_C = 1.0 mAdc, V_{CE} = 10 Vdc, f = 1.0 kHz)	2N3905 2N3906	h_{ie}	0.5 2.0	8.0 12	k ohms
Voltage Feedback Ratio (I_C = 1.0 mAdc, V_{CE} = 10 Vdc, f = 1.0 kHz)	2N3905 2N3906	h_{re}	0.1 0.1	5.0 10	X 10^{-4}
Small-Signal Current Gain (I_C = 1.0 mAdc, V_{CE} = 10 Vdc, f = 1.0 kHz)	2N3905 2N3906	h_{fe}	50 100	200 400	–
Output Admittance (I_C = 1.0 mAdc, V_{CE} = 10 Vdc, f = 1.0 kHz)	2N3905 2N3906	h_{oe}	1.0 3.0	40 60	μmhos
Noise Figure (I_C = 100 μAdc, V_{CE} = 5.0 Vdc, R_S = 1.0 k ohm, f = 1.0 kHz)	2N3905 2N3906	NF	– –	5.0 4.0	dB

SWITCHING CHARACTERISTICS

			Symbol	Min	Max	Unit
Delay Time	(V_{CC} = 3.0 Vdc, V_{BE} = 0.5 Vdc		t_d	–	35	ns
Rise Time	I_C = 10 mAdc, I_{B1} = 1.0 mAdc)		t_r	–	35	ns
Storage Time		2N3905 2N3906	t_s	– –	200 225	ns
Fall Time	(V_{CC} = 3.0 Vdc, I_C = 10 mAdc, I_{B1} = I_{B2} = 1.0 mAdc)	2N3905 2N3906	t_f	– –	60 75	ns

(1) Pulse Width ≤ 300 μs, Duty Cycle ≤ 2.0%.

FIGURE 1 – DELAY AND RISE TIME
EQUIVALENT TEST CIRCUIT

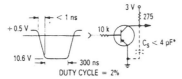

FIGURE 2 – STORAGE AND FALL TIME
EQUIVALENT TEST CIRCUIT

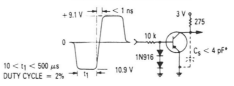

*Total shunt capacitance of test jig and connectors

TRANSIENT CHARACTERISTICS
—— T_J = 25°C --- T_J = 125°C

FIGURE 3 – CAPACITANCE

FIGURE 4 – CHARGE DATA

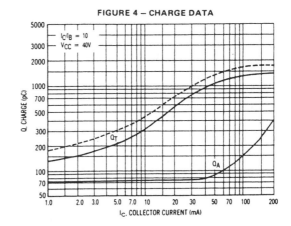

Silicon Controlled Rectifiers
Reverse Blocking Triode Thyristors

2N1595
thru
2N1599

SCRs
1.6 AMPERES RMS
50 thru 400 VOLTS

These devices are glassivated planar construction designed for gating operation in mA/μA signal or detection circuits.

- Low-Level Gate Characteristics — I_{GT} = 10 mA (Max) @ 25°C
- Low Holding Current — I_H = 5 mA (Typ) @ 25°C
- Glass-to-Metal Bond for Maximum Hermetic Seal

CASE 79-04
(TO-205AD)
STYLE 3

*MAXIMUM RATINGS (T_J = 125°C unless otherwise noted, R_{GC} = 1 kΩ.)

Rating		Symbol	Value	Unit
Repetitive Peak Reverse Blocking Voltage, Note 1	2N1595 2N1596 2N1597 2N1599	V_{RRM}	50 100 200 400	Volts
Repetitive Peak Forward Blocking Voltage, Note 1	2N1595 2N1596 2N1597 2N1599	V_{DRM}	50 100 200 400	Volts
RMS On-State Current (All Conduction Angles)		$I_{T(RMS)}$	1.6	Amps
Peak Non-Repetitive Surge Current (One Cycle, 60 Hz, T_J = −65 to +125°C)		I_{TSM}	15	Amps
Peak Gate Power		P_{GM}	0.1	Watt
Average Gate Power		$P_{G(AV)}$	0.01	Watt
Peak Gate Current		I_{GM}	0.1	Amp
Peak Gate Voltage — Forward Reverse		V_{GFM} V_{GRM}	10 10	Volts
Operating Junction Temperature Range		T_J	−65 to +125	°C
Storage Temperature Range		T_{stg}	−65 to +150	°C

*Indicates JEDEC Registered Data.

Note 1. V_{DRM} or V_{RRM} for all types can be applied on a continuous DC basis without incurring damage.

2N1595 thru 2N1599

ELECTRICAL CHARACTERISTICS (T_C = 25°C unless otherwise noted.)

Characteristic	Symbol	Min	Typ	Max	Unit
*Peak Forward or Reverse Blocking Current (Rated V_{DRM} or V_{RRM}) T_J = 25°C T_J = 125°C	I_{DRM}, I_{RRM}	— —	— —	10 6	μA mA
*Peak On-State Voltage (I_F = 1 Adc, Pulsed, 1 ms (Max), Duty Cycle ≤ 1%)	V_{TM}	—	1.1	2	Volts
*Gate Trigger Current (Continuous dc) (V_D = 7 V, R_L = 12 Ohms)	I_{GT}	—	2	10	mA
*Gate Trigger Voltage (Continuous dc) (V_D = 7 V, R_L = 12 Ohms) (V_D = 7 V, R_L = 12 Ohms, T_J = 125°C)	V_{GT}	— 0.2	0.7 —	3 —	Volts
Reverse Gate Current (V_{GK} = 10 V)	I_{GR}	—	17	—	mA
Holding Current (V_D = 7 V)	I_H	—	5	—	mA
Turn-On Time (I_{GT} – 10 mA, I_F = 1 A) (I_{GT} = 20 mA, I_F = 1 A)	t_{gt}	— —	0.8 0.6	— —	μs
Turn-Off Time (I_F = 1 A, I_R = 1 A, dv/dt = 20 V/μs, T_J = 125°C)	t_q	—	10	—	μs

*Indicates JEDEC Registered Data.

CURRENT DERATING

FIGURE 1 – CASE TEMPERATURE REFERENCE

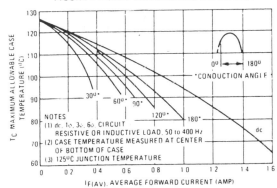

FIGURE 2 – AMBIENT TEMPERATURE REFERENCE

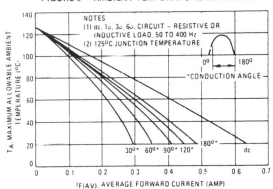

Triacs
Silicon Bidirectional Triode Thyristors

... designed primarily for industrial and military applications for the control of ac loads in applications such as light dimmers, power supplies, heating controls, motor controls, welding equipment and power switching systems; or wherever full-wave, silicon gate controlled solid-state devices are needed.

- Glass Passivated Junctions and Center Gate Fire
- Isolated Stud for Ease of Assembly
- Gate Triggering Guaranteed In All 4 Quadrants

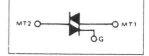

**2N5444
thru
2N5446
MAC5441
thru
MAC5443**

**TRIACs
40 AMPERES RMS
200 thru 600 VOLTS**

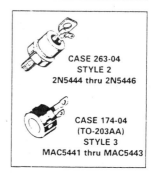

MT2 ○——————○ MT1
○ G

CASE 263-04
STYLE 2
2N5444 thru 2N5446

CASE 174-04
(TO-203AA)
STYLE 3
MAC5441 thru MAC5443

MAXIMUM RATINGS

Rating	Symbol	Value	Unit
*Peak Repetitive Off-State Voltage (T_J = −65 to +110°C) 1/2 Sine Wave 50 to 60 Hz, Gate Open	V_{DRM}		Volts
*Peak Principal Voltage MAC5441, 2N5444 MAC5442, 2N5445 MAC5443, 2N5446		200 400 600	
*RMS On-State Current (T_C per Figure 2) (T_C = +100°C) Full Sine Wave, 50 to 60 Hz	$I_{T(RMS)}$	40 20	Amps
*Peak Non-Repetitive Surge Current (One Full Cycle of surge current at 60 Hz, preceded and followed by a 40 A RMS current, T_C = 100°C)	I_{TSM}	300	Amps
*Peak Gate Power (Pulse Width = 10 μs Max)	P_{GM}	40	Watts
*Average Gate Power	$P_{G(AV)}$	0.75	Watt
*Peak Gate Current (10 μs Max)	I_{GM}	4	Amps
*Peak Gate Voltage	V_{GM}	30	Volts
*Operating Junction Temperature Range	T_J	−65 to +110	°C
*Storage Temperature Range	T_{stg}	−65 to +150	°C
*Stud Torque	—	30	in. lb.

THERMAL CHARACTERISTICS

Characteristic	Symbol	Max	Unit
*Thermal Resistance, Junction to Case MAC5441, MAC5442, MAC5443 2N5444, 2N5445, 2N5446	$R_{\theta JC}$	0.8 0.9	°C/W

*Indicates JEDEC Registered Data.

2N5444 thru 2N5446 ● MAC5441 thru MAC5443

ELECTRICAL CHARACTERISTICS (T_C = 25°C, and either polarity of MT2 to MT1 voltage, unless otherwise noted.)

Characteristic	Symbol	Min	Typ	Max	Unit
*Peak Forward or Reverse Blocking Current (Rated V_{DRM} or V_{RRM}) T_C = 25°C T_C = 100°C	I_{DRM}, I_{RRM}	— —	— 0.5	10 4	μA mA
*Peak On-State Voltage (I_{TM} = 56 A Peak, Pulse Width ≤ 1 ms, Duty Cycle ≤ 2%)	V_{TM}	—	1.65	1.85	Volts
Gate Trigger Current (Continuous dc), Note 1 (Main Terminal Voltage = 12 Vdc, R_L = 50 Ohms) MT2(+), G(+) MT2(+), G(−) MT2(−), G(−) MT2(−), G(+) *MT2(+), G(+); MT2(−), G(−) T_C = −65°C *MT2(+), G(−); MT2(−), G(+) T_C = −65°C	I_{GT}	 — — — — — —	 — — — — — —	 70 70 70 100 125 240	mA
*Gate Trigger Voltage (Continuous dc) (Main Terminal Voltage = 12 Vdc, R_L = 50 Ohms) MT2(+), G(+) MT2(+), G(−) MT2(−), G(−) MT2(−), G(+) *All Quadrants, T_C = −65°C *Main Terminal Voltage = Rated V_{DRM} = R_L = 10 k ohms, T_C = 100°C	V_{GT}	 — — — — — 0.2	 — — — — — —	 2 2 2 2.5 3.4 —	Volts
*Holding Current (Main Terminal Voltage = 12 Vdc, Gate Open) (Initiating Current = 150 mA) T_C = 25°C *T_C = −65°C	I_H	 — —	 — —	 70 100	mA
*Turn-On Time (Main Terminal Voltage = Rated V_{DRM}, I_{TM} = 56 A, Gate Source Voltage = 12 V, R_S = 12 Ohms, Rise Time = 0.1 μs, Pulse Width = 2 μs)	t_{gt}	—	1	2	μs
*Critical Rate-of-Rise of Commutation Voltage (Rated V_{DRM}, I_{TM} = 40 A, Commutating di/dt = 22 A·ms, gate energized) T_C = 70°C MAC5441, MAC5442, MAC5443 = 65°C 2N5444, 2N5445, 2N5446	dv/dt(c)	 5 5 5	 30 30 30	 — — —	V/μs
Critical Rate-of-Rise of Off State Voltage (Rated V_{DRM}, Exponential Voltage Rise, Gate Open, T_C = 110°C) MAC5441, 2N5444 MAC5442, 2N5445 MAC5443, 2N5446	dv/dt	 50 30 20			V/μs

*Indicates JEDEC Registered Data for 2N5541 thru 2N5446.

Note 1. All voltage polarities referenced to main terminal 1.

Chapter 1

2. $A = 4\phi = 67°$
3. 74Ω
10. $A_v = 2.424$ $R_i = -1.545$ kΩ
11. $I_L = v_i/2220$
15. $-4.78V \leq b \leq 4.78V; -1.43V \leq a \leq 1.43V$
18. $R_4 - R_1 = 500\Omega$
22. **a.** $v_{o(DM)} = 15.9$ mV; $v_{o(CM)} = 7.2$ μV
    b. $v_{o(DM)} = 143.3$ mV; $v_{o(CM)} = 0.8$ μV
28. **a.** $v_o(t) = 26.7(1 - 0.888e^{-1000t})$
    b. $t = 0^+$
30. $v_o = -1.472\ v_i \pm 12.9$ V
37. 74 dB

39. $v_o = R_1 \left(\dfrac{1}{R_1} + \dfrac{1}{R_2} + \dfrac{1}{R_v} \right) (v_{i1} - v_{i2})$

40. 79 Ω

Chapter 2

2. 9.75 nA; 6.84 mA
3. 0.721; 93.7 pA; 0.1 mA
6. 36 mV; 57.1 mV
13. 5.1 A
14. **a.** 0 A; 0 V
    b. 1.28 mA; 1.54 V
17. 1 mA; 5.2 Ω; $v_o(t) = 0.94 + 960 \times 10^{-6} \sin \omega_o t$ V
20. 52Ω; 0.583 V

21. **a.** $v_o = -\eta V_t \ln\left(\dfrac{v_s}{I_s R_s} + 1 \right)$

    b. $v_o = R_f I_s \left(e^{-v_s/\eta V_t} - 1 \right)$

32. **a.** -2 μA $\quad$ $v_{d1} = -14.94$ V $\quad$ $v_{d2} = -57.13$ mV
    b. -3 μA $\quad$ $v_{d1} = -30$ V $\quad$ $v_{d2} = -19.99$ V
35. $4.167\ \Omega < R < 5.556\ \Omega$; $P \approx 6$ W (various design choices possible)

Chapter 3

2. $-I_E \approx I_C = 6$ mA; $V_{BE} \approx 0.68$ V; $V_{CE} \approx 5.1$ V
5. $I_B = 40$ μA; $I_C = 9$ mA; $V_{BE} = 0.69$ V; $V_{CE} = 4.1$ V
10. $I_B = 15.53$ μA $I_C = 1.165$ mA
12. 302.1 kΩ
13. 3.62 V
16. $R_E = 294\ \Omega$; $I_C = 11.1$ mA; $V_{CE} = 1.7$ V; $V_{BE} = 0.71$ V
17. $R_{B2} = 8.45$ kΩ; $V_{CE} = -3$ V; $I_B = -13.3$ μA.
18. 2.78 V; 4.477 V
19. 40 gates
21. 25 gates
24. NAND gate; 13.7 V & 0.2V; 94 gates; 6.5 V & -6.3 V; 69.7 W

Chapter 4

1. **a.** -0.92 V
   b. 2.68 V
   c. -0.8 V
3. $R_s = 965\ \Omega$, $R_D = 6.04$ kΩ
5. $R_S = 287\ \Omega$; $R_D = 3$ kΩ
9. $I_D = 5.6$ mA; $V_{DS} = 7.16$ V; $V_{GS} = 7.16$ V; $V_{GD} = 0$ V
13. $R_S = 1.69$ kΩ; $R_D = 1.3$ kΩ
17. -2.75 mA; -3.5 V
18. -1.6 mA; -12.1 V
25. 459.4 Ω; 187.8 Ω
29. 1.29 V
33. $R_{off} \approx 500$ GΩ; $R_{on} \approx 17\ \Omega$
35. 0.772 mA; 6.36 V

Chapter 5

4. **a.** 160; 2.093 kΩ; 16.67 μS
   b. 160; 8.32 kΩ; 4.17 μS
6. **a.** $I_c = 1.99$ mA $V_{CE} = 1.52$ V
   b. -2.33; 13.26 kΩ; 1.5 kΩ
7. $R_E = 4.12$ kΩ; $h_{fe} = 200$; $h_{ie} = 2.6$ kΩ; $A_{vs} = -34.8$; $R_{in} = 2.53$ kΩ; $R_o = 453\ \Omega$
9. $R_B = 649$ kΩ; $h_{ie} = 2.35$ kΩ; $A_{vs} = 0.99$; $R_{in} = 143$ kΩ; $R_{out} = 12.8\ \Omega$

10. $A_{vs} = \dfrac{h_{fe}R_C}{h_{ie}}$; $R_{in} = R_E \left\| \dfrac{h_{ie}}{1 + h_{fe}} \right.$; $R_{out} = 12.8\ \Omega$

17. 103; 16 Ω
22. -17
25. $A_{vs} = -6.1$; $R_{in} = 510$ kΩ; $R_{out} = 3$ kΩ
31. $A_{vs} = 0.92$; $R_{in} = 765$ kΩ; $R_{out} = 340\ \Omega$
32. -4.04
36. 158 kΩ & 66.5 kΩ; 0.98

Chapter 6

1. $I_{B1} = 7.05$ μA; $I_{C1} = 1.55$ mA; $V_{CE1} = 4.9$ V
   $I_{B2} = 5.28$ μA; $I_{C2} = 950$ μA; $V_{CE2} = 5.4$ V
   $A_{vs} = -72.3$; $R_{in} = 8.4$ kΩ; $R_{out} = 53.3\ \Omega$
2. 1.477; 8.614; 83.26 Ω; 8.65 Ω
8. $A_{vs} = 185.4$; $R_{in} = 510.051$ kΩ; $R_{out} = 2.62$ kΩ
9. $R_{B1} = 270$ kΩ; $R_{B2} = 51$ kΩ
   $I_C = 1$ mA; $V_{CE} = 3.8$ V; $I_D = 960$ μA; $V_{GS} = -3.4$ V
   $A_{vs} = 2025$; $R_{in} = 4.64$ kΩ
10. 4.974 mA, 220 Ω; $h_{fe} = 40,000$, $h_{ie} = 209$ kΩ, $h_{oe} = 14.21$ μS; 0.435
15. 3.63 V; -31.8
16. $I_{B1} = 1.2$ μA; $I_{C1} = 216$ μA; $V_{CE1} = 6.4$ V
    $I_{B2} = 217$ μA; $I_{C2} = 39.1$ mA; $V_{CE2} = 7.1$ V;
    $A_{vs} = 104.8$

18. $I_C = 1$ mA; $V_{CE} = 4.13$ V; $I_D = 1$ mA; $V_{GS} = -2.2$ V; $A_{vs} = -4.1$

19. -82.5; 56.97 dB

21. $i_L = 0.01742(v_{i1} - v_{i2})$; $i_L = 80 \times 10^{-6} \dfrac{(v_{i1} + v_{i2})}{2}$

23. $R_1 = 10.5$ kΩ; $R_2 = 562$ Ω

Chapter 7

3. 7.32 kΩ & 38.3 kΩ; 7.68%

5. ± 8 V; -0.5 V $< v_o <$ 4 V

8. ± 7.454 V; 18.36%

11. 4.6 Vp-p

12. 19.3 V; 30.7 W; 23.3 W; 75.9%

13. a. -11.1 V $\leq V_o \leq$ 11.3 V
   b. Same as (a)

16. 68.32%; 32.09%

18. 2.8V; ± 30V; 5.05 kΩ; 57%

27. -19.7 dB, -23.3 dB; 12.3%

37. 500 mW; 1 W; 0.07°C/mW

Chapter 8

2. 0.019; 45.22

5. two series stages with global feedback, $f = 0.005$

10. 3.4 μV, 27%; 0.92%

14. 3.52 kΩ; 67.9 Ω

18. 10.5 kΩ

27. 1.54 kΩ; 27.5% increase in R_{in}

36. series-shunt; 4.249977318; 0.05% difference

Chapter 9

6. 0.112, 0.3303, 0.532, 0.7071, 0.8467, 0.9439, 0.9937

8. 5th order filter

10. first order low-pass, pole at $1/RC$; pole at 0, zero at $1/RC$

17. filter stages are never identical: change to 4th order Butterworth filter

19. sixth order filter

33. 100 pF

34. 5 MΩ; 833 kΩ

36. $C_2 = 50$ pF

40. 223 kHz

41. 71 kHz

Chapter 10

2. $5 \to 13.98$ dB; 80 kHz

5. 32.9 Hz; 31.69 Hz; 33.023 Hz

8. 1.873 krad/s, 9.78 rad/s, 35.46 rad/s; 1.873 krad/s is dominant; 298 Hz

9. 1.8 Hz; -4.5

11. 0.96; 17.3 Hz

13. 4 Hz; -3.1

17. ≈ 39 μF; 7.03; 0.83 Hz

22. $g_m = 94.21$ mA/V, $r_\pi = 1.486$ kΩ, $r_0 = 54.3$ kΩ, $r_b = \Omega$, C_μ 3 pF, $C_\pi = 121.9$ pF; 3.7 Hz, 100 Mhz.

24. 4.6k

26. 0.49 Hz, 49 Mhz; $A_{v0} = 21.7$ dB

32. 3.67 kΩ; 280 kHz, 0 Hz

37. Q_1 & Q_2—$I_C = 2.906$ mA, $V_{CE} = 6.11$ V, Q_3—$I_C = 5.86$ mA, $V_{CE} = 0.7$ V, Q_4—$I_C = 5.86$ mA, $V_{CE} = 14.3$V; 175; 4.85 MHz, 0 Hz.

Chapter 11

2. 48.97 kHz; 16 μs; 6.98 μs

4. 50.32 Hz; 16%; 13.6; 3.17 Hz

7. 0.00577; 2.45; 99 kHz, 23 Hz; 102

10. 5Hz, 1.4 MHz; 88.7 kHz; 89kHz, 0.34% error

18. 30 Hz, 16 Mhz; SPICE: 27 Hz, 14 MHz—11–14% error

22. ≈ 10 MHz; $f = 0.00045$ (max); 13 dB

Chapter 12

3. -18.4; $0.133/(RC)$

4. $0.039/(RC)$

5. -18.4; $0.19/(RC)$

8. $R_S = 2.7$ kΩ; $R = 8.2$ kΩ

INDEX